D0772795

Modern Machine Shop's
Guide to Threads, Threading and Threaded Fastners

By
Woodrow Chapman

Hanser Gardner Publications
Cincinnati, Ohio

Library of Congress Cataloging-in-Publication Data

Chapman, Woodrow W. (Woodrow Wilson), 1944-
  Modern machine shops guide to threads, threading, and threaded fasteners / By Woodrow Chapman.
      p. cm.
    ISBN 1-56990-359-X
1. Bolts and nuts--Design and construction. I. Title.
    TJ1330.C37 2004
    621.8'82--dc22

                                                                                        2004004400

While the advice and information in this book are believed to be true, accurate, and reliable, neither the author nor the publisher can accept any legal responsibility for any errors, omissions, or damages that may arise out of the use of this advice and information. The author and publisher make no warranty of any kind, expressed or implied, with regard to the material contained in this work.

A **MODERN MACHINE SHOP** book

Published by

Gardner Publications, Metalworking's Premier Publisher

www.mmsonline.com

Hanser Gardner Publications

6915 Valley Avenue

Cincinnati, OH 45244-3029

www.hansergardner.com

Preface

Since the publication of *Modern Machine Shop's Handbook for the Metalworking Industries* two years ago, we have had several inquiries and suggestions about offering a number of selections from the handbook in shorter, focused volumes. This book, which covers the production and use of fasteners used in engineering and manufacturing, is the third and final volume of these more concise editions, printed in an enlarged format in order to enhance readability.

The text, tables, and diagrams contained in this book cover thread systems, threading methods, and threaded fasteners and their capabilities, and special attention has been given to the wide variety of machining operations and tools utilized in the creation of threaded fasteners. An important addition to this book that is not contained in the parent *Handbook* is a discussion on aircraft fasteners (including rivets, which are sometimes substituted for threaded fasteners) that are employed in several industries. Every effort has been made to provide current, useful, and practical knowledge that an engineer, designer, or machinist normally consults in order to select a suitable threading operation or fastener for a particular application.

Finally, a word of thanks to the companies and associations listed on the following page whose cooperation and contributions have made this book possible. Every effort has been made to acknowledge the copyrighted material supplied by each contributor within the text, and any lapses of identification are regretted.

ANSI
American National Standards Institute
11 West 42nd Street
New York, NY 10036
www.ansi.org

ASME
American Society of Mechanical Engineers
345 East 47th Street
New York, NY 10017
www.asme.org

Alvord-Polk Tool
125 Gearhart Street
Millersburg, PA 17061
www.alvordpolk.com

American Fasteners Technologies Corp.
Nine Frontier Drive
Gibsonia, PA 15044-7999
www.americanfastener.com

Brown & Sharpe
Precision Park
200 Frenchtown Road
North Kingston, RI 02852
www.brownandsharpe.com

Brubaker Tool Corp.
200 Front Street
Millersburg, PA 17061
www.brubakertool.com

Cleveland Twist Drill
PO Box 2578
Augusta, GA 30903-2578
www.gfii.com

Darling Bolt Co.
2941 E. 10 Mile Road
Warren, MI 48090-2035
www.darlingbolt.com

DoAll Co.
254 N. Laurel Avenue
Des Plaines, IL 60016-4398
www.doall.com

Fastbolt Corp.
200 Louis Street
South Hackensack, NJ 07606
www.fastboltcorp.com

Fastcut Tool Corporation
31800 Sherman
Madison Heights, MI 48071-1463
www.fastcut.com

Giddings & Lewis
142 Doty Street
Fond du Lac, WI 54936-0590
www.giddings.com

Greenfield Industries, Inc.
PO Box 2587
Augusta, GA 30903-2578
www.gfii.com

Hanson Whitney Company
725 Marshall Phelps Road
Windsor, CT 06095
www.hansonwhitney.com

Hassay Savage Company
10 Industrial Boulevard
Turners Falls, MA 01376-2251
www.hassay-savage.com

Kennametal Inc.
1600 Technology Way
Latrobe, PA 15650
www.kennametal.com

Landis Threading Systems
360 South Church Street
Waynesboro, PA 17268-2659
www.landisthreadingsystems.com

Maryland Metrics
PO Box 261
Owens Mills, MD 21117
www.mdmetric.com

Metric & Multistandard Components Corp.
120 Old Saw Mill River Road
Hawthorne, NY 10532
www.metricmcc.com

Miniature Thread Specialists (MTS)
2005 W. Acoma Blvd., Unit C
Lake Havasu City, AZ 86403
www.minithread.com

New England Tap Corp.
175 Cocasset Street
Foxboro, MA 02035
www.netap.com

OSG Tap & Die, Inc.
676 East Fullerton Avenue
Glendale Heights, IL 60139
www.osg-sossner.com

Precision Twist Drill
One Precision Plaza
Crystal Lake, IL 60039-9000
www.precisiontwistdrill.com

Quadrant Engineering Plastic Products
2120 Fairmont Avenue
Reading, PA 19612-4235
www.quadrantepp.com

Renishaw Inc.
623 Cooper Court
Schaumburg, IL 60173
www.renishaw.com

SPS Technologies/Unbrako
301 Highland Avenue
Jenkintown, PA 19046
www.spstech.com

Surburban Tool
4141 N. Atlantic Blvd.
Auburn Hills, MI 48326
www.subtool.com

Taylor & Jones Limited
17 Southgate Honley
Huddersfield HD7 2NU
England
www.taylorandjones.co.uk

Tinnerman Fasteners
Engineered Components LLC
1060 West 130th Street
Brunswick, OH 44212
www.tinnerman.com

Unbrako/SPS Technologies, Inc.
4444 Lee Road
Cleveland, OH 44128
www.spstech.com/unbrako

Vermont Gage Company
6 Brooklyn Street
Swanton, VT 05488
www.vtgage.com

Weldon Tool Company
6030 Carey Drive
Valley View, OH 44125-4218
www.weldonmachinetool.com

West Coast Lockwasher (WCL)
16730 East Johnson Drive
Industry, CA 91744
www.wclco.com

The Yankee Corporation
Box 1255 RD#2
Fairfax, VT 05454
www.yankeereamer.com

THREAD SYSTEMS

TIGHTENING AND TENSIONING THREADED FASTENERS

THREADING OPERATIONS

Screw Threads

Unified inch and ISO metric threads

The origin of the parallel helical thread has been lost to time, but what is known of its history suggests that early versions took on a wide variety of forms. The Englishman Sir Joseph Whitworth, in the early 1840s, is generally credited with being the first to standardize a thread: the 55° radiused root thread that bore his name became the British Standard thread and remained so for over a century. Other countries, of course, went their own way. Just three years after the introduction of the Whitworth thread, an American, William Sellers, contributed a new design with a 60° thread. Sellers found the 55° angle difficult to gauge, and his thread design was eventually accepted as the American Standard. To further confuse the issue, an international conference convened in Switzerland in 1898 and announced the S.I. (Systeme International) Metric thread. Various countries throughout Europe nationalized this thread, making it the standard on the Continent.

In 1948, in the interest of promoting standardized manufacturing—which had become a nagging problem for Allied efforts during the Second World War—the United States, Canada, and the United Kingdom agreed to adopt a new thread, the Unified screw thread. The Unified Standard thread is the result, defined by American Standard B1.1-1949, which identifies a thread configuration that is essentially the same as the 60° American National thread that preceded it. (There is still an "AN" series of fasteners available, but the abbreviation stands for the first letters of Air Force–Navy rather than American National, and AN fasteners use standard Unified threads.) Changes are most evident in the permitted allowances and tolerances for individual sizes, and these modifications were introduced to provide greater fatigue strength, root clearance, and ease of assembly. In addition, the new Unified Standard (as it came to be known) and the ISO metric thread also have the same basic configuration, but differ in diameters and pitch lengths for standard sizes. In instances where the tolerances for the two threads overlap, ISO metric and Unified inch threads are compatible, as shown in **Table 1a** which lists the six potentially compatible sizes. ISO metric and Unified size threads should never be combined, but this listing provides evidence of the basic compatibility of the two thread forms. Either series satisfies the need for a general purpose screw thread system for most applications. Allmetal Screw Products notes that there are 319 thread sizes in the Unified Thread Standard ranging in diameter from 0.06" (1.5 mm) to 6" (152 mm), and in pitches from 80 threads per inch (0.3175 mm pitch) to 4 tpi (6.35 mm pitch). Each is available in at least two classes of fit. There are even more metric sizes, 332 in all, ranging in diameter from 1 mm (0.03937") to 300 mm (11.81"), and in pitches from 0.2 mm (127 tpi) to 6 mm (4.23 tpi). Again, each size is available in more than a single tolerance grade.

Converting inch pitches to metric dimensions can be confusing, and **Table 1b** provides conversions for all popular metric pitches in millimeters to pitch and threads per inch values in inches.

Thread definitions

All threads share fundamental characteristics that, when defined, precisely identify all relevant dimensions of its shape. The following definitions were provided by Cleveland Twist Drill Co. *Figure 1* illustrates important screw thread nomenclature.

General terms relating to screw threads

Complete Thread. The complete or full form thread is the cross section of a threaded length having full form at both crest and root.

Incomplete Thread. An incomplete thread is a threaded profile having either crests

Table 1a. Potentially Compatible Unified and ISO Metric Threads. *(Source, Allmetal Screw Products.)*

Unified Thread Size	ISO Metric Thread Size
1–72 UNF Class 3A	1.8 × 0.35 (Coarse Pitch)
2–56 UNC Class 3A	2.2 × 0.45 (Coarse Pitch)
3–56 UNF Class 3A	2.5 × 0.45 (Coarse Pitch)
8–36 UNF Class 3A	4.0 × 0.70 (Coarse Pitch)
10–32 UNF Class 3A	5.0 × 0.80 (Coarse Pitch)
5/16–20 UN Class 3A	8.0 × 1.25 (Coarse Pitch)

Table 1b. Millimeters Pitch to Threads Per Inch Metric Conversion. *(Source, New England Tap Corp.)*

Pitch mm	Pitch inch	Threads per inch	Basic Height*	Pitch mm	Pitch inch	Threads per inch	Basic Height*
.25	0.00984	101.6000	0.00639	1.25	0.04921	20.3211	0.03196
.30	0.01181	84.6668	0.00767	1.50	0.05906	16.9316	0.03836
.35	0.01378	72.5689	0.00895	1.75	0.06890	14.5138	0.04475
.40	0.01575	63.4921	0.01023	2.00	0.07874	12.7000	0.05114
.45	0.01772	56.4334	0.01151	2.50	0.09843	10.1595	0.06393
.50	0.01969	50.8000	0.01279	3.00	0.11811	8.4667	0.07671
.60	0.02362	42.3370	0.01534	3.50	0.13780	7.2569	0.08950
.70	0.02756	36.2845	0.01790	4.00	0.15748	6.3500	0.10229
.75	0.02953	33.8639	0.01918	4.50	0.17717	5.6443	0.11508
.80	0.03150	31.7460	0.02046	5.00	0.19685	5.0800	0.12758
.90	0.03543	28.2247	0.02301	6.00	0.23622	4.2333	0.15344
1.00	0.03937	25.4000	0.02557	mm × 0.03937 = inch equiv. inch × 25.4 = mm equiv.			

* Calculated as twice the addendum, in inches.

or roots, or both crests and roots, not fully formed, resulting from their intersection with the cylindrical or end surface of the work or the vanish cone. It may occur at either end of the thread.

NOTE: Formerly in pipe thread terminology this was referred to as "the imperfect thread" but that is no longer considered desirable.

Lead Thread. The lead thread is the portion of the incomplete thread that is fully formed at root but not fully formed at crest which occurs at the entering end of either external or internal threads.

Vanish Thread (Partial Thread, Washout Thread or Thread Run-Out). A vanish thread is that part of the thread which is not fully formed at the root. It is produced by the chamfer at the starting end of the threading tool.

Effective Thread. The effective (or useful) thread includes the complete thread and those portions of the incomplete thread which are fully formed at the root but not at the crest (in taper pipe threads this includes the so-called black crest threads), thus excluding the vanish thread.

Total Thread. The total thread includes the complete and all of the incomplete thread, thus including the vanish thread.

Classes of Threads. Classes of threads are distinguished from each other by the amounts of tolerance or tolerance and allowance specified.

Thread Series. Thread series are groups of diameter-pitch combinations distinguished from each other by the number of threads per inch applied to specific diameters.

Single and Multiple Start Threads. A single thread is a thread made by forming one groove around a cylinder or inside a hole. Most threads are single threads. Double, triple, or other multiple threads have two, three, or more grooves formed around a cylinder or inside a hole.

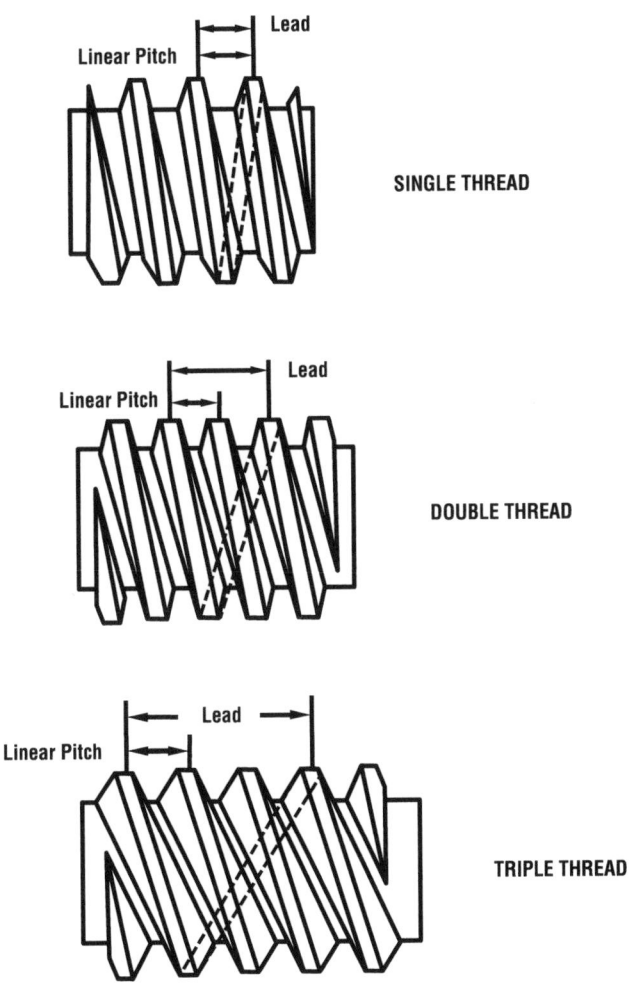

Screw Thread Forms

Terms relating to size and fit

Nominal Size. The nominal size is the designation that is used for the purpose of general information.

Basic Size. The basic size is the size from which the limits of size are derived by the application of allowances and tolerances.

Reference Size. A reference size is a size without tolerance used only for information purposes and does not govern manufacturing or inspection operations.

Design Size. The design size is the basic size with allowance applied, from which the limits of size are derived by the application of tolerances. If there is no allowance, the design size is the same as the basic size.

Actual Size. An actual size is a measured size.

Limits of Size. The limits of size are the applicable maximum and minimum sizes.

Maximum Material Limit. A maximum material limit is the limit of size that provides the maximum amount of material for the part. Normally it is the maximum limit of size of an external dimension or the minimum limit of size of an internal dimension.

Minimal Material Limit. A minimum material limit is the limit of size that provides the minimum amount of material for the part. Normally it is the minimum limit of size of an external dimension or the maximum limit of size of an internal dimension.

NOTE: Examples of exceptions are: an exterior corner radius where the maximum radius is the minimum material limit and the minimum radius is the maximum material limit.

Allowance. An allowance is a prescribed difference between the maximum material limits of mating parts. It is the minimum clearance (positive allowance) or maximum interference (negative allowance) between such parts.

Tolerance. A tolerance is the total permissible variation of a size. The tolerance is the difference between the limits of size.

Unilateral Tolerance. A unilateral tolerance is a tolerance in which variation is permitted only in one direction from the design size.

Bilateral Tolerance. A bilateral tolerance is a tolerance in which variation is permitted in both directions from the design size.

Terms relating to geometrical elements of screw threads

Axis of Thread. The axis of a thread is the axis of its pitch cylinder or cone.

Major Cylinder. The major cylinder is one that would bound the crests of an external straight thread or the roots of an internal straight thread.

Major Cone. The major cone is one that would bound the crests of an external taper thread or the roots of an internal taper thread.

Pitch Cylinder. The pitch cylinder is one of such diameter and location of its axis that its surface would pass through a straight thread in such a manner as to make the widths of the thread ridge and the thread groove equal and, therefore, is located equidistantly between the sharp major and minor cylinders of a given thread form. On a theoretically perfect thread these widths are equal to one-half of the basic pitch.

Pitch Cone. The pitch cone is one of such apex angle and location of its vertex and axis that its surface would pass through a taper thread in such a manner as to make the widths of the thread ridge and the thread groove equal and, therefore, is located equidistantly between the sharp major and minor cones of a given thread form. On a theoretically perfect taper thread these widths are equal to one-half of the basic pitch.

Minor Cylinder. The minor cylinder is one that would bound the roots of an external straight thread or the crests of an internal straight thread.

Minor Cone. The minor cone is one that would bound the roots of an external taper thread or the crests of an internal taper thread.

Pitch Line. The pitch line is a generator of the cylinder or cone specified in the definitions of pitch cylinder and pitch cone.

Form of Thread. The form of thread is its profile in an axial plane for a length of one pitch of the complete thread.

Basic Form of Thread. The basic form of a thread is the theoretical profile of the thread for a length of one pitch in an axial plane, from which the design forms of the threads for both the external and internal threads are developed.

Design Forms of Thread. The design forms for a thread are the maximum material forms permitted for the external and internal threads. In practice, however, the forms of roots are indeterminate roundings not encroaching on the maximum material form of the mating thread when assembled.

Terms relating to dimensions of screw threads

Pitch. The pitch of a thread having uniform spacing is the distance, measured parallel to its axis, between corresponding points on adjacent thread forms in the same side of the axis. The basic pitch is equal to the lead divided by the number of thread starts. See Table 4, No. 1.

Lead. When a threaded part is rotated about its axis with respect to a fixed mating, the lead is the axial distance moved by the part in relation to the amount of angular rotation. Lead is commonly specified as the distance to be moved in one complete rotation. It is necessary to distinguish measurement of lead from measurement of pitch, as uniformity of pitch measurements does not assure uniformity of lead. Variations in either lead or pitch cause the functional diameter of a thread to differ from the pitch diameter. See Table 4, No. 41.

Helix Variation. Helix variation of a thread is a wavy deviation from true helical advancement. The "helical path" includes the helix with its superimposed variation and is measured either as the maximum deviation from the true helix or as the "cumulative pitch." The cumulative pitch is the distance measured parallel to the axis of the thread between corresponding points on any two thread forms whether or not they are in the same axial plane.

Lead Angle. On a straight thread, the lead angle is the angle made by the helix of the thread at the pitch line with a plane perpendicular to the axis. On a taper thread, the lead angle at a given axial position is the angle made by the conical spiral of the thread, with the plane perpendicular to the axis, at the pitch line. See Table 4, No. 42.

Helix Angle. On a straight thread, the helix angle is the angle made by the helix of the thread at the pitch line with the axis. On a taper thread, the helix angle at a given axial position is the angle made by the conical spiral of the thread with the axis at the pitch line. The helix angle is the complement of the lead angle. See Table 4, No. 43.

NOTE: The helix angle was formerly defined in accordance with the present definition of lead angle.

Thickness of Thread Ridge. The thickness of thread ridge is the distance between the flanks of one thread ridge, normally measured parallel to the axis at the specified pitch radius. The thickness of thread ridge may be specified and measured parallel to the axis at any other specified radius.

NOTE: The pitch radius is equal to one-half of the pitch diameter.

Width of Thread Groove. The width of the thread groove is the distance between the flanks of adjacent thread ridges normally measured parallel to the axis at the specified pitch radius. The width of thread groove may be specified and measured parallel to the axis at any other specified radius.

Height of Thread. The height (or depth) of thread is the distance, measured radially between the major and minor cylinders or cones, respectively.

NOTE: In American practice the height of thread is often expressed as a percentage of three-fourths of the height of the fundamental triangle.

Addendum. The addendum of an external thread is the radial distance between the major and pitch cylinders or cones, respectively. The addendum of an internal thread is the radial distance between the minor and pitch cylinders or cones, respectively. See Table 4, No. 16 and No. 20.

Dedendum. The dedendum of an external thread is the radial distance between the pitch and minor cylinders or cones, respectively. The dedendum of an internal thread is the radial distance between the major and pitch cylinders or cones, respectively. See Table 4, No. 16–19.

Crest. The crest is the outermost surface of the thread form that joins the thread flanks. This is the farthest point on the thread from the cylinder or cone from which the thread projects. Hence, it is the point of the major diameter on external threads, and the minor diameter on internal threads. See Table 4, No. 34–36.

Crest Truncation. The crest truncation of a thread is the radial distance between the sharp crest (crest apex) and the cylinder or cone that would bound the crest. See Table 4, No. 20, 31, and 32.

Root. The root is the innermost surface of the thread form that joins the flanks, immediately adjacent to the cylinder or cone from which the thread projects. It is the point of the minor diameter on an external thread and the major diameter of an internal thread. See Table 4, No. 25–30.

Root Truncation. The root truncation of a thread is the radial distance between the sharp root (root apex) and the cylinder or cone that would bound the root. See Table 4, No. 20, 32, and 33.

Pitch Diameter. On a straight thread, the pitch diameter is the diameter of the pitch cylinder. On a taper thread, the pitch diameter at a given position on the thread axis is the diameter of the pitch cone at that position. On a single start thread of perfect form and lead, it is also the length between intercepts of a line that is perpendicular to the thread axis and intersects thread flanks on opposite sides of the thread axis. See Table 4, No. 12–15.

NOTE: When the crest of a thread is truncated beyond the pitch line, the pitch diameter, pitch cylinder, or pitch cone would be based on a theoretical extension of the thread flanks.

NOTE: Pitch diameter on the buttress casing thread is defined by the American Petroleum Institute, in API Standard 5B, as being midway between the major and minor diameters.

Major Diameter. On a straight thread, the major diameter is the diameter of the major cylinder. See Table 4, No. 3–5.

Minor Diameter. On a straight thread, the minor diameter is the diameter of the minor cylinder. See Table 4, No. 6–11.

Thread Groove Diameter (Simple Effective Diameter). On a straight thread, the thread groove diameter is the diameter of the coaxial cylinder, the surface of which would pass through the thread profiles at such points as to make the width of the thread groove equal to one-half of the basic pitch. It is the diameter yielded by measuring over or under cylinders (wires) or spheres (balls) inserted in the thread groove on opposite sides of the axis and computing the thread groove diameter as thus defined. On a taper thread, the thread groove diameter is the diameter at a given position on the thread axis of the coaxial cone, the surface of which would pass through the thread profiles at such points as to make the width of the thread groove (measured parallel to the axis) equal to one-half of the basic pitch. It is the diameter yielded by measuring over or under cylinders (wires) or spheres (balls) inserted in the thread groove on opposite sides of the axis and computing the thread groove diameter as thus defined.

Thread Ridge Diameter. On a straight thread, the thread ridge diameter is the diameter of the coaxial cylinder, the surface of which would pass through the thread profiles at such points as to make the thickness of the thread ridge equal to one-half of the basic pitch. On a taper thread, the thread ridge diameter is the diameter at a given position on the thread axis of the coaxial cone, the surface of which would pass through the thread profiles at such points as to make the thickness of the thread ridge (measured parallel to the axis) equal to one-half of the basic pitch.

Functional (Virtual) Diameter. The functional diameter of an external or internal thread is the pitch diameter of the enveloping thread of perfect pitch, lead, and flank angles, having full depth of engagement but clear at crests and roots, and of a specified length of engagement. It may be derived by adding to the pitch diameter in the case of an external thread, or subtracting from the pitch diameter in the case of an internal thread, the cumulative effects of deviations from specified profile, including variations in lead and flank angle over a specified length of engagement. The effects of taper, out-of-roundness, and surface defects may be positive or negative on either external or internal threads. (A perfect GO thread plug or ring gage, having a pitch diameter equal to that specified for the maximum material limit and having clearance at crest and root, is the enveloping thread corresponding to that limit.)

NOTE: Also called the Virtual Diameter, Effective Size, or Virtual Effective Diameter.

Length of Complete Thread. The length of complete thread is the axial length of a part where the thread section has full form at both crest and root; that is, the vanish threads are not included. However, on commercial fasteners where there are unfilled crests at the start of rolled threads or a chamfer at the start of a thread, not exceeding two pitches in length, this is traditionally included in the specified thread length.

NOTE: When designing threaded products, it is necessary to take cognizance of: (1) such permissible length of chamfer and (2) the first threads which by virtue of gaging practice may exceed the product limits and which may be included within the length of complete thread. However, when the application is such as to require a minimum or maximum number, or length, of complete threads, the specification shall so state. Similar specification is required for a definite length of engagement.

Length of Thread Engagement. The length of thread engagement of two mating threads is the axial distance over which two mating threads are designed to contact.

Depth of Thread Engagement. The depth (or height) of thread engagement between two coaxially assembled mating threads is the radial distance by which their thread forms overlap each other.

Unified threads by class

Unified threads have five fit classifications. The classes are intended to reflect the amount of tolerance specified. It should be remembered that these classes are entirely based on fit tolerances, and they should never be used as a determination of the grade or strength of an individual fastener. In all classes, the letter A applies to external threads, and the letter B applies to internal threads.

Class 1A and *1B* are very rarely used, although they are retained in the standards for Unified threads. Fasteners conforming to Class 1 are intended for conditions of low stress. Their loose fit is intended to allow for ease in assembly, even when the threads are slightly damaged or dirty. Only Class 1 external threads have an allowance.

Class 2A and *2B* are used for most applications. The specified allowance is intended to minimize galling and seizing in high cycle assembly, and to accommodate electroplated finishes. The 2A maximum diameters apply to an unplated part or to a part prior to plating, whereas the basic diameters (the 2A maximum diameter plus allowance) apply to a part

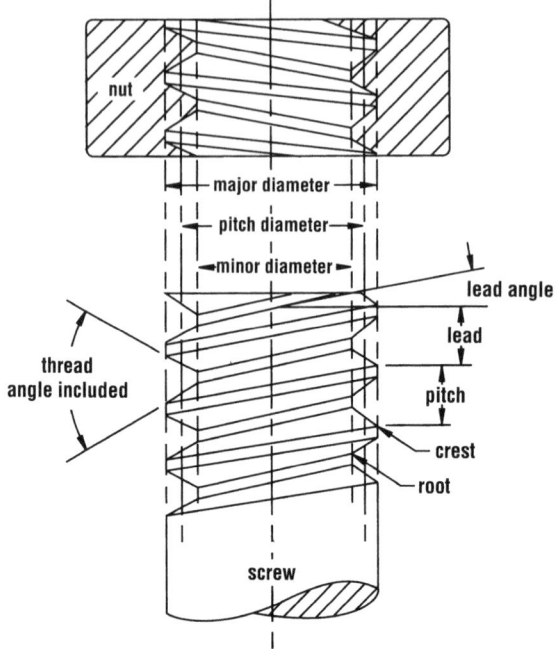

Figure 1. Basic screw thread dimensions. (Source, Kennametal, Inc.)

after plating. The minimum diameters of class 2B threads, whether or not plated or coated, are basic, affording no allowance or clearance in assembly at maximum material limits. This class is widely used for non-critical assembly with medium stress levels. Class 2A thread major diameter tolerances are equal to approximately twice their pitch diameter tolerances. The use of Class 2 threads provides for a small clearance between maximum material parts, except when the external thread is plated. Plated parts should be gaged with basic size GO gages.

Class 3A and *3B* are used where a close tolerance fit is important. Most aerospace fasteners are manufactured to Class 3 specifications. These threads require high quality assembly and inspection equipment. The maximum diameters of Class 3A threads and the minimum diameters of class 3B threads, whether or not plated or coated, are basic. This means that no allowance or clearance exists for assembly of maximum material components. However, since the tolerances on GO gages are within the limits of size of the product, the gages will assure a slight clearance when the fastener is made to maximum material limits. Note that thread formulas are generally based on specifications for Class 3 threads. For Class 1 and Class 2, the allowance stated in the tables must be subtracted.

Class 4 was designed for American National threads, and is no longer in use. Class 4 external threads have a negative allowance on the pitch diameter only.

Class 5 is an interference fit, designed for permanent installations. *Figure 2* illustrates the Class 5 thread at maximum and minimum material condition. Interference threads are intended to develop sufficient breakloose torques to prevent loosening of externally threaded fasteners. This is accomplished by precisely sizing the external thread larger than the internal thread so that, when torqued, either elastic compression or plastic movement of material causes both the internal and external components to become identical in

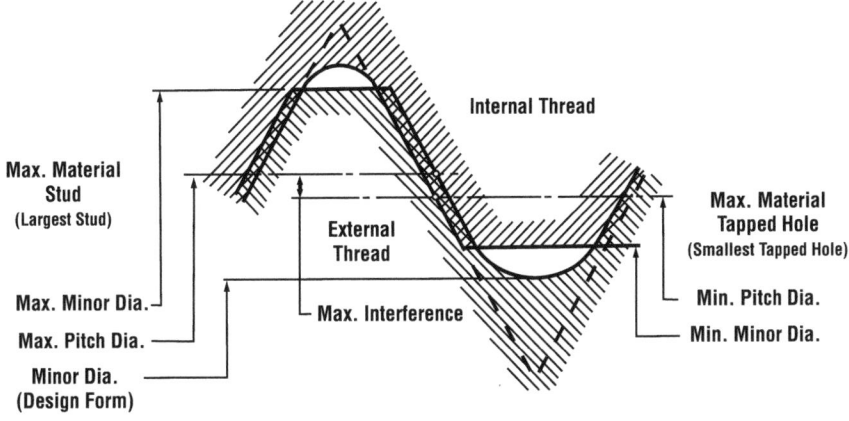

MAXIMUM INTERFERENCE

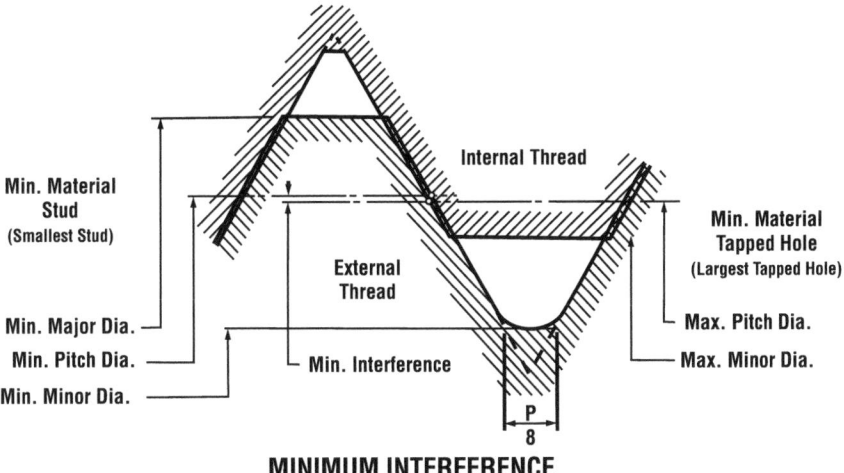

MINIMUM INTERFERENCE

*Figure 2. Class 5 thread form. (Source, ANSI B1.12-1972 as published by
the American Society of Mechanical Engineers.)*

size. Class 5 fits are most often used for installing studs in hard materials or in similar
situations where the mating of the components is intended to be permanent. Dimensions
for Class 5 interference fit threads are given in **Table 10.** The subclass categories for
Class 5 threads are as follows.

NC-5 HF   Ferrous material external threads for driving in hard ferrous material
           with hardness greater than 160 HB. The intended length of engagement
           should equal 1 $1/4$ times the diameter.

NC-5 CSF   Ferrous material external threads for driving in copper alloy and soft
           ferrous material with hardness of 160 HB or less. The intended length of
           engagement should equal 1 $1/4$ times the diameter.

NC-5 ONF   Ferrous material external threads for driving in non-ferrous materials
           other than copper alloys. The intended length of engagement should
           equal 2 $1/2$ times the diameter.

NC-5 IF   All ferrous material internal threads.

NC-5 INF   All non-ferrous material internal threads.

Designations for Unified threads

In addition to class, Unified threads are segmented by design, and identified by a two to four letter symbol intended to recognize the characteristics of the thread. Basic dimensions and specifications for UN and UNR thread forms are shown in *Figure 3*. The basic symbol designations are as follows.

UN   Unified constant-pitch thread.*
UNC   Unified coarse thread.*
UNF   Unified fine thread.*
UNEF   Unified extra-fine thread.*
UNS   Unified threads of special diameters, pitches, or lengths of engagement.*
UNR   Unified constant-pitch thread with a $0.108P$ to $0.144P$ controlled root radius.*
UNRC   Unified coarse thread with a $0.108P$ to $0.144P$ controlled root radius.*
UNRF   Unified fine thread with a $0.108P$ to $0.144P$ controlled root radius.*
UNREF   Unified extra-fine thread with a $0.108P$ to $0.144P$ controlled root radius.*
UNJ   Unified constant-pitch thread with a $0.15011P$ to a $0.18042P$ controlled root radius.**
UNJC   Unified coarse thread with a $0.15011P$ to a $0.18042P$ controlled root radius.**
UNJF   Unified fine thread with a $0.15011P$ to a $0.18042P$ controlled root radius.**
UNJEF   Unified extra-fine thread with a $0.15011P$ to a $0.18042P$ controlled root radius.**
UNM   Unified miniature thread.***

*For complete specifications, consult ANSI B1.1 Standard.
**For complete specifications, consult ANSI B1.15 Standard.
***For complete specifications, consult ANSI B1.10 Standard.

While UN and UNR threads are essentially identical, the root contour specified for external UN threads may be either flat (equal to $0.125P$) or rounded. For UN external threads, the nominal minor diameter is measured from the height of the flat root of the thread. The root contour (radius) for external UNR threads (both fine and course) is specified as continuous, and at a distance of not less than $0.625H$ below the basic major diameter. Radiused root contours (as opposed to flat) improve strength and fatigue life of the fastener. The majority of available socket headed cap screws are manufactured to UNR specifications. UNS and UNRS threads are intended for use only when standard series threads fail to conform to design standards.

The coarse thread forms, UNC and UNRC, are adaptable to rapid assembly and to conditions where corrosion or thread damage is likely to be encountered. Coarse threads are commonly specified in situations where the female thread is weaker than the male thread (for example, when threading steel bolts into aluminum castings). The coarse thread allows for the female thread to have maximum thread area and minimal minor diameter.

Fine thread forms, UNF and UNRF, provide more tensile strength than comparably sized coarse threads. Fine threads are specified where strength and fatigue resistance are important considerations, and in situations where the length of engagement is short. UNEF and UNREF forms enhance the positive characteristics of the basic fine thread, but require greater precision and increased time in assembly.

The limits of size for external and internal UN, UNF, UNC, UNS, and UNEF thread forms are provided in **Table 5** (external threads) and **Table 6** (internal threads). UNR form thread dimensions are identical to UN dimensions, with the exception of the minor

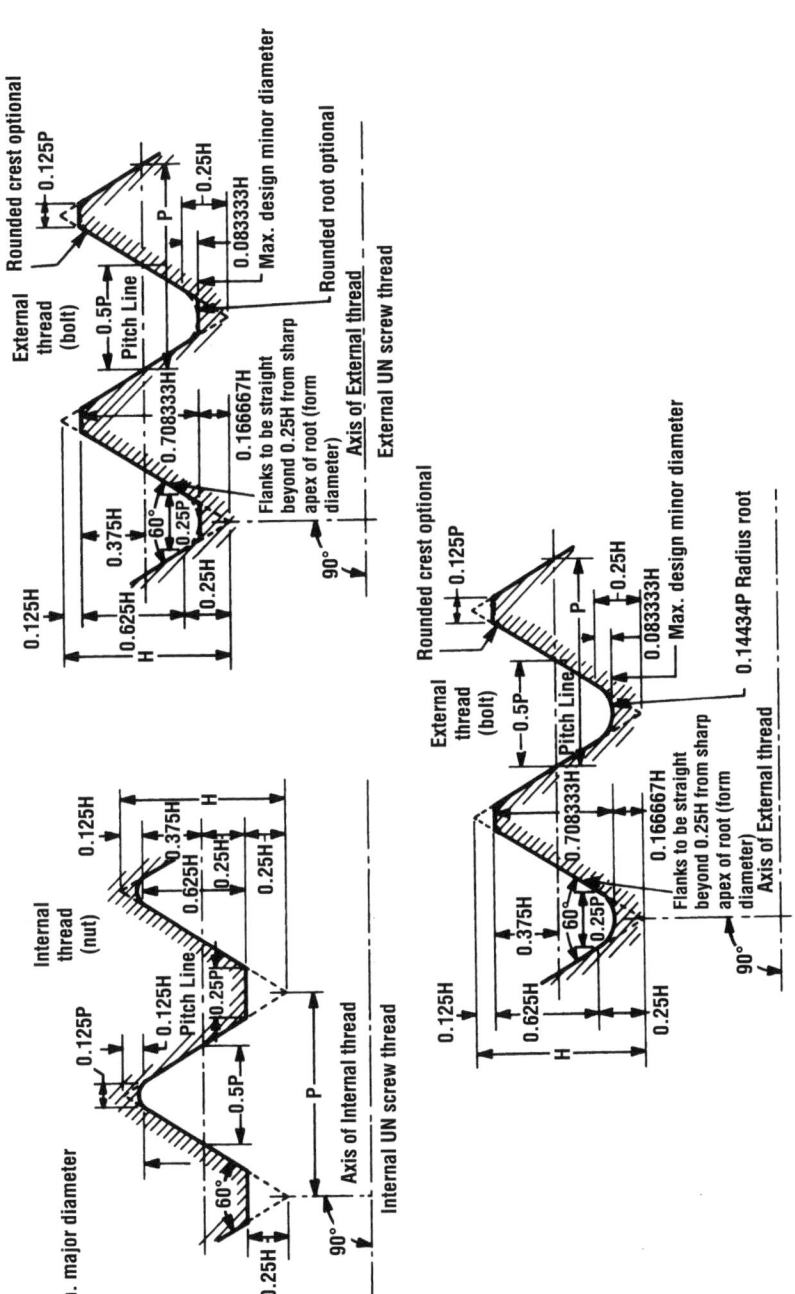

Figure 3. UN and UNR thread forms.

diameter of UNR external threads. To determine the minor diameter for an external UNR thread, use the constant for d_l from **Table 2**.

UNJ threads were originally developed to meet aerospace requirements. The radius, larger than specified for UNR threads, was designed to provide maximum fatigue strength resulting primarily from their increased minor diameter. Bolts made to UNJ specifications normally have tensile strength of 160,000 psi or more. The length of engagement for UNJC, UNJF, and 8UNJ threads is equal to the basic major diameter. For UNJEF, 12UNJ, and 16UNJ threads, the length of engagement is equal to 9 pitches. See **Table 7** (external threads) and **Table 8** (internal threads) for specifications for UNJ threads. *Figure 4* illustrates the UNJ thread form.

UNM threads are discussed later in this section. See **Table 11** and **Table 12** for formulas and dimensions.

When designating a specific Unified fastener, the nominal size is stated first, usually as a fraction but in some cases as a decimal, and followed by the number of threads per inch, the series symbol, and the class symbol. If a number in parenthesis follows the class symbol, it is a reference to the gaging system number. A left-hand thread is denoted by "LH" at the end. An example of a Unified thread designation would be $^1/_2$–16 UN-2A. A designation of the same thread including a left-hand thread indicator and a gaging system number would be $^1/_2$–16 UN-2A-LH(21).

Gaging systems

UN, UNR, UNJ, M, and MJ are inspected by one of three gaging systems, depending on the engineering requirement of the threaded product. Selection of the appropriate inspection method is based on the features of most importance for the intended use of the thread. Consideration is given to such features as form, fit, function, and fabrication of the threaded product. Guidance for selection of an inspection method from the three standard gaging systems is given in FED-STD-H28/20B as follows.

System 21. System 21 provides for interchangeable assembly with functional size control at the maximum material limits within the length of standard gaging elements, and also control of characteristics identified as NOT GO functional diameters or as HI (internal) and LO (external) functional diameters. These functional gages provide some control at the minimum material limit when there is little variation in thread form characteristics such as lead, flank angle, taper, and roundness. System 21 is appropriate for use under one or more of the following conditions.

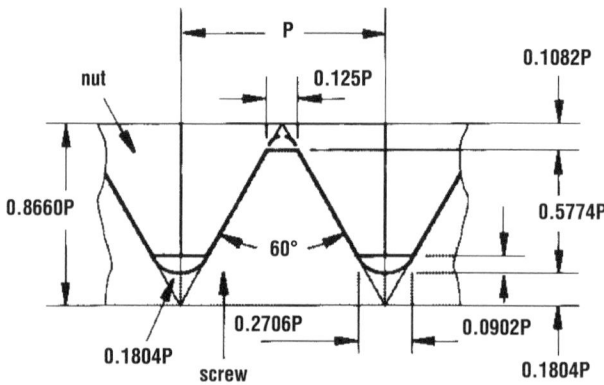

Figure 4. UNJ thread form. (Source, Kennametal, Inc.)

A) Where the threads of the product do not need specific mechanical strength properties, or where mechanical strength requirements are not specific for the product threads by either material strength and dimensional limits or by testing strength of the threads.

B) The threaded product has the mechanical properties specified, mechanical property testing of the threads is required, and the testing requires that the screw threads be subject to shear and beam loading by matching threads. If the threads have a locking element incorporated, locking torque values and tests must be specified and run on matching threads inspected by either System 22 or System 23.

C) For standard off-the-shelf general application fasteners when considered acceptable.

D) Internal thread is less than 0.190" (5 mm) nominal size.

System 22. System 22 provides for interchangeable assembly with functional size control at the maximum material limits within the length of standard gaging elements, and also control of the minimum material size limits over the length of the full thread. Other thread characteristics such as lead, flank angle, taper, and roundness variations are confined within these limits with no specific control of their magnitude. For UNJ and MJ external threads, control is also provided for the thread root radius and rounded root minor diameter. System 22 is recommended when none of the qualifying conditions for using either System 21 or System 23 is applicable.

System 23. System 23 provides for interchangeable assembly with functional size control at the maximum material limits within the length of standard gaging elements, and also control of the minimum material size limits over the length of the full thread. The magnitude of other thread characteristics such as lead, flank angle, taper, and roundness are further controlled within these limits. For UNJ and MJ external threads, control is also provided for the thread root radius and rounded root minor diameter. System 23 is recommended for use under one or more of the following conditions.

A) When thread element control is required to determine the extent of deviation in any of the elements of the thread, normally special applications.

B) For threaded product used in research investigations and testing to determine the effect that a specific thread element variation has on the attributes of the threaded product or the attributes of a threaded product's application.

C) The conductive of investigation and testing in analysis of thread failures.

Basic Unified thread dimensions

The basic pitch diameter (internal thread, specified as D_2 minimum) and the minor diameters of both external and internal Unified threads are all directly related to the basic major diameter of the internal thread (D). The basic major diameter is the exact equivalent of the decimal size of the fastener. For example, a $3/8$" nut has a basic major diameter of exactly 0.3750". These dimensions are constant for all Unified threads of the same pitch. For instance, from **Table 2** it can be seen that for all Unified threads, regardless of diameter, with 8 tpi, the basic (minimum) pitch diameter (D_2 min.) of the internal thread is equal to the basic major diameter (D) minus 0.0812". Also on the table, it can be seen that the minor diameter of an 8 tpi Class 3A external thread (d_1 basic) is equal to the basic major diameter minus 0.1488". Finally, from the same table, the minor diameter of an 8 tpi internal thread (D_1 min) is found to be equal to the basic major diameter minus 0.1353".

Theoretically, a 1 tpi internal Unified thread would have a minimum pitch diameter equal to the basic major diameter (D) minus 0.6496". The following formula, based on the constant 0.6496", can be used to find the minimum internal pitch diameter (D_2 min.) for any unified thread. Also, it should be noted that D_2 min. is *always equal* to the maximum pitch diameter for external (d_2 max.) Class 3A threads. For Class 1 and 2 threads, subtract the allowance for the specific thread as given in Table 5.

$$D_2 \text{ min.} = D - (0.6496 \div \text{tpi})$$

Similarly, the minor diameter (d_1) of a 1 tpi external Unified thread would be equal to the basic internal major diameter (D) minus 1.1908". Therefore, the following formula can be used to find d_1 for Class 3A threads. For other classes, subtract the allowance. It should be noted that this formula provides d_1 for UNR rounded radius threads (preferred).

$$d_1 = D - (1.1908 \div \text{tpi})$$

Finally, D_1, the internal thread minor diameter, for a 1 tpi nut would be equal to D minus 1.0824". Using this number, D_1 for any unified thread can be computed with the following formula.

$$D_1 = D - (1.0824 \div \text{tpi})$$

Table 2 can be used for quickly determining D_2 min., d_1, and D_1. To use the table, find the tpi in the left column, then subtract the number in any of the other columns from the basic internal major diameter. For example, to find D_1 for a 1 $^1/_{16}$–12 thread, subtract

Table 2. Constants for Determining Basic Dimensions for Unified Threads. (Subtract the constant from the major diameter of the internal thread.)

TPI	D_2 min.	d_1	D_1
4	0.1624	0.2977	0.2706
4.5	0.1444	0.2646	0.2405
5	0.1299	0.2382	0.2165
6	0.1083	0.1985	0.1804
7	0.0928	0.1701	0.1546
8	0.0812	0.1488	0.1353
9	0.0722	0.1323	0.1203
10	0.0650	0.1191	0.1082
11	0.0591	0.1082	0.0984
12	0.0541	0.0992	0.0902
13	0.0500	0.0916	0.0833
14	0.0464	0.0850	0.0773
16	0.0406	0.0744	0.0676
18	0.0361	0.0662	0.0601
20	0.0325	0.0595	0.0541
24	0.0271	0.0496	0.0451
27	0.0241	0.0441	0.0401
28	0.0232	0.0425	0.0386
32	0.0203	0.0372	0.0338
36	0.0180	0.0331	0.0301
40	0.0162	0.0298	0.0271
44	0.0148	0.0271	0.0246
48	0.0135	0.0248	0.0226

0.0902 from the major internal diameter (1.0625 − 0.0902 = 0.9723"). If the desired tpi is not on the table, use the universal formulas above.

Formulas for computing other dimensions are found in **Table 4**.

Unified threads by series

Unified threads are available in sizes that reflect the physical specifications of the individual thread forms. Larger diameters, naturally, tend to have fewer threads per inch, while smaller diameters have more. The "constant pitch" thread series, 4-UN/R, 6-UN/R, 8-UN/R, 8-UNJ, 12-UN/R, 12-UNJ, 16-UN/R, 16-UNJ, 20-UN/R, 28-UN/R, and 32-UN/R are commonly referred to as 8-thread series, 12-thread series, etc. The 8-, 12-, and 16-thread series are the preferred constant pitch sizes. The UN/UNR 8-series was designed for securing high-temperature, high-pressure joints, and is commonly used instead of UNC threads for applications where sizes in excess of one-inch are required. The UN/UNR 12-thread series was designed for use on boilers, but is commonly used in place of UNF threads for applications where diameters in excess of 1.5" are required. The UN/UNR 16-thread series is applied where fine-pitch fasteners 1 $^3/_4$" or larger are required. Available diameters, broken down by form and tpi, are shown in **Table 3**.

Table 3. Size/Pitch Combinations for Unified Threads. (See Note.)

Size*	Basic Dia.	4 UN	6 UN	8 UN/UNJ	12 UN/UNJ	16 UN/UNJ	20 UN	28 UN	32 UN	UNC/UNJC	UNF/ENJF	UNEF/UNJEF	UNS	NC-5**
0	0.0600										80			
1	0.0730									64	72			
2	0.0860									56	64			
3	0.0990									48	56			
4	0.1120									40	48			
5	0.1250									40	44			
6	0.1380								C	32	40			
8	0.1640								C	32	36			
10	0.1900								F	24	32		28, 36, 40, 48. 56	
12	0.1260							F	EF	24	28	32	36, 40,48, 56	
1/4	0.2500						C	F	EF	20	28	32	24, 27, 36, 40, 48, 56	20
5/16	0.3125						▮	▮	EF	18	24	32	27, 36, 40, 48	18
3/8	0.3750					C	▮	▮	EF	16	24	32	18, 27, 36, 40	16
7/16	0.4375						F	EF	▮	14	20	28	18, 24, 27	14
1/2	0.5000						F	EF	▮	13	20	28	12, 14, 18, 24, 27	13
9/16	0.5625				C		▮	▮		12	18	24	14, 27	12
5/8	0.6250				▮		▮	▮		11	18	24	14, 27	11
11/16	0.6875				▮		▮	▮				24		
3/4	0.7500				F	EF	▮	▮		10	16	20	14, 18, 24, 27	10
13/16	0.8125					EF	▮	▮				20		
7/8	0.8750					EF	▮	▮		9	14	20	10, 18, 24, 27	9
15/16	0.9375					EF	▮	▮				20		

(Continued)

Table 3. *(Continued)* **Size/Pitch Combinations for Unified Threads.** (See Note.)

Size*	Basic Dia.	4 UN	6 UN	8 UN/UNJ	12 UN/UNJ	16 UN/UNJ	20 UN	28 UN	32 UN	UNC/UNJC	UNF/ENJF	UNEF/UNJEF	UNS	NC-5**
1	1.0000			C	F		EF			8	12	20	10, 14, 18, 24, 27	8
1 1/16	1.0625											18		
1 1/8	1.1250				F					7	12	18	10, 14, 24	7
1 3/16	1.1875											18		
1 1/4	1.2500				F					7	12	18	10, 14, 24	7
1 5/16	1.3125											18		
1 3/8	1.3750		C		F					6	12	18	10, 14	6
1 7/16	1.4375											18		
1 1/2	1.5000		C		F					6	12	18	10, 14, 24	6
1 9/16	1.5625											18		
1 5/8	1.6250											18	10, 14, 24	
1 11/16	1.6875											18		
1 3/4	1.7500									5			10, 14, 18	
1 13/16	1.8125													
1 7/8	1.8750												10, 14, 18	
1 15/16	1.9375													
2	2.0000									4.5			10, 14, 18	
2 1/16	2.0625												16	
2 1/8	2.1250													
2 3/16	2.1875												16	
2 1/4	2.2500									4.5			14, 18	
2 5/16	2.3125												16	
2 3/8	2.3750													
2 7/16	2.4375												16	
2 1/2	2.5000	C								4			10, 14, 18	
2 5/8	2.6250													
2 3/4	2.7500	C								4			10, 14, 18	
2 7/8	2.8750													
3	3.0000	C								4			10, 14, 18	
3 1/8	3.1250													
3 1/4	3.2500	C								4			10, 14, 18	
3 3/8	3.3750													
3 1/2	3.5000	C								4			10, 14, 18	
3 5/8	3.6250													
3 3/4	3.7500	C								4			10, 14, 18	
3 7/8	3.8750													
4	4.0000	C								4			10, 14	

(Continued)

Table 3. *(Continued)* Size/Pitch Combinations for Unified Threads. (See Note.)

Size*	Basic Dia.	4 UN	6 UN	8 UN/UNJ	12 UN/UNJ	16 UN/UNJ	20 UN	28 UN	32 UN	UNC/UNJC	UNF/ENJF	UNEF/UNJEF	UNS	NC-5**
4 1/8	4.1250	■	■	■	■	■								
4 1/4	4.2500	■	■	■	■	■							10, 14	
4 3/8	4.3750	■	■	■	■	■								
4 1/2	4.5000	■	■	■	■	■							10, 14	
4 5/8	4.6250	■	■	■	■	■								
4 3/4	4.7500	■	■	■	■	■							10, 14	
4 7/8	4.8750	■	■	■	■	■								
5	5.0000	■	■	■	■	■							10, 14	
5 1/8	5.1250	■	■	■	■	■								
5 1/4	5.2500	■	■	■	■	■							10, 14	
5 3/8	5.3750	■	■	■	■	■								
5 1/2	5.5000	■	■	■	■	■							10, 14	
5 5/8	5.6250	■	■	■	■	■								
5 3/4	5.7500	■	■	■	■	■							10, 14	
5 7/8	5.8750	■	■	■	■	■								
6	6.0000	■	■	■	■	■							10, 14	

*Primary sizes are in normal type. Secondary sizes are in italics.
Note: A "C" in a box means that the constant pitch thread of that diameter is coarse. "F" designates fine, and "EF" extra fine. Dark boxes indicate availability. A blank box means that the thread combination is not available. A number in a box indicates the tpi of the thread of that diameter.

Formulas for 60° threads

Because of the similarities of the ISO Metric thread and the Unified thread, their dimensions can be determined with shared equations. Traditionally, metric thread equations have been stated as a fractional expression of the full height of the thread (the fundamental triangle) and Unified equations have been expressed as a decimal percentage of the pitch. Where applicable, both forms are given in **Table 4**.

Table 4. Basic Formulas for Calculating Unified and ISO 68 (1973) Thread Dimensions. *(See Notes Below.)*

	Symbol	Dimension	Equation	Remarks
1.	P	Pitch (number of threads per unit of length).	$P = 1 \div$ threads per inch, or $P = 1 \div$ threads per mm	To convert pitch in inches to metric pitch, multiply by 25.4. To convert metric pitch to pitch in inches, multiply by 0.03937.
2.	H	Height of fundamental triangle (sharp V-thread).	$H = 0.866025404P$ or $H = P \times \cos 30°$	The thread height if the crest and root were sharp, rather than having a slight flat.
3.	D	Major dia., int. thread.		Also called Basic Major Diameter. It is equal to the basic thread designation.

(Continued)

Table 4. *(Continued)* **Basic Formulas for Calculating Unified and ISO 68 (1973) Thread Dimensions.** (See Notes Below.)

	Symbol	Dimension	Equation	Remarks
4.	d (max)	Major dia. (max.), ext. thread.	See Table 5	For Class 3A, equal to the basic thread designation. For Class 1A and 2A, equal to the basic thread dimension minus the allowance.
5.	d (min)	Major dia. (min.), ext. thread.	$d \text{ (min)} = d \text{ (max)} - 0.06 \times \sqrt[3]{P^2}$	This equation will not provide 100% accurate results for all thread classes. See Table 5 for exact dimensions.
6.	D_1 (min)	Minor dia. (min.), int. thread.	$D_1 \text{ (min)} = D - 1.082532P$	The major diameter minus the depth of thread. Corresponds to recommended tap drill size. Also see Table 2.
7.	d_1 (max)	Minor dia. (max.), ext. thread.	$d_1 \text{ (max)} = d - 1.226869P$	See Note [1] below.
8.	D_1 (max)	Minor dia. (max.), int. thread.	$D_1 \text{ (max)} = D_1 \text{ (min)} + [0.05 \times \sqrt[3]{P^2} + 0.03P \div D] - 0.002$	See Note [2] below.
9.	d_1 (min)	Minor dia. (min.), ext. thread.	$d_1 \text{ (min)} = d_2 \text{ (min)} - 0.56580P$	See Note [1] below.
10.	d_1 UNR	Minor dia., ext. UNR thread.	See Table 2	See Note [1] below.
11.	D_1 UNJ	Minor dia., int. UNJ and MJ thread.	$D_1 = D - 0.974279P$	Controlled root radius thread.
12.	D_2 (min)	Pitch dia. (min.), int. thread.	$D_2 \text{ (min)} = D - 0.649519P$	Often estimated as $D - 2 \times h_{as}$. Also see Table 2.
13.	d_2 (max)	Pitch dia. (max.), ext. thread.	$d_2 \text{ (max)} = d \text{ (max)} - 0.649519P$	Often estimated as $d - 2 \times h_{dn}$.
14.	D_2 (max)	Pitch dia. (max.), int. thread.	See Tables	
15.	d_2 (min)	Pitch dia. (min.), ext. thread.	See Tables	
16.	$h_{as}*$ $h_{dn}*$	Addendum, ext. thread. Dedendum, ext. and int. thread. UN, UNR (max.)	h_{as} or $h_{dn} = 3/8 \times H$, or h_{as} or $h_{dn} = 0.3247595P$	Twice $h_{as} = 3/4 \times H$, or $0.649519P$. One-half $h_{as} = 3H \div 16$, or $0.16238P$.
17.	h_{ds}	Dedendum, ext. thread. UNR (min.)	$h_{ds} = H/3$, or $h_{ds} = 0.288867P$	UNR only.
18.	h_{ds}	Dedendum, ext. thread. UNJ (max.)	$h_{ds} = 0.28290P$	UNJ only.
19.	h_{ds}	Dedendum, ext. thread. UNJ (min.)	$h_{ds} = 7/24 \times H$, or $h_{ds} = 0.252588P$	UNJ only.
20.	$h_{an}*$ $f_{cn}*$ $f_{rs}*$	Addendum, int. thread. Truncation, int. thread crest. Truncation, ext. thread root.	h_{an}, f_{cn}, or $f_{rs} = H/4$, or h_{an}, f_{cn}, or $f_{rs} = 0.216506351P$	f_{cn} is the radial distance from the apex of the fundamental triangle to the flat at the crest of the internal thread. Does not apply to UNJ and MJ threads. f_{rs} is the radial distance from the apex of the fundamental triangle to the flat at the root of the external thread.

(Continued)

Table 4. *(Continued)* **Basic Formulas for Calculating Unified and ISO 68 (1973) Thread Dimensions.** (See Notes Below.)

	Symbol	Dimension	Equation	Remarks
21.	h_n* h_e* h_s* UN	Height of int. thread. Depth of thread engagement. Height of ext. UN and ISO metric thread.	h_n or $h_e = 5/8 \times H$, or h_n or $h_e = 0.541265877P$	Does not apply to UNJ and MJ threads. See separate listing. Twice $h_n = 5/4 \times H$, or $1.082532P$.
22.	h_s* UNR	Height of ext. UNR thread.	$h_s = 0.59539P$	UNR only.
23.	h_n* UNJ h_e* UNJ	Height of int. UNJ thread. Depth of UNJ thread engagement.	h_n or $h_e = 9H \div 16$, or h_n or $h_e = 0.48714P$	UNJ and MJ only.
24.	h_s* UNJ	Height of ext. UNJ thread.	$h_s = 2H \div 3$, or $h_s = 0.57735P$	UNJ and MJ only.
25.	r (max)	Root radius (max.), ext. thread.	$r = H \div 6$, or $r = 0.14434P$	Does not apply to UNJ and MJ threads.
26.	r (min)	Root radius/flat (min.), ext. thread.	$r = 0.125P$	Does not apply to UNR, UNJ and MJ threads.
27.	r UNR	Root radius (min.), ext. UNR thread.	$r = H \div 8$, or $r = 0.10825P$	UNR only.
28.	r UNR	Root radius (max.), ext. UNR thread.	$r = H \div 6$, or $r = 0.14434P$	UNR only.
29.	r UNJ	Root radius (min.), ext UNJ and MJ only.	$r = 0.15011P$	UNJ and MJ only.
30.	r UNJ	Root radius (max.), UNJ and MJ only.	$r = 5H \div 24$, or $r = 0.18042P$	Also height from sharp "V" to ext. thread root. UNJ and MJ only.
31.	$f_{cn}*$ UNJ	Truncation, int. thread crest, UNJ thread.	$f_{cn} = 5H \div 16$, or $f_{cn} = 0.27063P$	UNJ and MJ only.
32.	$f_{cs}*$ $f_{rn}*$	Truncation, ext. thread crest. Truncation, int. thread root.	f_{cs} or $f_{rn} = H \div 8$, or f_{cs} or $f_{rn} = 0.10825$	f_{cs} is the radial distance from the apex of the fundamental triangle to flat at the crest of the external thread. f_{rn} is the radial distance from the apex of the fundamental triangle to the flat at the root of the internal thread. Does not apply to UNR.
33.	$f_{rn}*$ UNR	Truncation, int. thread root, UNR.	$f_{rn} = 0.16238P$	UNR only.
34.	$F_{cs}*$ $F_{rn}*$	Flat, crest of ext. thread. Flat, root of int. thread.	F_{cs} or $F_{rn} = P \div 8$, or F_{cs}, or $F_{rn} = 0.125P$	
35.	$F_{cn}*$	Flat, int. thread crest.	$F_{cn} = 0.25P$	Does not apply to UNJ and MJ threads.
36.	$F_{cn}*$ UNJ	Flat, int. thread crest.	$F_{cn} = 5P \div 16$, or $F_{cn} = 0.3125P$	UNJ and MJ only.
37.	NA	See remarks.	$11/12 \times H$, or $0.79386P$	Use to calculate the difference between D (max.) and D_2 (max.).

(Continued)

Table 4. *(Continued)* **Basic Formulas for Calculating Unified and ISO 68 (1973) Thread Dimensions.**
(See Notes Below.)

	Symbol	Dimension	Equation	Remarks
38.	NA	See remarks.	$0.711325H$, or $0.616025P$	Use to calculate the difference between the pitch dia. (min.) and minor dia. (min.) of external thread for $0.125P$ root radius.
39.	NA	See remarks. For UNJ and MJ only.	$0.6533H$, or $0.56580P$	Use to calculate the difference between the pitch dia. (min.) and minor dia. (min.) of external UNJ and MJ thread.
40.	NA	See remarks.	$H/2$, or $0.433013P$	Use to calculate the difference between the pitch dia. (max.) and minor dia. (max.) of external thread, and between the pitch dia. (min.) and minor dia. (min.) of internal thread.
41.	L *	Lead.	$1 \div 1/P$	$1/P$ is the number of turns per unit of length (threads per inch).
42.	λ *	Lead angle.	$\tan \lambda = L \div (\pi \times \text{pitch dia.})$	
43.	ψ *	Helix angle.	$\cot \psi = L \div (\pi \times \text{pitch dia.})$	Sometimes called the rake angle.
44.	NA	Area of minor dia., ext. thread.	$0.7854 \times d_1{}^2$	$\pi \div 4 = 0.7854$. Expressed in in.^{2} or mm^2.
45.	A_s	Tensile stress area.	$A_s = \pi \times ([d_2 \div 2] - [3H \div 16])^2$, or $A_s = 0.7854 \times (d - [0.9743 \div 1/P])^2$	$\pi \div 4 = 0.7854$. $1/P$ = number of threads per inch. Expressed in in.^{2} or mm^2.
46.	LE	Length of thread engagement.	$LE = 4\,A_s \div (\pi \times d_{2\,BASIC})$	Based on combined shear failure of external and internal threads. See Note [3] below.
47.	AS_n	Shear area, internal threads.	$AS_n = \pi \times D_{2\,BASIC} \times (3LE \div 4)$	For use only when d is greater than 0.250 inch. See Note [4] below.
48.	AS_s	Shear area, external threads.	$AS_s = \pi \times d_{2\,BASIC} \times (5/8 \times LE)$	See Note [4] below.

Unless otherwise noted, formulas are for basic UN/ISO 68 thread form. Dimensions derived from these formulas for ISO 68 and M Profile threads do not compensate for tolerance grade variations.

* Symbol is traditional.

Note [1]: This specification has been stated in various ways in ANSI/ASME and Federal Standards. In FED-STD-H28, a single specification is provided for d_1 that is approximately the average of d_1 (max) and d_1 (min). These figures are given in the tables in this Handbook. In ANSI B1.1-1989, the specification given is stated as being for "UNR Minor Diameter (max)" which can be obtained by using the constant for d_1 that is specified in Table 2.

Note [2]: For 13 threads per inch or finer, this tolerance will not exceed $0.259809P$, nor shall it be less than $0.135315P$. For coarser threads, the tolerance will be equal to $0.120P$. This equation will not return values equal to the stated maximum and minimum figures provided in the tables, but they will be very close and will not exceed the values stated in the tables.

Note [3]: For a hollow part, subtract $0.7854\,d_h{}^2$ from A_s. d_h is the hole diameter on the externally threaded part.

Note [4]: Formulas given for AS_n and AS_s are simplified and are based on empirical data that vary from the minimum material shear areas.

Unified miniature threads

UNM threads are intended for use as general purpose fastening screws in watches, instruments, and miniature mechanisms. The diameter of UNM threads ranges from 0.30 mm to 1.40 mm (0.0118 to 0.0551 in.). Therefore, this thread form supplements the standard Unified series that begins at 0.060 in. The fourteen sizes are systematically distributed to provide a uniformly proportioned selection throughout the range. Primary sizes are recommended, but for restrictive conditions, secondary sizes may be specified. Since the standard is specified in metric units and then converted to inch units, the metric specifications take preference. Formulas for UNM threads are found in **Table 11**, and specifications are provided in **Table 12**. The thread form is illustrated in *Figure 5*. It should be noted that the maximum material limits given in Table 12 apply to both uncoated and coated threads. It is essential that measurable coatings remain with the stated dimensions since no allowance is provided between the limits of the internal and external threads. *(Text continued on p. 70)*

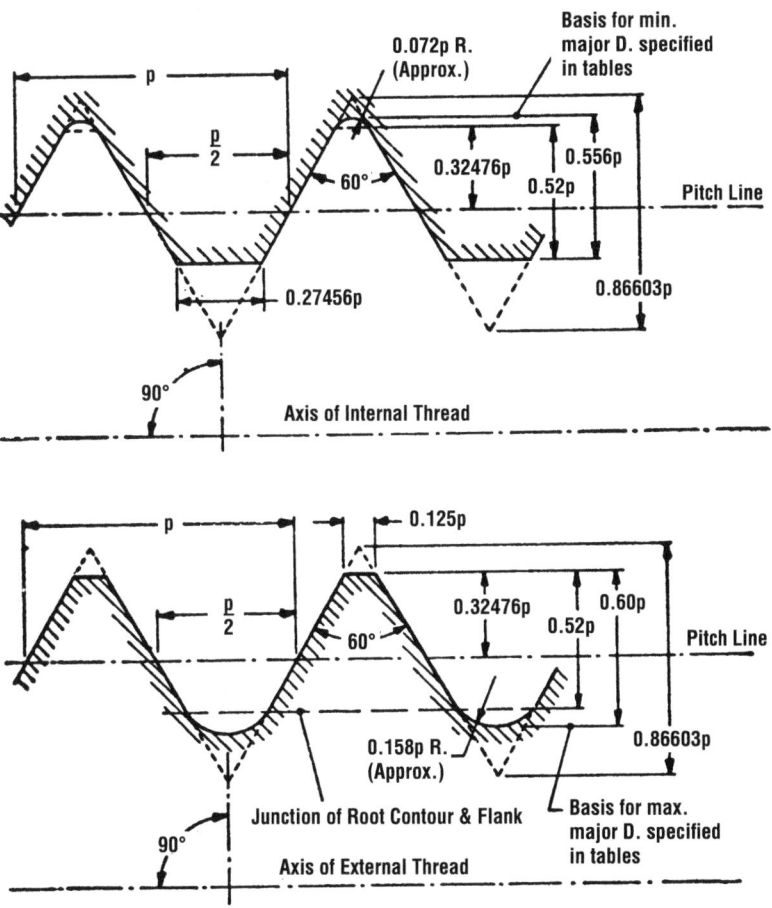

Figure 5. UN miniature thread form.

Table 5. Unified Screw Threads. External Dimensions. Classes 1A, 2A, and 3A.

Size, TPI, and Series	Class	Allowance	Major Diameter		Pitch Diameter		Minor Dia.
			max	min	max	min	
0-80 UNF	2A	.0005	.0595	.0563	.0514	.0496	.0442
	3A	-	.0600	.0568	.0519	.0506	.0447
1-64 UNC	2A	.0006	.0724	.0686	.0623	.0603	.0532
	3A	-	.0730	.0692	.0629	.0614	.0538
1-72 UNF	2A	.0006	.0724	.0689	.0634	.0615	.0554
	3A	-	.0730	.0695	.0640	.0626	.0560
2-56 UNC	2A	.0006	.0854	.0813	.0738	.0717	.0635
	3A	-	.0860	.0819	.0744	.0728	.0641
2-64 UNF	2A	.0006	.0854	.0816	.0753	.0733	.0662
	3A	-	.0860	.0822	.0759	.0744	.0668
3-48 UNC	2A	.0007	.0983	.0938	.0848	.0825	.0727
	3A	-	.0990	.0945	.0855	.0838	.0734
3-56 UNF	2A	.0007	.0983	.0942	.0867	.0845	.0764
	3A	-	.0990	.0949	.0874	.0858	.0771
4-40 UNC	2A	.0008	.1112	.1061	.0950	.0925	.0805
	3A	-	.1120	.1069	.0958	.0939	.0813
4-48 UNF	2A	.0007	.1113	.1068	.0978	.0954	.0857
	3A	-	.1120	.1075	.0985	.0967	.0864
5-40 UNC	2A	.0008	.1242	.1191	.1080	.1054	.0935
	3A	-	.1250	.1199	.1088	.1069	.0943
5-44 UNF	2A	.0007	.1243	.1195	.1095	.1070	.0964
	3A	-	.1250	.1202	.1102	.1083	.0971
6-32 UNC	2A	.0008	.1372	.1312	.1169	.1141	.0989
	3A	-	.1380	.1320	.1177	.1156	.0997
6-40 UNF	2A	.0008	.1372	.1321	.1210	.1184	.1065
	3A	-	.1380	.1329	.1218	.1198	.1073
8-32 UNC	2A	.0009	.1631	.1571	.1428	.1399	.1248
	3A	-	.1640	.1580	.1437	.1415	.1257
8-36 UNF	2A	.0008	.1632	.1577	.1452	.1424	.1291
	3A	-	.1640	.1585	.1460	.1439	.1299
10-24 UNC	2A	.0010	.1890	.1818	.1619	.1586	.1379
	3A	-	.1900	.1828	.1629	.1604	.1389
10-28 UNS	2A	.0010	.1890	.1825	.1658	.1625	.1452
10-32 UNF	2A	.0009	.1891	.1831	.1688	.1658	.1508
	3A	-	.1900	.1840	.1697	.1674	.1517
10-36 UNS	2A	.0009	.1891	.1836	.1711	.1681	.1550
10-40 UNS	2A	.0009	.1891	.1840	.1729	.1700	.1584
10-48 UNS	2A	.0008	.1892	.1847	.1757	.1731	.1636
10-56 UNS	2A	.0007	.1893	.1852	.1777	.1752	.1674
12-24 UNC	2A	.0010	.2150	.2078	.1879	.1845	.1639
	3A	-	.2160	.2088	.1889	.1863	.1649
12-28 UNF	2A	.0010	.2150	.2085	.1918	.1886	.1712
	3A	-	.2160	.2095	.1928	.1904	.1722
12-32 UNEF	2A	.0009	.2151	.2091	.1948	.1917	.1768
	3A	-	.2160	.2100	.1957	.1933	.1777

(Continued)

Table 6. Unified Screw Threads. Internal Dimensions. Classes 1B, 2B, and 3B.

Size, TPI, and Series	Class	Minor Diameter		Pitch Diameter		Major Dia.	Tap Drill*
		min	max	min	max		
0-80 UNF	2B	.0465	.0514	.0519	.0542	.0600	56
	3B	.0465	.0514	.0519	.0536	.0600	
1-64 UNC	2B	.0561	.0623	.0629	.0655	.0730	54
	3B	.0561	.0623	.0629	.0648	.0730	
1-72 UNF	2B	.0580	.0635	.0640	.0665	.0730	1.5mm
	3B	.0580	.0635	.0640	.0659	.0730	
2-56 UNC	2B	.0667	.0737	.0744	.0772	.0860	51
	3B	.0667	.0737	.0744	.0765	.0860	
2-64 UNF	2B	.0691	.0753	.0759	.0786	.0860	50
	3B	.0691	.0753	.0759	.0779	.0860	
3-48 UNC	2B	.0764	.0845	.0855	.0885	.0990	48
	3B	.0764	.0845	.0855	.0877	.0990	
3-56 UNF	2B	.0797	.0865	.0874	.0902	.0990	46
	3B	.0797	.0865	.0874	.0895	.0990	
4-40 UNC	2B	.0849	.0939	.0958	.0991	.1120	44
	3B	.0849	.0939	.0958	.0982	.1120	
4-48 UNF	2B	.0894	.0968	.0985	.1016	.1120	2.3mm
	3B	.0894	.0968	.0985	.1008	.1120	
5-40 UNC	2B	.0979	.1062	.1088	.1121	.1250	39
	3B	.0979	.1062	.1088	.1113	.1250	
5-44 UNF	2B	.1004	.1079	.1102	.1134	.1250	38
	3B	.1004	.1079	.1102	.1126	.1250	
6-32 UNC	2B	.104	.114	.1177	.1214	.1380	36
	3B	.1040	.1140	.1177	.1204	.1380	
6-40 UNF	2B	.111	.119	.1218	.1252	.1380	33
	3B	.1110	.1186	.1218	.1243	.1380	
8-32 UNC	2B	.130	.139	.1437	.1475	.1640	3.4mm
	3B	.1300	.1389	.1437	.1465	.1640	
8-36 UNF	2B	.134	.142	.1460	.1496	.1640	29
	3B	.1340	1416	.1460	.1487	.1640	
10-24 UNC	2B	.145	.156	.1629	.1672	.1900	3.7mm
	3B	.1450	.1555	.1629	.1661	.1900	
10-28 UNS	2B	.151	.160	.1668	.1711	.1900	
10-32 UNF	2B	.156	.164	.1697	.1736	.1900	5/32
	3B	.1560	.1641	.1697	.1726	.1900	
10-36 UNS	2B	.160	.166	.1720	.1759	.1900	
10-40 UNS	2B	.163	.169	.1738	.1775	.1900	
10-48 UNS	2B	.167	.172	.1765	.1799	.1900	
10-56 UNS	2B	.171	.175	.1784	.1816	.1900	
12-24 UNC	2B	.171	.181	.1889	.1933	.2160	17
	3B	.1710	.1807	.1889	.1922	.2160	
12-28 UNF	2B	.177	.186	.1928	.1970	.2160	15
	3B	.1770	.1857	.1928	.1959	.2160	
12-32 UNEF	2B	.182	.190	.1957	.1998	.2160	14
	3B	.1820	.1895	.1957	.1988	.2160	

(Continued)

Table 5. *(Continued)* **Unified Screw Threads. External Dimensions. Classes 1A, 2A, and 3A.**

Size, TPI, and Series	Class	Allowance	Major Diameter max	Major Diameter min	Pitch Diameter max	Pitch Diameter min	Minor Dia.
12-36 UNS	2A	.0009	.2151	.2096	.1971	.1941	.1810
12-40 UNS	2A	.0009	.2151	.2100	.1989	.1960	.1844
12-48 UNS	2A	.0008	.2152	.2107	.2017	.1991	.1896
12-56 UNS	2A	.0007	.2153	.2112	.2037	.2012	.1934
1/4-20 UNC	1A	.0011	.2489	.2367	.2164	.2108	.1876
	2A	.0011	.2489	.2408	.2164	.2127	.1876
	3A	-	.2500	.2419	.2175	.2147	.1887
1/4-24 UNS	2A	.0011	.2489	.2417	.2218	.2181	.1978
1/4-27 UNS	2A	.0010	.2490	.2423	.2249	.2214	.2036
1/4-28 UNF	1A	.0010	.2490	.2392	.2258	.2208	.2052
	2A	.0010	.2490	.2425	.2258	.2225	.2052
	3A	-	.2500	.2435	.2268	.2243	.2062
1/4-32 UNEF	2A	.0010	.2490	.2430	.2287	.2255	.2107
	3A	-	.2500	.2440	.2297	.2273	.2117
1/4-36 UNS	2A	.0009	.2491	.2436	.2311	.2280	.2150
1/4-40 UNS	2A	.0009	.2491	.2440	.2329	.2300	.2184
1/4-48 UNS	2A	.0008	.2492	.2447	.2357	.2330	.2236
1/4-56 UNS	2A	.0008	.2492	.2451	.2376	.2350	.2273
5/16-18 UNC	1A	.0012	.3113	.2982	.2752	.2691	.2431
	2A	.0012	.3113	.3026	.2752	.2712	.2431
	3A	-	.3125	.3038	.2764	.2734	.2443
5/16-20 UN	2A	.0012	.3113	.3032	.2788	.2748	.2500
	3A	-	.3125	.3044	.2800	.2770	.2512
5/16-24 UNF	1A	.0011	.3114	.3006	.2843	.2788	.2603
	2A	.0011	.3114	.3042	.2843	.2806	.2603
	3A	-	.3125	.3053	.2854	.2827	.2614
5/16-27 UNS	2A	.0010	.3115	.3048	.2874	.2839	.2661
5/16-28 UN	2A	.0010	.3115	.3050	.2883	.2849	.2677
	3A	-	.3125	.3060	.2893	.2867	.2687
5/16-32 UNEF	2A	.0010	.3115	.3055	.2912	.2880	.2732
	3A	-	.3125	.3065	.2922	.2898	.2742
5/16-36 UNS	2A	.0009	.3116	.3061	.2936	.2905	.2775
5/16-40 UNS	2A	.0009	.3116	.3065	.2954	.2925	.2809
5/16-48 UNS	2A	.0008	.3117	.3072	.2982	.2955	.2861
3/8-16 UNC	1A	.0013	.3737	.3595	.3331	.3266	.2970
	2A	.0013	.3737	.3643	.3331	.3287	.2970
	3A	-	.3750	.3656	.3344	.3311	.2983
3/8-18 UNS	2A	.0013	.3737	.3650	.3376	.3333	.3055
3/8-20 UN	2A	.0012	.3738	.3657	.3413	.3372	.3125
	3A	-	.3750	.3669	.3425	.3394	.3137
3/8-24 UNF	1A	.0011	.3739	.3631	.3468	.3411	.3228
	2A	.0011	.3739	.3667	.3468	.3430	.3228
	3A	-	.3750	.3678	.3479	.3450	.3239
3/8-27 UNS	2A	.0011	.3739	.3672	.3498	.3462	.3285

(Continued)

Table 6. *(Continued)* **Unified Screw Threads. Internal Dimensions. Classes 1B, 2B, and 3B.**

Size, TPI, and Series	Class	Minor Diameter		Pitch Diameter		Major Dia.	Tap Drill*
		min	max	min	max		
12-36 UNS	2B	.186	.192	.1980	.2019	.2160	
12-40 UNS	2B	.189	.195	.1998	.2035	.2160	
12-48 UNS	2B	.193	.198	.2025	.2059	.2160	
12-56 UNS	2B	.197	.201	.2044	.2076	.2160	
1/4-20 UNC	1B	.196	.207	.2175	.2248	.2500	9
	2B	.196	.207	.2175	.2223	.2500	
	3B	.1960	.2067	.2175	.2211	.2500	
1/4-24 UNS	2B	.205	.215	.2229	.2277	.2500	
1/4-27 UNS	2B	.210	.219	.2259	.2304	.2500	
1/4-28 UNF	1B	.211	.220	.2268	.2333	.2500	5.4mm
	2B	.211	.220	.2268	.2311	.2500	
	3B	.2110	.2190	.2268	.2300	.2500	
1/4-32 UNEF	2B	.216	.224	.2297	.2339	.2500	7/32
	3B	.2160	.2229	.2297	.2328	.2500	
1/4-36 UNS	2B	.220	.226	.2320	.2360	.2500	2
1/4-40 UNS	2B	.223	.229	.2338	.2376	.2500	
1/4-48 UNS	2B	.227	.232	.2365	.2401	.2500	
1/4-56 UNS	2B	.231	.235	.2384	.2417	.2500	
5/16-18 UNC	1B	.252	.265	.2764	.2843	.3125	F
	2B	.252	.265	.2764	.2817	.3125	
	3B	.2520	.2630	.2764	.2803	.3125	
5/16-20 UN	2B	.258	.270	.2800	.2852	.3125	G
	3B	.2580	.2662	.2800	.2839	.3125	
5/16-24 UNF	1B	.267	.277	.2854	.2925	.3125	H
	2B	.267	.277	.2854	.2902	.3125	
	3B	.2670	.2754	.2854	.2890	.3125	
5/16-27 UNS	2B	.272	.281	.2884	.2929	.3125	
5/16-28 UN	2B	.274	.282	.2893	.2937	.3125	J
	3B	.2740	.2801	.2893	.2926	.3125	
5/16-32 UNEF	2B	.279	.286	.2922	.2964	.3125	K
	3B	.2790	.2847	.2922	.2953	.3125	
5/16-36 UNS	2B	.282	.289	.2945	.2985	.3125	7.25mm
5/16-40 UNS	2B	.285	.291	.2963	.3001	.3125	
5/16-48 UNS	2B	.290	.295	.2990	.3026	.3125	
3/8-16 UNC	1B	.307	.321	.3344	.3429	.3750	5/16
	2B	.307	.321	.3344	.3401	.3750	
	3B	.3070	.3182	.3344	.3387	.3750	
3/8-18 UNS	2B	.315	.328	.3389	.3445	.3750	
3/8-20 UN	2B	.321	.332	.3425	.3479	.3750	P
	3B	.3210	.3287	.3425	.3465	.3750	
3/8-24 UNF	1B	.330	.340	.3479	.3553	.3750	Q
	2B	.330	.340	.3479	.3528	.3750	
	3B	.3300	.3372	.3479	.3516	.3750	
3/8-27 UNS	2B	.335	.344	.3509	.3556	.3750	

(Continued)

Table 5. *(Continued)* **Unified Screw Threads. External Dimensions. Classes 1A, 2A, and 3A.**

Size, TPI, and Series	Class	Allowance	Major Diameter		Pitch Diameter		Minor Dia.
			max	min	max	min	
3/8-28 UN	2A	.0011	.3739	.3674	.3507	.3471	.3301
	3A	-	.3750	.3685	.3518	.3491	.3312
3/8-32 UNEF	2A	.0010	.3740	.3680	.3537	.3503	.3357
	3A	-	.3750	.3690	.3547	.3522	.3367
3/8-36 UNS	2A	.0010	.3740	.3685	.3560	.3528	.3399
3/8-40 UNS	2A	.0009	.3741	.3690	.3579	.3548	.3434
7/16-14 UNC	1A	.0014	.4361	.4206	.3897	.3826	.3485
	2A	.0014	.4361	.4258	.3897	.3850	.3485
	3A	-	.4375	.4272	.3911	.3876	.3499
7/16-16 UN	2A	.0014	.4361	.4267	.3955	.3909	.3594
	3A	-	.4375	.4281	.3969	.3935	.3608
7/16-18 UNS	2A	.0013	.4362	.4275	.4001	.3958	.3680
7/16-20 UNF	1A	.0013	.4362	.4240	.4037	.3975	.3749
	2A	.0013	.4362	.4281	.4037	.3995	.3749
	3A	-	.4375	.4294	.4050	.4019	.3762
7/16-24 UNS	2A	.0011	.4364	.4292	.4093	.4055	.3853
7/16-27 UNS	2A	.0011	.4364	.4297	.4123	.4087	.3910
7/16-28 UNEF	2A	.0011	.4364	.4299	.4132	.4096	.3926
	3A	-	.4375	.4310	.4143	.4116	.3937
7/16-32 UN	2A	.0010	.4365	.4305	.4162	.4128	.3982
	3A	-	.4375	.4315	.4172	.4147	.3992
1/2-12 UNS	2A	.0016	.4984	.4870	.4443	.4389	.3962
	3A	-	.5000	.4886	.4459	.4419	.3978
1/2-13 UNC	1A	.0015	.4985	.4822	.4485	.4411	.4041
	2A	.0015	.4985	.4876	.4485	.4435	.4041
	3A	-	.5000	.4891	.4500	.4463	.4056
1/2-14 INS	2A	.0015	.4985	.4882	.4521	.4471	.4109
1/2-16 UN	2A	.0014	.4986	.4892	.4580	.4533	.4219
	3A	-	.5000	.4906	.4594	.4559	.4233
1/2-18 UNS	2A	.0013	.4987	.4900	.4626	.4582	.4305
1/2-20 UNF	1A	.0013	.4987	.4865	.4662	.4598	.4374
	2A	.0013	.4987	.4906	.4662	.4619	.4374
	3A	-	.5000	.4919	.4675	.4643	.4387
1/2-24 UNS	2A	.0012	.4988	.4916	.4717	.4678	.4477
1/2-27 UNS	2A	.0011	.4989	.4922	.4748	.4711	.4535
1/2-28 UNEF	2A	.0011	.4989	.4924	.4757	.4720	.4551
	3A	-	.5000	.4935	.4768	.4740	.4562
1/2-32 UN	2A	.0011	.4990	.4930	.4787	.4752	.4607
	3A	-	.5000	.4940	.4797	.4771	.4617
9/16-12 UNC	1A	.0016	.5609	.5437	.5068	.4990	.4587
	2A	.0016	.5609	.5495	.5068	.5016	.4587
	3A	-	.5625	.5511	.5084	.5045	.4603
9/16-14 UNS	2A	.0015	.5610	.5507	.5146	.5096	.4734
9/16-16 UN	2A	.0014	.5611	.5517	.5205	.5158	.4844
	3A	-	.5625	.5531	.5219	.5184	.4858

(Continued)

Table 6. *(Continued)* Unified Screw Threads. Internal Dimensions. Classes 1B, 2B, and 3B.

Size, TPI, and Series	Class	Minor Diameter		Pitch Diameter		Major Dia.	Tap Drill*
		min	max	min	max		
3/8-28 UN	2B	.336	.345	.3518	.3564	.3750	R
	3B	.3360	.3426	.3518	.3553	.3750	
3/8-32 UNEF	2B	.341	.349	.3547	.3591	.3750	11/32
	3B	.3410	.3469	.3547	.3580	.3750	
3/8-36 UNS	2B	.345	.352	.3570	.3612	.3750	S
3/8-40 UNS	2B	.348	.354	.3588	.3628	.3750	
7/16-14 UNC	1B	.360	.376	.3911	.4003	.4375	9.3mm
	2B	.360	.376	.3911	.3972	.4375	
	3B	.3600	.3717	.3911	.3957	.4375	
7/16-16 UN	2B	.370	.384	.3969	.4028	.4375	3/8
	3B	.3700	.3783	.3969	.4009	.4375	
7/16-18 UNS	2B	.377	.390	.4014	.4070	.4375	
7/16-20 UNF	1B	.383	.395	.4050	.4131	.4375	W
	2B	.383	.395	.4050	.4104	.4375	
	3B	.3830	.3916	.4050	.4091	.4375	
7/16-24 UNS	2B	.392	.402	.4104	.4153	.4375	
7/16-27 UNS	2B	.397	.406	.4134	.4181	.4375	
7/16-28 UNEF	2B	.399	.407	.4143	.4189	.4375	Y
	3B	.3990	.4051	.4143	.4178	.4375	
7/16-32 UN	2B	.404	.411	.4172	.4216	.4375	Y
	3B	.4040	.4094	.4172	.4205	.4375	
1/2-12 UNS	2B	.410	.428	.4459	.4529	.5000	Z
	3B	.4100	.4223	.4459	.4511	.5000	
1/2-13 UNC	1B	.417	.434	.4500	.4597	.5000	27/64
	2B	.417	.434	.4500	.4565	.5000	
	3B	.4170	.4284	.4500	.4548	.5000	
1/2-14 INS	2B	.423	.438	.4536	.4601	.5000	
1/2-16 UN	2B	.432	.446	.4594	.4655	.5000	7/16
	3B	.4320	.4408	.4594	.4640	.5000	
1/2-18 UNS	2B	.440	.453	.4639	.4697	.5000	
1/2-20 UNF	1B	.446	.457	.4675	.4759	.5000	11.4mm
	2B	.446	.457	.4675	.4731	.5000	
	3B	.4460	.4537	.4675	.4717	.5000	
1/2-24 UNS	2B	.455	.465	.4729	.4780	.5000	
1/2-27 UNS	2B	.460	.469	.4759	.4807	.5000	
1/2-28 UNEF	2B	.461	.470	.4768	.4816	.5000	15/32
	3B	.4610	.4676	.4768	.4804	.5000	
1/2-32 UN	2B	.466	.474	.4797	.4842	.5000	15/32
	3B	.4660	.4719	.4797	.4831	.5000	
9/16-12 UNC	1B	.472	.490	.5084	.5186	.5625	15/32
	2B	.472	.490	.5084	.5152	.5625	
	3B	.4720	.4843	.5084	.5135	.5625	
9/16-14 UNS	2B	.485	.501	.5161	.5226	.5625	
9/16-16 UN	2B	.495	.509	.5219	.5280	.5625	1/2
	3B	.4950	.5033	.5219	.5265	.5625	

(Continued)

Table 5. *(Continued)* **Unified Screw Threads. External Dimensions. Classes 1A, 2A, and 3A.**

Size, TPI, and Series	Class	Allowance	Major Diameter		Pitch Diameter		Minor Dia.
			max	min	max	min	
9/16-18 UNF	1A	.0014	.5611	.5480	.5250	.5182	.4929
	2A	.0014	.5611	.5524	.5250	.5205	.4929
	3A	-	.5625	.5538	.5264	.5230	.4943
9/16-20 UN	2A	.0013	.5612	.5531	.5287	.5245	.4999
	3A	-	.5625	.5544	.5300	.5268	.5012
9/16-24 UNEF	2A	.0012	.5613	.5541	.5342	.5303	.5102
	3A	-	.5625	.5553	.5354	.5325	.5114
9/16-27 UNS	2A	.0011	.5614	.5547	.5373	.5336	.5160
9/16-28 UN	2A	.0011	.5614	.5549	.5382	.5345	.5176
	3A	-	.5625	.5560	.5393	.5365	.5187
9/16-32	2A	.0010	.5615	.5555	.5412	.5377	.5232
	3A	-	.5625	.5565	.5422	.5396	.5242
5/8-11 UNC	1A	.0016	.6234	.6052	.5644	.5561	.5119
	2A	.0016	.6234	.6113	.5644	.5589	.5119
	3A	-	.6250	.6129	.5660	.5619	.5135
5/8-12 UN	2A	.0016	.6234	.6120	.5693	.5639	.5212
	3A	-	.6250	.6136	.5709	.5688	.5228
5/8-14 UNS	2A	.0015	.6235	.6132	.5771	.5720	.5359
5/8-16 UN	2A	.0014	.6236	.6142	.5830	.5782	.5469
	3A	-	.6250	.6156	.5844	.5808	.5483
5/8-18 UNF	1A	.0014	.6236	.6105	.5875	.5805	.5554
	2A	.0014	.6236	.6149	.5875	.5828	.5554
	3A	-	.6250	.6163	.5889	.5854	.5568
5/8-20 UN	2A	.0013	.6237	.6156	.5912	.5869	.5624
	3A	-	.6250	.6169	.5925	.5893	.5637
5/8-24 UNEF	2A	.0012	.6238	.6166	.5967	.5927	.5727
	3A	-	.6250	.6178	.5979	.5949	.5739
5/8-27 UNS	2A	.0011	.6239	.6172	.5998	.5960	.5785
5/8-28 UN	2A	.0011	.6239	.6174	.6007	.5969	.5801
	3A	-	.6250	.6185	.6018	.5990	.5812
5/8-32 UN	2A	.0011	.6239	.6179	.6036	.6000	.5856
	3A	-	.6250	.6190	.6047	.6020	.5867
11/16-12 UN	2A	.0016	.6859	.6745	.6318	.6264	.5837
	3A	-	.6875	.6761	.6334	.6293	.5853
11/16-16 UN	2A	.0014	.6861	.6767	.6455	.6407	.6094
	3A	-	.6875	.6781	.6469	.6433	.6108
11/16-20 UN	2A	.0013	.6862	.6781	.6537	.6494	.6249
	3A	-	.6875	.6794	.6550	.6518	.6262
11/16-24 UNEF	2A	.0012	.6863	.6791	.6592	.6552	.6352
	3A	-	.6875	.6803	.6604	.6574	.6364
11/16-28 UN	2A	.0011	.6864	.6799	.6632	.6594	.6426
	3A	-	.6875	.6810	.6643	.6615	.6437
11/16-32 UN	2A	.0011	.6864	.6804	.6661	.6625	.6481
	3A	-	.6875	.6815	.6672	.6645	.6492

(Continued)

Table 6. *(Continued)* **Unified Screw Threads. Internal Dimensions. Classes 1B, 2B, and 3B.**

Size, TPI, and Series	Class	Minor Diameter		Pitch Diameter		Major Dia.	Tap Drill*
		min	max	min	max		
9/16-18 UNF	1B	.502	.515	.5264	.5253	.5625	1/2
	2B	.502	.515	.5264	.5323	.5625	
	3B	.5020	.5106	.5264	.5308	.5625	
9/16-20 UN	2B	.508	.520	.5300	.5355	.5625	33/64
	3B	.5080	.5162	.5300	.5341	.5625	
9/16-24 UNEF	2B	.517	.527	.5354	.5405	.5625	33/64
	3B	.5170	.5244	.5354	.5392	.5625	
9/16-27 UNS	2B	.522	.531	.5384	.5432	.5625	
9/16-28 UN	2B	.524	.532	.5393	.5441	.5625	17/32
	3B	.5240	.5301	.5393	.5429	.5625	
9/16-32	2B	.529	.536	.5422	.5467	.5625	19/32
	3B	.5290	.5344	.5422	.5456	.5625	
5/8-11 UNC	1B	.527	.546	.5660	.5767	.6250	17/32
	2B	.527	.546	.5660	.5732	.6250	
	3B	.5270	.5391	.5660	.5714	.6250	
5/8-12 UN	2B	.535	.553	.5709	.5780	.6250	35/64
	3B	.5350	.5463	.5709	.5762	.6250	
5/8-14 UNS	2B	.548	.564	.5786	.5852	.6250	
5/8-16 UN	2B	.557	.571	.5844	.5906	.6250	9/16
	3B	.5570	.5658	.5844	.5890	.6250	
5/8-18 UNF	1B	.565	.578	.5889	.5980	.6250	9/16
	2B	.565	.578	.5889	.5949	.6250	
	3B	.5650	.5730	.5889	.5934	.6250	
5/8-20 UN	2B	.571	.582	.5925	.5981	.6250	37/64
	3B	.5710	.5787	.5925	.5967	.6250	
5/8-24 UNEF	2B	.580	.590	.5979	.6031	.6250	37/64
	3B	.5800	.5869	.5979	.6018	.6250	
5/8-27 UNS	2B	.585	.594	.6009	.6059	.6250	
5/8-28 UN	2B	.586	.595	.6018	.6067	.6250	19/32
	3B	.5860	.5936	.6018	.6055	.6250	
5/8-32 UN	2B	.591	.599	.6047	.6093	.6250	19/32
	3B	.5910	.5969	.6047	.6082	.6250	
11/16-12 UN	2B	.597	.615	.6334	.6405	.6875	19/32
	3B	.5970	.6085	.6334	.6387	.6875	
11/16-16 UN	2B	.620	.634	.6469	.6531	.6875	5/8
	3B	.6200	.6283	.6469	.6515	.6875	
11/16-20 UN	2B	.633	.645	.6550	.6606	.6875	41/64
	3B	.6330	.6412	.6550	.6592	.6875	
11/16-24 UNEF	2B	.642	.652	.6604	.6656	.6875	41/64
	3B	.6420	.6494	.6604	.6643	.6875	
11/16-28 UN	2B	.649	.657	.6643	.6692	.6875	21/32
	3B	.6490	.6551	.6643	.6680	.6875	
11/16-32 UN	2B	.654	.661	.6672	.6718	.6875	21/32
	3B	.6540	.6594	.6672	.6707	.6875	

(Continued)

Table 5. *(Continued)* **Unified Screw Threads. External Dimensions. Classes 1A, 2A, and 3A.**

Size, TPI, and Series	Class	Allowance	Major Diameter		Pitch Diameter		Minor Dia.
			max	min	max	min	
3/4-10 UNC	1A	.0018	.7482	.7288	.6832	.6744	.6255
	2A	.0018	.7482	.7353	.6832	.6773	.6773
	3A	-	.7500	.7371	.6850	.6806	.6273
3/4-12 UN	2A	.0017	.7483	.7369	.6942	.6887	.6461
	3A	-	.7500	.7386	.6959	.6918	.6478
3/4-14 UNS	2A	.0015	.7485	.7382	.7021	.6970	.6609
3/4-16 UNF	1A	.0015	.7485	.7343	.7079	.7004	.6718
	2A	.0015	.7485	.7391	.7079	.7029	.6718
	3A	-	.7500	.7406	.7094	.7056	.6733
3/4-18 UNS	2A	.0014	.7486	.7399	.7125	.7079	.6804
3/4-20 UNEF	2A	.0013	.7487	.7406	.7162	.7118	.6874
	3A	-	.7500	.7419	.7175	.7142	.6887
3/4-24 UNS	2A	.0012	.7488	.7416	.7217	.7176	.6977
3/4-27 UNS	2A	.0012	.7488	.7421	.7247	.7208	.7034
3/4-28 UN	2A	.0012	.7488	.7423	.7256	.7218	.7050
	3A	-	.7500	.7435	.7268	.7239	.7062
3/4-32 UN	2A	.0011	.7489	.7429	.7286	.7250	.7106
	3A	-	.7500	.7440	.7297	.7270	.7117
13/16-12 UN	2A	.0017	.8108	.7994	.7567	.7512	.7086
	3A	-	.8125	.8011	.7584	.7543	.7103
13/16-16 UN	2A	.0015	.8110	.8016	.7704	.7655	.7343
	3A	-	.8125	.8031	.7719	.7683	.7358
13/16-20 UNEF	2A	.0013	.8112	.8031	.7787	.7743	.7498
	3A	-	.8125	.8044	.7800	.7767	.7512
13/16-28 UN	2A	.0012	.8113	.8048	.7881	.7843	.7675
	3A	-	.8125	.8060	.7893	.7864	.7687
13/16-32 UN	2A	.0011	.8114	.8054	.7911	.7875	.7731
	3A	-	.8125	.8065	.7922	.7895	.7742
7/8-9 UNC	1A	.0019	.8731	.8523	.8009	.7914	7368
	2A	.0019	.8731	.8592	.8009	.7946	.7368
	3A	-	.8750	.8611	.8028	.7981	.7387
7/8-10 UNS	2A	.0018	.8732	.8603	.8082	.8022	.7505
7/8-12 UN	2A	.0017	.8733	.8619	.8192	.8137	.7711
	3A	-	.8750	.8636	.8209	.8168	.7728
7/8-14 UNF	1A	.0016	.8734	.8579	.8270	.8189	.7858
	2A	.0016	.8734	.8631	.8270	.8216	.7858
	3A	-	.8750	.8647	.8286	.8245	.7874
7/8-16 UN	2A	.0015	.8735	.8641	.8329	.8280	.7968
	3A	-	.8750	.8656	.8344	.8308	.7983
7/8-18 UNS	2A	.0014	.8736	.8649	.8375	.8329	.8054
7/8-20 UNEF	2A	.0013	.8737	.8656	.8412	.8368	.8124
	3A	-	.8750	.8669	.8425	.8392	.8137
7/8-24 UNS	2A	.0012	.8738	.8666	.8467	.8426	.8227
7/8-27 UNS	2A	.0012	.8738	.8671	.8497	.8458	.8284

(Continued)

Table 6. *(Continued)* **Unified Screw Threads. Internal Dimensions. Classes 1B, 2B, and 3B.**

Size, TPI, and Series	Class	Minor Diameter		Pitch Diameter		Major Dia.	Tap Drill*
		min	max	min	max		
3/4-10 UNC	1B	.642	.663	.6850	.6965	.7500	41/64
	2B	.642	.663	.6850	.6927	.7500	
	3B	.6420	.6545	.6850	.6907	.7500	
3/4-12 UN	2B	.660	.678	.6959	.7031	.7500	21/32
	3B	.6600	.6707	.6959	.7013	.7500	
3/4-14 UNS	2B	.673	.688	.7036	.7103	.7500	
3/4-16 UNF	1B	.682	.696	.7094	.7192	.7500	11/16
	2B	.682	.696	.7094	.7159	.7500	
	3B	.6820	.6908	.7094	.7143	.7500	
3/4-18 UNS	2B	.690	.703	.7139	.7199	.7500	
3/4-20 UNEF	2B	.696	.707	.7175	.7232	.7500	45/64
	3B	.6960	.7037	.7175	.7218	.7500	
3/4-24 UNS	2B	.705	.715	.7229	.7282	.7500	
3/4-27 UNS	2B	.710	.719	.7259	.7310	.7500	
3/4-28 UN	2B	.711	.720	.7268	.7318	.7500	23/32
	3B	.7110	.7176	.7268	.7305	.7500	
3/4-32 UN	2B	.716	.724	.7297	.7344	.7500	23/32
	3B	.7160	.7219	.7297	.7333	.7500	
13/16-12 UN	2B	.722	.740	.7584	.7656	.8125	47/64
	3B	.7220	.7329	.7584	.7638	.8125	
13/16-16 UN	2B	.745	.759	.7719	.7782	.8125	3/4
	3B	.7450	.7533	.7719	.7766	.8125	
13/16-20 UNEF	2B	.758	.770	.7800	.7857	.8125	49/64
	3B	.7580	.7662	.7800	.7843	.8125	
13/16-28 UN	2B	.774	.782	.7893	.7943	.8125	25/32
	3B	.7740	.7801	.7893	.7930	.8125	
13/16-32 UN	2B	.779	.786	.7922	.7969	.8125	25/32
	3B	.7790	.7844	.7922	.7958	.8125	
7/8-9 UNC	1B	.755	.778	.8028	.8151	.8750	49/64
	2B	.755	.778	.8028	.8110	.8750	
	3B	.7550	.7681	.8028	.8089	.8750	
7/8-10 UNS	2B	.767	.788	.8100	.8178	.8750	
7/8-12 UN	2B	.785	.803	.8209	.8281	.8750	25/32
	3B	.7850	.7948	.8209	.8263	.8750	
7/8-14 UNF	1B	.798	.814	.8286	.8392	.8750	51/64
	2B	.798	.814	.8286	.8356	.8750	
	3B	.7980	8068	.8286	.8339	.8750	
7/8-16 UN	2B	.807	.821	.8344	.8407	.8750	13/16
	3B	.8070	.8158	.8344	.8391	.8750	
7/8-18 UNS	2B	.815	.828	.8389	.8449	.8750	
7/8-20 UNEF	2B	.821	.832	.8425	.8482	.8750	53/64
	3B	.8210	.8287	.8425	.8468	.8750	
7/8-24 UNS	2B	.830	.840	.8479	.8532	.8750	
7/8-27 UNS	2B	.835	.844	.8509	.8560	.8750	

(Continued)

Table 5. *(Continued)* Unified Screw Threads. External Dimensions. Classes 1A, 2A, and 3A.

Size, TPI, and Series	Class	Allowance	Major Diameter max	Major Diameter min	Pitch Diameter max	Pitch Diameter min	Minor Dia.
7/8-28 UN	2A	.0012	.8738	.8673	.8506	.8468	.8300
	3A	-	.8750	.8685	.8518	.8489	.8312
7/8-32 UN	2A	.0011	.8739	.8679	.8536	.8500	.8356
	3A	-	.8750	.8690	.8547	.8520	.8367
15/16-12 UN	2A	.0017	.9358	.9244	.8817	.8760	.8336
	2B	-	.9375	.9261	.8834	.8793	.8353
15/16-16 UN	2A	.0015	.9360	.9266	.8954	.8904	.8593
	3A	-	.9375	.9281	.8969	.8932	.8608
15/16-20 UNEF	2A	.0014	.9361	.9280	.9036	.8991	.8748
	3A	-	.9375	.9294	.9050	.9016	.8762
15/16-28 UN	2A	.0012	.9363	.9298	.9131	.9091	.8925
	3A	-	.9375	.9310	.9143	.9113	.8937
15/16-32 UN	2A	.0011	.9364	.9304	.9161	.9123	.8981
	3A	-	.9375	.9315	.9172	.9144	.8992
1-8 UNC	1A	.0020	.9980	.9755	.9168	.9067	.8446
	2A	.0020	.9980	.9830	.9168	.9100	.8446
	3A	-	1.0000	.9850	.9188	.9137	.8466
1-10 UNS	2A	.0018	.9982	.9853	.9332	.9270	.8755
1-12 UNF	1A	.0018	.9982	.9810	.9441	.9353	.8960
	2A	.0018	.9982	.9868	.9441	.9382	.8960
	3A	-	1.0000	.9886	.9459	.9415	.8978
1-14 UNS	1A	.0017	.9983	.9828	.9519	.9435	.9107
	2A	.0017	.9983	.9880	.9519	.9463	.9107
	3A	-	1.0000	.9897	.9536	.9494	.9124
1-16 UN	2A	.0015	.9985	.9891	.9579	.9529	.9218
	3A	-	1.0000	.9906	.9594	.9557	.9233
1-18 UNS	2A	.0014	.9986	.9899	.9625	.9578	.9304
1-20 UNEF	2A	.0014	.9986	.9905	.9661	.9616	.9373
	3A	-	1.0000	.9919	.9675	.9641	.9387
1-24 UNS	2A	.0013	.9987	.9915	.9716	.9674	.9476
1-27 UNS	2A	.0012	.9988	.9921	.9747	.9707	.9534
1-28 UN	2A	.0012	.9988	.9923	.9756	.9716	.9550
	3A	-	1.0000	.9935	.9768	.9738	.9562
1-32 UN	2A	.0011	.9989	.9929	.9786	.9748	.9606
	3A	-	1.0000	.9940	.9797	.9769	.9617
1 1/16-8 UN	2A	.0020	1.0605	1.0455	.9793	.9725	.9071
	3A	-	1.0625	1.0475	.9812	.9762	.9091
1 1/16-12 UN	2A	.0017	1.0608	1.0494	1.0067	1.0010	.9586
	3A	-	1.0625	1.0511	1.0084	1.0042	.9603
1 1/16-16 UN	2A	.0015	1.0610	1.0516	1.0204	1.0154	.9843
	3A	-	1.0625	1.0531	1.0219	1.0182	.9858
1 1/16-18 UNEF	2A	.0014	1.0611	1.0524	1.0250	1.0203	.9929
	3A	-	1.0625	1.0538	1.0264	1.0228	.9943
1 1/16-20 UN	2A	.0014	1.0611	1.0530	1.0286	1.0241	.9998
	3A	-	1.0625	1.0544	1.0300	1.0266	1.0012

(Continued)

Table 6. *(Continued)* **Unified Screw Threads. Internal Dimensions. Classes 1B, 2B, and 3B.**

Size, TPI, and Series	Class	Minor Diameter min	Minor Diameter max	Pitch Diameter min	Pitch Diameter max	Major Dia.	Tap Drill*
7/8-28 UN	2B	.836	.845	.8518	.8568	.8750	27/32
	3B	.8360	.8426	.8518	.8555	.8750	
7/8-32 UN	2B	.841	.849	.8547	.8594	.8750	27/32
	3B	.8410	.8469	.8547	.8583	.8750	
15/16-12 UN	2B	.847	.865	.8834	.8908	.9375	27/32
	3B	.8470	.8575	.8834	.8889	.9375	
15/16-16 UN	2B	.870	.884	.8969	.9034	.9375	7/8
	3B	.8700	.8783	.8969	.9018	.9375	
15/16-20 UNEF	2B	.883	.895	.9050	.9109	.9375	57/64
	3B	.8830	.8912	.9050	.9094	.9375	
15/16-28 UN	2B	.899	.907	.9143	.9195	.9375	29/32
	3B	.8990	.9051	.9143	.9182	.9375	
15/16-32 UN	2B	.904	.911	.9172	.9221	.9375	29/32
	3B	.9040	.9094	.9172	.9209	.9375	
1-8 UNC	1B	.865	.890	.9188	.9320	1.0000	55/64
	2B	.865	.890	.9188	.9276	1.0000	
	3B	.8650	.8797	.9188	.9254	1.0000	
1-10 UNS	2B	.892	.913	.9350	.9430	1.0000	
1-12 UNF	1B	.910	.928	.9459	.9573	1.0000	29/32
	2B	.910	.928	.9459	.9535	1.0000	
	3B	.9100	.9198	.9459	.9516	1.0000	
1-14 UNS	1B	.923	.938	.9536	.9645	1.0000	59/64
	2B	.923	.938	.9536	.9609	1.0000	
	3B	.9230	.9315	.9536	.9590	1.0000	
1-16 UN	2B	.932	.946	.9594	.9659	1.0000	15/16
	3B	.9320	.9408	.9594	.9643	1.0000	
1-18 UNS	2B	.940	.953	.9639	.9701	1.0000	
1-20 UNEF	2B	.946	.957	.9675	.9734	1.0000	61/64
	3B	.9460	.9537	.9675	.9719	1.0000	
1-24 UNS	2B	.955	.965	.9729	.9784	1.0000	
1-27 UNS	2B	.960	.969	.9759	.9811	1.0000	
1-28 UN	2B	.961	.970	.9768	.9820	1.0000	31/32
	3B	.9610	.9676	.9768	.9807	1.0000	
1-32 UN	2B	.966	.974	.9797	.9846	1.0000	31/32
	3B	.966	.9719	.9797	.9834	1.0000	
1 1/16-8 UN	2B	.927	.952	.9813	.9902	1.0625	59/64
	3B	.9270	.9422	.9813	.9880	1.0625	
1 1/16-12 UN	2B	.972	.990	1.0084	1.0158	1.0625	31/32
	3B	.9720	.9823	1.0084	1.0139	1.0625	
1 1/16-16 UN	2B	.995	1.009	1.0219	1.0284	1.0625	1
	3B	.9950	1.0033	1.0219	1.0268	1.0625	
1 1/16-18 UNEF	2B	1.002	1.015	1.0264	1.0326	1.0625	1
	3B	1.0020	1.0105	1.0264	1.0310	1.0625	
1 1/16-20 UN	2B	1.008	1.020	1.0300	1.0359	1.0625	1 1/64
	3B	1.0080	1.0162	1.0300	1.0344	1.0625	

(Continued)

Table 5. *(Continued)* Unified Screw Threads. External Dimensions. Classes 1A, 2A, and 3A.

Size, TPI, and Series	Class	Allowance	Major Diameter max	Major Diameter min	Pitch Diameter max	Pitch Diameter min	Minor Dia.
1 1/16-28 UN	2A	.0012	1.0613	1.0548	1.0381	1.0341	1.0175
	3A	-	1.0625	1.0560	1.0393	1.0363	1.0187
1 1/8-7 UNC	1A	.0022	1.1228	1.0982	1.0300	1.0191	.9475
	2A	.0022	1.1228	1.1064	1.0300	1.0228	.9475
	3A	-	1.1250	1.1086	1.0322	1.0268	.9497
1 1/8-8 UN	2A	.0021	1.1229	1.1079	1.0417	1.0348	.9695
	3A	-	1.1250	1.1100	1.0438	1.0386	.9716
1 1/8-10 UNS	2A	.0018	1.1232	1.1103	1.0582	1.0520	1.0005
1 1/8-12 UNF	1A	.0018	1.1232	1.1060	1.0691	1.0601	1.0210
	2A	.0018	1.1232	1.1118	1.0691	1.0631	1.0210
	3A	-	1.1250	1.1136	1.0709	1.0664	1.0228
1 1/8-14 UNS	2A	.0016	1.1234	1.1131	1.0770	1.0717	1.0358
1 1/8-16 UN	2A	.0015	1.1235	1.1141	1.0829	1.0779	1.0468
	3A	-	1.1250	1.1156	1.0844	1.0807	1.0483
1 1/8-18 UNEF	2A	.0014	1.1236	1.1149	1.0875	1.0828	1.0554
	3A	-	1.1250	1.1163	1.0889	1.0853	1.0568
1 1/8-20 UN	2A	.0014	1.1236	1.1155	1.0911	1.0866	1.0623
	3A	-	1.1250	1.1169	1.0925	1.0891	1.0637
1 1/8-24 UNS	2A	.0013	1.1237	1.1165	1.0966	1.0924	1.0726
1 1/8-28 UN	2A	.0012	1.1238	1.1173	1.1006	1.0966	1.0800
	3A	-	1.1250	1.1185	1.1018	1.0988	1.0812
1 3/16-8 UN	2A	.0021	1.1854	1.1704	1.1042	1.0972	1.0320
	3A	-	1.1875	1.1725	1.1063	1.1011	1.0341
1 3/16-12 UN	2A	.0017	1.1858	1.1744	1.1317	1.1259	1.0836
	3A	-	1.1875	1.1761	1.1334	1.1291	1.0853
1 3/16-16 UN	2A	.0015	1.1860	1.1766	1.1454	1.1403	1.1093
	3A	-	1.1875	1.1781	1.1469	1.1431	1.1108
1 3/16-18 UNEF	2A	.0015	1.1860	1.1773	1.1499	1.1450	1.1178
	3A	-	1.1875	1.1788	1.1514	1.1478	1.1193
1 3/16-20 UN	2A	.0014	1.1861	1.1780	1.1536	1.1489	1.1248
	3A	-	1.1875	1.1794	1.1550	1.1515	1.1262
1 3/16-28 UN	2A	.0012	1.1863	1.1798	1.1631	1.1590	1.1425
	3A	-	1.1875	1.1810	1.1643	1.1612	1.1437
1 1/4-7 UNC	1A	.0022	1.2478	1.2232	1.1550	1.1439	1.0725
	2A	.0022	1.2478	1.2314	1.1550	1.1476	1.0725
	3A	-	1.2500	1.2336	1.1572	1.1517	1.0747
1 1/4-8 UN	2A	.0021	1.2479	1.2329	1.1667	1.1579	1.0945
	3A	-	1.2500	1.2350	1.1688	1.1635	1.0966
1 1/4-10 UNS	2A	.0019	1.2481	1.2352	1.1831	1.1768	1.1254
1 1/4-12 UNF	1A	.0018	1.2481	1.2310	1.1941	1.1849	1.1460
	2A	.0018	1.2482	1.2368	1.1941	1.1879	1.1460
	3A	-	1.2500	1.2386	1.1959	1.1913	1.1478
1 1/4-14 UNS	2A	.0016	1.2484	1.2381	1.2020	1.1966	1.1608
1 1/4-16 UN	2A	.0015	1.2485	1.2391	1.2079	1.2028	1.1718
	3A	-	1.2500	1.2406	1.2094	1.2056	1.1733

(Continued)

Table 6. *(Continued)* **Unified Screw Threads. Internal Dimensions. Classes 1B, 2B, and 3B.**

Size, TPI, and Series	Class	Minor Diameter		Pitch Diameter		Major Dia.	Tap Drill*
		min	max	min	max		
1 1/16-28 UN	2B	1.024	1.032	1.0393	1.0445	1.0625	1 1/32
	3B	1.0240	1.0301	1.0393	1.0432	1.0625	
1 1/8-7 UNC	1B	.970	.998	1.0322	1.0463	1.1250	31/32
	2B	.970	.998	1.0322	1.0416	1.1250	
	3B	.9700	.9875	1.0322	1.0393	1.1250	
1 1/8-8 UN	2B	.990	1.015	1.0438	1.0528	1.1250	1
	3B	.990	1.0047	1.0438	1.0505	1.1250	
1 1/8-10 UNS	2B	1.017	1.0380	1.0600	1.0680	1.1250	
1 1/8-12 UNF	1B	1.035	1.053	1.0709	1.0826	1.1250	1 1/32
	2B	1.035	1.053	1.0709	1.0787	1.1250	
	3B	1.0350	1.0448	1.0709	1.0768	1.1250	
1 1/8-14 UNS	2B	1.048	1.064	1.0786	1.0855	1.1250	
1 1/8-16 UN	2B	1.057	1.071	1.0844	1.0909	1.1250	1 1/16
	3B	1.0570	1.0658	1.0844	1.0893	1.1250	
1 1/8-18 UNEF	2B	1.065	1.078	1.0889	1.0951	1.1250	1 1/16
	3B	1.0650	1.0730	1.0889	1.0935	1.1250	
1 1/8-20 UN	2B	1.071	1.082	1.0925	1.0984	1.1250	1 5/64
	3B	1.0710	1.0787	1.0925	1.0969	1.1250	
1 1/8-24 UNS	2B	1.080	1.090	1.0979	1.1034	1.1250	
1 1/8-28 UN	2B	1.086	1.095	1.1018	1.1070	1.1250	1 3/32
	3B	1.0860	1.0926	1.1018	1.1057	1.1250	
1 3/16-8 UN	2B	1.052	1.077	1.1063	1.1154	1.1875	1 1/16
	3B	1.0520	1.0672	1.1063	1.1131	1.1875	
1 3/16-12 UN	2B	1.097	1.115	1.1334	1.1409	1.1875	1 3/32
	3B	1.0970	1.1073	1.1334	1.1390	1.1875	
1 3/16-16 UN	2B	1.120	1.134	1.1469	1.1535	1.1875	1 1/8
	3B	1.1200	1.1283	1.1469	1.1519	1.1875	
1 3/16-18 UNEF	2B	1.127	1.140	1.1514	1.1577	1.1875	1 1/8
	3B	1.1270	1.1355	1.1514	1.1561	1.1875	
1 3/16-20 UN	2B	1.133	1.145	1.1550	1.1611	1.1875	1 9/64
	3B	1.1330	1.1412	1.1550	1.1595	1.1875	
1 3/16-28 UN	2B	1.149	1.157	1.1643	1.1696	1.1875	1 5/32
	3B	1.1490	1.1551	1.1643	1.1683	1.1875	
1 1/4-7 UNC	1B	1.095	1.123	1.1572	1.1716	1.2500	1 3/32
	2B	1.095	1.123	1.1572	1.1668	1.2500	
	3B	1.0950	1.1125	1.1572	1.1644	1.2500	
1 1/4-8 UN	2B	1.115	1.140	1.1688	1.1780	1.2500	1 1/8
	3B	1.1150	1.1297	1.1688	1.1757	1.2500	
1 1/4-10 UNS	2B	1.142	1.163	1.1850	1.1932	1.1250	
1 1/4-12 UNF	1B	1.160	1.178	1.1959	1.2079	1.2500	1 5/32
	2B	1.160	1.178	1.1959	1.2039	1.2500	
	3B	1.1600	1.1698	1.1959	1.2019	1.2500	
1 1/4-14 UNS	2B	1.173	1.188	1.2036	1.2106	1.2500	
1 1/4-16 UN	2B	1.182	1.196	1.2094	1.2160	1.2500	1 3/16
	3B	1.1820	1.1908	1.2094	1.2144	1.2500	

(Continued)

Table 5. *(Continued)* Unified Screw Threads. External Dimensions. Classes 1A, 2A, and 3A.

Size, TPI, and Series	Class	Allowance	Major Diameter max	Major Diameter min	Pitch Diameter max	Pitch Diameter min	Minor Dia.
1 1/4-18 UNEF	2A	.0015	1.2485	1.2398	1.2124	1.2075	1.1803
	3A	-	1.2500	1.2413	1.2139	1.2103	1.1818
1 1/4-20 UN	2A	.0014	1.2486	1.2405	1.2161	1.2114	1.1873
	3A	-	1.2500	1.2419	1.2175	1.2140	1.1887
1 1/4-24 UNS	2A	.0013	1.2487	1.2415	1.2216	1.2173	1.1976
1 1/4-28 UN	2A	.0012	1.2488	1.2423	1.2256	1.2215	1.2050
	3A	-	1.2500	1.2435	1.2268	1.2237	1.2062
1 5/16-8 UN	2A	.0021	1.3104	1.2954	1.2292	1.2221	1.1570
	3A	-	1.3125	1.2975	1.2313	1.2260	1.1591
1 5/16-12 UN	2A	.0017	1.3108	1.2994	1.2567	1.2509	1.2086
	3A	-	1.3125	1.3011	1.2584	1.2541	1.2103
1 5/16-16 UN	2A	.0015	1.3110	1.3016	1.2704	1.2653	1.2343
	3A	-	1.3125	1.3031	1.2719	1.2681	1.2358
1 5/16-18 UNEF	2A	.0015	1.3110	1.3023	1.2749	1.2700	1.2428
	3A	-	1.3125	1.3038	1.2764	1.2728	1.2443
1 5/16-20 UN	2A	.0014	1.3111	1.3030	1.2786	1.2739	1.2498
	3A	-	1.3125	1.3044	1.2800	1.2765	1.2512
1 5/16-28 UN	2A	.0012	1.3113	1.3048	1.2881	1.2840	1.2675
	3A	-	1.3125	1.3060	1.2893	1.2862	1.2687
1 3/8-6 UNC	1A	.0024	1.3726	1.3453	1.2643	1.2523	1.1681
	2A	.0024	1.3726	1.3544	1.2643	1.2563	1.1681
	3A	-	1.3750	1.3568	1.2667	1.2607	1.1705
1 3/8-8 UN	2A	.0022	1.3728	1.3578	1.2916	1.2844	1.2194
	3A	-	1.3750	1.3600	1.2938	1.2884	1.2216
1 3/8-10 UNS	2A	.0019	1.3731	1.3602	1.3081	1.3018	1.2504
1 3/8-12 UNF	1A	.0019	1.3731	1.3559	1.3190	1.3096	1.2709
	2A	.0019	1.3731	1.3617	1.3190	1.3127	1.2709
	3A	-	1.3750	1.3636	1.3209	1.3162	1.2728
1 3/8-14 UNS	2A	.0016	1.3734	1.3631	1.3270	1.3216	1.2858
1 3/8-16 UN	2A	.0015	1.3735	1.3641	1.3329	1.3278	1.2968
	3A	-	1.3750	1.3656	1.3344	1.3306	1.2983
1 3/8-18 UNEF	2A	.0015	1.3735	1.3648	1.3374	1.3325	1.3053
	3A	-	1.3750	1.3663	1.3389	1.3353	1.3068
1 3/8-20 UN	2A	.0014	1.3736	1.3655	1.3411	1.3364	1.3123
	3A	-	1.3750	1.3669	1.3425	1.3390	1.3137
1 3/8-28 UN	2A	.0012	1.3738	1.3673	1.3506	1.3465	1.3300
	3A	-	1.3750	1.3685	1.3518	1.3487	1.3312
1 7/16-6 UN	2A	.0024	1.4351	1.4169	1.3268	1.3188	1.2306
	3A	-	1.4375	1.4193	1.3292	1.3232	1.2330
1 7/16-8 UN	2A	.0022	1.4353	1.4203	1.3541	1.3469	1.2819
	3A	-	1.4375	1.4225	1.3563	1.3509	1.2841
1 7/16-12 UN	2A	.0018	1.4357	1.4243	1.3816	1.3757	1.3335
	3A	-	1.4375	1.4261	1.3834	1.3790	1.3353
1 7/16-16 UN	2A	.0016	1.4359	1.4265	1.3953	1.3901	1.3592
	3A	-	1.4375	1.4281	1.3969	1.3930	1.3608

(Continued)

Table 6. *(Continued)* **Unified Screw Threads. Internal Dimensions. Classes 1B, 2B, and 3B.**

Size, TPI, and Series	Class	Minor Diameter		Pitch Diameter		Major Dia.	Tap Drill*
		min	max	min	max		
1 1/4-18 UNEF	2B	1.190	1.203	1.2139	1.2202	1.2500	1 3/16
	3B	1.1900	1.1980	1.2139	1.2186	1.2500	
1 1/4-20 UN	2B	1.196	1.207	1.2175	1.2236	1.2500	1 13/64
	3B	1.1960	1.2037	1.2175	1.2220	1.2500	
1 1/4-24 UNS	2B	1.205	1.215	1.2229	1.2285	1.2500	
1 1/4-28 UN	2B	1.211	1.220	1.2268	1.2321	1.2500	1 7/32
	3B	1.2110	1.2176	1.2268	1.2308	1.2500	
1 5/16-8 UN	2B	1.177	1.202	1.2313	1.2405	1.3125	1 11/64
	3B	1.1770	1.1922	1.2313	1.2382	1.3125	
1 5/16-12 UN	2B	1.222	1.240	1.2584	1.2659	1.3125	1 7/32
	3B	1.2220	1.2323	1.2584	1.2640	1.3125	
1 5/16-16 UN	2B	1.245	1.259	1.2719	1.2785	1.3125	1 1/4
	3B	1.2450	1.2533	1.2719	1.2769	1.3125	
1 5/16-18 UNEF	2B	1.252	1.265	1.2764	1.2827	1.3125	1 1/4
	3B	1.2520	1.2605	1.2764	1.2811	1.3125	
1 5/16-20 UN	2B	1.258	1.270	1.2800	1.2861	1.3125	1 17/64
	3B	1.2580	1.2662	1.2800	1.2845	1.3125	
1 5/16-28 UN	2B	1.274	1.282	1.2893	1.2946	1.3125	1 9/32
	3B	1.2740	1.2801	1.2893	1.2933	1.3125	
1 3/8-6 UNC	1B	1.195	1.225	1.2667	1.2822	1.3750	1 13/16
	2B	1.195	1.225	1.2667	1.2771	1.3750	
	3B	1.1950	1.2146	1.2667	1.2745	1.3750	
1 3/8-8 UN	2B	1.240	1.265	1.2938	1.3031	1.3750	1 15/64
	3B	1.2400	1.2547	1.2938	1.3008	1.3750	
1 3/8-10 UNS	2B	1.267	1.288	1.3100	1.3182	1.3750	
1 3/8-12 UNF	1B	1.285	1.303	1.3209	1.3332	1.3750	1 9/32
	2B	1.285	1.303	1.3209	1.3291	1.3750	
	3B	1.2850	1.2948	1.3209	1.3270	1.3750	
1 3/8-14 UNS	2B	1.298	1.314	1.3286	1.3356	1.3750	
1 3/8-16 UN	2B	1.307	1.321	1.3344	1.3410	1.3750	1 5/16
	3B	1.3070	1.3158	1.3344	1.3394	1.3750	
1 3/8-18 UNEF	2B	1.315	1.328	1.3389	1.3452	1.3750	1 5/16
	3B	1.3150	1.3230	1.3389	1.3436	1.3750	
1 3/8-20 UN	2B	1.321	1.332	1.3425	1.3486	1.3750	1 21/64
	3B	1.3210	1.3287	1.3425	1.3470	1.3750	
1 3/8-28 UN	2B	1.336	1.345	1.3518	1.3571	1.3750	1 11/32
	3B	1.3360	1.3426	1.3518	1.3558	1.3750	
1 7/16-6 UN	2B	1.257	1.288	1.3292	1.3396	1.4375	1 17/64
	3B	1.2570	1.2771	1.3292	1.3370	1.4375	
1 7/16-8 UN	2B	1.302	1.327	1.3563	1.3657	1.4375	1 19/64
	3B	1.3020	1.3172	1.3563	1.3634	1.4375	
1 7/16-12 UN	2B	1.347	1.365	1.3834	1.3910	1.4375	1 11/32
	3B	1.3470	1.3573	1.3834	1.3891	1.4375	
1 7/16-16 UN	2B	1.370	1.384	1.3969	1.4037	1.4375	1 3/8
	3B	1.3700	1.3783	1.3969	1.4020	1.4375	

(Continued)

Table 5. *(Continued)* **Unified Screw Threads. External Dimensions. Classes 1A, 2A, and 3A.**

Size, TPI, and Series	Class	Allowance	Major Diameter max	Major Diameter min	Pitch Diameter max	Pitch Diameter min	Minor Dia.
1 7/16-18 UNEF	2A	.0015	1.4360	1.4273	1.3999	1.3949	1.3678
	3A	-	1.4375	1.4288	1.4014	1.3977	1.3693
1 7/16-20 UN	2A	.0014	1.4361	1.4280	1.4036	1.3988	1.3748
	3A	-	1.4375	1.4294	1.4050	1.4014	1.3762
1 7/16-28 UN	2A	.0013	1.4362	1.4297	1.4130	1.4088	1.3924
	3A	-	1.4375	1.4310	1.4143	1.4112	1.3937
1 1/2-6 UNC	1A	.0024	1.4976	1.4703	1.3893	1.3772	1.2931
	2A	.0024	1.4976	1.4794	1.3893	1.3812	1.2931
	3A	-	1.5000	1.4818	1.3917	1.3856	1.2955
1 1/2-8 UN	2A	.0022	1.4978	1.4828	1.4166	1.4093	1.3444
	3A	-	1.5000	1.4850	1.4188	1.4133	1.3466
1 1/2-10 UNS	2A	.0019	1.4981	1.4852	1.4331	1.4267	1.3754
1 1/2-12 UNF	1A	.0019	1.4981	1.4809	1.4440	1.4344	1.3959
	2A	.0019	1.4981	1.4867	1.4440	1.4376	1.3959
	3A	-	1.5000	1.4886	1.4459	1.4411	1.3978
1 1/2-14 UNS	2A	.0017	1.4983	1.4880	1.4519	1.4464	1.4107
1 1/2-16 UN	2A	.0016	1.4984	1.4890	1.4578	1.4526	1.4217
	3A	-	1.5000	1.4906	1.4594	1.4555	1.4233
1 1/2-18 UNEF	2A	.0015	1.4985	1.4898	1.4264	1.4574	1.4303
	3A	-	1.5000	1.4913	1.4639	1.4602	1.4318
1 1/2-20 UN	2A	.0014	1.4986	1.4905	1.4661	1.4613	1.4373
	3A	-	1.5000	1.4919	1.4675	1.4639	1.4387
1 1/2-24 UNS	2A	.0013	1.4987	1.4915	1.4716	1.4672	1.4476
1 1/2-28 UN	2A	.0013	1.4987	1.4922	1.4755	1.4713	1.4549
	3A	-	1.5000	1.4935	1.4768	1.4737	1.4562
1 9/16-6 UN	2A	.0024	1.5601	1.5419	1.4518	1.4436	1.3556
	3A	-	1.5625	1.5443	1.4542	1.4481	1.3580
1 9/16-8 UN	2A	.0022	1.5603	1.5453	1.4791	1.4717	1.4069
	3A	-	1.5625	1.5475	1.4813	1.4758	1.4091
1 9/16-12 UN	2A	.0018	1.5607	1.5493	1.5066	1.5007	1.4585
	3A	-	1.5625	1.5511	1.5084	1.5040	1.4603
1 9/16-16 UN	2A	.0016	1.5609	1.5515	1.5203	1.5151	1.4842
	3A	-	1.5625	1.5531	1.5219	1.5180	1.4858
1 9/16-18 UNEF	2A	.0015	1.5610	1.5523	1.5249	1.5199	1.4928
	3A	-	1.5625	1.5538	1.5264	1.5227	1.4943
1 9/16-20 UN	2A	.0014	1.5611	1.5530	1.5286	1.5238	1.4998
	3A	-	1.5625	1.5544	1.5300	1.5264	1.5012
1 5/8-6 UN	2A	.0025	1.6225	1.6043	1.5142	1.5060	1.4180
	3A	-	1.6250	1.6068	1.5167	1.5105	1.4205
1 5/8-8 UN	2A	.0022	1.6228	1.6078	1.5416	1.5342	1.4694
	3A	-	1.6250	1.6100	1.5438	1.5382	1.4716
1 5/8-10 UNS	2A	.0019	1.6231	1.6102	1.5581	1.5517	1.5004
1 5/8-12 UN	2A	.0018	1.6232	1.6118	1.5691	1.5632	1.5210
	3A	-	1.6250	1.6136	1.5709	1.5665	1.5228
1 5/8-14 UNS	2A	.0017	1.6233	1.6130	1.5769	1.5714	1.5357

(Continued)

Table 6. *(Continued)* Unified Screw Threads. Internal Dimensions. Classes 1B, 2B, and 3B.

Size, TPI, and Series	Class	Minor Diameter min	Minor Diameter max	Pitch Diameter min	Pitch Diameter max	Major Dia.	Tap Drill*
1 7/16-18 UNEF	2B	1.377	1.390	1.4014	1.4079	1.4375	1 3/8
	3B	1.3770	1.3855	1.4014	1.4062	1.4375	
1 7/16-20 UN	2B	1.383	1.395	1.4050	1.4112	1.4375	1 25/64
	3B	1.3830	1.3912	1.4050	1.4096	1.4375	
1 7/16-28 UN	2B	1.399	1.407	1.4143	1.4198	1.4375	1 13/32
	3B	1.3990	1.4051	1.4143	1.4184	1.4375	
1 1/2-6 UNC	1B	1.320	1.350	1.3917	1.4075	1.5000	1 5/16
	2B	1.320	1.350	1.3917	1.4022	1.5000	
	3B	1.3200	1.3396	1.3917	1.3996	1.5000	
1 1/2-8 UN	2B	1.365	1.390	1.4188	1.4283	1.5000	1 23/64
	3B	1.3650	1.3797	1.4188	1.4259	1.5000	
1 1/2-10 UNS	2B	1.392	1.413	1.4350	1.4433	1.5000	
1 1/2-12 UNF	1B	1.410	1.428	1.4459	1.4584	1.5000	1 13/32
	2B	1.410	1.428	1.4459	1.4542	1.5000	
	3B	1.4100	1.4198	1.4459	1.4522	1.5000	
1 1/2-14 UNS	2B	1.423	1.438	1.4536	1.4608	1.5000	
1 1/2-16 UN	2B	1.432	1.446	1.4594	1.4662	1.5000	1 7/16
	3B	1.4320	1.4408	1.4594	1.4645	1.5000	
1 1/2-18 UNEF	2B	1.440	1.452	1.4639	1.4704	1.5000	1 7/16
	3B	1.4400	1.4480	1.4639	1.4687	1.5000	
1 1/2-20 UN	2B	1.446	1.457	1.4675	1.4737	1.5000	1 29/64
	3B	1.4460	1.4537	1.4675	1.4721	1.5000	
1 1/2-24 UNS	2B	1.455	1.465	1.4729	1.4787	1.5000	
1 1/2-28 UN	2B	1.461	1.470	1.4768	1.4823	1.5000	1 15/32
	3B	1.4610	1.4676	1.4768	1.4809	1.5000	
1 9/16-6 UN	2B	1.382	1.413	1.4542	1.4648	1.5625	1 25/64
	3B	1.3820	1.4021	1.4542	1.4622	1.5625	
1 9/16-8 UN	2B	1.427	1.452	1.4813	1.4909	1.5625	1 27/64
	3B	1.4270	1.4422	1.4813	1.4885	1.5625	
1 9/16-12 UN	2B	1.472	1.490	1.5084	1.5160	1.5625	1 15/32
	3B	1.4720	1.4823	1.5084	1.5141	1.5625	
1 9/16-16 UN	2B	1.495	1.509	1.5219	1.5287	1.5625	1 1/2
	3B	1.4950	1.5033	1.5219	1.5270	1.5625	
1 9/16-18 UNEF	2B	1.502	1.515	1.5264	1.5329	1.5625	1 1/2
	3B	1.5020	1.5105	1.5264	1.5312	1.5625	
1 9/16-20 UN	2B	1.508	1.520	1.5300	1.5362	1.5625	1 33/64
	3B	1.5080	1.5162	1.5300	1.5346	1.5625	
1 5/8-6 UN	2B	1.445	1.475	1.5167	1.5274	1.6250	1 29/64
	3B	1.4450	1.4646	1.5167	1.5247	1.6250	
1 5/8-8 UN	2B	1.490	1.515	1.5438	1.5535	1.6250	1 31/64
	3B	1.4900	1.5047	1.5438	1.5510	1.6250	
1 5/8-10 UNS	2B	1.517	1.538	1.5600	1.5683	1.6250	
1 5/8-12 UN	2B	1.535	1.553	1.5709	1.5785	1.6250	1 17/32
	3B	1.5350	1.5448	1.5709	1.5766	1.6250	
1 5/8-14 UNS	2B	1.548	1.564	1.5786	1.5858	1.6250	

(Continued)

Table 5. *(Continued)* **Unified Screw Threads. External Dimensions. Classes 1A, 2A, and 3A.**

Size, TPI, and Series	Class	Allowance	Major Diameter max	Major Diameter min	Pitch Diameter max	Pitch Diameter min	Minor Dia.
1 5/8-16 UN	2A	.0016	1.6234	1.6140	1.5828	1.5776	1.5467
	3A	-	1.6250	1.6156	1.5844	1.5805	1.5483
1 5/8-18 UNEF	2A	.0015	1.6235	1.6148	1.5874	1.5824	1.5553
	3A	-	1.6250	1.6163	1.5889	1.5852	1.5568
1 5/8-20 UN	2A	.0014	1.6236	1.6155	1.5911	1.5863	1.5623
	3A	-	1.6250	1.6169	1.5925	1.5889	1.5637
1 5/8-24 UNS	2A	.0013	1.6237	1.6165	1.5966	1.5922	1.5726
1 11/16-6 UN	2A	.0025	1.6850	1.6668	1.5767	1.5684	1.4805
	3A	-	1.6875	1.6693	1.5792	1.5730	1.4830
1 11/16-8 UN	2A	.0022	1.6853	1.6703	1.6041	1.5966	1.5319
	3A	-	1.6875	1.6725	1.6063	1.6007	1.5341
1 11/16-12 UN	2A	.0018	1.6857	1.6743	1.6316	1.6256	1.5835
	3A	-	1.6875	1.6761	1.6334	1.6289	1.5853
1 11/16-16 UN	2A	.0016	1.6859	1.6765	1.6453	1.6400	1.6092
	3A	-	1.6875	1.6781	1.6469	1.6429	1.6108
1 11/16-18 UNEF	2A	.0015	1.6860	1.6773	1.6499	1.6448	1.6178
	3A	-	1.6875	1.6788	1.6514	1.6476	1.6193
1 11/16-20 UN	2A	.0015	1.6860	1.6779	1.6535	1.6487	1.6247
	3A	-	1.6875	1.6794	1.6550	1.6514	1.6262
1 3/4-5 UNC	1A	.0027	1.7473	1.7165	1.6174	1.6040	1.5019
	2A	.0027	1.7473	1.7268	1.6174	1.6085	1.5019
	3A	-	1.7500	1.7295	1.6201	1.6134	1.5046
1 3/4-6 UN	2A	.0025	1.7475	1.7293	1.6392	1.6309	1.5430
	3A	-	1.7500	1.7318	1.6417	1.6354	1.5455
1 3/4-8 UN	2A	.0023	1.7477	1.7327	1.6665	1.6590	1.5943
	3A	-	1.7500	1.7350	1.6688	1.6632	1.5966
1 3/4-10 UNS	2A	.0019	1.7481	1.7352	1.6831	1.6766	1.6254
1 3/4-12 UN	2A	.0019	1.7482	1.7368	1.6941	1.6881	1.6460
	3A	-	1.7500	1.7386	1.6959	1.6914	1.6478
1 3/4-14 UNS	2A	.0017	1.7483	1.7380	1.7019	1.6963	1.6607
1 3/4-16 UN	2A	.0016	1.7484	1.7390	1.7078	1.7025	1.6717
	3A	-	1.7500	1.7406	1.7094	1.7054	1.6733
1 3/4-18 UNS	2A	.0015	1.7485	1.7398	1.7124	1.7073	1.6803
1 3/4-20 UN	2A	.0015	1.7485	1.7404	1.7160	1.7112	1.6872
	3A	-	1.7500	1.7419	7.7175	1.7139	1.6887
1 13/16-6 UN	2A	.0025	1.8100	1.7918	1.7017	1.6933	1.6055
	3A	-	1.8125	1.7943	1.7042	1.6979	1.6080
1 13/16-8 UN	2A	.0023	1.8102	1.7952	1.7290	1.7214	1.6568
	3A	-	1.8125	1.7975	1.7313	1.7256	1.6591
1 13/16-12 UN	2A	.0018	1.8107	1.7933	1.7566	1.7506	1.7085
	3A	-	1.8125	1.8011	1.7584	1.7539	1.7103
1 13/16-16 UN	2A	.0016	1.8109	1.8015	1.7703	1.7650	1.7342
	3A	-	1.8125	1.8031	1.7719	1.7679	1.7358
1 13/16-20 UN	2A	.0015	1.8110	1.8029	1.7785	1.7737	1.7497
	3A	-	1.8125	1.8044	1.7800	1.7764	1.7512

(Continued)

Table 6. *(Continued)* **Unified Screw Threads. Internal Dimensions. Classes 1B, 2B, and 3B.**

Size, TPI, and Series	Class	Minor Diameter		Pitch Diameter		Major Dia.	Tap Drill*
		min	max	min	max		
1 5/8-16 UN	2B	1.557	1.571	1.5844	1.5912	1.6250	1 9/16
	3B	1.5570	1.5658	1.5844	1.5895	1.6250	
1 5/8-18 UNEF	2B	1.565	1.578	1.5889	1.5954	1.6250	1 9/16
	3B	1.5650	1.5730	1.5889	1.5937	1.6250	
1 5/8-20 UN	2B	1.571	1.582	1.5925	1.5987	1.6250	1 37/64
	3B	1.5710	1.5787	1.5925	1.5971	1.6250	
1 5/8-24 UNS	2B	1.580	1.590	1.5979	1.6037	1.6250	
1 11/16-6 UN	2B	1.507	1.538	1.5792	1.5900	1.6875	1 1/2
	3B	1.5070	1.5271	1.5792	1.5873	1.6875	
1 11/16-8 UN	2B	1.552	1.577	1.6063	1.6160	1.6875	1 9/16
	3B	1.5520	1.5672	1.6063	1.6136	1.6875	
1 11/16-12 UN	2B	1.597	1.615	1.6334	1.6412	1.6875	1 19/32
	3B	1.5970	1.6070	1.6334	1.6392	1.6875	
1 11/16-16 UN	2B	1.620	1.634	1.6469	1.6538	1.6875	1 5/8
	3B	1.6200	1.6283	1.6469	1.6521	1.6875	
1 11/16-18 UNEF	2B	1.627	1.640	1.6514	1.6580	1.6875	1 5/8
	3B	1.6270	1.6355	1.6514	1.6563	1.6875	
1 11/16-20 UN	2B	1.633	1.645	1.6550	1.6613	1.6875	1 41/64
	3B	1.6330	1.6412	1.6550	1.6597	1.6875	
1 3/4-5 UNC	1B	1.534	1.568	1.6201	1.6375	1.7500	1 17/32
	2B	1.534	1.568	1.6201	1.6317	1.7500	
	3B	1.5340	1.5575	1.6201	1.6288	1.7500	
1 3/4-6 UN	2B	1.570	1.600	1.6417	1.6525	1.7500	1 9/16
	3B	1.5700	1.5896	1.6417	1.6498	1.7500	
1 3/4-8 UN	2B	1.615	1.640	1.6688	1.6786	1.7500	1 39/64
	3B	1.6150	1.6297	1.6688	1.6762	1.7500	
1 3/4-10 UNS	2B	1.642	1.663	1.6850	1.6934	1.7500	
1 3/4-12 UN	2B	1.660	1.678	1.6959	1.7037	1.7500	1 21/32
	3B	1.6600	1.6698	1.6959	1.7017	1.7500	
1 3/4-14 UNS	2B	1.673	1.688	1.7036	1.7109	1.7500	
1 3/4-16 UN	2B	1.682	1.696	1.7094	1.7163	1.7500	1 11/16
	3B	1.6820	1.6908	1.7094	1.7146	1.7500	
1 3/4-18 UNS	2B	1.690	1.703	1.7139	1.7205	1.7500	
1 3/4-20 UN	2B	1.696	1.707	1.7175	1.7338	1.7500	1 45/64
	3B	1.6960	1.7037	1.7175	1.7222	1.7500	
1 13/16-6 UN	2B	1.632	1.663	1.7042	1.7151	1.8125	1 45/64
	3B	1.6320	1.6521	1.7042	1.7124	1.8125	
1 13/16-8 UN	2B	1.677	1.702	1.7313	1.7412	1.8125	1 3/4
	3B	1.6770	1.6922	1.7313	1.7387	1.8125	
1 13/16-12 UN	2B	1.722	1.740	1.7584	1.7662	1.8125	1 23/32
	3B	1.7220	1.7320	1.7584	1.7642	1.8125	
1 13/16-16 UN	2B	1.745	1.759	1.7719	1.7788	1.8125	1 3/4
	3B	1.7450	1.7533	1.7719	1.7771	1.8125	
1 13/16-20 UN	2B	1.758	1.770	1.7800	1.7863	1.8125	1 49/64
	3B	1.7580	1.7662	1.7800	1.7847	1.8125	

(Continued)

Table 5. *(Continued)* Unified Screw Threads. External Dimensions. Classes 1A, 2A, and 3A.

Size, TPI, and Series	Class	Allowance	Major Diameter		Pitch Diameter		Minor Dia.
			max	min	max	min	
1 7/8-6 UN	2A	.0025	1.8725	1.8543	1.7642	1.7558	1.6680
	3A	-	1.8750	1.8568	1.7667	1.7604	1.6705
1 7/8-8 UN	2A	.0023	1.8727	1.8577	1.7915	1.7838	1.7193
	3A	-	1.8750	1.8600	1.7938	1.7881	1.7216
1 7/8-10 UNS	2A	.0019	1.8731	1.8602	1.8081	1.8016	1.7504
1 7/8-12 UN	2A	.0018	1.8732	1.8618	1.8191	1.8131	1.7710
	3A	-	1.8750	1.8636	1.8209	1.8164	1.7728
1 7/8-14 UNS	2A	.0017	1.8733	1.8630	1.8269	1.8213	1.7857
1 7/8-16 UN	2A	.0016	1.8734	1.8640	1.8328	1.8275	1.7967
	3A	-	1.8750	1.8656	1.8344	1.8304	1.7983
1 7/8-18 INS	2A	.0015	1.8735	1.8648	1.8374	1.8323	1.8053
1 7/8-20 UN	2A	.0015	1.8735	1.8654	1.8410	1.8362	1.8122
	3A	-	1.8750	1.8669	1.8425	1.8389	1.8137
1 15/16-6 UN	2A	.0026	1.9349	1.9167	1.8266	1.8181	1.7304
	3A	-	1.9375	1.9193	1.8292	1.8228	1.7330
1 15/16-8 UN	2A	.0023	1.9352	1.9202	1.8540	1.8463	1.7818
	3A	-	1.9375	1.9225	1.8563	1.8505	1.7841
1 15/16 -12 UN	2A	.0018	1.9357	1.9243	1.8816	1.8755	1.8335
	3A	-	1.9375	1.9261	1.8834	1.8789	1.8353
1 15/16-16 UN	2A	.0016	1.9359	1.9265	1.8953	1.8899	1.8592
	3A	-	1.9375	1.9281	1.8969	1.8929	1.8608
1 15/16-20 UN	2A	.0015	1.9360	1.9279	1.9035	1.8986	1.8747
	3A	-	1.9375	1.9294	1.9050	1.9013	1.8762
2-4 1/2 UNC	1A	.0029	1.9971	1.9641	1.8528	1.8385	1.7245
	2A	.0029	1.9971	1.9751	1.8528	1.8433	1.7245
	3A	-	2.0000	1.9780	1.8557	1.8486	1.7274
2-6 UN	2A	.0026	1.9974	1.9792	1.8891	1.8805	1.7929
	3A	-	2.0000	1.9818	1.8917	1.8853	1.7955
2-8 UN	2A	.0023	1.9977	1.9827	1.9165	1.9087	1.8443
	3A	-	2.0000	1.9850	1.9188	1.9130	1.8466
2-10 UNS	2A	.0020	1.9980	1.9851	1.9330	1.9265	1.8753
2-12 UN	2A	.0018	1.9982	1.9868	1.9441	1.9380	1.8960
	3A	-	2.0000	1.9886	1.9459	1.9414	1.8978
2-14 UNS	2A	.0017	1.9983	1.9880	1.9519	1.9462	1.9107
2-16 UN	2A	.0016	1.9984	1.9890	1.9578	1.9524	1.9217
	3A	-	2.0000	1.9906	1.9594	1.9554	1.9233
2-18 UNS	2A	.0015	1.9985	1.9898	1.9624	1.9573	1.9303
2-10 UN	2A	.0015	1.9985	1.9904	1.9660	1.9611	1.9372
	3A	-	2.0000	1.9919	1.9675	1.9638	1.9387
2 1/16-16 UNS	2A	.0016	2.0609	2.0515	2.0203	2.0149	1.9842
	3A	-	2.0625	2.0531	2.0219	2.0179	1.9858
2 1/8-6 UN	2A	.0026	2.1224	2.1042	2.0141	2.0054	1.9179
	3A	-	2.1250	2.1068	2.0167	2.0102	1.9205
2 1/8-8 UN	2A	.0024	2.1226	2.1076	2.0414	2.0335	1.9692
	3A	-	2.1250	2.1100	2.0438	2.0379	1.9716

(Continued)

Table 6. *(Continued)* **Unified Screw Threads. Internal Dimensions. Classes 1B, 2B, and 3B.**

Size, TPI, and Series	Class	Minor Diameter		Pitch Diameter		Major Dia.	Tap Drill*
		min	max	min	max		
1 7/8-6 UN	2B	1.695	1.725	1.7667	1.7777	1.8750	1 45/64
	3B	1.6950	1.7146	1.7667	1.7749	1.8750	
1 7/8-8 UN	2B	1.740	1.765	1.7938	1.8038	1.8750	1 3/4
	3B	1.7400	1.7547	1.7938	1.8013	1.8750	
1 7/8-10 UNS	2B	1.767	1.788	1.8100	1.8184	1.8750	
1 7/8-12 UN	2B	1.785	1.803	1.8209	1.8287	1.8750	1 25/32
	3B	1.7850	1.7948	1.8209	1.8267	1.8750	
1 7/8-14 UNS	2B	1.798	1.814	1.8286	1.8359	1.8750	
1 7/8-16 UN	2B	1.807	1.821	1.8344	1.8413	1.8750	1 13/16
	3B	1.8070	1.8158	1.8344	1.8396	1.8750	
1 7/8-18 UNS	2B	1.815	1.828	1.8389	1.8455	1.8750	
1 7/8-20 UN	2B	1.821	1.832	1.8425	1.8488	1.8750	1 53/64
	3B	1.8210	1.8287	1.8425	1.8472	1.8750	
1 15/16-6 UN	2B	1.757	1.788	1.8292	1.8403	1.9375	1 49/64
	3B	1.7570	1.7771	1.8292	1.8375	1.9375	
1 15/16-8 UN	2B	1.802	1.827	1.8563	1.8663	1.9375	1 51/64
	3B	1.8020	1.8172	1.8563	1.8638	1.9375	
1 15/16-12 UN	2B	1.847	1.865	1.8834	1.8913	1.9375	1 27/32
	3B	1.8470	1.8570	1.8834	1.8893	1.9375	
1 15/16-16 UN	2B	1.870	1.884	1.8969	1.9039	1.9375	1 7/8
	3B	1.8700	1.8783	1.8969	1.9021	1.9375	
1 15/16-20 UN	2B	1.883	1.895	1.9050	1.9114	1.9375	1 57/64
	3B	1.8830	1.8912	1.9050	1.9098	1.9375	
2-4 1/2 UNC	1B	1.759	1.795	1.8557	1.8743	2.0000	1 25/32
	2B	1.759	1.795	1.8557	1.8681	2.0000	
	3B	1.7590	1.7861	1.8557	1.8650	2.0000	
2-6 UN	2B	1.820	1.850	1.8917	1.9028	2.0000	1 53/64
	3B	1.8200	1.8396	1.8917	1.9000	2.0000	
2-8 UN	2B	1.865	1.890	1.9188	1.9289	2.0000	1 7/8
	3B	1.8650	1.8797	1.9188	1.9264	2.0000	
2-10 UNS	2B	1.892	1.913	1.9350	1.9435	2.0000	
2-12 UN	2B	1.910	1.928	1.9459	1.9538	2.0000	1 29/32
	3B	1.9100	1.9198	1.9459	1.9518	2.0000	
2-14 UNS	2B	1.923	1.938	1.9536	1.9610	2.0000	
2-16 UN	2B	1.932	1.946	1.9594	1.9664	2.0000	1 15/16
	3B	1.9320	1.9408	1.9594	1.9646	2.0000	
2-18 UNS	2B	1.940	1.953	1.9639	1.9706	2.0000	
2-20 UN	2B	1.946	1.957	1.9675	1.9739	2.0000	1 61/64
	3B	1.9460	1.9537	1.9675	1.9723	2.0000	
2 1/16-16 UNS	2B	1.995	2.009	2.0219	2.0289	2.0625	2
	3B	1.9950	2.0033	2.0219	2.0271	2.0625	
2 1/8-6 UN	2B	1.945	1.975	2.0167	2.0280	2.1250	1 61/64
	3B	1.9450	1.9646	2.0167	2.0251	2.1250	
2 1/8-8 UN	2B	1.990	2.015	2.0438	2.0540	2.1250	2
	3B	1.9900	2.0047	2.0438	2.0515	2.1250	

(Continued)

Table 5. *(Continued)* **Unified Screw Threads. External Dimensions. Classes 1A, 2A, and 3A.**

Size, TPI, and Series	Class	Allowance	Major Diameter max	Major Diameter min	Pitch Diameter max	Pitch Diameter min	Minor Dia.
2 1/8-12 UN	2A	.0018	2.1232	2.1118	2.0691	2.0630	2.0210
	3A	-	2.1250	2.1136	2.0709	2.0664	2.0228
2 1/8-16 UN	2A	.0016	2.1234	2.1140	2.0828	2.0774	2.0467
	3A	-	2.1250	2.1156	2.0844	2.0803	2.0483
2 1/8-20 UN	3A	.0015	2.1235	2.1154	2.0910	2.0861	2.0622
	3A	-	2.1250	2.1169	2.0925	2.0888	2.0637
2 3/16-16 UNS	2A	.0016	2.1859	2.1765	2.1453	2.1399	2.1092
	3A	-	2.1875	2.1781	2.1469	2.1428	2.1108
2 1/4-4 1/2 UNC	1A	.0029	2.2471	2.2141	2.1028	2.0882	1.9745
	2A	.0029	2.2471	2.2251	2.1028	2.0931	1.9745
	3A	-	2.2500	2.2280	2.1057	2.0984	1.9774
2 1/4-6 UN	2A	.0026	2.2474	2.2292	2.1391	2.1303	2.0429
	3A	-	2.2500	2.2318	2.1417	2.1351	2.0455
2 1/4-8 UN	2A	.0024	2.2476	2.2326	2.1664	2.1584	2.0942
	3A	-	2.2500	2.2350	2.1688	2.1628	2.0966
2 1/4-10 UNS	2A	.0020	2.2480	2.2351	2.1830	2.1765	2.1253
2 1/4-12 UN	2A	.0018	2.2482	2.2368	2.1941	2.1880	2.1460
	3A	-	2.2500	2.2386	2.1959	2.1914	2.1478
2 1/4-14 UNS	2A	.0017	2.2483	2.2380	2.2019	2.1962	2.1607
2 1/4-16 UN	2A	.0016	2.2484	2.2390	2.2078	2.2024	2.1717
	3A	-	2.2500	2.2406	2.2094	2.2053	2.1733
2 1/4-18 UNS	2A	.0015	2.2485	2.2398	2.2124	2.2073	2.1803
2 1/4-20 UN	2A	.0015	2.2485	2.2404	2.2160	2.2111	2.1872
	3A	-	2.2500	2.2419	2.2175	2.2137	2.1887
2 5/16-16 UNS	2A	.0017	2.3108	2.3014	2.2702	2.2647	2.2341
	3A	-	2.3125	2.3031	2.2719	2.2678	2.2358
2 3/8-6 UN	2A	.0027	2.3723	2.3541	2.2640	2.2551	2.1678
	3A	-	2.3750	2.3568	2.2667	2.2601	2.1705
2 3/8-8 UN	2A	.0024	2.3726	2.3576	2.2914	2.2833	2.2192
	3A	-	2.3750	2.3600	2.2938	2.2878	2.2216
2 3/8-12 UN	2A	.0019	2.3731	2.3617	2.3190	2.3128	2.2709
	3A	-	2.3750	2.3636	2.3209	2.3163	2.2728
2 3/8-16 UN	2A	.0017	2.3733	2.3639	2.3327	2.3272	2.2966
	3A	-	2.3750	2.3656	2.3344	2.3303	2.2983
2 3/8-20 UN	2A	.0015	2.3735	2.3654	2.3410	2.3359	2.3122
	3A	-	2.3750	2.3669	2.3425	2.3387	2.3137
2 7/16-16 UNS	2A	.0017	2.4358	2.4264	2.3952	2.3897	2.3591
	3A	-	2.4375	2.4281	2.3969	2.3928	2.3608
2 1/2-4 UNC	1A	.0031	2.4969	2.4612	2.3345	2.3190	2.1902
	2A	.0031	2.4969	2.4731	2.3345	2.3241	2.1902
	3A	-	2.5000	2.4762	2.3376	2.3298	2.1933
2 1/2-6 UN	2A	.0027	2.4973	2.4791	2.3890	2.3800	2.2928
	3A	-	2.5000	2.4818	2.3917	2.3850	2.2955
2 1/2-8 UN	2A	.0024	2.4976	2.4826	2.4164	2.4082	2.3442
	3A	-	2.5000	2.4850	2.4188	2.4127	2.3466

(Continued)

Table 6. *(Continued)* Unified Screw Threads. Internal Dimensions. Classes 1B, 2B, and 3B.

Size, TPI, and Series	Class	Minor Diameter		Pitch Diameter		Major Dia.	Tap Drill*
		min	max	min	max		
2 1/8-12 UN	2B	2.035	2.053	2.0709	2.0788	2.1250	2 1/32
	3B	2.0350	2.0448	2.0709	2.0768	2.1250	
2 1/8-16 UN	2B	2.057	2.071	2.0844	2.0914	2.1250	2 1/16
	3B	2.0570	2.0658	2.0844	2.0896	2.1250	
2 1/8-20 UN	3B	2.071	2.082	2.0925	2.0989	2.1250	2 1/16
	3B	2.0710	2.0787	2.0925	2.0973	2.1250	
2 3/16-16 UNS	2B	2.120	2.134	2.1469	2.1539	2.1875	2 1/8
	3B	2.1200	2.1283	2.1469	2.1521	2.1875	
2 1/4-4 1/2 UNC	1B	2.009	2.045	2.1057	2.1247	2.2500	2
	2B	2.009	2.045	2.1057	2.1183	2.2500	
	3B	2.0090	2.0361	2.1057	2.1152	2.2500	
2 1/4-6 UN	2B	2.070	2.100	2.1417	2.1531	2.2500	2 1/16
	3B	2.0700	2.0896	2.1417	2.1502	2.2500	
2 1/4-8 UN	2B	2.115	2.140	2.1688	2.1792	2.2500	2 1/8
	3B	2.1150	2.1297	2.1688	2.1766	2.2500	
2 1/4-10 UNS	2B	2.142	2.163	2.1850	2.1935	2.2500	
2 1/4-12 UN	2B	2.160	2.178	2.1959	2.2038	2.2500	2 5/32
	3B	2.1600	2.1698	2.1959	2.2018	2.2500	
2 1/4-14 UNS	2B	2.173	2.188	2.2036	2.2110	2.2500	
2 1/4-16 UN	2B	2.182	2.196	2.0294	2.2164	2.2500	2 3/16
	3B	2.1820	2.1908	2.0294	2.2146	2.2500	
2 1/4-18 UNS	2B	2.190	2.203	2.2139	2.2206	2.2500	
2 1/4-20 UN	2B	2.196	2.207	2.2175	2.2239	2.2500	2 3/16
	3B	2.1960	2.2037	2.2175	2.2223	2.2500	
2 5/16-16 UNS	2B	2.245	2.259	2.2719	2.2791	2.3125	2 1/4
	3B	2.2450	2.2533	2.2719	2.2773	2.3125	
2 3/8-6 UN	2B	2.195	2.226	2.2667	2.2782	2.3750	2 3/16
	3B	2.1950	2.2146	2.2667	2.2753	2.3750	
2 3/8-8 UN	2B	2.240	2.265	2.2938	2.3043	2.3750	2 1/4
	3B	2.2400	2.2552	2.2938	2.3017	2.3750	
2 3/8-12 UN	2B	2.285	2.303	2.3209	2.3290	2.3750	58mm
	3B	2.2850	2.2948	2.3209	2.3269	2.3750	
2 3/8-16 UN	2B	2.307	2.321	2.3344	2.3416	2.3750	2 5/16
	3B	2.3070	2.3158	2.3344	2.3398	2.3750	
2 3/8-20 UN	2B	2.321	2.332	2.3425	2.3491	2.3750	2 5/16
	3B	2.3210	2.3287	2.3425	2.3475	2.3750	
2 7/16-16 UNS	2B	2.370	2.384	2.3969	2.4041	2.4375	2 3/8
	3B	2.3700	2.3783	2.3969	2.4023	2.4375	
2 1/2-4 UNC	1B	2.229	2.267	2.3376	2.3578	2.5000	2 7/32
	2B	2.229	2.267	2.3376	2.3511	2.5000	
	3B	2.2290	2.2594	2.3376	2.3477	2.5000	
2 1/2-6 UN	2B	2.320	2.350	2.3917	2.4033	2.5000	2 5/16
	3B	2.3200	2.3396	2.3917	2.4004	2.5000	
2 1/2-8 UN	2B	2.365	2.390	2.4188	2.4294	2.5000	2 3/8
	3B	2.3650	2.3797	2.4188	2.4268	2.5000	

(Continued)

Table 5. *(Continued)* **Unified Screw Threads. External Dimensions. Classes 1A, 2A, and 3A.**

Size, TPI, and Series	Class	Allowance	Major Diameter max	Major Diameter min	Pitch Diameter max	Pitch Diameter min	Minor Dia.
2 1/2-10 UNS	2A	.0020	2.4980	2.4851	2.4330	2.4263	2.3753
2 1/2-12 UN	2A	.0019	2.4981	2.4867	2.4440	2.4378	2.3959
	3A	-	2.5000	2.4886	2.4459	2.4413	2.3978
2 1/2-14 UNS	2A	.0017	2.4983	2.4880	2.4519	2.4461	2.4107
2 1/2-16 UN	2A	.0017	2.4983	2.4889	2.4577	2.4522	2.4216
	3A	-	2.5000	2.4906	2.4594	2.4553	2.4233
2 1/2-18 UNS	2A	.0016	2.4984	2.4897	2.4623	2.4570	2.4302
2 1/2-20 UN	2A	.0015	2.4985	2.4904	2.4660	2.4609	2.4372
	3A	-	2.5000	2.4919	2.4675	2.4637	2.4387
2 5/8-6 UN	2A	.0027	2.6223	2.6041	2.5140	2.5050	2.4178
	3A	-	2.6250	2.6068	2.5167	2.5099	2.4205
2 5/8-8 UN	2A	.0025	2.6225	2.6075	2.5413	2.5331	2.4691
	3A	-	2.6250	2.6100	2.5438	2.5376	2.4716
2 5/8-12 UN	2A	.0019	2.6231	2.6117	2.5690	2.5628	2.5209
	3A	-	2.6250	2.6136	2.5709	2.5663	2.5228
2 5/8-16 UN	2A	.0017	2.6233	2.6139	2.5827	2.5772	2.5466
	3A	-	2.6250	2.6156	2.5844	2.5803	2.5483
2 5/8-20 UN	2A	.0015	2.6235	2.6154	2.5910	2.5859	2.5622
	3A	-	2.6250	2.6169	2.5925	2.5887	2.5637
2 3/4-4 UNC	1A	.0032	2.7468	2.7111	2.5844	2.5686	2.4401
	2A	.0032	2.7468	2.7230	2.5844	2.5739	2.4401
	3A	-	2.7500	2.7262	2.5876	2.5797	2.4433
2 3/4-6 UN	2A	.0027	2.7473	2.7291	2.6390	2.6299	2.5428
	3A	-	2.7500	2.7318	2.6417	2.6349	2.5455
2 3/4-8 UN	2A	.0025	2.7475	2.7325	2.6663	2.6580	2.5941
	3A	-	2.7500	2.7350	2.6688	2.6625	2.5966
2 3/4-10 UNS	2A	.0020	2.7480	2.7351	2.6830	2.6763	2.6253
2 3/4-12 UN	2A	.0019	2.7481	2.7367	2.6940	2.6878	2.6459
	3A	-	2.7500	2.7386	2.6959	2.6913	2.6478
2 3/4-14 UNS	2A	.0017	2.7483	2.7380	2.7019	2.6961	2.6607
2 3/4-16 UN	2A	.0017	2.7483	2.7389	2.7077	2.7022	2.6716
	3A	-	2.7500	2.7406	2.7094	2.7053	2.6733
2 3/4-18 UNS	2A	.0016	2.7484	2.7397	2.7123	2.7070	2.6802
2 3/4-20 UN	2A	.0015	2.7485	2.7404	2.7160	2.7109	2.6872
	3A	-	2.7500	2.7419	2.7175	2.7137	2.6887
2 7/8-6 UN	2A	.0028	2.8722	2.8540	2.7639	2.7547	2.6677
	3A	-	2.8750	2.8568	2.7667	2.7598	2.6705
2 7/8-8 UN	2A	.0025	2.8725	2.8575	2.7913	2.7829	2.7191
	3A	-	2.8750	2.8600	2.7938	2.7875	2.7216
2 7/8-12 UN	2A	.0019	2.8731	2.8617	2.8190	2.8127	2.7709
	3A	-	2.8750	2.8636	2.8209	2.8162	2.7728
2 7/8-16 UN	2A	.0017	2.8733	2.8639	2.8327	2.8271	2.7966
	3A	-	2.8750	2.8656	2.8344	2.8302	2.7983
2 7/8-20 UN	2A	.0016	2.8734	2.8653	2.8409	2.8357	2.8121
	3A	-	2.8750	2.8669	2.8425	2.8386	2.8137

(Continued)

Table 6. *(Continued)* **Unified Screw Threads. Internal Dimensions. Classes 1B, 2B, and 3B.**

Size, TPI, and Series	Class	Minor Diameter min	Minor Diameter max	Pitch Diameter min	Pitch Diameter max	Major Dia.	Tap Drill*
2 1/2-10 UNS	2B	2.392	2.413	2.4350	2.4437	2.5000	
2 1/2-12 UN	2B	2.41	2.428	2.4459	2.4540	2.5000	2 13/32
	3B	2.410	2.4198	2.4459	2.4519	2.5000	
2 1/2-14 UNS	2B	2.423	2.438	2.4536	2.4612	2.5000	
2 1/2-16 UN	2B	2.432	2.466	2.4594	2.4666	2.5000	2 7/16
	3B	2.4320	2.4408	2.4594	2.4648	2.5000	
2 1/2-18 UNS	2B	2.440	2.453	2.4639	2.4708	2.5000	
2 1/2-20 UN	2B	2.446	2.457	2.4675	2.4741	2.5000	2 1/16
	3B	2.4460	2.4537	2.4675	2.4725	2.5000	
2 5/8-6 UN	2B	2.445	2.475	2.5167	2.5285	2.6250	2 7/16
	3B	2.4450	2.4646	2.5167	2.5255	2.6250	
2 5/8-8 UN	2B	2.490	2.515	2.5438	2.5545	2.6250	2 1/2
	3B	2.4900	2.5052	2.5438	2.5518	2.6250	
2 5/8-12 UN	2B	2.535	2.553	2.5709	2.5790	2.6250	2 17/32
	3B	2.5350	2.5448	2.5709	2.5769	2.6250	
2 5/8-16 UN	2B	2.557	2.571	2.5844	2.5916	2.6250	2 9/16
	3B	2.5570	2.5658	2.5844	2.5898	2.6250	
2 5/8-20 UN	2B	2.571	2.582	2.5925	2.5991	2.6250	2 9/16
	3B	2.5710	2.5787	2.5925	2.5975	2.6250	
2 3/4-4 UNC	1B	2.479	2.517	2.5876	2.6082	2.7500	2 1/2
	2B	2.479	2.517	2.5876	2.6013	2.7500	
	3B	2.4790	2.5094	2.5876	2.5979	2.7500	
2 3/4-6 UN	2B	2.570	2.600	2.6417	2.6536	2.7500	2 9/16
	3B	2.5700	2.5896	2.6417	2.6506	2.7500	
2 3/4-8 UN	2B	2.615	2.640	2.6688	2.6796	2.7500	2 5/8
	3B	2.6150	2.6297	2.6688	2.6769	2.7500	
2 3/4-10 UNS	2B	2.642	2.663	2.6850	2.6937	2.7500	
2 3/4-12 UN	2B	2.660	2.678	2.6959	2.7040	2.7500	2 21/32
	3B	2.6600	2.6698	2.6959	2.7019	2.7500	
2 3/4-14 UNS	2B	2.673	2.688	2.7036	2.7112	2.7500	
2 3/4-16 UN	2B	2.682	2.696	2.7094	2.7166	2.7500	2 11/16
	3B	2.6820	2.6908	2.7094	2.7148	2.7500	
2 3/4-18 UNS	2B	2.690	2.703	2.7139	2.7208	2.7500	
2 3/4-20 UN	2B	2.696	2.707	2.7175	2.7241	2.7500	2 11/16
	3B	2.6960	2.7037	2.7175	2.7225	2.7500	
2 7/8-6 UN	2B	2.695	2.725	2.7667	2.7787	2.8750	2 11/16
	3B	2.6950	2.7146	2.7667	2.7757	2.8750	
2 7/8-8 UN	2B	2.740	2.765	2.7938	2.8048	2.8750	2 3/4
	3B	2.7400	2.7552	2.7938	2.8020	2.8750	
2 7/8-12 UN	2B	2.785	2.803	2.8209	2.8291	2.8750	2 25/32
	3B	2.7850	2.7948	2.8209	2.8271	2.8750	
2 7/8-16 UN	2B	2.807	2.821	2.8344	2.8417	2.8750	2 13/16
	3B	2.8070	2.8158	2.8344	2.8399	2.8750	
2 7/8-20 UN	2B	2.821	2.832	2.8425	2.8493	2.8750	2 13/16
	3B	2.8210	2.8287	2.8425	2.8476	2.8750	

(Continued)

Table 5. *(Continued)* Unified Screw Threads. External Dimensions. Classes 1A, 2A, and 3A.

Size, TPI, and Series	Class	Allowance	Major Diameter		Pitch Diameter		Minor Dia.
			max	min	max	min	
3-4 UNC	1A	.0032	2.9968	2.9611	2.8344	2.8183	2.6901
	2A	.0032	2.9968	2.9730	2.8344	2.8237	2.6901
	3A	-	3.0000	2.9762	2.8376	2.8296	2.6933
3-6 UN	2A	.0028	2.9972	2.9790	2.8889	2.8796	2.7927
	3A	-	3.0000	2.9818	2.8917	2.8847	2.7955
3-8 UN	2A	.0026	2.9974	2.9824	2.9162	2.9077	2.8440
	3A	-	3.0000	2.9850	2.9188	2.9124	2.8466
3-10 UNS	2A	.0020	2.9980	2.9851	2.9330	2.9262	2.8753
3-12 UN	2A	.0019	2.9981	2.9867	2.9440	2.9377	2.8959
	3A	-	3.0000	2.9886	2.9459	2.9412	2.8978
3-14 UNS	2A	.0018	2.9982	2.9879	2.9518	2.9459	2.9106
3-16 UN	2A	.0017	2.9983	2.9889	2.9577	2.9521	2.9216
	3A	-	3.0000	2.9906	2.9594	2.9552	2.9233
3-18 UNS	2A	.0016	2.9984	2.9897	2.9623	2.9569	2.9302
3-20 UN	2A	.0016	2.9984	2.9903	2.9659	2.9607	2.9371
	3A	-	3.0000	2.9919	2.9675	2.9636	2.9387
3 1/8-6 UN	2A	.0028	3.1222	3.1040	3.0139	3.0045	2.9177
	3A	-	3.1250	3.1068	3.0167	3.0097	2.9205
3 1/8-8 UN	2A	.0026	3.1224	3.1074	3.0412	3.0326	2.9690
	3A	-	3.1250	3.1100	3.0438	3.0374	2.9716
3 1/8-12 UN	2A	.0019	3.1231	3.1117	3.0690	3.0627	3.0209
	3A	-	3.1250	3.1136	3.0709	3.0662	3.0228
3 1/8-16 UN	2A	.0017	3.1233	3.1139	3.0827	3.0771	3.0466
	3A	-	3.1250	3.1156	3.0844	3.0802	3.0483
3 1/4-4 UNC	1A	.0033	3.2467	3.2110	3.0843	3.0680	2.9400
	2A	.0033	3.2467	3.2229	3.0843	3.0734	2.9400
	3A	-	3.2500	3.2262	3.0876	3.0794	2.9433
3 1/4-6 UN	2A	.0028	3.2472	3.2290	3.1389	3.1294	3.0427
	3A	-	3.2500	3.2318	3.1417	3.1346	3.0455
3 1/4-8 UN	2A	.0026	3.2474	3.2324	3.1662	3.1575	3.0940
	3A	-	3.2500	3.2350	3.1688	3.1623	3.0966
3 1/4-10 UNS	2A	.0020	3.2480	3.2351	3.1830	3.1762	3.1253
3 1/4-12 UN	2A	.0019	3.2481	3.2367	3.1940	3.1877	3.1459
	3A	-	3.2500	3.2386	3.1959	3.1912	3.1478
3 1/4-14 UNS	2A	.0018	3.2482	3.2379	3.2018	3.1959	3.1606
3 1/4-16 UN	2A	.0017	3.2483	3.2389	3.2077	3.2021	3.1716
	3A	-	3.2500	3.2406	3.2094	3.2052	3.1733
3 1/4-18 UNS	2A	.0016	3.2484	3.2397	3.2123	3.2069	3.1802
3 3/8-6 UN	2A	.0029	3.3721	3.3539	3.2638	3.2543	3.1676
	3A	-	3.3750	3.3568	3.2667	3.2595	3.1705
3 3/8-8 UN	2A	.0016	3.3724	3.3574	3.2912	3.2824	3.2190
	3A	-	3.3750	3.3600	3.2938	3.2876	3.2216

(Continued)

Table 6. *(Continued)* **Unified Screw Threads. Internal Dimensions. Classes 1B, 2B, and 3B.**

Size, TPI, and Series	Class	Minor Diameter		Pitch Diameter		Major Dia.	Tap Drill*
		min	max	min	max		
3-4 UNC	1B	2.729	2.767	2.8376	2.8585	3.0000	2 3/4
	2B	2.729	2.767	2.8376	2.8515	3.0000	
	3B	2.7290	2.7954	2.8376	2.8480	3.0000	
3-6 UN	2B	2.820	2.850	2.8917	2.9038	3.0000	2 13/16
	3B	2.8200	2.8396	2.8917	2.9008	3.0000	
3-8 UN	2B	2.865	2.890	2.9188	2.9299	3.0000	2 7/8
	3B	2.8650	2.8797	2.9188	2.9271	3.0000	
3-10 UNS	2B	2.892	2.913	2.9350	2.9439	3.0000	
3-12 UN	2B	2.910	2.928	2.9459	2.9541	3.0000	74mm
	3B	2.9100	2.9198	2.9459	2.9521	3.0000	
3-14 UNS	2B	2.923	2.938	2.9536	2.9613	3.0000	
3-16 UN	2B	2.932	2.946	2.9594	2.9667	3.0000	2 15/16
	3B	2.9320	2.9408	2.9594	2.9649	3.0000	
3-18 UNS	2B	2.940	2.953	2.9639	2.9709	3.0000	
3-20 UN	2B	2.946	2.957	2.9675	2.9743	3.0000	2 15/16
	3B	2.9460	2.9537	2.9675	2.9726	3.0000	
3 1/8-6 UN	2B	2.945	2.975	3.0167	3.0289	3.1250	
	3B	2.9450	2.9646	3.0167	3.0259	3.1250	
3 1/8-8 UN	2B	2.990	3.015	3.0438	3.0550	3.1250	
	3B	2.9900	3.0052	3.0438	3.0522	3.1250	
3 1/8-12 UN	2B	3.035	3.053	3.0709	3.0791	3.1250	
	3B	3.0350	3.0448	3.0709	3.0771	3.1250	
3 1/8-16 UN	2B	3.057	3.071	3.0844	3.0917	3.1250	
	3B	3.0570	3.0658	3.0844	3.0899	3.1250	
3 1/4-4 UNC	1B	2.979	3.017	3.0876	3.1088	3.2500	
	2B	2.979	3.017	3.0876	3.1017	3.2500	
	3B	2.9790	3.0094	3.0876	3.0982	3.2500	Reaming Recommended
3 1/4-6 UN	2B	3.070	3.100	3.1417	3.1540	3.2500	
	3B	3.0700	3.0896	3.1417	3.1509	3.2500	
3 1/4-8 UN	2B	3.115	3.140	3.1688	3.1801	3.2500	
	3B	3.1150	3.1297	3.1688	3.1772	3.2500	
3 1/4-10 UNS	2B	3.142	3.163	3.1850	3.1939	3.2500	
3 1/4-12 UN	2B	3.160	3.178	3.1959	3.2041	3.2500	
	3B	3.1600	3.1698	3.1959	3.2021	3.2500	
3 1/4-14 UNS	2B	3.173	3.188	3.2036	3.2113	3.2500	
3 1/4-16 UN	2B	3.182	3.196	3.2094	3.2167	3.2500	
	3B	3.1820	3.1908	3.2094	3.2149	3.2500	
3 1/4-18 UNS	2B	3.190	3.203	3.2139	3.2209	3.2500	
3 3/8-6 UN	2B	3.195	3.225	3.2667	3.2791	3.3750	
	3B	3.1950	3.2146	3.2667	3.2760	3.3750	
3 3/8-8 UN	2B	3.240	3.265	3.2938	3.3052	3.3750	
	3B	3.2400	3.2552	3.2938	3.3023	3.3750	

(Continued)

Table 5. *(Continued)* Unified Screw Threads. External Dimensions. Classes 1A, 2A, and 3A.

Size, TPI, and Series	Class	Allowance	Major Diameter max	Major Diameter min	Pitch Diameter max	Pitch Diameter min	Minor Dia.
3 3/8-12 UN	2A	.0019	3.3731	3.3617	3.3190	3.3126	3.2709
	3A	-	3.3750	3.3636	3.3209	3.3161	3.2728
3 3/8-16 UN	2A	.0017	3.3733	3.3639	3.3327	3.3269	3.2966
	3A	-	3.3750	3.3656	3.3344	3.3301	3.2983
3 1/2-4 UNC	1A	.0033	3.4967	3.4610	3.3343	3.3177	3.1900
	2A	.0033	3.4967	3.4729	3.3343	3.3233	3.1900
	3A	-	3.5000	3.4762	3.3376	3.3293	3.1933
3 1/2-6 UN	2A	.0029	3.4971	3.4789	3.3888	3.3792	3.2926
	3A	-	3.5000	3.4818	3.3917	3.3845	3.2955
3 1/2-8 UN	2A	.0026	3.4974	3.4824	3.4162	3.4074	3.4440
	3A	-	3.5000	3.4850	3.4188	3.4122	3.3466
3 1/2-10 UNS	2A	.0021	3.4979	3.4850	3.4329	3.4260	3.3752
3 1/2-12 UN	2A	.0019	3.4981	3.4867	3.4440	3.4376	3.3959
	3A	-	3.5000	3.4886	3.4459	3.4411	3.3978
3 1/2-14 UNS	2A	.0018	3.4982	3.4879	3.4518	3.4457	3.4106
3 1/2-16 UN	2A	.0017	3.4983	3.4889	3.4577	3.4519	3.4216
	3A	-	3.5000	3.4906	3.4594	3.4551	3.4233
3 1/2-18 UNS	2A	.0017	3.4983	3.4896	3.4622	3.4567	3.4301
3 5/8-6 UN	2A	.0029	3.6221	3.6039	3.5138	3.5041	3.4176
	3A	-	3.6250	3.6068	3.5167	3.5094	3.4205
3 5/8-8 UN	2A	.0027	3.6223	3.6073	3.5411	3.5322	3.4689
	3A	-	3.6250	3.6100	3.5438	3.5371	3.4716
3 5/8-12 UN	2A	.0019	3.6231	3.6117	3.5690	3.5626	3.5209
	3A	-	3.6250	3.6136	3.5709	3.5661	3.5228
3 5/8-16 UN	2A	.0017	3.6233	3.6139	3.5827	3.5769	3.5466
	3A	-	3.6250	3.6156	3.5844	3.5801	3.5483
3 3/4-4 UNC	1A	.0034	3.7466	3.7109	3.5842	3.5674	3.4399
	2A	.0034	3.7466	3.7228	3.5842	3.5730	3.4399
	3A	-	3.7500	3.7262	3.5876	3.5792	3.4433
3 3/4-6 UN	2A	.0029	3.7471	3.7289	3.6388	3.6290	3.5426
	3A	-	3.7500	3.7318	3.6417	3.6344	3.5455
3 3/4-8 UN	2A	.0027	3.7473	3.7323	3.6661	3.6571	3.5939
	3A	-	3.7500	3.7350	3.6688	3.6621	3.5966
3 3/4-10 UNS	2A	.0021	3.7479	3.7350	3.6829	3.6760	3.3652
3 3/4-12 UN	2A	.0019	3.7481	3.7367	3.6940	3.6876	3.6459
	3A	-	3.7500	3.7386	3.6959	3.6911	3.6478
3 3/4-14 UNS	2A	.0018	3.7482	3.7379	3.7018	3.6957	3.6606
3 3/4-16 UN	2A	.0017	3.7483	3.7389	3.7077	3.7019	3.6716
	3A	-	3.7500	3.7406	3.7094	3.7051	3.6733
3 3/4-18 UNS	2A	.0017	3.7483	3.7396	3.7122	3.7067	3.6801
3 7/8-6 UN	2A	.0030	3.8720	3.8538	3.7637	3.7538	3.6675
	3A		3.8750	3.8568	3.7667	3.7593	3.6705
3 7/8-8 UN	2A	.0027	3.8723	3.8573	3.7911	3.7820	3.7189
	3A	-	3.8750	3.8600	3.7938	3.7870	3.7216

(Continued)

Table 6. *(Continued)* **Unified Screw Threads. Internal Dimensions. Classes 1B, 2B, and 3B.**

Size, TPI, and Series	Class	Minor Diameter		Pitch Diameter		Major Dia.	Tap Drill*
		min	max	min	max		
3 3/8-12 UN	2B	3.285	3.303	3.3209	3.3293	3.3750	
	3B	3.2850	3.2948	3.3209	3.3272	3.3750	
3 3/8-16 UN	2B	3.307	3.321	3.3344	3.3419	3.3750	
	3B	3.3070	3.3158	3.3344	3.3400	3.3750	
3 1/2-4 UNC	1B	3.229	3.267	3.3376	3.3591	3.5000	
	2B	3.229	3.267	3.3376	3.3519	3.5000	
	3B	3.2290	3.2594	3.3376	3.3484	3.5000	
3 1/2-6 UN	2B	3.320	3.350	3.3917	3.4042	3.5000	
	3B	3.3200	3.3396	3.3917	3.4011	3.5000	
3 1/2-8 UN	2B	3.365	3.390	3.4188	3.4303	3.5000	
	3B	3.3650	3.3797	3.4188	3.4274	3.5000	
3 1/2-10 UNS	2B	3.392	3.413	3.4350	3.4440	3.5000	
3 1/2-12 UN	2B	3.410	3.428	3.4459	3.4543	3.5000	
	3B	3.4100	3.4198	3.4459	3.4522	3.5000	
3 1/2-14 UNS	2B	3.423	3.438	3.4536	3.4615	3.5000	
3 1/2-16 UN	2B	3.432	3.446	3.4594	3.4669	3.5000	
	3B	3.4320	3.4408	3.4594	3.4650	3.5000	
3 1/2-18 UNS	2B	3.440	3.453	3.4639	3.4711	3.5000	
3 5/8-6 UN	2B	3.445	3.475	3.5167	3.5293	3.6250	
	3B	3.4450	3.4646	3.5167	3.5262	3.6250	
3 5/8-8 UN	2B	3.490	3.515	3.5438	3.5554	3.6250	
	3B	3.4900	3.5052	3.5438	3.5525	3.6250	
3 5/8-12 UN	2B	3.535	3.553	3.5709	3.5793	3.6250	
	3B	3.5350	3.5448	3.5709	3.5772	3.6250	
3 5/8-16 UN	2B	3.557	3.571	3.5844	3.5919	3.6250	
	3B	3.5570	3.5658	3.5844	3.5900	3.6250	
3 3/4-4 UNC	1B	3.479	3.517	3.5876	3.6094	3.7500	
	2B	3.479	3.517	3.5876	3.6021	3.7500	
	3B	3.4790	3.5094	3.5876	3.5985	3.7500	
3 3/4-6 UN	2B	3.570	3.600	3.6417	3.6544	3.7500	
	3B	3.5700	3.5896	3.6417	3.6512	3.7500	
3 3/4-8 UN	2B	3.615	3.640	3.6688	3.6805	3.7500	
	3B	3.6150	3.6297	3.6688	3.6776	3.7500	
3 3/4-10 UNS	2B	3.642	3.663	3.6850	3.6940	3.7500	
3 3/4-12 UN	2B	3.660	3.678	3.6959	3.7043	3.7500	
	3B	3.6600	3.6698	3.6959	3.7022	3.7500	
3 3/4-14 UNS	2B	3.673	3.688	3.7036	3.7115	3.7500	
3 3/4-16 UN	2B	3.682	3.696	3.7094	3.7169	3.7500	
	3B	3.6820	3.6908	3.7094	3.7150	3.7500	
3 3/4-18 UNS	2B	3.690	3.703	3.7139	3.7211	3.7500	
3 7/8-6 UN	2B	3.695	3.725	3.7667	3.7795	3.8750	
	3B	3.6950	3.7146	3.7667	3.7763	3.8750	
3 7/8-8 UN	2B	3.740	3.765	3.7938	3.8056	3.8750	
	3B	3.7400	3.7552	3.7938	3.8026	3.8750	

Reaming Recommended

(Continued)

Table 5. *(Continued)* **Unified Screw Threads. External Dimensions. Classes 1A, 2A, and 3A.**

Size, TPI, and Series	Class	Allowance	Major Diameter max	Major Diameter min	Pitch Diameter max	Pitch Diameter min	Minor Dia.
3 7/8-12 UN	2A	.0020	3.8730	3.8616	3.8189	3.8124	3.7708
	3A	-	3.8750	3.8636	3.8209	3.8160	3.7728
3 7/8-16 UN	2A	.0018	3.8732	3.8638	3.8326	3.8267	3.7965
	3A	-	3.8750	3.8656	3.8344	3.8300	3.7983
4-4 UNC	1A	.0034	3.9966	3.9609	3.8342	3.8172	3.6899
	2A	.0034	3.9966	3.9728	3.8342	3.8229	3.6899
	3A	.0034	4.0000	3.9762	3.8376	3.8291	3.6933
4-8 UN	2A	.0027	3.9973	3.9823	3.9161	3.9070	3.8439
	3A	.0000	4.0000	3.9850	3.9188	3.9120	3.8466
4-12 UN	2A	.0020	3.9980	3.9866	3.9439	3.9374	3.8958
	3A	.0000	4.0000	3.9886	3.9459	3.9410	3.8978
4-16 UN	2A	.0018	3.9982	3.9888	3.9576	3.9517	3.9215
	3A	.0000	4.0000	3.9906	3.9594	3.9550	3.9233
4 1/4-8 UN	2A	.0028	4.2472	4.2322	4.1660	4.1567	4.0938
	3A	.0000	4.2500	4.2350	4.1688	4.1618	4.0966
4 1/4-12 UN	2A	.0020	4.2480	4.2366	4.1939	4.1874	4.1458
	3A	.0000	4.2500	4.2386	4.1959	4.1910	4.1478
4 1/4-16 UN	2A	.0018	4.2482	4.2388	4.2076	4.2017	4.1715
	3A	.0000	4.2500	4.2406	4.2094	4.2050	4.1735
4 1/2-8 UN	2A	.0028	4.4972	4.4822	4.4160	4.4066	4.3438
	3A	.0000	4.5000	4.4850	4.4188	4.4117	4.3466
4 1/2-12 UN	2A	.0020	4.4980	4.4866	4.4439	4.4374	4.3958
	3A	.0000	4.5000	4.4886	4.4459	4.4410	4.3978
4 1/2-16 UN	2A	.0018	4.4982	4.4888	4.4576	4.4517	4.4215
	3A	.0000	4.5000	4.4906	4.4594	4.4550	4.4233
4 3/4-8 UN	2A	.0029	4.7471	4.7321	4.6659	4.6464	4.5937
	3A	.0000	4.7500	4.7350	4.6688	4.5616	4.5966
4 3/4-12 UN	2A	.0020	4.7480	4.7366	4.6939	4.6872	4.6458
	3A	.0000	4.7500	4.7386	4.6959	4.6909	4.6478
4 3/4-16 UN	2A	.0018	4.7482	4.7388	4.7076	4.7015	4.5715
	3A	.0000	4.7500	4.7406	4.7094	4.7049	4.6733
5-8 UN	2A	.0029	4.9971	4.9821	4.9159	4.9062	4.8137
	3A	.0000	5.0000	4.9850	4.9188	4.9116	4.8466
5-12 UN	2A	.0020	4.9980	4.9866	4.9439	4.9372	4.8958
	3A	.0000	5.0000	4.9886	4.9459	4.9409	1.8978
5-16 UN	2A	.0018	4.9982	4.9888	4.9576	4.9515	4.9215
	3A	.0000	5.0000	4.9906	4.9594	4.9549	4.9233
5 1/4-8 UN	2A	.0029	5.2471	5.2321	5.1659	5.1561	5.0937
	3A	.0000	5.2500	5.2350	5.1688	5.1615	5.0966
5 1/4-12 UN	2A	.0020	5.2480	5.2366	5.1939	5.1872	5.1458
	3A	.0000	5.2500	5.2386	5.1959	5.1909	5.1478
5 1/4-16 UN	2A	.0018	5.2482	5.2388	5.2076	5.2015	5.1715
	3A	.0000	5.2500	5.2406	5.2094	5.2049	5.1733
5 1/2-8 UN	2A	.0030	5.4970	5.4820	5.4158	5.4059	5.3436
	3A	.0000	5.5000	5.4850	5.4188	5.4114	5.3466

(Continued)

Table 6. *(Continued)* **Unified Screw Threads. Internal Dimensions. Classes 1B, 2B, and 3B.**

Size, TPI, and Series	Class	Minor Diameter min	Minor Diameter max	Pitch Diameter min	Pitch Diameter max	Major Dia.	Tap Drill*
3 7/8-12 UN	2B	3.785	3.803	3.8209	3.8294	3.8750	
	3B	3.7850	3.7948	3.8209	3.8273	3.8750	
3 7/8-16 UN	2B	3.807	3.821	3.8344	3.8420	3.8750	
	3B	3.8070	3.8158	3.8344	3.8401	3.8750	
4-4 UNC	1B	3.729	3.767	3.8376	3.8597	4.000	
	2B	3.729	3.767	3.8376	3.8523	4.000	
	3B	3.7290	3.7594	3.8376	3.8487	4.000	
4-8 UN	2B	3.865	3.890	3.9188	3.9307	4.000	
	3B	3.8650	3.8797	3.9188	3.9277	4.000	
4-12 UN	2B	3.910	3.928	3.9459	3.9544	4.000	
	3B	3.9100	3.9198	3.9459	3.9523	4.000	
4-16 UN	2B	3.932	3.946	3.9594	3.9670	4.000	
	3B	3.9320	3.9408	3.9594	3.9651	4.000	
4 1/4-8 UN	2A	4.115	4.140	4.1688	4.1809	4.2500	
	3A	4.1150	4.1297	4.1688	4.1778	4.2500	
4 1/4-12 UN	2A	4.160	4.178	4.1959	4.2044	4.2500	
	3A	4.1600	4.1689	4.1959	4.2023	4.2500	
4 1/4-16 UN	2A	4.182	4.196	4.2094	4.2170	4.2500	
	3A	4.1820	4.1908	4.2094	4.2151	4.2500	
4 1/2-8 UN	2A	4.365	4.390	4.4188	4.4310	4.5000	
	3A	4.3650	4.3797	4.4188	4.4280	4.5000	Reaming Recommended
4 1/2-12 UN	2A	4.410	4.420	4.4459	4.4544	4.5000	
	3A	4.4100	4.4198	4.4459	4.4523	4.5000	
4 1/2-16 UN	2A	4.432	4.446	4.4594	4.4670	4.5000	
	3A	4.4320	4.4408	4.4594	4.4651	4.5000	
4 3/4-8 UN	2A	4.615	4.640	4.6688	4.6812	4.7500	
	3A	4.6150	4.6297	4.6688	4.6781	4.7500	
4 3/4-12 UN	2A	4.660	4.678	4.6959	4.7046	4.7500	
	3A	4.6600	4.6698	4.6959	4.7025	4.7500	
4 3/4-16 UN	2A	4.682	4.696	4.7094	4.7133	4.7500	
	3A	4.6820	4.6908	4.7094	4.7153	4.7500	
5-8 UN	2A	4.865	4.890	4.9188	4.9314	5.0000	
	3A	4.8650	4.8797	4.9188	4.9282	5.0000	
5-12 UN	2A	4.910	4.928	4.9459	4.9546	5.0000	
	3A	4.9100	4.9198	4.9459	4.9525	5.0000	
5-16 UN	2A	4.932	4.946	4.9594	4.9673	5.0000	
	3A	4.9320	4.9408	4.9594	4.9653	5.0000	
5 1/4-8 UN	2A	5.115	5.140	5.1688	5.1815	5.2500	
	3A	5.1150	5.1297	5.1688	5.1783	5.2500	
5 1/4-12 UN	2A	5.160	5.178	5.1959	5.2046	5.2500	
	3A	5.1600	5.1698	5.1959	5.2025	5.2500	
5 1/4-16 UN	2A	5.182	5.196	5.2094	5.2173	5.2500	
	3A	5.1820	5.1908	5.2094	5.2153	5.2500	
5 1/2-8 UN	2A	5.365	5.390	5.4188	5.4317	5.5000	
	3A	5.3650	5.3797	5.4188	5.4285	5.5000	

(Continued)

Table 5. *(Continued)* **Unified Screw Threads. External Dimensions. Classes 1A, 2A, and 3A.**

Size, TPI, and Series	Class	Allowance	Major Diameter		Pitch Diameter		Minor Dia.
			max	min	max	min	
5 1/2-12 UN	2A	.0020	5.4980	5.4866	5.4439	5.4372	5.3958
	3A	.0000	5.5000	5.4886	5.4459	5.4409	5.3978
5 1/2-16 UN	2A	.0018	5.4982	5.4888	5.4576	5.4515	5.4215
	3A	.0000	5.5000	5.4906	5.4594	5.4549	5.4233
5 3/4-8 UN	2A	.0030	5.7470	5.7320	5.6658	5.6558	5.5936
	3A	.0000	5.7500	5.7350	5.6688	5.6613	5.5966
5 3/4-12 UN	2A	.0021	5.7479	5.7365	5.6938	5.5869	5.6457
	3A	.0000	5.7500	5.7386	5.6959	5.6907	5.6478
5 3/4-16 UN	2A	.0019	5.7481	5.7386	5.7075	5.7013	5.6714
	3A	.0000	5.7500	5.7406	5.7094	5.7047	5.6733
6-8 UN	2A	.0030	5.9970	5.9820	5.9158	5.9056	5.8436
	3A	.0000	6.0000	5.9850	5.9188	5.9112	5.8466
6-12 UN	2A	.0021	5.9979	5.9865	5.9438	5.9369	5.8957
	3A	.0000	6.0000	5.9886	5.9459	5.9407	5.8978
6-16 IN	2A	.0019	5.9981	5.9887	5.9575	5.9513	5.9214
	3A	.0000	6.0000	5.9906	5.9594	5.9547	5.9233

Notes: Source, National Bureau of Standards H28 and TC 9-524. Dimensions are generally in accordance with ANSI B1.1-1974 (See Note 2, Table 4). UNS series threads are to be used only if standard series threads do not meet requirements. Class 2A threads with an additive finish increase by the allowance to Class 3A dimensions.

Table 6. *(Conclusion)* **Unified Screw Threads. Internal Dimensions. Classes 1B, 2B, and 3B.**

Size, TPI, and Series	Class	Minor Diameter		Pitch Diameter		Major Dia.	Tap Drill*
		min	max	min	max		
5 1/2-12 UN	2A	5.410	5.428	5.4459	5.4546	5.5000	
	3A	5.4100	5.4198	5.4459	5.4525	5.5000	
5 1/2-16 UN	2A	5.432	5.446	5.4594	5.4673	5.5000	
	3A	5.4320	5.4408	5.4594	5.4653	5.5000	
5 3/4-8 UN	2A	5.615	5.640	5.6688	5.6818	5.7500	Reaming Recommended
	3A	5.6150	5.6297	5.6688	5.6786	5.7500	
5 3/4-12 UN	2A	5.660	5.678	5.6959	5.7049	5.7500	
	3A	5.6600	5.6698	5.6959	5.7026	5.7500	
5 3/4-16 UN	2A	5.682	5.696	5.7094	5.7175	5.7500	
	3A	5.6820	5.6908	5.7094	5.7155	5.7500	
6-8 UN	2A	5.865	5.890	5.9188	5.9320	6.0000	
	3A	5.8650	5.8797	5.9188	5.9287	6.0000	
6-12 UN	2A	5.910	5.928	5.9459	5.9549	6.0000	
	3A	5.9100	5.9198	5.9459	5.9526	6.0000	
6-16 IN	2A	5.932	5.946	5.9594	5.9675	6.0000	
	3A	5.9320	5.9408	5.9594	5.9655	6.0000	

* Tap drill recommendations are for maximum percent of thread engagement (Source, FED-STD-H28/2B). See Tapping Section for alternative sizes. Reaming is suggested for larger sizes.

Notes: Source, National Bureau of Standards H28 and TC 9-524. Dimensions are generally in accordance with ANSI B1.1-1974 (See Note 2, Table 4). UNS series threads are to be used only if standard series threads do not meet requirements.

Table 7. UNJ Series External Threads, Class 3A. Secondary Sizes in italics. *(Source, MIL-S-8879C.)*

Size, TPI, and Series	Major Diameter		Pitch Diameter		Minor Diameter		Root Radius	
	max	min	max	min	max	min	max	min
0-80 UNJF	0.0600	0.0568	0.0519	0.0506	0.0456	0.0435	0.0023	0.0019
1-64 UNJC	0.0730	0.0692	0.0629	0.0614	0.0550	0.0526	0.0028	0.0023
1-72 UNJF	0.0730	0.0695	0.0640	0.0626	0.0570	0.0547	0.0025	0.0021
2-56 UNJC	0.0860	0.0819	0.0744	0.0728	0.0654	0.0627	0.0032	0.0027
2-64 UNJF	0.0860	0.0822	0.0759	0.0744	0.0680	0.0656	0.0028	0.0023
3-48 UNJC	0.0990	0.0945	0.0855	0.0838	0.0750	0.0720	0.0038	0.0031
3-56 UNJF	0.0990	0.0949	0.0874	0.0858	0.0784	0.0757	0.0032	0.0027
4-40 UNJC	0.1120	0.1069	0.0958	0.0939	0.0832	0.0798	0.0045	0.0038
4-48 UNJF	0.1120	0.1075	0.0985	0.0967	0.0880	0.0849	0.0038	0.0031
5-40 UNJC	0.1250	0.1199	0.1088	0.1069	0.0962	0.0928	0.0045	0.0038
5-44 UNJF	0.1250	0.1202	0.1102	0.1083	0.0987	0.0954	0.0041	0.0034
6-32 UNJC	0.1380	0.1320	0.1177	0.1156	0.1019	0.0979	0.0056	0.0047
6-40 UNJF	0.1380	0.1329	0.1218	0.1198	0.1092	0.1057	0.0045	0.0038
8-32 UNJC	0.1640	0.1580	0.1437	0.1415	0.1279	0.1238	0.0056	0.0047
8-36 UNJF	0.1640	0.1585	0.1460	0.1439	0.1320	0.1282	0.0050	0.0042
10-24 UNJC	0.1900	0.1828	0.1629	0.1604	0.1418	0.1368	0.0075	0.0063
10-32 UNJF	0.1900	0.1840	0.1697	0.1674	0.1539	0.1497	0.0056	0.0047
12-24 UNJC	0.2160	0.2088	0.1889	0.1863	0.1678	0.1627	0.0075	0.0063
12-28 UNJF	0.2160	0.2095	0.1928	0.1904	0.1748	0.1702	0.0064	0.0054
12-32 UNJEF	0.2160	0.2100	0.1957	0.1933	0.1799	0.1756	0.0056	0.0047
1/4-20 UNJC	0.2500	0.2419	0.2175	0.2147	0.1922	0.1864	0.0090	0.0075
1/4-28 UNJF	0.2500	0.2435	0.2268	0.2243	0.2088	0.2041	0.0064	0.0054
1/4-32 UNJEF	0.2500	0.2440	0.2297	0.2273	0.2139	0.2096	0.0056	0.0047
5/16-18 UNJC	0.3125	0.3038	0.2764	0.2734	0.2483	0.2420	0.0100	0.0083
5/16-24 UNJF	0.3125	0.3053	0.2854	0.2827	0.2644	0.2591	0.0075	0.0063
5/16-32 UNJEF	0.3125	0.3065	0.2922	0.2898	0.2764	0.2721	0.0056	0.0047
3/8-16 UNJC	0.3750	0.3656	0.3344	0.3311	0.3028	0.2957	0.0113	0.0094
3/8-24 UNJF	0.3750	0.3678	0.3479	0.3450	0.3268	0.3214	0.0075	0.0063
3/8-32 UNJEF	0.3750	0.3690	0.3547	0.3522	0.3389	0.3345	0.0056	0.0047
7/16-14 UNJC	0.4375	0.4272	0.3911	0.3876	0.3550	0.3472	0.0129	0.0107
7/16-16 UNJ	0.4375	0.4281	0.3969	0.3935	0.3653	0.3581	0.0113	0.0094
7/16-20 UNJF	0.4375	0.4294	0.4050	0.4019	0.3797	0.3736	0.0090	0.0075
7/16-28 UNJEF	0.4375	0.4310	0.4143	0.4116	0.3963	0.3914	0.0064	0.0054
1/2-13 UNJC	0.5000	0.4891	0.4500	0.4463	0.4111	0.4028	0.0139	0.0115
1/2-16 UNJ	0.5000	0.4906	0.4594	0.4559	0.4278	0.4205	0.0113	0.0094
1/2-20 UNJF	0.5000	0.4919	0.4675	0.4643	0.4422	0.4360	0.0090	0.0075
1/2-28 UNJEF	0.5000	0.4935	0.4768	0.4740	0.4588	0.4538	0.0064	0.0054
9/16-12 UNJC	0.5625	0.5511	0.5084	0.5045	0.4663	0.4574	0.0150	0.0125
9/16-16 UNJ	0.5625	0.5531	0.5219	0.5184	0.4903	0.4830	0.0113	0.0094
9/16-18 UNJF	0.5625	0.5538	0.5264	0.5230	0.4983	0.4916	0.0100	0.0083
9/16-24 UNJEF	0.5625	0.5553	0.5354	0.5325	0.5144	0.5089	0.0075	0.0063
5/8-11 UNJC	0.6250	0.6129	0.5660	0.5619	0.5201	0.5105	0.0164	0.0136
5/8-12 UNJ	0.6250	0.6136	0.5709	0.5668	0.5288	0.5196	0.0150	0.0125
5/8-16UNJ	0.6250	0.6156	0.5844	0.5808	0.5528	0.5454	0.0113	0.0094
5/8-18 UNJF	0.6250	0.6163	0.5889	0.5854	0.5608	0.5540	0.0100	0.0083
5/8-24 UNJEF	0.6250	0.6178	0.5979	0.5949	0.5768	0.5713	0.0075	0.0063
11/16-12 UNJ	0.6875	0.6761	0.6334	0.6293	0.5913	0.5822	0.0150	0.0125
11/16-16 UNJ	0.6875	0.6781	0.6469	0.6433	0.6153	0.6079	0.0113	0.0094
11/16-24 UNJEF	0.6875	0.6803	0.6604	0.6574	0.6394	0.6338	0.0075	0.0063

(Continued)

Table 8. UNJ Series Internal Threads, Class 3B. Secondary Sizes in italics. *(Source, MIL-S-8879C.)*

Size, TPI, and Series	Minor Diameter		Pitch Diameter		Major Diameter
	min	max	min	max	min
0-80 UNJF	0.0479	0.0511	0.0519	0.0536	0.0600
1-64 UNJC	0.0578	0.0619	0.0629	0.0648	0.0730
1-72 UNJF	0.0595	0.0631	0.0640	0.0659	0.0730
2-56 UNJC	0.0686	0.0732	0.0744	0.0765	0.0860
2-64 UNJF	0.0708	0.0749	0.0759	0.0779	0.0860
3-48 UNJC	0.0787	0.0841	0.0855	0.0877	0.0990
3-56 UNJF	0.0816	0.0862	0.0874	0.0895	0.0990
4-40 UNJC	0.0877	0.0942	0.0958	0.0982	0.1120
4-48 UNJF	0.0917	0.0971	0.0985	0.1008	0.1120
5-40 UNJC	0.1007	0.1072	0.1088	0.1113	0.1250
5-44 UNJF	0.1029	0.1088	0.1102	0.1126	0.1250
6-32 UNJC	0.1076	0.1157	0.1177	0.1204	0.1380
6-40 UNJF	0.1137	0.1202	0.1218	0.1243	0.1380
8-32 UNJC	0.1336	0.1417	0.1437	0.1465	0.1640
8-36 UNJF	0.1370	0.1442	0.1460	0.1487	0.1640
10-24 UNJC	0.1494	0.1600	0.1629	0.1661	0.1900
10-32 UNJF	0.1596	0.1675	0.1697	0.1726	0.1900
12-24 UNJC	0.1754	0.1852	0.1889	0.1922	0.2160
12-28 UNJF	0.1812	0.1896	0.1928	0.1959	0.2160
12-32 UNJEF	0.1856	0.1929	0.1957	0.1988	0.2160
1/4-20 UNJC	0.2013	0.2121	0.2175	0.2211	0.2500
1/4-28 UNJF	0.2152	0.2229	0.2268	0.2300	0.2500
1/4-32 UNJEF	0.2196	0.2263	0.2297	0.2328	0.2500
5/16-18 UNJC	0.2584	0.2690	0.2764	0.2803	0.3125
5/16-24 UNJF	0.2719	0.2799	0.2854	0.2890	0.3725
5/16-32 UNJEF	0.2820	0.2880	0.2922	0.2953	0.3125
3/8-16 UNJC	0.3142	0.3251	0.3344	0.3387	0.3750
3/8-24 UNJF	0.3344	0.3418	0.3479	0.3516	0.3750
3/8-32 UNJEF	0.3446	0.3501	0.3547	0.3580	0.3750
7/16-14 UNJC	0.3680	0.3795	0.3911	0.3957	0.4375
7/16-16 UNJ	0.3767	0.3869	0.3969	0.4014	0.4375
7/16-20 UNJF	0.3888	0.3970	0.4050	0.4091	0.4375
7/16-28 UNJEF	0.4027	0.4086	0.4143	0.4178	0.4375
1/2-13 UNJC	0.4251	0.4368	0.4500	0.4548	0.5000
1/2-16 UNJ	0.4392	0.4488	0.4594	0.4640	0.5000
1/2-20 UNJF	0.4513	0.4591	0.4675	0.4717	0.5000
1/2-28 UNJEF	0.4652	0.4708	0.4768	0.4804	0.5000
9/16-12 UNJC	0.4814	0.4914	0.5084	0.5135	0.5625
9/16-16 UNJ	0.5017	0.5109	0.5219	0.5265	0.5625
9/16-18 UNJF	0.5084	0.5166	0.5264	0.5308	0.5625
9/16-24 UNJEF	0.5219	0.5281	0.5354	0.5392	0.5625
5/8-11 UNJC	0.5365	0.5474	0.5660	0.5714	0.6250
5/8-12 UNJ	0.5439	0.5539	0.5709	0.5762	0.6250
5/8-16UNJ	0.5642	0.5731	0.5844	0.5890	0.6250
5/8-18 UNJF	0.5709	0.5788	0.5889	0.5934	0.6250
5/8-24 UNJEF	0.5844	0.5904	0.5979	0.6018	0.6250
11/16-12 UNJ	0.6064	0.6164	0.6334	0.6387	0.6875
11/16-16 UNJ	0.6267	0.6353	0.6469	0.6515	0.6875
11/16-24 UNJEF	0.6469	0.6547	0.6604	0.6643	0.6875

(Continued)

Table 7. *(Continued)* **UNJ Series External Threads, Class 3A.** Secondary Sizes in italics. *(Source, MIL-S-8879C.)*

Size, TPI, and Series	Major Diameter max	Major Diameter min	Pitch Diameter max	Pitch Diameter min	Minor Diameter max	Minor Diameter min	Root Radius max	Root Radius min
3/4-10 UNJC	0.7500	0.7371	0.6850	0.6806	0.6345	0.6240	0.0180	0.0150
3/4-12 UNJ	0.7500	0.7386	0.6959	0.6918	0.6538	0.6446	0.0150	0.0125
3/4-16 UNJF	0.7500	0.7406	0.7094	0.7056	0.6778	0.6702	0.0113	0.0094
3/4-20 UNJEF	0.7500	0.7419	0.7175	0.7142	0.6922	0.6859	0.0090	0.0075
13/16-12 UNJ	0.8125	0.8011	0.7584	0.7543	0.7163	0.7072	0.0150	0.0125
13/16-16 UNJ	0.8125	0.8031	0.7719	0.7683	0.7403	0.7329	0.0113	0.0094
13/16-20 UNJEF	0.8125	0.8044	0.7800	0.7767	0.7547	0.7484	0.0090	0.0075
7/8-9 UNJC	0.8750	0.8611	0.8028	0.7981	0.7467	0.7352	0.0200	0.0167
7/8-12 UNJ	0.8750	0.8636	0.8209	0.8168	0.7788	0.7696	0.0150	0.0125
7/8-14 UNJF	0.8750	0.8647	0.8286	0.8245	0.7925	0.7841	0.0129	0.0107
7/8-16 UNJ	0.8750	0.8656	0.8344	0.8308	0.8028	0.7954	0.0113	0.0094
7/8-20 UNJEF	0.8750	0.8669	0.8425	0.8392	0.8172	0.8109	0.0090	0.0075
15/16-12 UNJ	0.9375	0.9261	0.8834	0.8793	0.8413	0.8320	0.0150	0.0125
15/16-16 UNJ	0.9375	0.9281	0.8969	0.8932	0.8653	0.8578	0.0113	0.0094
15/16-20 UNJEF	0.9375	0.9294	0.9050	0.9016	0.8797	0.8733	0.0090	0.0075
1-8 UNJC	1.0000	0.9850	0.9188	0.9137	0.8556	0.8430	0.0226	0.0188
1-12 UNJF	1.0000	0.9886	0.9459	0.9415	0.9038	0.8944	0.0150	0.0125
1-16 UNJ	1.0000	0.9906	0.9594	0.9557	0.9278	0.9203	0.0113	0.0094
1-20 UNJEF	1.0000	0.9919	0.9675	0.9641	0.9422	0.9358	0.0090	0.0075
1 1/16-8 UNJ	1.0625	1.0475	09813	0.9762	0.9182	0.9055	0.0226	0.0188
1 1/16-12 UNJ	1.0625	1.0511	1.0084	1.0042	0.9663	0.9570	0.0150	0.0125
1 1/16-16 UNJ	1.0625	1.0531	1.0219	1.0182	0.9903	0.9828	0.0113	0.0094
1 1/16-18 UNJEF	1.0625	1.0538	1.0264	1.0228	0.9983	0.9914	0.0100	0.0083
1 1/8-7 UNJC	1.1250	1.1086	1.0322	1.0268	0.9600	0.9460	0.0258	0.0214
1 1/8-8 UNJ	1.1250	1.1100	1.0438	1.0386	0.9806	0.9661	0.0226	0.0188
1 1/8-12 UNJF	1.1250	1.1136	1.0709	1.0664	1.0288	1.0192	0.0150	0.0125
1 1/8-16 UNJ	1.1250	1.1156	1.0844	1.0807	1.0528	1.0453	0.0113	0.0094
1 1/8-18 UNJEF	1.1250	1.1163	1.0889	1.0853	1.0608	1.0539	0.0100	0.0083
1 3/16-8 UNJ	1.1875	1.1725	1.1063	1.1011	1.0432	1.0304	0.0226	0.0188
1 3/16-12 UNJ	1.1875	1.1761	1.1334	1.1291	1.0913	1.0820	0.0150	0.0125
1 3/16-16 UNJ	1.1875	1.1781	1.1469	1.1431	1.1153	1.1077	0.0113	0.0094
1 3/16-18 UNJEF	1.1875	1.1788	1.1514	1.1478	1.1233	1.1164	0.0100	0.0083
1 1/4-7 UNJC	1.2500	1.2336	1.1572	1.1517	1.0850	1.0709	0.0258	0.0214
1 1/4-8 UNJ	1.2500	1.2350	1.1688	1.1635	1.1056	1.0928	0.0226	0.0188
1 1/4-12 UNJF	1.2500	1.2386	1.1959	1.1913	1.1538	1.1442	0.0150	0.0125
1 1/4-16 UNJ	1.2500	1.2406	1.2094	1.2056	1.1778	1.1702	0.0113	0.0094
1 1/4-18 UNJEF	1.2500	1.2413	1.2139	1.2103	1.1858	1.1789	0.0100	0.0083
1 5/16-8 UNJ	1.3125	1.2975	1.2313	1.2260	1.1682	1.1553	0.0226	0.0188
1 5/16-12 UNJ	1.3125	1.3011	1.2584	1.2541	1.2163	1.2070	0.0150	0.0125
1 5/16-16 UNJ	1.3125	1.3031	1.2719	1.2681	1.2403	1.2327	0.0113	0.0094
1 5/16-18 UNJEF	1.3125	1.3038	1.2764	1.2728	1.2483	1.2414	0.0100	0.0083
1 3/8-6 UNJC	1.3750	1.3568	1.2667	1.2607	1.1825	1.1664	0.0301	0.0250
1 3/8-8 UNJ	1.3750	1.3600	1.2938	1.2884	1.2306	1.2177	0.0226	0.0188
1 3/8-12 UNJF	1.3750	1.3636	1.3209	1.3162	1.2788	1.2690	0.0150	0.0125
1 3/8-16 UNJ	1.3750	1.3656	1.3344	1.3306	1.3028	1.2952	0.0113	0.0094
1 3/8-18 UNJEF	1.3750	1.3663	1.3389	1.3353	1.3108	1.3039	0.0100	0.0083
1 7/16-8 UNJ	1.4375	1.4225	1.3563	1.3509	1.2932	1.2802	0.0226	0.0188
1 7/16-12 UNJ	1.4375	1.4261	1.3834	1.3790	1.3413	1.3318	0.0150	0.0125
1 7/16-16 UNJ	1.4375	1.4281	1.3969	1.3930	1.3653	1.3576	0.0113	0.0094
1 7/16-18 UNJEF	1.4375	1.4288	1.4014	1.3977	1.3733	1.3663	0.0100	0.0083

(Continued)

Table 8. *(Continued)* **UNJ Series Internal Threads, Class 3B.** Secondary Sizes in italics. *(Source, MIL-S-8879C.)*

Size, TPI, and Series	Minor Diameter		Pitch Diameter		Major Diameter
	min	max	min	max	min
3/4-10 UNJC	0.6526	0.6646	0.6850	0.6907	0.7500
3/4-12 UNJ	0.6689	0.6789	0.6959	0.7013	0.7500
3/4-16 UNJF	0.6892	0.6977	0.7094	0.7143	0.7500
3/4-20 UNJEF	0.7013	0.7081	0.7175	0.7218	0.7500
13/16-12 UNJ	0.7314	0.7414	0.7584	0.7638	0.8125
13/16-16 UNJ	0.7517	0.7602	0.7719	0.7766	0.8125
13/16-20 UNJEF	0.7638	0.7706	0.7800	0.7843	0.8125
7/8-9 UNJC	0.7668	0.7801	0.8028	0.8089	0.8750
7/8-12 UNJ	0.7939	0.8039	0.8209	0.8263	0.8750
7/8-14 UNJF	0.8055	0.8152	0.8286	0.8339	0.8750
7/8-16 UNJ	0.8142	0.8227	0.8344	0.8391	0.8750
7/8-20 UNJEF	0.8263	0.8331	0.8425	0.8468	0.8750
15/16-12 UNJ	0.8564	0.8664	0.8834	0.8889	0.9375
15/16-16 UNJ	0.8767	0.8852	0.8969	0.9018	0.9375
15/16-20 UNJEF	0.8888	0.8956	0.9050	0.9094	0.9375
1-8 UNJC	0.8783	0.8933	0.9188	0.9254	1.0000
1-12 UNJF	0.9189	0.9298	0.9459	0.9516	1.0000
1-16 UNJ	0.9392	0.9477	0.9594	0.9643	1.0000
1-20 UNJEF	0.9513	0.9581	0.9675	0.9719	1.0000
1 1/16-8 UNJ	0.9408	0.9558	0.9813	0.9880	1.0625
1 1/16-12 UNJ	0.9814	0.9914	1.0084	1.0139	1.0625
1 1/16-16 UNJ	1.0017	1.0102	1.0219	1.0268	1.0625
1 1/16-18 UNJEF	1.0084	1.0159	1.0264	1.0310	1.0625
1 1/8-7 UNJC	0.9859	1.0030	1.0322	1.0393	1.1250
1 1/8-8 UNJ	1.0033	1.0183	1.0438	1.0505	1.1250
1 1/8-12 UNJF	1.0439	1.0539	1.0709	1.0768	1.1250
1 1/8-16 UNJ	1.0642	1.0727	1.0844	1.0893	1.1250
1 1/8-18 UNJEF	1.0709	1.0784	1.0889	1.0935	1.1250
1 3/16-8 UNJ	1.0658	1.0808	1.1063	1.1131	1.1875
1 3/16-12 UNJ	1.1064	1.1164	1.1334	1.1390	1.1875
1 3/16-16 UNJ	1.1267	1.1352	1.1469	1.1519	1.1875
1 3/16-18 UNJEF	1.1334	1.1409	1.1514	1.1561	1.1875
1 1/4-7 UNJC	1.1109	1.1280	1.1572	1.1644	1.2500
1 1/4-8 UNJ	1.1283	1.1433	1.1688	1.1757	1.2500
1 1/4-12 UNJF	1.1689	1.1789	1.1959	1.2019	1.2500
1 1/4-16 UNJ	1.1892	1.1977	1.2094	1.2144	1.2500
1 1/4-18 UNJEF	1.1959	1.2034	1.2139	1.2186	1.2500
1 5/16-8 UNJ	1.1908	1.2058	1.2313	1.2382	1.3125
1 5/16-12 UNJ	1.2314	1.2414	1.2584	1.2640	1.3125
1 5/16-16 UNJ	1.2517	1.2602	1.2719	1.2769	1.3125
1 5/16-18 UNJEF	1.2584	1.2659	1.2764	1.2811	1.3125
1 3/8-6 UNJC	1.2127	1.2327	1.2667	1.2745	1.3750
1 3/8-8 UNJ	1.2533	1.2683	1.2938	1.3008	1.3750
1 3/8-12 UNJF	1.2939	1.3039	1.3209	1.3270	1.3750
1 3/8-16 UNJ	1.3142	1.3227	1.3344	1.3394	1.3750
1 3/8-18 UNJEF	1.3209	1.3284	1.3389	1.3436	1.3750
1 7/16-8 UNJ	1.3158	1.3308	1.3563	1.3634	1.4375
1 7/16-12 UNJ	1.3564	1.3664	1.3834	1.3891	1.4375
1 7/16-16 UNJ	1.3767	1.3852	1.3969	1.4020	1.4375
1 7/16-18 UNJEF	1.3834	1.3909	1.4014	1.4062	1.4375

(Continued)

Table 7. *(Continued)* **UNJ Series External Threads, Class 3A.** Secondary Sizes in italics. *(Source, MIL-S-8879C.)*

Size, TPI, and Series	Major Diameter		Pitch Diameter		Minor Diameter		Root Radius	
	max	min	max	min	max	min	max	min
1 1/2-6 UNJC	1.5000	1.4818	1.3917	1.3856	1.3075	1.2913	0.0301	0.0250
1 1/2-8 UNJ	1.5000	1.4850	1.4188	1.4133	1.3556	1.3426	0.0226	0.0188
1 1/2-12 UNJF	1.5000	1.4886	1.4459	1.4411	1.4038	1.3940	0.0150	0.0125
1 1/2-16 UNJ	1.5000	1.4906	1.4594	1.4555	1.4278	1.4201	0.0113	0.0094
1 1/2-18 UNJEF	1.5000	1.4913	1.4639	1.4602	1.4358	1.4288	0.0100	0.0083
1 9/16-8 UNJ	1.5625	1.5475	1.4813	1.4758	1.4182	1.4051	0.0226	0.0188
1 9/16-12 UNJ	1.5625	1.5511	1.5084	1.5040	1.4663	1.4568	0.0150	0.0125
1 9/16-16 UNJ	1.5625	1.5531	1.5219	1.5180	1.4903	1.4826	0.0113	0.0094
1 9/16-18 UNJEF	1.5625	1.5538	1.5264	1.5227	1.4983	1.4913	0.0100	0.0083
1 5/8-8 UNJ	1.6250	1.6100	1.5438	1.5382	1.4806	1.4675	0.0226	0.0188
1 5/8-12 UNJ	1.6250	1.6136	1.5709	1.5665	1.5288	1.5194	0.0150	0.0125
1 5/8-16 UNJ	1.6250	1.6156	1.5844	1.5805	1.5528	1.5451	0.0113	0.0094
1 5/8-18 UNJEF	1.6250	1.6163	1.5889	1.5852	1.5608	1.5538	0.0100	0.0083
1 11/16-8 UNJ	1.6875	1.6725	1.6063	1.6007	1.5432	1.5300	0.0226	0.0188
1 11/16-12 UNJ	1.6875	1.6761	1.6334	1.6289	1.5913	1.5818	0.0150	0.0125
1 11/16-16 UNJ	1.6875	1.6781	1.6469	1.6429	1.6153	1.6075	0.0113	0.0094
1 11/16-18 UNJEF	1.6875	1.6788	1.6514	1.6476	1.6233	1.6162	0.0100	0.0083
1 3/4-5 UNJC	1.7500	1.7295	1.6201	1.6134	1.5191	1.5002	0.0361	0.0300
1 3/4-8 UNJ	1.7500	1.7350	1.6688	1.6632	1.6056	1.5924	0.0226	0.0188
1 3/4-12 UNJ	1.7500	1.7386	1.6959	1.6914	1.6538	1.6442	0.0150	0.0125
1 3/4-16 UNJ	1.7500	1.7406	1.7094	1.7054	1.6778	1.6700	0.0113	0.0094
1 13/16-8 UNJ	1.8125	1.7975	1.7313	1.7256	1.6682	1.6549	0.0226	0.0188
1 13/16-12 UNJ	1.8125	1.8011	1.7584	1.7539	1.7163	1.7068	0.0150	0.0125
1 13/16-16 UNJ	1.8125	1.8031	1.7719	1.7679	1.7403	1.7325	0.0113	0.0094
1 7/8-8 UNJ	1.8750	1.8600	1.7938	1.7881	1.7306	1.7174	0.0226	0.0188
1 7/8-12 UNJ	1.8750	1.8636	1.8209	1.8164	1.7788	1.7692	0.0150	0.0125
1 7/8-16 UNJ	1.8750	1.8656	1.8344	1.8304	1.8028	1.7950	0.0113	0.0094
1 15/16-8 UNJ	1.9375	1.9225	1.8563	1.8505	1.7932	1.7798	0.0226	0.0188
1 15/16-12 UNJ	1.9375	1.9261	1.8834	1.8789	1.8413	1.8318	0.0150	0.0125
1 15/16-16 UNJ	1.9375	1.9281	1.8969	1.8929	1.8653	1.8575	0.0113	0.0094
2-4 1/2 UNJC	2.0000	1.9780	1.8557	1.8486	1.7434	1.7229	0.0401	0.0334
2-8 UNJ	2.0000	1.9850	1.9188	1.9130	1.8556	1.8423	0.0226	0.0188
2-12 UNJ	2.0000	1.9886	1.9459	1.9414	1.9038	1.8942	0.0150	0.0125
2-16 UNJ	2.0000	1.9906	1.9594	1.9554	1.9278	1.9200	0.0113	0.0094
2 1/8-8 UNJ	2.1250	2.1100	2.0438	2.0379	1.9806	1.9672	0.0226	0.0188
2 1/8-12 UNJ	2.1250	2.1136	2.0709	2.0664	2.0288	2.0192	0.0150	0.0125
2 1/8-16 UNJ	2.1250	2.1156	2.0844	2.0803	2.0528	2.0450	0.0113	0.0094
2 1/4-4 1/2 UNJC	2.2500	2.2280	2.1057	2.0984	1.9934	1.9727	0.0401	0.0334
2 1/4-8 UNJ	2.2500	2.2350	2.1688	2.1628	2.1056	2.0921	0.0226	0.0188
2 1/4-12 UNJ	2.2500	2.2386	2.1959	2.1914	2.1538	2.1442	0.0150	0.0125
2 1/4-16 UNJ	2.2500	2.2406	2.2094	2.2053	2.1778	2.1700	0.0113	0.0094
2 3/8-8 UNJ	2.3750	2.3600	2.2938	2.2878	2.2306	2.2171	0.0226	0.0188
2 3/8-12 UNJ	2.3750	2.3636	2.3209	2.3163	2.2788	2.2692	0.0150	0.0125
2 3/8-16 UNJ	2.3750	2.3656	2.3344	2.3303	2.3028	2.2949	0.0113	0.0094
2 1/2-4 UNJC	2.5000	2.4762	2.3376	2.3298	2.2113	2.1884	0.0401	0.0375
2 1/2-8 UNJ	2.5000	2.4850	2.4188	2.4127	2.3556	2.3420	0.0226	0.0188
2 1/2-12 UNJ	2.5000	2.4886	2.4459	2.4413	2.4038	2.3942	0.0150	0.0125
2 1/2-16 UNJ	2.5000	2.4906	2.4594	2.4553	2.4278	2.4199	0.0113	0.0094

(Continued)

Table 8. *(Continued)* **UNJ Series Internal Threads, Class 3B.** Secondary Sizes in italics. *(Source, MIL-S-8879C.)*

Size, TPI, and Series	Minor Diameter		Pitch Diameter		Major Diameter
	min	max	min	max	min
1 1/2-6 UNJC	1.3377	1.3577	1.3917	1.3996	1.5000
1 1/2-8 UNJ	1.3783	1.3933	1.4188	1.4259	1.5000
1 1/2-12 UNJF	1.1489	1.4289	1.4459	1.4522	1.5000
1 1/2-16 UNJ	1.4392	1.4477	1.4594	1.4645	1.5000
1 1/2-18 UNJEF	1.4459	1.4534	1.4639	1.4687	1.5000
1 9/16-8 UNJ	1.4408	1.4558	1.4813	1.4885	1.5625
1 9/16-12 UNJ	1.4814	1.4914	1.5084	1.5141	1.5625
1 9/16-16 UNJ	1.5017	1.5102	1.5219	1.5270	1.5625
1 9/16-18 UNJEF	1.5084	1.5159	1.5264	1.5312	1.5625
1 5/8-8 UNJ	1.5033	1.5183	1.5438	1.5510	1.6250
1 5/8-12 UNJ	1.5439	1.5539	1.5709	1.5766	1.6250
1 5/8-16 UNJ	1.5642	1.5727	1.5844	1.5895	1.6250
1 5/8-18 UNJEF	1.5709	1.5784	1.5889	1.5937	1.6250
1 11/16-8 UNJ	1.5658	1.5808	1.6063	1.6136	1.6875
1 11/16-12 UNJ	1.6064	1.6164	1.6334	1.6392	1.6875
1 11/16-16 UNJ	1.6267	1.6352	1.6469	1.6521	1.6875
1 11/16-18 UNJEF	1.6334	1.6409	1.6514	1.6563	1.6875
1 3/4-5 UNJC	1.5552	1.5792	1.6201	1.6288	1.7500
1 3/4-8 UNJ	1.6283	1.6433	1.6688	1.6762	1.7500
1 3/4-12 UNJ	1.6689	1.6789	1.6959	1.7017	1.7500
1 3/4-16 UNJ	1.6892	1.6977	1.7094	1.7146	1.7500
1 13/16-8 UNJ	1.6908	1.7058	1.7313	1.7387	1.8125
1 13/16-12 UNJ	1.7314	1.7414	1.7584	1.7642	1.8125
1 13/16-16 UNJ	1.7517	1.7602	1.7719	1.7771	1.8125
1 7/8-8 UNJ	1.7533	1.7683	1.7938	1.8013	1.8750
1 7/8-12 UNJ	1.7939	1.8039	1.8209	1.8267	1.8750
1 7/8-16 UNJ	1.8142	1.8227	1.8344	1.8396	1.8750
1 15/16-8 UNJ	1.8158	1.8308	1.8563	1.8638	1.9375
1 15/16-12 UNJ	1.8564	1.8664	1.8834	1.8893	1.9375
1 15/16-16 UNJ	1.8767	1.8852	1.8969	1.9021	1.9375
2-4 1/2 UNJC	1.7835	1.8102	1.8557	1.8650	2.0000
2-8 UNJ	1.8783	1.8933	1.9188	1.9264	2.0000
2-12 UNJ	1.9189	1.9289	1.9459	1.9518	2.0000
2-16 UNJ	1.9392	1.9477	1.9594	1.9646	2.0000
2 1/8-8 UNJ	2.0033	2.0183	2.0438	2.0515	2.1250
2 1/8-12 UNJ	2.0439	2.0539	2.0709	2.0768	2.1250
2 1/8-16 UNJ	2.0642	2.0727	2.0844	2.0896	2.1250
2 1/4-4 1/2 UNJC	2.0335	2.0602	2.1057	2.1152	2.2500
2 1/4-8 UNJ	2.1283	2.1433	2.1688	2.1766	2.2500
2 1/4-12 UNJ	2.1689	2.1789	2.1959	2.2018	2.2500
2 1/4-16 UNJ	2.1892	2.1977	2.2094	2.2146	2.2500
2 3/8-8 UNJ	2.2533	2.2683	2.2938	2.3017	2.3750
2 3/8-12 UNJ	2.2939	2.3039	2.3209	2.3269	2.3750
2 3/8-16 UNJ	2.3142	2.3227	2.3344	2.3398	2.3750
2 1/2-4 UNJC	2.2565	2.2865	2.3376	2.3477	2.5000
2 1/2-8 UNJ	2.3783	2.3933	2.4188	2.4268	2.5000
2 1/2-12 UNJ	2.4189	2.4289	2.4459	2.4519	2.5000
2 1/2-16 UNJ	2.4392	2.4477	2.4594	2.4648	2.5000

(Continued)

Table 7. *(Continued)* **UNJ Series External Threads, Class 3A.** Secondary Sizes in italics. *(Source, MIL-S-8879C.)*

Size, TPI, and Series	Major Diameter		Pitch Diameter		Minor Diameter		Root Radius	
	max	min	max	min	max	min	max	min
2 5/8-8 UNJ	2.6250	2.6100	2.5438	2.5376	2.4806	2.4669	0.0226	0.0188
2 5/8-12 UNJ	2.6250	2.6136	2.5709	2.5663	2.5288	2.5192	0.0150	0.0125
2 5/8-16 UNJ	2.6250	2.6156	2.5844	2.5803	2.5528	2.5449	0.0113	0.0094
2 3/4-4 UNJC	2.7500	2.7262	2.5876	2.5797	2.4613	2.4382	0.0401	0.0375
2 3/4-8 UNJ	2.7500	2.7350	2.6688	2.6625	2.6056	2.5918	0.0226	0.0188
2 3/4-12 UNJ	2.7500	2.7386	2.6959	2.6913	2.6538	2.6442	0.0150	0.0125
2 3/4-16 UNJ	2.7500	2.7406	2.7094	2.7053	2.6778	2.6699	0.0113	0.0094
2 7/8-8 UNJ	2.8750	2.8600	2.7938	2.7875	2.7306	2.7168	0.0226	0.0188
2 7/8-12 UNJ	2.8750	2.8636	2.8209	2.8162	2.7788	2.7690	0.0150	0.0125
2 7/8-16 UNJ	2.8750	2.8656	2.8344	2.8302	2.8028	2.7948	0.0113	0.0094
3-4 UNJC	3.0000	2.9762	2.8376	2.8296	2.7113	2.6882	0.0451	0.0375
3-8 UNJ	3.0000	2.9850	2.9188	2.9124	2.8556	2.8417	0.0226	0.0188
3-12 UNJ	3.0000	2.9886	2.9459	2.9412	2.9038	2.8940	0.0150	0.0125
3-16 UNJ	3.0000	2.9906	2.9594	2.9552	2.9278	2.9198	0.0113	0.0094
3 1/8-8 UNJ	3.1250	3.1100	3.0438	3.0374	2.9806	2.9667	0.0226	0.0188
3 1/8-12 UNJ	3.1250	3.1136	3.0709	3.0662	3.0288	3.0190	0.0150	0.0125
3 1/8-16 UNJ	3.1250	3.1156	3.0844	3.0802	3.0528	3.0448	0.0113	0.0094
3 1/4-4 UNJC	3.2500	3.2262	3.0876	3.0794	2.9613	2.9380	0.0451	0.0375
3 1/4-8 UNJ	3.2500	3.2350	3.1688	3.1623	3.1056	3.0916	0.0226	0.0188
3 1/4-12 UNJ	3.2500	3.2386	3.1959	3.1912	3.1538	3.1440	0.0150	0.0125
3 1/4-16 UNJ	3.2500	3.2406	3.2094	3.2052	3.1778	3.1698	0.0113	0.0094
3 3/8-8 UNJ	3.3750	3.3600	3.2938	3.2872	3.2306	3.2176	0.0226	0.0188
3 3/8-12 UNJ	3.3750	3.3636	3.3209	3.3161	3.2788	3.2690	0.0150	0.0125
3 3/8-16 UNJ	3.3750	3.3656	3.3344	3.3301	3.3028	3.2947	0.0113	0.0094
3 1/2-4 UNJC	3.5000	3.4762	3.3376	3.3293	3.2113	3.1878	0.0451	0.0375
3 1/2-8 UNJ	3.5000	3.4850	3.4188	3.4122	3.3556	3.3415	0.0226	0.0188
3 1/2-12 UNJ	3.5000	3.4886	3.4459	3.4411	3.4038	3.3940	0.0150	0.0125
3 1/2-16 UNJ	3.5000	3.4909	3.4594	3.4551	3.4278	3.4197	0.0113	0.0094
3 5/8-8 UNJ	3.6250	3.6100	3.5438	3.5371	3.4806	3.4664	0.0226	0.0188
3 5/8-12 UNJ	3.6250	3.6136	3.5709	3.5661	3.5288	3.5190	0.0150	0.0125
3 5/8-16 UNJ	3.6250	3.6156	3.5844	3.5801	3.5528	3.5447	0.0113	0.0094
3 3/4-4 UNJC	3.7500	3.7262	3.5876	3.5792	3.4613	3.4378	0.0451	0.0375
3 3/4-8 UNJ	3.7500	3.7350	3.6688	3.6621	3.6056	3.5914	0.0226	0.0188
3 3/4-12 UNJ	3.7500	3.7386	3.6959	3.6911	3.6538	3.6440	0.0150	0.0125
3 3/4-16 UNJ	3.7500	3.7406	3.7094	3.7051	3.6778	3.6697	0.0113	0.0094
3 7/8-8 UNJ	3.8750	3.8600	3.7938	3.7870	3.7306	3.7163	0.0226	0.0188
3 7/8-12 UNJ	3.8750	3.8636	3.8209	3.8160	3.7788	3.7688	0.0150	0.0125
3 7/8-16 UNJ	3.8750	3.8656	3.8344	3.8300	3.8028	3.7946	0.0113	0.0094
4-4 UNJC	4.0000	3.9762	3.8376	3.8291	3.7113	3.6876	0.0451	0.0375
4-8 UNJ	4.0000	3.9850	3.9188	3.9120	3.8556	3.8413	0.0226	0.0188
4-12 UNJ	4.0000	3.9886	3.9459	3.9410	3.9038	3.8938	0.0150	0.0125
4-16 UNJ	4.0000	3.9906	3.9594	3.9550	3.9278	3.9196	0.0113	0.0094
4 1/8-12 UNJ	4.1250	4.1136	4.0709	4.0660	4.0288	4.0188	0.0150	0.0125
4 1/8-16 UNJ	4.1250	4.1156	4.0844	4.0800	4.0528	4.0446	0.0113	0.0094
4 1/4-12 UNJ	4.2500	4.2386	4.1959	4.1910	4.1538	4.1438	0.0150	0.0125
4 1/4-16 UNJ	4.2500	4.2406	4.2094	4.2050	4.1778	4.1696	0.0113	0.0094
4 3/8-12 UNJ	4.3750	4.3636	4.3209	4.3160	4.2788	4.2688	0.0150	0.0125
4 3/8-16 UNJ	4.3750	4.3656	4.3344	4.3300	4.3028	4.2946	0.0113	0.0094

(Continued)

Table 8. *(Continued)* **UNJ Series Internal Threads, Class 3B.** Secondary Sizes in italics. *(Source, MIL-S-8879C.)*

Size, TPI, and Series	Minor Diameter min	Minor Diameter max	Pitch Diameter min	Pitch Diameter max	Major Diameter min
2 5/8-8 UNJ	2.5033	2.5183	2.5438	2.5518	2.6250
2 5/8-12 UNJ	2.5439	2.5539	2.5709	2.5769	2.6250
2 5/8-16 UNJ	2.5642	2.5727	2.5844	2.5898	2.6250
2 3/4-4 UNJC	2.5065	2.5365	2.5876	2.5979	2.7500
2 3/4-8 UNJ	2.6283	2.6433	2.6688	2.6769	2.7500
2 3/4-12 UNJ	2.6689	2.6789	2.6959	2.7019	2.7500
2 3/4-16 UNJ	2.6892	2.6977	2.7094	2.7148	2.7500
2 7/8-8 UNJ	2.7533	2.7683	2.7938	2.8020	2.8750
2 7/8-12 UNJ	2.7939	2.8039	2.8209	2.8271	2.8750
2 7/8-16 UNJ	2.8142	2.8227	2.8344	2.8399	2.8750
3-4 UNJC	2.7565	2.7865	2.8376	2.8480	3.0000
3-8 UNJ	2.8783	2.8933	2.9188	2.9271	3.0000
3-12 UNJ	2.9189	2.9289	2.9459	2.9521	3.0000
3-16 UNJ	2.9392	2.9477	2.9594	2.9649	3.0000
3 1/8-8 UNJ	3.0033	3.0183	3.0438	3.0522	3.1250
3 1/8-12 UNJ	3.0439	3.0539	3.0709	3.0771	3.1250
3 1/8-16 UNJ	3.0642	3.0727	3.0844	3.0899	3.1250
3 1/4-4 UNJC	3.0065	3.0365	3.0876	3.0982	3.2500
3 1/4-8 UNJ	3.1283	3.1433	3.1688	3.1773	3.2500
3 1/4-12 UNJ	3.1689	3.1789	3.1959	3.2021	3.2500
3 1/4-16 UNJ	3.1892	3.1977	3.2094	3.2149	3.2500
3 3/8-8 UNJ	3.2533	3.2683	3.2938	3.3023	3.3750
3 3/8-12 UNJ	3.2939	3.3039	3.3209	3.3272	3.3750
3 3/8-16 UNJ	3.3142	3.3227	3.3344	3.3400	3.3750
3 1/2-4 UNJC	3.2565	3.2865	3.3376	3.3484	3.5000
3 1/2-8 UNJ	3.3783	3.3933	3.4188	3.4274	3.5000
3 1/2-12 UNJ	3.4189	3.4289	3.4459	3.4522	3.5000
3 1/2-16 UNJ	3.4392	3.4477	3.4594	3.4650	3.5000
3 5/8-8 UNJ	3.5033	3.5183	3.5438	3.5525	3.6250
3 5/8-12 UNJ	3.5439	3.5539	3.5709	3.5772	3.6250
3 5/8-16 UNJ	3.5642	3.5727	3.5844	3.5900	3.6250
3 3/4-4 UNJC	3.5065	3.5365	3.5876	3.5985	3.7500
3 3/4-8 UNJ	3.6283	3.6433	3.6688	3.6776	3.7500
3 3/4-12 UNJ	3.6689	3.6789	3.6959	3.7022	3.7500
3 3/4-16 UNJ	3.6892	3.6977	3.7094	3.7150	3.7500
3 7/8-8 UNJ	3.7533	3.7683	3.7938	3.8026	3.8750
3 7/8-12 UNJ	3.7939	3.8039	3.8209	3.8273	3.8750
3 7/8-16 UNJ	3.8142	3.8227	3.8344	3.8401	3.8750
4-4 UNJC	3.7565	3.7865	3.8376	3.8487	4.0000
4-8 UNJ	3.8783	3.8933	3.9188	3.9277	4.0000
4-12 UNJ	3.9189	3.9289	3.9459	3.9523	4.0000
4-16 UNJ	3.9392	3.9477	3.9594	3.9651	4.0000
4 1/8-12 UNJ	4.0439	4.0539	4.0709	4.0773	4.1250
4 1/8-16 UNJ	4.0642	4.0727	4.0844	4.0901	4.1250
4 1/4-12 UNJ	4.1689	4.1789	4.1959	4.2023	4.2500
4 1/4-16 UNJ	4.1892	4.1977	4.2094	4.2151	4.2500
4 3/8-12 UNJ	4.2939	4.3039	4.3209	4.3273	4.3750
4 3/8-16 UNJ	4.3142	4.3227	4.3344	4.3401	4.3750

(Continued)

Table 7. *(Continued)* **UNJ Series External Threads, Class 3A.** Secondary Sizes in italics. *(Source, MIL-S-8879C.)*

Size, TPI, and Series	Major Diameter		Pitch Diameter		Minor Diameter		Root Radius	
	max	min	max	min	max	min	max	min
4 1/2-12 UNJ	4.5000	4.4886	4.4459	4.4410	4.4038	4.3938	0.0150	0.0125
4 1/2-16 UNJ	4.5000	4.4906	4.4594	4.4550	4.4278	4.4196	0.0113	0.0094
4 5/8-12 UNJ	4.6250	4.6136	4.5709	4.5659	4.5288	4.5188	0.0150	0.0125
4 5/8-16 UNJ	4.6250	4.6156	4.5844	4.5799	4.5528	4.5445	0.0113	0.0094
4 3/4-12 UNJ	4.7500	4.7386	4.6959	4.6909	4.6538	4.6438	0.0150	0.0125
4 3/4-16 UNJ	4.7500	4.7406	4.7094	4.7049	4.6778	4.6695	0.0113	0.0094
4 7/8-12 UNJ	4.8750	4.8636	4.8209	4.8159	4.7788	4.7688	0.0150	0.0125
4 7/8-16 UNJ	4.8750	4.8656	4.8344	4.8299	4.8028	4.7945	0.0113	0.0094
5-12 UNJ	5.0000	4.9886	4.9459	4.9409	4.9038	4.8938	0.0150	0.0125
5-16 UNJ	5.0000	4.9906	4.9594	4.9549	4.9278	4.9195	0.0113	0.0094
5 1/8-12 UNJ	5.1250	5.1136	5.0709	5.0659	5.0288	5.0188	0.0150	0.0125
5 1/8-16 UNJ	5.1250	5.1156	5.0844	5.0799	5.0528	5.0445	0.0113	0.0094
5 1/4-12 UNJ	5.2500	5.2386	5.1959	5.1909	5.1538	5.1438	0.0150	0.0125
5 1/4-16 UNJ	5.2500	5.2406	5.2094	5.2049	5.1778	5.1695	0.0113	0.0094
5 3/8-12 UNJ	5.3750	5.3636	5.3209	5.3159	5.2788	5.2688	0.0150	0.0125
5 3/8-16 UNJ	5.3750	5.3656	5.3344	5.3299	5.3028	5.2945	0.0113	0.0094
5 1/2-12 UNJ	5.5000	5.4886	5.4459	5.4409	5.4038	5.3938	0.0150	0.0125
5 1/2-16 UNJ	5.5000	5.4906	5.4594	5.4549	5.4278	5.4195	0.0113	0.0094
5 5/8-12 UNJ	5.6250	5.6136	5.5709	5.5657	5.5288	5.5186	0.0150	0.0125
5 5/8-16 UNJ	5.6250	5.6156	5.5844	5.5797	5.5528	5.5443	0.0113	0.0094
5 3/4-12 UNJ	5.7500	5.7386	5.6959	5.6907	5.6538	5.6436	0.0150	0.0125
5 3/4-16 UNJ	5.7500	5.7406	5.7094	5.7047	5.6778	5.6693	0.0113	0.0094
5 7/8-12 UNJ	5.8750	5.8636	5.8209	5.8157	5.7788	5.7686	0.0150	0.0125
5 7/8-16 UNJ	5.8750	5.8656	5.8344	5.8297	5.8028	5.7943	0.0113	0.0094
6-12 UNJ	6.0000	5.9886	5.9459	5.9407	5.9038	5.8936	0.0150	0.0125
6-16 UNJ	6.0000	5.9906	5.9594	5.9547	5.9278	5.9193	0.0113	0.0094

All dimensions in inches. Source, MIL-S-8879C.

Table 8. *(Continued)* **UNJ Series Internal Threads, Class 3B.** Secondary Sizes in italics. *(Source, MIL-S-8879C.)*

Size, TPI, and Series	Minor Diameter		Pitch Diameter		Major Diameter
	min	max	min	max	min
4 1/2-12 UNJ	4.4189	4.4289	4.4459	4.4523	4.5000
4 1/2-16 UNJ	4.4392	4.4477	4.4594	4.4651	4.5000
4 5/8-12 UNJ	4.5439	4.5539	4.5709	4.5775	4.6250
4 5/8-16 UNJ	4.5642	4.5727	4.5844	4.5903	4.6250
4 3/4-12 UNJ	4.6689	4.6789	4.6959	4.7025	4.7500
4 3/4-16 UNJ	4.6892	4.6977	4.7094	4.7153	4.7500
4 7/8-12 UNJ	4.7939	4.8039	4.8209	4.8275	4.8750
4 7/8-16 UNJ	4.8142	4.8227	4.8344	4.8403	4.8750
5-12 UNJ	4.9189	4.9289	4.9459	4.9525	5.0000
5-16 UNJ	4.9392	4.9477	4.9594	4.9653	5.0000
5 1/8-12 UNJ	5.0439	5.0539	5.0709	5.0775	5.1250
5 1/8-16 UNJ	5.0642	5.0727	5.0844	5.0903	5.1250
5 1/4-12 UNJ	5.1689	5.1789	5.1959	5.2025	5.2500
5 1/4-16 UNJ	5.1892	5.1977	5.2094	5.2153	5.2500
5 3/8-12 UNJ	5.2939	5.3039	5.3209	5.3275	5.3750
5 3/8-16 UNJ	5.3142	5.3227	5.3344	5.3403	5.3750
5 1/2-12 UNJ	5.4189	5.4289	5.4459	5.4525	5.5000
5 1/2-16 UNJ	5.4392	5.4477	5.4594	5.4653	5.5000
5 5/8-12 UNJ	5.5439	5.5539	5.5709	5.5776	5.6250
5 5/8-16 UNJ	5.5642	5.5727	5.5844	5.5905	5.6250
5 3/4-12 UNJ	5.6689	5.6789	5.6959	5.7026	5.7500
5 3/4-16 UNJ	5.6892	5.6977	5.7094	5.7155	5.7500
5 7/8-12 UNJ	5.7939	5.8039	5.8209	5.8276	5.8750
5 7/8-16 UNJ	5.8142	5.8227	5.8344	5.8405	5.8750
6-12 UNJ	5.9189	5.9289	5.9459	5.9526	6.0000
6-16 UNJ	5.9392	5.9477	5.9594	5.9655	6.0000

All dimensions in inches. Source, MIL-S-8879C.

Table 9. Thread Form for UN, UNR, and UNJ Threads. *(Source, Cleveland Twist Drill .)*

TPI	Pitch P	Height of External Thread UN, UNR	Height of External Thread UNJ	Addendum UN, UNR, UNJ	Crest Flat UN, UNR, UNJ	Root Flat (Basic Profile) UN	Root Radii UN Max.	Root Radii UN Min.	Root Radii UNJ Max.	Root Radii UNJ Min.
		0.70833H	0.66667H	0.375H	0.125P	0.250P	0.14434P	0.10825P	0.18042P	0.15011P
80	0.125	.0077	.0072	.0041	.0016	.0031	.0018	.0014	.0023	.0019
72	.0139	.0085	.0080	.0045	.0017	.0035	.0020	.0015	.0025	.0021
64	.0156	.0096	.0090	.0051	.0020	.0039	.0023	.0017	.0028	.0023
56	.0179	.0110	.0103	.0058	.0022	.0045	.0026	.0019	.0032	.0027
48	.0208	.0128	.0120	.0068	.0026	.0052	.0030	.0023	.0038	.0031
44	.0227	.0139	.0131	.0074	.0028	.0057	.0033	.0025	.0041	.0034
40	.0250	.0153	.0144	.0081	.0031	.0062	.0036	.0027	.0045	.0038
36	.0278	.0170	.0160	.0090	.0035	.0069	.0040	.0030	.0050	.0042
32	.0312	.0192	.0180	.0101	.0039	.0078	.0045	.0034	.0056	.0047
28	.0357	.0219	.0206	.0116	.0045	.0089	.0052	.0039	.0064	.0054
27	.0370	.0227	.0214	.0120	.0046	.0093	.0053	.0040	.0067	.0056
24	.0417	.0256	.0241	.0135	.0052	.0104	.0060	.0045	.0075	.0063
20	.0500	.0307	.0289	.0162	.0062	.0125	.0072	.0054	.0090	.0075
18	.0556	.0341	.0321	.0180	.0069	.0139	.0080	.0060	.0100	.0088
16	.0625	.0383	.0361	.0203	.0078	.0156	.0090	.0068	.0113	.0094
14	.0714	.0438	.0412	.0232	.0089	.0179	.0103	.0077	.0129	.0107
13	.0770	.0472	.0444	.0250	.0096	.0193	.0111	.0083	.0139	.0115
12	.0833	.0511	.0481	.0271	.0104	.0208	.0120	.0090	.0150	.0125
11-1/2	.08696	.0533	.0502	.0282	.0109	.0217	.0126	.0094	.0157	.0130
11	.0909	.0558	.0525	.0295	.0114	.0227	.0131	.0098	.0164	.0136
10	.100	.0613	.0577	.0325	.0125	.0250	.0144	.0108	.0180	.0150
9	.1111	.0682	.0642	.0361	.0139	.0278	.0160	.0120	.0201	.0167
8	.125	.0767	.0722	.0406	.0156	.0312	.0180	.0135	.0226	.0188
7	.1429	.0876	.0825	.0464	.0179	.0357	.0206	.0155	.0258	.0214
6	.1667	.1022	.0962	.0541	.0208	.0417	.0241	.0180	.0301	.0250
5	.200	.1227	.1155	.0650	.0250	.0500	.0239	.0216	.0361	.0300
4-1/2	.225	.1363	.1283	.0722	.0278	.0556	.0321	.0241	.0401	.0334
4	.250	.1534	.1443	.0812	.0312	.0625	.0361	.0271	.0451	.0375

All dimensions in inches.

Table 10. Unified Class 5 Interference-Fit Thread Dimensions.

	External Threads						
Class	NC-5 HF		NC-5 CSF NC-5 ONF		All Classes-Diameters		
Size	Major Dia.		Major Dia.		Pitch		Minor
	Max.	Min.	Max.	Min.	Max.	Min.	Max.
1/4-20	.2470	.2408	.2470	.2408	.2230	.2204	.1932
5/16-18	.3080	.3020	.3090	.3030	.2829	.2799	.2508
3/8-16	.3690	.3626	.3710	.3646	.3414	.3382	.3053
7/16-14	.4305	.4233	.4330	.4258	.3991	.3955	.3579
1/2-13	.4920	.4846	.4950	.4876	.4584	.4547	.4140
9/16-12	.5540	.5460	.5580	.5495	.5176	.5136	.4695
5/8-11	.6140	.6056	.6195	.6111	.5758	.5716	.5233
3/4-10	.7360	.7270	.7440	.7350	.6955	.6910	.6378
7/8-9	.8600	.8502	.8685	.8587	.8144	.8095	.7503
1-8	.9835	.9727	.9935	.9827	.9316	.9262	.8594
1 1/8-7	1.1070	1.0952	1.1180	1.1062	1.0465	1.0406	.9640
1 1/4-7	1.2320	1.2200	1.2430	1.2312	1.1715	1.1656	1.0890
1 3/8-6	1.3560	1.3410	1.3680	1.3538	1.2839	1.2768	1.1877
1 1/2-6	1.4810	1.4670	1.4930	1.4788	1.4089	1.4018	1.3127

	Internal Threads						
Class	NC-5 IF			NC-5 INF			Both Classes
Size	Minor Dia.		Tap Drill	Minor Dia.		Tap Drill	Pitch Dia.
	Min.	Max.		Min.	Max.		Min.	Max.	Major Min.

Class / Size	Minor Dia. Min.	Minor Dia. Max.	Tap Drill	Minor Dia. Min.	Minor Dia. Max.	Tap Drill	Pitch Dia. Min.	Pitch Dia. Max.	Major Min.
1/4-20	.1960	.2060	.2031	.1960	.2060	.2031	.2175	.2201	.2532
5/16-18	.2520	.2630	.2610	.2520	.2630	.2610	.2764	.2794	.3161
3/8-16	.3070	.3180	.3160	.3070	.3180	.3160	.3344	.3376	.3790
7/16-14	.3740	.3810	.3750	.3600	.3720	.3680	.3911	.3947	.4421
1/2-13	.4310	.4400	.4331	.4170	.4290	.4219	.4500	.4537	.5050
9/16-12	.4880	.4970	.4921	.4720	.4850	.4844	.5084	.5124	.5679
5/8-11	.5440	.5540	.5469	.5270	.5400	.5313	.5660	.5702	.6309
3/4-10	.6670	.6780	.6719	.6420	.6550	.6496	.6850	.6895	.7565
7/8-9	.7770	.7890	.7812	.7550	.7690	.7656	.8028	.8077	.8822
1-8	.8900	.9040	.8906	.8650	.8800	.8750	.9188	.9242	1.0081
1 1/8-7	1.0000	1.0150	1.0000	.9700	.9860	.9844	1.0322	1.0381	1.1343
1 1/4-7	1.1250	1.1400	1.1250	1.0950	1.1110	1.1094	1.1572	1.1631	1.2593
1 3/8-6	1.2290	1.2470	1.2344	1.1950	1.2130	1.2031	1.2667	1.2738	1.3858
1 1/2-6	1.3540	1.3720	1.3594	1.3200	1.3380	1.3281	1.3917	1.3988	1.5108

All dimensions in inches. For descriptions of each class of interference-fit thread, refer to "Unified threads by class" earlier in this section. Source, ASA B1.12-1972 as published by the American Society of Mechanical Engineers.

Table 11. Basic Formulas for Calculating UNM Thread Dimensions and Tolerances.

Symbol	Dimension	Equation
P	Pitch (number of threads per unit of length).	$P = 1 \div$ threads per inch, or $P = 1 \div$ threads per mm
H	Height of fundamental triangle.	$H = 0.866025404P$
D	Major dia. (basic).	
D	Major dia., int. thread.	$D + 0.072P$
d	Minor dia., ext. thread.	$D - 1.20P$
D_1	Minor dia., int. thread.	$D - 1.04P$
E	Pitch dia. (basic).	$D - 0.64952P$
h	Thread height (basic).	$0.556P$
h_c	Depth of engagement.	$0.52P$
r_n	Int. root radius.	$0.072P$
r_s	Ext. root radius.	$0.158P$

Other dimensions in accordance with formulas for UN threads in Table 4.

Tolerances on Design Dimensions*

External Thread (−)	Dimension	Internal Thread (+)
$0.12P + 0.006$	Major Diameter	$0.168P + 0.008$
$0.08P + 0.008$	Pitch Diameter	$0.08P + 0.008$
$0.16P + 0.008$	Minor Diameter	$0.32P + 0.012$

*Tolerances are applied to the dimensions specified in Table 12.

Table 12. Unified Miniature Thread Series. *(Source, FED-STD-H28/5.)*

UNM - External Threads							
Size (External)	Pitch mm	Major Dia. mm		Pitch Dia. mm		Minor Dia. mm	
		Max.	Min.	Max.	Min.	Max.	Min.
0.30 UNM	0.080	0.300	0.284	0.248	0.234	0.204	0.183
0.35 UNM	0.090	0.350	0.333	0.292	0.277	0.242	0.220
0.40 UNM	0.100	0.400	0.382	0.335	0.319	0.280	0.256
0.45 UNM	0.100	0.450	0.432	0.385	0.369	0.330	0.306
0.50 UNM	0.125	0.500	0.479	0.419	0.401	0.350	0.322
0.55 UNM	0.125	0.550	0.529	0.469	0.451	0.400	0.372
0.60 UNM	0.150	0.600	0.576	0.503	0.483	0.420	0.388
0.70 UNM	0.175	0.700	0.673	0.586	0.564	0.490	0.454
0.80 UNM	0.200	0.800	0.770	0.670	0.646	0.560	0.520
0.90 UNM	0.225	0.900	0.867	0.754	0.728	0.630	0.586
1.00 UNM	0.250	1.000	0.964	0.838	0.810	0.700	0.652
1.10 UNM	0.250	1.100	1.064	0.938	0.910	0.800	0.752
1.20 UNM	0.250	1.200	1.164	1.038	1.010	0.900	0.852
1.40 UNM	0.300	1.400	1.358	1.205	1.173	1.040	0.984
Size (External)	Pitch tpi	Major Dia. inch		Pitch Dia. inch		Minor Dia. inch	
		Max.	Min.	Max.	Min.	Max.	Min.
0.30 UNM	318	0.0118	0.0112	0.0098	0.0092	0.0080	0.0072
0.35 UNM	282	0.0138	0.0131	0.0115	0.0109	0.0095	0.0086

(Continued)

Table 12. *(Continued)* **Unified Miniature Thread Series.** *(Source, FED-STD-H28/5.)*

UNM - External Threads							
Size (External)	Pitch tpi	Major Dia. inch		Pitch Dia. inch		Minor Dia. inch	
		Max.	Min.	Max.	Min.	Max.	Min.
0.40 UNM	254	0.0157	0.0150	0.0132	0.0126	0.0110	0.0101
0.45 UNM	254	0.0177	0.0170	0.0152	0.0145	0.0130	0.0120
0.50 UNM	203	0.0197	0.0189	0.0165	0.0158	0.0138	0.0127
0.55 UNM	203	0.0217	0.0208	0.0185	0.0177	0.0157	0.0146
0.60 UNM	169	0.0236	0.0227	0.0198	0.0190	0.0165	0.0153
0.70 UNM	145	0.0276	0.0265	0.0231	0.0222	0.0193	0.0179
0.80 UNM	127	0.0315	0.0303	0.0264	0.0254	0.0220	0.0205
0.90 UNM	113	0.0354	0.0341	0.0297	0.0287	0.0248	0.0231
1.00 UNM	102	0.0394	0.0380	0.0330	0.0319	0.0276	0.0257
1.10 UNM	102	0.0433	0.0419	0.0369	0.0358	0.0315	0.0296
1.20 UNM	102	0.0472	0.00458	0.0409	0.0397	0.0354	0.0335
1.40 UNM	85	0.0551	0.0535	0.0474	0.0462	0.0409	0.0387

UNM - Internal Threads							
Size (Internal)	Pitch mm	Minor Dia. mm		Pitch Dia. mm		Major Dia. mm	
		Min.	Max.	Min.	Max.	Min.	Max.
0.30 UNM	0.080	0.217	0.254	0.248	0.262	0.306	0.327
0.35 UNM	0.090	0.256	0.297	0.292	0.307	0.356	0.380
0.40 UNM	0.100	0.296	0.340	0.335	0.351	0.407	0.432
0.45 UNM	0.100	0.346	0.390	0.385	0.401	0.457	0.482
0.50 UNM	0.125	0.370	0.422	0.419	0.437	0.509	0.538
0.55 UNM	0.125	0.420	0.472	0.469	0.487	0.559	0.588
0.60 UNM	0.150	0.444	0.504	0.503	0.523	0.611	0.644
0.70 UNM	0.175	0.518	0.586	0.586	0.608	0.713	0.750
0.80 UNM	0.200	0.592	0.668	0.670	0.694	0.814	0.856
0.90 UNM	0.225	0.666	0.750	0.754	0.780	0.916	0.962
1.00 UNM	0.250	0.740	0.832	0.838	0.866	1.018	1.068
1.10 UNM	0.250	0.840	0.932	0.938	0.966	1.118	1.168
1.20 UNM	0.250	0.940	1.032	1.038	1.066	1.218	1.268
1.40 UNM	0.300	1.088	1.196	1.205	1.237	1.422	1.480
Size (Internal)	Pitch tpi	Minor Dia. inch		Pitch Dia. inch		Major Dia. inch	
		Min.	Max.	Min.	Max.	Min.	Max.
0.30 UNM	318	0.0085	0.0100	0.0098	0.0104	0.0120	0.0129
0.35 UNM	282	0.0101	0.0117	0.0115	0.0121	0.0140	0.0149
0.40 UNM	254	0.0117	0.0134	0.0132	0.0138	0.0160	0.0170
0.45 UNM	254	0.0136	0.0154	0.0152	0.0158	0.0180	0.0190
0.50 UNM	203	0.0146	0.0166	0.0165	0.0172	0.0200	0.0212
0.55 UNM	203	0.0165	0.0186	0.0185	0.0192	0.0220	0.0231
0.60 UNM	169	0.0175	0.0198	0.0198	0.0206	0.0240	0.0254

(Continued)

Table 12. *(Continued)* **Unified Miniature Thread Series.** *(Source, FED-STD-H28/5.)*

Size (Internal)	Pitch tpi	Minor Dia. inch		Pitch Dia. inch		Major Dia. inch	
		Min.	Max.	Min.	Max.	Min.	Max.
0.70 UNM	145	0.0204	0.0231	0.0231	0.0240	0.0281	0.0295
0.80 UNM	127	0.0233	0.0263	0.0264	0.0273	0.0321	0.0337
0.90 UNM	113	0.0262	0.0295	0.0297	0.0307	0.0361	0.0379
1.00 UNM	102	0.0291	0.0327	0.0330	0.0341	0.0401	0.0420
1.10 UNM	102	0.0331	0.0367	0.0369	0.0380	0.0440	0.0460
1.20 UNM	102	0.0370	0.0406	0.0409	0.0420	0.0480	0.0499
1.40 UNM	85	0.0428	0.0471	0.0474	0.0487	0.0560	0.0583

All dimensions are provided in both millimeters and inches, as stated in FED-STD-H28/5.
Preferred sizes are in normal type. Secondary sizes are in italics.

60° Stub thread

The stub version of the 60° thread was actually designed for use as an Acme style thread for producing transverse movement. Shear strength is higher than with the Acme form, but load transmission efficiency is lower. The angle between the flanks of the thread is 60°, and the threads are truncated top and bottom. The width of flat at the screw crest is equal to 0.250P and the flat at the root equals 0.227P. Threads have a basic height of 0.433P, a basic thickness at pitch line of 0.50P, and are symmetrical about a line perpendicular to the axis of the screw. A typical thread designation is 1 $^1/_2$–9 SPL 60° FORM, followed by limits of size. *Figure 6* illustrates the thread form. FED-STD-H28/19 contains information on specifications, allowances, and tolerances.

ISO Metric threads

As described earlier, the ISO Metric and ANSI/ASME M Profile thread shares the basic profile of the Unified inch screw thread. The tolerances in this system have a heritage that

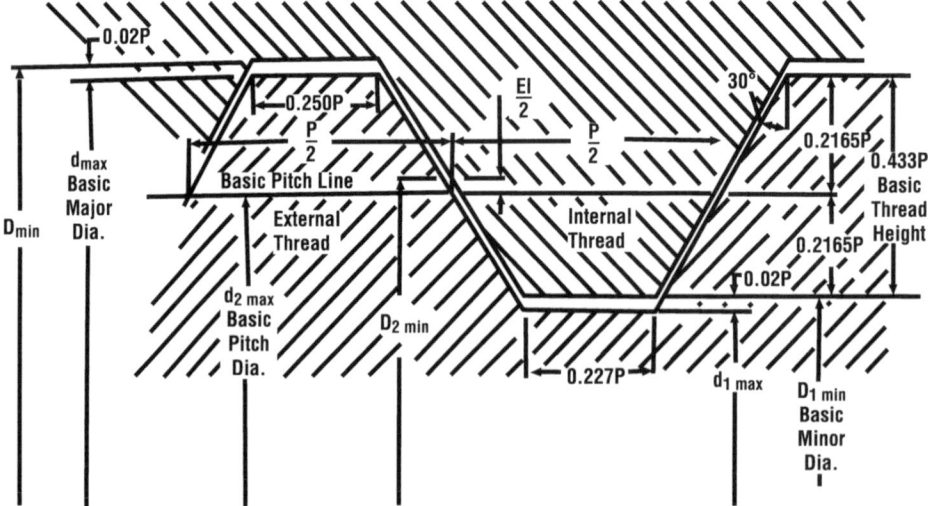

Figure 6. 60° Stub thread form.

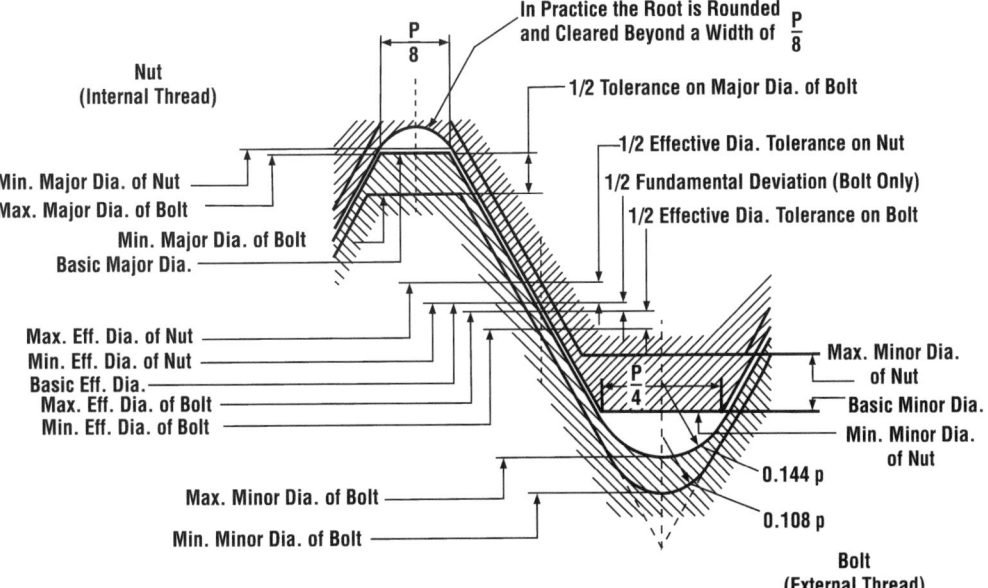

Figure 7. ISO Metric/M Profile thread form.

can be traced to the ISO system of limits and fits, and thread designations are intended to provide full tolerance information about a thread, much as the Class 1–5 system identifies the basic fit of Unified threads. The metric identification system first identifies the diameter in millimeters, then places a multi sign before the pitch. A dash then separates the thread dimensions from the tolerance information. Tolerance grades are numbers, and the number is followed by a tolerance position letter. A typical metric thread designation would be M10 × 1.5-6g. To indicate a gaging system number and a left hand thread, the designation would be M10 × 1.5-6g – LH (21). *Figure 7* illustrates the ISO Metric thread form. See the heading "Gaging systems" earlier in this section for details about System 21, System 22, and System 23.

The ISO Metric screw thread tolerance system provides for tolerance grades and tolerance positions (allowances) for the pitch diameter and crest diameter. For internal threads, the tolerance position letter is capitalized, and for external threads it is lower case. Internal threads have five grades of tolerance, 4,5,6,7,8. Grade 6 is the most commonly used grade, and lower numbers have tighter tolerances. There are two tolerance position letters for internal threads, H and G. "H" indicates that the minimum internal thread is at the basic pitch diameter size, and it is preferred for general use. "G" indicates that the minimum internal thread is above the pitch diameter, thereby creating an allowance. External threads have seven tolerance grades, 3,4,5,6,7,8,9, with the tolerance increasing with the numbers. "6" is the most commonly used grade. There are three tolerance position letters for external threads, h, g, and e, with the first two being the most widely used. "h" indicates no allowance, "g" provides a small allowance, and "e" has a larger allowance that does not find many applications. Metric threads can have two tolerance references. For instance, a thread designated 4g6g, has two tolerance grade and two tolerance position references. This thread will have its pitch diameter toleranced to 4g and its major diameter will have a 6g tolerance. The first reference is always to pitch diameter, and the second refers to the major diameter of the external thread of the minor diameter of the internal thread. Two factors determine the selection of a suitable tolerance class: the length of thread engagement, and the quality requirement. **Table 13** provides preferred ISO metric tolerance classes.

Since tolerance Grade 6 is by far the most common thread, the other tolerance grades are commonly measured in relation to Grade 6. The approximate ratios of other grades to Grade 6 are as follows: 3 = 0.5; 4 = 0.63; 5 = 0.8; 7 = 1.25; 8 = 1.6; and 9 = 2.0. The tolerances for Grade 6 threads are given in **Table 14** and **Table 15**. To find the approximate tolerance for other grades, multiply the appropriate Grade 6 tolerance by the ratio for the desired tolerance. For example, to find the external major diameter (d) tolerance for a Grade 4 thread with a pitch of 1.25, consult Table 15. From the table, it will be seen that the Grade 6 tolerance is 0.212 mm. To find the Grade 4 tolerance, multiple by the ratio given above (0.63): 0.212 × 0.63 = 0.133.

Metric grade tolerances can be derived from the tables of dimensions for 6H and 6g threads (**Table 16A** for coarse external threads, **Table 16B** for coarse internal threads, **Table 17** for fine external threads, **Table 18** for fine internal threads), **Table 19** for coarse pitch tolerance class 4h/5H, and **Table 20** for fine pitch tolerance class 4h/5H. In consulting the tables it will be seen that the difference between the maximum and minimum specifications for all of the four dimensions specified in Table 14 and Table 15 (internal pitch diameter, external pitch diameter, internal minor diameter, and external major diameter) are equal to the tolerances specified in these tables. It should be noted that although fine threads enhance static strength and in some instances can be used without locking mechanisms because they are less prone to vibration, coarse threads are recommended for most applications.

Table 13. Preferred Tolerance Classes for Metric Screw Threads. *(Source, OSG Tap and Die, Inc.)*

Quality Requirement	Tolerance Position "e" (Large Allowance)			External Thread					
				Tolerance Position "g" (Small Allowance)			Tolerance Position "h" (No Allowance)		
	Length of Thread Engagement			Length of Thread Engagement			Length of Thread Engagement		
Fine Close Fit							3h-4h	4h	5h-4h
Medium General Purpose		6e	7e-6e	5g-6g	6g	7g-6g	5h-6h	6h	7h-6h
Coarse Difficult Manufacturing Applications					8g	9g-8g			
Quality Requirement	Internal Thread								
	Tolerance Position "G" (Small Allowance)			Tolerance Position "H" (No Allowance)					
	Length of Thread Engagement			Length of Thread Engagement					
Fine Close Fit				4H		5H		6H	
Medium General Purpose	5G	6G	7G	5H		6H		7H	
Coarse Difficult Manufacturing Applications		7G	8G			7H		8H	

Note: Tolerance Position "e" is not to be applied to pitches finer than 0.5 mm. Tolerance Classes 6g and 6H are for commercial screw, bolt, and nut threads.

Therefore, it is suggested that coarse thread fasteners be employed unless a particular application requires the functional properties of fine threads. The Metric position allowance (also called the fundamental deviation) can also be derived from the tables. As noted above there is no allowance for positions H and h. The allowance for 6g is given in Table 16A, and explained in the notes in the bottom of the table. G allowances are mirror images of g allowances. As can be seen in Table 16, the allowance for any diameter "g" thread with a 1.25 pitch will be 0.028. That number is derived from the basic formula at the bottom of the table. For any "G" thread, the allowance will be the exact same, but a positive (in other words, the allowance for the 'g' thread will be expressed as –0.028, and for the 'G' thread it will be +0.028).

Generally, M profile 6g/6H metric designation can be equated to a Class 2A/2B Unified designation, and 4h/5H will provide a close fit similar to Class 3A/3B. British metric threads are defined in British Standard 3643, which is basically in accordance with the ISO and ANSI/ASME standards. For dimensions, consult Tables 16–20.

The **MJ profile** is used where 3A/3B or UNJ threads are required. MJ threads should be used where efficient use of material (maximum loads with minimum material), fatigue life, and stress levels take priority over cost considerations. FED-STD-H28/21B states that "The MJ profile provides a mandatory-controlled rounded root, for external thread, with an increased minor diameter. The rounded roots greatly reduce the concentration of stress, hence increase the fatigue life of the part. External threads produced by rolling after heat treatment and with rounded roots more than double part fatigue life. Its large core diameter improves the static tensile strength." As can be seen in Table 4, many specifications for MJ threads can be calculated with the same formulas as used for determining UNJ dimensions. For full specifications of MJ threads, consult ANSI B1.21M. See **Table 21** for additional information on the M and MJ thread forms.

Table 14. Internal (D_2) and External (d_2) Pitch Diameter Tolerances for Grade 6 Metric Thread.

Major Diameter		Pitch	D_2 Tolerance	d_2 Tolerance	Major Diameter		Pitch	D_2 Tolerance	d_2 Tolerance
Over	To				Over	To			
1.5	2.8	0.2	–	0.050	22.4	45	1	0.170	0.125
		0.25	–	0.056			1.5	0.200	0.150
		0.35	0.085	0.063			2	0.224	0.170
		0.4	0.090	0.067			3	0.265	0.200
		0.45	0.095	0.071			3.5	0.280	0.212
							4	0.300	0.224
							4.5	0.315	0.236
2.8	5.6	0.35	0.090	0.067	45	90	1.5	0.212	0.160
		0.5	0.100	0.075			2	0.236	0.180
		0.6	0.112	0.085			3	0.280	0.212
		0.7	0.118	0.090			4	0.315	0.236
		0.75	0.118	0.090			5	0.335	0.250
		0.8	0.125	0.095			5.5	0.355	0.265
							6	0.375	0.280
5.6	11.2	0.75	0.132	0.100	90	180	2	0.250	0.190
		1	0.150	0.112			3	0.300	0.224
		1.25	0.160	0.118			4	0.335	0.250
		1.5	0.180	0.132			6	0.400	0.300
11.2	22.4	1	0.160	0.118	180	355	3	0.335	0.250
		1.25	0.180	0.132			4	0.375	0.280
		1.5	0.190	0.140			6	0.425	0.315
		1.75	0.200	0.150					
		2	0.212	0.160					
		2.5	0.224	0.170					

All dimensions in millimeters.

Table 15. Internal Minor Diameter (D_1) and External Major Diameter (d) Tolerances for Grade 6 Metric Thread.

Pitch	D_1 Tolerance	d Tolerance	Pitch	D_1 Tolerance	d Tolerance
0.2	–	0.056	1.25	0.265	0.212
0.25	–	0.067	1.5	0.300	0.236
0.3	0.085	0.075	1.75	0.335	0.265
0.35	0.100	0.085	2	0.375	0.280
0.4	0.112	0.095	2.5	0.450	0.335
0.45	0.125	0.100	3	0.500	0.375
0.5	0.140	0.106	3.5	0.560	0.425
0.6	0.160	0.125	4	0.600	0.475
0.7	0.180	0.140	4.5	0.670	0.500
0.75	0.190	0.140	5	0.710	0.530
0.8	0.200	0.150	5.5	0.750	0.560
1	0.236	0.180	6	0.800	0.600

All dimensions in millimeters.

Table 16A. External Metric Thread—Coarse Pitch (Medium Fit)—Tolerance Class 6g.

Nominal Diameter[1] (d_{nom})	Pitch	Major Dia. (d)		Pitch Dia. (d_2)		Minor Dia. (d_1)		Allowance*
		max	min	max	min	max	min	
1	0.25	1.000	0.933	0.838	0.785	0.693	0.622	-
1.1		1.100	1.033	0.938	0.885	0.793	0.722	-
1.2		1.200	1.133	1.038	0.985	0.893	0.822	-
1.4	0.30	1.400	1.325	1.205	1.149	1.032	0.954	-
1.6	0.35	1.581	1.496	1.354	1.291	1.151	1.063	0.019
1.8		1.781	1.696	1.554	1.491	1.251	1.263	
2	0.40	1.981	1.886	1.721	1.654	1.490	1.394	
2.2	0.45	2.180	2.080	1.888	1.817	1.628	1.525	0.020
2.5		2.480	2.380	2.188	2.117	1.928	1.825	
3	0.50	2.980	2.874	2.655	2.580	2.367	2.256	
3.5	0.60	3.479	3.354	3.089	3.004	2.742	2.614	0.021
4	0.70	3.978	3.838	3.523	3.433	3.119	2.979	0.022
4.5	0.75	4.478	4.338	3.991	3.901	3.558	3.414	
5	0.80	4.976	4.826	4.456	4.361	3.994	3.841	0.024
6	1.00	5.974	5.794	5.325	5.212	4.747	4.563	0.026
7		6.974	6.794	6.324	6.212	5.747	5.563	
8	1.25	7.972	7.760	7.160	7.042	6.439	6.231	0.028
9		8.972	8.760	8.160	8.042	7.439	7.231	
10	1.50	9.968	9.732	8.994	8.862	8.127	7.887	0.032
11		10.968	10.732	9.994	9.862	9.127	8.887	

(Continued)

Table 16A. *(Continued)* **External Metric Thread—Coarse Pitch (Medium Fit)—Tolerance Class 6g.**

Nominal Diameter[1] (d_{nom})	Pitch	Major Dia. (d)		Pitch Dia. (d_2)		Minor Dia. (d_1)		Allowance*
		max	min	max	min	max	min	
12	1.75	11.966	11.701	10.829	10.679	9.819	9.543	0.034
14	2.00	13.962	13.682	12.663	12.503	11.508	11.204	0.038
16		15.962	15.682	14.663	14.503	13.508	13.204	
18		17.958	17.623	16.334	16.164	14.891	14.541	
20	2.50	19.958	19.623	18.334	18.164	16.891	16.541	0.042
22		21.958	21.623	20.334	20.164	18.891	18.541	
24	3.00	23.952	23.577	22.003	21.803	20.271	19.855	0.048
27		26.952	26.577	25.003	24.803	23.271	22.855	
30	3.50	29.947	29.522	27.674	27.462	25.553	25.189	0.053
33		32.947	32.522	30.674	30.462	28.553	28.189	
36	4.00	35.940	35.465	33.342	33.118	31.033	30.521	0.060
39		38.940	38.465	36.342	36.118	34.033	33.521	
42	4.50	41.937	41.437	39.014	38.778	36.418	35.856	0.063
45		44.937	44.437	42.014	41.778	39.416	38.856	
48	5.00	47.929	47.399	44.681	44.431	41.794	41.184	0.071
52		51.929	51.399	48.681	48.431	45.794	45.184	
56	5.50	55.925	55.365	52.353	52.088	49.177	48.516	0.075
60		59.925	59.365	56.353	56.088	53.177	52.516	

[1] The diameters listed to the left of the column are 1st choice, in the middle 2nd choice, and to the right 3rd choice. All sizes are 6g tolerances except those for M1 and M1.4 diameters. These diameters 6h threads and the figures given have been calculated on that basis. *The external thread allowance, or upper deviation, is derived from the following formulas: d_{nom} - d_{max}. All dimensions in millimeters. (Source, Landis Threading Systems)

Table 16B. Internal Metric Thread—Coarse Pitch (Medium Fit)—Tolerance Class 6H.

Nominal Diameter[1] (D_{min})	Pitch	Minor Dia. (D_1)		Pitch Dia. (D_2)		D_2 (min) Minus D_1 (min)	Tap Drill Size*
		max	min	max	min		
1		0.785	0.729	0.894	0.838	0.109	.75 mm
1.1	0.25	0.885	0.829	0.994	0.938	0.109	.85 mm
1.2		0.985	0.929	1.094	1.038	0.109	.95 mm
1.4	0.30	1.160	1.075	1.280	1.205	0.130	1.1 mm
1.6	0.35	1.321	1.221	1.458	1.373	0.152	1.25 mm
1.8		1.521	1.421	1.650	1.573	0.152	1.45 mm
2	0.40	1.679	1.567	1.830	1.740	0.173	1.6 mm
2.2	0.45	1.838	1.713	2.003	1.908	0.195	1.75 mm
2.5		2.138	2.013	2.303	2.208	0.195	2.05 mm
3.0	0.50	2.599	2.459	2.775	2.675	0.216	2.5 mm
3.5	0.60	3.010	2.850	3.222	3.110	0.260	2.9 mm
4	0.70	3.422	3.242	3.663	3.545	0.303	3.3 mm
4.5	0.75	3.878	3.688	4.131	4.013	0.325	#26

(Continued)

Table 16B. *(Continued)* **Internal Metric Thread—Coarse Pitch (Medium Fit)—Tolerance Class 6H.**

Nominal Diameter[1] (D_{min})	Pitch	Minor Dia. (D_1)		Pitch Dia. (D_2)		D_2 (min) Minus D_1 (min)	Tap Drill Size*
		max	min	max	min		
5	0.80	4.334	4.134	4.605	4.480	0.346	#19
6	1.00	5.153	4.917	5.500	5.350	0.433	5 mm
7		6.153	5.917	6.500	6.350	0.433	6 mm
8	1.25	6.912	6.647	7.348	7.188	0.541	17/64"
9		7.912	7.647	8.348	8.188	0.541	7.75 mm
10	1.50	8.676	8.376	9.206	9.026	0.650	8.5 mm
11		9.676	9.376	10.206	10.026	0.650	9.5 mm
12	1.75	10.441	10.106	11.063	10.863	0.757	Y
14	2.00	12.210	11.835	12.913	12.701	0.866	12 mm
16		14.210	13.835	14.913	14.701	0.866	14 mm
18	2.50	15.744	15.294	16.660	16.376	1.082	39/64"
20		17.744	17.294	18.600	18.376	1.082	17.5 mm
22		19.744	19.294	20.600	20.376	1.082	49/64"
24	3.00	21.252	20.752	22.316	22.051	1.299	21 mm
27		24.252	23.752	25.316	25.051	1.299	24 mm
30	3.50	26.771	26.211	28.007	27.727	1.516	26.5 mm
33		29.771	29.211	31.007	30.727	1.516	29.5 mm
36	4.00	32.270	31.670	33.702	33.402	1.732	32 mm
39		35.270	34.670	36.702	36.402	1.732	35 mm
42	4.50	37.799	37.129	39.392	39.077	1.948	1 15/32"
45		40.799	40.129	42.392	42.077	1.948	40.5 mm
48	5.00	43.297	42.587	45.087	44.752	2.165	1 11/16"
52		47.297	46.587	49.087	48.752	2.165	47 mm
56	5.50	50.796	50.046	52.783	52.428	2.382	50.5 mm
60		54.796	54.046	56.783	56.428	2.382	54.5 mm

[1] The diameters listed to the left of the column are 1st choice, in the middle 2nd choice, and to the right 3rd choice. *See Tapping Section for alternative sizes. The pitch diameter tolerance, not given in the table, is equal to the difference between the maximum and minimum pitch diameters: $D_{2\ max}$ - $D_{2\ min}$. All dimensions in millimeters. (Source, Landis Threading Systems.)

Table 17. External Metric Thread—Fine Pitch (Medium Fit)—Tolerance Class 6g (Preferred Sizes Only).

Nominal Diameter[1] (d_{nom})	Pitch	Major Dia. (d)		Pitch Dia. (d_2)		Minor Dia. (d_1)		Allowance*
		max	min	max	min	max	min	
1	0.20	1.000	0.964	0.870	0.840	0.754	0.710	-
1.2		1.200	1.164	1.070	1.040	0.954	0.910	-
1.6		1.583	1.527	1.453	1.403	1.338	1.274	0.17
2	0.25	1.982	1.915	1.820	1.764	1.675	1.601	0.18
2.5	0.35	2.481	2.396	2.254	2.191	2.052	1.964	0.19
3		2.981	2.896	2.754	2.687	2.552	2.460	

(Continued)

Table 17. *(Continued)* **External Metric Thread—Fine Pitch (Medium Fit)—Tolerance Class 6g (Preferred Sizes Only).**

Nominal Diameter[1] (d_{nom})	Pitch	Major Dia. (d)		Pitch Dia. (d_2)		Minor Dia. (d_1)		Allowance*
		max	min	max	min	max	min	
4	0.50	3.980	3.874	3.655	3.580	3.367	3.256	0.20
5		4.980	4.874	4.655	4.580	4.367	4.256	
6	0.75	5.978	5.838	5.491	5.391	5.058	4.904	0.22
8		7.978	7.838	7.491	7.391	7.058	6.904	
10		9.978	9.838	9.491	9.391	9.058	8.904	
8	1.00	7.974	7.794	7.324	7.212	6.747	6.563	0.26
10		9.974	9.794	9.324	9.212	8.747	8.563	
12		11.974	11.794	11.324	11.206	10.747	10.557	
16		16.974	15.794	15.324	15.206	14.747	14.557	
20		10.974	10.794	19.324	19.206	19.747	18.557	
24		23.974	23.794	23.324	23.199	22.747	22.550	
30		29.974	29.794	29.324	29.199	28.747	28.550	
10	1.25	9.972	9.760	9.160	9.042	8.438	8.230	0.28
12		11.972	11.760	11.160	11.028	10.438	10.216	
12	1.50	11.968	11.732	10.994	10.854	10.128	9.880	0.32
16		15.968	15.732	14.994	14.854	14.128	13.880	
20		19.968	19.732	18.994	18.854	18.128	17.880	
24		23.968	23.732	22.994	22.844	22.128	21.870	
30		29.968	29.732	28.994	28.844	28.128	27.870	
36		35.968	35.732	34.994	34.844	34.128	33.870	
42		41.968	41.732	40.994	40.844	40.128	39.870	
48		47.968	47.732	46.994	46.834	46.128	45.860	
56		55.968	55.732	54.994	54.834	54.128	53.860	
20	2.00	19.962	19.682	18.663	18.503	17.508	17.204	0.38
24		23.962	23.682	22.663	22.493	21.508	21.194	
30		29.962	29.682	28.663	28.493	27.508	27.194	
36		35.962	35.682	34.663	34.493	33.508	33.194	
42		41.962	41.682	40.663	40.493	39.508	39.194	
48		47.962	47.682	46.663	46.483	45.508	45.184	
56		55.962	55.682	54.663	54.483	53.508	53.184	
30	3.00	29.952	29.577	28.003	27.803	26.271	25.855	0.48
36		35.952	35.577	34.003	33.803	32.271	31.855	
42		41.952	41.577	40.003	39.803	38.271	37.855	
48		47.952	47.577	46.003	45.791	44.271	43.843	
56		55.952	55.577	54.003	53.791	52.271	51.843	
42	4.00	41.940	41.465	39.342	39.118	37.033	36.521	0.60
48		47.940	47.465	45.342	45.106	43.033	42.509	
56		55.940	55.465	53.342	53.106	51.033	50.509	

This Table contains dimensions for preferred nominal diameters through 56mm only. All sizes are to 6g tolerances except M1-M4 0.2 pitch, which are to 4h tolerances. *The external thread allowance, or upper deviation, is derived from the following formulas: d_{nom} - d_{max}. All dimensions in millimeters. (Source, Landis Threading Systems.)

Table 18. Internal Metric Thread—Fine Pitch (Medium Fit)—Tolerance Class 6H (Preferred Sizes Only).

Nominal Diameter[1] (D_{min})	Pitch	Minor Dia. (D_1)		Pitch Dia. (D_2)		D_2 (min) Minus D_1 (min)	Tap Drill Size*
		max	min	max	min		
1		0.821	0.783	0.910	0.870	0.087	.80 mm
1.2	0.20	1.021	0.983	1.110	1.070	0.087	1 mm
1.6		1.421	1.383	1.512	1.470	0.087	1.4 mm
2	0.25	1.785	1.729	1.886	1.838	0.109	1.75 mm
2.5	0.35	2.221	2.121	2.358	2.273	0.152	2.15 mm
3		2.721	2.621	2.863	2.773	0.152	2.6 mm
4	0.50	3.599	3.459	3.775	3.675	0.216	3.5 mm
5		4.599	4.459	4.775	4.675	0.216	4.5 mm
6		5.378	5.188	5.645	5.513	0.325	5.25 mm
8	0.75	7.378	7.188	7.645	7.513	0.325	7.25 mm
10		9.378	9.188	9.645	9.513	0.325	#64
8		7.153	6.917	7.500	7.350	0.433	7 mm
10		9.153	8.917	9.500	9.350	0.433	9 mm
12		11.153	10.917	11.510	11.350	0.433	11 mm
16	1.00	15.153	14.917	15.510	15.350	0.433	15 mm
20		19.153	18.917	19.510	19.350	0.433	19 mm
24		23.153	22.917	23.520	23.350	0.433	23 mm
30		29.153	28.917	29.520	29.350	0.433	29 mm
10	1.25	8.912	8.647	9.348	9.188	0.541	8.75 mm
12		10.912	10.647	11.368	11.188	0.541	27/64"
12		10.676	10.376	11.216	11.026	0.650	10.5 mm
16		14.676	14.376	15.216	15.026	0.650	14.5 mm
20		18.676	18.376	19.216	19.026	0.650	18.5 mm
24		22.676	22.376	23.226	23.026	0.650	22.5 mm
30	1.50	28.676	28.376	29.226	29.026	0.650	28.5 mm
36		34.676	34.376	35.225	35.026	0.650	34.5 mm
42		40.676	40.376	41.226	41.026	0.650	40.5 mm
48		46.676	46.376	47.238	47.026	0.650	46.5 mm
56		54.676	54.376	55.238	55.026	0.650	54.5 mm
20		18.210	17.835	18.913	18.701	0.866	18 mm
24		22.210	21.835	22.925	22.701	0.866	22 mm
30		28.210	27.835	28.925	28.701	0.866	28 mm
36	2.00	34.210	33.835	34.925	34.701	0.866	34 mm
42		40.210	39.835	40.925	40.701	0.866	40 mm
48		46.210	45.835	46.937	46.701	0.866	46 mm
56		54.210	53.835	54.937	54.701	0.866	2 1/8"

(Continued)

Table 18. *(Continued)* **Internal Metric Thread—Fine Pitch (Medium Fit)—Tolerance Class 6H (Preferred Sizes Only).**

Nominal Diameter[1] (D_{min})	Pitch	Minor Dia. (D_1)		Pitch Dia. (D_2)		D_2 (min) Minus D_1 (min)	Tap Drill Size*
		max	min	max	min		
30		27.252	26.752	28.316	28.051	1.299	27mm
36		33.252	32.752	34.316	34.051	1.299	33mm
42	3.00	39.252	38.752	40.316	40.051	1.299	39mm
48		45.252	44.752	46.331	46.051	1.299	45mm
56		53.252	52.752	54.331	54.051	1.299	53mm
42		38.270	37.670	39.702	39.402	1.732	1 1/2"
48	4.00	44.270	43.670	45.717	45.402	1.732	44mm
56		52.270	51.670	53.717	53.402	1.732	52mm

* See tapping Section for alternative sizes. This Table contains dimensions for preferred nominal diameters through 56mm only. All sizes are to 6g tolerances except M1-M1.6 with 0.2 pitch, which are to 4h tolerances. The pitch diameter tolerance, not given in this table, is equal to the difference between the maximum and minimum pitch diameters: D_{2max} - D_{2min}. All dimensions in millimeters. (Source, Landis Threading Systems.)

Table 19. Metric Thread—Coarse Pitch (Close Fit)—Tolerance Class 4h/5H.

Nominal Diameter[1]	Pitch	External Threads					Internal Threads			
		Major Dia.[2]	Pitch Dia. (d_2)		Minor Dia. (d_1)		Pitch Dia. (D_2)		Minor Dia. (D_1)	
		min	max	min	max	min	max	min	max	min
1		0.958	0.838	0.804	0.693	0.641	0.894	0.838	0.785	0.729
1.1	0.25	1.058	0.938	0.904	0.793	0.741	0.994	0.938	0.885	0.829
1.2		1.158	1.038	1.004	0.893	0.841	1.094	1.038	0.985	0.929
1.4	0.30	1.352	1.205	1.169	1.032	0.974	1.265	1.205	1.142	1.075
1.6	0.35	1.547	1.373	1.333	1.170	1.105	1.440	1.373	1.301	1.221
1.8		1.747	1.573	1.533	1.370	1.305	1.640	1.573	1.501	1.421
2	0.40	1.940	1.740	1.698	1.509	1.438	1.811	1.740	1.657	1.567
2.2	0.45	2.137	1.908	1.863	1.648	1.571	1.983	1.908	1.813	1.713
2.5		2.437	2.208	2.163	1.948	1.871	2.283	2.208	2.113	2.013
3	0.50	2.933	2.675	2.627	2.387	2.303	2.755	2.675	2.571	2.459
3.5	0.60	3.420	3.110	3.047	2.764	2.668	3.200	3.110	2.975	2.850
4	0.70	3.910	3.545	3.489	3.141	3.035	3.640	3.545	3.382	3.242
4.5	0.75	4.410	4.013	3.953	3.580	3.470	4.108	4.013	3.838	3.688
5	0.80	4.905	4.480	4.420	4.109	3.901	4.580	4.480	4.294	4.134
6	1.00	5.888	5.350	5.279	4.773	4.630	5.468	5.350	5.107	4.917
7		6.888	6.350	6.279	5.773	5.630	6.468	6.350	6.107	5.917
8	1.25	7.868	7.188	7.113	6.466	6.301	7.313	7.188	6.859	6.647
9		8.868	8.188	8.113	7.466	7.301	8.313	8.188	7.859	7.647
10	1.50	9.850	9.026	8.941	8.160	7.967	9.166	9.026	8.612	8.376
11		10.850	10.026	9.941	9.160	8.967	10.166	10.026	9.612	9.376
12	1.75	11.830	10.863	10.768	9.853	9.632	11.023	10.863	10.371	10.106

(Continued)

Table 19. *(Continued)* Metric Thread—Coarse Pitch (Close Fit)—Tolerance Class 4h/5H.

Nominal Diameter[1]	Pitch	External Threads							Internal Threads			
		Major Dia.[2]	Pitch Dia. (d₂)		Minor Dia. (d₁)			Pitch Dia. (D₂)		Minor Dia. (D₁)	
		min	max	min	max	min		max	min	max	min
14	2.00	13.820	12.701	12.601	11.546	11.302		12.871	12.701	12.135	11.835
16		15.820	14.701	14.601	13.546	13.302		14.871	14.701	14.135	13.835
18	2.50	17.788	16.376	16.270	14.933	14.647		16.556	16.376	15.649	15.294
20		19.788	18.376	18.270	16.933	16.647		18.556	18.376	17.649	17.294
22		21.788	20.375	20.270	18.933	18.647		20.556	20.376	19.649	19.294
24	3.00	23.764	22.051	21.925	20.319	19.978		22.263	22.051	21.152	20.752
27		26.764	25.051	24.926	23.319	22.978		25.263	25.051	24.152	23.752
30	3.50	29.735	27.727	27.595	25.706	25.322		27.951	27.727	26.661	26.211
33		32.735	30.727	30.595	28.706	28.322		30.951	30.727	29.661	29.211
36	4.00	35.700	33.402	33.262	31.093	30.665		33.638	33.402	32.145	31.670
39		38.700	36.402	36.262	34.093	33.665		36.638	36.402	35.145	34.670
42	4.50	41.685	39.077	38.927	36.479	36.004		39.327	39.077	37.659	37.129
45		44.685	42.007	41.927	39.479	39.004		42.327	42.077	40.659	40.129
48	5.00	47.665	44.753	44.592	41.866	41.345		45.017	44.752	43.147	42.587
52		51.665	48.752	48.592	45.866	45.345		49.017	48.752	47.147	46.587
56	5.50	55.645	52.428	52.258	49.252	48.685		52.708	52.428	50.646	50.046
60		59.645	56.428	56.258	53.252	52.685		56.708	56.428	56.646	54.046
64	6.00	63.625	60.106	59.923	56.639	56.026		60.403	60.103	58.135	57.505
68		67.625	64.103	63.923	60.639	60.026		64.403	64.103	62.135	61.505

[1] Secondary size nominal diameters are in italics. [2] Maximum major diameter (d) of the external thread, and the minimum major diameter of the internal thread (D), are equal to the nominal diameter shown in the first column. All dimensions in millimeters. (Source, Landis Threading Systems)

Table 20. Metric Thread—Fine Pitch (Close Fit)—Tolerance Class 4h/5H (Preferred Sizes Only).

Nominal Diameter[1]	Pitch	External Threads							Internal Threads			
		Major Dia.[2]	Pitch Dia. (d₂)		Minor Dia. (d₁)			Pitch Dia. (D₂)		Minor Dia. (D₁)	
		min	max	min	max	min		max	min	max	min
1	0.20	0.964	0.870	0.840	0.754	0.710		-	-	-	-
1.2		1.164	1.070	1.040	0.945	0.910		-	-	-	-
1.6		1.564	1.470	1.438	1.354	1.308		-	-	-	-
2	0.25	1.958	1.838	1.802	1.693	1.639		1.898	1.838	1.785	1.729
2.5	0.35	2.447	2.273	2.233	2.070	2.005		2.340	2.273	2.201	2.121
3		2.947	2.773	2.731	2.570	2.503		2.844	2.773	2.701	2.621
4	0.50	3.933	3.675	3.627	3.387	3.303		3.755	3.675	3.571	3.459
5		4.933	4.675	4.627	4.387	4.303		4.755	4.675	4.571	4.459
6		5.910	5.513	5.453	5.080	4.966		5.619	5.513	5.338	5.188
8	0.75	7.910	7.513	7.453	7.080	6.966		7.619	7.513	7.338	7.188
10		9.910	9.513	9.453	9.080	8.966		9.619	9.513	9.338	9.188

(Continued)

Table 20. *(Continued)* Metric Thread—Fine Pitch (Close Fit)—Tolerance Class 4h/5H (Preferred Sizes Only).

Nominal Diameter[1]	Pitch	External Threads					Internal Threads			
		Major Dia.[2]	Pitch Dia. (d_2)		Minor Dia. (d_1)		Pitch Dia. (D_2)		Minor Dia. (D_1)	
		min	max	min	max	min	max	min	max	min
8		7.888	7.350	7.279	6.773	6.630	7.468	7.350	7.107	6.917
10		9.888	9.350	9.279	8.773	8.630	9.468	9.350	9.107	8.917
12		11.888	11.350	11.275	10.773	10.630	11.468	11.350	11.107	10.917
16	1.00	15.888	15.350	15.275	14.773	14.630	15.468	15.350	15.107	14.917
20		19.888	19.350	19.275	18.773	18.630	19.468	19.350	19.107	18.917
24		23.888	23.350	23.275	22.773	22.630	23.468	23.350	23.107	22.917
30		29.888	29.350	29.275	28.773	28.630	29.468	29.350	29.107	28.917
10	1.25	9.868	9.188	9.113	8.467	8.302	9.313	9.188	8.859	8.647
12		11.868	11.188	11.103	10.467	10.292	11.328	11.188	10.859	10.647
12		11.850	11.026	10.936	10.159	9.961	11.176	11.026	10.612	10.376
16		15.850	15.026	14.936	14.159	13.961	15.176	15.026	14.612	14.376
20		19.850	19.026	18.936	18.159	17.961	19.176	19.026	18.612	18.376
24		23.850	23.026	22.936	22.159	21.956	23.186	23.026	22.612	22.376
30	1.50	29.850	29.026	28.931	28.159	27.956	29.186	29.025	28.612	28.376
36		35.850	35.026	34.931	34.159	33.956	35.186	35.026	34.612	34.376
42		41.850	41.026	40.931	40.159	39.956	41.186	41.026	40.612	40.376
48		47.850	47.026	46.926	46.159	45.951	47.196	47.026	46.612	46.376
56		55.850	55.026	54.926	54.159	53.951	55.196	55.026	54.612	54.376
64		63.850	63.026	62.926	62.159	61.951	63.196	63.026	62.612	62.376
20		19.820	18.701	18.601	17.546	17.302	18.871	18.701	18.135	17.835
24		23.820	22.701	22.595	21.546	21.296	22.881	22.701	22.134	21.835
30		29.820	28.701	28.595	27.546	27.296	28.881	28.701	28.135	27.835
36		35.820	34.701	34.595	33.546	33.296	34.881	34.701	34.135	33.835
42	2.00	41.820	40.701	40.595	39.546	39.296	40.881	40.701	40.135	39.835
48		47.820	46.701	46.589	45.546	45.290	46.891	46.701	46.135	45.835
56		55.820	54.701	54.589	53.546	53.290	54.891	54.701	54.135	53.835
64		63.820	62.701	62.589	61.546	61.290	62.891	62.701	62.135	61.835
30		29.764	28.051	27.926	26.319	25.978	28.263	28.051	27.152	26.752
36		35.764	340.51	33.926	32.319	31.978	34.263	34.051	33.152	32.752
42		41.764	40.051	39.926	38.319	37.978	40.263	40.051	39.152	38.752
48	3.00	47.764	46.051	45.919	44.319	43.971	46.275	46.051	45.152	44.752
56		55.764	54.051	53.919	52.319	52.971	54.275	54.051	53.152	52.752
64		63.764	62.051	61.919	60.319	59.971	62.275	62.051	61.152	60.752
42		41.700	39.402	39.262	37.093	36.665	39.638	39.402	38.145	37.670
48		47.700	45.402	45.252	43.093	42.655	45.652	45.402	44.145	43.670
56	4.00	55.700	53.402	53.252	51.093	50.655	53.652	53.402	52.145	51.670
64		63.700	61.402	61.252	59.093	58.655	61.652	61.402	60.145	59.670

[1] Nominal diameters on this Table are preferred sizes. [2] Both the maximum major diameter (d) of the external thread and the minimum major diameter of the internal thread (D) are equal to the nominal diameter shown in the first column. This Table contains dimensions for preferred nominal diameters through 64mm only. All dimensions in millimeters. (Source, Landis Threading Systems)

Table 21. Thread Form for M and MJ Threads. *(Source, Cleveland Twist Drill.)*

Pitch P	Height of External Thread		Addendum M, MJ	Root Flat (Basic Profile) M	Root Radii		
					M (Min.)	MJ	
	M	MJ				Min.	Max.
	0.69717H	0.66667H	0.375H	0.250P	0.125P	0.18042P	0.15011P
0.2	.1208	.1155	.0650	.0500	.0250	.0361	.0300
0.25	.1509	.1443	.0812	.0625	.0313	.0451	.0375
0.3	.1811	.1732	.0974	.0750	.0375	.0541	.0450
0.35	.2113	.2021	.1137	.0788	.0438	.0631	.0525
0.4	.2415	.2309	.1299	.1000	.0500	.0722	.0600
0.45	.2717	.2598	.1461	.1125	.0563	.0812	.0675
0.5	.3019	.2887	.1624	.1250	.0625	.0902	.0751
0.6	.3623	.3464	.1949	.1500	.0750	.1083	.0901
0.7	.4226	.4041	.2273	.1750	.0875	.1263	.1051
0.75	.4528	.4330	.2436	.1875	.0938	.1353	.1126
0.8	.4830	.4619	.2598	.2000	.1000	.1443	.1201
1	.6038	.5774	.3248	.2500	.1250	.1804	.1501
1.25	.7547	.7217	.4059	.3125	.1563	.2255	.1876
1.5	.9057	.8660	.4871	.3750	.1875	.2706	.2252
1.75	1.0566	1.0104	.5683	.4375	.2188	.3157	.2627
2	1.2075	1.1547	.6495	.5000	.2500	.3608	.3002
2.5	1.5094	1.4434	.8119	.6250	.3125	.4511	.3753
3	1.8113	1.7321	.9743	.7500	.3750	.5413	.4503
3.5	2.1132	2.0207	1.1367	.8750	.4375	.6315	.5254
4	2.4151	2.3094	1.2990	1.0000	.5000	.7217	.6004
4.5	2.7170	2.5981	1.4614	1.1250	.5625	.8119	.6755
5	3.0188	2.8868	1.6238	1.2500	.6250	.9021	.7506
5.5	3.3207	3.1754	1.7862	1.3750	.6875	.9923	.8256
6	3.6226	3.4641	1.9486	1.5000	.7500	1.0825	.9007
8	4.8301	4.6188	2.5981	2.0000	1.0000	1.4434	1.2009

All dimensions in millimeters.

B.S.F., B.S.W. Whitworth, and other British thread forms

Although the Whitworth thread form had been superceded by the Unified thread and the British Standard metric thread in Great Britain, its widespread usage over time assures a continued demand for replacing worn or damaged fasteners. There are two designations, B.S.W. (British Standard Whitworth) which indicates the coarse series, and B.S.F., the fine series. All Whitworth threads have a 55° included angle, a radius equal to 0.137329P, and a thread height equal to 0.640327P. The height of the sharp V thread is equal to 0.96049P. The Whitworth thread form is shown in *Figure 8*, and basic dimensions are given in **Table 22** (coarse threads) and **Table 23** (fine threads). It should be noted that, by reducing the maximum bolt limits below basic size for medium and free fit dimensions, an allowance is created for plating.

The British Standard Taper Pipe (BSTP) thread for pressure tight joints and the British

(Text continued on p. 89)

Table 22. British Standard Whitworth (B.S.W. - Coarse) Thread Dimensions. *(Source, Landis Threading Systems.)*

Size and TPI	Fit	Major Diameter			Effective Diameter			Minor Diameter			Drill Size
		Max.	Min.	Plated	Max.	Min.	Plated	Max.	Min.	Plated	
1/8-40	Close	0.1250	0.1215	-	0.1090	0.1071	-	0.0930	0.0890	-	2.50mm
	Med.	0.1238	0.1193	0.1250	0.1078	0.1049	0.1090	0.0918	0.0857	0.0930	
	Free	0.1238	0.1179	0.1250	0.1078	0.1035	0.1090	0.0918	0.0843	0.0930	
3/16-24	Close	0.1875	0.1832	-	0.1608	0.1585	-	0.1341	0.1291	-	3.70mm
	Med.	0.1863	0.1808	0.1875	0.1596	0.1561	0.1608	0.1329	0.1253	0.1341	
	Free	0.1863	0.1791	0.1875	0.1596	0.1544	0.1608	0.1329	0.1236	0.1341	
1/4-20	Close	0.2500	0.2452	-	0.2180	0.2154	-	0.1860	0.1805	-	5.10mm
	Med.	0.2488	0.2427	0.2500	0.2168	0.2129	0.2180	0.1848	0.1764	0.1860	
	Free	0.2488	0.2408	0.2500	0.2168	0.2110	0.2180	0.1848	0.1764	0.1860	
5/16-18	Close	0.3125	0.3073	-	0.2769	0.2741	-	0.2413	0.2354	-	6.50mm
	Med.	0.3112	0.3046	0.3125	0.2756	0.2714	0.2769	0.2400	0.2311	0.2413	
	Free	0.3112	0.3025	0.3125	0.2756	0.2693	0.2769	0.2400	0.2290	0.2413	
3/8-16	Close	0.3750	0.3695	-	0.3350	0.3320	-	0.2950	0.2888	-	7.90mm
	Med.	0.3736	0.3666	0.3750	0.3336	0.3291	0.3350	0.2936	0.2841	0.2950	
	Free	0.3736	0.3646	0.3750	0.3336	0.3268	0.3350	0.2936	0.2818	0.2950	
7/16-14	Close	0.4375	0.4316	-	0.3918	0.3886	-	0.3461	0.3394	-	9.20mm
	Med.	0.4360	0.4285	0.4375	0.3903	0.3855	0.3918	0.3446	0.3345	0.3461	
	Free	0.4360	0.4260	0.4375	0.3903	0.3830	0.3918	0.3446	0.3320	0.3461	
1/2-12	Close	0.5000	0.4937	-	0.4466	0.4432	-	0.3932	0.3860	-	10.40mm
	Med.	0.4985	0.4904	0.5000	0.4451	0.4399	0.4466	0.3917	0.3807	0.3932	
	Free	0.4985	0.4879	0.5000	0.4451	0.4374	0.4466	0.3917	0.3782	0.3932	
5/8-11	Close	0.6250	0.6183	-	0.5668	0.5631	-	0.5086	0.5010	-	13.40mm
	Med.	0.6233	0.6147	0.6250	0.5651	0.5595	0.5668	0.5069	0.4953	0.5086	
	Free	0.3233	0.6119	0.6250	0.5651	0.5567	0.5668	0.5069	0.4925	0.5086	
3/4-10	Close	0.7500	0.7428	-	0.6860	0.6820	-	0.6220	0.6139	-	16.25mm
	Med.	0.7482	0.7390	0.7500	0.6842	0.6782	0.6860	0.6202	0.6079	0.6220	
	Free	0.7482	0.7360	0.7500	0.6842	0.6752	0.6860	0.6202	0.6049	0.6220	

(Continued)

An allowance is provided on bolts in the "Medium" and "Free" classes by reducing the maximum bolt limits below basic size. This reduction allows for plating, the plating falling within the allowance. It also facilitates easy assembly of maximum metal bolts and nuts.

Table 22. (Continued) British Standard Whitworth (B.S.W. - Coarse) Thread Dimensions. (Source, Landis Threading Systems.)

Size and TPI	Fit	Major Diameter			Effective Diameter			Minor Diameter			Drill Size
		Max.	Min.	Plated	Max.	Min.	Plated	Max.	Min.	Plated	
7/8-9	Close	0.8750	0.8674	–	0.8039	0.7996	–	0.7328	0.7242	–	19.25mm
	Med.	0.8750	0.8653	–	0.8039	0.7975	–	0.7328	0.7197	–	
	Free	0.8750	0.8621	–	0.8039	0.7943	–	0.7328	0.7165	–	
1-8	Close	1.0000	0.9920	–	0.9200	0.9155	–	0.8400	0.8309	–	22.00mm
	Med.	1.0000	0.9897	–	0.9200	0.9132	–	0.8400	0.8261	–	
	Free	1.0000	0.9863	–	0.9200	0.9098	–	0.8400	0.8227	–	
1 1/8-7	Close	1.1250	1.1164	–	1.0335	1.0287	–	0.9420	0.9323	–	24.50mm
	Med.	1.1250	1.1140	–	1.0335	1.0263	–	0.9420	0.9272	–	
	Free	1.1250	1.1105	–	1.0335	1.0228	–	0.9420	0.9237	–	
1 1/4-7	Close	1.2500	1.2413	–	1.1585	1.1536	–	1.0670	1.0572	–	27.25mm
	Med.	1.2500	1.2388	–	1.1585	1.1511	–	1.0670	1.0520	–	
	Free	1.2500	1.2351	–	1.1585	1.1474	–	1.0670	1.0483	–	
1 1/2-6	Close	1.5000	1.4906	–	1.3933	1.3880	–	1.2866	1.2760	–	33.50mm
	Med.	1.5000	1.4879	–	1.3933	1.3853	–	1.2866	1.2704	–	
	Free	1.5000	1.4839	–	1.3933	1.3813	–	1.2866	1.2664	–	
1 3/4-5	Close	1.7500	1.7398	–	1.6219	1.6162	–	1.4938	1.4823	–	38.50mm
	Med.	1.7500	1.7369	–	1.6219	1.6133	–	1.4938	1.4763	–	
	Free	1.7500	1.7326	–	1.6219	1.6090	–	1.4938	1.4720	–	
2-4.5	Close	2.0000	1.9892	–	1.8577	1.8516	–	1.7154	1.7032	–	44.50
	Med.	2.0000	1.9862	–	1.8577	1.8486	–	1.7154	1.6969	–	
	Free	2.0000	1.9816	–	1.8577	1.8440	–	1.7154	1.6923	–	
2 1/4-4	Close	2.2500	2.2368	–	2.0899	2.0838	–	1.9298	1.9169	–	50.00mm
	Med.	2.2500	2.2354	–	2.0899	2.0803	–	1.9298	1.9102	–	
	Free	2.2500	2.2306	–	2.0899	2.0755	–	1.9298	1.9054	–	
2 1/2-4	Close	2.5000	2.4884	–	2.3399	2.3333	–	2.1798	2.1667	–	56.00mm
	Med.	2.5000	2.4850	–	2.3399	2.3299	–	2.1798	2.1598	–	
	Free	2.5000	2.4801	–	2.3399	2.3250	–	2.1798	2.1549	–	

An allowance is provided on bolts in the "Medium" and "Free" classes by reducing the maximum bolt limits below basic size. This reduction allows for plating, the plating falling within the allowance. It also facilitates easy assembly of maximum metal bolts and nuts.

(Continued)

Table 22. (Continued) **British Standard Whitworth (B.S.W. - Coarse) Thread Dimensions.** (Source, Landis Threading Systems.)

Size and TPI	Fit	Major Diameter			Effective Diameter			Minor Diameter			Drill Size
		Max.	Min.	Plated	Max.	Min.	Plated	Max.	Min.	Plated	
2 3/4-3.5	Close	2.7500	2.7377	-	2.5670	2.5600	-	2.3840	2.3701	-	61.50mm
	Med.	2.7500	2.7343	-	2.5670	2.5566	-	2.3840	2.3629	-	
	Free	2.7500	2.7290	-	2.5670	2.5513	-	2.3840	2.3576	-	
3-3.5	Close	3.0000	2.9875	-	2.8170	2.8098	-	2.6340	2.6199	-	68.00mm
	Med.	3.0000	2.9839	-	2.8170	2.8062	-	2.6340	2.6125	-	
	Free	3.0000	2.9786	-	2.8170	2.8009	-	2.6340	2.6072	-	
3 1/2-3.25	Close	3.5000	3.4868	-	3.3030	3.2954	-	3.1060	3.0912	-	73.75mm
	Med.	3.5000	3.4830	-	3.3030	3.2916	-	3.1060	3.0835	-	
	Free	3.5000	3.4773	-	3.3030	3.2859	-	3.1060	3.0778	-	
3 3/4-3	Close	3.7500	3.7364	-	3.5366	3.5287	-	3.3232	3.3078	-	85.50mm
	Med.	3.7500	3.7324	-	3.5366	3.5248	-	3.3232	3.2998	-	
	Free	3.7500	3.7265	-	3.5366	3.5189	-	3.3232	3.2939	-	
4-3	Close	4.0000	3.9862	-	3.7866	3.7786	-	3.5732	3.5577	-	92.00mm
	Med.	4.0000	3.9822	-	3.7866	3.7745	-	3.5732	3.5496	-	
	Free	4.0000	3.9761	-	3.7866	3.7685	-	3.5732	3.5436	-	
4 1/2-2.875	Close	4.5000	4.4857	-	4.2773	4.2689	-	4.0546	4.0385	-	104.20mm
	Med.	4.5000	4.4815	-	4.2773	4.2647	-	4.0546	4.0302	-	
	Free	4.5000	4.4752	-	4.2773	4.2584	-	4.0546	4.0239	-	
5-2.75	Close	5.0000	4.9852	-	4.7672	4.7584	-	4.5344	4.5178	-	116.50mm
	Med.	5.0000	4.9808	-	4.7672	4.7541	-	4.5344	4.5092	-	
	Free	5.0000	4.9743	-	4.7672	4.7475	-	4.5344	4.5026	-	
5 1/2-2.875	Close	5.5000	5.4847	-	5.2561	5.2470	-	5.0122	4.9951	-	128.50mm
	Med.	5.5000	5.4802	-	5.2561	5.2424	-	5.0122	4.9862	-	
	Free	5.5000	5.4733	-	5.2561	5.2356	-	5.0122	4.9794	-	
6-2.5	Close	6.0000	5.9842	-	5.7439	5.7345	-	5.4878	5.4701	-	141.00mm
	Med.	6.0000	5.9795	-	5.7439	5.7298	-	5.4878	5.4601	-	
	Free	6.0000	5.9725	-	5.7439	5.7227	-	5.4878	5.4539	-	

An allowance is provided on bolts in the "Medium" and "Free" classes by reducing the maximum bolt limits below basic size. This reduction allows for plating, the plating falling within the allowance. It also facilitates easy assembly of maximum metal bolts and nuts.

Table 23. British Standard Fine (B.S.F.) Thread Dimensions. *(Source, Landis Threading Systems.)*

Size and TPI	Fit	Major Diameter			Effective Diameter			Minor Diameter			Drill Size
		Max.	Min.	Plated	Max.	Min.	Plated	Max.	Min.	Plated	
3/16-32	Close	0.1875	0.1835	-	0.1675	0.1653	-	0.1475	0.1430	-	4mm
	Med.	0.1864	0.1813	0.1875	0.1664	0.1631	0.1675	0.1464	0.1396	0.1475	
	Free	0.1864	0.1796	0.1875	0.1664	0.1614	0.1675	0.1464	0.1379	0.1475	
7/32-28	Close	0.2188	0.2145	-	0.1959	0.1935	-	0.1730	0.1681	-	4.60mm
	Med.	0.2177	0.2122	0.2188	0.1948	0.1912	0.1959	0.1719	0.1645	0.1730	
	Free	0.2177	0.2105	0.2188	0.1948	0.1895	0.1959	0.1719	0.1628	0.1730	
1/4-26	Close	0.2500	0.2455	-	0.2254	0.2229	-	0.2008	0.1958	-	5.30mm
	Med.	0.2489	0.2432	0.2500	0.2243	0.2206	0.2254	0.1997	0.1921	0.2008	
	Free	0.2489	0.2413	0.2500	0.2243	0.2187	0.2254	0.1997	0.1902	0.2008	
5/16-22	Close	0.3125	0.3077	-	0.2834	0.2807	-	0.2543	0.2488	-	6.80mm
	Med.	0.3113	0.3051	0.3125	0.2822	0.2781	0.2834	0.2531	0.2447	0.2543	
	Free	0.3113	0.3030	0.3125	0.2822	0.2760	0.2834	0.2531	0.2426	0.2543	
3/8-20	Close	0.3750	0.3699	-	0.3430	0.3401	-	0.3110	0.3052	-	8.30mm
	Med.	0.3737	0.3671	0.3750	0.3417	0.3373	0.3430	0.3097	0.3008	0.3110	
	Free	0.3737	0.3649	0.3750	0.3417	0.3351	0.3430	0.3097	0.2986	0.3110	
7/16-18	Close	0.4375	0.4320	-	0.4019	0.3988	-	0.3663	0.3601	-	9.70mm
	Med.	0.4361	0.4290	0.4375	0.4005	0.3958	0.4019	0.3649	0.3555	0.3663	
	Free	0.4361	0.4267	0.4375	0.4005	0.3935	0.4019	0.3649	0.3532	0.3663	
1/2-16	Close	0.5000	0.4942	-	0.4600	0.4567	-	0.4200	0.4135	-	11.10mm
	Med.	0.4985	0.4910	0.5000	0.4585	0.4535	0.4600	0.4158	0.4085	0.4200	
	Free	0.4985	0.4886	0.5000	0.4585	0.4511	0.4600	0.4185	0.4061	0.4200	
9/16-16	Close	0.5625	0.5566	-	0.5225	0.5191	-	0.4825	0.4759	-	12.70mm
	Med.	0.5610	0.5533	0.5625	0.5210	0.5158	0.5225	0.4810	0.4708	0.825	
	Free	0.5610	0.5508	0.5625	0.5210	0.5133	0.5225	0.4810	0.4683	0.4825	
5/8-14	Close	0.6250	0.6187	-	0.5793	0.5757	-	0.5336	0.5265	-	14.00mm
	Med.	0.6234	0.6153	0.6250	0.5777	0.5723	0.5793	0.5320	0.5213	0.5336	
	Free	0.6234	0.6126	0.6250	0.5777	0.5696	0.5793	0.5320	0.5186	0.5336	

An allowance is provided on bolts in the "Medium" and "Free" classes by reducing the maximum bolt limits below basic size. This reduction allows for plating, the plating falling within the allowance. It also facilitates easy assembly of maximum metal bolts and nuts.

(Continued)

Table 23. (*Continued*) **British Standard Fine (B.S.F.) Thread Dimensions.** (*Source, Landis Threading Systems.*)

Size and TPI	Fit	Major Diameter			Effective Diameter			Minor Diameter			Drill Size
		Max.	Min.	Plated	Max.	Min.	Plated	Max.	Min.	Plated	
3/4-12	Close	0.7500	0.7432	-	0.6966	0.6927	-	0.6432	0.6355	-	-
	Med.	0.7482	0.7394	0.7500	0.6948	0.6889	0.6966	0.6414	0.6297	0.6432	16.75mm
	Free	0.7482	0.7365	0.7500	0.6948	0.6860	0.6966	0.6414	0.6268	0.6432	
7/8-11	Close	0.8750	0.8678	-	0.8168	0.8126	-	0.7586	0.7505	-	
	Med.	0.8750	0.8658	-	0.8168	0.8106	-	0.7586	0.7464	-	19.75mm
	Free	0.8750	0.8627	-	0.8168	0.8075	-	0.7586	0.7433	-	
1-10	Close	1.0000	0.9924	-	0.9360	0.9316	-	0.8720	0.8635	-	
	Med.	1.0000	0.9902	-	0.9360	0.9294	-	0.8720	0.8591	-	22.75mm
	Free	1.0000	0.9869	-	0.9360	0.9261	-	0.8720	0.8558	-	
1 1/8-9	Close	1.1250	1.1171	-	1.0539	1.0493	-	0.9828	0.9739	-	
	Med.	1.1250	1.1148	-	1.0539	1.0470	-	0.9828	0.9692	-	25.50mm
	Free	1.1250	1.1113	-	1.0539	1.0435	-	0.9828	0.9657	-	
1 1/4-9	Close	1.2500	1.2419	-	1.1789	1.1741	-	1.1078	1.0987	-	
	Med.	1.2500	1.2395	-	1.1789	1.1717	-	1.1078	1.0939	-	28.50mm
	Free	1.2500	1.2359	-	1.1789	1.1681	-	1.1078	1.0903	-	
1 1/2-8	Close	1.5000	1.4913	-	1.4200	1.4148	-	1.3400	1.3302	-	
	Med.	1.5000	1.4888	-	1.4200	1.4123	-	1.3400	1.3252	-	34.50mm
	Free	1.5000	1.4849	-	1.4200	1.4084	-	1.3400	1.3213	-	
1 3/4-7	Close	1.7500	1.7407	-	1.6585	1.6530	-	1.5670	1.5566	-	
	Med.	1.7500	1.7380	-	1.6585	1.6502	-	1.5670	1.5512	-	40.50mm
	Free	1.7500	1.7338	-	1.6585	1.6461	-	1.5670	1.5470	-	
2-7	Close	2.0000	1.9905	-	1.9085	1.9027	-	1.8170	1.8063	-	
	Med.	2.0000	1.9876	-	1.9085	1.8998	-	1.8170	0.8008	-	47.00mm
	Free	2.0000	1.9832	-	1.9085	1.8955	-	1.8170	1.7965	-	
2 1/4-6	Close	2.2500	2.2398	-	2.1433	2.1372	-	2.0366	2.0252	-	
	Med.	2.2500	2.2368	-	2.1433	2.1341	-	2.0366	2.0193	-	53.00mm
	Free	2.2500	2.2322	-	2.1433	2.1296	-	2.0366	2.0147	-	

An allowance is provided on bolts in the "Medium" and "Free" classes by reducing the maximum bolt limits below basic size. This reduction allows for plating, the plating falling within the allowance. It also facilitates easy assembly of maximum metal bolts and nuts.

(*Continued*)

Table 23. (Continued) British Standard Fine (B.S.F.) Thread Dimensions. (Source, Landis Threading Systems.)

Size and TPI	Fit	Major Diameter			Effective Diameter			Minor Diameter			Drill Size
		Max.	Min.	Plated	Max.	Min.	Plated	Max.	Min.	Plated	
2 1/2-6	Close	2.2500	2.4896	-	2.3933	2.3870	-	2.2866	2.2750	-	59.00
	Med.	2.2500	2.4864	-	2.3933	2.3838	-	2.2866	2.2689	-	
	Free	2.2500	2.4817	-	2.3933	2.3791	-	2.2866	2.2642	-	
2 3/4-6	Close	2.7500	2.7394	-	2.6433	2.6368	-	2.5366	2.5247	-	-
	Med.	2.7500	2.7361	-	2.6433	2.6355	-	2.5366	2.5186	-	
	Free	2.7500	2.7312	-	2.6433	2.6286	-	2.5366	2.5137	-	
3-5	Close	3.0000	2.9887	-	2.8719	2.8650	-	2.7438	2.7311	-	-
	Med.	3.0000	2.9852	-	2.8719	2.8616	-	2.7438	2.7245	-	
	Free	3.0000	2.9801	-	2.8719	2.8564	-	2.7438	2.7194	-	

An allowance is provided on bolts in the "Medium" and "Free" classes by reducing the maximum bolt limits below basic size. This reduction allows for plating, the plating falling within the allowance. It also facilitates easy assembly of maximum metal bolts and nuts.

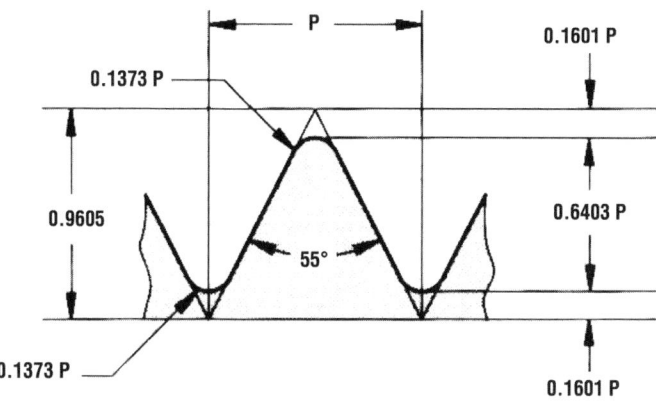

Figure 8. Whitworth (B.S.W.) thread form.

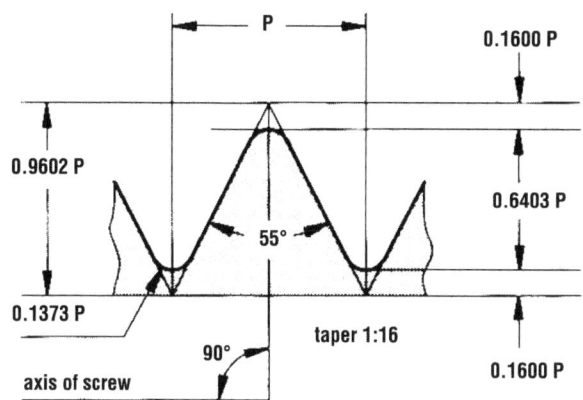

Figure 9. British Standard Taper Pipe form.

Standard Parallel Pipe (BSPP) for non-pressure type joints are also based on the Whitworth form (see *Figure 9*). Both BSTP and BSPP threads share major, pitch (effective), and minor diameters. Full specifications for BSTP threads are given in **Table 24**. (The Japanese parallel pipe thread–JIS PF–shares dimensions with the BSPP thread.)

The British Association (B.A.) Miniature Thread's origin dates back to the mid-1880s, and it was widely used for screws in instruments and clocks. It has an included angle of 47.5°, a radius equal to $0.2617P$, and a thread height equal to $0.6000P$. The height of the sharp V thread is equal to $1.1363365P$. Tolerances are based on two classes of fit (close and normal), and provide an allowance for plating sizes 0 through 10, and are given in **Table 25**. Dimensions for the B.A. thread are given in **Table 26**.

Table 24. Specifications for British Standard Taper Pipe (BSTP) Thread for Pressure Fit. *(Source, Landis Threading Systems.)*

Nominal Size		1/8	1/4	3/8	1/2	3/4	1	1 1/4	1 1/2	2	2 1/2	3
Pipe O.D.		13/32	17/32	11/16	27/32	1 1/16	1 11/32	1 11/16	1 29/32	2 3/8	3	3 1/2
TPI		28	19	19	14	14	11	11	11	11	11	11
Pitch		0.03571	0.05263	0.05263	0.07143	0.07143	0.09091	0.09091	0.09091	0.09091	0.09091	0.09091
Thread Height		0.0229	0.0337	0.0337	0.0457	0.0457	0.0582	0.0582	0.0582	0.0582	0.0582	0.0582
Basic Dia. at Gage Plane	Major Gage Dia.	0.3830	0.5180	0.6560	0.8250	1.0410	1.3090	1.6500	1.8820	2.3470	2.9600	3.4600
	Pitch Dia.	0.3601	0.4843	0.6223	0.7793	0.9953	1.2508	1.5918	1.8238	2.2888	2.9018	3.4018
	Minor Dia.	0.3372	0.4506	0.5886	0.7336	0.9496	1.1926	1.5336	1.7656	2.2306	2.8436	3.3436
Gage Length	Basic - Turns	4 3/8	4 1/2	4 3/4	4 1/2	5 1/4	4 1/2	5 1/2	5 1/2	6 7/8	7 9/16	8 15/16
	Basic - Inches	0.1563	0.2367	0.2500	0.3214	0.3750	0.4091	0.5000	0.5000	0.6250	0.6875	0.8125
	Tol. (+ and -) Turns	1	1	1	1	1	1	1	1	1	1	1 1/2
	Tol. (+ and -) Inches	0.0357	0.0526	0.0526	0.0714	0.0714	0.0909	0.0909	0.0909	0.0909	0.1364	0.1364
	Max. - Turns	5 3/8	5 1/2	5 3/4	5 1/2	6 1/4	5 1/2	6 1/2	6 1/2	7 7/8	9 1/16	10 7/16
	Max - Inches	0.1920	0.2893	0.3026	0.3928	0.4464	0.5000	0.5909	0.5909	0.7159	0.8239	0.9489
	Min. - Turns	3 3/8	3 1/2	3 3/4	3 1/2	4 1/2	3 1/2	4 1/2	4 1/2	5 7/8	6 1/16	7 7/16
	Min. - Inches	0.1206	0.1841	0.1974	0.2500	0.3036	0.3182	0.4091	0.4091	0.5341	0.5511	0.6761
Length of Useful Thread at Pipe End (Minimum)	For Basic Gage - Turns	7 1/8	7 1/4	7 1/2	7 1/4	8	7 1/4	8 1/4	8 1/4	10 1/8	11 9/16	12 15/16
	For Basic Gage - Inches	0.2545	0.3814	0.3947	0.5178	0.5714	0.6591	0.7500	0.7500	0.9204	1.0511	1.1761
	For Max Gage - Turns	8 1/8	8 1/4	8 1/2	8 1/4	9	8 1/4	9 1/4	9 1/4	11 1/8	13 1/16	14 7/16
	For Max Gage - Inches	0.2902	0.4340	0.4473	0.5892	0.6428	0.7500	0.8409	0.8409	1.0113	1.1875	1.3125
	For Min Gage - Turns	6 1/8	6 1/4	6 1/2	6 1/4	7	6 1/4	7 1/4	7 1/4	9 1/8	10 1/16	11 7/16
	For Min Gage - Inches	0.2188	0.3288	0.3421	0.4464	0.5000	0.5682	0.6591	0.6591	0.8295	0.9147	0.0397

(Continued)

Table 24. *(Continued)* Specifications for British Standard Taper Pipe (BSTP) Thread for Pressure Fit. *(Source, Landis Threading Systems.)*

Nominal Size		3 1/2	4	5	6	7	8	9	10	11	12
					Sizes 3 1/2 - 12"						
Pipe O.D.		4	4 1/2	5 1/2	6 1/2	7 1/2	8 1/2	9 1/2	10 5/8	11 5/8	12 5/8
TPI		11	11	11	11	10	10	10	10	8	8
Pitch		0.09091	0.09091	0.09091	0.09091	0.1000	0.1000	0.1000	0.1000	0.1250	0.1250
Thread Height		0.0582	0.0582	0.0582	0.0582	0.0640	0.0640	0.0640	0.0640	0.8000	0.8000
Basic Dia. at Gage Plane	Major Gage Dia.	3.9500	4.4500	5.4500	6.4500	7.4500	8.4500	9.4500	10.4500	11.4500	12.4500
	Pitch Dia.	3.8918	4.3918	5.3918	6.3918	7.3860	8.3860	9.3860	10.3860	11.3700	12.3700
	Minor Dia.	3.8336	4.3336	5.3336	6.3336	7.3220	8.3220	9.3220	10.3220	11.2900	12.2900
Gage Length	Basic - Turns	9 5/8	11	12 3/8	12 3/8	13 3/4	15	15	16 1/4	13	13
	Basic - Inches	0.8750	1.000	1.1250	1.1250	1.3750	1.5000	1.5000	1.6250	1.6250	1.6250
	Tol. (+ and -) Turns	1 1/2	1 1/2	1 1/2	1 1/2	2	2	2	2	2	2
	Tol. (+ and -) Inches	0.1364	0.1364	0.1364	0.1364	0.2000	0.2000	0.2000	0.2000	0.2000	0.2000
	Max. - Turns	11 1/8	12 1/2	13 7/8	13 7/8	15 3/4	17	17	18 1/4	15	15
	Max - Inches	0.0114	1.1364	1.2614	1.2614	1.5750	1.7000	1.7000	1.8250	1.8750	1.8750
	Min. - Turns	8 1/8	9 1/2	10 7/8	10 7/8	11 3/4	13	13	14 1/4	11	11
	Min. - Inches	0.7386	0.8636	0.9886	0.9886	1.1750	1.3000	1.3000	1.4250	1.3750	1.3750
Length of Useful Thread at Pipe End (Minimum)	For Basic Gage -Turns	13 5/8	15 1/2	17 3/8	17 3/8	19 1/4	20 1/2	20 1/2	21 3/4	18 1/2	18 1/2
	For Basic Gage - Inches	1.2386	1.4091	1.5795	1.5795	1.9250	2.0500	2.0500	2.1750	2.3125	2.3125
	For Max Gage - Turns	15 1/8	17	18 7/8	18 7/8	21 1/4	22 1/2	22 ½	23 3/4	20 1/2	20 1/2
	For Max Gage - Inches	1.3750	1.5455	1.7519	1.7519	2.1250	2.2500	2.2500	2.3750	2.5625	2.5625
	For Min Gage - Turns	12 1/8	14	15 7/8	15 7/8	17 1/4	18 1/2	18 ½	19 3/4	16 1/2	16 1/2
	For Min Gage - Inches	1.1022	1.2727	1.4431	1.4431	1.7250	1.8500	1.8500	1.9750	2.0625	2.0625

Table 25. Tolerances for B.A. Screw Threads.

Thread and Fit		Pitch Dia.	Minor Dia.	Major Dia.
Ext. Thread (minus)	Close Fit	$0.08P + 0.02$mm	$0.16P + 0.04$mm	$0.15P$
	Normal Fit	$0.10P + 0.025$mm	$0.020P + 0.05$mm	Size $0–10 = 0.20P$
				Size $11–16 = 0.025P$
Int. Thread (+)	All	$0.12P + 0.03$mm	$0.375P$	–

Table 26. B.A. Standard (Miniature) Thread Dimensions.

Size #	Major Dia. mm	Pitch mm	TPI	Pitch Dia. mm	Minor Dia. mm	Thread Height mm	Tap Drill mm
0	6.000	1.0000	25.40	5.400	4.800	0.600	5.00
1	5.300	0.9000	28.22	4.760	4.220	0.540	4.40
2	4.700	0.8100	31.35	4.215	3.730	0.485	3.90
3	4.100	0.7300	34.79	3.660	3.220	0.440	3.40
4	3.600	0.6600	38.48	3.205	2.810	0.395	2.95
5	3.200	0.5900	43.05	2.845	2.490	0.355	2.60
6	2.800	0.5300	47.92	2.480	2.160	0.320	2.30
7	2.500	0.4800	52.92	2.210	1.920	0.290	2.00
8	2.200	0.4300	59.07	1.940	1.680	0.260	1.80
9	1.900	0.3900	65.12	1.665	1.430	0.235	1.50
10	1.700	0.3500	72.57	1.490	1.280	0.210	1.35
11	1.500	0.3100	38.93	1.315	1.130	0.185	1.20
12	1.300	0.2800	90.71	1.130	0.960	0.170	1.00
13	1.200	0.2500	101.60	1.050	0.900	0.150	0.95
14	1.000	0.2300	110.40	0.860	0.720	0.140	0.75
15	0.900	0.2100	121.00	0.775	0.650	0.125	–
16	0.790	0.1900	134.00	0.675	0.560	0.115	–

Acme screw threads

The form for Acme threads dates back to the last decade of the 19th century, when the form was designed to replace square and other configured thread forms used for the purpose of producing transverse motion on mechanical mechanisms. Today, their use has expanded to a wide variety of applications wherever heavy loads must be transmitted. There are three standardized forms of Acme threads: General Purpose, Centralizing, and Stub, and each has an included angle of 29°. The maximum major diameter of the external thread is basic and equal to the nominal major diameter. The minimum pitch diameter of the internal thread is basic and equal to the basic major diameter minus the basic height of the thread ($0.5P$). The basic minor diameter is equal to the basic major diameter minus twice the basic thread height. The minimum minor diameter of the internal General Purpose thread is basic. The minimum minor diameter of the internal Centralizing thread is $0.1P$ above basic. Tolerances specified are applicable to lengths of engagement not exceeding twice the nominal diameter. The descriptions in this section are from FED-STD-H28/12, FED-STD-H28/13. ANSI/ASME B1.5 and ANSI/ASME B1.8 also provide dimensional data for Acme threads. The original Acme thread standards were issued in 1952 as ASA B1.5-1952 and ASA B1.8-1952. The form remains basically unaltered from the initial standard.

There are three classes of *General Purpose* threads (2G, 3G, 4G, with the G standing for General Purpose), each with clearances on all diameters for free movement so that they may be used in assemblies with the internal thread rigidly fixed and the external thread moving perpendicular to its axis limited by its bearing or bearings. Class 2G is preferred, but either of the other classes should be selected if excessive backlash or endplay is desired. The three classes of *Centralizing* threads (2C, 3C, and 4C, with the C standing for Centralizing) have a limited clearance on the major diameters of the internal and external threads, so that a bearing at the major diameter maintains approximate alignment of the thread axis and prevents wedging on the flanks of the thread. Previously, classes 5C and 6C were recognized, but they are no longer recommended. Class 2C is preferred; and while selecting one of the other classes can reduce excessive backlash or endplay, the result may be that the tolerances will be difficult to maintain. For both General Purpose and Centralizing threads, it is recommended that external and internal threads of the same class be used together. Centralizing threads are intended to prevent wedging on the flanks of the thread. The *Stub* Acme thread originated in the early 1900s, and its use is generally confined to those unusual applications where a coarse pitch thread of shallow depth is required due to mechanical or metallurgical considerations. Class 2G, general-purpose tolerances and allowances, are recommended for Stub Acme threads, with 3G and 4G being alternative selections where excessive backlash and endplay are encountered (see the following section).

Acme threads are designated by nominal diameter expressed as a fraction or a decimal, tpi, and class. A typical designation would be 7/16–12 ACME-2G. The same thread, with a left-hand thread, would be designated 7/16–12 Acme-2G-LH. Stub Acme threads do not utilize a class and the designation would read 7/16–12 Stub Acme, or 7/16–12 Stub Acme-LH.

All classes of General Purpose external and internal threads may be used interchangeably. The requirement for a Centralizing fit is that the sum of the major diameter tolerance plus the major diameter allowance on the internal thread, and the major diameter tolerance on the external thread shall equal or be less than the pitch diameter allowance on the external thread. A class 2C external thread that has a larger pitch diameter allowance than either

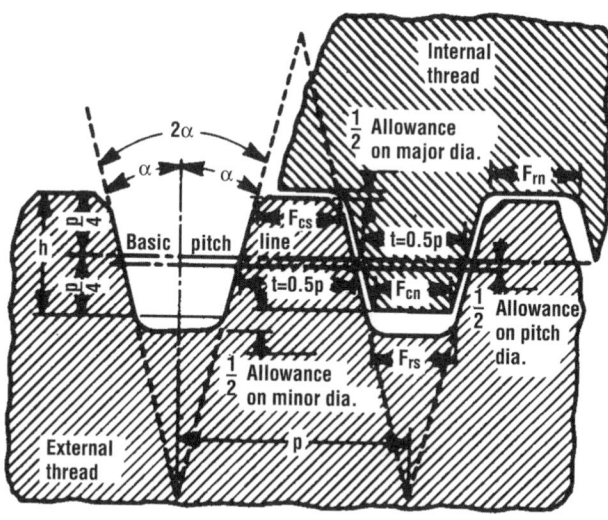

Figure 10. General Purpose Acme thread form.

a class 3C or 4C external thread can be used interchangeably with classes 2C, 3C, or 4C internal threads and fulfill this requirement. Similarly, a class 3C external thread can be used interchangeably with class 3C and 4C internal threads, but only a class 4C internal thread can be used with a class 4C external thread. *Figure 10* shows the General Purpose form, *Figure 11* the Centralizing form, and the Stub form is illustrated in *Figure 12*.

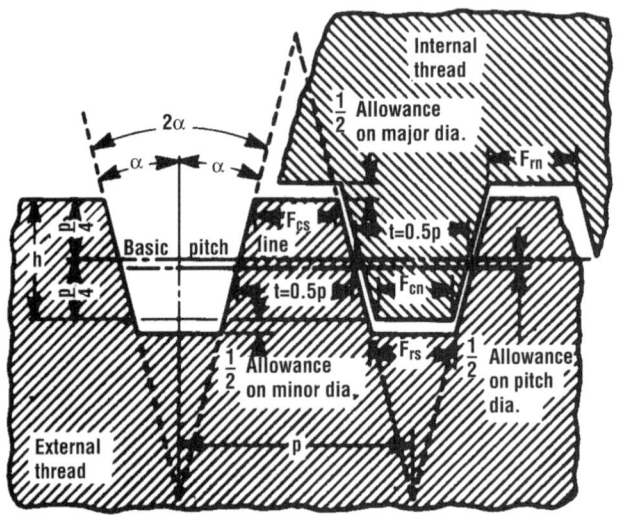

Figure 11. Centralizing Acme thread form.

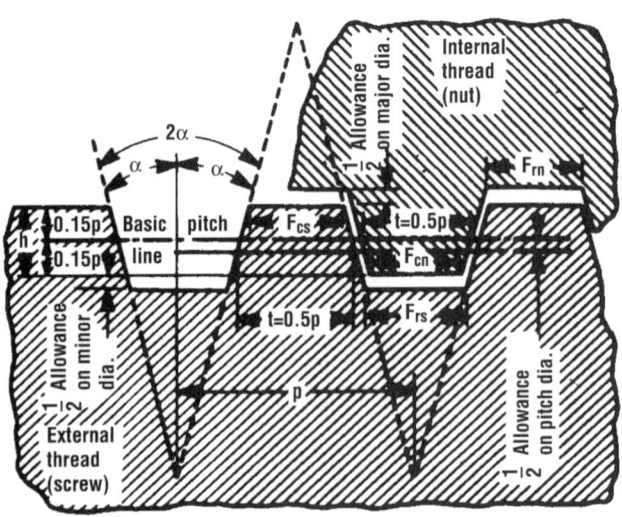

Figure 12. Stub Acme thread form.

Table 27. Basic Formulas for Calculating Acme Thread Dimensions.

Dimension	General Purpose Acme Thread	Centralizing Acme Thread	Stub Acme Threads *	Remarks
Basic Height of Thread	$H = 0.5P$	$H = 0.5P$	$H = 0.3P$	See Note [1]
Basic Thickness of Thread**	$t = 0.5P$	$t = 0.5P$	$t = 0.5P$	See Note [2]
Basic Width of Flat of Crest of Internal Thread	$F_{cn} = 0.3707P$	$F_{cn} = 0.3707P + 0.259 \times$ minor dia. allowance, int. thread	$F_{cn} = 0.4224P + 0.259 \times$ minor dia. allowance, int. thread	See Table 28 for pitch diameter allowances.
Basic Width of Flat of Crest of External Thread	$F_{cs} = 0.3707P - 0.259 \times$ pitch dia. allowance, ext. thread	$F_{cs} = 0.3707P - 0.259 \times$ pitch dia. allowance, ext. thread	$F_{cn} = 0.4224P + 0.259 \times$ pitch dia. allowance, ext. thread	
Basic Flat at Root of Internal Thread	$F_{rn} = 0.0307P - 0.259 \times$ major dia. allowance, int. thread	$F_{rn} = 0.0307P - 0.259 \times$ major dia. allowance, int. thread	$F_{rn} = 0.4224P - 0.259 \times$ major dia. allowance, int. thread	
Basic Flat at Root of External Thread	$F_{rs} = 0.0307P - 0.259 \times$ (minor dia. allowance, ext. thread − pitch dia. allowance, ext. thread)	$F_{rs} = 0.0307P - 0.259 \times$ (minor dia. allowance, ext. thread − pitch dia. allowance, ext. thread)	$F_{rs} = 0.4224P - 0.259 \times$ (minor dia. allowance, ext. thread − pitch dia. allowance, ext. thread)	See Table 28 for pitch diameter allowances.

* Basic Stub Acme form.
** The basic thickness of the thread measured at a diameter smaller by one-half the pitch than the basic major diameter (i.e., at the pitch diameter line).
Note [1]: For Modified Form 1 Stub Acme, $H = 0.375P$. For Modified Form 2 Stub Acme, $H = 0.025P$.
Note [2]: For Modified Form 1 Stub Acme, $F_{cn} = 0.4030P$. For Modified Form 2 Stub Acme, $F_{cs} = 0.4353P$.

Allowances, tolerances, and dimensions for Acme threads

For General Purpose Acme threads and Stub Acme threads, the allowance (minimum clearance) at major and minor diameters is established as follows: a minimal diametrical clearance is provided at the minor diameter of all external threads by establishing the maximum minor diameter of 0.020 in. below the basic minor diameter for 10 tpi and coarser, and 0.010 in. below the basic minor diameter for finer pitches. A minimal diametrical clearance at the major diameter is obtained by establishing the minimum major diameter of the internal thread 0.020 in. above the basic minor diameter for 10 tpi and coarser, and 0.010 in. above the basic minor diameter for finer pitches. General Purpose Acme external threads may have the crest corners chamfered at an angle of 45° with the axis, to a maximum depth of $0.667P$. This corresponds to a maximum width of chamfer flat of $0.945P$.

For Centralizing Acme threads, a minimal diametrical clearance is provided at the minor diameter of all external threads by establishing the maximum minor diameter 0.020 in. below the basic minor for 10 tpi and coarser, and 0.010 in. for finer pitches. A minimum diametrical clearance for the fillet is provided at the minor diameter by establishing the minimum minor diameter of the internal thread $0.1P$ greater than the basic minor diameter. A minimum diametrical clearance at the major diameter is obtained by establishing the minimum major diameter of the internal thread at $0.001\sqrt{D}$ above the basic major diameter. Centralizing Acme external threads have the crest corners chamfered to an angle of 45° with the axis, to a minimum depth of $0.05P$ and a maximum depth of $0.667P$, corresponding to a minimum width of chamfer flat of $0.0707P$ and a maximum width of $0.0945P$. External threads may also have a fillet at the minor diameter not greater than $0.1P$. Internal threads may have a fillet in the major diameter not greater than $0.06P$.

Pitch diameter allowances for General Purpose and Centralizing Acme Threads are given in **Table 28**. Dimensions and tolerances for General Purpose Acme Threads are given in **Tables 29–31**, for Centralizing Acme Threads see **Tables 30–32**, and for Stub Acme Threads see **Table 33**.

Table 28. Pitch Diameter Allowances for Acme Threads. *(Source, FED-STD H28/12.)*

Nominal Size Range		Pitch Diameter Allowances on External Threads, General Purpose, Centralizing, and Stub Acme*		
Above	To and Including	Class 2G and 2C and Stub Acme $0.008\sqrt{D}$	Class 3G and 3C $0.006\sqrt{D}$	Class 4G and 4C $0.004\sqrt{D}$
0	3/16	0.0024	0.0018	0.0012
3/16	5/16	0.0040	0.0030	0.0020
5/16	7/16	0.0049	0.0037	0.0024
7/16	9/16	0.0057	0.0042	0.0028
9/16	11/16	0.0063	0.0047	0.0032
11/16	13/16	0.0069	0.0052	0.0035
13/16	15/16	0.0075	0.0056	0.0037
15/16	1 1/16	0.0080	0.0060	0.0040
1 1/16	1 3/16	0.0085	0.0064	0.0042
1 3/16	1 5/16	0.0089	0.0067	0.0045
1 5/16	1 7/16	0.0094	0.0070	0.0047

(Continued)

Table 28. *(Continued)* **Pitch Diameter Allowances for Acme Threads.** *(Source, FED-STD H28/12.)*

Nominal Size Range		Pitch Diameter Allowances on External Threads, General Purpose, Centralizing, and Stub Acme*		
Above	To and Including	Class 2G and 2C and Stub Acme $0.008\sqrt{D}$	Class 3G and 3C $0.006\sqrt{D}$	Class 4G and 4C $0.004\sqrt{D}$
1 7/16	1 9/16	0.0098	0.0073	0.0049
1 9/16	1 7/8	0.0105	0.0079	0.0052
1 7/8	2 1/8	0.0113	0.0085	0.0057
2 1/8	2 3/8	0.0120	0.0090	0.0060
2 3/8	2 5/8	0.0126	0.0095	0.0063
2 5/8	2 7/8	0.0133	0.0099	0.0066
2 7/8	3 1/4	0.0140	0.0105	0.0070
3 1/4	3 3/4	0.0150	0.0112	0.0075
3 3/4	4 1/4	0.0160	0.0120	0.0080
4 1/4	4 3/4	0.0170	0.0127	0.0085
4 3/4	5 1/2	0.0181	0.0136	0.0091

All dimensions in inches.
*An increase in 10% of the allowance is recommended for each inch, or fraction thereof, that the length of engagement exceeds two diameters.

Modified square thread

Another example of an Acme type thread is the modified square thread that was designed for translation applications and for resisting axial loads where only small radial loads are permitted. The angle between the flanks of the thread is 10°, and the threads are truncated top and bottom. The thread had a basic height of $0.5P$, a basic thickness at the pitch line of $0.5P$, and is symmetrical about a line perpendicular to the axis of the external thread. The 10° angle results in a thread that is practically the equivalent of a square thread, and it is therefore capable of being produced economically. A typical Modified Square thread designation would be 1 ³/₄–6 SPL 10° FORM. *Figure 13* shows the thread form. FED-STD-H28/19A provides additional information on dimensions, allowances, and tolerances.

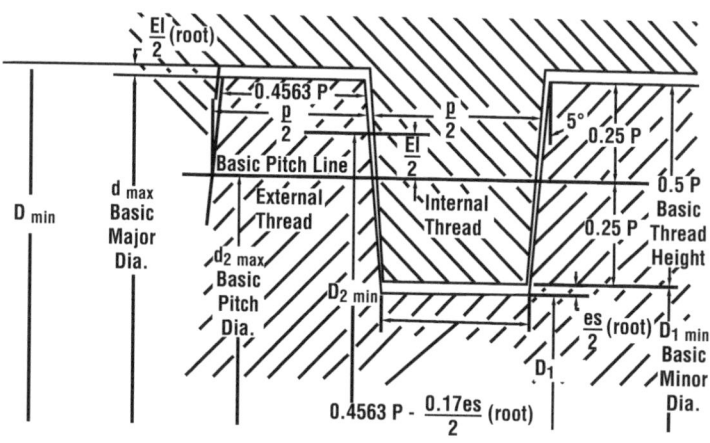

Figure 13. Modified Square thread form.

Table 29. Dimensions and Tolerances for Class 2G General Purpose Acme Screw Threads.
(Source, Landis Threading Systems.)

Nominal Size and TPI	External Threads - Class 2G								
	Major Diameter			Minor Diameter			Pitch Diameter		
	Max.	Min.	Tol.	Max.	Min.	Max.	Min.	Tol.
1/4-16	0.2500	0.2450	0.0050	0.1775	0.1618	0.2148	0.2043	0.0105
5/16-14	0.3125	0.3075	0.0050	0.2311	0.2140	0.2728	0.2614	0.0114
3/8-12	0.3750	0.3700	0.0050	0.2817	0.2632	0.3284	0.3161	0.0123
7/16-12	0.4375	0.4325	0.0050	0.3442	0.3253	0.3909	0.3783	0.0126
1/2-10	0.5000	0.4950	0.0050	0.3800	0.3594	0.4443	0.4306	0.0137
5/8-8	0.6250	0.6188	0.0062	0.4800	0.4570	0.5562	0.5408	0.0154
3/4-6	0.7500	0.7417	0.0083	0.5633	0.5371	0.6598	0.6424	0.0174
7/8-6	0.8750	0.8667	0.0083	0.6883	0.6615	0.7842	0.7663	0.0179
1-5	1.0000	0.9900	0.0100	0.7800	0.7509	0.8920	0.8726	0.0194
1 1/8-5	1.1250	1.1150	0.0100	0.9050	0.8753	1.0165	0.9967	0.0198
1 1/4-5	1.2500	1.2400	0.0100	1.0300	0.9998	1.1411	1.1210	0.0201
1 3/8-4	1.3750	1.3625	0.0125	1.1050	1.0719	1.2406	1.2185	0.0220
1 1/2-4	1.5000	1.4875	0.0125	1.2300	1.1965	1.3652	1.3429	0.0223
1 3/4-4	1.7500	1.7375	0.0125	1.4800	1.4456	1.6145	1.5916	0.0229
2-4	2.0000	1.9875	0.0125	1.7300	1.6948	1.8637	1.8402	0.0235
2 1/4-3	2.2500	2.2333	0.0167	1.8967	1.8572	2.0713	2.0450	0.0263
2 1/2-3	2.5000	2.4833	0.0167	2.1467	2.1065	2.3207	2.2939	0.0268
2 3/4-3	2.7500	2.7333	0.0167	2.3967	2.3558	2.5700	2.5427	0.0273
3-2	3.0000	2.9750	0.2500	2.4800	2.4326	2.7360	2.7044	0.0316
3 1/2-2	3.5000	3.4750	0.2500	2.9800	2.9314	3.2350	3.2026	0.0324
4-2	4.0000	3.9750	0.2500	3.4800	3.4302	3.7340	3.7008	0.0332
4 1/2-2	4.5000	4.4750	0.2500	3.9800	3.9201	4.2330	4.1991	0.0339
5-2	5.0000	4.9750	0.2500	4.4800	4.4281	4.7319	4.6973	0.0346

Nominal Size and TPI	Internal Threads - Class 2G								
	Major Diameter			Minor Diameter			Pitch Diameter		
	Min.	Max.	Tol.	Min.	Max.	Tol.	Min.	Max.	Tol.
1/4-16	0.2600	0.2700	0.0100	0.1875	0.1925	0.0050	0.2188	0.2293	0.0105
5/16-14	0.3225	0.3325	0.0100	0.2411	0.2461	0.0050	0.2768	0.2882	0.0114
3/8-12	0.3850	0.3950	0.0100	0.2917	0.2967	0.0050	0.3333	0.3456	0.0123
7/16-12	0.4475	0.4575	0.0100	0.3542	0.3592	0.0050	0.3958	0.4084	0.0126
1/2-10	0.5200	0.5400	0.0200	0.4000	0.4050	0.0050	0.4500	0.4637	0.0137
5/8-8	0.6450	0.6650	0.0200	0.5000	0.5062	0.0062	0.5625	0.5779	0.0154
3/4-6	0.7700	0.7900	0.0200	0.5833	0.5916	0.0083	0.6667	0.6841	0.0174
7/8-6	0.8950	0.9150	0.0200	0.7083	0.7166	0.0083	0.7917	0.8096	0.0179
1-5	1.0200	1.0400	0.0200	0.8000	0.8100	0.0100	0.9000	0.9194	0.0194
1 1/8-5	1.1450	1.1650	0.0200	0.9250	0.9350	0.0100	1.0250	1.0448	0.0198
1 1/4-5	1.2700	1.2900	0.0200	1.0500	1.0600	0.0100	1.1500	1.1701	0.0201

(Continued)

Table 29. *(Continued)* **Dimensions and Tolerances for Class 2G General Purpose Acme Screw Threads.**
(Source, Landis Threading Systems.)

Nominal Size and TPI	Major Diameter			Minor Diameter			Pitch Diameter		
	Min.	Max.	Tol.	Min.	Max.	Tol.	Min.	Max.	Tol.
				Internal Threads - Class 2G					
1 3/8-4	1.3950	1.4150	0.0200	1.1250	1.1375	0.0125	1.2500	1.2720	0.0220
1 1/2-4	1.5200	1.5400	0.0200	1.2500	1.2625	0.0125	1.3750	1.3973	0.0223
1 3/4-4	1.7700	1.7900	0.0200	1.5000	1.5125	0.0125	1.6250	1.6479	0.0229
2-4	2.0200	2.0400	0.0200	1.7500	1.7625	0.0125	1.8750	1.8985	0.0235
2 1/4-3	2.2700	2.2900	0.0200	1.9167	1.9334	0.0167	2.0833	2.1096	0.0263
2 1/2-3	2.5200	2.5400	0.0200	2.1667	2.1834	0.0167	2.3333	2.3601	0.0268
2 3/4-3	2.7700	2.7900	0.0200	2.4167	2.4334	0.0167	2.5833	2.6106	0.0273
3-2	3.0200	3.0400	0.0200	2.5000	2.5250	0.0250	2.7500	2.7816	0.0316
3 1/2-2	3.5200	3.5400	0.0200	3.0000	3.0250	0.0250	3.2500	3.2824	0.0324
4-2	4.0200	4.0400	0.0200	3.5000	3.5250	0.0250	3.7500	3.7832	0.0332
4 1/2-2	4.5200	4.5400	0.0200	4.0000	4.0250	0.0250	4.2500	4.2839	0.0339
5-2	5.0200	5.0400	0.0200	4.5000	4.5250	0.0250	4.7500	4.7846	0.0346

All dimensions in inches.

Table 30. Dimensions and Tolerances for Class 3G General Purpose Acme Screw Threads.
(Source, Landis Threading Systems.)

Nominal Size and TPI	Major Diameter			Minor Diameter			Pitch Diameter		
	Max.	Min.	Tol.	Max.	Min.	Tol.	Max.	Min.	Tol.
				External Threads - Class 3G					
1/4-16	0.2500	0.2450	0.0050	0.1775	0.1702		0.2158	0.2109	0.0049
5/16-14	0.3125	0.3075	0.0050	0.2311	0.2231		0.2738	0.2685	0.0053
3/8-12	0.3750	0.3700	0.0050	0.2817	0.2730		0.3296	0.3238	0.0058
7/16-12	0.4375	0.4325	0.0050	0.3442	0.3354		0.3921	0.3862	0.0059
1/2-10	0.5000	0.4950	0.0050	0.3800	0.3704		0.4458	0.4394	0.0064
5/8-8	0.6250	0.6188	0.0062	0.4800	0.4693		0.5578	0.5506	0.0072
3/4-6	0.7500	0.7417	0.0083	0.5633	0.5511		0.6615	0.6534	0.0081
7/8-6	0.8750	0.8667	0.0083	0.6883	0.6758		0.7861	0.7778	0.0083
1-5	1.0000	0.9900	0.0100	0.7800	0.7664		0.8940	0.8849	0.0091
1 1/8-5	1.1250	1.1150	0.0100	0.9050	0.8912		1.0186	1.0094	0.0092
1 1/4-5	1.2500	1.2400	0.0100	1.0300	1.0159		1.1433	1.1339	0.0094
1 3/8-4	1.3750	1.3625	0.0125	1.1050	1.0896		1.2430	1.2327	0.0103
1 1/2-4	1.5000	1.4875	0.0125	1.2300	1.2144		1.3677	1.3573	0.0104
1 3/4-4	1.7500	1.7375	0.0125	1.4800	1.4640		1.6171	1.6064	0.0107
2-4	2.0000	1.9875	0.0125	1.7300	1.7136		1.8665	1.8555	0.0110
2 1/4-3	2.2500	2.2333	0.0167	1.8967	1.8783		2.0743	2.0620	0.0123
2 1/2-3	2.5000	2.4833	0.0167	2.1467	2.1279		2.3238	2.3113	0.0125
2 3/4-3	2.7500	2.7333	0.0167	2.3967	2.3776		2.5734	2.5607	0.0127

(Continued)

Table 30. *(Continued)* **Dimensions and Tolerances for Class 3G General Purpose Acme Screw Threads.** *(Source, Landis Threading Systems.)*

Nominal Size and TPI	Major Diameter			Minor Diameter			Pitch Diameter		
	Max.	**Min.**	**Tol.**	**Max.**	**Min.**		**Max.**	**Min.**	**Tol.**
						External Threads - Class 3G			
3-2	3.0000	2.9750	0.2500	2.4800	2.4579		2.7395	2.7248	0.0147
3 1/2-2	3.5000	3.4750	0.2500	2.9800	2.9574		3.2388	3.2237	0.0151
4-2	4.0000	3.9750	0.2500	3.4800	3.4568		3.7380	3.7225	0.0155
4 1/2-2	4.5000	4.4750	0.2500	3.9800	3.9563		4.2373	4.2215	0.0158
5-2	5.0000	4.9750	0.2500	4.4800	4.4558		4.7364	4.7202	0.0162

Internal Threads - Class 3G

Nominal Size and TPI	Major Diameter			Minor Diameter			Pitch Diameter		
	Min.	**Max.**	**Tol.**	**Min.**	**Max.**	**Tol.**	**Min.**	**Max.**	**Tol.**
1/4-16	0.2600	0.2700	0.0100	0.1875	0.1925	0.0050	0.2188	0.2237	0.0049
5/16-14	0.3225	0.3325	0.0100	0.2411	0.2461	0.0050	0.2768	0.2821	0.0053
3/8-12	0.3850	0.3950	0.0100	0.2917	0.2967	0.0050	0.3333	0.3391	0.0058
7/16-12	0.4475	0.4575	0.0100	0.3542	0.3592	0.0050	0.3958	0.4017	0.0059
1/2-10	0.5200	0.5400	0.0200	0.4000	0.4050	0.0050	0.4500	0.4564	0.0064
5/8-8	0.6450	0.6650	0.0200	0.5000	0.5062	0.0062	0.5625	0.5697	0.0072
3/4-6	0.7700	0.7900	0.0200	0.5833	0.5916	0.0083	0.6667	0.6748	0.0081
7/8-6	0.8950	0.9150	0.0200	0.7083	0.7166	0.0083	0.7917	0.8000	0.0083
1-5	1.0200	1.0400	0.0200	0.8000	0.8100	0.0100	0.9000	0.9091	0.0091
1 1/8-5	1.1450	1.1650	0.0200	0.9250	0.9350	0.0100	1.0250	1.0342	0.0092
1 1/4-5	1.2700	1.2900	0.0200	1.0500	1.0600	0.0100	1.1500	1.1594	0.0094
1 3/8-4	1.3950	1.4150	0.0200	1.1250	1.1375	0.0125	1.2500	1.2603	0.0103
1 1/2-4	1.5200	1.5400	0.0200	1.2500	1.2625	0.0125	1.3750	1.3854	0.0104
1 3/4-4	1.7700	1.7900	0.0200	1.5000	1.5125	0.0125	1.6250	1.6357	0.0107
2-4	2.0200	2.0400	0.0200	1.7500	1.7625	0.0125	1.8750	1.8860	0.0110
2 1/4-3	2.2700	2.2900	0.0200	1.9167	1.9334	0.0167	2.0833	2.0956	0.0123
2 1/2-3	2.5200	2.5400	0.0200	2.1667	2.1834	0.0167	2.3333	2.3458	0.0125
2 3/4-3	2.7700	2.7900	0.0200	2.4167	2.4334	0.0167	2.5833	2.5960	0.1270
3-2	3.0200	3.0400	0.0200	2.5000	2.5250	0.0250	2.7500	2.7647	0.1470
3 1/2-2	3.5200	3.5400	0.0200	3.0000	3.0250	0.0250	3.2500	3.2651	0.1510
4-2	4.0200	4.0400	0.0200	3.5000	3.5250	0.0250	3.7500	3.7655	0.1550
4 1/2-2	4.5200	4.5400	0.0200	4.0000	4.0250	0.0250	4.2500	4.2658	0.1580
5-2	5.0200	5.0400	0.0200	4.5000	4.5250	0.0250	4.7500	4.7662	0.1620

All dimensions in inches.

Table 31. Dimensions and Tolerances for Class 4G General Purpose Acme Screw Threads.
(Source, Landis Threading Systems.)

Nominal Size and TPI	External Threads - Class 4G							
	Major Diameter			Minor Diameter		Pitch Diameter		
	Max.	Min.	Tol.	Max.	Min.	Max.	Min.	Tol.
1/4-16	0.2500	0.2450	0.0050	0.1775	0.1722	0.2168	0.2133	0.0035
5/16-14	0.3125	0.3075	0.0050	0.2311	0.2254	0.2748	0.2710	0.0038
3/8-12	0.3750	0.3700	0.0050	0.2817	0.2755	0.3309	0.3268	0.0041
7/16-12	0.4375	0.4325	0.0050	0.3442	0.3379	0.3934	0.3892	0.0042
1/2-10	0.5000	0.4950	0.0050	0.3800	0.3731	0.4472	0.4426	0.0046
5/8-8	0.6250	0.6188	0.0062	0.4800	0.4723	0.5593	0.5542	0.0051
3/4-6	0.7500	0.7417	0.0083	0.5633	0.5546	0.6632	0.6574	0.0058
7/8-6	0.8750	0.8667	0.0083	0.6883	0.6794	0.7880	0.7820	0.0060
1-5	1.0000	0.9900	0.0100	0.7800	0.7703	0.8960	0.8895	0.0065
1 1/8-5	1.1250	1.1150	0.0100	0.9050	0.8951	1.0208	1.0142	0.0066
1 1/4-5	1.2500	1.2400	0.0100	1.0300	1.0199	1.1455	1.1388	0.0067
1 3/8-4	1.3750	1.3625	0.0125	1.1050	1.0940	1.2453	1.2380	0.0073
1 1/2-4	1.5000	1.4875	0.0125	1.2300	1.2188	1.3701	1.3627	0.0074
1 3/4-4	1.7500	1.7375	0.0125	1.4800	1.4685	1.6198	1.6122	0.0076
2-4	2.0000	1.9875	0.0125	1.7300	1.7183	1.8693	1.8615	0.0078
2 1/4-3	2.2500	2.2333	0.0167	1.8967	1.8835	2.0773	2.0685	0.0088
2 1/2-3	2.5000	2.4833	0.0167	2.1467	2.1333	2.3270	2.3181	0.0089
2 3/4-3	2.7500	2.7333	0.0167	2.3967	2.3831	2.5767	2.5676	0.0091
3-2	3.0000	2.9750	0.2500	2.4800	2.4642	2.7430	2.7325	0.0105
3 1/2-2	3.5000	3.4750	0.2500	2.9800	2.9638	3.2425	3.2317	0.0108
4-2	4.0000	3.9750	0.2500	3.4800	3.4634	3.7420	3.7309	0.0111
4 1/2-2	4.5000	4.4750	0.2500	3.9800	3.9631	4.2415	4.2302	0.0113
5-2	5.0000	4.9750	0.2500	4.4800	4.4627	4.7409	4.7294	0.0115

Nominal Size and TPI	Internal Threads - Class 4G							
	Major Diameter			Minor Diameter		Pitch Diameter		
	Min.	Max.	Tol.	Min.	Max.	Tol.	Min.	Max.	Tol.
1/4-16	0.2600	0.2700	0.0100	0.1875	0.1925	0.0050	0.2188	0.2223	0.0035
5/16-14	0.3225	0.3325	0.0100	0.2411	0.2461	0.0050	0.2768	0.2806	0.0038
3/8-12	0.3850	0.3950	0.0100	0.2917	0.2967	0.0050	0.3333	0.3374	0.0041
7/16-12	0.4475	0.4575	0.0100	0.3542	0.3592	0.0050	0.3958	0.4000	0.0042
1/2-10	0.5200	0.5400	0.0200	0.4000	0.4050	0.0050	0.4500	0.4546	0.0046
5/8-8	0.6450	0.6650	0.0200	0.5000	0.5062	0.0062	0.5625	0.5676	0.0051
3/4-6	0.7700	0.7900	0.0200	0.5833	0.5916	0.0083	0.6667	0.6725	0.0058
7/8-6	0.8950	0.9150	0.0200	0.7083	0.7166	0.0083	0.7917	0.7977	0.0060
1-5	1.0200	1.0400	0.0200	0.8000	0.8100	0.0100	0.9000	0.9065	0.0065
1 1/8-5	1.1450	1.1650	0.0200	0.9250	0.9350	0.0100	1.0250	1.0316	0.0066
1 1/4-5	1.2700	1.2900	0.0200	1.0500	1.0600	0.0100	1.1500	1.1567	0.0067

(Continued)

Table 31. *(Continued)* **Dimensions and Tolerances for Class 4G General Purpose Acme Screw Threads.**
(Source, Landis Threading Systems.)

Nominal Size and TPI	Major Diameter			Minor Diameter			Pitch Diameter		
	Min.	Max.	Tol.	Min.	Max.	Tol.	Min.	Max.	Tol.
1 3/8-4	1.3950	1.4150	0.0200	1.1250	1.1375	0.0125	1.2500	1.2573	0.0073
1 1/2-4	1.5200	1.5400	0.0200	1.2500	1.2625	0.0125	1.3750	1.3824	0.0074
1 3/4-4	1.7700	1.7900	0.0200	1.5000	1.5125	0.0125	1.6250	1.6326	0.0076
2-4	2.0200	2.0400	0.0200	1.7500	1.7625	0.0125	1.8750	1.8828	0.0078
2 1/4-3	2.2700	2.2900	0.0200	1.9167	1.9334	0.0167	2.0833	2.0921	0.0088
2 1/2-3	2.5200	2.5400	0.0200	2.1667	2.1834	0.0167	2.3333	2.3422	0.0089
2 3/4-3	2.7700	2.7900	0.0200	2.4167	2.4334	0.0167	2.5833	2.5924	0.0091
3-2	3.0200	3.0400	0.0200	2.5000	2.5250	0.0250	2.7500	2.7605	0.0105
3 1/2-2	3.5200	3.5400	0.0200	3.0000	3.0250	0.0250	3.2500	3.2608	0.0108
4-2	4.0200	4.0400	0.0200	3.5000	3.5250	0.0250	3.7500	3.7611	0.0111
4 1/2-2	4.5200	4.5400	0.0200	4.0000	4.0250	0.0250	4.2500	4.2613	0.0113
5-2	5.0200	5.0400	0.0200	4.5000	4.5250	0.0250	4.7500	4.7615	0.0115

All dimensions in inches.

Table 32. Dimensions and Tolerances for Class 2C Centralizing Acme Screw Threads.
(Source, Landis Threading Systems.)

Nominal Size and TPI	Major Diameter			Minor Diameter			Pitch Diameter		
	Max.	Min.	Tol.	Max.	Min.	Tol.	Max.	Min.	Tol.
1/2-10	0.500	0.4975	0.0025	0.3800	0.3594	0.0137	0.4443	0.4306	0.0137
5/8-8	0.6250	0.6222	0.0028	0.4800	0.4570	0.0154	0.5562	0.5408	0.0154
3/4-6	0.7500	0.7470	0.0030	0.5633	0.5371	0.0174	0.6598	0.6424	0.0174
7/8-6	0.8750	0.8717	0.0033	0.6883	0.6615	0.0179	0.7842	0.7663	0.0179
1-5	1.0000	0.9965	0.0035	0.7800	0.7509	0.0194	0.8920	0.8726	0.0194
1 1/8-5	1.1250	1.1213	0.0037	0.9050	0.8753	0.0198	1.0165	0.9967	0.0198
1 1/4-5	1.1250	1.2461	0.0039	1.0300	0.9998	0.0201	1.1411	1.1210	0.0201
1 3/8-4	1.3750	1.3709	0.0041	1.1050	1.0719	0.0220	1.2406	1.2185	0.0220
1 1/2-4	1.5000	1.4957	0.0043	1.2300	1.1965	0.0223	1.3652	1.3429	0.0223
1 3/4-4	1.7500	1.7454	0.0046	1.4800	1.4456	0.0229	1.6145	1.5916	0.0229
2-4	2.0000	1.9951	0.0049	1.7300	1.6948	0.0235	1.8637	1.8402	0.0235
2 1/4-3	2.2500	2.2448	0.0052	1.8967	1.8572	0.0263	2.0713	2.0450	0.0263
2 1/2-3	2.5000	2.4945	0.0055	2.1467	2.1065	0.0268	2.3207	2.2939	0.0268
2 3/4-3	2.7500	2.7442	0.0058	2.3967	2.3558	0.0273	2.5700	2.5427	0.0273
3-2	3.0000	2.9939	0.0061	2.4800	2.4326	0.0316	2.7360	2.7044	0.0316
3 1/2-2	3.5000	3.4935	0.0065	2.9800	2.9314	0.0324	3.2350	3.2026	0.0324
4-2	4.0000	3.9930	0.0070	3.4800	3.4302	0.0332	3.7340	3.7008	0.0332
4 1/2-2	4.5000	4.4926	0.0074	3.9800	3.9201	0.0339	4.2330	4.1991	0.0339
5-2	5.0000	4.9922	0.0078	4.4800	4.4281	0.0346	4.7319	4.6973	0.0346

(Continued)

Table 32. *(Continued)* **Dimensions and Tolerances for Class 2C Centralizing Acme Screw Threads.**
(Source, Landis Threading Systems.)

Nominal Size and TPI	Major Diameter			Minor Diameter			Pitch Diameter		
	Min.	Max.	Tol.	Min.	Max.	Tol.	Min.	Max.	Tol.
				Internal Threads - Class 2C					
1/2-10	0.5007	0.5032	0.0025	0.4100	0.4150	0.0050	0.4500	0.4637	0.0137
5/8-8	0.6258	0.6286	0.0026	0.5125	0.5187	0.0062	0.5625	0.5779	0.0154
3/4-6	0.7509	0.7539	0.0030	0.6000	0.6083	0.0083	0.6667	0.6841	0.0174
7/8-6	0.8759	0.8792	0.0033	0.7250	0.7333	0.0083	0.7917	0.8096	0.0179
1-5	1.0010	1.0045	0.0035	0.8200	0.8300	0.0100	0.9000	0.9194	0.0194
1 1/8-5	1.1261	1.1298	0.0037	0.9450	0.9550	0.0100	1.0250	1.0448	0.0198
1 1/4-5	1.2511	1.2550	0.0039	1.0700	1.0800	0.0100	1.1500	1.1701	0.0201
1 3/8-4	1.3762	1.3803	0.0041	1.1500	1.1625	0.0125	1.2500	1.2720	0.0220
1 1/2-4	1.5012	1.5055	0.0043	1.2750	1.2875	0.0125	1.3750	1.3973	0.0223
1 3/4-4	1.7513	1.7559	0.0046	1.5250	1.5375	0.0125	1.6250	1.6479	0.0229
2-4	2.0014	2.0063	0.0049	1.7750	1.7875	0.0125	1.8750	1.8985	0.0235
2 1/4-3	2.2515	2.2567	0.0052	1.9500	1.9667	0.0167	2.0833	2.1096	0.0263
2 1/2-3	2.5016	2.5071	0.0055	2.2000	2.2167	0.0167	2.3333	2.3601	0.0268
2 3/4-3	2.7517	2.7575	0.0058	2.4500	2.4667	0.0167	2.5833	2.6106	0.0273
3-2	3.0017	3.0078	0.0061	2.5500	2.5750	0.0250	2.7500	2.7816	0.0316
3 1/2-2	3.5019	3.5084	0.0065	3.0500	3.0750	0.0250	3.2500	3.2824	0.0324
4-2	4.0020	4.0090	0.0070	3.5500	3.5750	0.0250	3.7500	3.7832	0.0332
4 1/2-2	4.5021	4.5095	0.0074	4.0500	4.0750	0.0250	4.2500	4.2839	0.0339
5-2	5.0022	5.0100	0.0078	4.5500	4.5750	0.0250	4.7500	4.7846	0.0346

All dimensions in inches.

Table 33. Dimensions and Tolerances for Class 3C Centralizing Acme Screw Threads.
(Source, Landis Threading Systems.)

Nominal Size and TPI	Major Diameter			Minor Diameter			Pitch Diameter		
	Max.	Min.	Tol.	Max.	Min.	Tol.	Max.	Min.	Tol.
				External Threads - Class 3C					
1/2-10	0.500	0.4989	0.0011	0.3800	0.3594		0.4443	0.4306	0.0137
5/8-8	0.6250	0.6238	0.0012	0.4800	0.4570		0.5562	0.5408	0.0154
3/4-6	0.7500	0.7487	0.0013	0.5633	0.5371		0.6598	0.6424	0.0174
7/8-6	0.8750	0.8736	0.0014	0.6883	0.6615		0.7842	0.7663	0.0179
1-5	1.0000	0.9985	0.0015	0.7800	0.7509		0.8920	0.8726	0.0194
1 1/8-5	1.1250	1.1234	0.0016	0.9050	0.8753		1.0165	0.9967	0.0198
1 1/4-5	1.1250	1.2483	0.0017	1.0300	0.9998		1.1411	1.1210	0.0201
1 3/8-4	1.3750	1.3732	0.0018	1.1050	1.0719		1.2406	1.2185	0.0220
1 1/2-4	1.5000	1.4982	0.0019	1.2300	1.1965		1.3652	1.3429	0.0223
1 3/4-4	1.7500	1.7480	0.0020	1.4800	1.4456		1.6145	1.5916	0.0229
2-4	2.0000	1.9979	0.0021	1.7300	1.6948		1.8637	1.8402	0.0235
2 1/4-3	2.2500	2.2478	0.0022	1.8967	1.8572		2.0713	2.0450	0.0263

(Continued)

Table 33. *(Continued)* **Dimensions and Tolerances for Class 3C Centralizing Acme Screw Threads.**
(Source, Landis Threading Systems.)

External Threads - Class 3C								
Nominal Size and TPI	**Major Diameter**			**Minor Diameter**		**Pitch Diameter**		
	Max.	Min.	Tol.	Max.	Min.	Max.	Min.	Tol.
2 1/2-3	2.5000	2.4976	0.0024	2.1467	2.1065	2.3207	2.2939	0.0268
2 3/4-3	2.7500	2.7475	0.0025	2.3967	2.3558	2.5700	2.5427	0.0273
3-2	3.0000	2.9974	0.0026	2.4800	2.4326	2.7360	2.7044	0.0316
3 1/2-2	3.5000	3.4972	0.0028	2.9800	2.9314	3.2350	3.2026	0.0324
4-2	4.0000	3.9970	0.0030	3.4800	3.4302	3.7340	3.7008	0.0332
4 1/2-2	4.5000	4.4968	0.0032	3.9800	3.9201	4.2330	4.1991	0.0339
5-2	5.0000	4.9966	0.0034	4.4800	4.4281	4.7319	4.6973	0.0346
Internal Threads - Class 3C								
Nominal Size and TPI	**Major Diameter**			**Minor Diameter**		**Pitch Diameter**		
	Min.	Max.	Tol.	Min.	Max.	Tol.	Min.	Max.	Tol.
1/2-10	0.5007	0.5032	0.0025	0.4100	0.4150	0.0050	0.4500	0.4637	0.0137
5/8-8	0.6258	0.6286	0.0028	0.5125	0.5187	0.0062	0.5625	0.5779	0.0154
3/4-6	0.7509	0.7539	0.0030	0.6000	0.6083	0.0083	0.6667	0.6841	0.0174
7/8-6	0.8759	0.8792	0.0033	0.7250	0.7333	0.0083	0.7917	0.8096	0.0179
1-5	1.0010	1.0045	0.0035	0.8200	0.8300	0.0100	0.9000	0.9194	0.0194
1 1/8-5	1.1261	1.1298	0.0037	0.9450	0.9550	0.0100	1.0250	1.0448	0.0198
1 1/4-5	1.2511	1.2550	0.0039	1.0700	1.0800	0.0100	1.1500	1.1701	0.0201
1 3/8-4	1.3762	1.3803	0.0041	1.1500	1.1625	0.0125	1.2500	1.2720	0.0220
1 1/2-4	1.5012	1.5055	0.0043	1.2750	1.2875	0.0125	1.3750	1.3973	0.0223
1 3/4-4	1.7513	1.7559	0.0046	1.5250	1.5375	0.0125	1.6250	1.6479	0.0229
2-4	2.0014	2.0063	0.0049	1.7750	1.7875	0.0125	1.8750	1.8985	0.0235
2 1/4-3	2.2515	2.2567	0.0052	1.9500	1.9667	0.0167	2.0833	2.1096	0.0263
2 1/2-3	2.5016	2.5071	0.0055	2.2000	2.2167	0.0167	2.3333	2.3601	0.0268
2 3/4-3	2.7517	2.7575	0.0058	2.4500	2.4667	0.0167	2.5833	2.6106	0.0273
3-2	3.0017	3.0078	0.0061	2.5500	2.5750	0.0250	2.7500	2.7816	0.0316
3 1/2-2	3.5019	3.5084	0.0065	3.0500	3.0750	0.0250	3.2500	3.2824	0.0324
4-2	4.0020	4.0090	0.0070	3.5500	3.5750	0.0250	3.7500	3.7832	0.0332
4 1/2-2	4.5021	4.5095	0.0074	4.0500	4.0750	0.0250	4.2500	4.2839	0.0339
5-2	5.0022	5.0100	0.0078	4.5500	4.5750	0.0250	4.7500	4.7846	0.0346

All dimensions in inches.

Table 34. Dimensions and Tolerances for Class 4C Centralizing Acme Screw Threads.
(Source, Landis Threading Systems.)

Nominal Size and TPI	Major Diameter			Minor Diameter		Pitch Diameter		
External Threads - Class 4C								
	Max.	Min.	Tol.	Max.	Min.	Max.	Min.	Tol.
1/2-10	0.5000	0.4993	0.0007	0.3800	0.3594	0.4443	0.4306	0.0137
5/8-8	0.6250	0.6242	0.0008	0.4800	0.4570	0.5562	0.5408	0.0154
3/4-6	0.7500	0.7491	0.0009	0.5633	0.5371	0.6598	0.6424	0.0174
7/8-6	0.8750	0.8741	0.0009	0.6883	0.6615	0.7842	0.7663	0.0179
1-5	1.0000	0.9990	0.0010	0.7800	0.7509	0.8920	0.8726	0.0194
1 1/8-5	1.1250	1.1239	0.0011	0.9050	0.8753	1.0165	0.9967	0.0198
1 1/4-5	1.1250	1.2489	0.0011	1.0300	0.9998	1.1411	1.1210	0.0201
1 3/8-4	1.3750	1.3738	0.0012	1.1050	1.0719	1.2406	1.2185	0.0220
1 1/2-4	1.5000	1.4988	0.0012	1.2300	1.1965	1.3652	1.3429	0.0223
1 3/4-4	1.7500	1.7487	0.0013	1.4800	1.4456	1.6145	1.5916	0.0229
2-4	2.0000	1.9986	0.0014	1.7300	1.6948	1.8637	1.8402	0.0235
2 1/4-3	2.2500	2.2485	0.0015	1.8967	1.8572	2.0713	2.0450	0.0263
2 1/2-3	2.5000	2.4984	0.0016	2.1467	2.1065	2.3207	2.2939	0.0268
2 3/4-3	2.7500	2.7483	0.0017	2.3967	2.3558	2.5700	2.5427	0.0273
3-2	3.0000	2.9983	0.0017	2.4800	2.4326	2.7360	2.7044	0.0316
3 1/2-2	3.5000	3.4981	0.0019	2.9800	2.9314	3.2350	3.2026	0.0324
4-2	4.0000	3.9980	0.0020	3.4800	3.4302	3.7340	3.7008	0.0332
4 1/2-2	4.5000	4.4979	0.0021	3.9800	3.9201	4.2330	4.1991	0.0339
5-2	5.0000	4.9978	0.0022	4.4800	4.4281	4.7319	4.6973	0.0346
Internal Threads - Class 4C								
Nominal Size and TPI	Min.	Max.	Tol.	Min.	Max.	Tol.	Min.	Max.	Tol.
1/2-10	0.5007	0.5021	0.0014	0.4100	0.4150	0.0050	0.4500	0.4637	0.0137
5/8-8	0.6258	0.6274	0.0016	0.5125	0.5187	0.0062	0.5625	0.5779	0.0154
3/4-6	0.7509	0.7526	0.0017	0.6000	0.6083	0.0083	0.6667	0.6841	0.0174
7/8-6	0.8759	0.8778	0.0019	0.7250	0.7333	0.0083	0.7917	0.8096	0.0179
1-5	1.0010	1.0030	0.0020	0.8200	0.8300	0.0100	0.9000	0.9194	0.0194
1 1/8-5	1.1261	1.1282	0.0021	0.9450	0.9550	0.0100	1.0250	1.0448	0.0198
1 1/4-5	1.2511	1.2533	0.0022	1.0700	1.0800	0.0100	1.1500	1.1701	0.0201
1 3/8-4	1.3762	1.3785	0.0023	1.1500	1.1625	0.0125	1.2500	1.2720	0.0220
1 1/2-4	1.5012	1.5036	0.0024	1.2750	1.2875	0.0125	1.3750	1.3973	0.0223
1 3/4-4	1.7513	1.7539	0.0026	1.5250	1.5375	0.0125	1.6250	1.6479	0.0229
2-4	2.0014	2.0042	0.0028	1.7750	1.7875	0.0125	1.8750	1.8985	0.0235
2 1/4-3	2.2515	2.2545	0.0030	1.9500	1.9667	0.0167	2.0833	2.1096	0.0263
2 1/2-3	2.5016	2.5048	0.0032	2.2000	2.2167	0.0167	2.3333	2.3601	0.0268
2 3/4-3	2.7517	2.7550	0.0033	2.4500	2.4667	0.0167	2.5833	2.6106	0.0273
3-2	3.0017	3.0052	0.0035	2.5500	2.5750	0.0250	2.7500	2.7816	0.0316

(Continued)

Table 34. *(Continued)* **Dimensions and Tolerances for Class 4C Centralizing Acme Screw Threads.** *(Source, Landis Threading Systems.)*

Nominal Size and TPI	Major Diameter			Minor Diameter			Pitch Diameter		
	Min.	Max.	Tol.	Min.	Max.	Tol.	Min.	Max.	Tol.
3 1/2-2	3.5019	3.5056	0.0037	3.0500	3.0750	0.0250	3.2500	3.2824	0.0324
4-2	4.0020	4.0060	0.0040	3.5500	3.5750	0.0250	3.7500	3.7832	0.0332
4 1/2-2	4.5021	4.5063	0.0042	4.0500	4.0750	0.0250	4.2500	4.2839	0.0339
5-2	5.0022	5.0067	0.0045	4.5500	4.5750	0.0250	4.7500	4.7846	0.0346

Internal Threads - Class 4C

All dimensions in inches.

Table 35. Dimensions and Tolerances for Stub Acme Screw Threads. *(Source, FED-STD-H28/13.)*

External Threads - Stub Acme

Nominal Size and TPI	Major Diameter			Pitch Diameter			Minor Diameter	
	Max.	Min.	Tol.	Max.	Min.	Tol.	Max.	Min.
1/4-16	0.2500	0.2469	0.0050	0.2272	0.2167	0.0105	0.2024	0.1919
5/16-14	0.3125	0.3089	0.0050	0.2871	0.2757	0.0114	0.2597	0.2483
3/8-12	0.3750	0.3708	0.0050	0.3451	0.3328	0.0123	0.3150	0.3027
7/16-12	0.4375	0.4333	0.0050	0.4076	0.3950	0.0126	0.3775	0.3649
1/2-10	0.5000	0.4950	0.0050	0.4643	0.4506	0.0137	0.4200	0.4063
5/8-8	0.6250	0.6188	0.0062	0.5812	0.5658	0.0154	0.5300	0.5146
3/4-6	0.7500	0.7417	0.0083	0.6931	0.6757	0.0174	0.6300	0.6126
7/8-6	0.8750	0.8667	0.0083	0.8175	0.7996	0.0179	0.7550	0.7371
1-5	1.0000	0.9900	0.0100	0.9320	0.9126	0.0194	0.8600	0.8406
1 1/8-5	1.1250	1.1150	0.0100	1.0565	1.0367	0.0198	0.9850	0.9652
1 1/4-5	1.2500	1.2400	0.0100	1.1811	1.1610	0.0201	1.1100	1.0899
1 3/8-4	1.3750	1.3625	0.0125	1.2906	1.2686	0.0220	1.2050	1.1830
1 1/2-4	1.500	1.4875	0.0125	1.4152	1.3929	0.0223	1.3300	1.3077
1 3/4-4	1.7500	1.7375	0.0125	1.6645	1.6416	0.0229	1.5800	1.5571
2-4	2.0000	1.9875	0.0125	1.9137	1.8902	0.0235	1.8300	1.8065
2 1/4-3	2.2500	2.2333	0.0167	2.1380	2.1117	0.0263	2.0300	2.0037
2 1/2-3	2.500	2.4833	0.0167	2.3874	2.3606	0.0268	2.2800	2.2532
2 3/4-3	2.7500	2.7333	0.0167	2.6367	2.6094	0.0273	2.5300	2.5027
3-2	3.0000	2.9750	0.2500	2.8360	2.8044	0.0316	2.6800	2.6484
3 1/2-2	3.5000	3.4750	0.2500	3.3350	3.3026	0.0324	3.1800	3.1476
4-2	4.0000	3.9750	0.2500	3.8340	3.8008	0.0332	3.6800	3.6468
4 1/2-2	4.5000	4.4750	0.2500	4.3330	4.2991	0.0339	4.1800	4.1461
5-2	5.0000	4.9750	0.2500	4.8319	4.7973	0.0346	4.6800	4.6454

Internal Threads - Stub Acme

Nominal Size and TPI	Major Diameter			Pitch Diameter			Minor Diameter		
	Min.	Max.	Tol.	Min.	Max.	Tol.	Min.	Max.	Tol.
1/4-16	0.2600	0.2750	0.0100	0.2312	0.2417	0.0105	0.2125	0.2156	0.0050

(Continued)

Table 35. *(Continued)* Dimensions and Tolerances for Stub Acme Screw Threads. *(Source, FED-STD-H28/13.)*

Nominal Size and TPI	Major Diameter			Pitch Diameter			Minor Diameter		
	Min.	Max.	Tol.	Min.	Max.	Tol.	Min.	Max.	Tol.
5/16-14	0.3225	0.3339	0.0100	0.2911	0.3025	0.0114	0.2696	0.2732	0.0050
3/8-12	0.3850	0.3973	0.0100	0.3500	0.3623	0.0123	0.3250	0.3292	0.0050
7/16-12	0.4475	0.4601	0.0100	0.4125	0.4251	0.0126	0.3875	0.3917	0.0050
1/2-10	0.5200	0.5337	0.0200	0.4700	0.4837	0.0137	0.4400	0.4450	0.0050
5/8-8	0.6450	0.6604	0.0200	0.5875	0.6029	0.0154	0.5500	0.5562	0.0062
3/4-6	0.7700	0.7874	0.0200	0.7000	0.7174	0.0174	0.6500	0.6583	0.0083
7/8-6	0.8950	0.9129	0.0200	0.8250	0.8429	0.0179	0.7750	0.7833	0.0083
1-5	1.0200	1.0394	0.0200	0.9400	0.9594	0.0194	0.8800	0.8900	0.0100
1 1/8-5	1.1450	1.1648	0.0200	1.0650	1.0848	0.0198	1.0050	1.0150	0.0100
1 1/4-5	1.2700	1.2901	0.0200	1.1900	1.2101	0.0201	1.1300	1.1400	0.0100
1 3/8-4	1.3950	1.4170	0.0200	1.3000	1.3220	0.0220	1.2250	1.2375	0.0125
1 1/2-4	1.5200	1.5423	0.0200	1.4250	1.4473	0.0223	1.3500	1.3625	0.0125
1 3/4-4	1.7700	1.7929	0.0200	1.6750	1.6979	0.0229	1.6000	1.6125	0.0125
2-4	2.0200	2.0435	0.0200	1.9250	1.9485	0.0235	1.8500	1.8625	0.0125
2 1/4-3	2.2700	2.2963	0.0200	2.1500	2.1763	0.0263	2.0500	2.0667	0.0167
2 1/2-3	2.5200	2.5468	0.0200	2.4000	2.4268	0.0268	2.3000	2.3167	0.0167
2 3/4-3	2.7700	2.7973	0.0200	2.6500	2.6773	0.0273	2.5500	2.5667	0.0167
3-2	3.0200	3.0516	0.0200	2.8500	2.8816	0.0316	2.7000	2.7250	0.0250
3 1/2-2	3.5200	3.5524	0.0200	3.3500	3.3824	0.0324	3.2000	3.2250	0.0250
4-2	4.0200	4.0532	0.0200	3.8500	3.8832	0.0332	3.7000	3.7250	0.0250
4 1/2-2	4.5200	4.5539	0.0200	4.3500	4.3839	0.0339	4.2000	4.2250	0.0250
5-2	5.0200	5.0546	0.0200	4.8500	4.8846	0.0346	4.7000	4.7250	0.0250

Buttress screw thread

The American Buttress thread form is capable of withstanding exceptionally high unidirectional stress parallel to the axis of the thread. This strength is derived from the basic thread form, which provides a pressure flank at a nearly right angle to the force, fortified by a "buttress" leaning into the flank at an angle of 45°, commonly called a 7°/45° thread form. Buttress threads have a variety of applications where tubular members are screwed together, ranging from the common automobile hose clamp to the breech assemblies of heavy artillery. The initial standardization of the thread came with the issue of ASA B1.9 in 1953, which defined the thread profile and the dimensional formulas. The present standard is ANSI B1.9, which was issued in 1973.

The standard Buttress thread as shown in *Figure 14* has a load flank angle of 7°, measured in the axial plane. The height of the sharp V thread is $0.89064P$, the thread height is $0.66271P$, and the basic height of thread engagement is $0.6P$. At maximum material condition, the root radius equals $0.07141P$, and the root truncation is $0.08261P$. At minimum material condition, root radius is equal to $0.0357P$, and the root truncation is $0.0413P$. Crest width equals $0.16316P$, and crest truncation is $0.14532P$. **Table 36** provides basic dimensions for preferred sizes of American Buttress screw threads in maximum material condition. The dimensions in the table include the recommended allowance

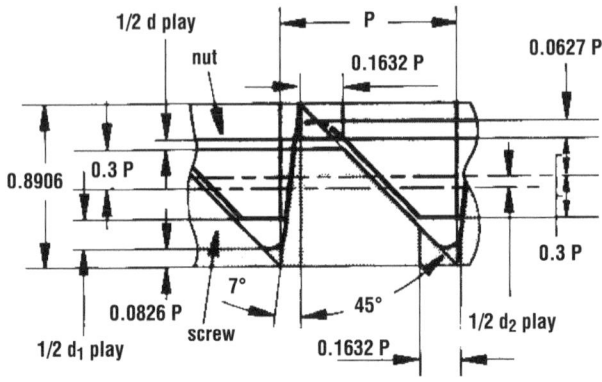

Figure 14. American Buttress 7°/45° thread form. (Source, Kennametal Inc.)

(clearance) specified for each diameter and pitch. Allowances are provided for external threads only. Internal threads are basic.

Pull-type Buttress threads are identified by their nominal diameter, stated as a decimal rather than a fraction, followed by the tpi and the letters BUTT. For push-type threads, the letters PUSH BUTT follow the diameter and tpi. The designation for a one and one-half inch, 10 tpi pull-type thread would be 1.5–10 BUTT. Class 2 Buttress threads are considered "Standard," and Class 3 are "Precision Grade." Both classes share identical allowances on their external threads, but Class 3 threads have slightly smaller tolerances (0.0019" on the smallest thread, and 0.0075" on the largest) on the major diameter of the external thread, the pitch diameter of both internal and external threads, and the minor diameter of the internal thread.

The British Standard 7°/45° Buttress thread shares its basic angular dimensions with the American form, but is not interchangeable. It has a basic thread height of 0.50586P and a root radius of 0.12055P. A second Standard British Buttress thread has a pressure flank angle of 0° and a flank angle of 52°. Known as a 0°/52°, it has a basic thread height of 0.49298P and a root radius of 0.09298P. Because of the 0° angle, this form offers economy of production. British Buttress threads are specified by British Standard B.S. 1657:1950.

Table 36. American Buttress Inch Screw Threads Basic Dimensions at Maximum Material Condition.

Nominal Size	TPI	External Thread			Internal Thread			Crest Width	Height of Thread	Height of Engagement
		Major Dia.	Pitch Dia.	Minor Dia.	Minor Dia.	Pitch Dia.	Major Dia.			
0.50	20	0.4963	0.4663	0.4300	0.4400	0.4700	0.5063	0.0082	0.0331	0.0300
0.50	16	0.4960	0.4585	0.4132	0.4250	0.4625	0.5078	0.0102	0.0414	0.0375
0.50	12	0.4956	0.4456	0.3852	0.4000	0.4500	0.5104	0.0136	0.0552	0.5000
0.65	20	0.6463	0.6163	0.5800	0.5900	0.6200	0.6563	0.0082	0.0331	0.0300
0.65	16	0.6460	0.6085	0.5632	0.5750	0.6125	0.6578	0.0102	0.0414	0.0375
0.65	12	0.6456	0.5956	0.5352	0.5500	0.6000	0.6604	0.0136	0.0552	0.0500
0.75	16	0.7458	0.7083	0.6630	0.6750	0.7125	0.7578	0.0102	0.0414	0.0375
0.75	12	0.7454	0.6954	0.6350	0.6500	0.7000	0.7604	0.0136	0.0552	0.0500
0.75	10	0.7451	0.6851	0.6126	0.6300	0.6900	0.7625	0.0163	0.0663	0.0600

(Continued)

Table 36. *(Continued)* American Buttress Inch Screw Threads Basic Dimensions at Maximum Material Condition.

| Nominal Size | TPI | External Thread | | | Internal Thread | | | Crest Width | Height of Thread | Height of Engagement |
		Major Dia.	Pitch Dia.	Minor Dia.	Minor Dia.	Pitch Dia.	Major Dia.			
0.875	16	0.8708	0.8333	0.7880	0.8000	0.8375	0.8828	0.0102	0.0414	0.0375
0.875	12	0.8704	0.8204	0.7600	0.7750	0.8250	0.8854	0.0136	0.0552	0.0500
0.875	10	0.8701	0.8101	0.7376	0.7550	0.8150	0.8875	0.0163	0.0663	0.0600
1.0	16	0.9958	0.9583	0.9130	0.9250	0.9625	1.0078	0.0102	0.0414	0.0375
1.0	12	0.9954	0.9454	0.8850	0.9000	0.9500	1.0104	0.0136	0.0552	0.0500
1.0	10	0.9951	0.9351	0.8626	0.8800	0.9400	1.0125	0.0163	0.0663	0.0600
1.25	12	1.2452	1.1952	1.1348	1.1500	1.2000	1.2604	0.0136	0.0552	0.0500
1.25	10	1.2449	1.1849	1.1124	1.1300	1.1900	1.2625	0.0163	0.0663	0.0600
1.25	8	1.2445	1.1695	1.0788	1.1000	1.1750	1.2657	0.0204	0.0828	0.0750
1.375	12	1.3702	1.3202	1.2598	1.2750	1.3250	1.3854	0.0136	0.0552	0.0500
1.375	10	1.3699	1.3099	1.2374	1.2550	1.3150	1.3875	0.0163	0.0663	0.0600
1.375	8	1.3695	1.2945	1.2038	1.2250	1.3000	1.3907	0.0204	0.0828	0.0750
1.50	12	1.4952	1.4452	1.3848	1.4000	1.4500	1.5104	0.0136	0.0552	0.0500
1.50	10	1.4949	1.4349	1.3624	1.3800	1.4400	1.5125	0.0163	0.0663	0.0600
1.50	8	1.4939	1.4189	1.3282	1.3500	1.4250	1.5157	0.0204	0.0828	0.0750
1.75	10	1.7447	1.6847	1.6122	1.6300	1.6900	1.7625	0.0163	0.0663	0.0600
1.75	8	1.7442	1.6692	1.5785	1.6000	1.6750	1.7657	0.0204	0.0828	0.0750
1.75	6	1.7436	1.6436	1.5227	1.5500	1.6500	1.7709	0.0272	0.1105	0.1000
2.0	10	1.9947	1.9347	1.8622	1.8800	1.9400	2.0125	0.0163	0.0663	0.0600
2.0	8	1.9942	1.9192	1.8285	1.8500	1.9250	2.0157	0.0204	0.0828	0.0750
2.0	6	1.9936	1.8936	1.7727	1.8000	1.9000	2.0209	0.0272	0.1105	0.1000
2.25	10	2.2447	2.1847	2.1122	2.1300	2.1900	2.2625	0.0163	0.0663	0.0600
2.25	8	2.2442	2.1692	2.0785	2.1000	2.1750	2.2657	0.0204	0.0828	0.0750
2.25	6	2.2436	2.1436	2.0227	2.0500	2.1500	2.2709	0.0272	0.1105	0.1000
2.50	10	2.4947	2.4347	2.3622	2.3800	2.4400	2.5125	0.0163	0.0663	0.0600
2.50	8	2.4942	2.4192	2.3285	2.3500	2.4250	2.5157	0.0204	0.0828	0.0750
2.50	6	2.4936	2.3936	2.2727	2.3000	2.4000	2.5209	0.0272	0.1105	0.1000
2.75	8	2.7439	2.6689	2.5782	2.6000	2.6750	2.7657	0.0204	0.0828	0.0750
2.75	6	2.7433	2.6433	2.5224	2.5500	2.6500	2.7709	0.0272	0.1105	0.1000
2.75	5	2.7429	2.6229	2.4778	2.5100	2.6300	2.7751	0.0326	0.1325	0.1200
3.0	8	2.9939	2.9189	2.8282	2.8500	2.9250	3.0157	0.0204	0.0828	0.0750
3.0	6	2.9933	2.8933	2.7724	2.8000	2.9000	3.0209	0.0272	0.1105	0.1000
3.0	5	2.9929	2.8729	2.7278	2.7600	2.8800	3.0251	0.0326	0.1325	0.1200
3.50	8	3.4939	3.4189	3.3282	3.3500	3.4250	3.5157	0.0204	0.0828	0.0750
3.50	6	3.4933	3.3933	3.2724	3.3000	3.4000	3.5209	0.0272	0.1105	0.1000
3.50	5	3.4929	3.3729	3.2278	3.2600	3.3800	3.5251	0.0326	0.1325	0.1200
4.0	8	3.9939	3.9189	3.8282	3.8500	3.9250	4.0157	0.0204	0.0828	0.0750
4.0	6	3.9933	3.8933	3.7724	3.8000	3.9000	4.0209	0.0272	0.1105	0.1000

(Continued)

Table 36. *(Continued)* **American Buttress Inch Screw Threads Basic Dimensions at Maximum Material Condition.**

Nominal Size	TPI	External Thread			Internal Thread			Crest Width	Height of Thread	Height of Engagement
		Major Dia.	Pitch Dia.	Minor Dia.	Minor Dia.	Pitch Dia.	Major Dia.			
4.0	5	3.9929	3.8729	3.7278	3.7600	3.8800	4.0251	0.0326	0.1325	0.1200
4.5	6	4.4930	4.3930	4.2721	4.3000	4.4000	4.5209	0.0272	0.1105	0.1000
4.5	5	4.4926	4.3726	4.2275	4.2600	4.3800	4.5251	0.0326	0.1325	0.1200
4.5	4	4.4920	4.3420	4.1607	4.2000	4.3500	4.5314	0.0408	0.1657	0.1500
5.0	6	4.9930	4.8930	4.7721	4.8000	4.9000	5.0209	0.0272	0.1105	0.1000
5.0	5	4.9926	4.8726	4.7275	4.7600	4.8800	5.0251	0.0326	0.1325	0.1200
5.0	4	4.9920	4.8420	4.6607	4.7000	4.8500	5.0314	0.0408	0.1657	0.1500
5.50	6	5.4930	5.3930	5.2721	5.3000	5.4000	5.5209	0.0272	0.1105	0.1000
5.50	5	5.4926	5.3726	5.2275	5.2600	5.3800	5.5251	0.0326	0.1325	0.1200
5.50	4	5.4920	5.3420	5.1607	5.2000	5.3500	5.5314	0.0408	0.1657	0.1500
6.0	6	5.9930	5.8930	5.7721	5.8000	5.9000	6.0209	0.0272	0.1105	0.1000
6.0	5	5.9926	5.8726	5.7275	5.7600	5.8800	6.0251	0.0326	0.1325	0.1200
6.0	4	5.9920	5.8420	5.6607	5.7000	5.8500	6.0314	0.0408	0.1657	0.1500
7.0	5	6.9922	6.8722	6.7271	6.7600	6.8800	7.0251	0.0326	0.1325	0.1200
7.0	4	6.9916	6.8416	6.6603	6.7000	6.8500	7.0314	0.0408	0.1657	0.1500
7.0	3	6.9907	6.7907	6.5489	6.6000	6.8000	7.0418	0.0544	0.2209	0.2000
8.0	5	7.9922	7.8722	7.7271	7.7600	7.8800	8.0251	0.0326	0.1325	0.1200
8.0	4	7.9916	7.8416	7.6603	7.7000	7.8500	8.0314	0.0408	0.1657	0.1500
8.0	3	7.9907	7.7907	7.5489	7.6000	7.8000	8.0418	0.0544	0.2209	0.2000
9.0	5	8.9922	8.8722	8.7271	8.7600	8.8800	9.0251	0.0326	0.1325	0.1200
9.0	4	8.9916	8.8416	8.6603	8.7000	8.8500	9.0314	0.0408	0.1657	0.1500
9.0	3	8.9907	8.7907	8.5489	8.6000	8.8000	9.0418	0.0544	0.2209	0.2000
10.0	5	9.9922	9.8722	9.7271	9.7600	9.8800	10.0251	0.0326	0.1325	0.1200
10.0	4	9.9916	9.8416	9.6603	9.7000	9.8500	10.0314	0.0408	0.1657	0.1500
10.0	3	9.9907	9.7907	9.5489	9.6000	9.8000	10.0418	0.0544	0.2209	0.2000
12.0	4	11.9911	11.8411	11.6598	11.7000	11.8500	12.0314	0.0408	0.1657	0.1500
12.0	3	11.9902	11.7902	11.5484	11.6000	11.8000	12.0418	0.0544	0.2209	0.2000
12.0	2.5	11.9896	11.7496	11.4594	11.5200	11.7600	12.0502	0.0653	0.2651	0.2400
14.0	4	13.9911	13.8411	13.6598	13.7000	13.8500	14.0314	0.0408	0.1657	0.1500
14.0	3	13.9902	13.7902	13.5484	13.6000	13.8000	14.0418	0.0544	0.2209	0.2000
14.0	2.5	13.9896	13.7496	13.4594	13.5200	13.7600	14.0502	0.0653	0.2651	0.2400
16.0	4	15.9911	15.8411	15.6598	15.7000	15.8500	16.0314	0.0408	0.1657	0.1500
16.0	3	15.9902	15.7902	15.5484	15.6000	15.8000	16.0418	0.0544	0.2209	0.2000
16.0	2.5	15.9896	15.7496	15.4594	15.5200	15.7600	16.0502	0.0653	0.2651	0.2400
18.0	3	17.9897	17.7897	17.5479	17.6000	17.8000	18.0418	0.0544	0.2209	0.2000
18.0	2.5	17.9891	17.7491	17.4589	17.5200	17.7600	18.0502	0.0653	0.2651	0.2400
18.0	2	17.9882	17.6882	17.3255	17.4000	17.7000	18.0627	0.0816	0.3314	0.3000
20.0	3	19.9897	19.7897	19.5479	19.6000	19.8000	20.0418	0.0544	0.2209	0.2000

(Continued)

Table 36. *(Continued)* American Buttress Inch Screw Threads Basic Dimensions at Maximum Material Condition.

Nominal Size	TPI	External Thread			Internal Thread			Crest Width	Height of Thread	Height of Engagement
		Major Dia.	Pitch Dia.	Minor Dia.	Minor Dia.	Pitch Dia.	Major Dia.			
20.0	2.5	19.9891	19.7491	19.4589	19.5200	19.7600	20.0502	0.0653	0.2651	0.2400
20.0	2	19.9882	19.6882	19.3255	19.4000	19.7000	20.0627	0.0816	0.3314	0.3000
22.0	3	21.9897	21.7897	21.5479	21.6000	21.8000	22.0418	0.0544	0.2209	0.2000
22.0	2.5	21.9891	21.7491	21.4589	21.5200	21.7600	22.0502	0.0653	0.2651	0.2400
22.0	2	21.9882	21.6882	21.3255	21.4000	21.7000	22.0627	0.0816	0.3314	0.3000
24.0	3	23.9897	23.7897	23.5479	23.6000	23.8000	24.0418	0.0544	0.2209	0.2000
24.0	2.5	23.9891	23.7491	23.4589	23.5200	23.7600	24.0502	0.0653	0.2651	0.2400
24.0	2	23.9882	23.6882	23.3255	23.4000	23.7000	24.0627	0.0816	0.3314	0.3000

All dimensions in inches.

Pipe threads

The modern American pipe form has been through several evolutions since it originated as the Briggs Standard in the last century. In 1913, the "Committee on the Standardization of Pipe Threads," jointly sponsored by the American Gas Association and the American Society of Mechanical Engineers, began six years of work that resulted in a standard that was published by the National Screw Thread Commission (NSTC). The NSTC continued work on the project, as did the AGA and the ASME. Since then, the standards have been reviewed and fine-tuned regularly, but the basic thread dimensions can be traced back to their earliest origins. There are two general categories of pipe threads: those that provide a pressure-tight fit, and those that produce a mechanical joint. Pressure fit threads may be designed to perform when assembled dry (NPTF for tapered fittings, NPSF for straight pipe fuel fittings, and NPSI for standard straight pipe fittings), or with a sealer (NPT series and NPSC for pipe couplings). Pipe threads that are used for joining, but are not pressure tight, are designed as parallel rather than taper threads. These include NPSM for free fitting mechanical joints, and NPSL for loose fitting mechanical joints. See Table 23 for specifications for the British Standard Taper Pipe Thread (BSTP).

The NPT (American National Standard Pipe Thread) form is the most widely used pipe thread. A taper is always present on the external thread, and in high pressure applications, the internal thread is also tapered. The basic form is shown in *Figures 15* and *16*. It utilizes a 60° included angle thread, and the height of the sharp V thread equals $0.866025P$. The basic thread height is $0.80000P$, and both the crest and root are truncated a minimum of $0.033P$. The maximum truncation is graduated by pitch as follows: 27 tpi = $0.0972P$; 18 tpi = $0.0882P$; 14 tpi = $0.0784P$; 11.5 tpi = $0.0725P$; and 8 tpi = $0.0624P$. The minimum width of the flat at the truncation $0.038P$. The maximum width of the flat at the truncation is graduated by pitch as follows: 27 tpi = $0.1107P$; 18 tpi = $0.1026P$; 14 tpi = $0.0896P$, 11.5 tpi = $0.0839P$, and 8 tpi = $0.0720P$. The taper on the NPT form is 0.75" per foot, producing a taper angle of 1°47'. The tolerance for the NPT thread is "plus or minus one turn (length of one pitch)." NPT thread specifications and formulas are contained in MIL-P-7105B (Aeronautical National Taper Pipe Thread Form, Symbol ANPT), ANSI B2.1-1968, and ANSI/ASME B1.20.1-1983. **Table 37** provides dimensions.

The NPSC thread is used on pipe couplings, and it exists only in an internal thread configuration. When assembled wrench-tight with an external NPT thread, a sealer must be used and the coupling should only be used for low-pressure applications. Dimensions for NPSC threads, which are the same form as NPT threads, are given in **Table 38**. ANSI

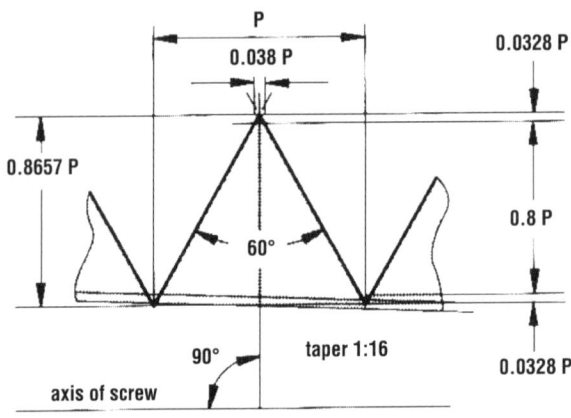

Figure 15. Basic NPT thread form.

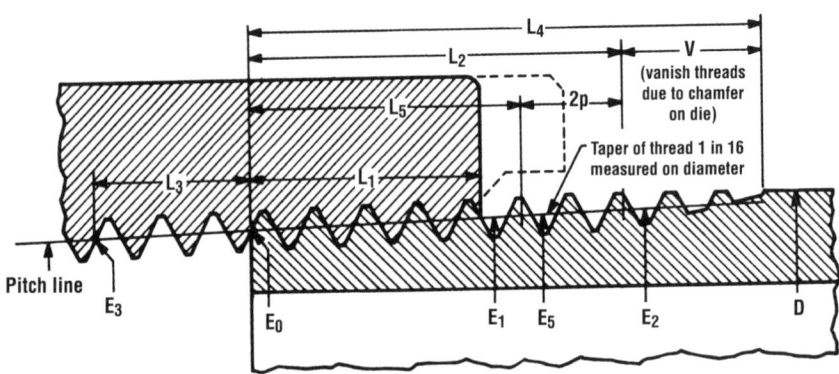

Figure 16. NPT assembly. Refer to Table 37 for dimensions.

B2.1-1968 and ANSI/ASME B1.20.1-1983 contain full information on the form.

NPSM pipe threads are used on free-fitting mechanical joints used for mechanical assembly in situations where there is no internal pressure. The form uses a 60° included angle thread with thread height (sharp V thread) equal to $0.866025P$. The internal thread height is truncated and is equal to $0.54126P$. The internal thread crest equals $0.2500P$. Tolerances are the same as Unified 2A/2B, and the minimum pitch diameter of the internal thread is equal to E_1 of the NPT thread form. **Table 39** provides dimensions for the NPSM thread form. NSPL pipe threads are intended for use in loose fitting mechanical joints with locknuts. Both NPSM and NPSL forms are covered by standards ANSI B2.2-1968 and ANSI/ASME B1.20.1-1983.

Dryseal pipe threads work through deformation when under load. Wrench-tight threads are "crushed," bringing the flanks into contact and thereby erecting a barrier to escaping fluids or gases. The dimensions of NPTF dryseal taper pipe threads are similar to those of NPT threads, except for the amount of truncation in the crest and root. **Table 40** provides formulas for determining both the truncation and the width of the flat at the crest and root. Although Class 1 and 2 NPTF threads are theoretically identical, Class 2 threads have been inspected to assure accuracy of the root and crest truncation. Therefore, when a pressure tight seal is critical, only Class 2 threads should be used. See **Table 41** for dimensions of

Table 37. American Standard Taper Pipe Threads NPT Basic Dimensions. (Source, Landis Threading Systems.)

Nominal Pipe Size		1/16	1/8	1/4	3/8	1/2	3/4	1	1 1/4	1 1/2	2	2 1/2	3
						Sizes 1/16 - 3"							
Pipe O.D.		0.3125	0.4050	0.5400	0.6750	0.8400	1.0500	1.3150	1.6600	1.9000	2.3750	2.8750	3.5000
TPI		27	27	18	18	14	14	11 1/2	11 1/2	11 1/2	11 1/2	8	8
Pitch P		0.03704	0.03704	0.05556	0.05556	0.07143	0.07143	0.08696	0.08696	0.08696	0.08696	0.12500	0.12500
E_0 *		0.27118	0.36351	0.47739	0.61201	0.75843	0.96768	1.21363	1.55713	1.79609	2.26902	2.71953	3.34062
Hand Tight Engagement	Inch L_1 *	0.1600	0.1615	0.2278	0.2400	0.3200	0.3390	0.4000	0.4200	0.4200	0.4360	0.682	0.766
	Threads	4.32	4.36	4.10	4.32	4.48	4.75	4.60	4.83	4.83	5.01	5.46	6.13
	E_1 *	0.28118	0.37360	0.49163	0.62701	0.77843	0.98887	1.23863	1.58338	1.82234	2.29627	2.76216	3.38850
Length of Full Thread	Inch L_2 *	0.2611	0.2639	0.4018	0.4078	0.5337	0.5457	0.6828	0.7068	0.7235	0.7565	1.1375	1.2000
	Threads	7.05	7.12	7.23	7.34	7.47	7.64	7.85	8.13	8.32	8.70	9.10	9.60
	E_2 *	0.28750	0.3800	0.50250	0.63750	0.79179	1.00179	1.25630	1.60130	1.84130	2.31630	2.79062	3.41562
Wrench Makeup Length, Internal Thread	Inch L_3 *	0.1111	0.1111	0.1667	0.1667	0.2143	0.2143	0.2609	0.2609	0.2609	0.2609	0.2500	0.2500
	Threads	3	3	3	3	3	3	3	3	3	3	2	2
	E_3 *	0.26424	0.35656	0.46697	0.60160	0.74504	0.95429	1.19733	1.54083	1.77978	2.25272	2.70391	3.32500
Vanish Threads	Inch	0.1285	0.1285	0.1928	0.1928	0.2478	0.2478	0.3017	0.3017	0.3017	0.3017	0.4337	0.4337
	Threads	3.47	3.47	3.47	3.47	3.47	3.47	3.47	3.47	3.47	3.47	3.47	3.47
Overall Length Ext. Threads	L_4 *	0.3896	0.3924	0.5946	0.6006	0.7815	0.7935	0.9845	1.0085	1.0252	1.0582	1.5712	1.6337
Nominal Perfect Ext Thread	Length L_5 *	0.1870	0.1898	0.2907	0.2967	0.3909	0.4029	0.5089	0.5329	0.5496	0.5826	0.8875	0.9500
	Dia.	0.28287	0.37537	0.49556	0.63056	0.78286	0.99286	1.24543	1.59043	1.83043	2.30543	2.77500	3.40000
Height of Thread		0.02963	0.2963	0.04444	0.04444	0.05714	0.05714	0.06957	0.06957	0.06957	0.06957	0.1000	0.1000
Increase in Dia. per Thread		0.00231	0.00231	0.00347	0.00347	0.00446	0.00446	0.00543	0.00543	0.00543	0.00543	0.00781	0.00781
Basic Minor Dia. at Small End		0.2416	0.3339	0.4329	0.5676	0.7013	0.9105	1.1441	1.4876	1.7265	2.1995	2.6195	3.2406

(Continued)

Table 37. *(Continued)* **American Standard Taper Pipe Threads NPT Basic Dimensions.** *(Source, Landis Threading Systems.)*

Nominal Pipe Size		3 1/2	4	5	6	8	10	12	14 OD	16 OD	18 OD	20 OD	24 OD
						Sizes 3 1/2 - 24 OD"						
Pipe O.D.		4.0000	4.5000	5.5630	6.6250	8.6250	10.7500	12.7500	14.0000	16.0000	18.0000	20.0000	24.0000
TPI		8	8	8	8	8	8	8	8	8	8	8	8
Pitch P		0.12500	0.12500	0.12500	0.12500	0.12500	0.12500	0.12500	0.12500	0.12500	0.12500	0.12500	0.12500
E_0 *		3.83750	4.33438	5.39073	6.44609	8.43359	10.54531	12.53281	13.77500	15.76250	17.7750	19.73750	23.71250
Hand Tight Engagement	Inch L_1 *	0.8210	0.8440	0.9370	0.9580	1.0630	1.2100	1.3600	1.5620	1.8120	2.0000	2.125	2.375
	Threads	6.57	6.75	7.50	7.66	8.50	9.68	10.88	12.50	14.50	16.00	17.00	19.00
	E_1 *	3.88881	4.38712	5.44929	6.50597	8.50003	10.62094	12.61781	13.87262	15.87575	17.8750	19.87031	23.86094
Length of Full Thread	Inch L_2 *	1.2500	1.3000	1.4063	1.5125	1.7125	1.9250	2.1250	2.2500	2.4500	2.6500	2.8500	3.2500
	Threads	10.00	10.40	11.25	12.10	13.70	15.40	17.00	18.90	19.60	21.20	22.80	26.00
	E_2 *	3.91562	4.41562	5.47862	6.54062	8.54062	10.66562	12.66562	13.91562	15.91562	17.91562	19.91562	23.91562
Wrench Makeup Length, Internal Thread	Inch L_3 *	0.2500	0.2500	0.2500	0.2500	0.2500	0.2500	0.2500	0.2500	0.2500	0.2500	0.2500	0.2500
	Threads	2	2	2	2	2	2	2	2	2	2	2	2
	E_3 *	3.82188	4.31875	5.37511	6.43047	8.41797	10.52969	12.51719	13.75938	15.74688	17.73438	19.72188	23.69688
Vanish Threads	Inch	0.4337	0.4337	0.4337	0.4337	0.4337	0.4337	0.4337	0.4337	0.4337	0.4337	0.4337	0.4337
	Threads	3.47	3.47	3.47	3.47	3.47	3.47	3.47	3.47	3.47	3.47	3.47	3.47
Overall Length Ext. Threads	L_4 *	1.6837	1.7337	1.8400	1.9462	2.1462	2.3587	2.5587	2.6837	2.8837	3.0837	3.2837	3.6837
Nominal Perfect Ext Thread	Length L_5 *	1.000	1.0500	1.1563	1.2625	1.4625	1.6750	1.8750	2.0000	2.2000	2.4000	2.6000	3.0000
	Dia.	3.90000	4.40000	5.46300	6.52500	8.52500	10.65000	12.65000	13.90000	15.90000	17.90000	19.90000	23.90000
Height of Thread		0.1000	0.1000	0.1000	0.1000	0.1000	0.1000	0.1000	0.1000	0.1000	0.1000	0.1000	0.1000
Increase in Dia. per Thread		0.00781	0.00781	0.00781	0.00781	0.00781	0.00781	0.00781	0.00781	0.00781	0.00781	0.00781	0.00781
Basic Minor Dia. at Small End		3.7375	4.2344	5.2907	6.3461	8.3336	10.4453	12.4328	13.6750	15.6625	17.6500	19.6375	23.6125

Notes: E_0 = The Pitch Diameter at the small end of the external thread = $D - (0.05D + 1.1)P$. E_1 = The Pitch Diameter at the large end of the Internal thread = $E_0 + 0.0625 L_1$. E_2 = Pitch Diameter at large end of the external tread = $E_1 + 0.0625 L_2$. E_3 = Pitch Diameter at small end of internal thread = $E_0 - 0.1875P$. L_1 = Normal engagement by hand between external and internal threads. L_2 = Effective length of external thread = $(0.80D + 6.8)P$. L_3 = Normal wrench makeup. Note that MIL-P-7105B differs from this chart by specifying the wrench makeup to be 3 threads for all sizes up to 3". L_4 = Overall length of the external thread. $L_1 + L_3$ = Overall length of the internal thread. L_5 = Nominal length of complete external thread = $L_2 - 2P$.

Table 38. NPSC Internal Pipe Coupling Thread Dimensions.

Nominal Pipe Size	TPI	Pipe O.D.	Minor Dia. (Min.)	Pitch Diameter Min.	Pitch Diameter Max.
1/8	27	0.405	0.340	0.3701	0.3771
1/4	18	0.540	0.442	0.4864	0.4968
3/8	18	0.675	0.577	0.6218	0.6322
1/2	14	0.840	0.715	0.7717	0.7851
3/4	14	1.050	0.925	0.9822	0.9956
1	11 1/2	1.315	1.161	1.2305	1.2468
1 1/4	11 1/2	1.660	1.506	1.5752	1.5915
1 1/2	11 1/2	1.900	1.745	1.8142	1.8305
2	11 1/2	2.375	2.219	2.2881	2.3044
2 1/2	8	2.875	2.650	2.7504	2.7739
3	8	3.500	3.277	3.3768	3.4002
3 1/2	8	4.000	3.777	3.8771	3.9005
4	8	4.500	4.275	4.3754	4.3988

All dimensions in inches.

Table 39. Dimensions for NPSM Threads for Free-Fitting Mechanical Joints.

Nominal Size	TPI	Pipe O.D.	Allowance	Class 2A External Thread Major Dia. Max.	Min.	Pitch Dia. Max.	Min.	Class 2B Internal Thread Minor Dia. Min.	Max.	Pitch Dia. Min.	Max.
1/8	27	0.405	0.0011	0.397	0.039	0.3725	0.3689	0.358	0.364	0.3736	0.3783
1/4	18	0.540	0.0013	0.526	0.517	0.4903	0.4859	0.468	0.481	0.4916	0.4974
3/8	18	0.675	0.0014	0.662	0.653	0.6256	0.6211	0.603	0.612	0.6270	0.6329
1/2	14	0.840	0.0015	0.823	0.813	0.7769	0.7718	0.747	0.759	0.7784	0.7851
3/4	14	1.050	0.0016	1.034	1.024	0.9873	0.9820	0.958	0.970	0.9889	0.9958
1	11.5	1.315	0.0017	1.293	1.281	1.2369	1.2311	1.201	1.211	1.2386	1.2462
1 1/4	11.5	1.660	0.0018	1.638	1.626	1.5816	1.5756	1.546	1.555	1.5834	1.5912
1 1/2	11.5	1.900	0.0018	1.877	1.865	1.8205	1.8144	1.785	1.794	1.8223	1.8302
2	11.5	2.375	0.0019	2.351	2.339	2.9444	2.2882	2.259	2.268	2.2963	2.3044
2 1/2	8	2.875	0.0022	2.841	2.826	2.7600	2.7526	2.708	2.727	2.7622	2.7720
3	8	3.500	0.0023	3.467	3.452	3.3862	3.3786	3.334	3.353	3.3885	3.3984
3 1/2	8	4.000	0.0023	3.968	3.953	3.8865	3.8788	3.835	3.848	3.8888	3.8988
4	8	4.500	0.0023	4.466	4.451	4.3848	4.3711	4.333	4.346	4.3871	4.3971
5	8	5.563	0.0024	5.528	5.513	5.4469	5.4390	5.395	5.408	5.4493	5.4598
6	8	6.625	0.0024	6.585	6.570	6.5036	6.4955	6.452	6.464	6.5060	6.5165

All dimensions in inches.

Table 40. Formulas for Determining Root and Crest Truncation on NPTF Threads.

TPI	Pitch P	Truncation				Equivalent Width of Flat			
		Crest		Root		Crest		Root	
		Max.	Min.	Max.	Min.	Max.	Min.	Max.	Min.
27	0.03704	0.094 P	0.047 P	0.140 P	0.094 P	0.108 P	0.054 P	0.162 P	0.108 P
18	0.05556	0.078 P	0.047 P	0.109 P	0.078 P	0.090 P	0.054 P	0.126 P	0.090 P
14	0.07143	0.060 P	0.036 P	0.085 P	0.060 P	0.070 P	0.042 P	0.098 P	0.070 P
11.5	0.08696	0.060 P	0.040 P	0.090 P	0.060 P	0.069 P	0.046 P	0.103 P	0.069 P
8	0.12500	0.055 P	0.042 P	0.076 P	0.055 P	0.064 P	0.048 P	0.088 P	0.064 P

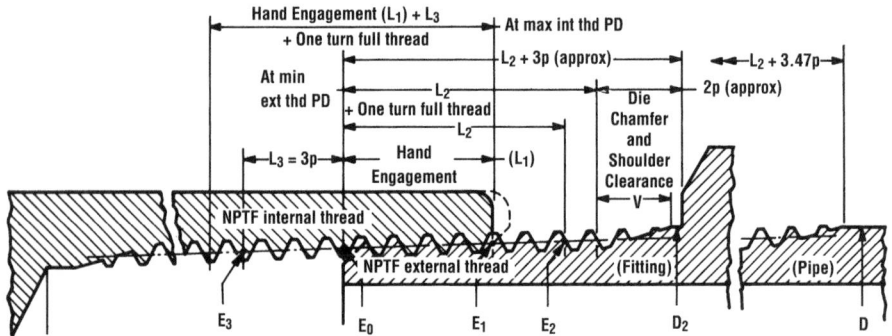

Figure 17. NPTF assembly. Refer to Table 38 for dimensions.

NPTF threads. The NPSF, or Standard Fuel Internal Straight Pipe Thread, is internal only. It is designed to be used with external NPTF threads and its major advantage is economy of manufacture because it is a straight rather than tapered form. It is not as efficient as the internal NPTF thread, but the form provides an effective seal by allowing for root and crest interference when assembled. *Figure 17* shows the NPTF form.

Although not technically a pipe thread, the NPSH and NH (National Hose coupling Threads and National Fire Hose Coupling Threads) form is used on threaded sections of hose couplings, valves, nozzles, and all other fittings used in direct connection with hose intended for fire protection or for domestic, industrial, and general service. NH threads are standard hose threads ranging in size from 0.75" to 6" created by cutting or rolling. NPSH threads are straight hose couplings that can be mated to NPT threads using a gasket for sealing. They are available in sizes from 0.50" to 4". Another form of the thread, the NHR, is for garden hose connections and is used on thin-walled materials that are formed to profile. The dimensions for these threads are given in FED-STD-H28/10. This is a 60° included angle thread with a flat at the crest and root equal to 0.125P. Thread height is equal to 0.649519P. Tolerances are as follows. The pitch diameter tolerance provided for a mating nipple (external thread) and coupling (internal thread) are the same. Pitch diameter tolerances include lead and angle variations. The tolerance on the major diameter is twice that of the pitch diameter. The tolerance on the minor diameter of the external thread is equal to the tolerance on the pitch diameter plus two-ninths of the basic thread height. The minimum minor diameter of the external thread is such as to result in a flat equal to one-third of the basic flat at the root when the pitch diameter of the external thread

Table 41. NPTF Dryseal Pipe Threads Basic Dimensions. *(Source, Landis Threading Systems.)*

Nominal Pipe Size	1/16	1/8	1/4	3/8	1/2	3/4	1	1 1/4	1 1/2	2	2 1/2	3
Pipe O.D.	0.3125	0.405	0.540	0.675	0.840	1.050	1.315	1.660	1.900	2.375	2.875	3.500
TPI	27	27	18	18	14	14	11 1/2	11 1/2	11 1/2	11 1/2	8	8
Pitch P	0.03704	0.03704	0.0556	0.0556	0.07143	0.07143	0.08696	0.08696	0.08696	0.08696	0.12500	0.12500
E_0 *	0.27118	0.36351	0.47739	0.61201	0.75843	0.96768	1.21363	1.55713	1.79606	2.26902	2.71953	3.34062
E_1	0.28118	0.37360	0.49160	0.62701	0.77843	0.98887	1.23863	1.58338	1.82234	2.29627	2.76216	3.38850
E_2 *	0.28750	0.38000	0.50250	0.63750	0.79179	1.00179	1.25630	1.60130	1.84130	2.31630	2.79062	3.41562
E_3 *	0.26424	0.35656	0.46697	0.60160	0.74504	0.95429	1.19733	1.54083	1.77978	2.25272	2.70391	3.32500
Hand Tight Engagement — Inch L_1	0.1600	0.1615	0.2278	0.2400	0.3200	0.3390	0.4000	0.4200	0.4200	0.4360	0.6820	0.7660
Hand Tight Engagement — Threads	4.32	4.36	4.10	4.32	4.48	4.75	4.60	4.83	4.83	5.01	5.46	6.13
Length of Full Thread — Inch L_2	0.2611	0.2639	0.4018	0.4078	0.5337	0.5457	0.6828	0.7068	0.7235	0.7565	1.1375	1.2000
Length of Full Thread — Threads	7.05	7.12	7.23	7.34	7.47	7.64	7.85	8.13	8.32	8.70	9.10	9.60
Vanish Threads — Inch	0.1139	0.1112	0.1607	0.1547	0.2163	0.2043	0.2547	0.2620	0.2765	0.2747	0.3781	0.3781
Vanish Threads — Threads	3.075	3.072	2.892	2.791	3.028	2.860	2.929	3.013	3.180	3.159	3.025	3.025
Shoulder Length	0.3750	0.3750	0.5625	0.5625	0.7500	0.7500	0.9375	0.9688	1.0000	1.0312	1.5156	1.5781
External Thread for Draw — Inch	0.1011	0.1024	0.1740	0.1678	0.2137	0.2067	0.2828	0.2868	0.3035	0.3205	0.4555	0.4340
External Thread for Draw — Threads	2.73	2.76	3.13	3.02	2.99	2.89	3.25	3.30	3.49	3.69	3.64	3.47
Basic Internal Full Thread Length — Inch	0.2711	0.2726	0.3945	0.4067	0.5343	0.5533	0.6609	0.6809	0.6809	0.6969	1.0570	1.1410
Basic Internal Full Thread Length — Threads	7.32	7.36	7.10	7.32	7.48	7.75	7.60	7.83	7.83	8.01	8.46	9.13
Outside Dia. of Fitting	0.315	0.407	0.546	0.681	0.850	1.060	1.327	1.672	1.912	2.387	2.893	3.518
Outside Dia. of Pipe	0.3125	0.405	0.540	0.675	0.840	1.050	1.315	1.660	1.900	2.375	2.875	3.500

*Notes: E_0 = The Pitch Diameter at the small end of the external thread = $D - (0.05D = 1.037P$. E_1 = The Pitch Diameter at the large end of the Internal thread = $E_0 + 0.0625\,L_1$. E_2 = Pitch Diameter at small end of internal thread. $L_2 = (0.8\,D + 5.8)P$. E_3 = Pitch Diameter at large end of the external thread.

is at its minimum value. The maximum minor diameter is basic. The tolerance on the major diameter of the internal thread is equal to the tolerance of the pitch diameter plus two-ninths of the basic thread height. The minimum major diameter of the internal thread is such as to result in a basic flat of $P/8$ $(0.125P)$ when the pitch diameter of the internal thread is at its minimum value. The maximum major diameter of the internal thread is that corresponding to a flat equal to one-third of the basic flat. The tolerance on the minor diameter of the internal thread is twice the tolerance on the pitch diameter of the internal thread. The minimum minor diameter of the internal thread is such as to result in a basic flat of $P/8$ at the crest when the pitch diameter of the internal thread is at its minimum value. See **Table 42** basic dimensions and *Figure 18* for the thread form.

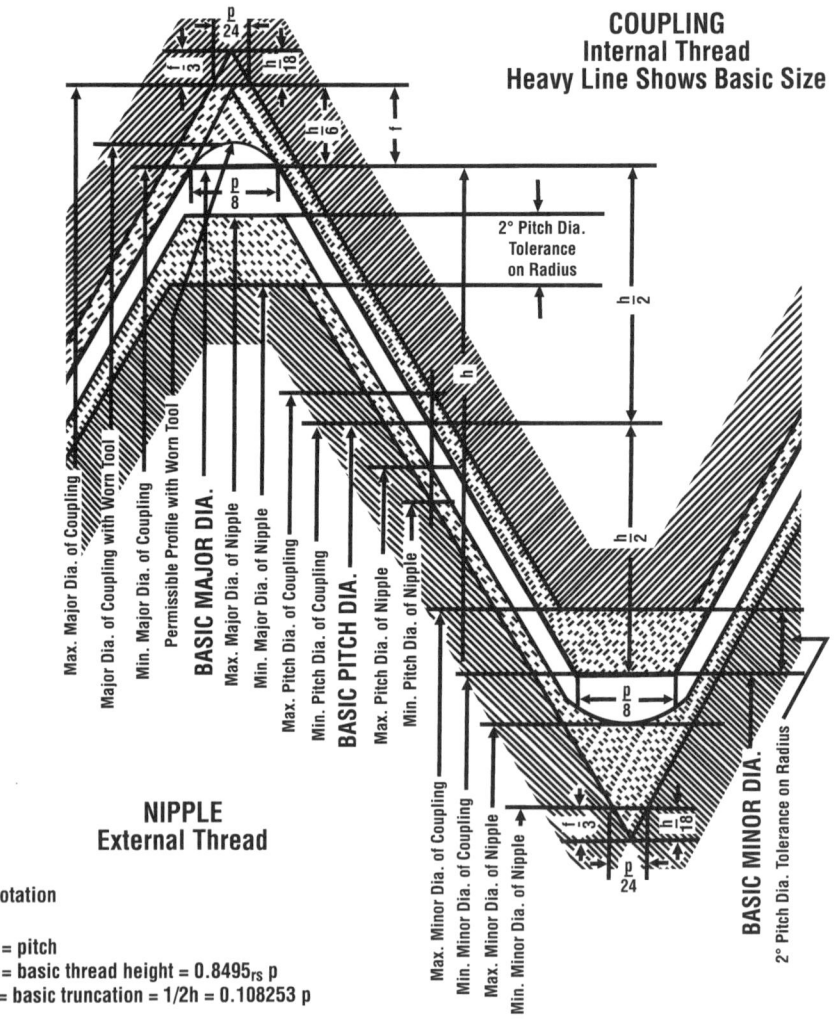

Figure 18. NPSH and NH thread form.

Table 42. Basic Dimensions of American National Hose Nipple and Coupling Threads, NPSH and NH.

Nominal Size*	TPI	Symbol	Service	Pitch	Allowance	External Thread Max.			Internal Thread Max.		
						Major Dia.	Pitch Dia.	Minor Dia.	Minor Dia.	Pitch Dia.	Major Dia.
1/2, 5/8, 3/4	11.5	.75-11.5 NHR	Garden Hose	0.08696	0.0100	1.0625	1.0060	0.9495	0.9595	1.0160	1.0725
3/4	8	.75-8 NH	Chemical	0.1250	0.0120	1.3750	1.2938	1.2126	1.2246	1.3058	1.3870
1 1/2	9	1.5-9 NH	Fire Hose	0.1111	0.0120	1.9900	1.9178	1.8457	1.8577	1.9298	2.0020
1/2	14	.5-14 NPSH	Steam, air, water and all other connections with standard pipe threads	0.07143	0.0075	0.8248	0.7784	0.7320	0.7395	0.7859	0.8323
3/4	14	.75-14 NSPH		0.07143	0.0075	1.0353	0.9889	0.9425	0.9500	0.9964	1.0428
1	11.5	1-11.5 NSPH		0.08696	0.0100	1.2951	1.2386	1.1821	1.1921	1.2486	1.3051
1 1/4	11.5	1.25-11.5 NSPH		0.08696	0.0100	1.6399	1.5834	1.5629	1.5369	1.5934	1.6499
1 1/2	11.5	1.5-11.5 NSPH		0.08696	0.0100	1.8788	1.8223	1.7658	1.7758	1.8323	1.8888
2	11.5	2-11 NSPH		0.08696	0.0100	2.3528	2.2963	2.2398	2.2498	2.3063	2.3628
2 1/2	7.5	2.5-7.5 NH		0.13333	0.0150	3.0686	2.9820	2.8954	2.9104	2.9970	3.0836
3	6	3-6 NH		0.16667	0.0150	3.6239	3.5156	3.4073	3.4223	3.5306	3.6389
3 1/2	6	3.5-6 NH		0.16667	0.0200	4.2439	4.1356	4.0273	4.0473	4.1556	4.2639
4	6	4-6 NH (SPL) *	Fire Hose	0.16667	0.0250	4.9082	4.7999	4.6916	4.7117	4.8200	4.9283
4	4	4-4 NH		0.2500	0.0250	5.0109	4.8485	4.6861	4.7111	4.8735	5.0359
4 1/2	4	4.5-4 NH		0.2500	0.0250	5.7609	5.5985	5.4361	5.4611	5.6235	5.7859
5	4	5-4 NH		0.2500	0.0250	6.2600	6.0976	5.9352	5.9602	6.1226	6.2850
6	4	6-4 NH		0.2500	0.0250	7.0250	6.8626	6.7002	6.7252	6.8876	7.0500

* 4-6 NH (SPL) thread is used extensively aboard ship by the Navy Dept. All dimensions in inches.

Tightening and Tensioning Threaded Fasteners

Bolted Joints

Axial Loaded Joints

(Federal Standard MIL-HDBK-60 contains information on calculating tension and torque in threaded fasteners. Much of the material in this section was derived from this standard and from FED-STD-H28/2B.) Applied preloads (or tension stress) for bolts, screws, and studs in *axial loaded* joints should be sufficient to assure that joint members remain in contact and in compression. Loss of joint compression can result in 1) leakage of pressurized fluids past compression gaskets, 2) loosening of the fastener, and 3) reduced fastener fatigue life. *Figure 1* shows that for a joint with no preload, the axial load on the joint members is transmitted directly to the fastener and the bolt load equals the joint load. In multiple bolted joints, each fastener will share a proportion of the joint load (shown as line OAB in Figure 1). However, when a preload (P_{B1} in the figure) is applied, the axial joint load is partially absorbed by reduction of the compression of the joint members. Therefore, bolt load increase is lower than in the nonpreloaded joint (shown as line $P_{B1}A$, with point A representing the separation of joint members with resulting loss of compression so that further loading beyond point A is direct along line AB). When an axial joint load in the range between lines P_a and P_b is applied to a nonpreloaded joint, the resulting bolt load will vary between P_{Ba} and P_{Bb}. When preload P_{B1} is applied, the resultant bolt load varies between $P_{Ba'}$ and $P_{Bb'}$. This condition considerably reduces cyclic load and results in enhanced fatigue life for the fastener. Preloading can increase fastener life from only a few hundred cycles to an essentially unlimited fatigue life. One laboratory study tested a bolt tightened to 1,420 pounds of tension (29,000 PSI of residual stress) by subjecting it to an alternating (tension and compression) load of 9,000 pounds. It experienced failure after 6,000 cycles. An identical bolt, tightened to 8,420 pounds of tension (80,000 PSI of residual stress), was cycled at the same 9,000 pound load for 4,650,000 cycles before failure.

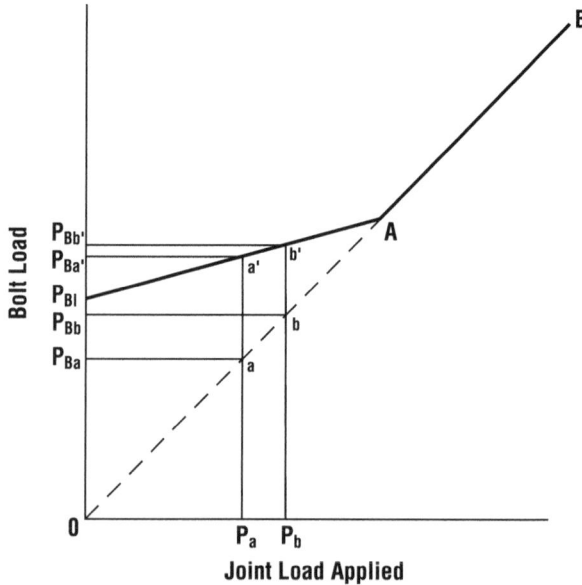

Figure 1. Bolt load in joint with applied axial load.

Shear Loaded Joints

Fasteners in *shear loaded* joints can be separated into two categories. First, in the case where the joint members are designed to slide, joint members transmit shear loads to the fasteners in the joint. The preload must be sufficient to keep the joint members in contact. Secondly, in the case where the joint members do not slide and there is no relative motion between the joint members, shear loads are transmitted within the joint by frictional force. Therefore, the preload must be sufficient to provide a frictional force greater than the applied shear forces. In both instances, shear stresses induced by preloading must be considered in the bolt design. The U.S. Navy Bureau of Aeronautics dictates the following fit tolerances for bolts loaded in shear.

Fit	Tolerance	Application
Class 1	± 0.001"	For use when one or two bolts are subjected to a reversal of loads in a critical joint assembly. Holes are to be reamed.
Class 2	± 0.002"	For use when there are more than two through primary structures, where vibration and reversal of loads are expected, or where joint rigidity is required. Holes are to be reamed.
Class 3	± 0.005 to ± 0.010"	For use where a large number of bolts not subjected to load reversals are used.

Combined Shear and Axial Loaded Joints

These joints are subjected to the sum of forces of axial loaded joints and shear loaded joints, and must be analyzed carefully to assure reliability.

Determining Preloads

Fasteners are normally designed for maximum utilization of material, and a maximum safe preload can usually be determined. However, if for standardization, logistical, or other reasons a lower strength fastener is replaced by one of higher strength, the preload values should not be increased. Preload should be based on calculated joint requirements. For joints subjected to cyclic loading and for joints using high strength fasteners where yield strain is near the point of fracture, preloads should be maintained below the yield point. Maximum preloads specified are generally within the following ranges.

50 – 80% of the maximum tensile ultimate strength,
75 – 90% of the minimum tensile yield strength or proof load, or
100% of the observed proportional limit or onset of yield.

For joints with primarily static loading using fasteners of ductile material, where yield strain is relatively far from the strain at fracture, preloads above yield are often used. Preload should not exceed the strength of the components (fastener head, juncture of head and shank, etc.), and there must be sufficient thread engagement to prevent stripping.

Torsion Load

When preloads are applied by turning nuts, bolts, or screws, a torsion load component is added to the desired axial bolt load. In fact, the highest load imposed on a fastener will be experienced during tightening. The combined loading increases the tensile stress on the fastener, a consideration often ignored by designers on the assumption that the majority of the torsional force rapidly dissipates after the driving force is removed. This may be true in joints tensioned near or beyond fastener yield, but for critical joints where fastener tension must be maintained below yield it is important to adjust the axial tension load requirements to include the effects of preload tension. For this adjustment, the combined tensile stress, f_{tc} (also known as "von Mises stress"), may be calculated from the following equation. (See **Table 1** for explanation of symbols for all equations found in this section.)

Table 1. Symbols Used in Equations in this Section.

A_B = Effective fastener* cross sectional area in in.^{2} or mm^2.
A_G = Gasket area in in.^{2} or mm^2.
A_J = Effective compressed area in the joint in in.^{2} or mm^2.
A_s = Tensile stress area in in.^{2} or mm^2.

d = External thread nominal diameter.
d_2 = External thread pitch diameter in inches.

E_B = Fastener* modulus of elasticity in PSI or MPa.
E_G = Gasket modulus of elasticity in PSI or MPa.
E_J = Joint material modulus of elasticity in PSI or MPa.
e = Coefficient of linear thermal expansion (in./in. % °F).

F = Maximum load stress in pounds or Newtons.
f_s = Shear stress caused by torsional load application in PSI or MPa.
f_t = Axial tensile stress applied in PSI or MPa.

L = Lead of thread helix in inches or millimeters.
L_B = Effective fastener* length in inches or millimeters.

n = Number of threads per inch.

P = Axial load applied to the joint in pounds or Newtons.
P (italicized) = Thread pitch in inch dimensions.
P_B = Axial fastener* load in pounds or Newtons.
P_G = Gasket compression load in pounds or Newtons.
P_J = Joint compression load in pounds or Newtons.

t = temperature in °F.
t_G = Gasket thickness in inches or millimeters.
t_0 = Operating temperature in °F.
t_J = Effective joint thickness in inches or millimeters.

Y = Proof load (stress) in PSI or MPa.

δ_B = Fastener* elongation in inches or millimeters.
δ_G = Gasket compression in inches or millimeters.
δ_J = Joint material compression in inches or millimeters.

μ = Coefficient of friction in threads.

Δ = Change caused by application of load "P."

θ = Amount of turn of a nut in degrees.

T (subscript) = Indicates applicability to total joint including gasket.
1 (subscript) = Indicates condition with applied joint fastener* preload only.

Relevant Conversions
1 PSI = 0.006894757 N/mm^2
1 ksi = 6.894757 N/mm^2
1 N/mm^2 = 145.0377 PSI
1 N/mm^2 = 0.1450377 ksi
1 MPa = 145.0377 PSI
1 N/mm^2 = 1 MPa
1 pound (force) = 4.448222 Newton
1 Newton = 0.2248089 pound (force)
1 in.^{2} = 645.16 mm^2
1 mm^2 = 0.001550003 in.^{2}

*The term "Fastener" as used in this section is interchangeable with threaded bolts, screws, and studs.

$$f_{tc} = \sqrt{f_t^2 + 3f_s^2}$$

A simple calculation for the amount of tensile stress on a fastener takes into consideration the force acting upon the fastener and its diameter.

$$f_t = F \div A_s$$

Maximum loading stresses for threaded fasteners are determined by their known proof yield capability and tensile stress area. Maximum load in pounds can be determined by the equation

$$F = Y \times A_s.$$

For single start Unified Threads and M Profile Metric threads, the following formula can be used to determine combined tensile stress. Tensile stress increase due to preload tension increases in significance as friction coefficients increase. (See **Table 2** for coefficient of friction values to be used in the following equations.)

$$f_{tc} = f_t \times \sqrt{1 + (3 \times [\{1.96 + 2.31\} \div \{1 - 0.325P \div d_2\} - 1.96]^2)}$$

For single start UNJ with a thread stress diameter equal to the pitch diameter, the following equation applies.

$$f_{tc} = f_t \times \sqrt{1 + (3 \times [\{0.637P \div d_2\} + 2.31 \ ]^2)}$$

Table 2. Coefficient of Friction (μ)Values for Threaded Fasteners.

Bolt/Nut Material	Lubricant	$\mu \pm 20\%$
Steel, Carbon and Low Alloy	Graphite in Petrolatum or Oil	0.07
Steel, Carbon and Low Alloy	Molybdenum Disulfide Grease	0.11
Steel, Carbon and Low Alloy	Machine Oil	0.15
Steel, Cadmium Plated	None	0.12
Steel, Zinc Plated	None	0.17
Steel/Bronze	None	0.15
Steel, Corrosion Resistant Steel or Nickel Base Alloys. Silver Plated Materials	None	0.14
Titanium/Steel	Graphite in Petrolatum	0.08
Titanium	Molybdenum Disulfide Grease	0.10

As a further example of the relationship of torque to lubrication, *Figure 2* is a chart based on the results obtained with a 9/16–18 steel bolt screwed into an aluminum casting. Without lubrication, a torque of 115 to 125 lb./in. was required to develop tension of 800 to 1,400 lb. When lubricated, torque values were just 75 to 85 lb./in. to develop tension in the range of 1,000 to 1,250 pounds. Torque values are affected in various ways by different types of lubricants.

Calculating Tensile Stress Area

Tests referenced in FED-STD-H28/2B have shown that externally threaded parts fail in tension at loads corresponding to those of unthreaded parts with diameters midway between the pitch diameter and minor diameter. The following formulas for calculation of tensile stress area provide stress area based upon a diameter approximately midway between the minimum pitch diameter and the minimum minor diameter. These formulas have been successfully applied to steel and other metals with ultimate strengths up to 180,000 PSI. In these formulas, A_s is tensile stress area, d_2 is the pitch diameter,

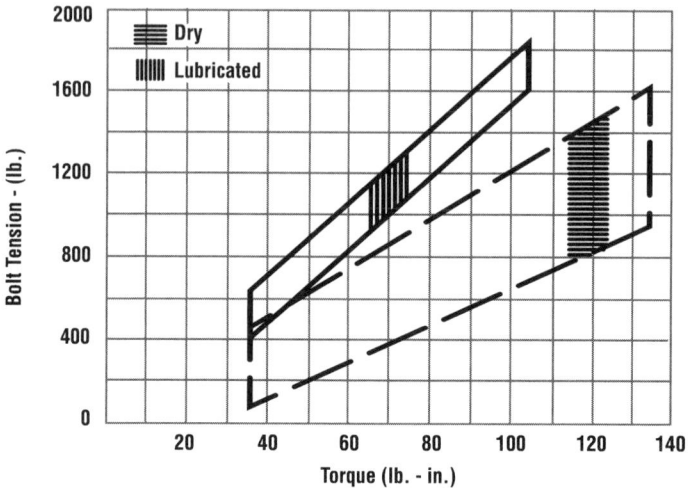

Figure 2. The effect of lubrication on the torque-tension relationship.(Source, Skidmore-Wilhiem Mfg. Co.)

d is major diameter, H is the height of the fundamental triangle, and n is the number of threads per inch $(1/P)$.

$$A_s = \pi \left([d_2 \div 2] - [3H \div 16]\right)^2$$
$$\text{or } A_s = 0.7854 \times (d - [0.9743 \div n])^2$$

The grade designation of a bolt or nut can be used as a guide in the selection. Federal Standard FF-B-584F gives mechanical and chemical properties of carbon and alloy steel bolts, Grades 1 – 8, as provided in **Table 3**. Ultimate tensile strengths for standard Unified threads are given in **Table 4**, and **Table 5** gives ultimate tensile strengths for UNJ 3A threads. Ultimate shear strengths for standard Unified threads are shown in **Table 6**. **Table 7** gives mechanical properties for bolts, studs, and hex cap screws for fasteners made of corrosion resistant steels, aluminum, and copper and nickel alloys. In *Figure 3*, tensile strength/tensile load relationships for various size fasteners are shown. **Table 8** gives mechanical properties of various nut grades. **Table 9**, provided by American Fastener Technologies Corp., contains ASTM, SAE, and ISO markings and mechanical properties for steel fasteners. Tensile stress areas of Unified Thread fasteners are given in **Table 10**. The three most commonly used fastener grades are Grade 2, Grade 5, and Grade 8. Grade 2 fasteners are made of mild low or medium carbon steel, Grade 5 fasteners are made of medium hard steel, and Grade 8 fasteners are made from medium carbon alloy steel, quenched and tempered. ASTM standards ASTM 193 and ASTM 194 should be consulted for full material and chemical properties of hex head cap screws, studs, bolts, and nuts made of steel and stainless steel. For further information on grades, see SAE standard SAE J429.

Tensile stress areas for coarse M Profile Metric threads can be found in **Table 11**, and **Table 12** gives mechanical and chemical properties for standard classes of M Profile Metric bolts, screws, and studs. Metric Class (or "Property Class") ratings for externally threaded fasteners contain two numbers separated by a period. The first number, which may be one or two digits, represents 1/100 of the nominal tensile strength of the fastener, expressed in Newtons/mm^2 (one N/mm^2 = 145.0377 PSI, and one PSI = 0.006894757 N/mm^2). The number following the period is equal to ten times the ratio between the fastener's nominal yield stress and its nominal tensile strength. Multiplying the two numbers provides a value roughly equivalent to 1/10 of the nominal yield strength in

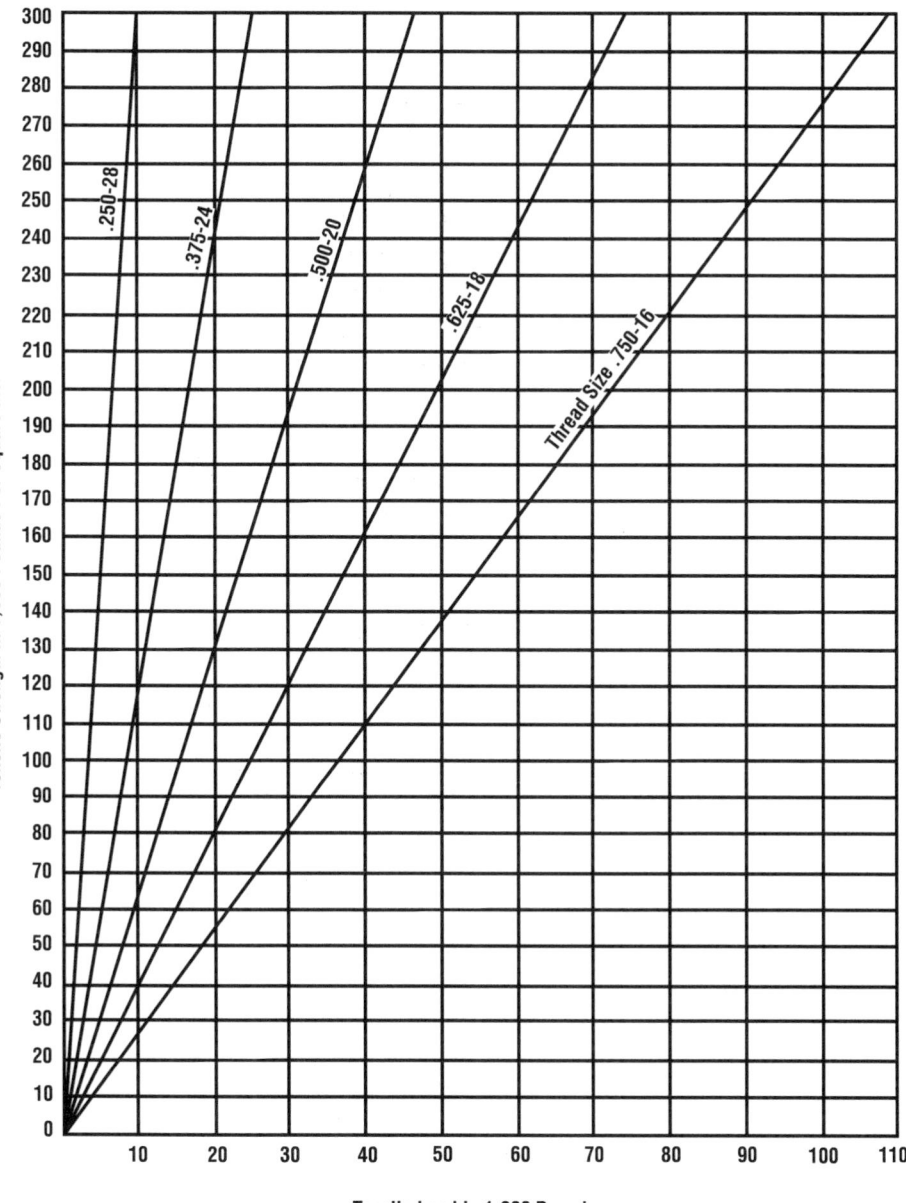

Figure 3. Tensile strength versus tensile load for various threaded fastener sizes. (Source, MIL-STD-1251A.)

N/mm². Additional information on external Metric fastener classes and properties can be found in ISO standard ISO 898. The four most commonly used classes are as follows. Class 4.8 indicates a low or medium carbon steel fastener commonly used for machine screws, machine bolts, and similar applications. Class 8.8 fasteners are usually hex cap screws and hex socket screws made of medium hard carbon steel, quenched and tempered. Class 10.9 is usually a quenched and tempered alloy steel used for hex cap screws. Finally,

(Text continued on p. 147)

Table 3. Properties of Carbon and Alloy Steel Bolts, Screws, and Studs. (Source, FF-B-548F.)

Mechanical Properties

Grade	Nominal Dia. (inches)	Full Size Fastener		Machine Test Specimen				Core Hardness Rockwell	
		Proof Load ksi	Tensile Strength Min. ksi	Yield Strength Min. ksi	Tensile Strength Min. ksi	Elongation %	Reduction in Area %	Min.	Max.
1	1/4" - 1.5"	33	60	36	60	18	35	B70	B100
2	1/4" - 3/4"	55	74	57	74	18	35	B80	B100
	Over 3/4" to 1.5"	33	60	36	60	18	35	B70	B100
4	1/4" - 1.5"	-	115	100	115	10	35	C22	C32
5	1/4" - 1"	85	120	92	120	14	35	C25	C34
	Over 1" to 1.5"	74	105	81	105	14	35	C19	C30
5.1	No. 6" - 3/8"	85	120	-	-	-	-	C25	C40
5.2	1/4" - 1"	85	120	92	120	14	35	C26	C36
7	1/4" - 1.5"	105	133	115	133	12	35	C28	C34
8	1/4" - 1.5"	120	150	130	150	12	35	C33	C39
8.1	1/4" - 1.5"	120	150	130	150	10	35	C32	C38
8.2	1/4" - 1"	120	150	130	150	10	35	C35	C42

Chemical Properties

Grade	Material & Treatment	Element %					
		Carbon Min.	Carbon Max.	Manganese Min.	Phosphorus Max.	Sulpher Max.	Boron Min.
1	Low or Medium Carbon Steel	-	0.55	-	0.048	0.058	-
2	Low or Medium Carbon Steel	-	0.28	-	0.048	0.058	-
4	Medium Carbon Cold Drawn Steel	-	0.55	-	0.048	0.058	-
5	Medium Carbon Steel, Quenched and Tempered	0.28	0.55	-	0.048	0.058	-

(Continued)

Table 3. *(Continued)* **Properties of Carbon and Alloy Steel Bolts, Screws, and Studs.** *(Source, FF-B-548F.)*

Grade	Material & Treatment	Chemical Properties							
		Element %							
		Carbon		Manganese	Phosphorus	Sulpher	Boron		
		Min.	Max.	Min.	Max.	Max.	Min.		
5.1	Low or Medium Carbon Steel, Quenched and Tempered	0.15	0.30	-	0.048	0.058	-		
5.2	Low Carbon Martensite Steel, Fully Killed, Fine Grain, Quenched and Tempered	0.15	0.25	0.74	0.048	0.058	0.0005		
7	Medium Carbon Alloy Steel, Quenched and Tempered	0.28	0.55	-	0.040	0.045	-		
8	Medium Carbon Alloy Steel, Quenched and Tempered	0.28	0.55	-	0.040	0.045	-		
8.1	Elevated Temperature Drawn Steel-Medium Carbon Alloy or SAE 1541 (formerly SAE 1041)	0.28	0.55	-	0.048	0.058	-		
8.2	Low Carbon Martensite Steel, Fully Killed, Fine Grain, Quenched and Tempered	0.15	0.25	0.74	0.048	0.058	0.0005		

Table 4. Ultimate Tensile Strengths of Class 2A and Class 3A Unified Threaded Fasteners. *(Source, MIL-HDBK-5H.)*

Fastener Size Dia. (inch)	See Note 3	Nominal Minor Area	Tensile Stress of Fastener (ksi) Ultimate Tensile Strength in Pounds (See Notes 1, 2)						
			55	62	62.5	125	140	160	180
0.112	4-40	0.0050896	280	316	318	636	713	814	916
0.138	6-32	0.0076821	423	476	480	960	1075	1225	1380
0.164	8-32	0.012233	673	758	765	1525	1710	1955	2200
0.190	10-32	0.018074	994	1120	1130	2255	2530	2890	3250
0.250	1/4-28	0.033394	1835	2070	2085	4170	4680	5340	6010
0.312	5/16-24	0.053666	2950	3325	3350	6710	7510	8590	9660
0.375	3/8-24	0.082397	4530	5110	5150	10300	11500	13150	14800
0.438	7/16-20	0.11115	6110	6890	6950	13850	15550	17750	20000
0.500	1/2-20	0.15116	8310	9370	9450	18900	21150	24150	27200
0.562	9/16-18	0.19190	10550	11900	11950	23950	26850	30700	34500
0.625	5/8-18	0.24349	13350	15100	15200	30400	34050	38950	43800
0.750	3/4-16	0.35605	19550	22050	22250	44500	49800	57000	64100
0.875	7/8-14	0.48695	26750	30150	30400	60900	68200	77900	87700
1.000	1-12	0.63307	34800	39250	39550	79100	88600	101000	114000
1.125	1-1/8-12	0.82162	45200	50900	51400	102500	115000	131500	147500
1.250	1-1/4-12	1.0347	56900	64200	64700	129000	144500	165500	186000
1.375	1-3/8-12	1.2724	70000	78900	79500	159000	178000	203500	229000
1.500	1-1/2-12	1.5345	84400	95100	95900	191500	214500	245500	276000

1) Values shown for 0.112, 0.138, and 0.164 diameters are for 2A threads. All other values are for 3A threads.
2) Nuts and fastener heads designed to develop the ultimate tensile strength of the fastener are required to develop the tabulated tension loads.
3) Fractional equivalent or number of threads per inch.
Nominal area dimensions are in inches.

Table 5. Ultimate Tensile Strength of UNJ Unified Threaded Fasteners. *(Source, MIL-HDBK-5H.)*

| Fastener Size | | Nominal Minor Area | Tensile Stress of Fastener (ksi) | | | |
Dia. (inch)	See Note 3		160	180	220	260
			Ultimate Tensile Strength in Pounds (See Notes 1, 2)			
0.112	4-40	0.0054367	869	979	1195	1410
0.138	6-32	0.0081553	1305	1465	1790	2120
0.164	8-32	0.012848	2055	2310	2825	3340
0.190	10-32	0.018602	2975	3345	4090	4840
0.250	1/4-28	0.034241	5480	6160	7530	8900
0.312	5/16-24	0.054905	8780	9880	12050	14250
0.375	3/8-24	0.083879	13400	15100	18450	21800
0.438	7/16-20	0.11323	18100	20350	24900	29400
0.500	1/2-20	0.15358	24550	27600	33750	39900
0.562	9/16-18	0.19502	31200	35100	42900	50700
0.625	5/8-18	0.24700	39500	44500	54300	64200
0.750	3/4-16	0.36082	57700	64900	79400	93800
0.875	7/8-14	0.49327	78900	88800	108500	128000
1.000	1-12	0.64156	102500	115500	141000	166500
1.125	1-1/8-12	0.83129	133000	149500	182500	216000
1.250	1-1/4-12	1.0456	167000	188000	230000	271500
1.375	1-3/8-12	1.2844	205500	231000	282500	333500
1.500	1-1/2-12	1.5477	247500	278500	340500	402000

1) Values are for 3A threads.
2) Nuts and fastener heads designed to develop the ultimate tensile strength of the fastener are required to develop the tabulated tension loads.
3) Fractional equivalent or number and threads per inch.
4) The tension fastener allowables above are based on the maximum minor diameter thread area for UNJ threads as defined in MIL-S-8879.
Maximum minor area is given in inches.

Table 6. Ultimate Shear Strength of Unified Threaded Fasteners. (Source, MIL-HDBK-5H.)

Fastener Dia. (inch)	Basic Shank Area	Shear Stress of Fastener (ksi)							
		Ultimate Single Shear Strength, in Pounds							
		35	38	75	90	108	125	145	156
0.112	0.0098520	345	374	739	887	1060	1230	1425	1535
0.125	0.012272	430	466	920	1105	1325	1530	1775	1910
0.138	0.014957	523	568	1120	1345	1615	1870	2165	2330
0.156	0.019175	671	729	1435	1725	2070	2395	2780	2990
0.164	0.021124	739	803	1580	1900	2280	2640	3060	3295
0.188	0.027612	966	1045	2070	2485	2980	3450	4005	4310
0.190	0.028353	992	1075	2125	2550	3060	3540	4110	4420
0.216	0.036644	1280	1390	2745	3295	3955	4580	5315	5720
0.219	0.037582	1315	1425	2815	3380	4060	4700	5445	5860
0.250	0.049087	1715	1865	3680	4420	5300	6140	7115	7660
0.312	0.076699	2680	2915	5750	6900	8280	9590	11100	11950
0.375	0.11045	3865	4200	8280	9935	11900	13800	16000	17200
0.438	0.15033	5260	5710	11250	13500	16200	18750	21750	23450
0.500	0.19635	6870	7460	14700	17650	21200	24500	28450	30600
0.562	0.24850	8700	9440	18600	22350	26800	31050	36000	38750
0.625	0.30680	10700	11650	23000	27600	33100	38350	44500	47900
0.750	0.44179	15450	16750	33100	39750	47700	55200	64000	68900
0.875	0.60132	21050	22850	45100	54100	64900	75200	87200	93800
1.000	0.78540	27450	29850	58900	70700	84800	98200	113500	122500
1.125	0.99402	34750	37750	74600	89500	107000	124000	144000	155000
1.250	1.2272	43000	46600	92000	110000	132500	153000	177500	191000
1.375	1.4849	52000	56400	111000	133500	160000	185500	215000	231500
1.500	1.7671	61800	67100	132500	159000	190500	220500	256000	275500

Basic shank area dimensions are in inches.

Table 7. Mechanical Properties for Stainless Steel and Nonferrous Fasteners.

| Grade | Description & Treatment (See Note 1) | Bolts, Screws & Studs | | | | | | Nuts | |
| | | Full Size Fastener | | Machine Test Specimen | | | | | |
		Yield Strength Min. ksi	Tensile Strength Min. ksi	Yield Strength Min. ksi	Tensile Strength Min. ksi	Elongation Min. %	Hardness Rockwell	Proof Load Stress	Hardness Rockwell
303A	Austenitic Stainless Steel, SA	30	75	30	75	20	B75	75	B75
304-A	Austenitic Stainless Steel, SA	30	75	30	75	20	B75	75	B75
304	Austenitic Stainless Steel, CW	50	90	45	85	20	B85	90	B85
304-SH	Austenitic Stainless Steel, SH	See Note 2	See Note 2	See Note 2	See Note 2	15	C25	See Note 2	C20
305-A	Austenitic Stainless Steel, SA	30	75	30	75	20	B70	75	B70
305	Austenitic Stainless Steel, CW	50	90	45	85	20	B85	90	B85
305-SH	Austenitic Stainless Steel, SH	See Note 2	See Note 2	See Note 2	See Note 2	15	C25	See Note 2	C20
316-A	Austenitic Stainless Steel, SA	30	75	30	75	20	B70	75	B70
316	Austenitic Stainless Steel, CW	50	90	45	85	20	B85	90	B85
316-SH	Austenitic Stainless Steel, SH	See Note 2	See Note 2	See Note 2	See Note 2	15	C25	See Note 2	C20
XM7-A	Austenitic Stainless Steel, SA	30	75	30	75	20	B70	75	B70
XM7	Austenitic Stainless Steel, CW	50	90	45	85	20	B85	90	B85

(Continued)

Table 7. *(Continued)* **Mechanical Properties for Stainless Steel and Nonferrous Fasteners.**

| | | Bolts, Screws & Studs | | | | | | | Nuts | |
| | | Full Size Fastener | | Machine Test Specimen | | | | | | |
Grade	Description & Treatment (See Note 1)	Yield Strength Min. ksi	Tensile Strength Min. ksi	Yield Strength Min. ksi	Tensile Strength Min. ksi	Elongation Min. %	Hardness Rockwell	Proof Load Stress	Hardness Rockwell
384-A	Austenitic Stainless Steel, SA	30	75	30	75	20	B70	75	B70
384	Austenitic Stainless Steel, CW	50	90	45	85	20	B85	90	B85
410-H	Martensitic Stainless Steel, HT	95	125	95	125	20	C22	125	C22
410-HT	Martensitic Stainless Steel, HT	135	180	135	180	12	C36	180	C36
416-H	Martensitic Stainless Steel, HT	95	125	95	125	20	C22	125	C22
416-HT	Martensitic Stainless Steel, HT	135	180	135	180	12	C36	180	C36
430	Ferritic Stainless Steel	40	70	40	70	20	B75	70	B75
464-HF	Naval Brass	15	52	14	50	25	B56	52	B56
464	Naval Brass	27	60	25	57	25	B65	60	B65
462	Naval Brass	27	52	24	50	20	B65	52	B65
642	Aluminum Bronze	35	72	35	72	15	B75	72	B75
630	Aluminum Bronze	50	105	50	105	10	B90	105	B90
614	Aluminum Bronze	40	75	40	75	30	B70	75	B70
510	Phosphor Bronze	35	60	35	60	15	B60	60	B60
675	Manganese Bronze	22	55	22	55	20	B60	55	B60

(Continued)

Table 7. *(Continued)* **Mechanical Properties for Stainless Steel and Nonferrous Fasteners.**

| Grade | Description & Treatment (See Note 1) | Bolts, Screws & Studs | | | | | | Nuts | |
| | | Full Size Fastener | | Machine Test Specimen | | | | | |
		Yield Strength Min. ksi	Tensile Strength Min. ksi	Yield Strength Min. ksi	Tensile Strength Min. ksi	Elongation Min. %	Hardness Rockwell	Proof Load Stress	Hardness Rockwell
655-HF	Silicon Bronze	20	52	18.5	50	20	B60	52	B60
655	Silicon Bronze	38	70	36	68	15	B75	70	B75
651	Silicon Bronze	45	75	42.5	72	8	B75	75	B75
661	Silicon Bronze	38	70	38	70	15	B75	70	B75
NICU-A-HF	Nickel-Copper Alloy A	25	70	25	70	20	B70	70	B70
NICU-A	Nickel-Copper Alloy A	40	80	40	80	20	B80	80	B80
NICU-B	Nickel-Copper Alloy B	40	80	40	80	20	B80	80	B80
NICU-K (7)	Nickel-Copper Aluminum Alloy	90	130	90	130	20	C24	130	C24
2024-T4	Aluminum Alloy	40	55	40	55	14	B70	55	B70
6061-T6	Aluminum Alloy	35	42	35	42	12	B50	42	B50

1) Abbreviations used for material treatment are: S.A. = Solution Annealed. CW = Cold Worked. SH = Strain Hardened. HT = Hardened and Tempered.
2) Austenitic stainless steel, strain hardened bolts, screws, studs, and nuts shall have the following strength per properties.

(Continued)

Table 7. *(Continued)* **Mechanical Properties for Stainless Steel and Nonferrous Fasteners.**

| Product Size | Bolts, Screws & Studs | | | | | Nuts |
| | Tested Full Size | | Machine Test Specimen | | | |
	Yield Strength Min. ksi	Tensile Strength Min. ksi	Yield Strength Min. ksi	Tensile Strength Min. ksi		Proof Load Stress ksi
To 5/8 inch	100	125	90	115		125
Over 5/8 to 1 inch	70	105	65	100		105
Over 1 to 1.5 inch	50	90	45	85		90

Source, American Fastener Technologies Corp., and MIL–S–1222H.

Table 8. Mechanical Properties for Nuts. (MIL-S-1222H.)

Material	Grade	Heat Treatment or Conditions	Nominal Size (inch)	Proof Stress for Hex Nut* (psi)	Rockwell Hardness
Carbon and Alloy Steels	2	-	1/4 and over	90,000	B80 min
	2H	Hardened and Tempered	1/4 and over	150,000	C24-38
	4, 7	Hardened and Tempered	1/4 and over	150,000	C24-38
	5	Hardened and Tempered	1/4 thru 1	120,000	C23-32
			Over 1	105,000	C19-32
	8	Hardened and Tempered	1/4 and over	150,000	C24-38
Corrosion Resistant Steels	303, 303 Se, 304, 305, 316, 321, 347, 384	Annealed	1/4 and over	75,000	B65-95
		Cold Worked	1/4 and over	90,000	B95 min
	410, 416, 416, Se, 431	Hardened	1/4 and over	125,000	C25-34
		Hardened and Tempered	1/4 and over	180,000	C38-47
	630	Annealed and Age Hardened	1/4 and over	135,000	C28-38
Aluminum Alloys	2024	T4	1/4 and over	55,000	B70 min
	6061	T6	1/4 and over	40,000	B40 min
	7075	T73	1/4 and over	52,000	B60 min
Copper Alloys	462, 464, 482	-	1/4 and over	55,000	-
	510, 544	-	1/4 and over	60,000	-
	632	-	1/4 and over	90,000	-
	655, 661	-	1/4 and over	65,000	-
	670, 675	-	1/4 and over	55,000	-
Nickel Alloys	400, 405	-	1/4 and over	80,000	B80 min
	500	Annealed and Age Hardened	1/4 and over	130,000	C24 min
Titanium Alloys	T7	Annealed	1/4 and over	120,000	C26 max
		Solution Treated and Aged	1/4 and over	140,000	C26 min

Table 9. ASTM, SAE, and ISO Grade Markings and Mechanical Properties for Steel Fasteners.
(Source, American Fasteners Technologies Corp.)

Identification Grade Mark	Specification	Fastener Description	Material	Nominal Size Range	Mechanical Properties (ksi)		
					Proof Load	Yield Strength (min)	Tensile Strength (min)
No Grade Mark	SAE J429 Grade 1	Bolts, Screws, Studs	Low or Medium Carbon Steel	1/4 thru 1 1/2	33	36	60
	ASTM A307 Grades A&B		Low Carbon Steel	1/4 thru 4	-	-	60
	SAE J429 Grade 2		Low or Medium Carbon Steel	1/4 thru 3/4	55	57	74
				3/4 thru 1 1/2	33	36	60

(Continued)

Table 9. *(Continued)* **ASTM, SAE, and ISO Grade Markings and Mechanical Properties for Steel Fasteners.**
(Source, American Fasteners Technologies Corp.)

Identification Grade Mark	Specification	Fastener Description	Material	Nominal Size Range	Proof Load	Yield Strength (min)	Tensile Strength (min)
No Grade Mark	SAE J429 Grade 4	Studs	Medium Carbon Cold Drawn Steel	1/4 thru 1 1/2	-	100	115
B5	ASTM A193 Grade B5	Bolts, Screws, Studs for High Temp. Service	AISI 501	1/4 thru 4	-	80	100
B6	ASTM A193 Grade B6		AISI 410			85	110
B7	ASTM A193 Grade B7		AISI 4140, 4142, or 4105	1/4 thru 2 1/2 Over 2 1/2 thru 4 Over 4 thru 7	- - -	105 95 75	125 115 100
B16	ASTM A193 Grade B16		CrMoVa Alloy Steel			105 95 85	125 115 100
B8	ASTM A193 Grade B8		AISI 304				
B8C	ASTM A193 Grade B8C		AISI 347	1/4 and larger	-	30	75
B8M	ASTM A193 Grade B8M		AISI 316				
B8T	ASTM A193 Grade B8T		AISI 321	1/4 and larger	-	30	75

(Continued)

Table 9. *(Continued)* **ASTM, SAE, and ISO Grade Markings and Mechanical Properties for Steel Fasteners.**
(Source, American Fasteners Technologies Corp.)

Identification Grade Mark	Specification	Fastener Description	Material	Nominal Size Range	Proof Load	Yield Strength (min)	Tensile Strength (min)
						Mechanical Properties (ksi)	
B8	ASTM A193 Grade B8	Bolts, Screws, Studs for High Temp. Service	AISI 304 Strain Hardened	1/4 thru 3/4 Over 3/4 thru 1 Over 1 thru 1 1/4 Over 1 1/4 thru 1 1/2		100 80 65 50	125 115 105 100
B8C	ASTM A193 Grade B8C		AISI 347 Strain Hardened		- - - -		
B8M	ASTM A193 Grade B8M		AISI 316 Strain Hardened			95 80 65 50	110 100 95 90
B8T	ASTM A193 Grade B8T		AISI 321 Strain Hardened			100 80 65 50	125 115 105 100
L7	ASTM A320 Grade L7	Bolts, Screws, Studs for Low Temp. Service	AISI 4140, 4142 or 4145	1/4 thru 2 1/2	-	105	125
L7A	ASTM A320 Grade L7A		AISI 4037				
L7B	ASTM A320 Grade L7B		AISI 4137				
L7C	ASTM A320 Grade L7C		AISI 8740				
L43	ASTM A320 Grade L43		AISI 4340	1/4 thru 4	-	105	125

(Continued)

Table 9. *(Continued)* **ASTM, SAE, and ISO Grade Markings and Mechanical Properties for Steel Fasteners.**
(Source, American Fasteners Technologies Corp.)

Identification Grade Mark	Specification	Fastener Description	Material	Nominal Size Range	Mechanical Properties (ksi)		
					Proof Load	Yield Strength (min)	Tensile Strength (min)
B8	ASTM A320 Grade B8		AISI 304				
B8C	ASTM A320 Grade B8C		AISI 347				
B8T	ASTM A320 Grade B8T		AISI 321	1/4 and larger	-	30	75
B8F	ASTM A320 Grade B8F		AISI 303 or 303Se				
B8M	ASTM A320 Grade B8M	Bolts, Screws, Studs for Low Temp. Service	AISI 316				
B8	ASTM A320 Grade B8		AISI 304				
B8C	ASTM A320 Grade B8C		AISI 347	1/4 thru 3/4	-	100	100
B8F	ASTM A320 Grade B8F		AISI 303 or 303Se	Over 3/4 thru 1	-	80	80
				Over 1 thru 1 1/4	-	65	65
B8M	ASTM A320 Grade B8M		AISI 316	Over 1 1/4 thru 1 1/2		50	50
B8T	ASTM A320 Grade B8T		AISI 321				

(Continued)

Table 9. *(Continued)* **ASTM, SAE, and ISO Grade Markings and Mechanical Properties for Steel Fasteners.**
(Source, American Fasteners Technologies Corp.)

Identification Grade Mark	Specification	Fastener Description	Material	Nominal Size Range	Mechanical Properties (ksi)		
					Proof Load	Yield Strength (min)	Tensile Strength (min)
	SAE J429 Grade 5	Bolts, Screws, Studs	Medium Carbon Steel, Quenched and Tempered	1/4 thru 1 Over 1 to 1 1/2	85 74	92 81	120 105
	ASTM A449			1/4 thru 1 Over 1 to 1 1/2 Over 1 1/2 thru 3	85 74 55	92 81 58	120 105 90
	SAE J429 Grade 5.1	Sems	Low or Medium Carbon Steel, Quenched and Tempered	No. 6 thru 3/8	85	-	120
	SAE J429 Grade 5.2	Bolts, Screws, Studs	Low Carbon Martensitic Steel, Quenched and Tempered	1/4 thru 1	85	92	120
A325	ASTM A325 Type 1	High Strength Structural Bolts	Medium Carbon Steel, Quenched and Tempered	1/2 thru 1 1 1/8 thru 1 1/2	85 74	92 81	120 105
A325	ASTM A325 Type 2		Low Carbon Martensitic Steel, Quenched and Tempered	1/2 thru 1	85	92	120
A325	ASTM A325 Type 3		Atmospheric Corrosion Resisting Steel, Quenched and Tempered	1/2 thru 1 1 1/8 thru 1 1/2	85 74	92 81	120 105
BB	ASTM A354 Grade BB	Bolts, Studs	Alloy Steel, Quenched and Tempered	1/4 thru 2 1/2 2 3/4 thru 4	80 75	83 78	105 100
BC	ASTM A354 Grade BC				105 95	109 99	125 115
	SAE J429 Grade 7	Bolts, Screws	Medium Carbon Alloy Steel, Quenched and Tempered [4]	1/4 thru 1 1/2	105	115	133

(Continued)

Table 9. *(Continued)* **ASTM, SAE, and ISO Grade Markings and Mechanical Properties for Steel Fasteners.** *(Source, American Fasteners Technologies Corp.)*

Identification Grade Mark	Specification	Fastener Description	Material	Nominal Size Range	Mechanical Properties (ksi)		
					Proof Load	Yield Strength (min)	Tensile Strength (min)
(hex, 6 radial lines)	SAE J429 Grade 8	Bolts, Screws, Studs	Medium Carbon Alloy Steel, Quenched and Tempered	1/4 thru 1 1/2	120	130	150
	ASTM A354 Grade BD		Alloy Steel, Quenched and Tempered [4]				
No Grade Mark	SAE J429 Grade 8.1	Studs	Medium Carbon Alloy or SAE 1041 Modified Elevated Temperature Drawn Steel	1/4 thru 1 1/2	120	130	150
A490	ASTM A490	High Strength Structural Bolts	Alloy Steel, Quenched and Tempered	1/2 thru 1 1/2	120	130	150 min. 170 max.
No Grade Mark	ISO R898 Class 4.6		Medium Carbon Steel, Quenched and Tempered		33	36	60
No Grade Mark	ISO R898 Class 5.8				55	57	74
8.8 or 88	ISO R898 Class 8.8	Bolts, Screws, Studs	Alloy Steel, Quenched and Tempered	All sizes thru 1 1/2	85	92	120
10.9 or 109	ISO R898 Class 10.9				120	130	150

(Continued)

Table 9. *(Continued)* **ASTM, SAE, and ISO Grade Markings and Mechanical Properties for Steel Fasteners.**
(Source, American Fasteners Technologies Corp.)

Identification Grade Mark	Specification	Fastener Description	Material	Nominal Size Range	Mechanical Properties (ksi)		
					Proof Load	Yield Strength (min)	Tensile Strength (min)
12.9 or 129	ISO 898 Class 12.9	Bolts, Screws, Studs	Alloy Steel, Quenched and Tempered	All sizes thru 1 1/2	140	175	200

Identification Grade Mark	Specification	Material	Nominal Size Range	Proof Load Stress (ksi)	Rockwell Hardness		See Note
					Min.	Max.	
No mark	ASTM A563-Grade 0	Carbon Steel	1/4 thru 1 1/2	69	B55	C32	3, 4
	ASTM A563-Grade A	Carbon Steel	1/4 thru 1 1/2	90	B68	C32	3, 4
	ASTM A563-Grade B	Carbon Steel	1/4 thru 1	120	B69	C32	3, 4
			Over 1 thru 1 1/2	105			
	ASTM A563-Grade C	Carbon Steel may be Quenched and Tempered	1/4 thru 4	144	B78	C38	5
	ASTM A563-Grade C3	Atmospheric Corrosion Resistant Steel may be Quenched and Tempered	1/4 thru 4	144	B78	C38	5, 9
	ASTM A563-Grade D	Carbon Steel may be Quenched and Tempered	1/4 thru 4	150	B84	C38	6
	ASTM A563-Grade DH	Carbon Steel, Quenched and Tempered	1/4 thru 4	175	C24	C38	6
	ASTM A563-Grade DH3	Atmospheric Corrosion Resistant Steel, Quenched and Tempered	1/4 thru 4	175	C24	C38	5, 9

(Continued)

Table 9. *(Continued)* **ASTM, SAE, and ISO Grade Markings and Mechanical Properties for Steel Fasteners.**
(Source, American Fasteners Technologies Corp.)

Identification Grade Mark	Specification	Material	Nominal Size Range	Proof Load Stress (ksi)	Rockwell Hardness Min.	Rockwell Hardness Max.	See Note
	ASTM A194-Grade 1	Carbon Steel	1/4 thru 4	130	B70	-	7
	ASTM A194-Grade 2	Medium Carbon Steel	1/4 thru 4	150	159	352	7, 8
	ASTM A194-Grade 2H	Medium Carbon Steel, Quenched and Tempered	1/4 thru 4	175	C24	C38	7
	ASTM A194-Grade 2HM	Medium Carbon Steel, Quenched and Tempered	1/4 thru 4	150	159	237	7, 8
	ASTM A194-Grade 4	Medium Carbon Alloy Steel, Quenched and Tempered	1/4 thru 4	175	C24	C38	7
	ASTM A194-Grade 7	Medium Carbon Alloy Steel, Quenched and Tempered	1/4 thru 4	175	C24	C38	7
	ASTM A194-Grade 7M	Medium Carbon Alloy Steel, Quenched and Tempered	1/4 thru 4	150	159	237	7
See Note 1, 2	See 10						

Notes:
1) In addition to the indicated grade marking, all grades, except A563 grades O, A and B, must be marked for manufacturer identification.
2) The markings shown for all grades of A194 nuts are for cold formed and hot forged nuts. When nuts are machined from bar stock the nut must be additionally marked with the letter 'B'.
3) Nuts are not required to be marked unless specified by the purchaser. When marked, the identification marking shall be the grade letter O, A, or B.
4) Properties shown are those of nonplated or noncoated coarse thread hex nuts.
5) Properties shown are those of coarse thread heavy hex bolts.
6) Properties shown are those of coarse thread heavy hex nuts.
7) Properties shown are those of coarse 8-pitch thread heavy hex nuts.
8) Hardnesses are Brinell Hardness Numbers.
9) The nut manufacturer, as an option, may add other markings to indicate the use of atmospheric corrosion resistant steel.
10) Specifications-ASTM A563-Carbon and Alloy Steel Nuts. ASTM A194/A194M-Carbon and Alloy Steel Nuts for bolts for high pressure and high temperature service.

Table 10. Tensile Stress Areas for Unified Thread Fasteners. *(Source, MIL-S-122H.)*

Nominal Dia. (inches)	Coarse Threads (UNC)		Fine Threads (UNF)		8-Thread Series (8UN)	
	Threads per inch	Stress Area (in.^{2})	Threads per inch	Stress Area (in.^{2})	Threads per inch	Stress Area (in.^{2})
1/4	20	0.0318	28	0.0364	-	-
5/16	18	0.0524	24	0.0580	-	-
3/8	16	0.0775	24	0.0878	-	-
7/16	14	0.1063	20	0.1187	-	-
1/2	13	0.1419	20	0.1599	-	-
9/16	12	0.182	18	0.203	-	-
5/8	11	0.226	18	0.256	-	-
3/4	10	0.334	16	0.373	-	-
7/8	9	0.462	14	0.509	-	-
1	8	0.606	12	0.663	8	0.606
1-1/8	7	0.763	12	0.856	8	0.790
1-1/4	7	0.969	12	1.073	8	1.000
1-3/8	6	1.155	12	1.315	8	1.233
1-1/2	6	1.405	12	1.581	8	1.492
1-5/8	-	-	-	-	8	1.78
1-3/4	5	1.90	-	-	8	2.08
1-7/8	-	-	-	-	8	2.41
2	4-1/2	2.50	-	-	8	2.77
2-1/4	4-1/2	3.25	-	-	8	3.56
2-1/2	4	4.00	-	-	8	4.44
2-3/4	4	4.93	-	-	8	5.43
3	4	5.97	-	-	8	6.51
3-1/4	4	7.10	-	-	8	7.69
3-1/2	4	8.33	-	-	8	8.96
3-3/4	4	9.66	-	-	8	10.34
4	4	11.08	-	-	8	11.81

Table 11. Tensile Stress Areas for M Profile Metric Fasteners - Tolerance Class 6g.

Basic Dia. mm	Coarse Series		Fine Series	
	Pitch	Stress Area mm^2	Pitch	Stress Area mm^2
1	.25	.4498	-	-
1.1	.25	.5766	-	-
1.2	.25	.7190	-	-
1.4	.30	.9635	-	-
1.6	.35	1.208	-	-
1.8	.35	1.629	-	-
2	.40	1.988	-	-

(Continued)

Table 11. *(Continued)* Tensile Stress Areas for M Profile Metric Fasteners - Tolerance Class 6g.

Basic Dia. mm	Coarse Series		Fine Series	
	Pitch	Stress Area mm²	Pitch	Stress Area mm²
2.2	.45	2.383	-	-
2.5	.45	3.274	-	-
3	.50	4.878	-	-
3.5	.60	6.579	-	-
4	.70	8.531	-	-
4.5	.75	11.030	-	-
5	.80	13.829	-	-
6	1.00	19.637	0.75	22.03
7	1.00	28.267	-	-
8	1.25	35.828	1.00	39.167
9	1.25	47.222	-	-
10	1.50	56.837	1.25	61.199
11	1.50	70.985	-	-
12	1.75	82.688	1.25	92.075
14	2.00	113.35	1.50	124.54
16	2.00	154.24	1.50	167.25
18	2.50	189.23	1.50	216.23
20	2.50	241.14	1.50	271.50
22	2.50	299.32	1.50	333.13
24	3.00	347.31	2.00	384.42
27	3.00	453.47	2.00	495.74
30	3.50	553.10	2.00	621.20
33	3.50	685.23	2.00	760.80
36	4.00	806.41	2.00	914.54
39	4.00	955.56	2.00	1082.41
42	4.50	1107.57	2.00	1264.42
45	4.50	1291.60	1.50	1492.51
48	5.00	1456.07	2.00	1670.85
52	5.00	1739.17	-	-
56	5.50	2008.27	2.00	2300.72
60	5.50	2338.56	1.50	2696.36
64	6.00	2648.87	2.00	3031.13
68	6.00	3026.33	-	-
72	6.00	3428.92	2.00	3862.06
80	6.00	4309.51	2.00	4793.53
90	6.00	5551.61	2.00	6099.23
100	6.00	6950.79	2.00	7562.02

Table 12. Mechanical and Chemical Properties of M Profile Carbon and Alloy Steel Bolts, Screws, and Studs.

Mechanical Properties

Class	Designation	4.6	4.8	5.6	5.8	6.6	6.8	6.9	8.8	10.9	12.9	14.9
	Formerly	4D	4S	5D	5S	6D	6S	6G	8G	10K	12K	-
	U.S. Grade	1	2	2	2	3	3	3	5	8	-	-
Brinell	Min.	110		140			170		225	280	330	390
	Max.	170		215			245		300	365	425	-
Rockwell B	Min.	63		78			88					
	Max.	88		97			102					
Rockwell C	Min.	-		-			-		18	27	34	40
	Max.	-		-			-		31	38	44	49
Yield Point	ksi Min.	45		56		76		76	91	128	153	180
Tensile Strength	ksi Min.	56		70		85	85	85.4	113.8	142.2	170	200
	ksi Max.	78		100		113	113	99.6	128	170.7	200	230

Chemical Properties

Class	Material & Treatment	Element %			
		Carbon		Phosphorus	Sulpher
		Min.	Max.	Max.	Max.
4.6	Carbon Steel	-	0.55	0.05	0.06
4.8	Carbon Steel	-	0.55	0.05	0.06
5.6	Carbon Steel	0.15	0.55	0.05	0.06
5.8	Carbon Steel	-	0.55	0.05	0.06
6.8	Carbon Steel	-	0.55	0.05	0.06
8.8	Carbon Steel with Additives (B, Mn, or Cr), Quenched and Tempered	0.15	0.40	0.035	0.035
	Carbon Steel, Quenched and Tempered	0.25	0.55	0.035	0.035

(Continued)

Table 12. *(Continued)* **Mechanical and Chemical Properties of M Profile Carbon and Alloy Steel Bolts, Screws, and Studs.**

Class	Material & Treatment	Chemical Properties			
		Element %			
		Carbon		Phosphorus	Sulpher
		Min.	Max.	Max.	Max.
10.9	Carbon Steel, Quenched and Tempered	0.25	0.55	0.035	0.035
	Carbon Steel with Additives (B, Mn, or Cr), Quenched and Tempered	0.20	0.55	0.035	0.035
	Alloy Steel, Quenched and Tempered	0.20	0.55	0.035	0.035
10.9*	Carbon Steel with Additives (B, Mn, or Cr), Quenched and Tempered	0.15	0.35	0.035	0.035
12.9	Alloy Steel, Quenched and Tempered	0.20	0.50	0.035	0.035

* Products to this specification have the Property Class underlined.

Table 13. Mechanical Properties of Steel Metric Nuts.

Nominal Size mm		Class 5		Class 8		Class 10		Class 12	
Over	Up To	Proof load N/mm²	Hardness Rockwell C	Proof load N/mm²	Hardness Rockwell C	Proof load N/mm²	Hardness Rockwell C	Proof load N/mm²	Hardness Rockwell C
-	4	520	30	800	30	1040	28-38	1150	31-38
4	7	580	30	810	30	1040	28-38	1150	31-38
7	10	590	30	830	30	1040	28-38	1160	31-38
10	16	610	30	840	30	1050	28-38	1190	31-38
16	39	630	30	920	38	1060	28-38	1200	31-38

Note: Hexagon nuts size M 5 and larger, and all nuts Class 8 and higher are marked either on the bearing surface or on the side on one flat. Left-hand thread nuts size M 6 and larger are marked either on the bearing surface with a left-hand arrow, or by a groove on the corners halfway up the nut height.

Class 12.9 fasteners are usually hex socket cap screws and set screws made of quenched and tempered alloy steel. Mechanical properties of internally threaded Metric nuts are presented in **Table 13**. Property Classes of nuts differ from bolts as they have only one designation number. ISO 898 Part 2 covers the properties of metric, and the values given in **Table 13** are in accord with that standard.

Preload Relaxation

When preloads are initially applied to a fastener in a joint, local yielding takes place due to excess bearing stress under nut and bolt heads, local high spots, rough surface finish, and lack of perfect squareness of the bearing surfaces. Also, loads are not distributed evenly on each thread in a joint, which may lead to thread deformation. These conditions can result in loss of the intended preload over time. As a general guideline, an allowance for 10% loss of preload may be anticipated when designing a joint. Increasing the resilience of a joint may increase its resistance to local yielding. If practical, a ratio of joint length to bolt diameter of four or higher is recommended. Use of through bolts, farside tapped holes, spacers, and washers are design options that can be used to increase the ratio. Over time, the prescribed preload may be reduced or completely lost due to vibration, temperature cycling, creep joint load, etc. Use of a thread-locking solution that prevents relative motion within a joint may solve these problems. **Table 14** provides operating parameters for four types of thread-locking liquids produced by the Loctite Corporation.

Table 14. Common Thread-locking Liquid Solutions. *(Source, Loctite Corp.)*

Product	Fastener Size	Break/Prevail Torque	Temperature Range	Recommended Uses	Key Specifications
Low Strength Thread-locker (Purple)	#2 to 1/4" 2.2 to 6 mm	62/27	−65° to 300°F −54° to 149°C	Set screws, adjustment screws, calibration screws meters and gages	Conforms to MIL-S-46163A
Removable Thread-locker (Blue)	1/4" to 3/4" 6 to 20 mm	115/53	−65° to 300°F −54° to 149°C	Machine tools and presses, pumps and compressors, mounting bolts, gear boxes	Conforms to MIL-S-46163A. NSF/ANSI 61 Approved
High Strength Thread-locker (Red)	3/8" to 1" 9.5 to 25 mm	85/250	−65° to 300°F −54° to 149°C	Heavy equipment, suspension bolts, motor and pump mounts, bearing cap bolts and studs	Conforms to MIL-S-46163A
Penetrating Thread-locker (Green)	#2 to 1/2" 2.2 to 12 mm	85/250	−65° to 300°F −54° to 149°C	Preassembled fasteners, instrumentation screws, electrical connectors	Conforms to MIL-S-46163A. NSF/ANSI 61 Approved

Bolted joint behavior under applied load

The following derivation of a general formula describing behavior of a bolted joint subjected to an applied axial load covers only the condition where the axial load (P in pounds or Newtons) applied to the joint is applied over the same effective joint thickness (t_J in inches or millimeters) that is subjected to the joint compression load (P_J in pounds or Newtons). In many instances, the actual loading planes reside within the joint, leading to a complex loading situation. The condition covered in this section predicts somewhat higher bolt tensions prior to joint separation than exists when loading planes are within the joint material and joint separation occurring at a higher applied joint load. Eccentric loading on a joint causes joint separation and at a lower load than axial loading, and is not covered by the following examples.

Preload Application

Tightening a fastener in a joint compresses the joint materials and puts the fastener in tension. The preload on the fastener (P_B in pounds or Newtons) is equal to the compressive load on the joint material (P_J) – see *Figure 4*. The Modulus of Elasticity (E_B), or Young's Modulus, is a ratio between the stress applied to a material and the elastic strain produced by the stress. For carbon and alloy steel, E_B = 28,500,000 PSI; for stainless steel, E_B = 27,600,000 PSI; and for aluminum alloys, E_B has a range of 9,900,000 to 10,300,000 PSI.

Stretch of a fastener under preload in this configuration is equal to
$$\delta_{B1} = (P_{B1} \times L_B) \div (A_B \times E_B).$$

Compression of the joint material under preload is equal to
$$\delta_{J1} = (P_{J1} \times t_J) \div (A_J \times E_J).$$

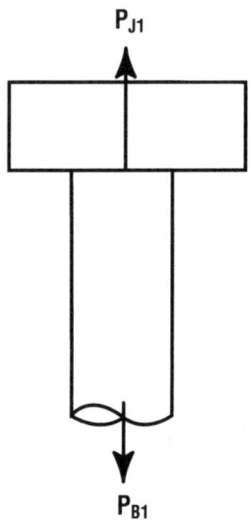

P_{J1}

P_{B1}

Figure 4. Fastener forces under preload.

Axial Load Applied and Joint Materials in Compression

When an axial load (P) is applied to a joint, fastener stretch is increased by $\Delta\delta_B$ and fastener load increases by ΔP_B. The increased (stretched) length of the fastener reduces the compression of the joint materials by $\Delta\delta_J$ and the compressive loading of the joint materials by ΔP_J. See *Figure 5*.

The increased amount of *stretch* of a fastener under preload in this configuration ($\Delta\delta_B$) and the amount of fastener load increase (ΔP_B) are equal to
$$\Delta\delta_B = (\Delta P_B \times L_B) \div (A_B \times E_B), \text{ and}$$
$$\Delta P_B = (\Delta\delta_B \times A_B \times E_B) \div L_B.$$

The reduction in the compression of joint materials ($\Delta\delta_J$) and the compressive loading (ΔP_J) are equal to
$$\Delta\delta_J = (\Delta P_J \times t_J) \div (A_J \times E_J), \text{ and}$$
$$\Delta P_J = (\Delta\delta_J \times A_J \times E_J) \div t_J.$$

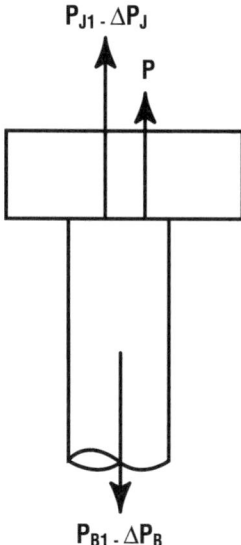

Figure 5. Forces inherent in a loaded joint fastener.

Therefore (see *Figure 5*), $P + P_{J1} - \Delta P_J = P_{B1} + \Delta P_B$. From *Figure 4* it can be seen that $P_{J1} = P_{B1}$. The amount of additional fastener stretch equals the reduction in joint material compression, i.e., $\Delta\delta_B = \Delta\delta_J = \Delta\delta$. Using these relationships and substituting the values for ΔP_B and ΔP_J, the following formulas can be derived.

$$\Delta P_B = (A_B \times E_B \div L_B) \div ([A_B \times E_B \div L_B] + [A_J \times E_J \div t_J]), \text{ and}$$
$$P_B = P_{B1} + \Delta P_B = P_{B1} + P \div (1 + [L_B \times A_J \times E_J \div t_J \times A_B \times E_B]).$$

Axial Load Applied, but Joint Separated

Whenever the applied axial load is sufficiently large to open the joint, there is no longer any load applied to the fastener by the joint materials. As shown in *Figure 6*, the load (P) becomes equal to the fastener load (P_B).

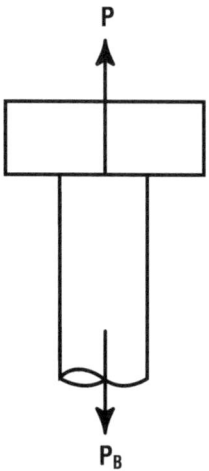

Figure 6. Forces applied to an open joint fastener.

Joint Separation

Referring to Figure 1, line $P_{B1}A$ is the condition discussed above in *Axial Load Supplied and Joint Materials in Compression*. Line AB in the same figure is the condition described immediately above in *Axial Load Applied but Joint Separated*. The point of joint separation, A, occurs when the applied load, P, is described by both conditions.

$$P = P_{B1} \times (1 + [t_J \times A_B \times E_B \div L_B \times A_J \times E_J]).$$

Gasketed Joints

When preload is applied to a joint, gasket compression load is equal to the load on the rest of the joint.

Gasket compression is

$$\delta_{G1} = P_{G1} \times t_G \div A_G \times E_G.$$

Total joint compression is

$$\delta_{JT1} = \delta_{J1} + \delta_{G1} = (P_{J1} \times t_J \div A_J \times E_J) + (P_{G1} \times t_G \div A_G \times E_G)$$
$$= P_{B1} \times ([t_J \div A_J \times E_J] + [t_G \div A_G \times E_G]) \text{ since } P_{J1} = P_{G1} = P_{B1}.$$

With the axial load, P, applied to the joint as in *Axial Load Applied and Joint Materials in Compression* (above) fastener loading remains unchanged, but compression in the total joint is reduced by $\Delta\delta_{JT}$ and compression load is reduced by ΔP_{JT}.

$$\Delta\delta_{JT} = \Delta P_{JT} ([t_J \div A_J \times E_J] + [t_G \div A_G \times E_G]), \text{ and}$$
$$\Delta P_{JT} = \Delta\delta_{JT} \div ([t_J \div A_J \times E_J] + [t_G \div A_G \times E_G]).$$

Therefore, $\Delta P_B = P \div (1 + K)$ where
$$K = ([L_B \div \{A_B \times E_B\}] \div [t_J \div \{A_J \times E_J\}] + [t_G \div A_G \times E_G]),$$
and $P_B = P_{B1} + \Delta P_B = P_{B1} + (P \div [1 + K]).$

It is important to apply approximately equal tension to each of the fasteners in a flange joint, especially if the joint contains a gasket. Unequal tension will result in leakage past the gasket. The recommended procedure is to tighten the fasteners in stages, following a pattern of tightening a pair of opposite fasteners, then a pair of opposite fasteners 90° away, followed by pairs of opposite fasteners in between as shown in *Figure 7*. This should be done in two or three steps in order to allow the shear stress to relax between steps. After tightening, it is advisable to retighten after a few hours to help assure that preload relaxation will not cause the joint to leak.

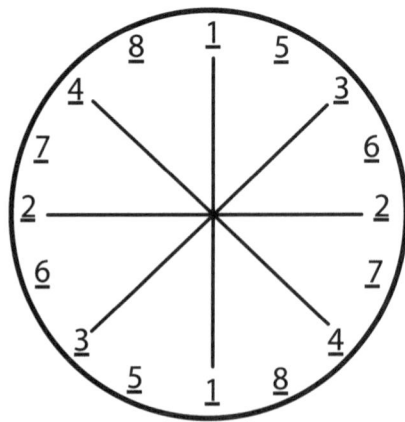

Figure 7. Correct tightening order for pairs of flange bolts.

Controlling bolt tension

There are numerous methods for controlling fastener tension, and **Table 15** provides the expected accuracy and relative costs for widely used preloading techniques as provided by SPS Technologies. Under ideal conditions, bolt elongation control using an ultrasonic detector can produce an accuracy as good as the strain gage method and at comparable cost. Tightening methods using power drive are similar in accuracy to manual methods, but costs are higher.

Table 15. Estimated Accuracy and Relative Costs for Manual Preloading Methods. *(Source, SPS Technologies.)*

Preload Method	Load Accuracy %	Relative Cost
Feel (Sensitivity to touch)	± 35	1
Torque Wrench	± 25	1.5
Turn of Nut	± 15	3
Preload Indicating Washer	± 10	7
Sensor Wrench (Computer controlled) —Below yield using turn of nut	± 15	8
—Yield point sensing	± 8	8
Bolt Elongation	± 3 – 5	15
Strain Gage	± 1	20

Elongation Measurement

Bolt elongation increases in the same proportion as stress increases. If both ends of a bolt are accessible, as in *Figure 8,* it is quite simple to measure the bolt with a micrometer before and after tension is applied to assure proper axial stress. The following formula applies:

$$\delta_B = (f_t \times L_B) \div E_B.$$

For more complex bolt geometry, elongation is equal to the sum of the elongations for each section, also taking into consideration traditional stresses in bolt head height and nut engagement length. Measurement by micrometer is most accurate when the bolt is threaded its entire length, or it has few threads in the bolt grip area so that elongation will be practically uniform throughout its length.

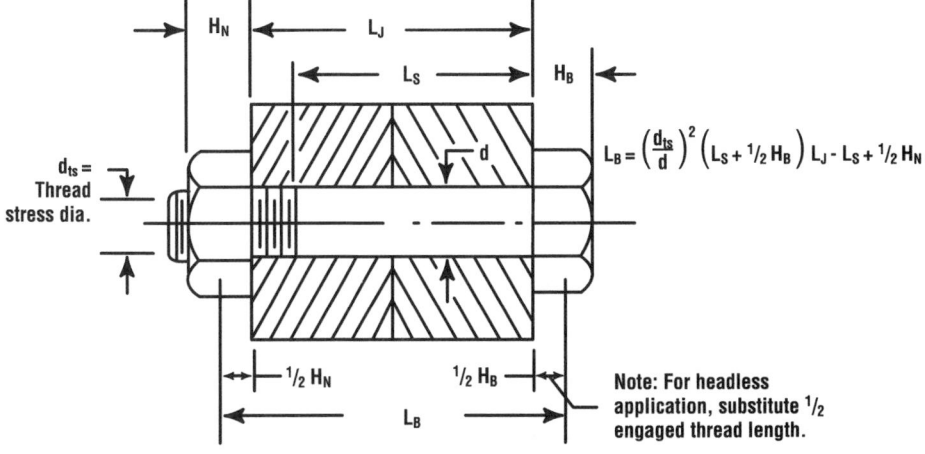

Figure 8. Effective length applicable in elongation formula.

In situations where both ends of a fastener are not accessible for measurement, it is sometimes possible to drill the fastener axially and take measurements with a micrometer depth gage as the fastener is tightened. *Figure 9* shows a cross section of a drilled fastener. A drilled fastener can also be used in another way to measure elongation. A pin placed through the length of the hole, and secured at the bottom, will move in relation to its reference surface as the fastener is tightened. If, for example, 2 mm of the pin protruded from the top of the fastener prior to tightening, the fastener will have elongated 1 mm over its length when the pin's protrusion is reduced to 1 mm during tightening.

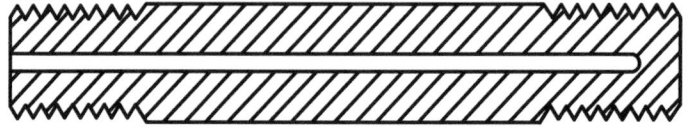

Figure 9. Axial hole in fastener for elongation measurement when one end is not accessible.

The *"turn of the nut" method* for measuring elongation uses the lead of the thread helix to determine the amount of elongation. Accuracy is affected by elastic deformation of the threads, by roughness of the bearing surfaces, and by the difficulty in establishing the starting point for measuring the angle. Generally, the nut is tightened to firmly seat the contacting surfaces and then it is loosened just enough to release bolt tension and twisting. This is the starting point for measuring the angle. Although nut turn angle will differ with bolt size, material, length, and thread lead, the following guidelines are useful for estimating the number of turns past snug for most bolt length/diameter relationships.

Bolt Length (diameters)	Amount of Rotation Past Snug
Length ≤ 4	One-third turn
$4 <$ Length ≤ 8	One-half turn
$8 <$ Length ≤ 12	Two-thirds turn

It should be kept in mind that unless the deformation of the nut and joint materials under load is negligible compared to that of the bolt, a significant portion of the nut rotation will be absorbed through means other than elongation of the bolt. In such instances, the following equation will not provide reliable results and the nut turn angle should be determined empirically with a tension measuring device. In most cases, the following equations can be used to determine the turn angle of the nut in degrees that will result in a given elongation.

$$\delta_B = (\theta \times L) \div 360 \quad \text{or} \quad \theta = 360 \times ([f_t \times L_B] \div [E \times L]).$$

The *ultrasonic method* of measuring elongation uses sound pulses that are generated from one end of a bolt and travel at the speed of sound through the bolt material and bounce off the other end and return to the point of origin. The time required to travel the bolt's length and return is measured to determine the length of the bolt. Using various inputs, the ultrasonic system can compute the stress, load, and elongation of a bolt at any time by comparing pulse travel time in loaded and unstressed conditions. The equipment requires a high degree of sophistication as the speed of sound varies with material, temperature, and stress level.

Strain Gages use a thin wire bonded to a bolt that stretches as the bolt elongates under load. The stretch causes a change in electrical resistance of the wire that can be measured and correlated with bolt load. Strain gage wires or groups of wires can be bonded either to the outside of a fastener or to the inside surface of a small axial hole drilled through the fastener. An additional mounting method uses the recess on the outside of a plastic tube that is bonded to the inside of a small axial hole in the bolt.

Temperature can also be used to induce tension and elongation. When a hot bolt and nut in a joint are air cooled, the bolt will shrink as it cools and tension will be developed. The following formula can be used to calculate the temperature necessary to induce the required axial tensile stress in a bolt when the stress is below the elastic limit.

$t = (f_t \div E_B \times e) + t_o.$

To provide bolt tension, the bolt and nut should be heated to a temperature slightly higher than calculated temperature to allow for cooling prior to nut tightening. The nut should be tightened snugly and the assembly should then be allowed to cool. Tension is induced by cooling. This method can only be applied where the bolt is accessible for heating and where the heating will not degrade joint materials. Accuracy is affected by the difficulty of controlling bolt temperature and by any significant deformity of the joint materials.

Temperature can also be used to control the expansion of a bolt. Using a thickness gage inserted under the nut, the elongation of the bolt can be measured as temperature is applied. When the desired length is achieved, the nut is tightened snugly. After cooling, the axial expansion in the bolt should be approximately equal to the desired stress level.

Torque control

Unified Threads

Suggested preloads and tightening torques for any size fastener can be calculated from the tables found in this section. The suggested preload, in most cases is 90% of the maximum clamping load for permanent connections, and 75% maximum clamping load for reusable connections. As shown earlier, the following equation provides maximum clamping load:

$F = Y \times A_s.$

Values for Y have been provided in Tables 3 and 9, and tensile stress areas for standard fasteners have been given in Tables 10 and 11. For materials without a given proof load (Y), the value can be estimated as 85% of the minimum yield strength. Once the value of 90% F or 75% F is known for a specific grade and size of fastener, the tightening torque (T) can be determined by multiplying F times the coefficient of friction (from Table 2) and the nominal fastener's nominal diameter.

$T = \%F \times \mu \times d.$

The following example shows how to estimate clamping force and desired tightening torque for a Grade 5, $^5/_8 - 11$ UNC bolt to be used for a permanent connection. From Table 3, it can be seen that the proof load for this fastener is 85,000 PSI, and from Table 10 it can be seen that the tensile stress area is 0.226 in.^{2}. Therefore, F = 85000 × 0.226 = 19,210 pounds. Since the connection is designed to be permanent, the desired clamping force will be 90% of the maximum clamping load, or 17,289 pounds. The desired torque can now be found by multiplying the clamping load times the coefficient of friction from Table 2 (15%), and the bolt diameter (0.6250). 17289 × .15 × .625 = 1621 lb in. or 135 lb ft.

Metric Threads

The equations above can be used, with metric measurements, to solve the following example for a Class 8.8, 16 × 2.00 fastener used in a reusable connection. From Table 9 it can be seen that the proof load is 85,000 PSI; 85000 × .006894757 = 586 MPa. The tensile stress area, from Table 11, for this fastener is 154.235 mm^2. To find F multiply,

586 × 154.235 = 90,382 Newtons. Next, multiply the clamping force by 75% to determine the amount of force desirable for a reusable connection. The desired clamping force is therefore 67,786 Newtons. From Table 2, obtain the coefficient of friction. The bolt diameter is 16 mm. Tightening torque can now be found with the following formula: 67786 × .15 × 16 = 162,686 N/mm or 162.69 Nm.

Table 16 provides clamping load and tightening torque recommendations for bolts, screws, and studs with Unified threads, and **Table 17** provides this information for M Profile metric threads. Seating torques for Unified thread socket screws are given in **Table 18** and **Table 19**, and **Table 20** provides these values for metric thread socket screws. The higher values in these tables represent 100% of the allowable load and torque and the lower values are 75% of the maximum. The values were obtained from the formulas discussed immediately above and assume a 0.15% coefficient of friction. Proof load values to determine the clamping load and tightening torque were obtained from Table 3 and Table 10. It should be noted that, depending on the source of the data, proof loads, tensile strength, and yield strength values will vary by as much as 10%. This can be attributed to independent testing on different samples that provide different results, to averaging several samples, and to discrepancies in the calibration of test equipment. The varying values for grades and classes is sufficient reason not to tension fasteners beyond 90% of their proof load rating.

UNJ Threads. As stated earlier, calculation of tensile stress is based on the diameter approximately midway between the minimum pitch diameter and the minimum minor diameter. However, for UNJ threads in accordance with MIL-S-8879, tensile stress area is considered to be at the basic pitch diameter. Therefore, tightening torque requirements for UNJ threaded fasteners is higher than for an equally stressed Unified threaded fastener in an equivalent joint. To convert the Unified torque to the appropriate UNJ torque requirement, use the following formula.

$$\text{UNJ}_{\text{Torque}} = \text{Unified}_{\text{Torque}} \times ([d_{\text{Basic}} \times n] \div [d_{\text{Basic}} \times n])^2.$$

Table 16. Clamp Load and Tightening Torque Values for Unified Thread Bolts, Screws, and Studs.
(See Notes for Dimensional Information.)

Nominal Size and tpi	Grade 2		Grade 5		Grade 8	
	Clamp Load Range [1]	Torque Range [2]	Clamp Load Range [1]	Torque Range [2]	Clamp Load Range [1]	Torque Range [2]
4-40	249 - 332	4 - 6	385 - 513	6 - 9	544 - 725	9 - 12
4-48	273 - 364	5 - 6	421 - 562	7 - 9	595 - 793	10 - 13
5-40	375 - 500	7 - 9	579 - 773	11 - 14	818 - 1091	15 - 20
5-44	342 - 457	6 - 9	529 - 706	10 - 13	747 - 996	14 - 19
6-32	371 - 495	8 - 10	574 - 765	12 - 16	810 - 1080	17 - 22
6-40	419 - 558	9 - 12	647 - 863	13 - 18	914 - 1218	19 - 25
8-32	578 - 770	14 - 19	893 - 1190	22 - 29	1260 - 1680	31 - 41
8-36	608 - 811	15 - 20	940 - 1253	23 - 31	1327 - 1769	33 - 44
10-24	722 - 963	21 - 27	1116 - 1488	32 - 42	1575 - 2100	45 - 60
10-32	825 - 1100	24 - 31	1275 - 1700	36 - 48	1800 - 2400	51 - 68
12-24	998 - 1331	32 - 43	1543 - 2057	50 - 67	2178 - 2904	71 - 94
12-28	1064 - 1419	34 - 46	1645 - 2193	53 - 71	2322 - 3096	75 - 100
1/4-20	1312 - 1749	4 - 5	2027 - 2703	6 - 8	2862 - 3816	9 - 12

(Continued)

Table 16. *(Continued)* Clamp Load and Tightening Torque Values for Unified Thread Bolts, Screws, and Studs. *(See Notes for Dimensional Information.)*

Nominal Size and tpi	Grade 2		Grade 5		Grade 8	
	Clamp Load Range [1]	Torque Range [2]	Clamp Load Range [1]	Torque Range [2]	Clamp Load Range [1]	Torque Range [2]
1/4-28	1502 - 2002	5 - 6	2321 - 3094	7 - 10	3276 - 4368	10 - 14
5/16-18	2162 - 2882	8 - 11	3341 - 4454	13 - 17	4716 - 6288	18 - 25
5/16-24	2393 - 3190	9 - 12	3698 - 4930	14 - 19	5220 - 6960	20 - 27
3/8-16	3197 - 4263	15 - 20	4941 - 6588	23 - 31	6975 - 9300	33 - 44
3/8-24	3622 - 4829	17 - 23	5597 - 7463	26 - 35	7902 - 10536	37 - 49
7/16-14	4385 - 5847	24 - 32	6777 - 9036	37 - 49	9567 - 12756	52 - 70
7/16-20	4896 - 6529	27 - 36	7567 - 10090	41 - 55	10683 - 14244	58 - 78
1/2-13	5853 - 7805	37 - 49	9046 - 12062	57 - 75	12771 - 17028	80 - 106
1/2-20	6596 - 8795	41 - 55	10194 - 13592	64 - 85	14391 - 19188	90 - 120
9/16-12	7508 - 10010	53 - 70	11603 - 15470	82 - 109	16380 - 21840	115 - 153
9/16-18	8374 - 11165	59 - 78	12941 - 17255	91 - 121	18270 - 24360	128 - 171
5/8-11	9323 - 12430	73 - 97	14408 - 19210	113 - 150	20340 - 27120	159 - 212
5/8-18	10560 - 14080	83 - 110	16320 - 21760	128 - 170	23040 - 30720	180 - 240
3/4-10	13778 - 18370	129 - 172	21293 - 28390	200 - 266	30060 - 40080	282 - 376
3/4-16	15386 - 20515	144 - 192	23779 - 31705	223 - 297	33570 - 44760	315 - 420
7/8-9	11435 - 15246	125 - 167	29453 - 39270	322 - 430	41580 - 55440	455 - 606
7/8-14	12598 - 16797	138 - 184	32449 - 43265	355 - 473	45810 - 61080	501 - 668
1-8	14999 - 19998	187 - 250	33633 - 44844	420 - 561	54540 - 72720	682 - 909
1-12	16409 - 21879	205 - 273	36797 - 49062	460 - 613	59670 -79560	746 - 995
1-1/8-7	18884 - 25179	266 - 354	42347 - 56462	595 - 794	68670 - 91560	966 - 1288
1-1/8-8	19553 - 26070	275 - 367	43845 - 58460	617 - 822	71100 - 94800	1000 - 1333
1-1/8-12	21186 -28248	298 - 397	47508 - 63344	668 - 891	77040 - 102720	1083 - 1445
1-1/4-7	23983 - 31977	375 - 500	53780 - 71706	840 - 1120	87210 - 116280	1363 - 1817
1-1/4-8	24750 - 33000	387 - 516	55500 - 74000	867 - 1156	90000 - 120000	1406 - 1875
1-1/4-12	26557 - 35409	415 - 553	59552 - 79402	930 - 1241	96570 - 128760	1509 - 2012
1-3/8-6	28586 - 38115	491 - 655	64103 - 85470	1102 - 1469	103950 - 138600	1787 - 2382
1-3/8-8	30517 - 40689	525 - 699	68432 - 91242	1176 - 1568	110970 - 147960	1907 - 2543
1-3/8-12	32546 - 43395	559 - 746	72983 - 97310	1254 - 1673	118350 - 157800	2034 - 2712
1-1/2-6	34774 - 46365	652 - 869	77978 - 103970	1462 - 1949	126450 - 168600	2371 - 3161
1-1/2-8	36927 - 49236	692 - 923	82806 - 110408	1553 - 2070	134280 - 179040	2518 - 3357
1-1/2-12	39130 - 52173	734 - 978	87746 - 116994	1645 - 2194	142290 - 189720	2668 - 3557
1-5/8-8	44055 - 58740	895 - 1193	98790 - 131720	2007 - 2676	160200 - 213600	3254 - 4339
1-3/4-5	47025 - 62700	1029 - 1372	105450 - 140600	2307 - 3076	171000 - 228000	3741 - 4988
1-3/4-8	51480 - 68640	1126 - 1502	115440 - 153920	2525 - 3367	187200 - 249600	4095 - 5460
1-7/8-8	59648 - 79530	1398 - 1864	133755 - 178340	3135 - 4180	216900 - 289200	5084 - 6778
2-4.5	61875 - 82500	1547 - 2063	138750 - 185000	3469 - 4625	225000 - 300000	5625 - 7500

Notes: 1) Lower value equals 75% of maximum clamping load in pounds. Higher value equals maximum (100%) clamping load in pounds.

2) For sizes through 12-28 (shaded on the Table), tightening torque is given in lb in. For sizes 1/4-20 and larger, tightening torques are given in lb ft. Lower value equals 75% of maximum tightening torque. Higher value equals maximum (100%) tightening torque.

Table 17. Clamp Load and Tightening Torque Values for M Profile Metric Bolts, Screws, and Studs.
(See Notes for Dimensional Information.)

Nominal Size and Pitch mm	Grade 4.8			Grade 8.8		
	Clamp Load Range [1]	Torque Range [2] Nm	Torque Range [3] lb ft	Clamp Load Range [1]	Torque Range [2] Nm	Torque Range [3] lb ft
1 × 0.25	105 - 139	0.02 - 0.02	0.14 - 0.19	202 - 270	0.03 - 0.04	0.27 - 0.36
1.1 × 0.25	134 - 179	0.02 - 0.03	0.20 - 0.26	259 - 346	0.04 - 0.06	0.38 - 0.51
1.2 × 0.25	167 - 223	0.03 - 0.04	0.27 - 0.36	324 - 431	0.06 - 0.08	0.52 - 0.69
1.4 × 0.3	224 - 299	0.05 - 0.06	0.42 - 0.56	434 - 578	0.09 - 0.12	0.81 - 1.1
1.6 × 0.35	281 - 374	0.07 - 0.09	0.60 - 0.80	544 - 725	0.13 - 0.17	1.2 - 1.5
1.8 × 0.35	379 - 505	0.10 - 0.14	0.9 - 1.2	733 - 977	0.20 - 0.26	1.8 - 2.3
2.0 × 0.4	462 - 616	0.14 - 0.18	1.2 - 1.6	895 - 1193	0.27 - 0.36	2.4 - 3.2
2.2 × 0.45	554 - 739	0.18 - 0.24	1.6 - 2.2	1072 - 1430	0.35 - 0.47	3.1 - 4.2
2.5 × 0.45	761 - 1015	0.29 - 0.38	2.5 - 3.4	1473 - 1964	0.55 - 0.74	4.9 - 7
3 × 0.5	1134 - 1512	0.51 - 0.68	4.5 - 6	2195 - 2927	0.99 - 1.3	9 - 12
3.5 × 0.6	1530 - 2039	0.80 - 1.1	7 - 9	2961 - 3947	1.6 - 2.1	14 - 18
4 × 0.7	1983 - 2645	1.2 - 1.6	11 - 14	3839 - 5119	2.3 - 3.1	20 - 27
4.5 × 0.75	2564 - 3419	1.7 - 2.3	15 - 20	4964 - 6618	3.4 - 4.5	30 - 40
5 × 0.8	3215 - 4287	2.4 - 3.2	1.8 - 2.4	6223 - 8297	4.7 - 6	3.4 - 4.6
6 × 1	4566 - 6087	4.1 - 5.5	3.0 - 4.0	8837 - 11782	8 - 11	6 - 8
6 × 0.75	5122 - 6829	4.6 - 6	3.4 - 4.5	9914 - 13218	9 - 12	7 - 9
7 × 1	6572 - 8763	7 - 9	5.1 - 7	12720 - 16960	13 - 18	10 - 13
8 × 1.25	8330 - 11107	10 - 13	7 - 10	16123 - 21497	19 - 26	14 - 19
8 × 1	9106 - 12142	11 - 15	8 - 11	17625 - 23500	21 - 28	16 - 21
9 × 1.25	10979 - 14639	15 - 20	11 - 15	21250 - 28333	29 - 38	21 - 28
10 × 1.5	13215 - 17619	20 - 26	15 - 19	25577 - 34102	38 - 51	28 - 38
10 × 1.25	14229 - 18972	21 - 28	16 - 21	27540 - 36719	41 - 55	30 - 41
11 × 1.5	16504 - 22005	27 - 36	20 - 27	31943 - 42591	53 - 70	39 - 52
12 × 1.75	19225 - 25633	35 - 46	26 - 34	37210 - 49613	67 - 89	49 - 66
12 × 1.25	21407 - 28543	39 - 51	28 - 38	41434 - 55245	75 - 99	55 - 73
14 × 2	26354 - 35139	55 - 74	41 - 54	51008 - 68010	107 - 143	79 - 105
14 × 1.5	28956 - 38607	61 - 81	45 - 60	56043 - 74724	118 - 157	87 - 116
16 × 2	35861 - 47814	86 - 115	63 - 85	69408 - 92544	167 - 222	123 - 164
16 × 1.5	38886 - 51848	93 - 124	69 - 92	75263 - 100350	181 - 241	133 - 178
18 × 2.5	43996 - 58661	119 - 158	88 - 117	85154 - 113538	230 - 307	170 - 226
18 × 1.5	50273 - 67031	136 - 181	100 - 133	97304 - 129738	263 - 350	194 - 258
20 × 2.5	56065 - 74753	168 - 224	124 - 165	108513 - 144684	326 - 434	240 - 320
20 × 1.5	63124 - 84165	189 - 252	140 - 186	122175 - 162900	367 - 489	270 - 360
22 × 1.5	69592 - 92789	230 - 306	169 - 226	134694 - 179592	444 - 593	328 - 437
22 × 1.5	77453 - 103270	256 - 341	189 - 251	149909 - 199878	495 - 660	365 - 486
24 × 3	80764 - 107685	291 - 388	214 - 286	156317 - 208422	563 - 750	415 - 553
24 × 2	89378 - 119170	322 - 429	237 - 316	172989 - 230652	623 - 830	459 - 612

(Continued)

Table 17. *(Continued)* **Clamp Load and Tightening Torque Values for M Profile Metric Bolts, Screws, and Studs.**
(See Notes for Dimensional Information.)

Nominal Size and Pitch mm	Grade 4.8			Grade 8.8		
	Clamp Load Range [1]	Torque Range [2] Nm	Torque Range [3] lb ft	Clamp Load Range [1]	Torque Range [2] Nm	Torque Range [3] lb ft
27 × 3	105432 - 140576	427 - 569	315 - 420	204062 - 272082	826 - 1102	610 - 813
27 × 2	115260 - 153679	467 - 622	344 - 459	223083 - 297444	903 - 1205	666 - 889
30 ×3.5	128596 - 171461	579 - 772	427 - 569	248895 - 331860	1120 - 1493	826 - 1101
30 × 2	144406 - 192541	650 - 866	479 - 639	279495 - 372660	1258 - 1677	928 - 1237
33 × 3.5	159316 - 212421	789 - 1051	582 - 776	308354 - 411138	1526 - 2035	1126 - 1501
33 × 2	176886 - 235848	876 - 1167	646 - 861	342360 - 456480	1695 - 2260	1250 - 1667
36 × 4	187490 - 249987	1012 - 1350	747 - 996	362885 - 483846	1960 - 2613	1445 - 1927
36 × 2	212631 - 283507	1148 - 1531	847 - 1129	411543 - 548724	2222 - 2963	1639 - 2185
39 × 4	222168 - 296224	1300 - 1733	959 - 1278	430002 - 573336	2516 - 3354	1855 - 2474
39 × 2	251660 - 335547	1472 - 1963	1086 - 1448	487085 - 649446	2849 - 3799	2102 - 2802

Nominal Size and Pitch mm	Grade 10.9			Grade 12.9		
	Clamp Load Range [1]	Torque Range [2] Nm	Torque Range [3] lb ft	Clamp Load Range [1]	Torque Range [2] Nm	Torque Range [3] lb ft
1 × 0.25	280 - 373	0.04 - 0.06	0.37 - 0.50	327 - 436	0.05 - 0.07	0.43 - 0.58
1.1 × 0.25	359 - 479	0.06 - 0.08	0.52 - 0.70	419 - 559	0.07 - 0.09	0.61 - 0.82
1.2 × 0.25	448 - 597	0.08 - 0.11	0.71 - 0.95	523 - 697	0.09 - 0.13	0.83 - 1.11
1.4 × 0.3	600 - 800	0.13 - 0.17	1.1 - 1.5	701 - 935	0.15 - 0.20	1.30 - 1.74
1.6 × 0.35	752 - 1003	0.18 - 0.24	1.6 - 2.1	879 - 1172	0.21 - 0.28	1.87 - 2.49
1.8 × 0.35	1014 - 1352	0.27 - 0.37	2.4 - 3.2	1185 - 1580	0.32 - 0.43	2.83 - 3.78
2.0 × 0.4	1238 - 1650	0.37 - 0.50	3.3 - 4.4	1446 - 1928	0.43 - 0.58	3.84 - 5.12
2.2 × 0.45	1483 - 1978	0.49 - 0.65	4.3 - 5.8	1734 - 2312	0.57 - 0.76	5.06 - 6.75
2.5 × 0.45	2038 - 2717	0.76 - 1.0	6.8 - 9.0	2382 - 3176	0.89 - 1.19	7.91 - 10.54
3 × 0.5	3037 - 4049	1.4 - 1.8	12 - 16	3549 - 4732	1.60 - 2.13	14.13 - 18.85
3.5 × 0.6	4095 - 5461	2.2 - 2.9	19 - 25	4786 - 6382	2.51 - 3.35	22.24 - 29.65
4 × 0.7	5311 - 7081	3.2 - 4.2	28 - 38	6206 - 8275	3.72 - 4.97	32.96 - 43.94
4.5 × 0.75	6866 - 9155	4.6 - 6	41 - 55	8024 - 10699	5.42 - 7.22	47.94 - 63.92
5 × 0.8	8609 - 11478	6 - 9	5 - 6	10061 - 13414	7.55 - 10.06	5.57 - 7.42
6 × 1	12224 - 16299	11 - 15	8 - 11	14286 - 19048	12.86 - 17.14	9.48 - 12.64
6 × 0.75	13714 - 18285	12 - 16	9 - 12	16027 - 21369	14.42 - 19.23	10.64 - 14.18
7 × 1	17596 - 23462	18 - 25	14 - 18	20564 - 27419	21.59 - 28.79	15.93 - 21.23
8 × 1.25	22303 - 29737	27 - 36	20 - 26	26065 - 34753	31.28 - 41.70	23.07 - 30.76
8 × 1	24381 - 32509	29 - 39	22 - 29	28494 - 37992	34.19 - 45.59	25.22 - 33.63
9 × 1.25	29396 - 39194	40 - 53	29 - 39	34354 - 45805	46.38 - 61.84	34.21 - 45.61
10 × 1.5	35381 - 47175	53 - 71	39 - 52	41349 - 55132	62.02 - 82.70	45.75 - 60.99
10 × 1.25	38096 - 50795	57 - 76	42 - 56	44522 - 59363	66.78 - 89.04	49.26 - 65.68
11 × 1.5	44188 - 58918	73 - 97	54 - 72	51642 - 68855	85.21 - 113.61	62.85 - 83.80
12 × 1.75	51473 - 68631	93 - 124	68 - 91	60156 - 80207	108.28 - 144.37	79.86 - 106.48
12 × 1.25	57317 - 76422	103 - 138	76 - 101	66985 - 89313	120.57 - 160.76	88.93 - 118.57

(Continued)

Table 17. *(Continued)* **Clamp Load and Tightening Torque Values for M Profile Metric Bolts, Screws, and Studs.** *(See Notes for Dimensional Information.)*

Nominal Size and Pitch mm	Grade 10.9			Grade 12.9		
	Clamp Load Range [1]	Torque Range [2] Nm	Torque Range [3] lb ft	Clamp Load Range [1]	Torque Range [2] Nm	Torque Range [3] lb ft
14 × 2	70560 - 94081	148 - 198	109 - 146	82462 - 109950	173.17 - 230.89	127.72 - 170.30
14 × 1.5	77526 - 103368	163 - 217	120 - 160	90603 - 120804	190.27 - 253.69	140.33 - 187.11
16 × 2	96014 - 128019	230 - 307	170 - 227	112210 - 149613	269.30 - 359.07	198.63 - 264.84
16 × 1.5	104113 - 138818	250 - 333	184 - 246	121674 - 162233	292.02 - 389.36	215.38 - 287.18
18 × 2.5	117796 - 157061	318 - 424	235 - 313	137665 - 183553	371.70 - 495.59	274.15 - 365.53
18 × 1.5	134603 - 179471	363 - 485	268 - 357	157307 - 209743	424.73 - 566.31	313.26 - 417.69
20 × 2.5	150110 - 200146	450 - 600	332 - 443	175429 - 233906	526.29 - 701.72	388.17 - 517.56
20 × 1.5	169009 - 225345	507 - 676	374 - 499	197516 - 263355	592.55 - 790.07	437.04 - 582.72
22 × 1.5	186327 - 248436	615 - 820	454 - 605	217755 - 290340	718.59 - 958.12	530.01 - 706.68
22 × 1.5	207373 - 276498	684 - 912	505 - 673	242352 - 323136	799.76 - 1066.35	589.87 - 786.50
24 × 3	216238 - 288317	778 - 1038	574 - 766	252712 - 336949	909.76 - 1213.02	671.01 - 894.67
24 × 2	239301 - 319069	861 - 1149	635 - 847	279666 - 372887	1006.80 - 1342.39	742.57 - 990.10
27 × 3	282285 - 376380	1143 - 1524	843 - 1124	329899 - 439866	1336.09 - 1781.46	985.45 - 1313.94
27 × 2	308598 - 411464	1250 - 1666	922 - 1229	360651 - 480868	1460.64 - 1947.51	1077.31 - 1436.41
30 ×3.5	344305 - 459073	1549 - 2066	1143 - 1524	402380 - 536507	1810.71 - 2414.28	1335.51 - 1780.68
30 × 2	386635 - 515513	1740 - 2320	1283 - 1711	451850 - 602467	2033.33 - 2711.10	1499.70 - 1999.61
33 × 3.5	426556 - 568741	2111 - 2815	1557 - 2076	498505 - 664673	2467.60 - 3290.13	1820.01 - 2426.68
33 × 2	473598 - 631464	2344 - 3126	1729 - 2305	553482 - 737976	2739.74 - 3652.98	2020.73 - 2694.30
36 × 4	501990 - 669320	2711 - 3614	1999 - 2666	586663 - 782218	3167.98 - 4223.98	2336.58 - 3115.44
36 × 2	569301 - 759068	3074 - 4099	2267 - 3023	665328 - 887104	3592.77 - 4790.36	2649.89 - 3533.19
39 × 4	594836 - 793115	3480 - 4640	2567 - 3422	695170 - 926893	4066.74 - 5422.33	2999.48 - 3999.30
39 × 2	673800 - 898400	3942 - 5256	2907 - 3876	787453 - 1049938	4606.60 - 6142.14	3397.66 - 4530.21

Notes: 1) Lower value equals 75% of maximum clamp load in Newtons. Higher value equals maximum (100%) clamp load in Newtons.

2) Lower value equals 75% of maximum tightening torque in Nm. Higher value equals maximum (100%) tightening torque.

3) For sizes through 4.5 × 0.75 (shaded in the Table), tightening torque is given in lb in. For sizes 5 × 0.8 and larger, tightening torques are given in lb ft. Lower value equals 75% of maximum tightening torque. Higher value equals maximum (100%) tightening torque.

Table 18. Recommended Seating Torques for Socket Head Cap and Flat Head Screws. *(See Notes.)*

Thread Size	Alloy Steel Socket Head Cap Screws[1]		Stainless Steel Socket Head Cap Screws[2]		Alloy Steel Flat Head Socket Screws[3]		Stainless Steel Flat Head Socket Screws[4]	
Nom	UNRC	UNRF	UNRC	UNRF	UNRC	UNRF	UNRC	UNRF
#0	-	3	-	1.3	-	1.5	-	1
#1	5	5	2	2.3	205	2.5	1.7	1.8
#2	7	8	3.8	4	4.5	4.5	2.8	3
#3	12	13	5.7	6	7	7	4.3	4.6
#4	18	19	8	9	8	8	6.0	6.6
#5	24	25	12	14	12	13	8.9	9.3

(Continued)

Table 18. *(Continued)* Recommended Seating Torques for Socket Head Cap and Flat Head Screws. *(See Notes.)*

Thread Size	Alloy Steel Socket Head Cap Screws[1]		Stainless Steel Socket Head Cap Screws[2]		Alloy Steel Flat Head Socket Screws[3]		Stainless Steel Flat Head Socket Screws[4]	
Nom	UNRC	UNRF	UNRC	UNRF	UNRC	UNRF	UNRC	UNRF
#6	34	36	15	17	15	17	11	12
#8	59	60	28	29	30	31	20	21
#10	77	91	40	45	40	45	30	34
1/4	200	240	95	110	100	110	71	81
5/16	425	475	170	190	200	220	123	136
3/8	750	850	300	345	350	400	218	247
7/16	1,200	1,350	485	545	560	625	349	388
1/2	1,850	2,150	750	850	850	1,000	532	600
9/16	2,500	2,700	920	1,050	1,200	1,360	767	856
5/8	3,400	3,820	1,270	4,450	1,700	1,900	1,060	1,200
3/4	6,000	6,800	2,260	2,520	3,000	3,200	1,880	2,100
7/8	8,400	9,120	3,790	4,180	5,000	5,400	3,030	3,340
1	12,500	13,200	5,690	6,230	7,200	7,600	4,550	5,000
1 1/8	14,900	16,600						
1 1/4	25,000	27,000						
1 3/8	33,000	35,000						
1 1/2	43,500	47,000						
1 3/4	71,500	82,500						
2	108,000	125,000						
2 1/4	155,000	186,000						
2 1/2	215,000	248,000						
2 3/4	290,000	330,000						
3	375,000	430,000						

Notes: Caution: Even when using the recommended seating torques in this Table, the induced loads obtained may vary by as much as ± 25%, depending on uncontrolled variables such as mating material, lubrication, surface finish, hardness, etc. Torque values are in inch pounds. To convert inch pounds (in.-lbs) to Newton meters (N-m), divide the values on this Table by 8.88. Seating torques given are for screws made to the specifications referenced, as follows.

1) Alloy Steel Socket Head Cap Screws: ASTM A574. Hardness: RC 38-43. Tensile Strength: 190,000 PSI thru $^1/_2$" size, 180,000 PSI over $^1/_2$" size. Yield Strength: 170,000 PSI thru $^1/_2$" size, 155,000 PSI over $^1/_2$" size.

2) Stainless Steel Socket Head Cap Screws: ASTM F837. Hardness: RC 33. Tensile Strength: 95,000 PSI. Yield Strength 30,000 PSI.

3) Alloy Steel Flat Head Socket Screws: ASTM F835. Hardness: RC 38-43. Tensile Strength: 160,000 PSI min.

4) Stainless Steel Flat Head Socket Screws: ASTM F879. Hardness: RC 33. Tensile Strength 90,000 PSI min.

Source, Unbrako/SPS Technologies, Inc.

Table 19. Recommended Seating Torques for Socket Button Head, Low Head Cap, Shoulder, and Set Screws. *(See Notes.)*

Thread Size	Alloy Steel Button Head Cap Screws[1]		Stainless Steel Button Head Cap Screws[2]		Low Head Cap Screws[3]	Socket Head Shoulder Screws[4]	Alloy Steel Socket Set Screws[5]	Stainless Steel Socket Set Screws[6]
Nom	UNRC	UNRF	UNRC	UNRF				
#0	-	1.5	-	1			1	0.4
#1	2.5	2.5	1.7	1.8			1.8	1.2
#2	4.5	4.5	2.8	3			1.8	1.2
#3	7	7	4.3	4.6			5	4
#4	8	8	6.0	6.6			5	4
#5	12	13	8.9	9.3			10	7
#6	15	17	11	12			10	7
#8	30	31	20	21	25		20	16
#10	40	45	30	34	35		36	26
1/4	100	110	71	81	80	45	87	70
5/16	200	220	123	136	157	112	165	130
3/8	350	400	218	247	278	230	290	230
7/16	560	625					430	340
1/2	850	1,000	532	600	226	388	620	500
9/16	1,200	1,360					620	500
5/8	1,700	1,900	1,060	1,200		990	1,325	980
3/4						1,975	2,400	1,700
7/8							3,600	3,000
1						3,490	5,000	4,000
1 1/8							7,200	5,600
1 1/4						5,610	9,600	7,700
1 3/8							9,600	7,700
1 1/2						12,000	11,300	9,100
1 3/4						16,000		
2						30,000		

Notes: Caution: Even when using the recommended seating torques in this Table, the induced loads obtained may vary by as much as ± 25%, depending on uncontrolled variables such as mating material, lubrication, surface finish, hardness, etc. Torque values are in inch pounds. To convert inch pounds (in.-lbs) to Newton meters (N-m), divide the values on this Table by 8.88. Seating torques given are for screws made to the specifications referenced, as follows.

1) Alloy Steel Button Head Socket Screws: ASTM F835. Hardness: RC 38-43. Tensile Strength: 160,000 PSI min.
2) Stainless Steel Button Head Socket Screws: ASTM F879. Hardness: RC 33. Tensile Strength: 90,000 PSI min.
3) Low Head Cap Screws: ASTM A574. Hardness: RC 38-43. Tensile Strength: 170,000 PSI min. Yield Strength: 150,000 PSI min.
4) For Socket Head Shoulder Screws, the nominal dimension given is for the shoulder diameter of the screw. ASTM A574. Hardness: RC 36-43. Tensile Strength: 160,000 PSI.
5) Alloy Steel Socket Set Screws: ASTM F912. Hardness: RC 45-53.
6) Stainless Steel Socket Set Screws: ASTM F880. Hardness: RC 33.

Source, Unbrako/SPS Technologies, Inc.

Table 20. Recommended Seating Torques for Metric Socket Head Screws. *(See Notes.)*

Thread Size[1]	Socket Head Cap Screws[2]		Flat Head Cap Screws[3]		Button Head Cap Screws[4]		Socket Head Shoulder Screws[5]		Low Head Cap Screws[6]		Socket Set Screws[7]	
Nom	N-m	in.-lbs	N-m	in.-lbs	N-m	in.-lbs	N-m	in.-lbs	N-m	in.-lbs	N-m	in.-lbs
M1.6	0.29	2.6									0.09	0.8
M2	0.60	5.3									0.21	1.8
M2.5	1.21	11									0.57	5
M3	2.1	19	1.2	11	1.2	11					0.92	8
M4	4.6	41	2.8	25	2.8	25			4.5	40	2.2	19
M5	9.5	85	5.5	50	5.5	50			8.5	75	4.0	35
M6	16	140	9.5	85	9.5	85	7	60	14.5	130	7.2	64
M8	39	350	24	210	24	210	12	105	35	310	17	150
M10	77	680	47	415	47	415	29	255	70	620	33	290
M12	135	1,200	82	725	82	725	57	500	120	1,060	54	480
M14	215	1,900					100	885				
M16	330	2,900	205	1,800	205	1,800	240	2,125	300	2,650	134	1,190
M20	650	5,700	400	3,550			470	4,160	575	5,100	237	2,100
M24	1,100	9,700	640	5,650							440	3,860
M30	2,250	19,900										
M36	3,850	34,100										
M42	6,270	19,900										
M48	8,560	75,800										

Notes: Caution: Even when using the recommended seating torques in this Table, the induced loads obtained may vary by as much as ± 25%, depending on uncontrolled variables such as mating material, lubrication, surface finish, hardness, etc. 1) All threads are Class 4g 6g. Seating torques given are for screws made to the specifications referenced, as follows. 2) Socket Head Cap Screws: ASTM A574M, DIN912 alloy steel. Hardness: RC 38-43. Tensile Stress: 1300 MPa thru M16 size, 1250 MPa over M16 size. Yield Stress: 1170 MPa thru M16 size, 1125 over M16 size. 3) Flat Head Cap Screws: ASTM F835M. Hardness: RC 38-43. Tensile Stress: 1040 MPa. Yield Stress: 945 MPa. 4) Button Head Cap Screws: ASTM F835M. Hardness: RC 38-43. Tensile Stress: 1040 MPa. Yield Stress: 945 MPa. 5) Socket Head Shoulder Screws: ASTM A574. Hardness: RC 36-43. Tensile Stress: 1100 MPa based on minimum thread neck area. Shear Stress: 660 MPa. 6) Low Head Cap Screws: ASTM A574M. Hardness: RC 33-39. Tensile Stress: 1040 MPa. Yield Stress: 940 MPa. 7) Socket Set Screws: ASTM F912M. Hardness: Hardness: RC 45-53. *Source, Unbrako/SPS Technologies, Inc.*

Dimensions for UNJ threads are included in the Screw Threads section of this book.

Bolt torque and axial tension relationship

Axial Load Torque (Figure 10)

In order to achieve a desired axial load, in a fastener, torque is applied. The torque must be sufficient to produce the axial load and also overcome friction in the threads and friction under the nut or bolt head (bearing surface). Axial load (P_B) is a component of the normal force developed between threads. The normal force component that is perpendicular to the thread's helix is the thread helix force ($P_N\lambda$), and the final component of this force is the torque load ($P_B \tan \lambda$). Assuming that the force is applied at the pitch diameter of the thread, the torque (T_1) required to develop the axial load is

$$T_1 = P_B \times \tan \lambda \times (d_2 \div 2).$$

As can be seen in *Figure 9*, $\tan \lambda = L \div (\pi \times d_2)$; therefore,

$$T_1 = (P_B \times L) \div 2\,\pi.$$

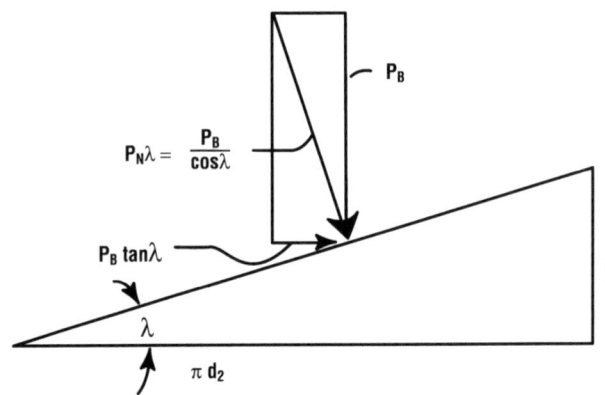

Figure 10. Thread helix forces.

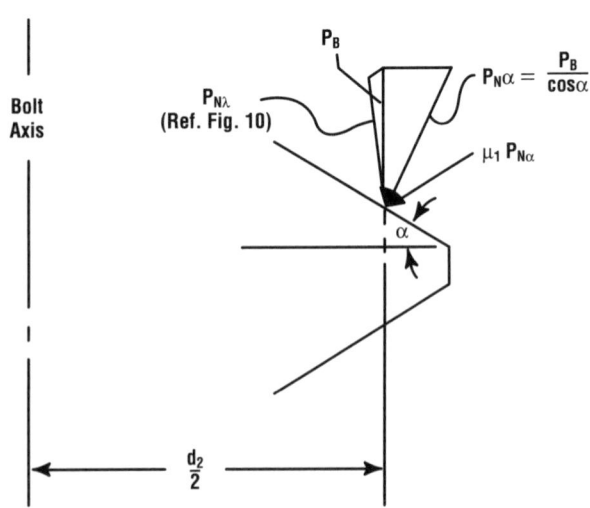

Figure 11. Thread friction force.

Thread Friction Torque (Figure 11)

As shown in the figure, the normal force component perpendicular to the thread flanks is $P_N\alpha$. With a coefficient of friction (μ_1) between the threads, the total friction load is equal to $\mu_1 \times P_N\alpha$, or $\mu_1 \times P_N\alpha \div \cos\alpha$. Assuming that the force is applied at the pitch diameter of the thread, the amount of torque required to overcome thread friction is

$$T_2 = (d_2 \times \mu_1 \times P_B) \times 2\cos\alpha.$$

Nut or Bolt Underhead Friction Torque (Figure 12)

Given a coefficient of friction of μ_2 between the bolt head washer face and the joint, the friction load equals $\mu_2 \times P_B$. Assuming that the force is applied midway between the nominal diameter (d) and the washer face diameter (b), the torque (T_3) necessary to overcome the nut or bolt underhead friction can be calculated as

$$T_3 = ([d + b] \div 4) \times (\mu_2 \times P_B).$$

Torque-Tension Relation

Total torque (T) required to develop axial bolt load (P_B) is equal to the sum of T_1, T_2, and T_3 above.

$$T = P_B ([L \div 2\pi] + [d_2 \times \mu_1 \div 2\cos\alpha] + [\{d + b \times \mu_2\} \div 4]).$$

For fasteners with 60° threads, $\mu = 30°$, and d_2 is approximately $0.92d$. If no loose washer is used under the rotated nut or bolt head, b = approximately $1.5d$.

$$T = P_B ([0.159 \times L] + [d \times \{\mu_1 + 0.625\,\mu_2\}]).$$

If, in addition to these conditions, μ_1 is approximately equal to μ_2, then $\mu_1 = \mu$, then $T = P_B \times ([0.159 \times L] + [0.625 \times \mu d]).$

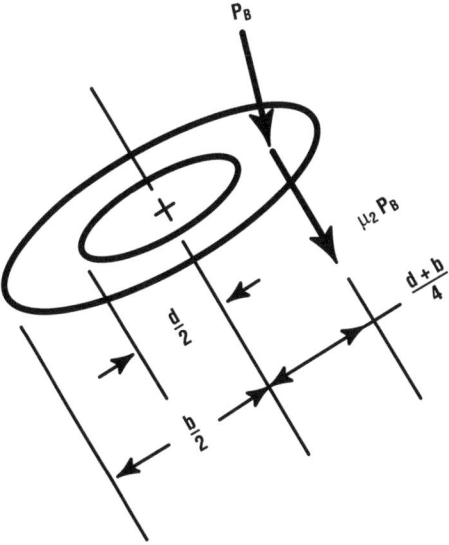

Figure 12. Nut or bolt head friction force.

Shear area

Minimum Material Condition
The geometric shear area of an internal thread at minimal material condition is equal to the area of that thread that is intersected by a cylinder with a diameter equal to the minimum major diameter of the mating external thread over the length of engagement (see *Figure 13*). Similarly, the geometric shear area of an external thread at minimum material condition is equal to the area of that thread that is intersected by a cylinder with a diameter equal to the maximum minor diameter of the mating internal thread (also shown in *Figure 13*).

Basic Size
The geometric shear area of an external thread at basic size is equal to the area of a basic size thread that is intersected by a cylinder with a diameter equal to the basic minor diameter. The geometric shear area of an internal thread at basic size is not normally used for calculation.

Formulas for Determining Shear Area
In order to determine stress areas with the formulas given below, it is first necessary to choose a length of engagement. Commonly, length of engagement, based on requirements, varies from a low of one-third of the major diameter of the internal thread to 1.5 times the major diameter of the internal thread. In these formulas, n = number of threads per inch; d is the minor diameter of the external thread, d_2 is the pitch diameter of the external thread; D_1 is the minor diameter of the internal thread; D_2 is the pitch diameter of the internal

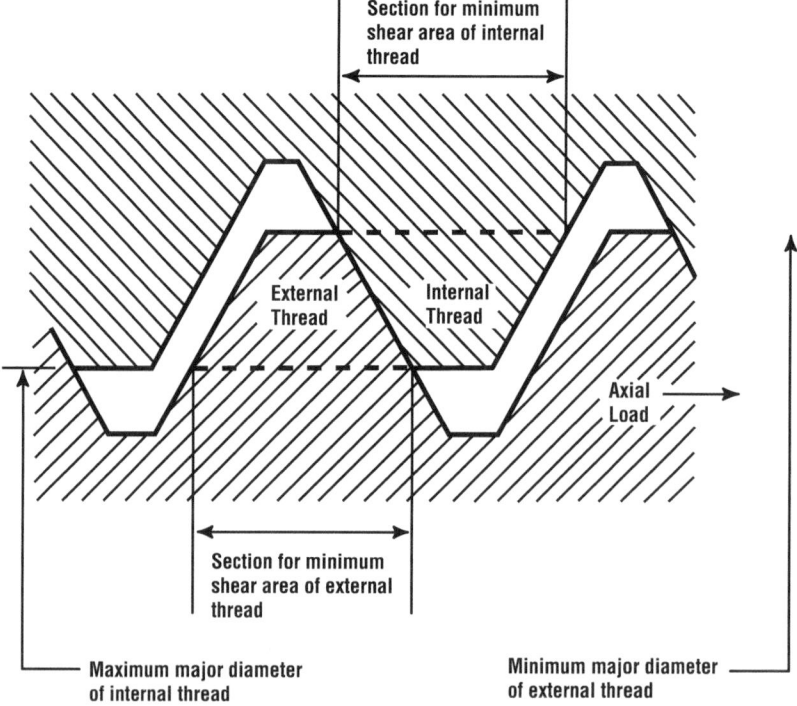

Figure 13. Shear areas at minimal material condition.

thread; Td, Td_2, TD_1, and TD_2 are the tolerances on d, d_2, D_1, and D_2, respectively; es is external thread allowance; AS_n is the shear area of the internal thread; AS_s is the shear area of the external thread; and LE is the length of engagement.

Shear area of internal threads when both external and internal threads are at minimal material condition:

$$AS_n \text{ (min)} = \pi \times n \times LE \times d_{min} \times ([1 \div 2n] + 0.57735 \times [d_{min} - D_{2\ max}]), \text{ or}$$
$$AS_n \text{ (min)} = \pi \times d_{min} \times (0.875 - 0.57735 \times n \times [Td + TD_2 + es] \times LE.$$

Shear area of internal thread (simplified), when d is 0.25" or greater:

$$AS_n = \pi \times D_{2\ basic} \times (3 \times LE \div 4).$$

Shear area of external threads when both external and internal threads are at minimal material condition:

$$AS_s \text{ (min)} = \pi \times n \times LE \times D_{1\ max} \times ([1 \div 2n + 0.57735] \times [d_{2\ min} - D_{1\ max}]), \text{ or}$$
$$AS_s \text{ (min)} = \pi \times D_{1\ max} ([0.75 - 0.57735 \times n] \times (TD_1 + Td_2 + es] \times LE.$$

Shear area of external thread (simplified):

$$AS_s = \pi \times d_{2\ basic} \times (^5/_8 \times LE).$$

Shear area of external threads when both external and internal threads are at basic size:

$$AS_s \text{ (max)} = \pi \times D_{1\ basic} \times (^3/_4 \times LE).$$

Shear area of combined failure:

$$AS = \pi \times D_{2\ basic} \times (LE \div 2).$$

The shear stress area ratio (R_1) can be found by $AS_{s\ max}$ to $AS_{n\ min}$.

Length of engagement

The length of engagement for UNC, UNF, and 8UN series threads, upon which their specified tolerances are based, is equal to basic major diameter. These tolerances are applicable for lengths of engagement for these threads of 1.0 to 1.5 times the basic major diameter. The length of engagement for UNEF, 12UN, and 16UN series threads, upon which their specified tolerances are based, is equal to 9 pitches. These tolerances are applicable for lengths of engagement for these threads of 5 to 15 pitches. For miniature thread series, all tolerances governing limits of size are based on pitch only and apply to lengths of engagement from 0.67 to 1.5 times the nominal diameter.

When mating internal and external threads are on parts manufactured from materials of approximately equal strength, failure will usually take place in both threads simultaneously. However, it is generally more economical if the externally threaded part should break, rather than either the internal or external part strip. In other words, the length of thread engagement (LE) should be sufficient to develop the full strength of the screw. Thus, the length of the internal thread and its dimensions (particularly its minor diameter) should be such that, taking into account a possible difference in strength of material of the internal and external threads, the externally threaded part will break before either the external or internal threads strip. Due to this situation, formulas for length of engagement are derived from shear formulas with tensile stress area (A_s) replaced by $2A_s$ because the required area in shear is twice the tensile stress area in order to fully develop the full strength of the externally threaded part (and $4A_s$ when calculating the combined full strength of both the external and internal threads). This relationship is based on experiments made by the National Bureau of Standards, in which it was found that for hot-rolled and cold-rolled steel, and brass screws and nuts, this factor varied from 1.7 to 2.0. The effect of combined stress is not often taken into consideration in calculating length of engagement because the added shear load affects both the tensile and shear stresses in approximately the same proportion.

To find length of engagement, based on shear of the external thread, requires tensile stress area and shear stress area values derived from formulas introduced above.

$$LE = 2 A_s \div AS_{s\ min}.$$

Length of engagement based on developing full tensile stress of the external thread with threads at basic size:

$$LE_{max} = 2A_s \div AS_{s\ max}.$$

Length of engagement based on combined shear failure of external and internal threads (assumes combination stress failure of both threads):

$$LE = 4A_s \pi \times d_{2\ basic}.$$

For a hollow part, subtract $0.7854d_h^2$ from A_s in the above formulas. d_h is the hole diameter on the externally threaded part.

Length of engagement based on shear of the internal thread is equal to the shear area stress ratio (R_1 – see above) and the ultimate material strength ratio (R_2), which is found by dividing the ultimate tensile strength of the internally threaded part by the ultimate tensile strength of the externally threaded part.

$$LE_{max} = R_1 \div R_2.$$

The above formulas for length of engagement yield approximate values because they are based in part on shear stress areas that are inexact due to nut dilation that varies with geometry, friction forces, and material properties.

Stripping strength of tapped holes

The Unbrako division of SPS Technologies has done research into obtaining minimum thread engagement based on applied load, material, type of thread, and bolt diameter. The results are reported in this section and shown graphically in *Figures 14 through 19*.

Knowledge of the thread stripping strength of tapped holes is necessary to develop full tensile strength of a bolt or, for that matter, the minimum engagement needed for any lesser load. Conversely, if only limited length of engagement is available, the data can help determine the maximum load that can be safely applied without stripping the threads of the tapped hole.

Attempts to compute lengths of engagement and related factors by formula have not been entirely satisfactory—mainly because of subtle differences between various materials. Therefore, strength data was empirically developed from a series of tensile tests of tapped specimens for seven commonly used metals including steel, aluminum, brass, and cast iron.

This design data is summarized in the charts shown in Figures 14 through 19, and covers a range of screw thread sizes from #0 to one-inch in diameter for both coarse and fine threads. Though developed from tests of Unbrako socket head cap screws having minimum tensile strength (depending on their diameter) from 190,000 to 180,000 PSI, these stripping strength values are valid for all other screws or bolts of equal or lower strength having a standard thread form. Data are based on static loading only.

In the test program, bolts threaded into tapped specimens of the metal under study were stressed in tension until the threads stripped. Load at which stripping occurred and the length of engagement of the specimen were noted. Conditions of the tests, all of which are met in a majority of industrial bolt applications, were as follows.

Tapped holes had a basic thread depth within the range of 65 to 80 percent. Threads of tapped holes were Class 2B or better.

Minimum amount of metal surrounding the tapped hole was 2.5 times the major diameter.

Test loads were applied slowly in tension to screws having standard Class 3A threads. (Data, though, will be equally applicable to Class 2A external threads as well.

Study of the test results revealed certain factors that greatly simplified the compilation of thread stripping strength data.

Stripping strengths are almost identical for loads applied either by pure tension or by screw torsion. These data are, therefore, equally valid for either condition of application.

Stripping strength values vary with diameter of screw. For a given load and material, larger diameter bolts require greater engagement.

Minimum length of engagement (as a percent of screw diameter) is a straight line function of load. This permits easy interpolation of test data for any intermediate load condition.

When engagement is plotted as a percentage of bolt diameter, it is apparent that stripping strengths for a wide range of screw sizes are close enough to be grouped in a single curve. Thus, in the accompanying charts, data for size #0 through #12 have been represented by a single set of curves.

With this data, it becomes a simple matter to determine stripping strengths and lengths of engagement for any condition of application, as can be seen from the four examples that follow.

Example 1. Calculate length of thread engagement to develop the minimum ultimate tensile strength (190,000 PSI) of a $1/2$-13 (National Coarse) Unbrako cap screw in cast iron having an ultimate shear strength of 30,000 PSI. *Figure 14* is for screw sizes from #0 through #10. *Figures 15 and 16* are for sizes from $1/4$ inch through $5/8$ inch. *Figures 17 and 18* are for sizes $3/4$ inch through 1 inch, and *Figure 19* is for screws over one inch in diameter. For this example, we can see from Figure 15 that a value of 1.4D is reached at the spot where the Cast Iron line intersects the ultimate tensile strength value of the screw. Multiplying the nominal bolt diameter (0.500 inch) by 1.4 gives a minimum length of engagement of 0.725 inch.

Example 2. Calculate the length of engagement for the above conditions if only 140,000 PSI is to be applied (which is the equivalent of using a bolt with a maximum tensile strength of 140,000 PSI). From Figure 15 obtain the value, which is 1.06D. Minimum length of engagement is therefore 1.06 × 0.5 = 0.530 inch.

Example 3. Suppose in Example 1 that minimum length of engagement to develop full tensile strength was not available because the thickness of the metal allowed a tapped hole only 0.600 inch deep. Hole depth in terms of bolt diameter equals 0.600 ÷ 0.500 = 1.20D. By working backwards in Figure 15, maximum hole load that can be carried is approximately 159,000 PSI.

Example 4. Suppose that the hole in Example 1 is in steel having an ultimate tensile shear strength of 65,000 PSI. Although there is no curve for this steel in Figure 15, a design value can be obtained by taking a point midway between the curves for the 80,000 PSI and 50,000 PSI steels that are listed. Under the conditions of the example, a length of engagement of either 0.825D or 0.413 inch will be obtained.

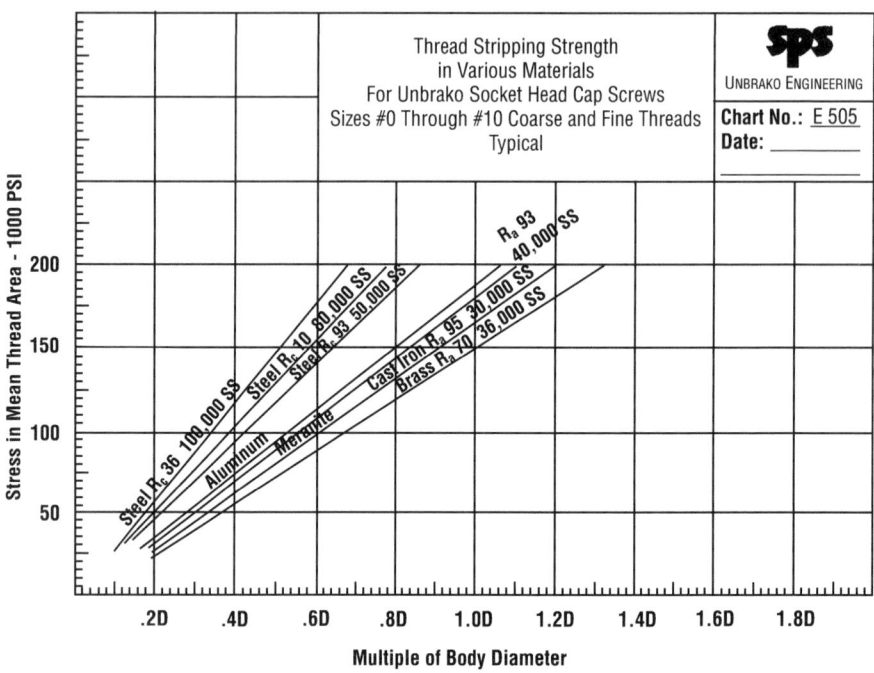

Figure 14. Thread stripping strength for nominal sizes #0 through #10.

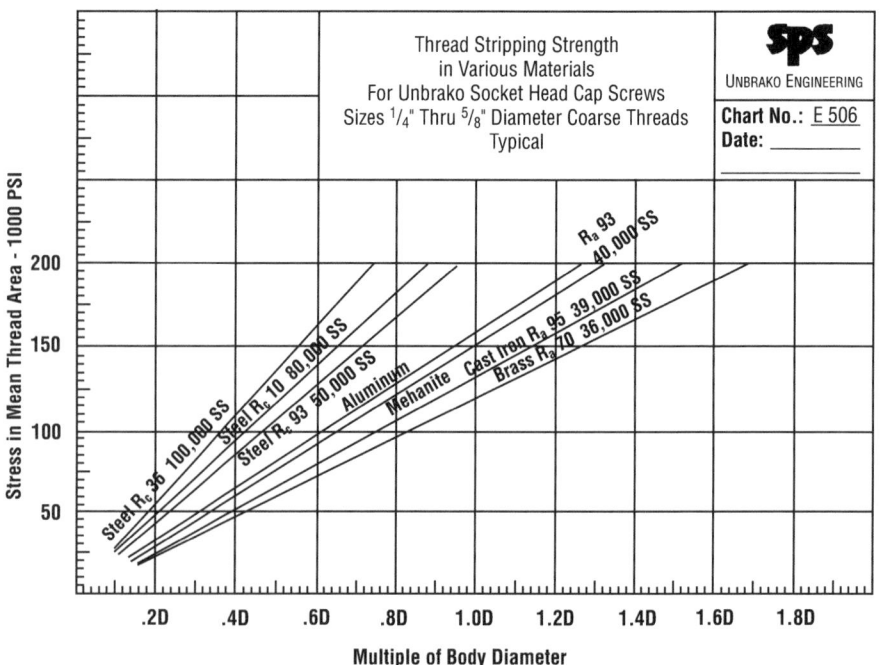

Figure 15. Thread stripping strength for nominal sizes $^1/_4$ inch through $^5/_8$ inch with coarse threads.

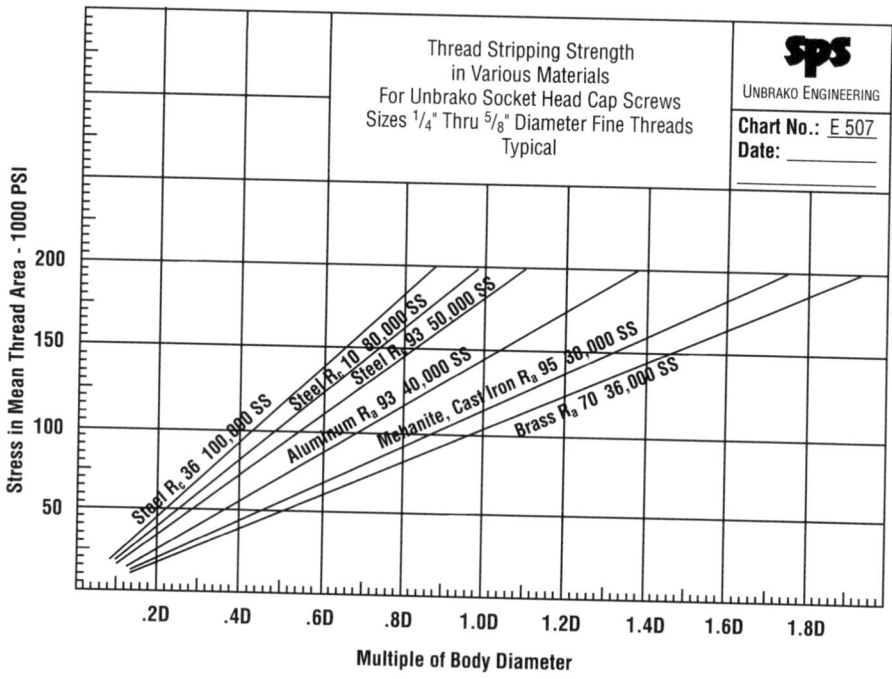

Figure 16. Thread stripping strength for nominal sizes ¹/₄ inch through ⁵/₈ inch with fine threads.

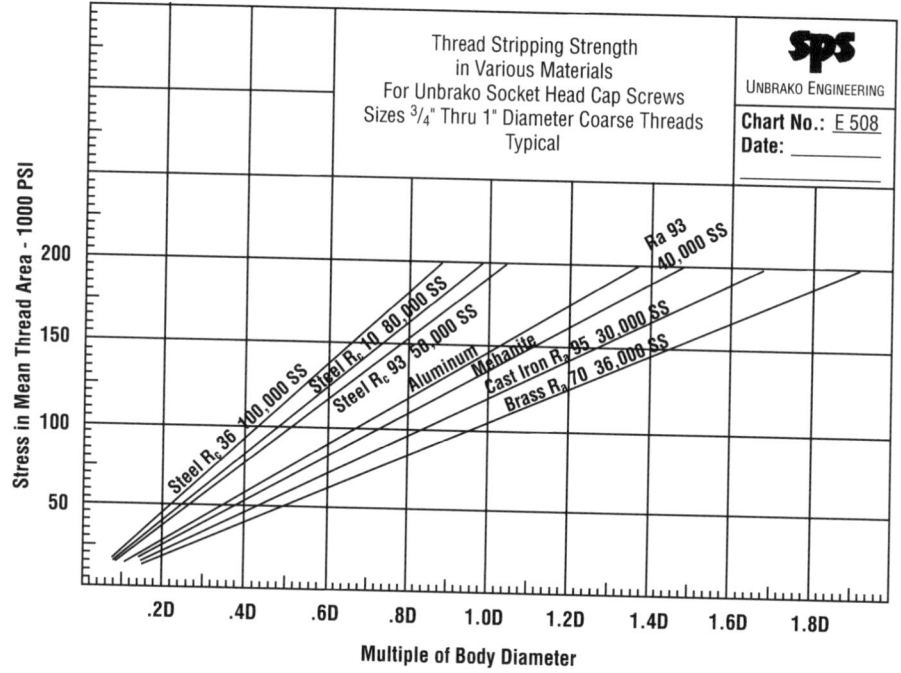

Figure 17. Thread stripping strength for nominal sizes ³/₄ inch through 1 inch with coarse threads.

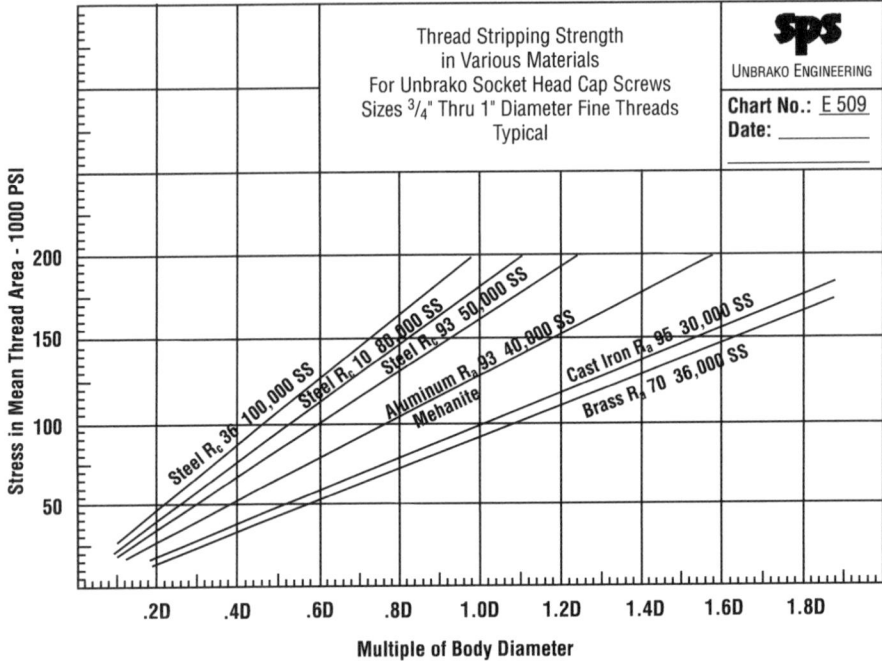

Figure 18. Thread stripping strength for nominal sizes ³/₄ inch through 1 inch with fine threads.

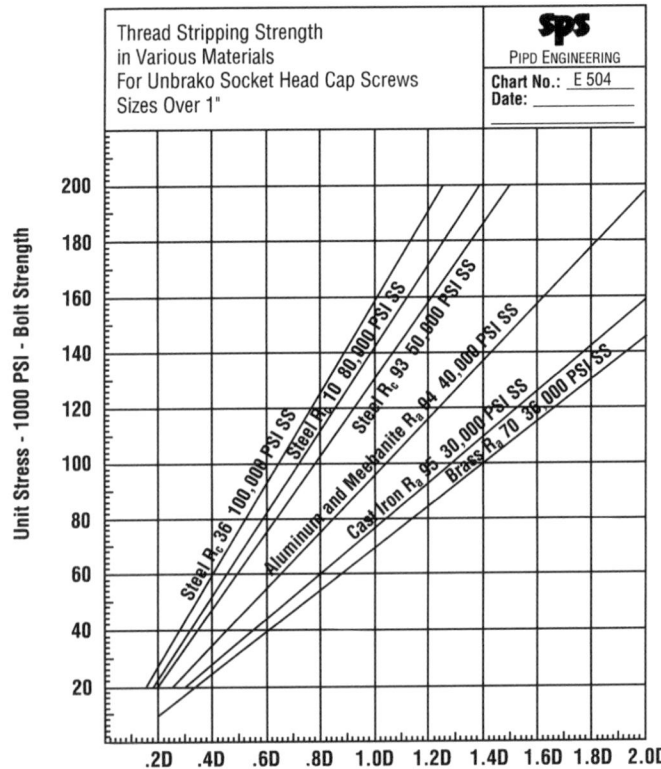

Figure 19. Thread stripping strength for nominal sizes over 1 inch.

Plating and coating threaded parts

Because they can change the diameters of a fastener, platings or coatings must be applied under strict quality regulations. The thickness of a coating can severely impact the torque and preload requirements, so special care should be taken to assure that coated fasteners remain with diameter limits. Plating and coating thickness can vary from as little as 0.00003" for gold flash to as much as 0.0005" for solder plate over nickel plate. However, most finishes range in thickness from 0.0002" to 0.0003". MIL-S-7742D provides the following guidelines for dimensioning threads for coating/plating.

Material Limits for Coated/Plated External Threads

When externally threaded parts are to be coated/plated, the threads should not be undercut more than 0.001" on all threads for which the specified pitch diameter tolerance does not exceed 0.0035". For threaded parts with a pitch diameter tolerance in excess of 0.0035", the specified minimum pitch diameter may be reduced by an amount equal to 0.3 times the pitch diameter tolerance, up to a maximum of 0.0015". To determine before coating/plating gaging limits for a uniformly coated/plated external thread, make the following reductions. Reduce the tabulated maximum pitch diameter by an amount equal to four times the maximum plating/coating thickness. Reduce the tabulated minimum pitch diameter, minimum major diameter, and minimum minor diameter by an amount equal to two times the minimum coating/plating thickness. Reduce the tabulated maximum major diameter and maximum minor diameter by an amount equal to two times the maximum coating/plating thickness. All thread elements must be in tolerance, as modified above, before coating/plating. Unless otherwise specified, coated Class 2A threads should meet the dimensional limits for Class 3A threads.

The following formulas, where d_2 is pitch diameter and t_c is coating thickness, can be used as a quick guide for predicting alteration in pitch diameter caused by coating/plating.

$$d_{2\ max}\ \text{before plating} = d_{2\ max}\ \text{after plating} - (t_{c\ max} \times 4)$$
$$d_{2\ min}\ \text{before plating} = d_{2\ min}\ \text{after plating} - (t_{c\ min} \times 4).$$

Material Limits for Coated/Plated Internal Threads

Internal threads may be overcut to allow for coating/plating thickness. Unless otherwise specified, before plating/coating gaging limits for a uniformly coated/plated internal thread should be increased as follows. Increase the tabulated minimum pitch diameter by an amount equal to four times the maximum coating/plating thickness. Increase the tabulated minimum major diameter and the minimum minor diameter by an amount equal to two times the maximum coating/plating thickness. Increase the tabulated maximum pitch diameter, maximum major diameter, and maximum minor diameter by an amount two times the minimum coating/plating thickness. All thread elements must be in tolerance, as modified above, before coating/plating.

Military specifications and requirements for fasteners

Mil-B-6812E *Military Specification, Bolts, Aircraft* provides stringent tensile strength requirements for bolts. As defined by the standard, the test specifies minimum ultimate tensile strength and minimum double shear strength for noncorrosion resistant steel bolts, corrosion resistant steel bolts, and aluminum alloy bolts. These strength requirements are shown in **Table 21**.

An important consideration for many fastener applications, especially the aerospace industry, is the strength to weight ratio of the fastener material. Strength to weight ratio is defined as the ratio of tensile strength to density. In applications where weight is an important

Table 21. Strength Requirements for Aircraft Bolts. *(Source, MIL-B-6812E.)*

Size	Ultimate Tensile Strength[1] (min. in pounds)				Double Shear Strength[2] (min. in pounds)	
		Steel		Al Alloy	Steel	Al Alloy
	Eyebolt	Hexagon Head Bolts			All Types of Bolts	
		Fine Thread[3]	Coarse Thread[3]	Fine Thread		
No. 6	-	-	-	-	2,120	1,080
No. 8	-	-	-	-	3,000	1,570
No. 10	1,150	2,210	1,800	-	4,250	2,092
1/4	2,450	4,080	3,360	1,100	7,360	3,650
5/16	3,910 (AN44)	6,500	5,660	2,030	11,500	5,700
5/16[4]	5,290 (AN45)	6,500	5,660	3,220	11,500	5,700
3/8	7,015	10,100	8,470	3,220	16,560	8,250
7/16	9,200	13,600	11,680	5,020	22,500	11,200
1/2	14,375	18,500	15,730	6,750	29,400	14,600
9/16	20,125	23,600	20,300	9,180	37,400	18,500
5/8	-	30,100	25,100	11,700	46,000	22,800
3/4	-	44,000	37,800	14,900	66,300	33,000
7/8	-	60,000	-	21,800	90,100	45,000
1	-	80,700	-	29,800	117,800	58,500
1-1/8	-	101,800	-	40,000	147,500	74,000
1-1/4	-	130,200	-	50,500	182,100	91,000

1) The values shown for the ultimate tensile strength are for minimum values and are based on:
    a. 125,000 PSI for noncorrosion-resistant and corrosion-resistant steel
    b. 62,000 PSI for aluminum alloy.
    The strength values shown for the eyebolts are based on the strength of the eye. The root area of the thread is the basis of calculation for the tensile strength of hexagon head bolts.
    Clevis bolts shall have tensile strengths equal to one-half of the requirements for hexagon-head bolts when used with AN320 or MS21083 nuts. Clevis bolts are intended primarily for use in shear applications.
2) Ultimate shear strengths are computed on the basis of 60 percent of the ultimate tensile strengths.
3) Class of thread is as specified on the applicable standard drawing.
4) Different from size $5/16$ above in the design of the eye section.

consideration, the strength to weight ratio becomes a factor (see **Table 22**).

All standard studs, bolts, hex cap screws, socket head screws, and nuts in diameters of $1/4$ inch and larger, "for critical applications where a high degree of reliability is required," that are procured by any branch of the military are required to meet the material specifications laid out in MIL-S-1222H *Military Specification, Studs, Bolts, Hex Cap Screws, Socket Head Cap Screws, and Nuts*. The grade, chemical, and identification requirements from that standard are shown in **Table 23**.

Washers

(The published informational catalogs of West Coast Lockwasher (WCL) Company were the source of much of the material in this section.)

Table 22. Strength to Weight Ratios for Various Fastener Materials. *(Source, ITT Harper.)*

Material	Tensile Strength ksi	Density lb./in.^{3}	Strength to Weight Ratio
Martensitic Stainless Steel (410, 416)	180	0.280	6.4
Aluminum (2002-T4)	60	0.098	6.1
Austenitic Stainless Steel (18-8) Strain Hardened	125	0.290	4.3
Titanium (Commercially Pure)	50	0.163	3.1
Nylon	12	0.041	2.9
Austenitic Stainless Steel (18-8) Annealed	80	0.290	2.8
Monel 400	80	0.319	2.5
Silicon Bronze	75	0.308	2.4
Brass	60	0.308	2.0
Mild Steel	50	0.282	1.8

Table 23. Material Grades, Chemical, and Identification Requirements for Fasteners. *(Source, MIL-S-1222H.)*

Material Type		Material Grade	Fastener Type	Chemical Requirement	I.D. Marking
Carbon and Alloy Steel		2	Stud, bolt, hex cap screw	SAE J 429	See Note 1
		5	Stud, bolt, hex cap screw	SAE J 429	See Note 2
		8	Stud, bolt, hex cap screw	SAE J 429	See Note 2
		2	Nut	SAE J 995	See Note 1
		5	Nut	SAE J 995	See Note 2
		8	Nut	SAE J 995	See Note 2
Alloy Steels		B7 See Note 5	Stud, bolt, hex cap screw	ASTM A 193	B7
		B16	Stud, bolt, hex cap screw	ASTM A 193	B16
		A574	Socket head cap screw	ASTM A 574	See Note 1
		4340	Socket head cap screw	ASTM A 574	4340
		L7	Stud, bolt, hex cap screw	ASTM A 320	L7
		L43	Stud, bolt, hex cap screw	ASTM A 320	L43
		2H	Nut	ASTM A 194	2H
		4	Nut	ASTM A 194	4
		7 See Note 9	Nut	ASTM A 194	7
Corrosion Resistant Steels	Austenitic	303	Stud, bolt, hex cap screw, socket head cap screw	ASTM A 194	303 See Notes 3 & 4
		303 Se See Note 6	Stud, bolt, hex cap screw, socket head cap screw	ASTM F 593	303 Se See Notes 3 & 4
		304	Stud, bolt, hex cap screw, socket head cap screw	ASTM F 593	304 See Notes 3 & 4
		305	Stud, bolt, hex cap screw, socket head cap screw	ASTM F 593	305 See Notes 3 & 4
		316	Stud, bolt, hex cap screw, socket head cap screw	ASTM F 593	316 See Notes 3 & 4

(Continued)

Table 23. *(Continued)* **Material Grades, Chemical, and Identification Requirements for Fasteners.**
(Source, MIL-S-1222H.)

Material Type		Material Grade	Fastener Type	Chemical Requirement	I.D. Marking
Corrosion Resistant Steels	Austenitic	321	Stud, bolt, hex cap screw, socket head cap screw	ASTM F 593	321 See Notes 3 & 4
		347	Stud, bolt, hex cap screw, socket head cap screw	ASTM F 593	347 See Notes 3 & 4
		384	Stud, bolt, hex cap screw, socket head cap screw	ASTM F 593	384 See Notes 3 & 4
	Martensitic	410 See Note 6	Stud, bolt, hex cap screw, socket head cap screw	ASTM F 593	410H, 410HT See Note 3
		416, 416 Se	Stud, bolt, hex cap screw, socket head cap screw	ASTM F 593	416H, 416HT, 416seH, 416SeHT See Note 3
		431	Stud, bolt, hex cap screw, socket head cap screw	ASTM F 593	431H, 431HT See Note 3
	Age Hardened	630	Stud, bolt, hex cap screw, socket head cap screw	ASTM F 593	630 See Note 3
	Austenitic	303	Nut	ASTM F 594	303 See Notes 3 & 4
		303 Se See Note 6	Nut	ASTM F 594	303Se See Notes 3 & 4
		304	Nut	ASTM F 594	304 See Notes 3 & 4
		305	Nut	ASTM F 594	305, See Notes 3 & 4
		384	Nut	ASTM F 594	384 See Notes 3 & 4
		316	Nut	ASTM F 594	316 See Notes 3 & 4
		321	Nut	ASTM F 594	321 See Notes 3 & 4
		347	Nut	ASTM F 594	347 See Notes 3 & 4
	Martensitic	410 See Note 6	Nut	ASTM F 594	410H, 410HT See Note 3
		416	Nut	ASTM F 594	416H, 416HT See Note 3
		416 Se	Nut	ASTM F 594	416seH, 416seHT See Note 3
		431	Nut	ASTM F 594	431H, 431HT See Note 3
	Age Hardened	630	Nut	ASTM F 594	630, See Note 3
Nickel Alloys	Ni-Cu	400	Stud, bolt, hex cap screw, socket head cap screw	QQ-N-281, class A	NC
	Ni-Cu-Al	500	Stud, bolt, hex cap screw, socket head cap screw	QQ-N-286, class A	K, See Note 8

(Continued)

Table 23. *(Continued)* **Material Grades, Chemical, and Identification Requirements for Fasteners.** *(Source, MIL-S-1222H.)*

Material Type		Material Grade	Fastener Type	Chemical Requirement	I.D. Marking
Nickel Alloys	Ni-Cu	400 See Note 7	Nut	QQ-N-281, class A	NC
		405	Nut	QQ-N-281, class A	NC-R
	Ni-Cu-Al	500	Nut	QQ-N-286, class A	K
Copper Alloys	N	462 See Note 6	Stud, bolt, hex cap screw	ASTM F 468	462, See Note 3
		464	Stud, bolt, hex cap screw	ASTM F 468	464, See Note 3
		482	Stud, bolt, hex cap screw	QQ-B-637	482, See Note 3
	P	510 See Note 6	Stud, bolt, hex cap screw	ASTM F 468	510, See Note 3
		544	Stud, bolt, hex cap screw	ASTM B 139	544, See Note 3
	Ni-Al Bronze	632	Stud, bolt, hex cap screw	QQ-C-465	632, See Note 3
	Silicon Bronze	655 See Note 6	Stud, bolt, hex cap screw	ASTM F 468	655, See Note 3
		661	Stud, bolt, hex cap screw	ASTM F 468	661, See Note 3
	M	670 See Note 6	Stud, bolt, hex cap screw	ASTM B 138	670, See Note 3
		675	Stud, bolt, hex cap screw	ASTM F 468	675, See Note 3
	N	462	Nut	ASTM F 467	462, See Note 3
		464	Nut	ASTM F 467	464, See Note 3
	P	510	Nut	ASTM F 467	510, See Note 3
	S	655	Nut	ASTM F 467	655, See Note 3
		661	Nut	ASTM F 467	661, See Note 3
	M	675	Nut	ASTM F 467	675, See Note 3
Titanium Alloy		T-7	Stud, bolt, hex cap screw	MIL-T-9047	T7
Aluminum Alloys		2024	Stud, bolt, hex cap screw	ASTM F 468	2024, See Note 3
		6061	Stud, bolt, hex cap screw	ASTM F 468	6061, See Note 3
		7075	Stud, bolt, hex cap screw	ASTM F 468	7075, See Note 3
		2024	Nut	ASTM F 467	2024, See Note 3
		6061	Nut	ASTM F 467	6061, See Note 3
		7075	Nut	ASTM F 467	7075, See Note 3

Notes:
1) Individual fastener marking to identify the material grade is not required.
2) Grade 5 shall be marked with three radial lines equally spaced for externally threaded fasteners and three circumferential dashes equally spaced for nuts. Grade 8 shall be marked with six radial lines equally spaced for externally threaded fasteners and six circumferential dashes equally spaced for nuts.
3) When specified each fastener shall be marked with the material grade.
4) In addition to the material grade, the marking shall include the heat treatment symbol "An" when the fastener is machined from annealed stock or when the fastener is annealed after forming, or the identification symbol "SH" when the fastener is cold headed and rolled from strain hardened stock thus acquiring a degree of cold work.
5) Suitable nuts for B7 externally threaded fasteners are grades 2H, 4, and 7, and suitable nuts for B16 externally threaded fasteners are grades 4 and 7.

(Continued)

Table 23. *(Continued)* **Material Grades, Chemical, and Identification Requirements for Fasteners.** *(Source, MIL-S-1222H.)*

6) The following groups of material are considered equivalent and interchangeable. When approved by the contracting activity, the contractor may select any alloy within each group:
   a. 303,303 Se, 304, 305, 384
   b. 32L, 347
   c. 410, 416, 416 Se, 431
   d. 462, 464, 482
   e. 510, 544
   f. 655, 661
   g. 670, 675
7) The following groups of material are considered equivalent and interchangeable. When approved by the contracting activity, the contractor may select any alloy within each group:
   a. 400, 405
8) Heat or lot number and the manufacturer's symbol are required) for fasteners $1/2$ inch in diameter and larger. Only the material symbol shall be required for fasteners less than $1/2$ inch in diameter. Numbers used for

Plain Flat Washers

Flat washers are intended to act as bearing surfaces that prevent bolt heads and nuts from damaging work surfaces. In addition, since their surface is known to be smooth and flat, they allow more accuracy in achieving installation torque or strain measurements. Several standards issuing organizations, in countries throughout the world, have specified preferred and alternative sizes for internal and external dimensions of flat washers. One manufacturer claims to keep over 7,000 different flat washers in inventory. In addition, it should be remembered that manufacturers of washers offer custom size washers in almost any conceivable dimension and material. For instance, one leading company offers sizes ranging from the smallest at 0.022" internal diameter and 0.040" outside diameter to the largest with dimensions of 2.515" internal diameter to 3.250" outside diameter. Each of the hundreds of available sizes is available in seventeen different thicknesses, eighteen different materials, and dozens of finishes. Typically these custom washers are available with the following tolerances: external diameter = +0.020" to –0.005"; internal diameter = ±0.010"; and thickness = ±0.003" for washers up to 0.032" thick, ±0.006" for washers from 0.033" to 0.062" thick, ±0.009" for washers from 0.063" to 0.093" thick; and ±0.010" for washers 0.094" thick and over. More exact tolerances are available on special order. Therefore, should a washer size be required that is not included in a printed standard of suggested or recommended sizes, that washer will almost certainly be available from one of several manufacturers or suppliers.

The inside diameter of conventional flat washers have three distinct profiles that are a result of the punch process that is used to produce the part. The punch, pushing through the washer, creates a rounded corner on the entry side. The center section has approximately parallel sides as the punch passes through, and the exit side is tapered resulting from punch breakout. The internal dimension of the washer is measured across the approximately parallel flats. The diameter of the breakout taper may exceed the specified internal diameter on the washer by an amount equal to 25% of the washer thickness (see *Figure 20*).

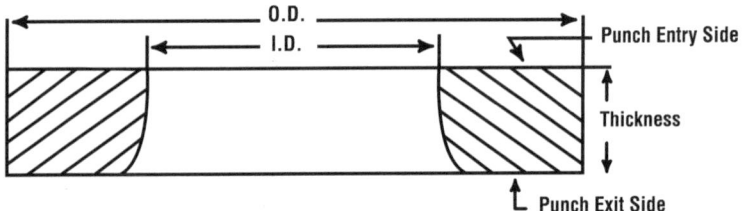

Figure 20. Profile view of a flat washer. (Source, West Coast Lockwasher Company.)

Helical Spring Lockwashers

Helical spring lockwashers, which are among the most widely used antiloosening assembly components available, have been specified in countless assembly applications because of their reliable performance. They enhance the security of general industrial assemblies by 1) providing greater bolt tension per unit of applied torque, 2) providing hardened bearing surfaces that create more uniform torque control, 3) allowing more uniform load distribution through controlled radii (section) cutoff, 4) providing protection against looseness resulting from vibration and corrosion, and 5) providing optimum locking performance in applications with hardened faying or bearing surfaces.

Helical spring lockwashers are slightly thicker on their inside diameter than on their outside diameter (see *Figure 21*). Therefore, the design thickness is, in theory, situated midway across the surface of the washer and may be determined with the formula

$$T = (t_o + t_i) \div 2$$

where T = specified thickness
t_o = outside thickness
t_i = inside thickness.

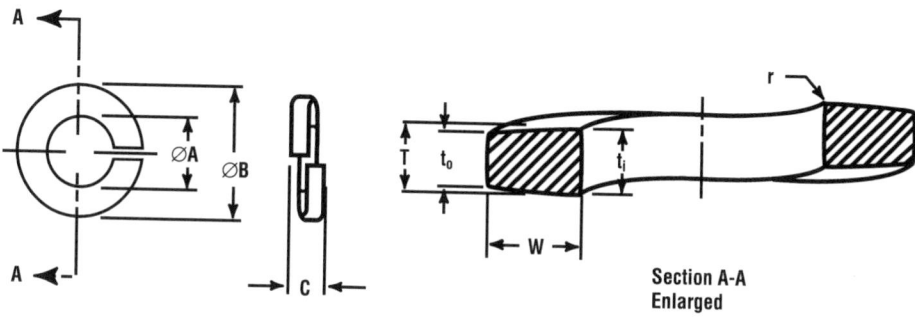

Figure 21. Sectional view of a helical spring lockwasher.

The "split" in the helical spring lockwasher absorbs initial driving torque and visually closes under nominal bolt loading. When tension in the assembly is reduced and loosening occurs, it provides resistance to the backoff rotation of the screw or nut. Helical washers are trapezoidal in section. After the single-coil spring closes to the flat condition, further loading results in additional deformation of the washer caused by a complex twisting of the trapezoidal section and a slight increase in the diameter of the washer under load (see *Figure 22*). The spring rate developed by the final deformation is very high and provides a reactive load that is equivalent to a significant increase in effective bolt length. Bolts stretch under load. The longer the effective length of a bolt, the more it can stretch. A

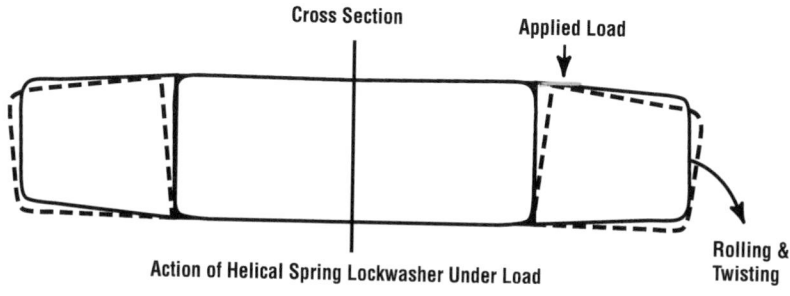

Figure 22. Action of a helical spring lockwasher under load. (Source, West Coast Lockwasher Company.)

hardened steel bolt, stressed at 60,000 PSI, will elongate approximately 0.002" per inch of effective length. Therefore, a long bolt can become a very effective spring that, like a spring when stretched, will attempt to return to its original length. This provides a clamping or tightening force to the assembly. Two terms are used to define and measure these forces: spring rate and effective bolt length.

Spring rate is the ratio of load to deflection in the loaded part. It is deflection related to load and is expressed in terms of the amount of load required to achieve specific levels of deflection. Optimum joint performance is achieved when the spring elements of the fastening system have a spring rate that is low enough to assure that any yielding of the joint members in compression will not significantly reduce the designed tensile stress of the fastener. To obtain this optimal condition, the conventional solution is to utilize the spring characteristics of long bolts, but an auxiliary spring element, such as a helical spring lockwasher, is an effective alternative.

Tests run at Lawrence Technological University have shown that the typical helical spring lockwasher exhibits a spring rate after flattening that is approximately 70% more effective than a flat washer of the same thickness. This means that the effective bolt length in the joint is increased by the thickness of the flattened helical spring lockwasher, plus the equivalent length provided by the spring rate derived from the visually flattened washer. Tested at 75% of the hardened bolt proof load, the equivalent bolt length is shown in **Table 24**. The total contribution of a helical spring lockwasher to the integrity of an assembly, in addition to the commonly recognized frictional resistance to backoff rotation, includes the reactive length added to the bolt by the washer thickness and by the tension of spring rate generated by its compression.

Testing helical spring lockwashers. In preparing a helical spring lockwasher for hardness testing, it should first be twisted to remove the helix so that it forms a near flat surface. Due to the trapezoidal section of the washer, both sides must be filed or ground flat to assure accurate readings (see *Figure 23*). Care must be taken to assure that surface temperature does not exceed 250° F (121° C) during this operation. Essential requirements

Table 24. Equivalent Bolt Length. *(Source, West Coast Lockwasher Company.)*

Washer Size (inches)	Additional Spring Rate Contributed by Compressed Washer	
	Regular Lockwasher	Heavy Lockwasher
3/8	65.6%	74.1%
7/16	68.6%	72.2%
1/2	73%	80.3%

Comparison of Washerless Joint to Joint with Lockwasher			
Joint without Washer		Joint with 3/8 inch Helical Spring Lockwasher	
Bolt head height	0.24"	Bolt head height	0.24"
Two assembled 1/8" plates	0.25"	Two assembled 1/8" plates	0.25"
Nut thickness	0.32"	Helical washer thickness	0.10"
		Nut thickness	0.32"
Total thickness	0.81"	**Total thickness**	0.91"
Half of head height (deflected)	0.12"	Half of head height (deflected)	0.12"
Total thickness of assembled plates	0.025"	Total thickness of assembled plates	0.025"
Half of nut thickness (deflected)	0.16"	Total thickness of washer	0.10"
		Additional spring rate contributed by compressed washer	0.656"
Equivalent bolt length	0.53"	Half of nut thickness (deflected)	0.16"
		Equivalent bolt length	1.286"

HARDNESS

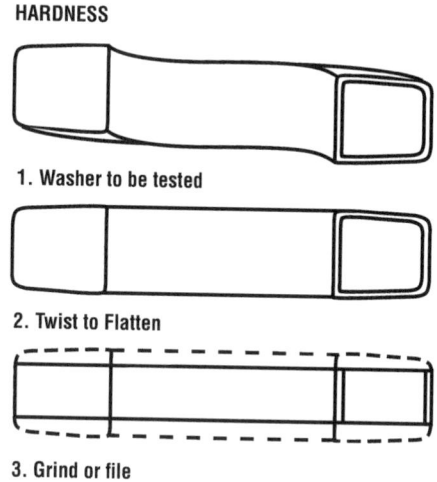

1. Washer to be tested

2. Twist to Flatten

3. Grind or file

*Figure 23. Steps required for preparing a helical spring lockwasher for testing.
(Source, West Coast Lockwasher Company.)*

of the Rockwell test are that the penetrator be perpendicular to the surface of the test piece, and that the test piece must not move, even in the slightest degree, when the test load is applied. Since one point of hardness represents a depth of only 0.00008", a movement of only 0.001" could cause an error of over 10 Rockwell numbers.

Penetration should be made in the center of the washer's flattened surface, and readings should not be taken too close together. If the indentation is made too close to the edge of the test washer, the material will yield, giving incorrect low readings. Also, since the area surrounding an indent will be cold worked, subsequent indentations made in these areas will give incorrect readings—usually higher than virgin material. The penetration test should be performed on only one side of the washer.

Applicable hardness requirements are as follows.

Material	HR$_C$	HV
Carbon Steel	38-46	372-458
Austenitic Stainless Steel	35-43	345-423
Monel K500	33-40	327-392
	HR$_B$	HV
Aluminum Alloy	75-97	137-222
Phosphor Bronze	90 min.	185 min.
Silicon Bronze	90 min.	185 min.

In addition, carbon steel spring washers, tested in accordance with ANSI B18.21.1, should meet the following limits for decarburization.

Dia. of Section of Washer	Max. Depth of Free Ferrite	Max. Total Affected Depth
Up to 0.140" incl.	0.002"	0.006"
0.140" to 0.250" incl.	0.003"	0.008"
0.250" to 0.375" incl.	0.004"	0.010"
0.375" to 0.500" incl.	0.006"	0.015"

Spring Washers

Spring washers are specifically designed to provide a compensating spring force and sustain a load or absorb a shock. Many design variations have evolved to best serve one or the other of these two basic functions or to optimize both functions in a single part within specific I.D./O/D. limits. Two principal factors continually increase the requirements for spring washers: 1) the continuing effort to downsize many end products, relative to both weight and cost, creates a need for small, multifunctional assembly components such as washers that support a load, span a hole, or both, while providing a compensating spring force; and 2) automated assembly requires some "play" or tolerance in the "fit" of components. Spring tension is needed to compensate for these tolerances.

Recognizing these two broad areas of influence, it can be stated that the more common applications for spring washers are 1) to take up "play" in assemblies due to cumulative tolerances, 2) to compensate for small dimensional changes in assembled components, 3) to eliminate end-play or rattles, 4) to maintain fastener tension, 5) to compensate for expansion and contraction or cold flow of material, and 6) to absorb intermittent shock loads and function as working springs capable of providing controlled reaction under dynamic loads.

Design considerations. Load and deflection are the key characteristics of a spring washer as it must be determined how much the washer will deflect under a given load and at what point it will flatten. These values are commonly expressed in Load/Deflection (L/D) curves with load (applied force) measured on one axis and washer deflection on the other. A typical L/D curve for a simple coil spring is shown in *Figure 24*. Point "A" on the curve represents a spring upon which no force has been applied and no deflection has taken place. The vertical axis (Load) is measured in increments of ten pounds of pressure, and the horizontal axis (Deflection) represents deflection in one-inch increments. It can be seen that with a load of 20 pounds, the spring deflects two inches (point "C"). At point "E", the spring is fully compressed. The graph represents a linear curve, or constant spring rate, where the sample spring deflects one inch for every increase of ten pounds applied to the spring. Most spring washers do not perform this consistently.

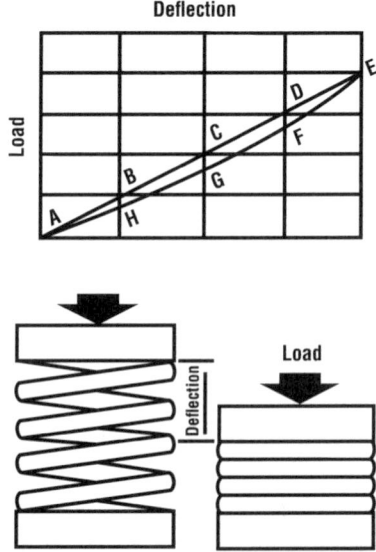

Figure 24. Load and deflection curve for a spring washer. (Source, West Coast Lockwasher Company.)

As the force is gradually released, it does not return to the full extension it had before deflection. This can be seen on the return at points "F," "G," and "H." At point "F," for example, at a deflection of three inches, the spring is supplying just under thirty pounds of reactive force. This results from the spring having used some of its stored energy to overcome friction caused by the bearing surface at both ends. Some stored energy is also lost through the increased temperature resulting from intermolecular friction caused by the initial deflection.

When considering washer design, the obvious relationship between washer thickness and load bearing characteristics should be kept in mind. Likewise, there is an inverse relationship between washer thickness and spring compensation or deflection. By varying either of these characteristics, the performance capabilities of a washer can be predicted by the designer.

Load/deflection characteristics. When specifying a spring washer, it is important to analyze both functional and physical requirements. In considering load requirements, two basic types (static and dynamic) must be recognized in addition to the amount of load. In a static load environment, the basic function of the spring washer is to retain load. In such an environment, the elastic load of the material may be exceeded. In a dynamic environment, the washer functions as a regularly flexing spring, and the elastic limits of the material must not be exceeded. Loading a washer beyond its yield strength will result in permanent distortion of its crown height. The type and magnitude of the load to which a spring washer will be subjected, and the reactive force it will be required to exert, are the primary factors that determine the type of spring washer best suited to a specific application. This range of deflection, or "spring travel," is an important element in spring washer design. Values for maximum load and maximum deflection for both single and multiple wave washers can be determined mathematically. The following equations, while useful as guidelines, can produce values that can vary up to 35% from actual test results. If load and deflection values are exceeded, a spring washer cannot perform within its elastic limits. Therefore, to avoid overstressing the washer, a general design rule when using these formulas is to select a washer that has twice the required deflection.

Estimating load (P) in pound-feet and deflection (f) in inches for single wave washers:

$$P = (S \times [D - d] \times t^2) \div 6 \times D$$
$$f = (S \times D^2) \div (6 \times [E \times t]).$$

Estimating load (P) in pound-feet and deflection (f) in inches for multiple wave washers:

$$P = (S \times N^2 \times t^2 \times [D - d]) \div (.75 \times [D + d])$$
$$f = (S \times 2D^2) \div (12\ E \times t \times N^2)$$

where: N = number of waves
         P = load in pound-feet
         E = modulus of elasticity in PSI (30,000,000 PSI for steel)
         t = material thickness in inches
         f = deflection in inches
         d = inside diameter in inches
         D = outside diameter in inches
         S = max. allowable stress in PSI (200,000 PSI for steel).

Space envelope. Wave washers and conical washers increase in diameter as they are compressed. Therefore, allowable inside and outside diameter limits are an important consideration when specifying spring washers. The overall space occupied by a spring washer can be described as a hollow cylinder as shown in *Figure 25*. This cylinder of space

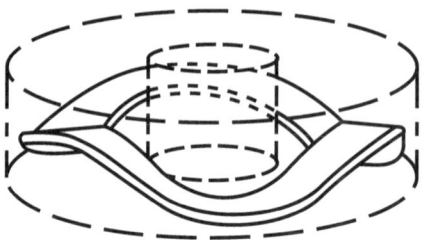

Figure 25. Space cylinder (envelope) for a spring washer. (Source, West Coast Lockwasher Company.)

must be recognized in assembly design considerations which may restrict the acceptable dimensions of the washer. Generally, the larger the washer outside diameter, the greater the load that can be supported and distributed.

Operational environment. The environment in which a spring washer operates can affect the anticipated performance of the washer in terms of load bearing and reactive characteristics. Temperature and exposure to corrosive agents are important environmental considerations, and washer materials specifications must respect these factors. The limitations imposed by even relatively low ambient temperatures on various spring washer materials are shown below. The indicated temperature limits can, of course, be exceeded, but this will result in increased relaxation of the washer under load.

Material	Recommended Operating Limit (°F)
SAE 1050 Steel	250°
SAE 1065 Steel	250°
425 Phosphor Bronze	225°
Beryllium Copper	225°
Spring Brass	150°

Belleville Conical Disc Spring Washers

Disc springs are among the most widely used tension generating washers. Common uses include spanning alignment holes, distributing bearing loads, and generating and sustaining the tension needed to hold assemblies together. Belleville disc spring washers (see *Figure 26*) combine high spring force for short movement with high-energy storage capacity. In true Belleville washers, the ratio of material thickness to rim width is held to about one in five, and crown height should not exceed 40% of the material thickness to assure that, when loaded to flat, yield strength is not exceeded and the washer will return to full crown height when compression is removed. However, commercial springs are not held to specific outside diameter/crown height/thickness ratio, and the crown height to thickness ratio is traditionally much greater for commercial disc springs. Therefore, when loaded to flat, their yield point may be exceeded. These washers are often used in applications where they function entirely within their elastic range. When loaded beyond their yield point, they will act in a consistent manner over a reduced crown height although they have, in effect, become deformed.

By carrying the thickness/crown height relationship, it is possible to meet a wide range of load/deflection requirements with disc spring washers. A crown height to thickness ratio ranging from 0.4 to 0.8, for example, produces a fairly constant spring rate (*Figure 27a*), while increasing the thickness ratio up to 1.4 will provide a positive rate of increase in the load up to 100% deflection (*Figure 27b*). Once a ratio of 1.4 is reached, the disc spring will show a constant load over a fairly large deflection, making it useful in applications where

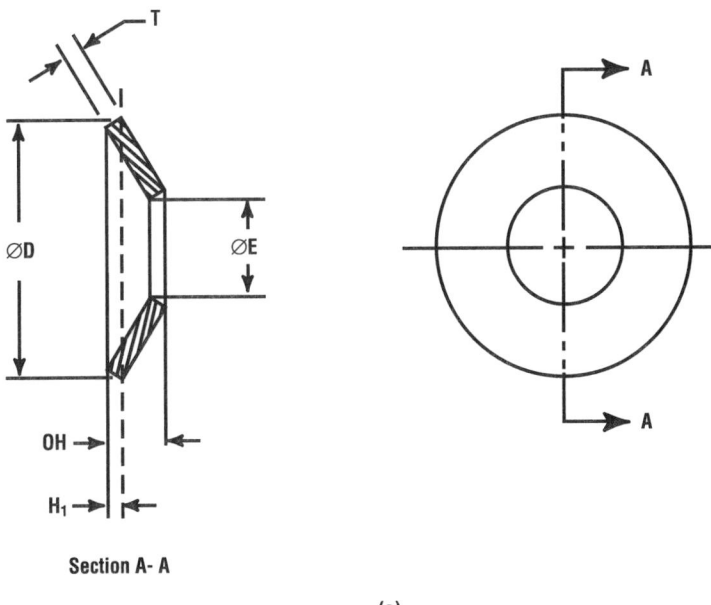

(a)

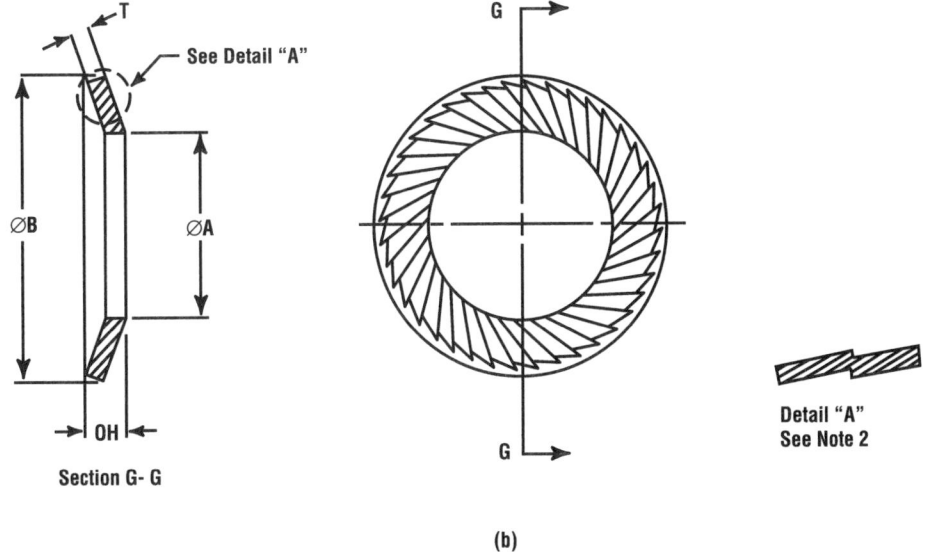

(b)

Figure 26. (a) Belleville spring washer in cross section. (b) Serrated Belleville spring washer. Serrated washers have sawtooth surfaces top and bottom with a rake angle of 30° +5°.

extreme wear conditions must be absorbed (*Figure 27c*). Finally, when the crown height to thickness ratio exceeds 1.4, yielding will occur (*Figure 27d*).

Disc springs can be stacked to enhance performance characteristics. When stacked in parallel (*Figure 28a*), load bearing characteristics are enhanced. When stacked in series (*Figure 28b*), greater travel can be achieved. Combination stacking, in both parallel and series, can be used to increase both load bearing and deflection.

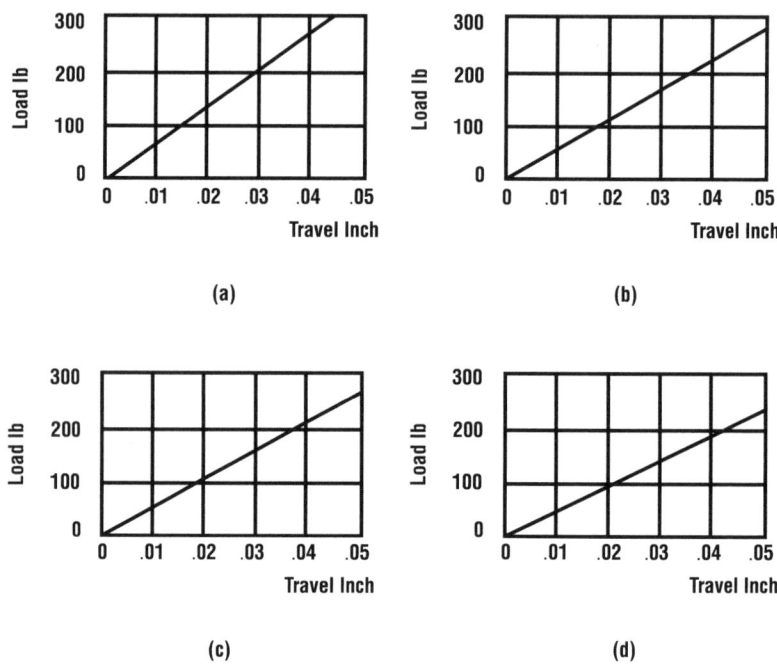

Figure 27. Effect of thickness/crown height ratio on disc springs (see text).
(Source, West Coast Lockwasher Company.)

Figure 28. (a) Belleville washers stacked in parallel. (b) Belleville washers stacked in series.

Taps and Tapping

Taps are among the most important and widely used tools in manufacturing. The selection of the ideal tap for a given operation is essential for assuring a strong, reliable threaded hole. Tapping can be done either by hand or machine, using one of the three basic varieties of taps: taper, plug, and bottoming. All three are identical in thread form and size, but differ in the chamfered (tapered) section at the point of the tap as follows.

Tapered taps are tapered for approximately the first seven to ten threads.

Plug taps are tapered for approximately three to five threads.

Bottoming taps are only tapered for one to two threads. Some manufacturers also offer a "semi-bottoming" tap that has two to two and one-half threads on the chamfer. This configuration is not universal.

Taps are also identified by a style designation as Style 1, Style 2, or Style 3 taps. Style refers to the tap diameter and physical characteristics of the tap. This identification method, which appears to have originated with the United States Cutting Tool Institute (and is used by ASME/ANSI in their Standard ASME/ANSI B94.9), identifies Styles as follows.

Style 1 taps are Machine Screw sizes #0–12 and Metric sizes 1.6 mm to 5 mm. These taps have external centers on the thread and shank ends that can be removed on the thread end for bottoming taps.

Style 2 taps are in Fractional sizes $1/4"$ – $3/8"$ (0.250 to 0.3750") and Metric sizes 6 mm to 10 mm. They may have an external center on the thread end that can be removed for bottoming taps, and a partial cone on the shank end that is about one-quarter of the diameter of the shank.

Style 3 taps are all taps above the Fractional size $3/8"$ and the Metric size 10 mm. They have internal centers on both the thread and shank ends.

Dimensions for all Style 1, Style 2, and Style 3 Screw, Fractional, and Metric sizes, plus applicable tolerances, are given in **Table 1**.

Tap nomenclature *(Figure 1)*

Angle of Thread – The angle included between the flanks of the thread measured in an axial plane.

Axis – The longitudinal centerline of the tap.

Back Taper – A slight axial relief on the thread of the tap that makes the pitch diameter of the thread near the shank somewhat smaller than that of the chamfered end.

Basic – The theoretical or nominal standard size from which all variations are made.

Chamfer – The tapering of the threads at the front end of each land of a tap by cutting away and relieving the crest of the first few teeth to distribute the cutting action over several teeth. When the tapering amounts to 7 to 10 threads, the tap is called a "taper" tap; 3 to 5 threads, a "plug" tap; and 1 to 2 threads, a "bottoming" tap.

Chamfer Angle – The angle formed between the chamfer and the axis of the tap, measured in an axial plane at the cutting edge.

Chamfer Relief – The gradual decrease in land height from cutting edge to heel on the chamfered portion, to provide clearance for the cutting action as the tap advances.

Core – The diameter of the body of the tap between the flutes.

Crest – The top surface joining the two flanks of a thread. The crest of an external thread is at its major diameter, while the crest of an internal thread is at its minor diameter.

Cutting Face – The leading side of the land in the direction of rotation for cutting on which the chip impinges.

Dryseal – A pipe threaded fuel connection for both external and internal application designed for use where the assembled product must withstand high fluid or gas pressures without the use of a sealing compound, or where a sealer is functionally objectionable.

Flutes – The longitudinal channels formed in a tap to create cutting edges on the thread profile and to provide chip spaces and cutting fluid passages.

Heel – The back edge of the land.

Height of Thread – The distance between the crest and the base of a thread, measured normal to the axis.

Helical Flute – A flute with uniform axial lead and constant helix in a helical path around the axis of a cylindrical tap.

Helix – The angle at which the thread is ground to give the tap its advance. Also used to describe the angle of the flutes in Spiral Flute taps.

Hook Face – A concave cutting face, usually specified either as Chordal Hook or Tangential Hook.

Chordal Hook – The angle between the chord passing through the root and crest of a thread form at the cutting face, and a radial line through the crest at the cutting edge.

Tangential Hook Angle – The angle between a line tangent to a hook cutting face at the cutting edge and a radial line to the same point.

Interrupted Thread – A tap having an odd number of lands, with every other tooth along with the thread helix removed.

Land – The threaded section of the tap between the flutes.

Lead – The distance a screw thread advances axially in one complete turn. On a single lead screw or tap, the lead and pitch are identical. On a double lead screw or tap, the lead is twice the pitch, etc.

Pitch – The distance from any point on a screw or tap thread to a corresponding point on the next thread, measured parallel to the axis. The pitch equals one divided by the number of threads per inch.

Pitch Diameter – On a straight thread, the diameter of an imaginary co-axial cylinder, the surface of which would pass through the thread profile at such points as to make equal the width of the threads and the width of the spaces cut by the surface of the cylinder. On a taper thread, the diameter at a given distance from a reference plane perpendicular to the axis of an imaginary co-axial cone, the surface of which would pass through the thread profile at such points as to make equal the width of the threads and the width of the spaces cut by the surface of the cone.

Rake – Any deviation of a straight cutting face of the tooth from a radial line. Positive Rake means that the crest of the cutting face is angularly advanced ahead of the balance of the face of the tooth. Negative Rake means that the same point is angularly behind the balance of the cutting face of the tooth. Zero Rake means that the cutting face is directly on the centerline.

Relief – Back taper is a form of relief. This usually equals .0005" to .001" per inch of thread. See *Thread Relief.*

Root – The bottom surface joining the flanks of two adjacent threads. The root of an external thread is at its minor diameter, while the root of an internal thread is at its major diameter.

Spiral Point (Chip Driver) – A supplementary angular fluting cut in the cutting face of the land at the chamfer end. It is slightly longer than the chamfer on the tap and of the opposite hand to that of rotation.

Thread Relief – The clearance produced by removal of metal from behind the cutting edge. When the thread angle is relieved from the heel to cutting edge, the tap is said to have "eccentric" relief. If relieved from heel for only a portion of land width, the tap is said to have "con-eccentric" relief (a portion of the thread is concentric before the relief begins). This can be $^1/_3$ concentric and $^2/_3$ eccentric.

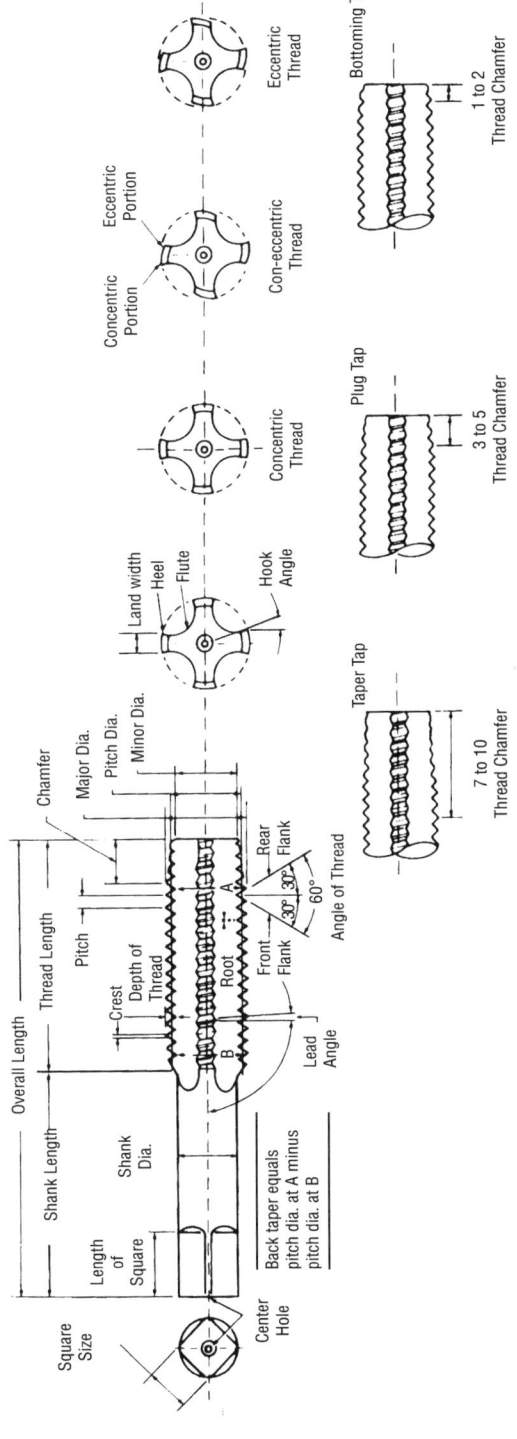

Figure 1. Tap terminology. (Source, Kennametal Inc.)

Types of taps

Standards for taps classify taps as having either "cut" or "ground" threads. Cut Thread Taps are created by turning and, for today's high accuracy and quality requirements, they cannot consistently produce threads to the same degree of excellence that can be achieved with Ground Thread Taps which are produced by grinding wheels to very exacting tolerances and have become the "standard" for taps. In fact, most manufacturers have abandoned Cut Thread Taps completely, and it is probable that it would be difficult to find one on the industrial market today. Even though Cut Thread Taps are uncommon, **Table 38** has been included for dimension limits for Unified threads produced with cut thread taps. Other than this table, all references and dimensions in this section refer to Ground Thread Taps. Thread limits and tolerances for Unified Thread and Metric Thread Ground Thread Taps are given in **Tables 35** through **37**. All Ground Thread Taps made to industry standards are marked with their nominal size, the tpi of the tap, and a symbol to identify the thread type (these symbols are identified in **Table 2**). In addition, left-hand threads are marked LH, and high-speed steel taps are marked HS.

Table 3 can be used for determining size and configuration availability of the first nine varieties of taps listed below. For example, #2 at the top of the Table refers to *Spiral Pointed Plug Taps*. From the Table, it can be seen that this tap is available in size 0-80 NF with an H2 Limit (see the section that immediately follows for a discussion of the H-Limit system). The "P" indicates that it is available only as a plug tap. The appearance of "TPB" in a box indicates that the tap size and type is available in all three configurations: taper, plug, and bottoming. The illustrations in this section were provided by Fastcut Tool Corporation.

1. Straight Fluted Taps. General purpose taps that can be used by hand or under power in both blind and through holes in most materials. The flutes retain chips during use, and for holes deeper than $1^1/_2$ times the diameter three flute taps should be used. Special versions designed to produce small, powdery chips are available for tapping cast iron, cast brass, and other brasses. They are available in all Machine Screw Sizes, Fractional Sizes through $1^1/_2$", and Metric sizes through M36, in taper, plug, and bottoming configurations. See Table 1 for dimensions, and Table 3 for size availability.

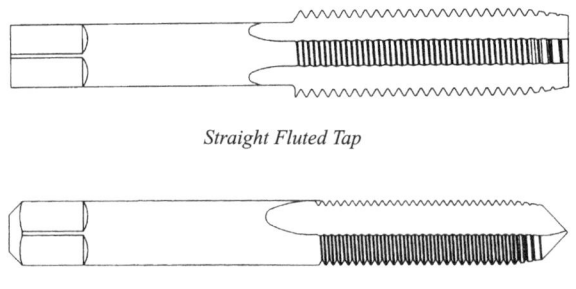

Straight Fluted Tap

Straight Fluted Tap

2. Spiral Pointed Plug Taps. Primarily intended for machine use, these taps can also be used by hand, although obtaining a true start in a hole can be difficult with any spiral pointed tap. Although designed for tapping through holes (and usually considered the first choice for this operation), they can be used in blind holes that are deep enough to allow for chip accumulation in the bottom of the hole. The spiral point forces chips ahead of the tap, which prevents clogging and recutting of the chips, and makes long holes, in excess of $1^1/_2$ diameters, possible to cut. Suitable for most materials, but especially good performance can be obtained in those with high ductility. Available in Machine Screw sizes, Fractional sizes through $^3/_4$", and Metric sizes through M20. See Table 1 for dimensions, and Table 3 for size availability.

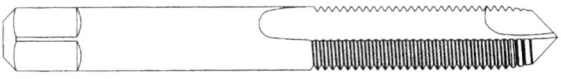

Spiral Pointed Plug Tap

3. Spiral Pointed Bottoming Taps. Short, blind holes that require full threads close to the hole's base can be cut with this tap. The notched front of the tap allows for additional area to accommodate cut chips; and the thicker chip, resulting from the short chamfer, tends to break up chips rather than leaving them long and stringy. These taps can be used by hand or under power in most materials, especially those with high ductility. Available in Machine Screw sizes and Fractional sizes through $^5/_{16}$". See Table 1 for dimensions, and Table 3 for size availability.

Spiral Pointed Bottoming Tap

4. Spiral Pointed Fluteless Plug Taps. Also called Spiral Point-only Taps, these taps are designed for tapping short through holes. The spiral point forces the chips ahead of the tap, which prevents clogging and recutting of the chips. They can be used by hand or under power in most materials. Available in Machine Screw sizes 4–12 and popular Fractional sizes to $^1/_2$". See Table 1 for dimensions, and Table 3 for size availability.

Spiral Pointed Fluteless Plug Tap

5. Spiral Fluted Taps. The helical flutes are the same hand of flute lead as the hand of cut. These taps can be used for most materials but are especially useful for tapping blind holes in mild steel and brass either by hand or under power. The spiral flutes draw the chips out of the hole, and also allow the tap to bridge a keyway or slot inside the hole without binding. The helix angle is 25° to 35°. Available in either plug or bottoming configurations in Machine Screw sizes 4–10 and most popular Fractional sizes through $^1/_2$". See Table 1 for dimensions, and Table 3 for size availability.

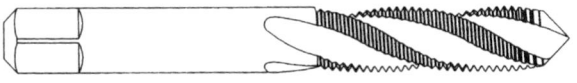

Spiral Fluted Tap

6. *Fast Spiral Fluted Taps.* Intended for tapping relatively deep blind holes, these taps have the same advantages as Spiral Fluted Taps but the helix angle is 45° to 60°. They perform best in mild steel, aluminum, magnesium, copper, and brass, and are available in both plug and bottoming configurations. Available in several Machine Screw sizes 3–12, popular Fractional sizes through $^1/_4$", and Metric sizes M3–M12. See Table 1 for dimensions, and Table 3 for size availability.

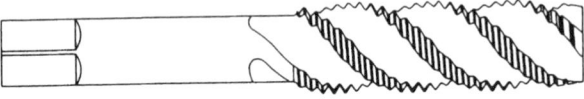

Fast Spiral Fluted Tap

7. *Screw Thread Insert Taps.* These taps produce threads with a pitch diameter that is slightly more than two thread heights larger than the basic pitch diameter, allowing for the installation of a screw thread insert. Inserts are primarily used in soft material prone to thread stripping, or to repair damaged threads. They can be used by hand or under power to produce either through or blind holes in most materials. They are available in plug and bottoming configurations in Machine Screw sizes 4–12 and Fractional sizes $^1/_4$"–$^1/_2$". See Table 1 for dimensions, Table 3 for size availability, and **Table 26** for recommended tap drill sizes for screw thread insert taps.

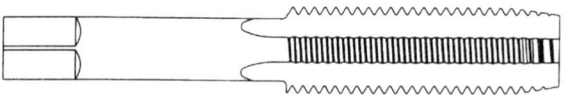

Screw Thread Insert Tap

8. *Spiral Pointed Screw Thread Insert Plug Taps.* The attributes of this tap are basically the same as the Screw Thread Insert Tap, but the spiral point, which forces chips ahead of the tap, allows for the tapping of long holes in excess of $1^1/_2$ times the diameter of the tap. Available in Machine Screw sizes 4–12 and Fractional sizes through $^1/_2$". See Table 1 for dimensions, Table 3 for size availability, and **Table 26** for recommended tap drill sizes for screw thread insert taps.

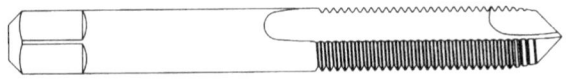

Spiral Pointed Screw Thread Insert Plug Tap

9. *Thread Forming Taps.* These taps do not cut threads, they cold form them by displacing material from the major diameter toward the minor diameter. They are extremely effective in ductile materials such as aluminum, copper, brass, leaded steels, low carbon steels, and stainless steels, but they will not function in materials that are brittle and have low ductility. Thread Forming Taps must be run under power and usually at speeds of 150% to 200% in excess of cutting taps. Because the tap displaces material above the mouth of the hole, countersinking before tapping is recommended. Also, since the hole diameter is actually reduced by the forming process, a larger tap drill is required for a forming tap than a cutting tap. See **Tables 18** through **21** for suggested drill sizes

for forming taps. These taps are available in Machine Screw sizes and Fractional sizes through $^3/_4$" in both plug and bottoming configurations. See Table 1 for dimensions, and Table 3 for size availability.

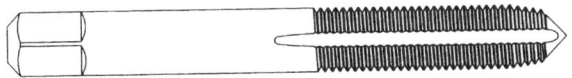

Thread Forming Taps

10. Pulley Taps. These specialty taps are used to tap threads into pulley hubs. The various lengths allow the machine spindle to clear the pulley sheave. See **Table 27** for dimensions.

Pulley Tap

11. Nut Taps. Designed for tapping nuts, these taps have a long chamfer that produces low chip loads. After tapping a nut, the tap is not reversed. Instead, the nuts pass onto the long shank and accumulate until the shank becomes loaded and the tap is removed and the nuts are dumped. See **Table 28** for dimensions.

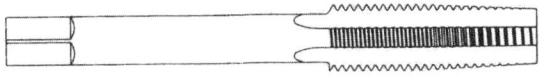

Nut Tap

12. Taper Pipe Taps. These taps are available for tapping NPT (Regular), NPTF (Dryseal), ANPT (Aeronautical), and PTF pipe threads. They are available in Continuous or Interrupted thread versions. Interrupted taps are designed to provide additional chip space, better coolant flow, and reduced drag. The thread size should be specified when ordering. See **Tables 29** through **31** and **Tables 34** and **35** for taper pipe tap specifications and information.

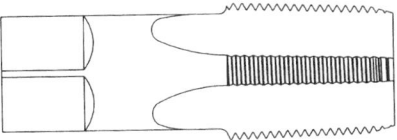

Taper Pipe Tap

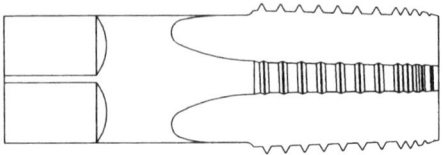

Interrupted Taper Pipe Tap

13. Straight Pipe Taps. These taps are available for tapping NPS (Regular) and NPSF (Dryseal) pipe threads. See **Tables 32**, **33**, and **34** for straight pipe tap specifications and information.

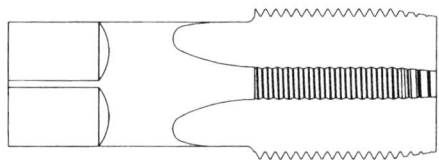

Straight Pipe Tap

14. Miniature Taps. Taps for cutting UNM threads are available in both thread forming and thread cutting varieties. Available tap sizes, recommended tap drill sizes, and information on their use can be found later in this "Taps and Tapping" section under the heading "Miniature taps." (Text continued on p. 203)

Table 1. Dimensions and Tolerances for Standard Ground Thread Hand Taps. *(Source, New England Tap Corp.)*

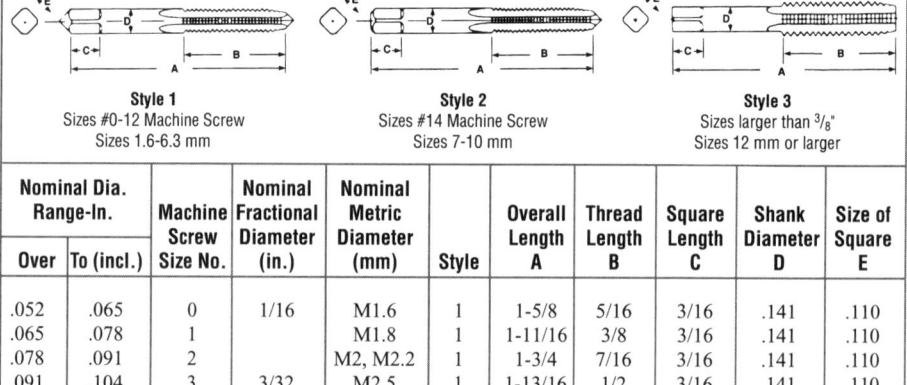

Style 1	Style 2	Style 3
Sizes #0-12 Machine Screw	Sizes #14 Machine Screw	Sizes larger than 3/8"
Sizes 1.6-6.3 mm	Sizes 7-10 mm	Sizes 12 mm or larger

Nominal Dia. Range-In.		Machine Screw Size No.	Nominal Fractional Diameter (in.)	Nominal Metric Diameter (mm)	Style	Overall Length A	Thread Length B	Square Length C	Shank Diameter D	Size of Square E
Over	To (incl.)									
.052	.065	0	1/16	M1.6	1	1-5/8	5/16	3/16	.141	.110
.065	.078	1		M1.8	1	1-11/16	3/8	3/16	.141	.110
.078	.091	2		M2, M2.2	1	1-3/4	7/16	3/16	.141	.110
.091	.104	3	3/32	M2.5	1	1-13/16	1/2	3/16	.141	.110
.104	.117	4			1	1-7/8	9/16	3/16	.141	.110
.117	.130	5	1/8	M3, M3.5	1	1-15/16	5/8	3/16	.141	.110
.130	.145	6		M3.5	1	2	11/16	3/16	.141	.110
.145	.171	8	5/32	M4	1	2-1/8	3/4	1/4	.168	.131
.171	.197	10	3/16	M4.5, M5	1	2-3/8	7/8	1/4	.194	.152
.197	.223	12	7/32		1	2-3/8	15/16	9/32	.220	.165

(Continued)

Table 1. *(Continued)* Dimensions and Tolerances for Standard Ground Thread Hand Taps.

Nominal Dia. Range-In.		Machine Screw Size No.	Nominal Fractional Diameter (in.)	Nominal Metric Diameter (mm)	Style	Overall Length A	Thread Length B	Square Length C	Shank Diameter D	Size of Square E
Over	To (incl.)									
.223	.260	14	1/4	M6, M6.3	2	2-1/2	1	5/16	.255	.191
.260	.323		5/16	M7, M8	2	2-23/32	1-1/8	3/8	.318	.238
.323	.395		3/8	M10	2	2-15/16	1-1/4	7/16	.381	.286
.395	.448		7/16		3	3-5/32	1-7/16	13/32	.323	.242
.448	.510		1/2	M12, M12.5	3	3-3/8	1-21/32	7/16	.367	.275
.510	.573		9/16	M14	3	3-19/32	1-21/32	1/2	.429	.322
.573	.635		5/8	M16	3	3-13/16	1-13/16	9/16	.480	.360
.635	.709		11/16	M18	3	4-1/32	1-13/16	5/8	.542	.406
.709	.760		3/4		3	4-1/4	2	11/16	.590	.442
.760	.823		13/16	M20	3	4-15/32	2	11/16	.652	.489
.823	.885		7/8	M22	3	4-11/16	2-7/32	3/4	.697	.523
.885	.948		15/16	M24	3	4-29/32	2-7/32	3/4	.760	.570
.948	1.010		1	M25	3	5-1/8	2-1/2	13/16	.800	.600
1.010	1.073		1-1/16	M27	3	5-1/8	2-1/2	7/8	.896	.672
1.073	1.135		1-1/8		3	5-7/16	2-9/16	7/8	.896	.672
1.135	1.198		1 3/16	M30	3	5 7/16	2 9/16	1	1.021	.766
1.198	1.260		1 1/4		3	5 3/4	2 9/16	1	1.021	.766
1.260	1.323		1 5/16	M33	3	5 3/4	2 9/16	1 1/16	1.108	.831
1.323	1.385		1 3/8		3	6 1/16	3	1 1/16	1.108	.831
1.385	1.448		1 7/16	M36	3	6 1/16	3	1 1/8	1.233	.925
1.448	1.510		1 1/2		3	6 3/8	3	1 1/8	1.233	.925

Tolerances

Element	Nominal Dia. Range Inches		Direction	Tolerance Inches Ground Thread
	Over	To (Incl.)		
Length Overall-A	.052	1.010	Plus or Minus	1/32
	1.010	1.510	Plus or Minus	1/16
Length of Thread-B	.052	.223	Plus or Minus	3/64
	.223	.510	Plus or Minus	1/16
	.510	1.510	Plus or Minus	3/32
Length of Square-C	.052	1.010	Plus or Minus	1/32
	1.010	1.510	Plus or Minus	1/16
Diameter of Shank-D	.052	.223	Minus	.0015
	.223	.635	Minus	.0015
	.635	1.010	Minus	.002
	1.010	1.510	Minus	.002
Size of Square-E	.052	.510	Minus	.004
	.510	1.010	Minus	.006
	1.010	1.510	Minus	.008

All dimensions in inches.

Table 2. Standard Tap Symbol Marking System. *(Source, Kennametal Inc.)*

Tap Marking	Thread Designations	Thread Series
ACME-C	ACME-C	Acme threads, centralizing
ACME-G	ACME-G	Acme threads, general purpose *(Also see STUB ACME)*
AMO	AMO	American Standard microscope objective threads
NPT	ANPT	Aeronautical National Form Taper pipe threads
BUTT	BUTT	Buttress Threads, pull type
PUSH-BUTT	PUSH-BUTT	Buttress Threads, push type
F-PTF	F-PTF	Dryseal fine taper pipe thread series
M	M	Metric Screw Threads—M Profile, with basic ISO 68 profile
MJ	MJ	Metric Screw Threads—MJ Profile, with rounded root of radius 0.1501 1P to 0.18042P
MJS	MJS	Metric Screw Threads—MJ Profile, special series
NC	NS5 IF	Class 5 interference fit Internal Threads entire ferrous material range
NC	NC5 INF	Class 5 interference fit Internal Threads entire nonferrous material range
NGO-RH or LH	NGO	National gas outlet threads
NGS	NGS	National gas straight threads
NGT	NGT	National gas taper threads *(Also see SGT)*
NH	NH	American Standard hose coupling threads of full form
NPS	NPSC	American Standard straight pipe threads in pipe couplings
NPSF	NPSF	Dryseal American Standard fuel internal straight pipe threads
NPSH	NPSH	American Standard straight hose coupling threads for joining to American Standard taper pipe threads
NPSI	NPSI	Dryseal American Standard intermediate internal straight pipe threads
NPSL	NPSL	American Standard straight pipe threads for loose fitting mechanical joints with locknuts
NPS	NPSM	American Standard straight pipe threads for free fitting mechanical joints for fixtures
NPT	NPT	American Standard taper pipe threads for general use
NPTF	NPTF	Dryseal American Standard taper pipe threads
NPTR	NPTR	American Standard taper pipe threads for railing joints
PTF SHORT	PTF—SAE SHORT	Dryseal SAE short taper pipe threads
PTF-SPL SHORT	PTF-SPL SHORT	Dryseal special short taper pipe threads
PTF-SPL EXTRA SHORT	PFT-SPL EXTRA SHORT	Dryseal special extra short taper pipe threads *(Also see SPL-PTF)*
M	S	ISO Miniature screw threads 0.25 to 1.4 mm inclusive
SGT	SGT	Special gas taper threads
SPL-PTF	SPL-PTF	Dryseal special taper pipe threads
STUB ACME	STUB ACME	Stub Acme threads
STUB ACME M1	STUB ACME M1	Stub Acme Modified Form 1
STUB ACME M2	STUB ACME M2	Stub Acme Modified Form 2

(Continued)

Table 2. *(Continued)* **Standard Tap Symbol Marking System.**

Tap Marking	Thread Designations	Thread Series
N	UN	Unified Inch Screw Thread, constant-pitch series
NC	UNC	Unified Inch Screw Thread, coarse pitch series
NF	UNF	Unified Inch Screw Thread, fine pitch series
NEF	UNEF	Unified Inch Screw Thread, extra-fine pitch series
N	UNJ	Unified Inch Screw Thread, constant-pitch series with rounded root of radius 0.15011P to 0.18042P, on external thread only
NC	UNJC	Unified Inch Screw Thread, coarse pitch series with rounded root of radius 0.15011P to 0.18042P. On external thread only Unified Inch Screw Thread, fine pitch series, with rounded root of radius 0.15011P to 0.18042P. On external thread only
NF	UNJF	Unified Inch Screw Thread, fine pitch series, with rounded root of radius 0.15011P. to 0.18042P. On external thread only
NEF	UNJEF	Unified Inch Screw Thread, extra-fine pitch series, with rounded root of radius 0.15011P to 0.18042P. On external thread only
N	UNR	Unified Inch Screw Thread, constant-pitch series, with rounded root of radius not less than 0.108P. (External thread only)
NC	UNRC	Unified Inch Screw Thread, course thread series, with rounded root of radius not less than 0.108P. (External thread only)
NF	UNRF	Unified Inch Screw Thread, fine pitch series, with rounded root of radius not less than 0.108P. (External thread only)
NEM	UNREF	Unified Inch Screw Thread, extra-fine pitch series, with rounded root of radius not less than 0.108P. (External thread only)
UNM	UNM	Unified miniature thread series
NS	UNS	Unified Inch Screw Thread, special diameter pitch, or length of engagement

Tap & Die Marking	Thread Designations	British Screw Threads
BA	BA	British Association
BSC	BSC	British Cycle
BSF	BSF	British Whitworth Fine
BSW	BSW	British Whitworth Course
BSPP (OLD)	BSPP	British Straight Pipe
BSPT (OLD)	BSPT	British Taper Pipe
WHIT	WHIT	British Whitworth Special

Tap Marking	Thread Designations	British Pipe Threads
G1/4 (BSPP)	G	British Internal Straight Pipe for Mechanical Joints
Rc1/4 (BSPT)	Rc	British Internal Taper Pipe for Pressure Tight Joints
Rp1/4 (BSPP)	Rp	British Internal Straight Pipe for Pressure Tight Joints

Die Marking	Thread Designations	British Pipe Threads
G1/4A (BSPP)	GxA	British External Straight Pipe for Mechanical Joints
G1/4B (BSPP)	GxB	British External Straight Pipe for Mechanical Joints
R1/4 (BSPT)	R	British External Taper Pipe for Pressure Tight Joints

Table 3. Commercially Available Ground Thread Tap Sizes for Taper, Plug, and Bottoming Taps. See Notes.

Machine Screw Sizes											
Size	Thread	H Limit	1.	2.	3.	4.	5.	6.	7.	8.	9.
0-80	NF	H1	TPB	P	B	-	-	-	-	-	-
0-80	NF	H2	PB	P	B	-	-	-	-	-	PB
0-80	NF	H3	-	-	-	-	-	-	-	-	PB
1-64	NC	H1	TPB	P	-	-	-	-	-	-	-
1-64	NC	H2	P	P	B	-	-	-	-	-	PB
1-72	NF	H1	TPB	P	-	-	-	-	-	-	-
1-72	NF	H2	PB	P	B	-	-	-	-	-	B
2-56	NC	H1	TPB	P	B	-	-	-	-	-	-
2-56	NC	H2	TPB	P	B	-	-	-	-	-	PB
2-56	NC	H3	-	-	-	-	-	-	-	-	PB
2-64	NF	H1	-	P	-	-	-	-	-	-	-
2-64	NF	H2	TPB	P	-	-	-	-	-	-	B
2-64	NF	H3	-	-	-	-	-	-	-	-	B
3-48	NC	H1	P	P	-	-	-	-	-	-	-
3-48	NC	H2	TPB	P	B	-	-	PB	-	-	PB
3-48	NC	H3	-	-	-	-	-	-	-	-	B
3-56	NF	H1	-	P	-	-	-	-	-	-	-
3-56	NF	H2	TPB	P	B	-	-	-	-	-	PB
3-56	NF	H3	-	-	-	-	-	-	-	-	B
4-40	NC	H1	TPB	P	B	-	-	-	PB	P	-
4-40	NC	H2	TPB	P	B	P	PB	PB	-	-	-
4-40	NC	H3	-	-	-	-	-	-	-	-	PB
4-40	NC	H5	-	-	-	-	-	-	-	-	PB
4-48	NF	H1	P	P	-	-	-	-	-	-	-
4-48	NF	H2	TPB	P	B	-	-	-	-	-	-
4-48	NF	H3	-	-	-	-	-	-	-	-	PB
4-48	NF	H5	-	-	-	-	-	-	-	-	PB
4-36	NS	H2	TPB	P	-	-	-	-	-	-	-
5-40	NC	H1	PB	P	-	-	-	-	PB	P	-
5-40	NC	H2	TPB	P	B	P	PB	PB	-	-	-
5-40	NC	H3	-	-	-	-	-	-	-	-	PB
5-40	NC	H5	-	-	-	-	-	-	-	-	PB
5-44	NF	H2	TPB	P	B	-	-	-	-	-	-

Note: The three column heads on left indicate Tap size, Fine or Course Unified Thread, and H Limit of Tap. The numbered column heads identify a specific tap style as follows: 1 = Straight Fluted Taps. 2 = Spiral Pointed Plug Taps. 3 = Spiral Pointed Bottoming Taps. 4 = Spiral Pointed Fluteless Plug Taps. 5 = Spiral Fluted Taps. 6 = Fast Spiral Fluted Taps. 7 = Screw Thread Insert Taps. 8 = Spiral Pointed Screw Thread Insert Plug Taps. 9 = Thread Forming Taps. A "T," "P," or "B" in any box indicates that the indicated tap is available in Taper, Plug, or Bottoming configurations. All taps are ground thread and commercially available in high-speed steel with wear-resistant coatings (see text). Check with manufacturer for number of flutes. See Table 1 for tap dimensions.

(Continued)

Table 3. *(Continued)* **Commercially Available Ground Thread Tap Sizes for Taper, Plug, and Bottoming Taps.** See Notes.

Size	Thread	H Limit	1.	2.	3.	4.	5.	6.	7.	8.	9.

Machine Screw Sizes											
Size	Thread	H Limit	1.	2.	3.	4.	5.	6.	7.	8.	9.
5-44	NF	H3	-	-	-	-	-	-	-	-	PB
5-44	NF	H5	-	-	-	-	-	-	-	-	PB
6-32	NC	H1	TPB	P	B	-	-	-	-	-	-
6-32	NC	H2	TPB	P	B	-	-	-	PB	P	-
6-32	NC	H3	TPB	P	B	P	PB	PB	PB	-	PB
6-32	NC	H5	-	-	-	-	-	-	-	-	PB
6-32	NC	H7	PB	P	B	-	-	-	-	-	PB
6-40	NF	H1	P	P	-	-	-	-	PB	P	-
6-40	NF	H2	TPB	P	B	-	-	-	-	-	-
6-40	NF	H3	-	-	-	-	-	-	-	-	PB
6-40	NF	H5	-	-	-	-	-	-	-	-	PB
8-32	NC	H1	TPB	P	B	-	-	-	-	-	-
8-32	NC	H2	TPB	P	B	-	-	-	PB	P	-
8-32	NC	H3	TPB	P	B	P	PB	PB	PB	-	PB
8-32	NC	H5	-	-	-	-	-	-	-	-	PB
8-32	NC	H7	PB	P	B	-	-	-	-	-	PB
8-36	NF	H1	P	P	-	-	-	-	-	-	-
8-36	NF	H2	TPB	P	B	-	-	-	-	-	-
8-36	NF	H3	-	-	-	-	-	-	-	-	PB
10-24	NC	H1	TPB	P	B	-	-	-	-	-	-
10-24	NC	H2	PB	P	B	-	-	-	PB	P	
10-24	NC	H2	TPB	-	-	-	-	-	-	-	-
10-24	NC	H3	TPB	P	B	P	PB	PB	PB	-	-
10-24	NC	H4	-	-	-	-	-	-	-	-	PB
10-24	NC	H6	-	-	-	-	-	-	-	-	PB
10-24	NC	H7	PB	P	-	-	-	-	-	-	-
10-32	NF	H1	PB	P	B	-	-	-	-	-	-
10-32	NF	H1	TPB	-	-	-	-	-	-	-	-
10-32	NF	H2	TPB	P	B	-	-	-	PB	P	-
10-32	NF	H3	TPB	P	B	P	PB	PB	PB	-	-
10-32	NC	H4	-	-	-	-	-	-	-	-	PB
10-32	NF	H6	-	-	-	-	-	-	-	-	PB
10-32	NF	H7	PB	P	-	-	-	-	-	-	-

Note: The three column heads on left indicate Tap size, Fine or Course Unified Thread, and H Limit of Tap. The numbered column heads identify a specific tap style as follows: 1 = Straight Fluted Taps. 2 = Spiral Pointed Plug Taps. 3 = Spiral Pointed Bottoming Taps. 4 = Spiral Pointed Fluteless Plug Taps. 5 = Spiral Fluted Taps. 6 = Fast Spiral Fluted Taps. 7 = Screw Thread Insert Taps. 8 = Spiral Pointed Screw Thread Insert Plug Taps. 9 = Thread Forming Taps. A "T," "P," or "B" in any box indicates that the indicated tap is available in Taper, Plug, or Bottoming configurations. All taps are ground thread and commercially available in high-speed steel with wear-resistant coatings (see text). Check with manufacturer for number of flutes. See Table 1 for tap dimensions.

(Continued)

Table 3. *(Continued)* Commercially Available Ground Thread Tap Sizes for Taper, Plug, and Bottoming Taps. See Notes.

Machine Screw Sizes											
Size	Thread	H Limit	1.	2.	3.	4.	5.	6.	7.	8.	9.
12-24	NC	H1	-	P	-	-	-	-	-	-	-
12-24	NC	H2	-	-	-	-	-	-	PB	P	-
12-24	NC	H3	TPB	P	B	P	-	PB	PB	-	-
12-24	NC	H4	-	-	-	-	-	-	-	-	PB
12-24	NC	H6	-	-	-	-	-	-	-	-	PB
12-28	NF	H1	P	-	-	-	-	-	-	-	-
12-28	NF	H3	TPB	P	-	-	-	-	-	-	-
12-28	NF	H4	-	-	-	-	-	-	-	-	PB
Fractional Sizes											
1/4-20	NC	H1	TPB	P	-	-	-	-	-	-	-
1/4-20	NC	H2	TPB	P	-	-	-	-	PB	P	-
1/4-20	NC	H3	TPB	P	B	P	PB	PB	PB	-	P
1/4-20	NC	H4	-	-	-	-	-	-	-	-	PB
1/4-20	NC	H5	PB	P	-	-	-	-	-	-	-
1/4-20	NC	H6	-	-	-	-	-	-	-	-	PB
1/4-28	NF	H1	PB	P	-	-	-	-	-	-	-
1/4-28	NF	H2	PB	P	-	-	-	-	PB	P	-
1/4-28	NF	H3	TPB	P	P	-	PB	PB	PB	-	-
1/4-28	NF	H4	PB	P	-	-	-	-	-	-	PB
1/4-28	NF	H6	-	-	-	-	-	-	-	-	PB
5/16-18	NC	H1	TPB	P	-	-	-	-	-	-	-
5/16-18	NC	H2	TPB	P	-	-	-	-	-	-	-
5/16-18	NC	H3	TPB	P	B	P	PB	PB	PB	P	-
5/16-18	NC	H5	PB	P	-	-	-	-	-	-	PB
5/16-18	NC	H7	-	-	-	-	-	-	-	-	PB
5/16-24	NF	H1	TPB	P	-	-	-	-	-	-	-
5/16-24	NF	H2	PB	P	-	-	-	-	PB	P	-
5/16-24	NF	H3	TPB	P	B	-	PB	PB	-	-	-
5/16-24	NF	H4	PB	P	-	-	-	-	-	-	. .
5/16-24	NF	H5	-	-	-	-	-	-	-	-	PB
5/16-24	NF	H7	-	-	-	-	-	-	-	-	PB
3/8-16	NC	H1	PB	P	-	-	-	-	-	-	-
3/8-16	NC	H2	TPB	P	-	-	-	-	-	-	-

Note: The three column heads on left indicate Tap size, Fine or Course Unified Thread, and H Limit of Tap. The numbered column heads identify a specific tap style as follows: 1 = Straight Fluted Taps. 2 = Spiral Pointed Plug Taps. 3 = Spiral Pointed Bottoming Taps. 4 = Spiral Pointed Fluteless Plug Taps. 5 = Spiral Fluted Taps. 6 = Fast Spiral Fluted Taps. 7 = Screw Thread Insert Taps. 8 = Spiral Pointed Screw Thread Insert Plug Taps. 9 = Thread Forming Taps. A "T," "P," or "B" in any box indicates that the indicated tap is available in Taper, Plug, or Bottoming configurations. All taps are ground thread and commercially available in high-speed steel with wear-resistant coatings (see text). Check with manufacturer for number of flutes. See Table 1 for tap dimensions.

(Continued)

Table 3. *(Continued)* **Commercially Available Ground Thread Tap Sizes for Taper, Plug, and Bottoming Taps.** See Notes.

			Fractional Sizes								
Size	**Thread**	**H Limit**	**1.**	**2.**	**3.**	**4.**	**5.**	**6.**	**7.**	**8.**	**9.**
3/8-16	NC	H5	PB	P	-	-	-	-	-	-	PB
3/8-16	NC	H7	-	-	-	-	-	-	-	-	PB
3/8-24	NF	H1	TPB	P	-	-	-	-	-	-	-
3/8-24	NF	H2	PB	P	-	-	-	-	PB	P	-
3/8-24	NF	H3	TPB	P	-	-	PB	PB	-	-	-
3/8-24	NF	H4	PB	P	-	-	-	-	-	-	-
3/8-24	NF	H5	-	-	-	-	-	-	-	-	PB
3/8-24	NF	H7	-	-	-	-	-	-	-	-	PB
7/16-14	NC	H1	PB	-	-	-	-	-	-	-	-
7/16-14	NC	H2	PB	P	-	-	-	-	-	-	-
7/16-14	NC	H3	TPB	P	-	-	PB	PB	PB	P	-
7/16-14	NC	H5	PB	P	-	-	-	-	-	-	PB
7/16-14	NC	H8	-	-	-	-	-	-	-	-	PB
7/16-20	NF	H1	PB	-	-	-	-	-	-	-	-
7/16-20	NF	H2	PB	P	-	-	-	-	-	-	-
7/16-20	NF	H3	TPB	P	-	-	PB	PB	PB	P	-
7/16-20	NF	H5	PB	P	-	-	-	-	-	-	PB
7/16-20	NF	H8	-	-	-	-	-	-	-	-	PB
1/2-13	NC	H1	PB	P	-	-	-	-	-	-	-
1/2-13	NC	H2	PB	P	-	-	-	-	-	-	-
1/2-13	NC	H3	TPB	P	-	P	PB	PB	PB	P	-
1/2-13	NC	H5	PB	P	-	-	-	-	-	-	PB
1/2-13	NC	H8	-	-	-	-	-	-	-	-	PB
1/2-20	NF	H1	TPB	P	-	-	-	-	-	-	-
1/2-20	NF	H2	PB	P	-	-	-	-	-	-	-
1/2-20	NF	H3	TPB	P	-	-	PB	PB	PB	P	-
1/2-20	NF	H5	PB	P	-	-	-	-	-	-	PB
1/2-20	NF	H8	-	-	-	-	-	-	-	-	PB
9/16-12	NC	H2	P	-	-	-	-	-	-	-	-
9/16-12	NC	H3	TPB	P	-	-	-	-	-	-	-
9/16-12	NC	H5	PB	-	-	-	-	-	-	-	-
9/16-12	NC	H7	-	-	-	-	-	-	-	-	PB
9/16-12	NC	H10	-	-	-	-	-	-	-	-	PB

Note: The three column heads on left indicate Tap size, Fine or Course Unified Thread, and H Limit of Tap. The numbered column heads identify a specific tap style as follows: 1 = Straight Fluted Taps. 2 = Spiral Pointed Plug Taps. 3 = Spiral Pointed Bottoming Taps. 4 = Spiral Pointed Fluteless Plug Taps. 5 = Spiral Fluted Taps. 6 = Fast Spiral Fluted Taps. 7 = Screw Thread Insert Taps. 8 = Spiral Pointed Screw Thread Insert Plug Taps. 9 = Thread Forming Taps. A "T," "P," or "B" in any box indicates that the indicated tap is available in Taper, Plug, or Bottoming configurations. All taps are ground thread and commercially available in high-speed steel with wear-resistant coatings (see text). Check with manufacturer for number of flutes. See Table 1 for tap dimensions.

(Continued)

Table 3. *(Continued)* Commercially Available Ground Thread Tap Sizes for Taper, Plug, and Bottoming Taps. See Notes.

Size	Thread	H Limit	1.	2.	3.	4.	5.	6.	7.	8.	9.
9/16-18	NF	H2	P	-	-	-	-	-	-	-	-
9/16-18	NF	H3	TPB	P	-	-	-	-	-	-	-
9/16-18	NF	H5	PB	-	-	-	-	-	-	-	-
9/16-18	NF	H7	-	-	-	-	-	-	-	-	PB
9/16-18	NF	H10	-	-	-	-	-	-	-	-	PB
5/8-11	NC	H1	P	-	-	-	-	-	-	-	-
5/8-11	NC	H2	P	-	-	-	-	-	-	-	-
5/8-11	NC	H3	TPB	P	-	-	-	-	-	-	-
5/8-11	NC	H5	PB	P	-	-	-	-	-	-	-
5/8-11	NC	H7	-	-	-	-	-	-	-	-	PB
5/8-11	NC	H10	-	-	-	-	-	-	-	-	PB
5/8-18	NF	H1	P	-	-	-	-	-	-	-	-
5/8-18	NF	H2	PB	-	-	-	-	-	-	-	-
5/8-18	NF	H3	TPB	P	-	-	-	-	-	-	-
5/8-18	NF	H5	PB	-	-	-	-	-	-	-	-
5/8-18	NF	H7	PB	-	-	-	-	-	-	-	PB
5/8-18	NF	H10	PB	-	-	-	-	-	-	-	PB
11/16-11	NS	H3	TPB	-	-	-	-	-	-	-	-
11/16-16	NS	H3	TPB	-	-	-	-	-	-	-	-
3/4-10	NC	H1	P	-	-	-	-	-	-	-	-
3/4-10	NC	H2	P	-	-	-	-	-	-	-	-
3/4-10	NC	H3	TPB	P	-	-	-	-	-	-	-
3/4-10	NC	H5	PB	P	-	-	-	-	-	-	-
3/4-10	NC	H7	-	-	-	-	-	-	-	-	PB
3/4-10	NC	H10	-	-	-	-	-	-	-	-	PB
3/4-16	NF	H1	P	-	-	-	-	-	-	-	-
3/4-16	NF	H2	P	-	-	-	-	-	-	-	-
3/4-16	NF	H3	TPB	P	-	-	-	-	-	-	-
3/4-16	NF	H5	PB	-	-	-	-	-	-	-	-
3/4-16	NF	H7	-	-	-	-	-	-	-	-	PB
3/4-16	NF	H10	-	-	-	-	-	-	-	-	PB
7/8-9	NC	H2	P	-	-	-	-	-	-	-	-
7/8-9	NC	H4	TPB	-	-	-	-	-	-	-	-

Note: The three column heads on left indicate Tap size, Fine or Course Unified Thread, and H Limit of Tap. The numbered column heads identify a specific tap style as follows: 1 = Straight Fluted Taps. 2 = Spiral Pointed Plug Taps. 3 = Spiral Pointed Bottoming Taps. 4 = Spiral Pointed Fluteless Plug Taps. 5 = Spiral Fluted Taps. 6 = Fast Spiral Fluted Taps. 7 = Screw Thread Insert Taps. 8 = Spiral Pointed Screw Thread Insert Plug Taps. 9 = Thread Forming Taps. A "T," "P," or "B" in any box indicates that the indicated tap is available in Taper, Plug, or Bottoming configurations. All taps are ground thread and commercially available in high-speed steel with wear-resistant coatings (see text). Check with manufacturer for number of flutes. See Table 1 for tap dimensions.

(Continued)

Table 3. *(Continued)* **Commercially Available Ground Thread Tap Sizes for Taper, Plug, and Bottoming Taps.** See Notes.

Fractional Sizes											
Size	Thread	H Limit	1.	2.	3.	4.	5.	6.	7.	8.	9.
7/8-9	NC	H4	TPB	-	-	-	-	-	-	-	-
7/8-9	NC	H6	P	-	-	-	-	-	-	-	-
7/8-14	NF	H2	P	-	-	-	-	-	-	-	-
7/8-14	NF	H4	TPB	-	-	-	-	-	-	-	-
7/8-14	NF	H6	P	-	-	-	-	-	-	-	-
1-8	NC	H2	P	-	-	-	-	-	-	-	-
1-8	NC	H4	TPB	-	-	-	-	-	-	-	-
1-8	NC	H6	P	-	-	-	-	-	-	-	-
1-12	NF	H4	TPB	-	-	-	-	-	-	-	-
1-14	NS	H2	P	-	-	-	-	-	-	-	-
1-14	NS	H4	TPB	-	-	-	-	-	-	-	-
1-14	NS	H6	TPB	-	-	-	-	-	-	-	-
1-1/8-7	NC	H4	TPB	-	-	-	-	-	-	-	-
1-1/8-12	NF	H4	TPB	-	-	-	-	-	-	-	-
1-1/4-7	NC	H4	TPB	-	-	-	-	-	-	-	-
1-1/4-12	NF	H4	TPB	-	-	-	-	-	-	-	-
1-3/8-6	NC	H4	TPB	-	-	-	-	-	-	-	-
1-3/8-12	NF	H4	TPB	-	-	-	-	-	-	-	-
1-1/2-6	NC	H4	TPB	-	-	-	-	-	-	-	-
1-1/2-12	NF	H4	TPB	-	-	-	-	-	-	-	-

Note: The three column heads on left indicate Tap size, Fine or Course Unified Thread, and H Limit of Tap. The numbered column heads identify a specific tap style as follows: 1 = Straight Fluted Taps. 2 = Spiral Pointed Plug Taps. 3 = Spiral Pointed Bottoming Taps. 4 = Spiral Pointed Fluteless Plug Taps. 5 = Spiral Fluted Taps. 6 = Fast Spiral Fluted Taps. 7 = Screw Thread Insert Taps. 8 = Spiral Pointed Screw Thread Insert Plug Taps. 9 = Thread Forming Taps. A "T," "P," or "B" in any box indicates that the indicated tap is available in Taper, Plug, or Bottoming configurations. All taps are ground thread and commercially available in high-speed steel with wear-resistant coatings (see text). Check with manufacturer for number of flutes. See Table 1 for tap dimensions.

Table 3. *(Continued)* **Commercially Available Ground Thread Tap Metric Sizes for Taper, Plug, and Bottoming Taps.** See Notes.

Metric Sizes									
Size	Pitch	Class of Fit	Limit	1.	2.	5.	6.	9.	10.
M1.6	0.35	6H	D3	TPB	P	-	-	-	-
M1.8	0.35	6H	D3	TPB	P	-	-	-	-
M2	0.40	6H	D3	TPB	P	-	-	-	-
M2.2	0.45	6H	D3	TPB	P	-	-	-	-
M2.5	0.45	6H	D3	TPB	P	-	-	-	-
M3	0.50	6H	D3	TPB	P	PB	PB	PB	-
M3.5	0.60	6H	D4	TPB	P	PB	PB	PB	-
M4	0.70	6H	D4	TPB	P	PB	PB	PB	-

(Continued)

Table 3. *(Continued)* **Commercially Available Ground Thread Tap Metric Sizes for Taper, Plug, and Bottoming Taps.** See Notes.

				Metric Sizes					
Size	**Pitch**	**Class of Fit**	**Limit**	**1.**	**2.**	**5.**	**6.**	**9.**	**10.**
M4.5	0.75	6H	D4	TPB	P	PB	PB	PB	-
M5	0.80	6H	D4	TPB	P	PB	PB	PB	-
M6	1.0	6H	D5	TPB	P	PB	PB	PB	-
M7	1.0	6H	D5	TPB	P	PB	PB	PB	-
M8	1.25	6H	D5	TPB	P	PB	PB	PB	-
M8	1.0	6H	D5	TPB	P	PB	PB	PB	-
M10	1.50	6H	D6	TPB	P	PB	PB	PB	-
M10	1.25	6H	D5	TPB	P	PB	PB	PB	-
M10	1.0	-	H3	-	-	-	-	-	P
M12	1.75	6H	D6	TPB	P	PB	PB	PB	-
M12	1.25	6H	D5	TPB	P	PB	PB	PB	-
M14	2.0	6H	D7	TPB	P	-	-	PB	-
M14	1.50	6H	D6	TPB	P	-	-	PB	-
M14	1.25	-	H4	-	-	-	-	-	P
M16	2.0	6H	D7	TPB	P	-	-	PB	-
M16	1.50	6H	D6	TPB	P	-	-	-	-
M18	2.50	6H	D7	TPB	P	-	-	-	-
M18	1.50	6H	D6	TPB	-	-	-	PB	-
M18	1.50	-	H4	-	-	-	-	-	P
M20	2.50	6H	D7	TPB	P	-	-	-	-
M20	1.50	6H	D6	TPB	-	-	-	-	-
M22	2.50	6H	D7	TPB	-	-	-	-	-
M22	1.50	6H	D6	TPB	-	-	-	-	-
M24	3.0	6H	D8	TPB	-	-	-	-	-
M24	2.0	6H	D7	TPB	-	-	-	-	-
M27	3.0	6H	D8	TPB	-	-	-	-	-
M27	2.0	6H	D7	TPB	-	-	-	-	-
M30	3.50	6H	D9	TPB	-	-	-	-	-
M30	2.0	6H	D7	TPB	-	-	-	-	-
M33	2.0	6H	D7	TPB	-	-	-	-	-
M36	4.0	6H	D9	TPB	-	-	-	-	-
M36	3.0	6H	D8	TPB	-	-	-	-	-
M39	4.0	6H	D9	TPB	-	-	-	-	-
M39	3.0	6H	D8	TPB	-	-	-	-	-

Note: The four column heads on left indicate Tap size, Pitch of Thread, Class of Metric Thread, and Limit of Tap. The numbered column heads identify a specific tap style as follows: 1 = Straight Fluted Taps. 2 = Spiral Pointed Plug Taps. 5 = Spiral Fluted Taps. 6 = Fast Spiral Fluted Taps. 9 = Thread Forming Taps. 10 = Spark Plug Tap. A "T," "P," or "B" in any box indicates that the indicated tap is available in Taper, Plug, or Bottoming configurations. All taps are ground thread and commercially available in high-speed steel with wear-resistant coatings (see text). Check with manufacturer for number of flutes. See Table 1 for tap dimensions.

Pitch diameter limits identification system

Inch System ground thread screw machine and fractional size taps that are made to conform to standards for pitch diameter limits are marked with the letter "G" (for ground thread), plus the letter "H" or "L" and a pitch diameter limit number. "H" stands for high, or above basic size, and "L" indicates low, or under basic size. For taps under one inch in diameter, the pitch diameter limits are the amount the pitch diameter of the tap is over or under basic pitch diameter divided by .0005 as follows.

L 4 = basic minus .0015 to basic minus .002" H 5 = basic plus .002 to basic plus .0025"
L 3 = basic minus .001 to basic minus .0015" H 6 = basic plus .0025 to basic plus .003"
L 2 = basic minus .0005 to basic minus .001" H 7 = basic plus .003 to basic plus .0035"
L 1 = basic to basic minus .0005" H 8 = basic plus .0035 to basic plus .004"
H 1 = basic to basic plus .0005" H 9 = basic plus .004 to basic plus .0045"
H 2 = basic plus .0005 to basic plus .001" H 10 = basic plus .0045 to basic plus .005"
H 3 = basic plus .001 to basic plus .0015" H 11 = basic plus .005 to basic plus .0055"
H 4 = basic plus .0015 to basic plus .002" H 12 = basic plus .0055 to basic plus .006"

For fractional size taps over one inch in diameter to one and one-half inches in diameter, the following pitch diameter limits apply.

H 4 = basic plus .001 to basic plus .002"
H 6 = basic plus .002 to basic plus .003"
H 8 = basic plus .003 to basic plus .004"

Table 4 gives pitch diameters for H–1 through H–3 limit taps for Machine Screw size ground thread taps, and **Table 5** provides H–1 through H–6 limits for Fractional size ground thread taps. Recommended taps for Class 2, 2B, 3, and 3B Machine Screw and Fractional size threads for ground thread taps are given in **Table 6**. Forming tap recommendations for Class 2, 2B, and 3B Machine Screw and Fractional size threads are found in **Table 22**.

Metric System ground thread taps use the letter "D" for taps where the pitch diameter is over basic, and the letters "DU" for taps that have pitch diameters under basic. The D and DU pitch diameter limits are the amount the pitch diameter of the tap is over or under the basic pitch diameter divided by .013 mm. Since .013 mm equals .0005118", D limits are essentially equal to H limits for inch system taps, and DU limits are roughly equivalent to L limits. Pitch diameter limits (in inches) for recommended metric taps are as follows.

D 3 = basic plus .0009" to basic plus .0015" D 7 = basic plus .0019" to basic plus .0035"
D 4 = basic plus .0012" to basic plus .002" D 8 = basic plus .0024" to basic plus .004"
D 5 = basic plus .0015" to basic plus .0025" D 9 = basic plus .0025" to basic plus .0045"
D 6 = basic plus .0018" to basic plus .003"

Table 7 gives pitch diameter limits for D–3 through D–9 limits for M Profile ground thread taps in both millimeters and inches, and **Table 8** shows recommended taps (with pitch diameter limits in both millimeters and inches) for Class 4H and 6H Metric Threads.

Calculating hole sizes for length of engagement and desired percentage of thread for ground thread cutting taps

Before drilling, reaming, and tapping a hole for a selected thread size, the percentage of thread engagement produced by the operation should be determined (see *Figure 2*). As the desired percentage of thread increases, the amount of strain on the tap's teeth increases, requiring more power and raising the risk of the tap breaking in the hole. Tap breakage is

(Text continued on p. 209)

Table 4. Machine Screw Size Taps—Ground Thread—Unified Screw Threads. *(Source, Kennametal Inc.)*

Size	Threads Per Inch			Major Diameter			Basic Pitch Dia.	Pitch Diameter Limits					
	UNC	UNF	UNS	Basic	Min.	Max.		H1 Limit		H2 Limit		H3 Limit	
								Min.	Max.	Min.	Max.	Min.	Max.
0	-	80	-	.0600	.0605	.0616	.0519	.0519	.0524	.0524	.0529	-	-
1	64	-	-	.0730	.0736	.0750	.0629	.0629	.0634	.0634	.0639	-	-
1	-	72	-	.0730	.0736	.0748	.0640	.0640	.0645	.0645	.0650	-	-
2	56	-	-	.0860	.0867	.0883	.0744	.0794	.0749	.0749	.0754	-	-
2	-	64	-	.0860	.0866	.0880	.0759	-	-	.0764	.0769	-	-
3	48	-	-	.0990	.0999	.1017	.0855	.0855	.0860	.0860	.0865	-	-
3	-	56	-	.0990	.0997	.1013	.0874	.0874	.0879	.0879	.0884	-	-
4	-	-	36	.1120	.1135	.1156	.0940	-	-	.0945	.0950	-	-
4	40	-	-	.1120	.1133	.1152	.0958	.0958	.0963	.0963	.0968	-	-
4	-	48	-	.1120	.1129	.1147	.0985	.0985	.0990	.0990	.0995	-	-
5	40	-	-	.1250	.1263	.1282	.1088	.1088	.1093	.1093	.1098	-	-
5	-	44	-	.1250	.1263	.1280	.1102	-	-	.1107	.1112	-	-
6	32	-	-	.1380	.1401	.1421	.1177	.1177	.1182	.1182	.1187	.1187	.1192
6	-	40	-	.1380	.1393	.1412	.1218	.1218	.1223	.1223	.1228	-	-
8	32	-	-	.1640	.1661	.1681	.1437	.1437	.1442	.1442	.1447	.1447	.1452
8	-	36	-	.1640	.1655	.1676	.1460	.1460	.1465	.1465	.1470	-	-
10	24	-	-	.1900	.1927	.1954	.1629	.1629	.1634	.1634	.1639	.1639	.1644
10	-	32	-	.1900	.1921	.1941	.1697	.1697	.1702	.1702	.1707	.1707	.1712
12	24	-	-	.2160	.2187	.2214	.1889	.1889	.1894	-	-	.1899	.1904
12	-	28	-	.2160	.2183	.2206	.1928	.1928	.1933	-	-	.1938	.1942

All dimensions in inches.

Table 5. Fractional Size Taps—Ground Thread—Unified Screw Threads. *(Source, Fastcut Tool Corp.)*

Size	Threads Per Inch			Major Diameter			Basic Pitch Dia.	Pitch Diameter Limits			
	UNC	UNF	UNS	Basic	Min.	Max.		H1 Limit		H2 Limit	
								Min.	Max.	Min.	Max.
1/4	20	-	-	0.2500	0.2533	0.2565	0.2175	0.2175	0.2180	0.2180	0.2185
1/4	-	28	-	0.2500	0.2533	0.2546	0.2268	0.2268	0.2273	0.2273	0.2278
5/16	18	-	-	0.3125	0.3161	0.3197	0.2764	0.2764	0.2769	0.2769	0.2774
5/16	-	24	-	0.3125	0.3152	0.3179	0.2854	0.2854	0.2859	0.2859	0.2864
3/8	16	-	-	0.3750	0.3790	0.3831	0.3344	0.3344	0.3349	0.3349	0.3354
3/8	-	24	-	0.3750	0.3777	0.3804	0.3479	0.3479	0.3484	0.3484	0.3489
7/16	14	-	-	0.4375	0.4422	0.4468	0.3911	0.3911	0.3916	0.3916	0.3921
7/16	-	20	-	0.4375	0.4408	0.4440	0.4050	0.4050	0.4055	0.4055	0.4060
1/2	13	-	-	0.5000	0.5050	0.5100	0.4500	0.4500	0.4505	0.4505	0.4510
1/2	-	20	-	0.5000	0.5033	0.5065	0.4675	0.4675	0.4680	0.4680	0.4685
9/16	12	-	-	0.5625	0.5679	0.5733	0.5084	0.5084	0.5089	0.5089	0.5094
9/16	-	18	-	0.5625	0.5661	0.5697	0.5264	0.5264	0.5269	0.5269	0.5274
5/8	11	-	-	0.6250	0.6309	0.6368	0.5660	0.5660	0.5665	0.5665	0.5670
5/8	-	18	-	0.6250	0.6286	0.6322	0.5889	0.5889	0.5894	0.5894	0.5899
11/16	-	-	11	0.6875	0.6934	0.6993	0.6285	-	-	-	-
11/16	-	-	16	0.6875	0.6915	0.6956	0.6469	-	-	-	-
3/4	10	-	-	0.7500	0.7565	0.7630	0.6850	0.6850	0.6855	0.6855	0.6860
3/4	-	16	-	0.7500	0.7540	0.7581	0.7094	0.7094	0.7099	0.7099	0.7104
7/8	9	-	-	0.8750	0.8822	0.8894	0.8028	0.8028	0.8033	0.8033	0.8038
7/8	-	14	-	0.8750	0.8797	0.8843	0.8286	0.8286	0.8291	0.8291	0.8296

(Continued)

Table 5. *(Continued)* **Fractional Size Taps—Ground Thread—Unified Screw Threads.** *(Source, Fastcut Tool Corp.)*

Size	Threads Per Inch			Major Diameter			Basic Pitch Dia.	Pitch Diameter Limits			
	UNC	UNF	UNS	Basic	Min.	Max.		H1 Limit		H2 Limits	
								Min.	Max.	Min.	Max.
1	8	-	-	1.0000	1.0081	1.0162	0.9188	0.9188	0.9193	0.9193	0.9198
1	-	12	-	1.0000	1.0054	1.0108	0.9459	-	-	-	-
1	-	-	14	1.0000	1.0047	1.0093	0.9536	-	-	0.9541	0.9546
1 1/8	7	-	-	1.1250	1.1343	1.1436	1.0322	-	-	-	-
1 1/8	-	12	-	1.1250	1.1304	1.1358	1.0709	-	-	-	-
1 1/4	7	-	-	1.2500	1.2593	1.2686	1.1572	-	-	-	-
1 1/4	-	12	-	1.2500	1.2554	1.2608	1.1959	-	-	-	-
1 3/8	6	-	-	1.3750	1.3859	1.3967	1.2667	-	-	-	-
1 3/8	-	12	-	1.3750	1.3804	1.3858	1.3209	-	-	-	-
1 1/2	6	-	-	1.5000	1.5109	1.5217	1.3917	-	-	-	-
1 1/2	-	12	-	1.5000	1.5054	1.5108	1.4459	-	-	-	-

Size	Threads Per Inch			Pitch Diameter Limits							
	UNC	UNF	UNS	H3 Limit		H4 Limit		H5 Limit		H6 Limit	
				Min.	Max.	Min.	Max.	Min.	Max.	Min.	Max.
1/4	20	-	-	0.2185	0.2190	-	-	0.2195	0.2200	-	-
1/4	-	28	-	0.2278	0.2283	0.2283	0.2288	-	-	-	-
5/16	18	-	-	0.2774	0.2779	-	-	0.2784	0.2789	-	-
5/16	-	24	-	0.2864	0.2869	0.2869	0.2874	-	-	-	-
3/8	16	-	-	0.3354	0.3359	-	-	0.3364	0.3369	-	-
3/8	-	24	-	0.3489	0.3494	0.3494	0.3499	-	-	-	-
7/16	14	-	-	0.3921	0.3926	-	-	0.3931	0.3936	-	-
7/16	-	20	-	0.4060	0.4065	-	-	0.4070	0.4075	-	-
1/2	13	-	-	0.4510	0.4515	-	-	0.4520	0.4525	-	-
1/2	-	20	-	0.4685	0.4690	-	-	0.4695	0.4700	-	-
9/16	12	-	-	0.5094	0.5099	-	-	0.5104	0.5109	-	-
9/16	-	18	-	0.5274	0.5279	-	-	0.5284	0.5289	-	-
5/8	11	-	-	0.5670	0.5675	-	-	0.5680	0.5685	-	-
5/8	-	18	-	0.5899	0.5904	-	-	0.5909	0.5914	-	-
11/16	-	-	11	0.6295	0.6300	-	-	-	-	-	-
11/16	-	-	16	0.6479	0.6484	-	-	-	-	-	-
3/4	10	-	-	0.6860	0.6865	-	-	0.6870	0.6875	-	-
3/4	-	16	-	0.7104	0.7109	-	-	0.7114	0.7119	-	-
7/8	9	-	-	-	-	0.8043	0.8048	-	-	0.8053	0.8058
7/8	-	14	-	-	-	0.8301	0.8306	-	-	0.8311	0.8316
1	8	-	-	-	-	0.9203	0.9208	-	-	0.9213	0.9218
1	-	12	-	-	-	0.9474	0.9479	-	-	-	-
1	-	-	14	-	-	0.9551	0.9556	-	-	0.9561	0.9566
1 1/8	7	-	-	-	-	1.0332	1.0342	-	-	-	-
1 1/8	-	12	-	-	-	1.0719	1.0729	-	-	-	-
1 1/4	7	-	-	-	-	1.1582	1.1592	-	-	-	-
1 1/4	-	12	-	-	-	1.1969	1.1979	-	-	-	-
1 3/8	6	-	-	-	-	1.2677	1.2687	-	-	-	-
1 3/8	-	12	-	-	-	1.3219	1.3229	-	-	-	-
1 1/2	6	-	-	-	-	1.3927	1.3937	-	-	-	-
1 1/2	-	12	-	-	-	1.4469	1.4479	-	-	-	-

All dimensions in inches.

Table 6. Tap Recommendations for Classes 2, 3, 2B, and 3B Unified Machine Screw and Fractional Size Screw Threads. *(Source, New England Tap Corp.)*

Size	Threads Per Inch		Recommended Tap for Class of Thread				Pitch Diameter Limits				
	UNC	UNF	Class 2	Class 3	Class 2B	Class 3B	Min. All Classes (Basic)	Max. Class 2	Max. Class 3	Max. Class 2B	Max. Class 3B
0	-	80	G H1	G H1	G H2	G H1	.0519	.0536	.0532	.0542	.0536
1	64	-	G H1	G H1	G H2	G H1	.0629	.0648	.0643	.0655	.0648
1	-	72	G H1	G H1	G H2	G H1	.0640	.0658	.0653	.0665	.0659
2	56	-	G H1	G H1	G H2	G H1	.0744	.0764	.0759	.0772	.0765
2	-	64	G H1	G H1	G H2	G H1	.0759	.0778	.0773	.0786	.0779
3	48	-	G H1	G H1	G H2	G H1	.0855	.0877	.0871	.0885	.0877
3	-	56	G H1	G H1	G H2	G H1	.0874	.0894	.0889	.0902	.0895
4	40	-	G H2	G H1	G H2	G H2	.0958	.0982	.0975	.0991	.0982
4	-	48	G H1	G H1	G H2	G H1	.0985	.1007	.1001	.1016	.1008
5	40	-	G H2	G H1	G H2	G H2	.1088	.1112	.1105	.1121	.1113
5	-	44	G H1	G H1	G H2	G H1	.1102	.1125	.1118	.1134	.1126
6	32	-	G H2	G H1	G H3	G H2	.1177	.1204	.1196	.1214	.1204
6	-	40	G H2	G H1	G H2	G H2	.1218	.1242	.1235	.1252	.1243
8	32	-	G H2	G H1	G H2	G H2	.1437	.1464	.1456	.1475	.1465
8	-	36	G H2	G H1	G H2	G H2	.1460	.1485	.1478	.1496	.1487
10	24	-	G H3	G H3	G H3	G H3	.1629	.1662	.1653	.1672	.1661
10	-	32	G H2	G H1	G H3	G H2	.1697	.1724	.1716	.1736	.1726
12	24	-	G H3	G H1	G H3	G H3	.1889	.1922	.1913	.1933	.1922
12	-	28	G H3	G H1	G H3	G H3	.1928	.1959	.1950	.1970	.1959
1/4	20	-	G H3	G H2	G H5	G H3	.2175	.2211	.2201	.2223	.2211
1/4	-	28	G H3	G H1	G H4	G H3	.2268	.2299	.2290	.2311	.2300
5/16	18	-	G H3	G H2	G H5	G H3	.2764	.2805	.2794	.2817	.2803
5/16	-	24	G H3	G H1	G H4	G H3	.2854	.2887	.2878	.2902	.2890
3/8	16	-	G H3	G H2	G H5	G H3	.3344	.3389	.3376	.3401	.3387
3/8	-	24	G H3	G H1	G H4	G H3	.3479	.3512	.3503	.3528	.3516
7/16	14	-	G H5	G H3	G H5	G H3	.3911	.3960	.3947	.3972	.3957
7/16	-	20	G H3	G H1	G H5	G H3	.4050	.4086	.4076	.4104	.4091
1/2	13	-	G H5	G H3	G H5	G H3	.4500	.4552	.4537	.4565	.4548
1/2	-	20	G H3	G H1	G H5	G H3	.4675	.4711	.4701	.4731	.4717
9/16	12	-	G H5	G H3	G H5	G H3	.5084	.5140	.5124	.5152	.5135
9/16	-	18	G H3	G H2	G H5	G H3	.5264	.5305	.5294	.5323	.5308
5/8	11	-	G H5	G H3	G H5	G H3	.5660	.5719	.5702	.5732	.5714
5/8	-	18	G H3	G H2	G H5	G H3	.5889	.5930	.5919	.5949	.5934
3/4	10	-	G H5	G H3	G H5	G H5	.6850	.6914	.6895	.6927	.6907
3/4	-	16	G H3	G H2	G H5	G H3	.7094	.7139	.7126	.7159	.7143
7/8	9	-	G H6	G H4	G H6	G H4	.8028	.8098	.8077	.8110	.8089

(Continued)

Table 6. *(Continued)* **Tap Recommendations for Classes 2, 3, 2B, and 3B Unified Machine Screw and Fractional Size Screw Threads.** *(Source, New England Tap Corp.)*

Size	Threads Per Inch		Recommended Tap for Class of Thread				Pitch Diameter Limits				
	UNC	UNF	Class 2	Class 3	Class 2B	Class 3B	Min. All Classes (Basic)	Max. Class 2	Max. Class 3	Max. Class 2B	Max. Class 3B
7/8	-	14	G H4	G H2	G H6	G H4	.8286	.8335	.8322	.8356	.8339
1	8	-	G H6	G H4	G H6	G H4	.9188	.9264	.9242	.9276	.9254
1	-	12	G H4	G H2	G H6	G H4	.9459	.9515	.9499	.9535	.9516
1	-	14 NS	G H4	G H2	G H6	G H4	.9536	.9585	.9572	.9609	.9590
1 1/8	7	-	G H8	G H4	G H8	G H4	1.0322	1.0407	1.0381	1.0416	1.0393
1 1/8	-	12	G H4	G H4	G H6	G H4	1.0709	1.0765	1.0749	1.0787	1.0768
1 1/4	7	-	G H8	G H4	G H8	G H4	1.1572	1.1657	1.1631	1.1668	1.1644
1 1/4	-	12	G H4	G H4	G H6	G H4	1.1959	1.2015	1.1999	1.2039	1.2019
1 3/8	6	-	G H8	G H4	G H8	G H4	1.2667	1.2768	1.2738	1.2771	1.2745
1 3/8	-	12	G H4	G H4	G H6	G H4	1.3209	1.3265	1.3249	1.3291	1.3270
1 1/2	6	-	G H8	G H4	G H8	G H4	1.3917	1.4018	1.3988	1.4022	1.3996
1 1/2	-	12	G H4	G H4	G H6	G H4	1.4459	1.4515	1.4499	1.4542	1.4522

The above recommended taps normally produce the Class of Thread indicated in average materials when used with reasonable care. However, if the tap specified does not give a satisfactory gage fit in the work, a choice of some other limit tap will be necessary. All dimensions in inches.

Table 7. Metric Size Taps Dimensions and Pitch Diameter Limits M Profile Thread. *(Source, Kennametal Inc.)*

	Dimensions									
	Millimeters						Inches			
Nom. Dia.	Pitch	Basic Major Dia.	Min. Major Dia.	Max. Major Dia.	Basic Pitch Dia.	Pitch	Basic Major Dia.	Min. Major Dia.	Max. Major Dia.	Basic Pitch Dia.
1.6	0.35	1.600	1.628	1.653	1.373	.01378	.06299	.06409	.06508	.05406
2	0.4	2.000	2.032	2.057	1.740	.01575	.07874	.08000	.08098	.06850
2.5	0.45	2.500	2.536	2.561	2.208	.01772	.09843	.09984	.10083	.08693
3	0.5	3.000	3.040	3.065	2.675	.01969	.11811	.11969	.12067	.10531
3.5	0.6	3.500	3.548	3.573	3.110	.02362	.13780	.13969	.14067	.12244
4	0.7	4.000	4.056	4.097	3.545	.02756	.15748	.15969	.16130	.13957
4.5	0.75	4.500	4.560	4.601	4.013	.02953	.17717	.17953	.18114	.15799
5	0.8	5.000	5.064	5.105	4.480	.03150	.19685	.19937	.20098	.17638
6	1	6.000	6.080	6.121	5.350	.03937	.23622	.23937	.24098	.21063
7	1	7.000	7.080	7.121	6.350	.03937	.27559	.27874	.28035	.25000
8	1.25	8.000	8.100	8.164	7.188	.04921	.31496	.31890	.32142	.28299
10	1.5	10.000	12.120	12.184	9.026	.05906	.39370	.39843	.40094	.35535
12	1.75	12.000	12.140	12.204	10.863	.06890	.47244	.47795	.48047	.42768
14	2	14.000	14.160	14.224	12.701	.07874	.55118	.55748	.56000	.50004
16	2	16.000	16.160	16.224	14.701	.07874	.62992	.63622	.63874	.57878
20	2.5	20.000	20.200	20.264	18.376	.09483	.78740	.79528	.79780	.72346
24	3	24.000	24.240	24.340	22.051	.11811	.94488	.95433	.95827	.86815
30	3.5	30.000	30.280	30.380	27.727	.13780	1.18110	1.19213	1.19606	1.09161
36	4	36.000	36.360	36.420	33.402	.15748	1.41732	1.42992	1.43386	1.31504

(Continued)

Table 7. *(Continued)* **Metric Size Taps Dimensions and Pitch Diameter Limits M Profile Thread.**
(Source, Kennametal Inc.)

Pitch Diameter Limits											
Nom. Dia. mm	Pitch	D3 Limits mm		D3 Limits inch		Nom. Dia. mm	Pitch	D6 Limits mm		D6 Limits inch	
		Min.	Max.	Min.	Max.			Min.	Max.	Min.	Max.
1.6	0.35	1.397	1.412	.05500	.05559	10	1.5	9.073	9.104	.35720	.35843
2	0.4	1.764	1.779	.06945	.07004	12	1.75	10.901	10.941	.42953	.43075
2.5	0.45	2.232	2.247	.08787	.08846			D7 Limits mm		D7 Limits inch	
3	0.5	2.699	2.714	.10626	.10685			Min.	Max.	Min.	Max.
		D4 Limits mm		D4 Limits inch		14	2	12.751	12.792	.50201	.50362
		Min.	Max.	Min.	Max.	16	2	14.751	14.792	.58075	.58236
3.5	0.6	3.142	3.162	.12370	.12449	20	2.5	18.426	18.467	.72543	.72705
4	0.7	3.577	3.597	.14083	.14161			D8 Limits mm		D8 Limits inch	
4.5	0.75	4.045	4.065	.15925	.16004			Min.	Max.	Min.	Max.
5	0.8	4.512	4.532	.17764	.17843	24	3	22.114	22.155	.87063	.87224
		D5 Limits mm		D5 Limits inch				D9 Limits mm		D9 Limits inch	
		Min.	Max.	Min.	Max.			Min.	Max.	Min.	Max.
6	1	5.390	5.415	.21220	.21319	30	3.5	27.792	27.844	1.09417	1.09622
7	1	6.390	6.415	.25157	.25256	36	4	33.467	33.519	1.31760	1.31965
8	1.25	7.222	7.253	.28433	.28555						

Lead Tolerance

A maximum lead deviation of ±0.013 mm (± 0.0005") within any two threads not more than 25.4 mm (1") apart is permitted.

Angle Tolerance

Pitch (mm)	Deviation of Half Angle
Over 0.25 to 2.5 Inclusive	±30'
Over 2.5 to 4 Inclusive	±25'
Over 4 to 6 Inclusive	±20'

Basic pitch diameter is the same as minimum pitch diameter of Internal Thread Class 6H—Table 21, ANSI B1.13M-1979.

Table 8. Tap Recommendations for Classes 4H and 6H Metric M Profile Screw Threads.
(Source, Fastcut Tool Corp.)

Threads per inch		Recommended Tap for Class of Thread		Pitch Diameter Limits for Class of Thread					
				Millimeters			Inches		
Nominal Diameter	Pitch	4H	6H	Min. All Classes (Basic)	Max. 4H	Max. 6H	Min. All Classes (Basic)	Max. 4H	Max. 6H
M1.6	0.35	D1	D3	1.373	1.426	1.458	.0541	.0561	.0574
M2	0.4	D1	D3	1.740	1.796	1.830	.0685	.0707	.0720
M2.5	0.45	D1	D3	2.208	2.268	2.303	.0869	.0893	.0907
M3	0.5	D1	D3	2.675	2.738	2.775	.1053	.1078	.1092
M3.5	0.6	D1	D4	3.110	3.181	3.222	.1224	.1252	.1268

(Continued)

Table 8. *(Continued)* Tap Recommendations for Classes 4H and 6H Metric M Profile Screw Threads. (Source, Fastcut Tool Corp.)

Threads per inch		Recommended Tap for Class of Thread		Pitch Diameter Limits for Class of Thread					
				Millimeters			Inches		
Nominal Diameter	Pitch	4H	6H	Min. All Classes (Basic)	Max. 4H	Max. 6H	Min. All Classes (Basic)	Max. 4H	Max. 6H
M4	0.7	D2	D4	3.545	3.620	3.663	.1396	.1425	.1442
M4.5	0.75	D2	D4	4.013	4.088	4.131	.1580	.1609	.1626
M5	0.8	D2	D4	4.480	4.560	4.605	.1764	.1795	.1813
M6	1	D3	D5	5.350	5.445	5.500	.2106	.2144	.2165
M6	0.75	D2	D3	5.513	5.598	5.645	.2170	.2204	.2222
M7	1	D3	D5	6.350	6.445	6.500	.2500	.2537	.2559
M7	0.75	D2	D4	6.513	6.598	6.645	.2564	.2598	.2616
M8	1.25	D3	D5	7.188	7.288	7.348	.2830	.2869	.2893
M8	1	D3	D5	7.350	7.445	7.500	.2894	.2931	.2953
M10	1.5	D3	D6	9.026	9.138	9.206	.3554	.3598	.3624
M10	1.25	D3	D5	9.188	9.288	9.348	.3617	.3657	.3680
M12	1.75	D3	D6	10.863	10.988	11.063	.4277	.4326	.4356
M12	1.25	D3	D5	11.188	11.300	11.368	.4405	.4449	.4476
M14	2	D3	D7	12.701	12.833	12.913	.5000	.5052	.5084
M14	1.5	D3	D6	13.026	13.144	13.216	.5128	.5175	.5203
M16	2	D4	D7	14.701	14.833	14.913	.5788	.5840	.5871
M16	1.5	D3	D6	15.026	15.144	15.216	.5916	.5962	.5990
M18	2.5	D4	D7	16.376	16.516	16.600	.6447	.6502	.6535
M18	1.5	D3	D6	17.026	17.144	17.216	.6703	.6750	.6778
M20	2.5	D4	D7	18.376	18.516	18.600	.7235	.7290	.7323
M20	1.5	D3	D5	19.026	19.144	19.216	.7490	.7537	.7565
M24	3	D4	D8	22.051	22.221	22.316	.8681	.8748	.8786
M24	1.5	D3	D5	23.026	23.151	23.226	.9065	.9114	.9144
M27	3	D5	D8	25.051	25.221	25.316	.9863	.9930	.9967
M27	2	D5	D7	25.701	25.841	25.925	1.0118	1.0174	1.0207
M30	3.5	D5	D9	27.727	27.907	28.007	1.0916	1.0987	1.1026
M30	2	D5	D7	28.701	28.841	28.925	1.1300	1.1355	1.1388
M33	3.5	D5	D9	30.727	30.907	31.007	1.2097	1.2168	1.2207
M33	2	D5	D7	31.701	31.841	31.925	1.2481	1.2536	1.2569
M36	4	D5	D9	33.402	33.592	33.702	1.3150	1.3225	1.3268
M36	2	D5	D7	34.701	34.841	34.925	1.3662	1.3717	1.3750

The above recommended taps normally produce the Class of Thread indicated in average materials when used with reasonable care. However, if the tap specified does not give a satisfactory gage fit in the work, a choice of some other limit tap will be necessary. D1 Limit to have minus .0005 tolerance.

infrequent if the diameter of the tap is over $1/2$" or if the length of thread to be tapped is less than $1/2$". In tougher materials, tapping a successful hole becomes increasingly difficult, which further discourages attempts to achieve high percentages of thread of 70% or more, especially when using small diameter taps to produce holes for UNC threads. Normally, in the production of threads, it is considered impractical to tap a thread unless its nominal diameter is greater than six times the basic thread height. Therefore, when the ratio of diameter (D) is greater than six times the basic thread height (h), the use of a larger diameter, a finer pitch of thread, or both should be considered.

As a rule, when both sides of the assembly are to be concentric, the difference between the minimum major diameter of the external thread and the maximum minor diameter of the internal thread should never be less than twice the addendum of the external thread (0.75H). Another way to calculate is that the sum of the major diameter tolerance and allowance (if any) of the external thread plus the minor diameter tolerance of the internal thread should never be more than 4/3 of the addendum of the external thread (0.5H). This will provide for a minimum safe overlap of 50 percent thread engagement. When one side of the assembly is to be eccentric, the difference between the maximum pitch diameter of the internal thread and the minimum pitch diameter of the external thread should not exceed the basic thread height (0.625H). Otherwise stated, the sum of the pitch diameter tolerances of both threads and the allowance, if any, should not be greater than 0.625H. This will provide for an eccentric assembly condition equal to half the basic thread height (0.312H) and zero minimum overlap on one side. If the results from the limits of size selected violate the above rules, the tolerances should be reduced by using a closer class of tolerance, assuming tolerance consistent with manufacturing possibility, or using a coarser pitch to increase the amount of overlap. The major diameter tolerance of the external thread or minor diameter tolerance of the internal thread should not be less than the pitch diameter allowance of the respective thread to maintain thread form. Also, it should be noted that if the tolerance on the minor diameter of the internal thread must necessarily be large, the major diameter of the external thread must be held close to the maximum major diameter and vice versa.

An additional important consideration when selecting the desired percentage of thread is hole depth and length of engagement. For short lengths of engagement, the hole diameter required prior to threading should be held near the minimum limit to maximize thread height for maximum joint strength. As length of engagement increases, the hole diameter should be increased for more economical tapping with less risk of tap breakage. The following recommended hole sizes before threading (from FED-STD-H28/2B) are suggested to permit economical tapping. It will be noted that the difference between limits in each range is the same and equal to half of the minor diameter tolerance. However, the minimum differences for thread sizes below $1/4$" are equal to the minor diameter tolerances (tolerance is the difference between specified maximum and minimum limits) for lengths of engagement to and including 0.33 diameter. For sizes $1/4$" and larger, with lengths of engagement exceeding 0.33 diameter, the minimum values between maximum and minimum hole sizes should never be less than 0.0040".

Table 9 gives minor diameter limits and corresponding percentages of thread for Class 1B and 2B Unified thread sizes from $1/4$–20 UNC through 1 $1/8$–16 UN. **Table 10** provides the same information for Class 1B and 2B sizes 1 $1/8$–18 UNEF through 3 $3/4$–4 UNC.

Length of Engagement	Minimum Hole Size	Maximum Hole Size
Up to and including 0.33 diameter	Minimum Minor Diameter	Minimum Minor Diameter plus $1/2$ Minor Diameter Tolerance
Above 0.33 through 0.67 diameter	Minimum Minor Diameter plus $1/4$ Minor Diameter Tolerance	Minimum Minor Diameter plus $3/4$ Minor Diameter Tolerance
Above 0.67 through 1.50 diameter	Minimum Minor Diameter plus $1/2$ Minor Diameter Tolerance	Maximum Minor Diameter (Minimum Minor Diameter plus Tolerance)
Above 1.5 through 3.0 diameter *	Minimum Minor Diameter plus $3/4$ Minor Diameter Tolerance	Maximum Minor Diameter plus $1/4$ Minor Diameter Tolerance *

* Recommended maximum hole size is outside standard minor diameter limits. Use of a minor diameter larger than standard will result in a reduction in shear area of the external threads of the mating part. If manufacturing process permits, maximum hole size before threading should be maintained at the high end of the standard minor diameter limits.

Table 11 gives this information for Class 3B Unified thread sizes 0–80 UNF through 1 $^1/_8$–16 UN. **Table 12** covers Class 3B sizes 1 $^1/_8$–18 UNEF through 3 $^3/_4$–4. These tables also list sizes of drills that may be expected to drill holes within or near the specified minor diameter limits. The resulting probable hole size and percentages of thread are tabulated. Since drills normally produce marginally oversize holes, the tabulated hole sizes have been derived from probable mean oversizes when drilling holes equal to 1.5 times drill diameter with measurement being taken at midpoint of the depth drilled. Due to the increase in amount oversize experienced with drills in excess of $^3/_4$", reaming of holes this size and larger is recommended. It should be noted that the Unified Thread profile allows for a maximum permissible percentage of thread of 83.3%, and that some of the indicated percentages on the Tables are in excess of this amount. This is either due to the allowances for drill sizes to cut oversize, or lack of availability of drills within specified minor diameter limits. A reading in excess of 83.3% therefore indicates that the drill size is smaller than the minimum minor diameter, and additional machining of the hole may be necessary to permit economical tapping.

Table 13 gives recommended hole size limits for Unified Class 1B and 2B threads for three lengths of engagement (see chart above), and **Table 14** provides the same suggested hole sizes for Unified Class 3B threads. For both Tables, refer to Tables 11 and 12 for basic minimum and minor diameter and percent of thread specifications. Hole size limits for diameter-pitch combinations for sizes over 1.00 inch not included on these Tables can be derived provided that there is a diameter-pitch combination in the Tables that: 1) has the same pitch and 2) has a diameter that is less by an integral amount than the diameter of the diameter-pitch combination for which hole size values are desired.

Example. To obtain the values for a 4.00–8UN–3B thread, add 2.00 to the values for the 2.00–8UN–3B thread given in the table. These values would then become 3.8650, 3.8722, 3.8684, 3.8759, 3.8722, 3.8797. The percentages of thread will remain unchanged.

Recommended tap drill sizes for Metric M Profile threads are given in **Table 15**, and **Table 16** provides pitch diameter limits for Class 4H and Class 6H threads. See **Table 17** for decimal inch equivalents of standard M Profile metric threads. *(Text continued p. 243)*

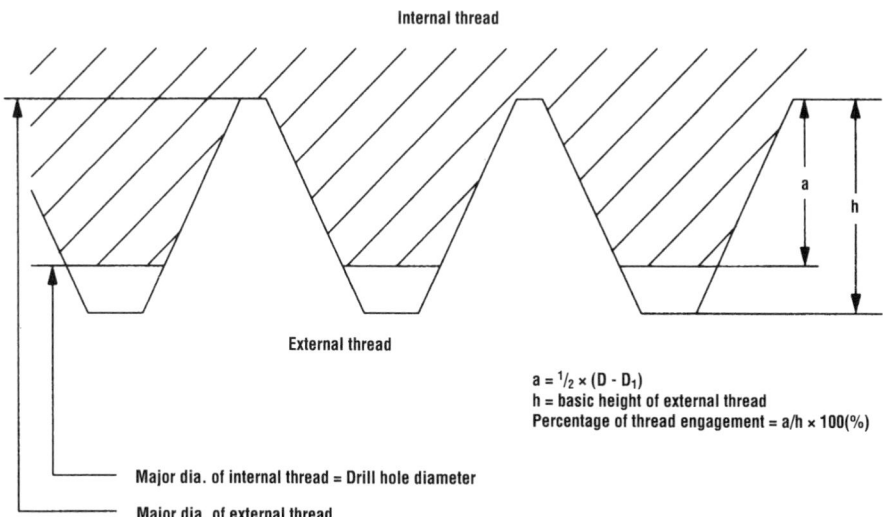

Internal thread

External thread

$a = ^1/_2 \times (D - D_1)$
h = basic height of external thread
Percentage of thread engagement = a/h × 100(%)

Major dia. of internal thread = Drill hole diameter

Major dia. of external thread

Figure 2. Percentage of thread engagement. External thread = basic profile. (Source, OSG Tap and Die, Inc.)

Table 9. Tap Drill Sizes and Percentage of Thread, Unified Threads Class 1B and 2B, Size $1/4$-20 UNC* Through 1 $1/8$-16 UN. (Source, FED-STD-H28/2B.)

Size, TPI, and Series*	Classes 1B and 2B Minor Dia.				Tap Drill and Percentage of Thread				
	D_1 (min.)	% of Thread*	D_1 (max.)	% of Thread*	Drill Size		% of Thread*	Probable Hole Size	% of Thread*
1/4-20 UNC	.196	83.1	.207	66.2	#9	.1960	83	.1998	77
					#8	.1990	79	.2028	73
					#7	.2010	75	.2048	70
					13/64	.2031	72	.2069	66
					#6	.2040	71	.2078	65
					#5	.2055	69	.2093	63
1/4-28 UNF	.211	84.1	.220	64.7	#3	.2130	80	.2168	72
					7/32	.2188	67	.2226	59
1/4-32 UNEF	.216	83.8	.224	64.0	7/32	.2188	77	.2226	67
					#2	.2210	71	.2248	62
1/4-36 UNS	.220	83.1	.226	66.5	#2	.2210	80	.2248	70
5/16-18 UNC	.252	83.8	.265	65.8	F	.2570	77	.2608	72
					G	.2610	71	.2651	66
5/16-20 UN	.258	83.9	.270	65.4	F	.2570	85	.2608	80
					G	.2610	79	.2651	73
					H	.2660	72	.2701	65
5/16-24 UNF	.267	84.1	.277	65.6	H	.2660	86	.2701	78
					I	.2720	75	.2761	67
					J	.2770	66	.2811	58
5/16-28 UN	.274	83.0	.282	65.7	J	.2770	77	.2811	68
					K	.2810	68	.2852	59
					9/32	.2812	67	.2854	58
5/16-32 UNEF	.279	82.5	.286	65.3	K	.2810	78	.2852	67
					9/32	.2812	77	.2854	67
5/16-36 UNS	.282	84.5	.289	65.1	7.25 mm	.2854	75	.2896	63
3/8-16 UNC	.307	83.8	.321	66.5	5/16	.3125	77	.3169	72
					O	.3160	73	.3204	67
3/8-20 UN	.321	83.1	.332	66.2	P	.3230	80	.3274	73
					Q	.3320	66	.3364	59
3/8-24 UNF	.330	83.1	.340	64.7	Q	.3320	79	.3364	71
					R	.3390	67	.3434	58
3/8-28 UN	.336	84.1	.345	64.7	R	.3390	78	.3434	68
					11/32	.3438	67	.3483	58
3/8-32 UNEF	.341	83.8	.349	64.0	11/32	.3438	77	.3483	66
					S	.3480	67	.3525	55
3/8-36 UNS	.345	83.1	.352	63.7	S	.3480	75	.3525	62
7/16-14 UNC	.360	83.5	.376	66.3	T	.3580	86	.3626	81
					23/64	.3594	84	.3640	79
7/16-16 UN	.370	83.1	.384	65.9	3/8	.3750	77	.3796	71
					V	.3770	75	.3816	69
7/16-20 UNF	.383	83.9	.395	65.4	W	.3860	79	.3906	72
					25/64	.3906	72	.3952	65
7/16-28 UNEF	.399	83.0	.407	65.7	Y	.4040	72	.4086	62

* For Machine Screw sizes, see **Table 11**: Tap Drill sizes and Percentage of Thread, Unified Threads Class 3B, Size 0-80 UNF Through 1 $1/8$-16 UN.

(Continued)

Table 9. *(Continued)* **Tap Drill Sizes and Percentage of Thread, Unified Threads Class 1B and 2B, Size ¹/₄-20 UNC* Through 1 ¹/₈-16 UN.** *(Source, FED-STD-H28/2B.)*

Size, TPI, and Series*	Classes 1B and 2B Minor Dia.				Tap Drill and Percentage of Thread				
	D_1 (min.)	% of Thread*	D_1 (max.)	% of Thread*	Drill Size		% of Thread*	Probable Hole Size	% of Thread*
7/16-32 UN	.404	82.5	.41	65.3	Y	.4040	83	.4086	71
					13/32	.4062	77	.4108	66
1/2-12 UNS	.410	83.1	.428	66.5	Z	.4130	80	.4177	76
					27/64	.4219	72	.4266	68
1/2-13 UNC	.417	83.1	.434	66.0	27/64	.4219	78	.4266	73
1/2-16 UN	.432	83.8	.446	66.5	7/16	.4375	77	.4422	71
1/2-20 UNF	.446	83.1	.457	66.2	29/64	.4531	72	.4578	65
1/2-28 UNEF	.461	84.1	.470	64.70	15/32	.4688	67	.4736	57
1/2-32 UN	.466	83.8	.474	64.0	15/32	.4688	77	.4736	65
9/16-12 UNC	.472	83.6	.490	67.0	15/32	.4688	87	.4736	82
					31/64	.4844	72	.4892	68
9/16-16 UN	.495	83.1	.509	65.9	1/2	.5000	77	.5048	71
					0.5062	.5062	69	.5110	63
9/16-18 UNF	.502	83.8	.515	65.8	1/2	.5000	87	.5048	80
					0.5062	.5062	78	.5110	71
9/16-20 UN	.508	83.9	.520	65.4	33/64	.5156	72	.5204	65
9/16-24 UNEF	.517	84.1	.527	65.6	33/64	.5156	87	.5204	78
					0.5203	.5203	78	.5251	69
9/16-28 UN	.524	83.0	.532	65.7	17/32	.5312	67	.5361	57
					0.5263	.5263	78	.5312	67
9/16-32 UN	.529	82.5	.536	65.3	17/32	.5312	77	.5361	65
5/8-11 UNC	.527	83.0	.546	66.9	17/32	.5312	79	.5361	75
5/8-12 UN	.535	83.1	.553	66.5	35/64	.5469	72	.5518	68
5/8-16 UN	.557	83.8	.571	66.5	9/16	.5625	77	.5674	71
					0.5687	.5687	69	.5736	63
5/8-18 UNF	.565	83.1	.578	65.1	9/16	.5625	87	.5674	80
					0.5687	.5687	78	.5736	71
5/8-20 UN	.571	83.1	.582	66.2	37/64	.5781	72	.5830	65
5/8-24 UNEF	.580	83.1	.590	64.7	37/64	.5781	87	.5830	78
					0.5828	.5828	78	.5877	69
5/8-28 UN	.586	84.1	.595	64.7	19/32	.5938	67	.5987	57
5/8-32 UN	.591	83.8	.599	64.0	19/32	.5938	77	.5987	65
11/16-12 UN	.597	83.6	.615	67.0	19/32	.5938	87	.5987	82
					39/64	.6094	72	.6143	68
11/16-16 UN	.620	83.1	.634	65.9	5/8	.6250	77	.6300	71
11/16-20 UN	.633	83.9	.645	65.4	41/64	.6406	72	.6456	65
11/16-24 UNEF	.642	84.1	.652	65.6	41/64	.6406	87	.6456	77
11/16-28 UN	.649	83.0	.657	65.7	21/32	.6562	67	.6612	57
11/16-32 UN	.654	82.5	.661	65.3	21/32	.6562	77	.6612	65

* For Machine Screw sizes, see **Table 11**: Tap Drill sizes and Percentage of Thread, Unified Threads Class 3B, Size 0-80 UNF Through 1 ¹/₈-16 UN.

(Continued)

Table 9. *(Continued)* **Tap Drill Sizes and Percentage of Thread, Unified Threads Class 1B and 2B, Size $1/4$-20 UNC* Through 1 $1/8$-16 UN.** *(Source, FED-STD-H28/2B.)*

Size, TPI, and Series*	Classes 1B and 2B Minor Dia.				Tap Drill and Percentage of Thread				
	D_1 (min.)	% of Thread*	D_1 (max.)	% of Thread*	Drill Size		% of Thread*	Probable Hole Size	% of Thread*
3/4-10 UNC	.642	83.1	.663	67.0	41/64	.6406	84	.6456	80
					21/32	.6562	72	.6612	68
3/4-12 UN	.660	83.1	.678	66.5	21/32	.6562	87	.6612	82
					43/64	.6719	72	.6769	68
3/4-16 UNF	.682	83.8	.696	66.5	11/16	.6875	77	.6925	71
3/4-20 UNEF	.696	83.1	.707	66.2	45/64	.7031	72	.7082	64
3/4-28 UN	.711	84.1	.720	64.7	23/32	.7188	67	.7239	56
3/4-32 UN	.716	83.8	.724	64.0	23/32	.7188	77	.7239	64
13/16-12 UN	.722	83.6	.740	67.0	47/64	.7344	72	.7395	67
13/16-16 UN	.745	83.1	.759	65.9	3/4	.7500	77	.7552	71
13/16-20 UNEF	.758	83.9	.770	65.4	49/64	.7656	72	.7708	64
13/16-28 UN	.774	83.0	.782	65.7	25/32	.7812	67	.7864	56
13/16-32 UN	.779	82.5	.786	65.3	25/32	.7812	77	.7864	64
7/8-9 UNC	.755	83.1	.778	67.2	49/64	.7656	76	.7708	72
7/8-12 UN	.785	83.1	.803	66.5	25/32	.7812	87	.7864	82
					54/64	.7969	72	.8021	67
7/8-14 UNF	.798	83.0	.814	65.7	57/64	.7969	84	.8021	67
					0.8024	.8024	78	.8076	73
					13/16	.8125	67	.8177	62
7/8-16 UN	.807	83.8	.821	66.5	13/16	.8125	77	.8178	70
7/8-20 UNEF	.821	83.1	.832	66.2	53/64	.8281	72	.8335	64
7/8-28 UN	.836	84.1	.845	64.7	27/32	.8438	67	.8493	55
7/8-32 UN	.841	83.8	.849	64.0	27/32	.8438	77	.8493	63
15/16-12 UN	.847	83.6	.865	67.0	27/32	.8438	87	.8493	81
					55/64	.8594	72	.8650	67
15/16-16 UN	.870	83.1	.884	65.9	7/8	.8750	77	.8807	70
15/16-20 UNEF	.883	83.9	.895	65.4	57/64	.8906	72	.8965	63
15/16-28 UN	.899	83.0	.907	65.7	29/32	.9062	67	.9122	55
15/16-32 UN	.904	82.5	.911	65.3	29/32	.9062	77	.9122	62
1-8 UNC	.865	83.1	.890	67.7	55/64	.8594	87	.8653	83
					7/8	.8750	77	.8809	73
1-12 UNF	.910	83.1	.928	66.5	29/32	.9062	87	.9122	81
					59/64	.9219	72	.9279	67
1-14 UNS	.923	83.0	.938	66.8	59/64	.9219	84	.9279	78
					0.9274	.9274	78	.9335	72
1-16 UN	.923	83.8	.946	66.5	15/16	.9375	77	.9437	69
1-20 UNEF	.946	83.1	.957	66.2	61/64	.9531	72	.9594	63
1-28 UN	.961	84.1	.970	64.7	31/32	.9688	67	.9753	53

* For Machine Screw sizes, see **Table 11**: Tap Drill sizes and Percentage of Thread, Unified Threads Class 3B, Size 0-80 UNF Through 1 $1/8$-16 UN.

(Continued)

Table 9. *(Continued)* Tap Drill Sizes and Percentage of Thread, Unified Threads Class 1B and 2B, Size $1/4$-20 UNC* Through 1 $1/8$-16 UN. *(Source, FED-STD-H28/2B.)*

Size, TPI, and Series*	Classes 1B and 2B Minor Dia.				Tap Drill and Percentage of Thread				
	D_1 (min.)	% of Thread*	D_1 (max.)	% of Thread*	Drill Size		% of Thread*	Probable Hole Size	% of Thread*
1-32 UN	.966	83.8	.974	64.0	31/32	.9688	77	.9753	61
1 1/16-8 UN	.927	83.4	.952	68.0	59/64	.9219	87	.9279	83
					0.9274	.9274	83	.9335	79
					15/16	.9375	77	.9437	73
1 1/16-12 UN	.927	83.6	.990	67.0	31/32	.9688	87	.9753	81
					63/64	.9844	72	.9911	66
1 1/16-16 UN	.995	83.1	1.009	65.9	1	1.0000	77	1.0069	68
1 1/16-18 UNEF	1.002	83.8	1.015	65.8	1	1.0000	87	1.0069	77
1 1/16-20 UN	1.008	83.9	1.020	65.4	1 1/64	1.0156	72	1.0226	61
1 1/16-28 UN	1.024	83.0	1.032	65.7	1 1/32	1.0312	67	1.0383	52
1 1/8-7 UNC	.970	83.5	.998	68.4	31/32	.9688	84	.9750	81
					63/64	.9844	76	.9911	72
1 1/8-8 UN	.990	83.1	1.015	67.7	1	1.0000	77	1.0069	73
1 1/8-12 UNF	1.035	83.1	1.053	66.5	1 1/32	1.0312	87	1.0383	80
					1 3/64	1.0469	72	1.0541	65
1 1/8-16 UN	1.057	83.8	1.071	66.5	1 1/16	1.0625	77	1.0699	68

* For Machine Screw sizes, see **Table 11**: Tap Drill sizes and Percentage of Thread, Unified Threads Class 3B, Size 0-80 UNF Through 1 $1/8$-16 UN.

Table 10. Tap Drill Sizes and Percentage of Thread, Unified Threads Class 1B and 2B, Size 1 $1/8$-18 UNEF Through 3 $3/4$-4 UNC. *(Source, FED-STD-H28/2B.)*

Size, TPI, and Series*	Classes 1B and 2B Minor Dia.				Tap Drill and Percentage of Thread		
	D_1 (min.)	% of Thread*	D_1 (max.)	% of Thread*	Drill Size		% of Thread*
1 1/8-18 UNEF	1.065	83.1	1.078	65.1	1 1/16	1.0625	87
					1 5/64	1.0781	65
1 1/8-20 UN	1.071	83.1	1.082	66.2	1 5/64	1.0781	72
1 1/8-28 UN	1.086	84.1	1.095	64.7	1 3/32	1.0938	67
1 3/16-8 UN	1.052	83.4	1.077	68.0	1 1/16	1.0625	77
1 3/16-12 UN	1.097	83.6	1.115	67.0	1 3/32	1.0938	87
1 3/16-16 UN	1.120	83.1	1.134	65.9	1 1/8	1.1250	77
1 3/16-18 UNEF	1.127	83.8	1.140	65.8	1 1/8	1.1250	87
					1 9/64	1.1406	65
1 3/16-20 UN	1.133	83.9	1.145	65.4	1 9/64	1.1406	72
1 3/16-28 UN	1.149	83.0	1.157	65.7	1 5/32	1.1562	67
1 1/8-7 UNC	1.095	83.5	1.123	68.4	1 3/32	1.0938	84
1 1/8-8 UN	1.115	83.1	1.140	67.7	1 1/8	1.1250	77
1 1/8-12 UNF	1.160	83.1	1.178	66.5	1 5/32	1.1562	87
					1 11/64	1.1719	72
1 1/8-16 UN	1.182	83.8	1.196	66.5	1 3/16	1.1875	77

(Continued)

Table 10. *(Continued)* Tap Drill Sizes and Percentage of Thread, Unified Threads Class 1B and 2B, Size 1 $^1/_8$-18 UNEF Through 3 $^3/_4$-4 UNC. *(Source, FED-STD-H28/2B.)*

Size, TPI, and Series*	Classes 1B and 2B Minor Dia.				Tap Drill and Percentage of Thread		
	D_1 (min.)	% of Thread*	D_1 (max.)	% of Thread*	Drill Size		% of Thread*
1 1/8-18 UNEF	1.190	83.1	1.203	65.1	1 3/16	1.1875	87
					1 13/64	1.2031	65
1 1/8-20 UN	1.196	83.1	1.207	66.2	1 13/64	1.2031	72
1 1/8-28 UN	1.211	84.1	1.220	64.7	1 7/32	1.2188	67
1 5/16-8 UN	1.177	83.4	1.202	68.0	1 11/64	1.1719	87
					1 3/16	1.1875	77
1 5/16-12 UN	1.222	83.6	1.240	67.0	1 7/32	1.2188	87
					1 15/64	1.2344	72
1 5/16-16 UN	1.245	83.1	1.259	65.9	1 1/4	1.2500	77
1 5/16-18 UNEF	1.252	83.8	1.265	65.8	1 1/4	1.2500	87
					1 17/64	1.2656	65
1 5/16-20 UN	1.258	83.9	1.270	65.4	1 17/64	1.2656	72
1 5/16-28 UN	1.274	83.0	1.282	65.7	1 9/32	1.2812	67
1 3/8-6 UNC	1.195	83.1	1.225	69.3	1 3/16	1.1875	87
					1 13/64	1.2031	79
					1 15/64	1.2188	72
1 3/8-8 UN	1.240	83.1	1.265	67.7	1 15/64	1.2344	87
					1 1/4	1.2500	77
1 3/8-12 UNF	1.285	83.1	1.303	66.5	1 9/32	1.2812	87
					1 19/64	1.2969	72
1 3/8-16 UN	1.307	83.8	1.321	66.5	1 5/16	1.3125	77
1 3/8-18 UNEF	1.315	83.1	1.328	65.1	1 5/16	1.3125	87
					1 21/64	1.3281	65
1 3/8-20 UN	1.321	83.1	1.332	66.2	1 21/64	1.3281	72
1 3/8-28 UN	1.336	84.1	1.345	64.7	1 11/32	1.3438	67
1 7/16-6 UN	1.257	83.4	1.288	69.1	1 17/64	1.2656	79
					1 9/32	1.2812	72
1 7/16-8 UN	1.302	83.4	1.327	68.0	1 19/64	1.2969	87
					1 5/16	1.3125	77
1 7/16-12 UN	1.347	83.6	1.365	67.0	1 11/32	1.3438	87
					1 23/64	1.3594	72
1 7/16-16 UN	1.370	83.1	1.384	65.9	1 3/8	1.3750	77
1 7/16-18 UNEF	1.377	83.8	1.390	65.8	1 3/8	1.3750	87
1 7/16-20 UN	1.383	83.9	1.395	65.4	1 25/64	1.3906	72
1 7/16-28 UN	1.399	83.0	1.407	65.7	1 13/32	1.4062	
1 1/2-6 UNC	1.320	83.1	1.350	69.3	1 5/16	1.3125	87
					1 21/64	1.3281	79
1 1/2-8 UN	1.365	83.1	1.390	67.7	1 23/64	1.3594	87
					1 3/8	1.3750	77
1 1/2-12 UNF	1.410	83.1	1.428	66.5	1 13/32	1.4062	87
					1 27/64	1.4219	72
1 1/2-16 UN	1.432	83.8	1.446	66.5	1 7/16	1.4375	77

(Continued)

Table 10. *(Continued)* Tap Drill Sizes and Percentage of Thread, Unified Threads Class 1B and 2B, Size 1 $^1/_8$-18 UNEF Through 3 $^3/_4$-4 UNC. *(Source, FED-STD-H28/2B.)*

Size, TPI, and Series*	Classes 1B and 2B Minor Dia.				Tap Drill and Percentage of Thread		
	D_1 (min.)	% of Thread*	D_1 (max.)	% of Thread*	Drill Size		% of Thread*
1 1/2-18 UNEF	1.440	83.1	1.452	66.5	1 7/16	1.4375	87
1 1/2-20 UN	1.446	83.1	1.457	66.2	1 29/64	1.4531	72
1 1/2-28 UN	1.461	84.1	1.470	64.7	1 15/32	1.4688	67
1 9/16-6 UN	1.382	83.4	1.413	69.1	1 25/64	1.3906	79
					1 13/32	1.4062	72
1 9/16-8 UN	1.427	83.4	1.452	68.0	1 27/64	1.4219	87
					1 7/16	1.4375	77
1 9/16-12 UN	1.472	83.6	1.490	67.0	1 15/32	1.4688	87
					1 31/64	1.4844	72
1 9/16-16 UN	1.495	83.1	1.509	65.9	1 1/2	1.5000	77
1 9/16-18 UNEF	1.502	83.8	1.515	65.8	1 1/2	1.5000	87
					1 33/64	1.5156	65
1 9/16-20 UN	1.508	83.9	1.520	65.4	1 33/64	1.5156	72
1 5/8-6 UN	1.445	83.1	1.475	69.3	1 29/64	1.4531	79
					1 15/32	1.4688	72
1 5/8-8 UN	1.490	83.1	1.515	67.7	1 31/64	1.4844	87
					1 1/2	1.5000	77
1 5/8-12 UN	1.535	83.1	1.553	66.5	1 17/32	1.5312	87
					1 35/64	1.5469	72
1 5/8-16 UN	1.557	83.8	1.571	66.5	1 9/16	1.5625	77
1 5/8-18 UNEF	1.565	83.1	1.578	65.1	1 9/16	1.5625	87
					1 37/64	1.5781	65
1 5/8-20 UN	1.571	83.1	1.582	66.2	1 37/64	1.5781	72
1 11/16-6 UN	1.507	83.4	1.538	69.1	1 1/2	1.5000	87
					1 33/64	1.5156	79
					1 17/32	1.5312	72
1 11/16-8 UN	1.552	83.4	1.577	68.0	1 9/16	1.5625	77
1 11/16-12 UN	1.597	83.6	1.615	67.0	1 19/32	1.5938	87
					1 39/64	1.6094	72
1 11/16-16 UN	1.620	83.1	1.634	65.9	1 5/8	1.6250	77
1 11/16-18 UNEF	1.627	83.8	1.640	65.8	1 5/8	1.6250	87
					1 41/64	1.6406	65
1 11/16-20 UN	1.633	83.9	1.645	65.4	1 41/64	1.6406	72
1 3/4-5 UNC	1.534	83.1	1.568	70.1	1 17/32	1.5312	84
					1 35/64	1.5469	78
1 3/4-6 UN	1.570	83.1	1.600	69.3	1 37/64	1.5781	79
					1 19/32	1.5938	72
1 3/4-8 UN	1.615	83.1	1.640	67.7	1 39/64	1.6094	87
					1 5/8	1.6250	77
					1 41/64	1.6406	67
1 3/4-12 UN	1.660	83.1	1.678	66.5	1 21/32	1.6562	87
					1 43/64	1.6719	72

(Continued)

Table 10. *(Continued)* **Tap Drill Sizes and Percentage of Thread, Unified Threads Class 1B and 2B, Size 1 1/8-18 UNEF Through 3 3/4-4 UNC.** *(Source, FED-STD-H28/2B.)*

Size, TPI, and Series*	Classes 1B and 2B Minor Dia.				Tap Drill and Percentage of Thread		
	D_1 (min.)	% of Thread*	D_1 (max.)	% of Thread*	Drill Size		% of Thread*
1 3/4-16 UN	1.682	83.8	1.696	66.5	1 11/16	1.6875	77
1 3/4-20 UN	1.696	83.1	1.707	66.2	1 45/64	1.7031	72
1 13/16-6 UN	1.632	83.4	1.663	69.1	1 5/8	1.6250	87
					1 41/64	1.6406	79
					1 21/32	1.6562	72
1 13/16-8 UN	1.677	83.4	1.702	68.0	1 43/64	1.6719	87
					1 11/16	1.6875	77
1 13/16-12 UN	1.722	83.6	1.740	67.0	1 23/32	1.7188	87
					1 47/64	1.7344	72
1 13/16-16 UN	1.745	83.1	1.759	65.9	1 3/4	1.7500	77
1 13/16-20 UN	1.758	83.9	1.770	65.4	1 49/64	1.7656	72
1 7/8-6 UN	1.695	83.1	1.725	69.3	1 45/64	1.7031	79
					1 23/32	1.7188	72
1 7/8-8 UN	1.740	83.1	1.765	67.7	1 3/4	1.7500	77
1 7/8-12 UN	1.785	83.1	1.803	66.5	1 25/32	1.7812	87
					1 51/64	1.7969	72
1 7/8-16 UN	1.807	83.8	1.821	66.5	1 13/16	1.8125	77
1 7/8-20 UN	1.821	83.1	1.832	66.2	1 53/64	1.8281	72
1 15/16-6 UN	1.757	83.4	1.788	69.1	1 49/64	1.7656	79
					1 25/32	1.7812	72
1 15/16-8 UN	1.802	83.4	1.827	68.0	1 51/64	1.7969	87
					1 13/16	1.8125	77
1 15/16-12 UN	1.847	83.6	1.865	67.0	1 27/32	1.8438	87
					1 55/64	1.8594	72
1 15/16-16 UN	1.870	83.1	1.884	65.9	1 7/8	1.8750	77
					1 25/32	1.7812	72
1 15/16-8 UN	1.802	83.4	1.827	68.0	1 51/64	1.7969	87
					1 13/16	1.8125	77
1 15/16-12 UN	1.847	83.6	1.865	67.0	1 27/32	1.8438	87
					1 55/64	1.8594	72
1 15/16-16 UN	1.870	83.1	1.884	65.9	1 7/8	1.8750	77
1 15/16-20 UN	1.883	83.9	1.895	65.4	1 57/64	1.8906	72
2-4.5 UNC	1.759	83.5	1.795	71.0	1 25/32	1.7812	76
2-6 UN	1.820	83.1	1.850	69.3	1 53/64	1.8281	79
					1 27/32	1.8438	72
2-8 UN	1.865	83.1	1.890	67.7	1 7/8	1.8750	77
2-12 UN	1.910	83.1	1.928	66.5	1 29/32	1.9062	87
					1 59/64	1.9219	72
2-16 UN	1.932	83.8	1.946	66.5	1 15/16	1.9375	77
2-20 UN	1.946	83.1	1.957	66.2	1 61/64	1.9531	72
2 1/16-16 UNS	1.995	83.1	2.009	65.9	2	2.0000	77

(Continued)

Table 10. *(Continued)* Tap Drill Sizes and Percentage of Thread, Unified Threads Class 1B and 2B, Size 1 $^1/_8$-18 UNEF Through 3 $^3/_4$-4 UNC. *(Source, FED-STD-H28/2B.)*

Size, TPI, and Series*	Classes 1B and 2B Minor Dia.				Tap Drill and Percentage of Thread		
	D_1 (min.)	% of Thread*	D_1 (max.)	% of Thread*	Drill Size		% of Thread*
2 1/8-6 UN	1.945	83.1	1.975	69.3	1 61/64	1.9531	79
					1 31/32	1.9688	72
2 1/8-8 UN	1.990	83.1	2.015	67.7	2	2.0000	77
2 1/8-12 UN	2.035	83.1	2.053	66.5	2 1/32	2.0312	87
2 1/8-16 UN	2.057	83.8	2.071	66.5	2 1/16	2.0625	77
2 1/8-20 UN	2.071	83.1	2.082	66.2	2 1/16	2.0625	96
2 3/16 UNS	2.120	83.1	2.134	65.9	2 1/8	2.1250	77
2 1/4-4.5 UNC	2.090	83.5	2.045	71.0	2	2.0000	87
					2 1/32	2.0312	76
2 1/4-6 UN	2.070	83.1	2.100	69.3	2 1/16	2.0625	87
2 1/4-8 UN	2.115	83.1	2.140	67.7	2 1/8	2.1250	77
2 1/4-12 UN	2.160	83.1	2.178	66.5	2 5/32	2.1562	87
2 1/4-16 UN	2.182	83.8	2.196	66.5	2 3/16	2.1875	77
2 1/4-20 UN	2.196	83.1	2.207	66.2	2 3/16	2.1875	96
2 5/16-16 UNS	2.245	83.1	2.259	65.9	2 1/4	2.2500	77
2 3/8-6 UN	2.195	83.1	2.226	68.8	2 3/16	2.1875	87
2 3/8-8 UN	2.240	83.1	2.265	67.7	2 1/4	2.2500	77
2 3/8-12 UN	2.285	83.1	2.303	66.5	58 mm	2.2835	85
2 3/8-16 UN	2.307	83.8	2.321	66.5	2 5/16	2.3125	77
2 3/8-20 UN	2.321	83.1	2.332	66.2	2 5/16	2.3125	96
2 7/16-16 UNS	2.370	83.1	2.384	65.9	2 3/8	2.3750	77
2 1/2-4 UNC	2.229	83.4	2.267	71.7	2 7/32	2.2188	87
					2 1/4	2.2500	77
2 1/2-6 UN	2.320	83.1	2.350	69.3	2 5/16	2.3125	87
2 1/2-8 UN	2.365	83.1	2.390	67.7	2 3/8	2.3750	77
2 1/2-12 UN	2.410	83.1	2.428	66.5	2 13/32	2.4062	87
2 1/2-16 UN	2.432	83.8	2.446	66.5	2 7/16	2.4375	77
2 1/2-20 UN	2.446	83.1	2.457	66.2	2 7/16	2.4375	96
2 5/8-4 UN	2.354	83.4	2.392	71.7	2 11/32	2.3438	87
					2 3/8	2.3750	77
2 5/8-6 UN	2.445	83.1	2.475	69.3	2 7/16	2.4375	87
2 5/8-8 UN	2.490	83.1	2.515	67.7	2 1/2	2.5000	77
2 5/8-12 UN	2.535	83.1	2.553	66.5	2 17/32	2.5312	87
2 5/8-16 UN	2.557	83.8	2.571	66.5	2 9/16	2.5625	77
2 5/8-20 UN	2.571	83.1	2.582	66.2	2 9/16	2.5625	96
2 3/4-4 UNC	2.479	83.4	2.517	71.7	2 1/2	2.5000	77
2 3/4-6 UN	2.570	83.1	2.600	69.3	2 9/16	2.5625	87
2 3/4-8 UN	2.615	83.1	2.640	67.7	2 5/8	2.6250	77

(Continued)

Table 10. *(Continued)* **Tap Drill Sizes and Percentage of Thread, Unified Threads Class 1B and 2B, Size 1 1/8-1 UNEF Through 3 3/4-4 UNC.** *(Source, FED-STD-H28/2B.)*

Size, TPI, and Series*	Classes 1B and 2B Minor Dia.				Tap Drill and Percentage of Thread		
	D_1 (min.)	% of Thread*	D_1 (max.)	% of Thread*	Drill Size		% of Thread*
2 3/4-12 UN	2.660	83.1	2.678	66.5	2 21/32	2.6562	87
2 3/4-16 UN	2.682	83.8	2.696	66.5	2 11/16	2.6875	77
2 3/4-20 UN	2.696	83.1	2.707	66.2	2 11/16	2.6875	96
2 7/8-4 UN	2.604	83.4	2.642	71.7	2 5/8	2.6250	77
2 7/8-6 UN	2.695	83.1	2.725	69.3	2 11/16	2.6875	87
2 7/8-8 UN	2.740	83.1	2.765	67.7	2 3/4	2.7500	77
2 7/8-12 UN	2.785	83.1	2.803	66.5	2 25/32	2.7812	87
2 7/8-16 UN	2.807	83.8	2.821	66.5	2 13/16	2.8125	77
2 7/8-20 UN	2.821	83.1	2.832	66.2	2 13/16	2.8125	96
3-4 UNC	2.729	83.4	2.767	71.7	2 3/4	2.7500	77
3-6 UN	2.820	83.1	2.850	69.3	2 13/16	2.8125	87
3-8 UN	2.865	83.1	2.890	67.7	2 7/8	2.8750	77
3-12 UN	2.910	83.1	2.928	66.5	74 mm	2.9134	80
3-16 UN	2.932	83.8	2.946	66.5	2 15/16	2.9375	77
3-20 UN	2.946	83.1	2.957	66.2	2 15/16	2.9375	96
3 1/4-4 UNC	2.979	83.4	3.017	71.7	3	3.0000	77
3 1/2-4 UNC	3.229	83.4	3.267	71.7	3 1/4	3.2500	77
3 3/4-4 UNC	3.479	83.4	3.517	71.7	3 1/2	3.5000	77

Table 11. Tap Drill Sizes and Percentage of Thread, Unified Threads Class 3B, Size 0-80 UNF Through 1 1/8-16 UN. *(Source, FED-STD-H28/2B.)*

Size, TPI, and Series*	Classes 3B Minor Dia.				Tap Drill and Percentage of Thread				
	D_1 (min.)	% of Thread*	D_1 (max.)	% of Thread*	Drill Size		% of Thread*	Probable Hole Size	% of Thread*
0-80 UNF	.0465	83.1	.0514	52.9	#56	.0465	83	.0480	74
					3/64	.0469	81	.0484	71
1-64 UNC	.0561	83.3	.0623	52.7	#54	.0550	89	.0565	81
					#53	.0595	67	.0610	59
1-72 UNF	.0580	83.1	.0635	52.7	#53	.0595	75	.0610	67
					1/16	.0625	58	.0640	50
2-56 UNC	.0667	83.2	.0737	53.0	#51	.0670	82	.0687	75
					#50	.0700	69	.0717	62
					#49	.0730	56	.0747	49
2-64 UNF	.0691	83.3	.0753	52.7	#50	.0700	79	.0717	70
					#49	.0730	64	.0747	56
3-48 UNC	.0764	83.5	.0845	53.6	#48	.0760	85	.0779	78
					5/64	.0781	77	.0800	70
					#47	.0785	76	.0804	69
					#46	.0810	67	.0829	60
					#45	.0820	63	.0839	56

(Continued)

Table 11. *(Continued)* Tap Drill Sizes and Percentage of Thread, Unified Threads Class 3B, Size 0-80 UNF Through 1 1/8-16 UN. *(Source, FED-STD-H28/2B.)*

Size, TPI, and Series*	Classes 3B Minor Dia.				Tap Drill and Percentage of Thread				
	D_1 (min.)	% of Thread*	D_1 (max.)	% of Thread*	Drill Size		% of Thread*	Probable Hole Size	% of Thread*
3-56 UNF	.0797	83.2	.0865	53.9	#46	.0810	78	.0829	69
					#45	.0820	73	.0839	65
					#44	.0860	56	.0879	48
4-40 UNC	.0849	83.4	.0939	55.7	#44	.0860	80	.0879	74
					#43	.0890	71	.0910	65
					#42	.0935	57	.0955	51
					3/32	.0938	56	.0958	50
4-48 UNF	.0894	83.5	.0968	56.2	#43	.0890	85	.0910	78
					#42	.0935	68	.0955	61
					3/32	.0938	67	.0958	60
					#41	.0960	59	.0980	52
5-40 UNC	.0979	83.4	.1062	57.9	#40	.0980	83	.1003	76
					#39	.0995	79	.1018	71
					#38	.1015	72	.1038	65
					#37	.1040	65	.1063	58
5-44 UNF	.1004	83.3	.1079	57.9	#38	.1015	80	.1038	72
					#37	.1040	71	.1063	63
					#36	.1065	63	.1088	55
6-32 UNC	.1040	83.8	.1140	59.1	#37	.1040	84	.1063	78
					#36	.1065	78	.1088	72
					7/64	.1094	70	.1120	64
					#35	.1100	69	.1126	63
					#34	.1110	67	.1136	60
					#33	.1130	62	.1156	55
6-40 UNF	.1110	83.1	.1186	59.7	#34	.1110	83	.1136	75
					#33	.1130	77	.1156	69
					#32	.1160	68	.1186	60
8-32 UNC	.1300	83.8	.1389	61.8	#29	.1360	69	.1389	62
8-36 UNF	.1340	83.1	.1416	62.1	#29	.1360	78	.1389	70
					#28	.1405	65	.1434	57
					9/64	.1406	65	.1435	57
10-24 UNC	.1450	83.1	.1555	63.7	#27	.1440	85	.1472	79
					#26	.1470	79	.1502	74
					#25	.1495	75	.1527	69
					#24	.1520	70	.1552	64
					#23	.1540	66	.1572	61
10-32 UNF	.1560	83.8	.1641	63.8	5/32	.1562	83	.1594	75
					#22	.1570	81	.1602	73
					#21	.1590	76	.1622	68
					#20	.1610	71	.1642	64
12-24 UNC	.1710	83.1	.1807	65.2	11/64	.1719	82	.1754	75
					#17	.1730	79	.1765	73
					#16	.1770	72	.1805	66
					#15	.1800	67	.1835	60
12-28 UNF	.1770	84.1	.1857	65.3	#16	.1770	84	.1805	77
					#15	.1800	78	.1835	70
					#14	.1820	73	.1855	66
					#13	.1850	67	.1885	59

(Continued)

Table 11. *(Continued)* **Tap Drill Sizes and Percentage of Thread, Unified Threads Class 3B, Size 0-80 UNF Through 1 1/8-16 UN.** *(Source, FED-STD-H28/2B.)*

Size, TPI, and Series*	Classes 3B Minor Dia.				Tap Drill and Percentage of Thread				
	D_1 (min.)	% of Thread*	D_1 (max.)	% of Thread*	Drill Size		% of Thread*	Probable Hole Size	% of Thread*
12-32 UNEF	.1820	83.8	.1895	65.3	#14	.1820	84	.1855	75
					#13	.1850	76	.1885	68
					3/16	.1875	70	.1910	62
					#12	.1890	67	.1925	58
1/4-20 UNC	.1960	83.1	.2067	66.7	#9	.1960	83	.1998	77
					#8	.1990	79	.2028	73
					#7	.2010	75	.2048	70
					13/64	.2031	72	.2069	66
					#6	.2040	71	.2078	65
					#5	.2055	69	.2093	63
1/4-28 UNF	.2110	84.1	.2190	66.8	#3	.2130	80	.2168	72
					7/32	.2188	67	.2226	59
1/4-32 UNEF	.2160	83.8	.2229	66.8	7/32	.2188	77	.2226	67
					#2	.2210	71	.2248	62
5/16-18 UNC	.2520	83.8	.2630	68.6	F	.2570	77	.2608	72
					G	.2610	71	.2651	66
5/16-20 UN	.2580	83.9	.2680	68.5	F	.2570	85	.2608	80
					G	.2610	79	.2651	73
					H	.2660	72	.2701	65
5/16-24 UNF	.2670	84.1	.2754	68.5	H	.2660	86	.2701	78
					I	.2720	75	.2761	67
5/16-28 UN	.2740	83.0	.2807	68.5	J	.2770	77	.2811	68
5/16-32 UNEF	.2790	82.5	.2847	68.5	K	.2810	78	.2852	67
					9/32	.2812	77	.2854	67
3/8-16 UNC	.3070	83.8	.3182	70.0	5/16	.3125	77	.3169	72
					O	.3160	73	.3204	67
3/8-20 UN	.3210	83.1	.3297	69.7	P	.3230	80	.3274	73
3/8-24 UNF	.3300	83.1	.3372	69.8	Q	.3320	79	.3364	71
3/8-28 UN	.3360	84.1	.3426	69.8	R	.3390	78	.3434	68
3/8-32 UNEF	.3410	83.8	.3469	69.2	11/32	.3438	77	.3483	68
7/16-14 UNC	.3600	83.5	.3717	70.9	T	.3580	86	.3626	81
					23/64	.3594	84	.3640	79
7/16-16 UN	.3700	83.1	.3800	70.8	3/8	.3750	77	.3796	71
					V	.3770	75	.3816	69
7/16-20 UNF	.3830	83.9	.3916	70.7	W	.3860	79	.3906	72
					25/64	.3906	72	.3952	65
7/16-28 UNEF	.3990	83.0	.4051	69.8	Y	.4040	72	.4086	62
7/16-32 UN	.4040	82.5	.4094	69.2	Y	.4040	83	.4086	71
					13/32	.4062	77	.4108	66
1/2-12 UNS	.4100	83.1	.4223	71.8	Z	.4130	80	.4177	76
					27/64	.4219	72	.4266	68
1/2-13 UNC	.4170	83.1	.4284	71.7	27/64	.4219	78	.4266	73
1/2-16 UN	.4320	83.8	.4419	71.6	7/16	.4375	77	.4422	71
1/2-20 UNF	.4460	83.1	.4537	71.3	29/64	.4531	72	.4578	65

(Continued)

Table 11. *(Continued)* **Tap Drill Sizes and Percentage of Thread, Unified Threads Class 3B, Size 0-80 UNF Through 1 1/8-16 UN.** *(Source, FED-STD-H28/2B.)*

Size, TPI, and Series*	Classes 3B Minor Dia.				Tap Drill and Percentage of Thread				
	D_1 (min.)	% of Thread*	D_1 (max.)	% of Thread*	Drill Size		% of Thread*	Probable Hole Size	% of Thread*
1/2-28 UNEF	.4610	84.1	.4676	69.8	11.8mm	.4646	76	.4693	66
1/2-32 UN	.4660	83.8	.4719	69.2	15/32	.4688	77	.4736	65
9/16-12 UNC	.4720	83.6	.4843	72.2	15/32	.4688	87	.4736	82
					31/64	.4844	72	.4892	68
9/16-16 UN	.4950	83.1	.5040	72.1	1/2	.5000	77	.5048	71
9/16-18 UNF	.5020	83.8	.5106	71.9	1/2	.5000	87	.5048	80
					0.5062	.5062	78	.5110	71
9/16-20 UN	.5080	83.9	.5162	71.3	33/64	.5156	72	.5204	65
9/16-24 UNEF	.5170	84.1	.5244	70.4	33/64	.5156	87	.5204	78
					0.5203	.5203	78	.5251	69
9/16-28 UN	.5240	83.0	.5301	69.8	0.5263	.5263	78	.5312	67
9/16-32 UN	.5290	82.5	.5344	69.2	17/32	.5312	77	.5361	65
5/8-11 UNC	.5270	83.0	.5391	72.7	17/32	.5312	79	.5361	75
5/8-12 UN	.5350	83.1	.5463	72.7	35/64	.5469	72	.5518	68
5/8-16 UN	.5570	83.8	.5662	72.4	9/16	.5625	77	.5674	71
5/8-18 UNF	.5650	83.1	.5730	72.1	9/16	.5625	87	.5674	80
					0.5687	.5687	78	.573	71
5/8-20 UN	.5710	83.1	.5787	71.3	37/64	.5781	72	.5830	65
5/8-24 UNEF	.5800	83.1	.5869	70.4	37/64	.5781	87	.5830	78
					0.5828	.5828	78	.5877	69
5/8-28 UN	.5860	84.1	.5926	69.8	0.5828	.5828	91	.5877	80
5/8-32 UN	.5910	83.8	.5969	69.2	19/32	.5938	77	.5987	65
11/16-12 UN	.5970	83.6	.6085	73.0	19/32	.5938	87	.5987	.82
11/16-16 UN	.6200	83.1	.6284	72.8	5/8	.6250	77	.6300	71
11/16-20 UN	.6330	83.9	.6412	71.3	41/64	.6406	72	.6456	65
11/16-24 UNEF	.6420	84.1	.6494	70.4	41/64	.6406	87	.6456	77
11/16-28 UN	.6490	83.0	.6551	69.8	16.5mm	.6496	82	.6546	71
11/16-32 UN	.6540	82.5	.6594	69.2	21/32	.6562	77	.6612	65
3/4-10 UNC	.6420	83.1	.6545	73.5	41/64	.6406	04	.6456	80
3/4-12 UN	.6600	83.1	.6707	73.3	21/32	.6562	87	.6612	82
3/4-16 UNF	.6820	83.8	.6908	72.9	11/16	.6875	77	.6925	71
3/4-20 UNEF	.6960	83.1	.7037	71.3	45/64	.7031	72	.7082	64
3/4-28 UN	.7110	84.1	.7176	69.8	18mm	.7087	89	.7138	78
3/4-32 UN	.7160	83.8	.7219	69.2	23/32	.7188	77	.7239	64
13/16-12 UN	.7220	83.6	.7329	73.5	18.5mm	.7283	78	.7334	73
13/16-16 UN	.7450	83.1	.7533	72.9	3/4	.7500	77	.7552	71
13/16-20 UNEF	.7580	83.9	.7662	71.3	49/64	.7656	72	.7708	64
13/16-28 UN	.7740	83.0	.7801	69.8	19.75mm	.7776	75	.7828	64
13/16-32 UN	.7790	82.5	.7844	69.2	25/32	.7812	77	.7864	64

(Continued)

Table 11. *(Continued)* Tap Drill Sizes and Percentage of Thread, Unified Threads Class 3B, Size 0-80 UNF Through 1 1/8-16 UN. *(Source, FED-STD-H28/2B.)*

Size, TPI, and Series*	Classes 3B Minor Dia.				Tap Drill and Percentage of Thread				
	D_1 (min.)	% of Thread*	D_1 (max.)	% of Thread*	Drill Size		% of Thread*	Probable Hole Size	% of Thread*
7/8-9 UNC	.7550	83.1	.7681	74.1	49/64	.7656	76	.7708	72
7/8-12 UN	.7850	83.1	.7952	73.7	25/32	.7812	87	.7864	82
7/8-14 UNF	.7980	83.0	.8068	73.5	51/64	.7969	84	.8021	79
					0.8024	.8024	78	.8076	73
7/8-16 UN	.8070	83.8	.8158	72.9	13/16	.8125	77	.8178	70
7/8-20 UNEF	.8210	83.1	.8287	71.3	53/64	.8281	72	.8335	64
7/8-28 UN	.8360	84.1	.8426	69.8	21.25mm	.8366	83	.8420	71
7/8-32 UN	.8410	83.8	.8469	69.2	27/32	.8438	77	.8493	63
15/16-12 UN	.8470	83.6	.8575	73.9	27/32	.8438	87	.8493	81
15/16-16 UN	.8700	83.1	.8783	72.9	7/8	.8750	77	.8807	70
15/16-20 UNEF	.8830	83.9	.8912	71.3	57/64	.8906	72	.8965	63
15/16-28 UN	.8990	83.0	.9051	69.8	22.75mm	.8957	90	.9017	77
15/16-32 UN	.9040	82.5	.9094	69.2	29/32	.9062	77	.9122	62
1.000-8 UNC	.8650	83.1	.8797	74.1	55/64	.8594	.87	.8653	83
					7/8	.8750	77	.8809	73
1.000-12 UNF	.9100	83.1	.9198	74.1	29/32	.9062	87	.9122	81
1.000-14 UNS	.9230	83.0	.9315	73.8	59/64	.9219	84	.9279	78
					0.9274	.9274	78	.9335	72
1.000-16 UN	.9320	83.8	.9408	72.9	15/16	.9375	77	.9437	69
1.000-20 UNEF	.9460	83.1	.9537	71.3	61/64	.9531	72	.9594	63
1.000-28 UN	.9610	84.1	.9676	69.8	24.5mm	.9645	77	.9709	63
1.000-32 UN	.9660	83.8	.9719	69.2	31/32	.9688	77	.9753	61
1 1/16-8 UN	.9270	83.4	.9422	74.1	59/64	.9219	87	.9279	83
					0.9274	.9274	83	.9335	79
					15/16	.9375	77	.9437	73
1 1/16-12 UN	.9720	83.6	.9823	74.1	31/32	.9688	87	.9753	81
1 1/16-16 UN	.9950	83.1	1.0033	72.9	1	1.0000	77	1.0069	68
1 1/16-18 UNEF	1.0020	83.8	1.0105	72.1	1	1.0000	87	1.0069	77
1 1/16-20 UN	1.0080	83.9	1.0162	71.3	1 1/64	1.0156	72	1.0226	61
1 1/16-28 UN	1.0240	83.0	1.0301	69.8	1 1/32	1.0312	67	1.0383	52
1 1/8-7 UNC	.9700	83.5	.9875	74.1	31/32	.9688	84	.9750	81
					63/64	.9844	76	.9911	72
1 1/8-8 UN	.9900	83.1	1.0047	74.1	1	1.0000	77	1.0069	73
1 1/8-12 UNF	1.0350	83.1	1.0448	74.1	1 1/32	1.0312	87	1.0383	80
1 1/8-16 UN	1.0570	83.8	1.0658	72.9	1 1/16	1.0625	77	1.0699	68

Table 12. Tap Drill Sizes and Percentage of Thread, Unified Threads Class 3B, Size 1 1/8-18 UNEF Through 3 3/4-4 UNC. (*Source, FED-STD-H28/2B.*)

Size, TPI, and Series*	Class 3B Minor Dia.				Tap Drill and Percentage of Thread		
	D_1 (min.)	% of Thread*	D_1 (max.)	% of Thread*	Drill Size		% of Thread*
1 1/8-18 UNEF	1.1650	83.1	1.0730	72.1	1 1/16	1.0625	87
1 1/8-20 UN	1.0710	83.1	1.0787	71.3	1 5/64	1.0781	72
1 1/8-28 UN	1.0860	84.1	1.0926	69.8	1 3/32	1.0938	67
1 3/16-8 UN	1.0520	83.4	1.0672	74.1	1 1/16	1.0625	77
1 3/16-12 UN	1.0970	83.6	1.1073	74.1	1 3/32	1.0938	87
1 3/16-16 UN	1.1200	83.1	1.1283	72.9	1 1/8	1.1250	77
1 3/16-18 UNEF	1.1270	83.8	1.1355	72.1	1 1/8	1.1250	87
1 3/16-20 UN	1.1330	83.9	1.1412	71.3	1 9/64	1.1406	72
1 3/16-28 UN	1.1490	83.0	1.1551	69.8	29.25mm	1.1516	77
1 1/4-7 UNC	1.0950	83.5	1.1125	74.1	1 3/32	1.0938	84
1 1/4-8 UN	1.1150	83.1	1.1297	74.1	1 1/8	1.1250	77
1 1/4-12 UNF	1.1600	83.1	1.1698	74.1	1 5/32	1.1562	87
1 1/4-16 UN	1.1820	83.8	1.1908	72.9	1 3/16	1.1875	77
1 1/4-18 UNEF	1.1900	83.1	1.1980	72.1	1 3/16	1.1875	87
1 1/4-20 UN	1.1960	83.1	1.2037	71.3	1 13/64	1.2031	72
1 1/4-28 UN	1.2110	84.1	1.2176	69.8	30.75mm	1.2106	85
1 5/16-8 UN	1.1770	83.4	1.1922	74.1	1 11/64 / 1 3/16	1.1719 / 1.1875	87 / 77
1 5/16-12 UN	1.2220	83.6	1.2323	74.1	1 7/32	1.2188	87
1 5/16-16 UN	1.2450	83.1	1.2533	72.9	1 1/4	1.2500	77
1 5/16-18 UNEF	1.2520	83.8	1.2605	72.1	1 1/4	1.2500	87
1 5/16-20 UN	1.2580	83.9	1.2662	71.3	1 17/64	1.2656	72
1 5/16-28 UN	1.2740	83.0	1.2801	69.8	32.5mm	1.2795	71
1 3/8-6 UNC	1.1950	83.1	1.2146	74.1	1 3/16 / 1 13/64	1.1875 / 1.2031	87 / 79
1 3/8-8 UN	1.2400	83.1	1.2547	74.1	1 15/64 / 1 1/4	1.2344 / 1.2500	87 / 77
1 3/8-12 UNF	1.2850	83.1	1.2948	74.1	1 9/32	1.2812	87
1 3/8-16 UN	1.3070	83.8	1.3158	72.9	1 5/16	1.3125	77
1 3/8-18 UNEF	1.3150	83.1	1.3230	72.1	1 5/16	1.3125	87
1 3/8-20 UN	1.3210	83.1	1.3287	71.3	1 21/64	1.3281	72
1 3/8-28 UN	1.3360	84.1	1.3426	69.8	34mm	1.3386	78
1 7/16-6 UN	1.2570	83.4	1.2771	74.1	1 17/64	1.2656	79
1 7/16-8 UN	1.3020	83.4	1.3172	74.1	1 19/64 / 1 5/16	1.2969 / 1.3125	87 / 77
1 7/16-12 UN	1.3470	83.6	1.3573	74.1	1 11/32	1.3438	87
1 7/16-16 UN	1.3700	83.1	1.3783	72.9	1 3/8	1.3750	77
1 7/16-18 UNEF	1.3770	83.8	1.3855	72.1	1 3/8	1.3750	87

(Continued)

Table 12. *(Continued)* Tap Drill Sizes and Percentage of Thread, Unified Threads Class 3B, Size 1 1/8-18 UNEF Through 3 3/4-4 UNC. *(Source, FED-STD-H28/2B.)*

Size, TPI, and Series*	Class 3B Minor Dia.				Tap Drill and Percentage of Thread		
	D_1 (min.)	% of Thread*	D_1 (max.)	% of Thread*	Drill Size		% of Thread*
1 7/16-20 UN	1.3830	83.9	1.3912	71.3	1 25/64	1.3906	72
1 7/16-28 UN	1.3990	83.0	1.4051	69.8	35.5mm	1.3976	86
1 1/2-6 UNC	1.3200	83.1	1.3396	74.1	1 5/16	1.3125	87
					1 21/64	1.3281	79
1 1/2-8 UN	1.3650	83.1	1.3797	74.1	1 23/64	1.3594	87
					1 3/8	1.3750	77
1 1/2-12 UNF	1.4100	83.1	1.4198	74.1	1 13/32	1.4062	87
1 1/2-16 UN	1.4320	83.8	1.4408	72.9	1 7/16	1.4375	77
1 1/2-18 UNEF	1.4400	83.1	1.4480	72.1	1 7/16	1.4375	87
1 1/2-20 UN	1.4460	83.1	1.4537	71.3	1 29/64	1.4531	72
1 1/2-28 UN	1.4610	84.1	1.4676	69.8	37mm	1.4567	93
1 9/16-6 UN	1.3820	83.4	1.4021	74.1	1 25/64	1.3906	79
1 9/16-8 UN	1.4270	83.4	1.4422	74.1	1 27/64	1.4219	87
					1 7/16	1.4375	77
1 9/16-12 UN	1.4720	83.6	1.4823	74.1	1 15/32	1.4688	87
1 9/16-16 UN	1.4950	83.1	1.5033	72.9	1 1/2	1.5000	77
1 9/16-18 UNEF	1.5020	83.8	1.5105	72.1	1 1/2	1.5000	87
1 9/16-20 UN	1.5080	83.9	1.5162	71.3	1 33/64	1.5156	72
1 5/8-6 UN	1.4450	83.1	1.4646	74.1	1 29/64	1.4531	79
1 5/8-8 UN	1.4900	83.1	1.5047	74.1	1 31/64	1.4844	87
					1 1/2	1.5000	77
1 5/8-12 UN	1.5350	83.1	1.5448	74.1	1 17/32	1.5312	87
1 5/8-16 UN	1.5570	83.8	1.5658	72.9	1 9/16	1.5625	77
1 5/8-18 UNEF	1.5650	83.1	1.5730	72.1	1 9/16	1.5625	87
1 5/8-20 UN	1.5710	83.1	1.5787	71.3	1 37/64	1.5781	72
1 11/16-6 UN	1.5070	83.4	1.5271	74.1	1 1/2	1.5000	87
					1 33/64	1.5156	79
1 11/16-8 UN	1.5520	83.4	1.5672	74.1	1 9/16	1.5625	77
1 11/16-12 UN	1.5970	83.6	1.6073	74.1	1 19/32	1.5938	87
1 11/16-16 UN	1.6200	83.1	1.6283	72.9	1 5/8	1.6250	77
1 11/16-18 UNEF	1.6270	83.8	1.6355	72.1	1 5/8	1.6250	87
1 11/16-20 UN	1.6330	83.9	1.6412	71.3	1 41/64	1.6406	72
1 3/4-5 UNC	1.5340	83.1	1.5575	74.1	1 17/32	1.5312	84
					1 35/64	1.5469	78
1 3/4-6 UN	1.5700	83.1	1.5896	74.1	1 9/16	1.5625	87
					1 37/64	1.5781	79
1 3/4-8 UN	1.6150	83.1	1.6297	74.1	1 39/64	1.6094	87
					1 5/8	1.6250	77
1 3/4-12 UN	1.6600	83.1	1.6698	74.1	1 21/32	1.6562	87

(Continued)

Table 12. *(Continued)* **Tap Drill Sizes and Percentage of Thread, Unified Threads Class 3B, Size 1 1/8-18 UNEF Through 3 3/4-4 UNC.** *(Source, FED-STD-H28/2B.)*

Size, TPI, and Series*	Class 3B Minor Dia.				Tap Drill and Percentage of Thread		
	D_1 (min.)	% of Thread*	D_1 (max.)	% of Thread*	Drill Size		% of Thread*
1 3/4-16 UN	1.6820	83.8	1.6908	72.9	1 11/16	1.6875	77
1 3/4-20 UN	1.6960	83.1	1.7037	71.3	1 45/64	1.7031	72
1 13/16-6 UN	1.6320	83.4	1.6521	74.1	1 5/8	1.6250	87
					1 41/64	1.6406	79
1 13/16-8 UN	1.6770	83.4	1.6922	74.1	1 43/64	1.6719	87
					1 11/16	1.6875	77
1 13/16-12 UN	1.7220	83.6	1.7323	74.1	1 23/32	1.7188	87
1 13/16-16 UN	1.7450	83.1	1.7533	72.9	1 3/4	1.7500	77
1 13/16-20 UN	1.7580	83.9	1.7662	71.3	1 49/64	1.7656	72
1 7/8-6 UN	1.6950	83.1	1.7146	74.1	1 45/64	1.7031	79
1 7/8-8 UN	1.7400	83.1	1.7547	74.1	1 3/4	1.7500	77
1 7/8-12 UN	1.7850	83.1	1.7948	74.1	1 25/32	1.7812	87
1 7/8-16 UN	1.8070	83.8	1.8158	72.9	1 13/16	1.8125	77
1 7/8-20 UN	1.8210	83.1	1.8287	71.3	1 53/64	1.8281	72
1 15/16-6 UN	1.7570	83.4	1.7771	74.1	1 49/64	1.7656	79
1 15/16-8 UN	1.8020	83.4	1.8172	74.1	1 51/64	1.7969	87
					1 13/16	1.8125	77
1 15/16-12 UN	1.8470	83.6	1.8573	74.1	1 27/32	1.8438	87
1 15/16-16 UN	1.8700	83.1	1.8783	72.9	1 7/8	1.8750	77
1 15/16-20 UN	1.8830	83.9	1.8912	71.3	1 57/64	1.8906	72
2-4.5 UNC	1.7590	83.5	1.7861	74.1	1 25/32	1.7812	76
2-6 UN	1.8200	83.1	1.8396	74.1	1 53/64	1.8281	79
2-8 UN	1.8650	83.1	1.8797	74.1	1 7/8	1.8750	77
2-12 UN	1.9100	83.1	1.9198	74.1	1 29/32	1.9062	87
2-16 UN	1.9320	83.8	1.9408	72.9	1 15/16	1.9375	77
2-20 UN	1.9460	83.1	1.9537	71.3	1 61/64	1.9531	72
2 1/16-16 UNS	1.9950	83.1	2.0033	72.9	2	2.0000	77
2 1/8-6 UN	1.9450	83.1	1.9646	74.1	1 61/64	1.9531	79
2 1/8-8 UN	1.9900	83.1	2.0047	74.1	2	2.0000	77
2 1/8-12 UN	2.0350	83.1	2.0448	74.1	2 1/32	2.0312	87
2 1/8-16 UN	2.0570	83.8	2.0658	72.9	2 1/16	2.0625	77
2 1/8-20 UN	2.0710	83.1	2.0787	71.3	2 1/16	2.0625	96
2 3/16-UNS	2.1200	83.1	2.1283	72.9	2 1/8	2.1250	77
2 1/4-4.5 UNC	2.0090	83.5	2.0361	74.1	2	2.0000	87
					2 1/32	2.0312	76
2 1/4-6 UN	2.0700	83.1	2.0896	74.1	2 1/16	2.0625	87
2 1/4-8 UN	2.1150	83.1	2.1297	74.1	2 1/8	2.1250	77
2 1/4-12 UN	2.1600	83.1	2.1698	74.1	2 5/32	2.1562	87

(Continued)

Table 12. *(Continued)* Tap Drill Sizes and Percentage of Thread, Unified Threads Class 3B, Size 1 1/8-18 UNEF Through 3 3/4-4 UNC. *(Source, FED-STD-H28/2B.)*

Size, TPI, and Series*	Class 3B Minor Dia.				Tap Drill and Percentage of Thread		
	D_1 (min.)	% of Thread*	D_1 (max.)	% of Thread*	Drill Size		% of Thread*
2 1/4-16 UN	2.1820	83.8	2.1908	72.9	2 3/16	2.1875	77
2 1/4-20 UN	2.1960	83.1	2.2037	71.3	2 3/16	2.1875	96
2 5/16-16 UNS	2.2450	83.1	2.2533	72.9	2 1/4	2.2500	77
2 3/8-6 UN	2.1950	83.1	2.2146	74.1	2 3/16	2.1875	87
2 3/8-8 UN	2.2400	83.1	2.2547	74.1	2 1/4	2.2500	77
2 3/8-12 UN	2.2850	83.1	2.2948	74.1	58mm	2.2835	85
2 3/8-16 UN	2.3070	83.8	2.3158	72.9	2 5/16	2.3125	77
2 3/8-20 UN	2.3210	83.1	2.3287	71.3	2 5/16	2.3125	96
2 7/16-16 UNS	2.3700	83.1	2.3783	72.9	2 3/8	2.3750	77
2 1/2-4 UNC	2.2290	83.4	2.2594	74.1	2 7/32 2 1/4	2.2188 2.2500	87 77
2 1/2-6 UN	2.3200	83.1	2.3396	74.1	2 5/16	2.3125	87
2 1/2-8 UN	2.3650	83.1	2.3797	74.1	2 3/8	2.3750	77
2 1/2-12 UN	2.4100	83.1	2.4198	74.1	2 13/32	2.4062	87
2 1/2-16 UN	2.4320	83.8	2.4408	72.9	2 7/16	2.4375	77
2 1/2-20 UN	2.4460	83.1	2.4537	71.3	2 7/16	2.4375	96
2 5/8-4 UN	2.3540	83.4	2.3844	74.1	2 11/32 2 3/8	2.3438 2.3750	87 77
2 5/8-6 UN	2.4450	83.1	2.4646	74.1	2 7/16	2.4375	87
2 5/8-8 UN	2.4900	83.1	2.5047	74.1	2 1/2	2.5000	77
2 5/8-12 UN	2.5350	83.1	2.5448	74.1	2 17/32	2.5312	87
2 5/8-16 UN	2.5570	83.8	2.5658	72.9	2 9/16	2.5625	77
2 5/8-20 UN	2.5710	83.1	2.5787	71.3	2 9/16	2.5625	96
2 3/4-4 UNC	2.4790	83.4	2.5094	74.1	2 1/2	2.5000	77
2 3/4-6 UN	2.5700	83.1	2.5896	74.1	2 9/16	2.5625	87
2 3/4-8 UN	2.6150	83.1	2.6297	74.1	2 5/8	2.6250	77
2 3/4-12 UN	2.6600	83.1	2.6698	74.1	2 21/32	2.6562	87
2 3/4-16 UN	2.6820	83.8	2.6908	72.9	2 11/16	2.6875	77
2 3/4-20 UN	2.6960	83.1	2.7037	71.3	2 11/16	2.6875	96
2 7/8-4 UN	2.6040	83.4	2.6344	74.1	2 5/8	2.6250	77
2 7/8-6 UN	2.6950	83.1	2.7146	74.1	2 11/16	2.6875	87
2 7/8-8 UN	2.7400	83.1	2.7547	74.1	2 3/4	2.7500	77
2 7/8-12 UN	2.7850	83.1	2.7948	74.1	2 25/32	2.7812	87
2 7/8-16 UN	2.8070	83.8	2.8158	72.9	2 13/16	2.8125	77

Table 13. Recommended Hole Size Limits Before Threading for Different Lengths of Engagement, Unified Class 1B and 2B Threads. *(Source, FED-STD-H28/2B.)*

Size, TPI, and Series	To and Including 0.33D		Above 0.33D to 0.67D		Above 0.67D through 1.5D	
	Min.	Max.	Min.	Max.	Min.	Max.
0-80 UNF	.0465	.0500	.0479	.0514	.0479	.0514
1-64 UNC	.0561	.0599	.0580	.0618	.0585	.0623
1-72 UNF	.0580	.0613	.0596	.0629	.0602	.0635
2-56 UNC	.0667	.0705	.0686	.0724	.0699	.0737
2-64 UNF	.0691	.0724	.0707	.0740	.0720	.0753
3-48 UNC	.0764	.0804	.0785	.0825	.0805	.0845
3-56 UNF	.0797	.0831	.0814	.0848	.0831	.0865
4-40 UNC	.0849	.0849	.0871	.0916	.0894	.0939
4-48 UNF	.0894	.0931	.0912	.0949	.0931	.0968
5-40 UNC	.0979	.1020	.1000	.1041	.1021	.1062
5-44 UNF	.1004	.1041	.1023	.1060	.1042	.1079
6-32 UNC	.104	.109	.106	.112	.109	.114
6-40 UNF	.111	.115	.113	.117	.115	.119
8-32 UNC	.130	.135	.132	.137	.134	.139
8-36 UNF	.134	.138	.136	.140	.138	.142
10-24 UNC	.145	.150	.147	.153	.150	.156
10-32 UNF	.156	.160	.158	.162	.160	.164
12-24 UNC	.171	.176	.173	.178	.176	.181
12-28 UNF	.177	.182	.179	.184	.181	.186
12-32 UNEF	.182	.186	.184	.188	.186	.190
1/4-20 UNC	.196	.202	.199	.204	.202	.207
1/4-28 UNF	.211	.216	.213	.218	.216	.220
1/4-32 UNEF	.216	.220	.218	.222	.220	.224
1/4-36 UNS	.220	.223	.221	.225	.222	.226
5/16-18 UNC	.252	.259	.256	.262	.259	.265
5/16-20 20UN	.258	.264	.261	.267	.264	.270
5/16-24 UNF	.267	.272	.270	.275	.272	.277
5/16-28 28UN	.274	.278	.276	.280	.278	.282
5/16-32 UNEF	.279	.282	.280	.284	.282	.286
5/16-36 UNS	.282	.286	.283	.287	.285	.289
3/8-16 UNC	.307	.314	.311	.318	.314	.321
3/8-20 20UN	.321	.327	.324	.330	.327	.332
3/8-24 UNF	.330	.335	.332	.337	.335	.340
3/8-28 28UN	.336	.340	.338	.343	.340	.345
3/8-32 UNEF	.341	.345	.343	.347	.345	.349
3/8-36 UNS	.345	.348	.346	.350	.348	.352
7/16-14 UNC	.360	.368	.364	.372	.368	.376
7/16-16 16UN	.370	.377	.373	.380	.377	.384

All dimensions in inches.

(Continued)

Table 13. *(Continued)* **Recommended Hole Size Limits Before Threading for Different Lengths of Engagement, Unified Class 1B and 2B Threads.** *(Source, FED-STD-H28/2B.)*

Size, TPI, and Series	To and Including 0.33D		Above 0.33D to 0.67D		Above 0.67D through 1.5D	
	Min.	Max.	Min.	Max.	Min.	Max.
7/16-20 UNF	.383	.389	.386	.392	.389	.395
7/16-28 UNEF	.399	.403	.401	.405	.403	.407
7/16-32 32UN	.404	.407	.405	.409	.407	.411
1/2-12 UNS	.410	.419	.414	.423	.419	.428
1/2-13 UNC	.417	.425	.421	.430	.425	.434
1/2-16 16UN	.432	.439	.436	.443	.439	.446
1/2-20 UNF	.446	.452	.449	.454	.452	.457
1/2-28 UNEF	.461	.466	.463	.468	.466	.470
1/2-32 32UN	.466	.470	.468	.472	.470	.474
9/16-12 UNC	.472	.481	.477	.486	.481	.490
9/16-16 16UN	.495	.502	.498	.505	.502	.509
9/16-18 UNF	.502	.509	.506	.512	.509	.515
9/16-20 20UN	.508	.514	.511	.517	.514	.520
9/16-24 UNEF	.517	.522	.520	.525	.522	.527
9/16-28 28UN	.524	.528	.526	.530	.528	.532
9/16-32 32UN	.529	.532	.530	.534	.532	.536
5/8-11 UNC	.527	.536	.532	.541	.536	.546
5/8-12 12UN	.535	.544	.539	.548	.544	.553
5/8-16 16UN	.557	.564	.561	.568	.564	.571
5/8-18UNF	.565	.571	.568	.574	.571	.578
5/8-20 20UN	.571	.577	.574	.580	.577	.582
5/8-24 UNEF	.580	.585	.582	.587	.585	.590
5/8-28 28UN	.586	.590	.588	.593	.590	.595
5/8-32 32UN	.591	.595	.593	.597	.595	.599
11/16-16-12 12UN	.597	.606	.602	.611	.606	.615
11/16-16-16UN	.620	.627	.623	.630	.627	.634
11/16-16-18UNS	.627	.634	.630	.637	.634	.640
11/16-20 20UN	.633	.639	.636	.642	.639	.645
11/16-24UNEF	.642	.647	.645	.650	.647	.652
11/16-28 28UN	.649	.653	.651	.655	.653	.657
11/16-32 32UN	.654	.657	.655	.659	.657	.661
3/4-10 UNC	.642	.652	.647	.658	.652	.663
3/4-12 12UN	.660	.669	.664	.673	.669	.678
3/4-16 UNF	.682	.689	.686	.693	.689	.696
3/4-18 UNS	.690	.696	.693	.699	.696	.703
3/4-20 UNEF	.696	.702	.699	.704	.702	.707
3/4-28 28UN	.711	.716	.713	.718	.716	.720
3/4-32 32UN	.716	.720	.718	.722	.720	.724

All dimensions in inches.   *(Continued)*

Table 13. *(Continued)* **Recommended Hole Size Limits Before Threading for Different Lengths of Engagement, Unified Class 1B and 2B Threads.** *(Source, FED-STD-H28/2B.)*

Size, TPI, and Series	To and Including 0.33D		Above 0.33D to 0.67D		Above 0.67D through 1.5D	
	Min.	Max.	Min.	Max.	Min.	Max.
13/16-12 12UN	.722	.731	.727	.736	.731	.740
13/16-16 16UN	.745	.752	.748	.755	.752	.795
13/16-18UNS	.752	.759	.756	.762	.759	.765
13/16-20UNEF	.758	.764	.761	.767	.764	.770
13/16-28 28UN	.774	.778	.776	.780	.778	.782
13/16-32 32UN	.779	.782	.780	.784	.782	.786
7/8-9UNC	.755	.766	.760	.772	.766	.778
7/8-12 12UN	.785	.794	.789	.798	.794	.803
7/8-14UNF	.798	.806	.802	.810	.806	.814
7/8-16 16UN	.807	.814	.811	.818	.814	.821
7/8-18UNS	.815	.821	.818	.824	.821	.828
7/8-20UNEF	.821	.827	.824	.830	.827	.832
7/8-28 28UN	.836	.840	.838	.843	.840	.845
7/8-32 32UN	.841	.845	.843	.847	.845	.849
15/16-12 12UN	.847	.856	.852	.861	.856	.865
15/16-16 16UN	.870	.877	.873	.880	.877	.884
15/16-20UNEF	.883	.889	.886	.892	.889	.895
15/16-28 28UN	.899	.903	.901	.905	.903	.907
15/16-32 32UN	.904	.907	.905	.909	.907	.911
1-8UNC	.865	.877	.871	.884	.877	.890
1-12UNF	.910	.919	.914	.923	.919	.928
1-14UNS	.923	.931	.927	.934	.931	.938
1-16 16UN	.932	.939	.936	.943	.939	.946
1-18UNS	.940	.946	.943	.949	.946	.953
1-20UNEF	.946	.952	.949	.954	.952	.957
1-28 28UN	.961	.966	.963	.968	.966	.970
1-32 32UN	.966	.970	.968	.972	.970	.974
1 1/16-8 8UN	.927	.940	.934	.946	.940	.952
1 1/16-12 12UN	.972	.981	.977	.986	.981	.990
1 1/16-14UNS	.985	.993	.989	.997	.993	1.001
1 1/16-16 16UN	.995	1.002	.998	1.005	1.002	1.009
1 1/16-18UNEF	1.002	1.009	1.006	1.012	1.009	1.015
1 1/16-20 20UN	1.008	1.014	1.011	1.017	1.014	1.020
1 1/16-28 28UN	1.024	1.028	1.026	1.030	1.028	1.032
1 1/8-7UNC	.970	.984	.977	.991	.984	.998
1 1/8 8UN	.990	1.002	.996	1.008	1.002	1.015
1 1/8-12UNF	1.035	1.044	1.039	1.048	1.044	1.053
1 1/8-16 16UN	1.057	1.064	1.061	1.068	1.064	1.071

All dimensions in inches.

(Continued)

Table 13. *(Continued)* **Recommended Hole Size Limits Before Threading for Different Lengths of Engagement, Unified Class 1B and 2B Threads.** *(Source, FED-STD-H28/2B.)*

Size, TPI, and Series	To and Including 0.33D		Above 0.33D to 0.67D		Above 0.67D through 1.5D	
	Min.	Max.	Min.	Max.	Min.	Max.
1 1/8-18UNEF	1.065	1.071	1.068	1.074	1.071	1.078
1 1/8-20 20UN	1.071	1.077	1.074	1.080	1.077	1.082
1 1/8-28 28UN	1.086	1.090	1.088	1.093	1.090	1.095
1 3/16-8 8UN	1.052	1.065	1.058	1.071	1.065	1.077
1 3/16-12 12UN	1.097	1.106	1.102	1.111	1.106	1.115
1 3/16-16 16UN	1.120	1.127	1.123	1.130	1.127	1.134
1 3/16-18UNEF	1.127	1.134	1.130	1.137	1.134	1.140
1 3/16-20 20UN	1.133	1.139	1.136	1.142	1.139	1.145
1 3/16-28 28UN	1.149	1.153	1.151	1.155	1.153	1.157
1 1/4-7UNC	1.095	1.109	1.102	1.116	1.109	1.123
1 1/4-8 8UN	1.115	1.127	1.121	1.134	1.127	1.140
1 1/4-12UNF	1.160	1.169	1.164	1.173	1.169	1.178
1 1/4-16 16UN	1.182	1.189	1.186	1.193	1.189	1.196
1 1/4-18UNEF	1.190	1.196	1.193	1.199	1.196	1.203
1 1/4-20 20UN	1.196	1.202	1.199	1.204	1.202	1.207
1 1/4-28 28UN	1.211	1.216	1.213	1.218	1.216	1.220
1 5/16-8 8UN	1.177	1.190	1.184	1.196	1.190	1.202
1 5/16-12 12UN	1.222	1.231	1.227	1.236	1.231	1.240
1 5/16-16 16UN	1.245	1.252	1.248	1.255	1.252	1.259
1 5/16-18 UNEF	1.252	1.259	1.256	1.262	1.259	1.265
1 5/16-20 20UN	1.258	1.264	1.261	1.267	1.264	1.270
1 5/16-28 28UN	1.274	1.278	1.276	1.280	1.278	1.282
1 3/8-6UNC	1.195	1.210	1.202	1.218	1.210	1.225
1 3/8-8 8UN	1.240	1.252	1.246	1.258	1.252	1.265
1 3/8-12UNF	1.285	1.294	1.289	1.298	1.294	1.303
1 3/8-16 16UN	1.307	1.314	1.311	1.318	1.314	1.321
1 3/8-18UNEF	1.315	1.321	1.318	1.324	1.321	1.328
1 3/8-20 20UN	1.321	1.327	1.324	1.330	1.327	1.332
1 3/8-28 28UN	1.336	1.340	1.338	1.343	1.340	1.345
1 7/16-6 6UN	1.257	1.272	1.265	1.280	1.272	1.288
1 7/16-8 8UN	1.302	1.315	1.308	1.321	1.315	1.327
1 7/16-12 12UN	1.347	1.356	1.352	1.361	1.356	1.365
1 7/16-16 16UN	1.370	1.377	1.373	1.380	1.377	1.384
1 7/16-18UNEF	1.377	1.384	1.380	1.387	1.384	1.390
1 7/16-20 20UN	1.383	1.389	1.386	1.392	1.389	1.395
1 7/16-28 28UN	1.399	1.403	1.401	1.405	1.403	1.407
1 1/2-6UNC	1.320	1.335	1.327	1.343	1.335	1.350
1 1/2-8 8UN	1.365	1.377	1.371	1.384	1.377	1.390

All dimensions in inches.

(Continued)

Table 13. *(Continued)* Recommended Hole Size Limits Before Threading for Different Lengths of Engagement, Unified Class 1B and 2B Threads. *(Source, FED-STD-H28/2B.)*

Size, TPI, and Series	To and Including 0.33*D*		Above 0.33*D* to 0.67*D*		Above 0.67*D* through 1.5*D*	
	Min.	Max.	Min.	Max.	Min.	Max.
1 1/2-12UNF	1.410	1.419	1.414	1.423	1.419	1.428
1 1/2-16 16UN	1.432	1.439	1.436	1.443	1.439	1.446
1 1/2-18UNEF	1.440	1.446	1.443	1.449	1.446	1.453
1 1/2-20 20UN	1.446	1.452	1.449	1.454	1.452	1.457
1 1/2-28 28UN	1.461	1.466	1.463	1.468	1.466	1.470
1 9/16-6 6UN	1.382	1.397	1.390	1.405	1.397	1.413
1 9/16-8 8UN	1.427	1.440	1.434	1.446	1.440	1.452
1 9/16-12 12UN	1.472	1.481	1.477	1.486	1.481	1.490
1 9/16-16 16UN	1.495	1.502	1.498	1.505	1.502	1.509
1 9/16-18UNEF	1.502	1.509	1.506	1.512	1.509	1.515
1 9/16-20 20UN	1.508	1.514	1.511	1.517	1.514	1.520
1 5/8-6 6UN	1.445	1.460	1.452	1.468	1.460	1.475
1 5/8-8 8UN	1.490	1.502	1.496	1.508	1.502	1.515
1 5/8-12 12UN	1.535	1.544	1.539	1.548	1.544	1.553
1 5/8-16 16UN	1.557	1.564	1.561	1.568	1.564	1.571
1 5/8-18UNEF	1.565	1.571	1.568	1.574	1.571	1.578
1 5/8-20 20UN	1.571	1.577	1.574	1.580	1.577	1.582
1 11/16-6 6UN	1.507	1.522	1.515	1.530	1.522	1.538
1 11/16-8 8UN	1.552	1.565	1.558	1.571	1.565	1.577
1 11/16-12 12UN	1.597	1.606	1.602	1.611	1.606	1.615
1 11/16-16 16UN	1.620	1.627	1.623	1.630	1.627	1.634
1 11/16-18UNEF	1.627	1.634	1.630	1.637	1.634	1.640
1 11/16-20 20UN	1.633	1.639	1.636	1.642	1.639	1.645
1 3/4-5UNC	1.534	1.550	1.542	1.559	1.550	1.568
1 3/4-6 6UN	1.570	1.585	1.577	1.593	1.585	1.600
1 3/4-8 8UN	1.615	1.627	1.6 21	1.634	1.627	1.640
1 3/4-12 12UN	1.660	1.669	1.664	1.673	1.669	1.678
1 3/4-16 16UN	1.682	1.689	1.686	1.693	1.689	1.696
1 3/4-20 20UN	1.696	1.702	1.699	1.704	1.702	1.707
1 13/16-6 6UN	1.632	1.647	1.640	1.655	1.647	1.663
1 13/16-8 8UN	1.677	1.690	1.684	1.696	1.690	1.702
1 13/16-12 12UN	1.722	1.731	1.727	1.736	1.731	1.740
1 13/16-16 16UN	1.745	1.752	1.748	1.755	1.752	1.759
1 13/16-20 20UN	1.7 58	1.764	1.761	1.767	1.764	1.770
1 7/8-6 6UN	1.695	1.710	1.702	1.718	1.710	1.725
1 7/8-8 8UN	1.740	1.752	1.746	1.758	1.752	1.765
1 7/8-12 12UN	1.785	1.794	1.789	1.798	1.794	1.803
1 7/8-16 16UN	1.807	1.814	1.811	1.818	1.814	1.821

All dimensions in inches. *(Continued)*

Table 13. *(Continued)* Recommended Hole Size Limits Before Threading for Different Lengths of Engagement, Unified Class 1B and 2B Threads. *(Source, FED-STD-H28/2B.)*

Size, TPI, and Series	To and Including 0.33D		Above 0.33D to 0.67D		Above 0.67D through 1.5D	
	Min.	Max.	Min.	Max.	Min.	Max.
1 7/8-20 20UN	1.821	1.827	1.824	1.830	1.827	1.832
1 15/16-6 6UN	1.757	1.772	1.765	1.780	1.772	1.788
1 15/16-8 8UN	1.802	1.815	1.808	1.821	1.815	1.827
1 15/16-12 12UN	1.847	1.856	1.852	1.861	1.856	1.865
1 15/16-16 16UN	1.870	1.877	1.873	1.880	1.877	1.884
1 15/16-20 20UN	1.883	1.889	1.886	1.892	1.889	1.895
2-4.5UNC	1.759	1.777	1.768	1.786	1.777	1.795
2-6 6UN	1.820	1.835	1.827	1.843	1.835	1.850
2-8 8UN	1.865	1.877	1.871	1.884	1.877	1.890
2-12 12UN	1.910	1.919	1.914	1.923	1.919	1.928
2-16 16UN	1.932	1.939	1.936	1.943	1.939	1.946
2-20 20UN	1.946	1.952	1.949	1.954	1.952	1.957
2 5/8-16UNS	1.995	2.002	1.998	2.005	2.002	2.009
2 1/8-6 6UN	1.945	1.960	1.952	1.968	1.960	1.975
2 1/8-8 8UN	1.990	2.002	1.996	2.008	2.002	2.015
2 1/8-12 12UN	2.035	2.044	2.039	2.048	2.044	2.053
2 1/8-16 16UN	2.057	2.064	2.061	2.068	2.064	2.071
2 1/8-20 20UN	2.071	2.077	2.074	2.080	2.077	2.082
2 3/16-16UNS	2.120	2.127	2.123	2.130	2.127	2.134
2 1/4-4.5UNC	2.009	2.027	2.018	2.036	2.027	2.045
2 1/4-6 6UN	2.070	2.085	2.077	2.093	2.085	2.100
2 1/2-4UNC	2.229	2.248	2.239	2.258	2.248	2.267
2 3/4-4UNC	2.479	2.498	2.489	2.508	2.498	2.517
3-4UNC	2.729	2.748	2.739	2.758	2.748	2.767
3 1/4-4UNC	2.979	2.998	2.989	3.008	2.998	3.017

All dimensions in inches.

Table 14. Recommended Hole Size Limits Before Threading for Different Lengths of Engagement, Unified Class 3B Threads. *(Source, FED-STD-H28/2B.)*

Size, TPI, and Series	To and Including 0.33D		Above 0.33D to 0.67D		Above 0.67D through 1.5D	
	Min.	Max.	Min.	Max.	Min.	Max.
0-80 UNF	.0465	.0500	.0479	.0514	.0479	.0514
1-64 UNC	.0561	.0599	.0580	.0618	.0585	.0623
1-72 UNF	.0580	.0613	.0596	.0629	.0602	.0635
2-56 UNC	.0667	.0705	.0686	.0724	.0699	.0737
2-64 UNF	.0691	.0724	.0707	.0740	.0720	.0753
3-48 UNC	.0764	.0804	.0785	.0825	.0805	.0845
3-56 UNF	.0797	.0831	.0814	.0848	.0831	.0865

All dimensions in inches.   *(Continued)*

Table 14. *(Continued)* **Recommended Hole Size Limits Before Threading for Different Lengths of Engagement, Unified Class 3B Threads.** *(Source, FED-STD-H28/2B.)*

Size, TPI, and Series	To and Including 0.33D		Above 0.33D to 0.67D		Above 0.67D through 1.5D	
	Min.	Max.	Min.	Max.	Min.	Max.
4-40 UNC	.0849	.0894	.0871	.0916	.0894	.0939
4-48 UNF	.0894	.0931	.0912	.0949	.0931	.0968
5-40 UNC	.0979	.1020	.1000	.1041	.1021	.1062
5-44 UNF	.1004	.1041	.1023	.1060	.1042	.1079
6-32 UNC	.1040	.1091	.1066	.1115	.1091	.1140
6-40 UNF	.1110	.1148	.1128	.1167	.1147	.1186
8-32 UNC	.1300	.1345	.1324	.1367	.1346	.1389
8-36 UNF	.1340	.1377	.1359	.1397	.1378	.1416
10-24 UNC	.1450	.1502	.1475	.1528	.1502	.1555
10-32 UNF	.1560	.1601	.1582	.1621	.1602	.1641
12-24 UNC	.1710	.1758	.1733	.1782	.1758	.1807
12-28 UNF	.1770	.1815	.1794	.1836	.1815	.1857
12-32 UNEF	.1820	.1858	.1841	.1877	.1859	.1895
1/4-20 UNC	.1960	.2013	.1986	.2040	.2013	.2067
1/4-28 UNF	.2110	.2152	.2131	.2171	.2150	.2190
1/4-32 UNEF	.2160	.2196	.2172	.2212	.2189	.2229
1/4-36 UNS	.2200	.2229	.2203	.2243	.2218	.2258
5/16-18 UNC	.2520	.2577	.2551	.2604	.2577	.2630
5/16-20 20UN	.2580	.2632	.2608	.2656	.2632	.2680
5/16-24 UNF	.2670	.2714	.2694	.2734	.2714	.2754
5/16-28 28UN	.2740	.2772	.2749	.2789	.2767	.2807
5/16-32 UNEF	.2790	.2817	.2792	.2832	.2807	.2847
5/16-36 UNS	.2820	.2850	.2823	.2863	.2837	.2877
3/8-16 UNC	.3070	.3127	.3101	.3155	.3128	.3182
3/8-20 20UN	.3210	.3253	.3231	.3275	.3253	.3297
3/8-24 UNF	.3300	.3336	.3314	.3354	.3332	.3372
3/8-28 28UN	.3360	.3395	.3370	.3410	.3386	.3426
3/8-32 UNEF	.3410	.3441	.3415	.3455	.3429	.3469
3/8-36 UNS	.3450	.3475	.3450	.3490	.3461	.3501
7/16-14 UNC	.3600	.3660	.3630	.3688	.3659	.3717
7/16-16 16UN	.3700	.3749	.3723	.3774	.3749	.3800
7/16-20 UNF	.3830	.3875	.3855	.3896	.3875	.3916
7/16-28 UNEF	.3990	.4020	.3995	.4035	.4011	.4051
7/16-32 32UN	.4040	.4066	.4040	.4080	.4054	.4094
1/2-12 UNS	.4100	.4161	.4129	.4192	.4160	.4223
1/2-13 UNC	.4170	.4225	.4196	.4254	.4226	.4284
1/2-16 16UN	.4320	.4371	.4347	.4395	.4371	.4419
1/2-20 UNF	.4460	.4498	.4477	.4517	.4497	.4537

All dimensions in inches.

(Continued)

Table 14. *(Continued)* **Recommended Hole Size Limits Before Threading for Different Lengths of Engagement, Unified Class 3B Threads.** *(Source, FED-STD-H28/2B.)*

Size, TPI, and Series	To and Including 0.33*D*		Above 0.33*D* to 0.67*D*		Above 0.67*D* through 1.5*D*	
	Min.	Max.	Min.	Max.	Min.	Max.
1/2-28 UNEF	.4610	.4645	.4620	.4660	.4636	.4676
1/2-32 32UN	.4660	.4691	.4665	.4705	.4679	.4719
9/16-12 UNC	.4720	.4783	.4753	.4813	.4783	.4843
9/16-16 16UN	.4950	.4994	.4971	.5017	.4994	.5040
9/16-18 UNF	.5020	.5065	.5045	.5086	.5065	.5106
9/16-20 20UN	.5080	.5123	.5102	.5142	.5122	.5162
9/16-24 UNEF	.5170	.5209	.5186	.5226	.5204	.5244
9/16-28 28UN	.5240	.5270	.5245	.5285	.5261	.5301
9/16-32 32UN	.5290	.5316	.5290	.5330	.5304	.5344
5/8-11 UNC	.5270	.5328	.5298	.5360	.5329	.5391
5/8-12 12UN	.5350	.5406	.5377	.5435	.5405	.5463
5/8-16 16UN	.5570	.5617	.5596	.5640	.5618	.5662
5/8-18 UNF	.5650	.5690	.5669	.5710	.5689	.5730
5/8-20 20UN	.5710	.5748	.5727	.5767	.5747	.5787
5/8-24 UNEF	.5800	.5834	.5811	.5851	.5829	.5869
5/8-28 28UN	.5860	.5895	.5870	.5910	.5886	.5926
5/8-32 32UN	.5910	.5941	.5915	.5955	.5929	.5969
11/16-12 12UN	.5970	.6029	.6001	.6057	.6029	.6085
11/16-16 16UN	.6200	.6241	.6219	.6262	.6241	.6284
11/16-18 UNS	.6270	.6315	.6294	.6335	.6314	.6355
11/16-20 20UN	.6330	.6373	.6352	.6392	.6372	.6412
11/16-24 UNEF	.6420	.6459	.6436	.6476	.6454	.6494
11/16-28 28UN	.6490	.6520	.6495	.6535	.6511	.6551
11/16-32 32UN	.6540	.6566	.6540	.6580	.6554	.6594
3/4-10 UNC	.6420	.6481	.6449	.6513	.6481	.6545
3/4-12 12UN	.6600	.6652	.6626	.6680	.6653	.6707
3/4-16 UNF	.6820	.6866	.6844	.6887	.6865	.6908
3/4-18 UNS	.6900	.6940	.6919	.6960	.6939	.6980
3/4-20 UNEF	.6960	.6998	.6977	.7017	.6997	.7037
3/4-28 28UN	.7110	.7145	.7120	.7160	.7136	.7176
3/4-32 32UN	.7160	.7191	.7165	.7205	.7179	.7219
13/16-12 12UN	.7220	.7276	.7250	.7303	.7276	.7329
13/16-16 16UN	.7450	.7491	.7469	.7512	.7490	.7533
13/16-18 UNS	.7520	.7565	.7544	.7585	.7564	.7605
13/16-20 UNEF	.7580	.7623	.7602	.7642	.7622	.7662
13/16-28 28UN	.7740	.7770	.7745	.7785	.7761	.7801
13/16-32 32UN	.7790	.7816	.7790	.7830	.7804	.7844
7/8-9 UNC	.7550	.7614	.7580	.7647	.7614	.7681

All dimensions in inches.   *(Continued)*

Table 14. *(Continued)* Recommended Hole Size Limits Before Threading for Different Lengths of Engagement, Unified Class 3B Threads. *(Source, FED-STD-H28/2B.)*

Size, TPI, and Series	To and Including 0.33D		Above 0.33D to 0.67D		Above 0.67D through 1.5D	
	Min.	Max.	Min.	Max.	Min.	Max.
7/8-12 12UN	.7850	.7900	.7874	.7926	.7900	.7952
7/8-14 UNF	.7980	.8022	.8000	.8045	.8023	.8068
7/8-16 16UN	.8070	.8116	.8094	.8137	.8115	.8158
7/8-18 UNS	.8150	.8190	.8169	.8210	.8189	.8230
7/8-20 UNEF	.8210	.8248	.8227	.8267	.8247	.8287
7/8-28 28UN	.8360	.8395	.8370	.8410	.8386	.8426
7/8-32 32UN	.8410	.8441	.8415	.8455	.8429	.8469
15/16-12 12UN	.8470	.8524	.8499	.8550	.8524	.8575
15/16-16 16UN	.8700	.8741	.8719	.8762	.8740	.8783
15/16-20 UNEF	.8830	.8873	.8852	.8892	.8872	.8912
15/16-28 28UN	.8990	.9020	.8995	.9035	.9011	.9051
15/16-32 32UN	.9040	.9066	.9040	.9080	.9054	.9094
1-8 UNC	.8650	.8722	.8684	.8759	.8722	.8797
1-12 UNF	.9100	.9148	.9123	.9173	.9148	.9198
1-14 UNS	.9230	.9271	.9249	.9293	.9271	.9315
1-16 16UN	.9320	.9366	.9344	.9387	.9365	.9408
1-18 UNS	.9400	.9440	.9419	.9460	.9439	.9480
1-20 UNEF	.9460	.9498	.9477	.9517	.9497	.9537
1-28 28UN	.9610	.9645	.9620	.9660	.9636	.9676
1-32 32UN	.9660	.9691	.9665	.9705	.9679	.9719
1 1/16-8 8UN	.9270	.9347	.9309	.9384	.9347	.9422
1 1/16-12 12UN	.9720	.9773	.9748	.9798	.9773	.9823
1 1/16-14 UNS	.9850	.9896	.9874	.9918	.9896	.9940
1 1/16-16 16UN	.9950	.9991	.9969	1.0012	.9990	1.0033
1 1/16-18 UNEF	1.0020	1.0065	1.0044	1.0085	1.0064	1.0105
1 1/16-20 20UN	1.0080	1.0123	1.0102	1.0142	1.0122	1.0162
1 1/16-28 28UN	1.0240	1.0270	1.0245	1.0285	1.0261	1.0301
1 1/8-7 UNC	.9700	.9790	.9747	.9833	.9789	.9875
1 1/8-8 8UN	.9900	.9972	.9934	1.0009	.9972	1.0047
1 1/8-12 UNF	1.0350	1.0398	1.0373	1.0423	1.0398	1.0448
1 1/8-16 16UN	1.0570	1.0616	1.0594	1.0637	1.0615	1.0658
1 1/8-18 UNEF	1.0650	1.0690	1.0669	1.0710	1.0689	1.0730
1 1/8-20 20UN	1.0710	1.0748	1.0727	1.0767	1.0747	1.0787
1 1/8-28 28UN	1.0860	1.0895	1.0870	1.0910	1.0886	1.0926
1 3/16-8 8UN	1.0520	1.0597	1.0559	1.0634	1.0597	1.0672
1 3/16-12 12UN	1.0970	1.1023	1.0998	1.1048	1.1023	1.1073
1 3/16-16 16UN	1.1200	1.1241	1.1219	1.1262	1.1240	1.1283
1 3/16-18 UNEF	1.1270	1.1315	1.1294	1.1335	1.1314	1.1355

All dimensions in inches.

(Continued)

Table 14. *(Continued)* Recommended Hole Size Limits Before Threading for Different Lengths of Engagement, Unified Class 3B Threads. *(Source, FED-STD-H28/2B.)*

Size, TPI, and Series	To and Including 0.33D		Above 0.33D to 0.67D		Above 0.67D through 1.5D	
	Min.	Max.	Min.	Max.	Min.	Max.
1 3/16-20 20UN	1.1330	1.1373	1.1352	1.1392	1.1372	1.1412
1 3/16-28 28UN	1.1490	1.1520	1.1495	1.1535	1.1511	1.1551
1 1/4-7 UNC	1.0950	1.1040	1.0997	1.1083	1.1039	1.1125
1 1/4-8 8UN	1.1150	1.1222	1.1184	1.1259	1.1222	1.1297
1 1/4-12 UNF	1.1600	1.1648	1.1623	1.1673	1.1648	1.1698
1 1/4-16 16UN	1.1820	1.1866	1.1844	1.1887	1.1865	1.1908
1 1/4-18 UNEF	1.1900	1.1940	1.1919	1.1960	1.939	1.1980
1 1/4-20 20UN	1.1960	1.1998	1.1977	1.2017	1.1997	1.2037
1 1/4-28 28UN	1.2110	1.2145	1.2120	1.2160	1.2136	1.2176
1 5/16-8 8UN	1.1770	1.1847	1.1809	1.1884	1.1847	1.1922
1 5/16-12 12UN	1.2220	1.2273	1.2248	1.2298	1.2273	1.2323
1 5/16-16 16UN	1.2450	1.2491	1.2469	1.2512	1.2490	1.2533
1 5/16-18 UNEF	1.2520	1.2565	1.2544	1.2585	1.2564	1.2605
1 5/16-20 20UN	1.2580	1.2623	1.2602	1.2642	1.2622	1.2662
1 5/16-28 28UN	1.2740	1.2770	1.2745	1.2785	1.2761	1.2801
1 3/8-6 UNC	1.1950	1.2046	1.1996	1.2096	1.2046	1.2146
1 3/8-8 8UN	1.2400	1.2472	1.2434	1.2509	1.2472	1.2547
1 3/8-12 UNF	1.2850	1.2898	1.2873	1.2923	1.2898	1.2948
1 3/8-16 16UN	1.3070	1.3116	1.3094	1.3137	1.3115	1.3158
1 3/8-18 UNEF	1.3150	1.3190	1.3169	1.3210	1.3189	1.3230
1 3/8-20 20UN	1.3210	1.3248	1.3227	1.3267	1.3247	1.3287
1 3/8-28 28UN	1.3360	1.3395	1.3370	1.3410	1.3386	1.3426
1 7/16-6 6UN	1.2570	1.2671	1.2621	1.2721	1.2671	1.2771
1 7/16-8 8UN	1.3020	1.3097	1.3059	1.3134	1.3097	1.3172
1 7/16-12 12UN	1.3470	1.3523	1.3498	1.3548	1.3523	1.3573
1 7/16-16 16UN	1.3700	1.3741	1.3719	1.3762	1.3740	1.3783
1 7/16-18 UNEF	1.3770	1.3815	1.3794	1.3835	1.3814	1.3855
1 7/16-20 20UN	1.3830	1.3873	1.3852	1.3892	1.3872	1.3912
1 7/16-28 28UN	1.3990	1.4020	1.3995	1.4035	1.4011	1.4051
1 1/2-6 UNC	1.3200	1.3296	1.3246	1.3346	1.3296	1.3396
1 1/2-8 8UN	1.3650	1.3722	1.3684	1.3759	1.3722	1.3797
1 1/2-12 UNF	1.4100	1.4148	1.4123	1.4173	1.4148	1.4198
1 1/2-16 16UN	1.4320	1.4366	1.4344	1.4387	1.4365	1.4408
1 1/2-18 UNEF	1.4400	1.4440	1.4419	1.4460	1.4439	1.4480
1 1/2-20 20UN	1.4460	1.4498	1.4477	1.4517	1.4497	1.4537
1 1/2-28 28UN	1.4610	1.4645	1.4620	1.4660	1.4636	1.4676
1 9/16-6 6UN	1.3820	1.3921	1.3871	1.3971	1.3921	1.4021
1 9/16-8 8UN	1.4270	1.4347	1.4309	1.4384	1.4347	1.4422

All dimensions in inches. *(Continued)*

Table 14. *(Continued)* **Recommended Hole Size Limits Before Threading for Different Lengths of Engagement, Unified Class 3B Threads.** *(Source, FED-STD-H28/2B.)*

Size, TPI, and Series	To and Including 0.33D		Above 0.33D to 0.67D		Above 0.67D through 1.5D	
	Min.	Max.	Min.	Max.	Min.	Max.
1 9/16-12 12UN	1.4720	1.4773	1.4748	1.4798	1.4773	1.4823
1 9/16-16 16UN	1.4950	1.4991	1.4969	1.5012	1.4990	1.5033
1 9/16-18 UNEF	1.5020	1.5065	1.5044	1.5085	1.5064	1.5105
1 9/16-20 20UN	1.5080	1.5123	1.5102	1.5142	1.5122	1.5162
1 5/8-6 6UN	1.4450	1.4546	1.4496	1.4596	1.4546	1.4646
1 5/8-8 8UN	1.4900	1.4972	1.4934	1.5009	1.4972	1.5047
1 5/8-12 12UN	1.5350	1.5398	1.5373	1.5423	1.5398	1.5448
1 5/8-16 16UN	1.5570	1.5616	1.5594	1.5637	1.5615	1.5658
1 5/8-18 UNEF	1.5650	1.5690	1.5669	1.5710	1.5689	1.5730
1 5/8-20 20UN	1.5710	1.5748	1.5727	1.5767	1.5747	1.5787
1 11/16-6 6UN	1.5070	1.5171	1.5121	1.5221	1.5171	1.5271
1 11/16-8 8UN	1.5520	1.5597	1.5559	1.5634	1.5597	1.5672
1 11/16-12 12UN	1.5970	1.6023	1.5998	1.6048	1.6023	1.6073
1 11/16-16 16UN	1.6200	1.6241	1.6219	1.6262	1.6240	1.6283
1 11/16-18 UNEF	1.6270	1.6315	1.6294	1.6335	1.6314	1.6355
1 11/16-20 20UN	1.6330	1.6373	1.6352	1.6392	1.6372	1.6412
1 3/4-5 UNC	1.5340	1.5455	1.5395	1.5515	1.5455	1.5575
1 3/4-6 6UN	1.5700	1.5796	1.5746	1.5846	1.5796	1.5896
1 3/4-8 8UN	1.6150	1.6222	1.6184	1.6259	1.6222	1.6297
1 3/4-12 12UN	1.6600	1.6648	1.6623	1.6673	1.6648	1.6698
1 3/4-16 16UN	1.6820	1.6866	1.6844	1.6887	1.6865	1.6908
1 3/4-20 20UN	1.6960	1.6998	1.6977	1.7017	1.6997	1.7037
1 13/16-6 6UN	1.6320	1.6421	1.6371	1.6471	1.6421	1.6521
1 13/16-8 8UN	1.6770	1.6847	1.6809	1.6884	1.6847	1.6922
1 13/16-12 12UN	1.7220	1.7273	1.7248	1.7298	1.7273	1.7323
1 13/16-16 16UN	1.7450	1.7491	1.7469	1.7512	1.7490	1.7533
1 13/16-20 20UN	1.7580	1.7623	1.7602	1.7642	1.7622	1.7662
1 7/8-6 6UN	1.6950	1.7046	1.6996	1.7096	1.7046	1.7146
1 7/8-8 8UN	1.7400	1.7472	1.7434	1.7509	1.7472	1.7547
1 7/8-12 12UN	1.7850	1.7898	1.7873	1.7923	1.7898	1.7948
1 7/8-16 16UN	1.8070	1.8116	1.8094	1.8137	1.8115	1.8158
1 7/8-20 20UN	1.8210	1.8248	1.8227	1.8267	1.8247	1.8287
1 15/16-6 6UN	1.7570	1.7671	1.7621	1.7721	1.7671	1.7771
1/15/16-8 8UN	1.8020	1.8097	1.8059	1.8134	1.8097	1.8172
1 15/16-12 12UN	1.8470	1.8523	1.8498	1.8548	1.8523	1.8573
1 15/16-16 16UN	1.8700	1.8741	1.8719	1.8762	1.8740	1.8783
1 15/16-20 20UN	1.8830	1.8873	1.8852	1.8892	1.8872	1.8912
2-4.5 UNC	1.7590	1.7727	1.7661	1.7794	1.7728	1.7861

All dimensions in inches.

(Continued)

Table 14. *(Continued)* Recommended Hole Size Limits Before Threading for Different Lengths of Engagement, Unified Class 3B Threads. *(Source, FED-STD-H28/2B.)*

Size, TPI, and Series	To and Including 0.33D		Above 0.33D to 0.67D		Above 0.67D through 1.5D	
	Min.	Max.	Min.	Max.	Min.	Max.
2-6 6UN	1.8200	1.8296	1.8246	1.8346	1.8296	1.8396
2-8 8UN	1.8650	1.8722	1.8684	1.8759	1.8722	1.8797
2-12 12UN	1.9100	1.9148	1.9123	1.9173	1.9148	1.9198
2-16 16UN	1.9320	1.9366	1.9344	1.9387	1.9365	1.9408
2-20 20UN	1.9460	1.9498	1.9477	1.9517	1.9497	1.9537
2 1/16-16 UNS	1.9950	1.9991	1.9969	2.0012	1.9990	2.0033
2 1/8-6 6UN	1.9450	1.9546	1.9496	1.9596	1.9546	1.9646
2 1/8-8 8UN	1.9900	1.9972	1.9934	2.0009	1.9972	2.0047
2 1/8-12 12UN	2.0350	2.0398	2.0373	2.0423	2.0398	2.0448
2 1/8-16 16UN	2.0570	2.0616	2.0594	2.0637	2.0615	2.0658
2 1/8-20 20UN	2.0710	2.0748	2.0727	2.0767	2.0747	2.0787
2 3/16-16 UNS	2.1200	2.1241	2.1219	2.1262	2.1240	2.1283
2 1/4-4.5 UNC	2.0090	2.0227	2.0161	2.0294	2.0228	2.0361
2 1/4-6 UN	2.0700	2.0796	2.0746	2.0846	2.0796	2.0896
2 1/2-4 UNC	2.2290	2.2444	2.2369	2.2519	2.2444	2.2594
2 3/4-4 UNC	2.4790	2.4944	2.4869	2.5019	2.4944	2.5094
3-4 UNC	2.7290	2.7444	2.7369	2.7519	2.7444	2.7594
3 1/4-4 UNC	2.9790	2.9944	2.9869	3.0019	2.9944	3.0094

All dimensions in inches.

Table 15. Drill Sizes for Metric Taps. *(Source, OSG Tap and Die, Inc.)*

Nom. size (mm)	Pitch (mm)	Basic Major Dia. (inches)	Tap Drill Size	Dec. Equiv. Tap Drill (inches)	Theo. % of Thread	Probable Oversize (inches)	Probable Hole Size (inches)	Prob. % of Thread
1.6	0.35	.0630	1.25mm	.0492	77	.0015	.0507	69
1.8	0.35	.0709	1.45mm	.0571	77	.0015	.0586	69
2	0.40	.0787	1/16	.0625	79	.0015	.0640	72
			1.60mm	.0630	77	.0017	.0647	69
			52	.0635	74	.0017	.0652	66
2.2	0.45	.0866	1.75mm	.0689	77	.0017	.0706	70
2.5	0.45	.0984	2.05mm	.0807	77	.0019	.0826	69
			46	.0810	76	.0019	.0829	67
			45	.0820	71	.0019	.0839	63
3	0.50	.1181	40	.0980	79	.0023	.1003	70
			2.5mm	.0984	77	.0023	.1007	68
			39	.0995	73	.0023	.1018	64

Note: Sizes with asterisk(*) are not standard drills.

(Continued)

Table 15. *(Continued)* **Drill Sizes for Metric Taps.** *(Source, OSG Tap and Die, Inc.)*

Nom. size (mm)	Pitch (mm)	Basic Major Dia. (inches)	Tap Drill Size	Dec. Equiv. Tap Drill (inches)	Theo. % of Thread	Probable Oversize (inches)	Probable Hole Size (inches)	Prob. % of Thread
3.5	0.60	.1378	33	.1130	81	.0026	.1156	72
			2.9mm	.1142	77	.0026	.1168	68
			32	.1160	71	.0026	.1186	63
4	0.70	.1574	3.25mm	.1280	82	.0029	.1309	74
			30	.1285	81	.0029	.1314	73
			3.3mm	.1299	77	.0029	.1328	69
4.5	0.75	.1772	3.7mm	.1457	82	.0032	.1489	74
			26	.1470	79	.0032	.1502	70
			3.75mm	.1476	77	.0032	.1508	69
			25	.1495	72	.0032	.1527	64
5	0.80	.1968	4.2mm	.1654	77	.0032	.1686	69
			19	.1660	75	.0032	.1692	68
6	1.00	.2362	10	.1935	84	.0038	.1973	76
			9	.1960	79	.0038	.1998	71
			5.0mm	.1968	77	.0038	.2006	70
			8	.1990	73	.0038	.2028	65
7	1.00	.2756	A	.2340	81	.0038	.2378	74
			15/64	.2344	81	.0038	.2382	73
			6.0mm	.2362	77	.0038	.2400	70
			B	.2380	74	.0038	.2418	66
8	1.25	.3150	6.7mm	.2638	80	.0041	.2679	74
			17/64	.2656	77	.0041	.2697	71
			6.75mm	.2657	77	.0041	.2698	71
			H	.2660	77	.0041	.2701	70
			6.8mm	.2677	74	.0041	.2718	68
8	1.00	.3150	7.0mm	.2756	77	.0041	.2797	69
			J	.2770	74	.0041	.2811	66
10	1.50	.3937	8.4mm	3307	82	.0044	.3351	76
			Q	.3320	80	.0044	.3364	75
			8.5mm	.3346	77	.0044	.3390	71
10	1.25	.3937	8.7mm	.3425	80	.0046	.3471	73
			11/32	.3438	78	.0046	.3483	71
			8.75mm	.3445	77	.0046	.3491	70
12	1.75	.4724	*10.25mm	.4035	77	.0047	.4082	72
			Y	.4040	76	.0047	.4087	71
			13/32	.4062	74	.0047	.4109	69
12	1.25	.4724	27/64*	.4219	79	.0047	.4266	72
			10.75mm	.4232	77	.0047	.4279	70
14	2	.5512	15/32	.4688	81	.0048	.4736	76
			12mm	.4724	77	.0048	.4772	72
14	1.5	.5512	12.5mm	.4921	77	.0048	.4969	71
16	2	.6299	35/64*	.5469	81	.0049	.5518	76
			14mm	.5512	77	.0049	.5561	72
16	1.5	.6299	*14.5mm	.5709	77	.0049	.5758	71
18	2.5	.7087	39/64*	.6094	78	.0050	.6144	74
			15.5mm	.6102	77	.0050	.6152	73
18	1.5	.7087	*16.5mm	.6496	77	.0050	.6546	70

Note: Sizes with asterisk(*) are not standard drills.   *(Continued)*

Table 15. *(Continued)* **Drill Sizes for Metric Taps.** *(Source, OSG Tap and Die, Inc.)*

Nom. size (mm)	Pitch (mm)	Basic Major Dia. (inches)	Tap Drill Size	Dec. Equiv. Tap Drill (inches)	Theo. % of Thread	Probable Oversize (inches)	Probable Hole Size (inches)	Prob. % of Thread
20	2.5	.7874	11/16* 17.5mm	.6875 .6890	78 77	.0050 .0052	.6925 .6942	74 73
20	1.5	.7874	*18.5mm	.7283	77	.0052	.7335	70
22	2.5	.8661	49/64* 19.5mm	.7656 .7677	79 77	.0052 .0052	.7708 .7729	75 73
22	1.5	.8661	*20.5mm	.8071	77	.0052	.8123	70
24	3	.9449	*21mm 53/64	.8268 .8281	77 76	.0059 .0059	.8327 .8340	73 72
24	2	.9449	*22mm	.8661	77	.0059	.8720	71
27	3	1.0630	15/16 *24mm	.9375 .9449	82 77	.0060 .0062	.9435 .9511	78 73
27	2	1.0630	*25mm 63/64	.9843 .9844	77 77	.0070 .0070	.9913 .9914	70 70
30	3.5	1.1811	*26.5mm 1-3/64	1.0433 1.0469	77 75			
30	2	1.1811	*28mm 1-7/64	1.1024 1.1094	77 70			
33	3.5	1.2992	1-5/32 *29.5mm 1-11/64	1.1562 1.1614 1.1719	80 77 71	Reaming Recommended		
33	2	1.2992	1-7/32 *31mm	1.2188 1.2205	79 77			
36	4	1.4173	1-1/4 *32mm	1.2500 1.2598	82 77			
36	3	1.4173	*33mm	1.2992	77			

Note: Sizes with asterisk(*) are not standard drills.

Table 16. Pitch Diameter Limits for Class 4H and 6H (Internal) M Profile Metric Threads.

Size (mm)	Pitch (mm)	Pitch Diameter Limits for Class of Thread		
		Min. (Basic)	Max. 4H	Max. 6H
M1.6	0.35	1.373	1.426	1.458
M1.8	0.35	1.573	1.626	1.658
M2	0.4	1.740	1.796	1.830
M2.2	0.45	1.908	1.968	2.003
M2.5	0.45	2.208	2.268	2.303
M3	0.5	2.675	2.738	2.775
M3.5	0.6	3.110	3.181	3.222
M4	0.7	3.545	3.620	3.663
M4.5	0.75	4.013	4.088	4.131
M5	0.8	4.480	4.560	4.605
M6	1	5.350	5.445	5.500
M7	1	6.350	6.445	6.500
M8	1.25	7.188	7.288	7.348
M8	1	7.350	7.445	7.500

All dimensions in millimeters.

(Continued)

Table 16. *(Continued)* Pitch Diameter Limits for Class 4H and 6H (Internal) M Profile Metric Threads.

Size (mm)	Pitch (mm)	Pitch Diameter Limits for Class of Thread		
		Min. (Basic)	Max. 4H	Max. 6H
M10	1.5	9.026	9.138	9.206
M10	1.25	9.188	9.288	9.348
M12	1.75	10.863	10.988	11.063
M12	1.25	11.188	11.300	11.368
M14	2	12.701	12.833	12.913
M14	1.5	13.026	13.144	13.216
M16	2	14.701	14.833	14.913
M16	1.5	15.026	15.144	15.216
M18	2.5	16.376	16.516	16.600
M18	1.5	17.026	17.144	17.216
M20	2.5	18.376	18.516	18.600
M20	1.5	19.026	19.144	19.216
M22	2.5	20.376	20.516	20.600
M22	1.5	21.026	21.144	21.216
M24	3	22.051	22.221	22.316
M24	2	22.701	22.841	22.925
M27	3	25.051	25.221	25.316
M27	2	25.701	25.841	25.925
M30	3.5	27.727	27.907	28.007
M30	2	28.701	28.841	28.925
M33	3.5	30.727	30.907	31.007
M33	2	31.701	31.841	31.925
M36	4	33.402	33.592	33.702
M36	3	34.051	34.221	34.316
M39	4	36.402	36.592	36.702
M39	3	37.051	37.221	37.316

All dimensions in millimeters.

Several different mathematical formulas have been advanced for predicting tap drill sizes for thread cutting taps that will result in a desired percentage of thread engagement. One thing that should always be kept in mind is that the drill will actually be forming the minor diameter of the thread. Therefore, under no circumstance should the drill diameter be in excess of the maximum minor diameter of the internal thread to be cut. Equation 1, which follows, is useful for finding the correct tap drill size for a desired percentage of thread, and Equation 2 can be used for determining the depth of thread that will be achieved with a given drill size. **One should always keep in mind** that the results predicted by these equations are based on a perfect hole, precisely the same size as the drill that produced it. Actual hole sizes will vary as a result of the condition of the drill, the machine doing the drilling, and the material being drilled. Actual percentage of thread engagement can only be determined by pin gaging the hole.

Equation 1) For ground thread taps only. Find the closest *tap drill size* for a desired percentage of thread for Unified and Metric M Profile threads.

Example for Unified Screw Threads:
>   Where: S is the size of the tap drill expressed in decimal inches
> D is the basic major diameter of the thread expressed in decimal inches

% is the desired percentage of thread engagement

tpi is the number of threads per inch.

$$S = D - ([.01299 \times \%] \div tpi)$$

Find the drill size needed for a $^{1}/_{4}$–20 tap to produce 75% of full thread.

.2500 – ([.01299 × 75] ÷ 20) = .20128" diameter hole.

Therefore, use a #7 (.2010") tap drill.

Example for Metric Screw Threads:

Where: *S* is the size of the tap drill expressed in millimeters

D is the basic major diameter of the thread expressed in millimeters

% is the desired percentage of thread engagement

P is the Pitch of the thread expressed in millimeters.

$$S = D - ([\% \times P] \div 76.98)$$

Find the drill size needed for an M5 × 0.80 tap to produce 75% of full thread.

5 – ([75 × .80] ÷ 76.98) = 4.22 mm

Therefore, use a #19 (4.216 mm / .1660") tap drill.

Equation 2) For ground thread taps only. Find the *percentage of thread engagement* that a specified tap drill will produce.

Example for Unified Screw Threads:

Where: % is the desired percentage of thread engagement

tpi is the number of threads per inch

D is the basic major diameter of the thread expressed in decimal inches

S is the size of the tap drill expressed in decimal inches.

$$\% = tpi \times ([D - S] \div .01299)$$

Find the percent of thread engagement that will result from using a #7 (.2010") drill to produce a hole for a $^{1}/_{4}$–20 tap.

20 × ([.2500 – .2010] ÷ .01299) = 75.44 %

Example for Metric Screw Threads:

Where: *S* is the size of the tap drill expressed in millimeters

D is the basic major diameter of the thread expressed in millimeters

% is the desired percentage of thread engagement

P is the pitch of the thread expressed in millimeters

$$\% = (76.980 \div P) \times (D - S)$$

Find the percent of thread engagement that will result from using a #19 (4.216 mm) drill to produce a hole for an M5 × 0.80 tap.

(76.980 ÷ .8) × (5 – 4.216) = 75.44 %

Calculating hole sizes for desired percentage of thread for forming taps

Also known as cold-forming taps, these taps do not remove metal by cutting. Instead, they displace material by plastic flow from the major diameter toward the minor diameter. Consequently, they do not produce chips, and they are extremely effective in ductile materials such as aluminum, copper, brass, leaded steels, low carbon steels, and stainless steels. However, they will not function in materials that are brittle and have low ductility. Threads created by these taps are stronger than those produced by cutting taps, and 65%

Table 17. Decimal Inch Equivalents of Standard M Profile Metric Thread Sizes.

Thread Designation	Dia. (Inches)	Pitch (Inches)	Thread Designation	Dia. (Inches)	Pitch (Inches)	Thread Designation	Dia. (Inches)	Pitch (Inches)
M1 × .25	.03937	.00984	M7 × 1.0	.27559	.03937	M18 × 2.5	.70866	.09483
M1.4 × .30	.05512	.01181	M8 × 1.25	.31496	.04921	M20 × 1.5	.78740	.05906
M1.6 × .35	.06299	.01378	M10 × 1.25	.39370	.04921	M20 × 2.5	.78740	.09483
M2 × .40	.07874	.01575	M10 × 1.5	.39370	.05906	M24 × 2	.94488	.07874
M2.5 × .45	.09843	.01772	M12 × 1.25	.47244	.04921	M24 × 3	.94488	.11811
M3 × .50	.11811	.01969	M12 × 1.75	.47244	.06890	M27 × 3	1.06299	.11811
M3.5 × .60	.13780	.02362	M14 × 1.5	.55118	.05906	M30 × 3.5	1.18110	.13780
M4 × .70	.15748	.02756	M14 × 2	.55118	.07874	M33 × 3.5	1.29921	.13780
M4.5 × .75	.17717	.02953	M16 × 1.5	.62992	.05906	M36 × 4	1.41732	.15748
M5 × .80	.19685	.03150	M16 × 2	.62992	.07874	mm × 0.03937 = inch equiv.		
M6 × 1.0	.23622	.03937	M18 × 1.5	.70866	.05906	inch × 25.4 = mm equiv.		

of thread is satisfactory for most applications. As can be seen in **Table 3**, thread forming taps for very coarse threads are not available. This is because of the difficulty in uniformly deforming the large volume of material required by coarse threads. Coarse threads down to 12 tpi can be reliably cut into most materials, but high ductility and low hardness are requirements for forming threads as coarse as 10 tpi.

Thread forming taps must be run under power and are usually effective when run at speeds of 150% to 200% in excess of thread cutting taps. Torque requirements are normally 125% more than with cutting taps, and power requirements can increase by 200% to 400%. Good lubrication and heat control can reduce these requirements. Because the tap displaces material above the mouth of the hole, countersinking before tapping is recommended. Also, since the hole diameter is actually reduced by the forming process, a larger tap drill is required for a forming tap than a cutting tap. Forming taps produce no chips, and tool life normally exceeds that of cutting taps. However, with forming taps, some imperfections can be expected at the minor diameter because the minor diameter, formed by displaced material that flows into the tap threads, is actually smaller than the diameter of the original drilled hole. With forming taps, the preferred tap drill size for a given thread designation will usually fall somewhere midway between the thread's specified major and minor diameters. These taps work best with fine pitches, and satisfactory threads coarser than 12 tpi are very difficult to achieve. **Tables 18** through **20** give recommended tap drill and forming tap sizes for Machine Screw and Fractional size threads, along with probable resulting percentage of thread. **Table 21** provides the same information for Metric M Profile threads. **Table 22** gives H limit forming tap recommendations for Machine Screw and Fractional size threads.

Since drilled holes for forming taps must be larger than holes drilled for cutting taps, the equation used for predicting percentage of full thread is unique to forming taps. Equation 3, which follows, is useful for finding the correct tap drill size for a desired percentage of thread, and Equation 4 can be used for determining the depth of thread that will be achieved with a given drill size. In all drilling operations, the true hole size will be affected by the condition of the drill and the material being machined. These equations can be used to predict percentage of thread for a given drill size, but the actual percentage of thread achieved can only be determined by pin gaging the drilled hole. The tables in this section are based on probable actual hole size and should be consulted for guidance in selecting a drill size for a particular thread.

Equation 3) For forming taps only. Find the closest *tap drill size* for a desired percentage of thread for Unified and Metric M Profile threads.

Example for Unified Screw Threads:

Where: S is the size of the tap drill expressed in decimal inches

D is the basic major diameter of the thread expressed in decimal inches

% is the desired percentage of thread engagement

tpi is the number of threads per inch.

$$S = D - ([.006799 \times \%] \div tpi)$$

Find the drill size needed for a $\frac{1}{4}$–20 tap to produce 70% of full thread.

$.2500 - ([.006799 \times .70] \div 20) = .2262"$ diameter hole.

Therefore, use a 5.75 mm (.2264") tap drill.

Example for Metric Screw Threads:

Where: S is the size of the tap drill expressed in millimeters

D is the basic major diameter of the thread expressed in millimeters

% is the desired percentage of thread engagement

P is the pitch of the thread expressed in millimeters.

$$S = D - ([\% \times P] \div 147.06$$

Find the drill size needed for an M5 × 0.80 tap to produce 70% of full thread.

$5 - ([70 \times .8] \div 147.06) = 4.619$ mm

Therefore, use a #14 (4.623 mm / .1820") tap drill.

Equation 4) For forming taps only. Find the *percentage of thread engagement* that a specified tap drill will produce.

Example for Unified Screw Threads:

Where: % is the desired percentage of thread engagement

tpi is the number of threads per inch

D is the basic major diameter of the thread expressed in decimal inches

S is the size of the tap drill expressed in decimal inches.

$$\% = tpi \times ([D - S] \div .006799)$$

Find the percent of thread engagement that will result from using a 5.75 mm drill to produce a hole for a $\frac{1}{4}$–20 tap.

5.75 mm $= .2264"$

$20 \times ([.2500 - .2264] \div .006799) = 69.42\ \%$

Example for Metric Screw Threads:

Where: % is the desired percentage of thread engagement

D is the basic major diameter of the thread expressed in millimeters

S is the size of the tap drill expressed in millimeters

P is the pitch of the thread expressed in millimeters.

$$\% = ([D - S] \times 147.06) \div P$$

Find the percent of thread engagement that will result from using a #14 (4.623 mm) drill to produce a hole for an M5 × 0.80 tap.

$([5 - 4.623] \times 147.06) \div .8 = 69.30\ \%$

(Text continued on p. 253)

Table 18. Forming Tap Drill Sizes for Class 2B Fit Unified Threads. *(Source, New England Tap Corp.)*

Size	UNC	UNF	75% Thread Theor. Size	75% Thread Near Drill	65% Thread Theor. Size	65% Thread Near Drill	55% Thread Theor. Size	55% Thread Near Drill
0		80	.0543	#54	.0549	#54	.0555	1.4mm
1	64		.0657	1.7mm	.0664	#51	.0672	#51
1		72	.0666	1.7mm	.0672	#51	.0679	#51
2	56		.0776	5/64	.0783	#47	.0792	#47
2		64	.0788	2mm	.0795	2.05mm	.0803	2.05mm
3	48		.0895	#43	.0904	2.3mm	.0914	2.3mm
3		56	.0905	2.3mm	.0913	2.3mm	.0922	2.35mm
4	40		.1000	#39	.1010	#38	.1023	2.6mm
4		48	.1021	2.6mm	.1030	2.6mm	.1040	#37
5	40		.1129	#33	.1140	2.9mm	.1152	2.9mm
5		44	.1141	2.9mm	.1149	2.9mm	.1161	#32
6	32		.1227	3.1mm	.1240	1/8	.1256	1/8
6		40	.1259	1/8	.1269	3.2mm	.1282	#30
8	32		.1486	#25	.1499	3.8mm	.1515	#24
8		36	.1504	#24	.1516	#24	.1530	#23
10	24		.1692	4.3mm	.1710	11/64	.1730	#17
10		32	.1746	4.4mm	.1760	#16	.1776	4.5mm
12	24		.1951	#9	.1969	5mm	.1990	#8
12		28	.1983	#8	.1998	5.1mm	.2016	#7
1/4	20		.2248	5.7mm	.2269	5.75mm	.2294	5.8mm
1/4		28	.2323	5.9mm	.2338	A	.2357	15/64
5/16	18		.2844	7.25mm	.2868	7.3mm	.2895	L
5/16		24	.2916	7.4mm	.2934	M	.2955	7.5mm
3/8	16		.3432	11/32	.3459	8.8mm	.3490	S
3/8		24	.3541	9mm	.3559	9mm	.3581	T
7/16	14		.4007	X	.4040	Y	.4076	13/32
7/16		20	.4122	Z	.4144	10.5mm	.4169	10.5mm
1/2	13		.4605	15/32	.4638	15/32	.4677	15/32
1/2		20	.4747	12mm	.4769	12mm	.4795	31/64
9/16	12		.5193	33/64	.5230	17/32	.5273	17/32
9/16		18	.5343	17/32	.5368	17/32	.5396	17/32
5/8	11		.5780	37/64	.5819	37/64	.5866	37/64
5/8		18	.5968	19/32	.5992	19/32	.6021	19/32
3/4	10		.6980	45/64	.7024	45/64	.7075	45/64
3/4		16	.7182	23/32	.7209	23/32	.7242	23/32
7/8	9		.818	13/16	.823	.823	.829	.823
7/8		14	.839	21.25mm	.843	27/32	.845	27/32

(Continued)

Table 18. *(Continued)* **Forming Tap Drill Sizes for Class 2B Fit Unified Threads.** *(Source, New England Tap Corp.)*

Tap			75% Thread		65% Thread		55% Thread	
Size	UNC	UNF	Theor. Size	Near Drill	Theor. Size	Near Drill	Theor. Size	Near Drill
1	8		.935	15/16	.942	15/16	.948	15/16
1		12	.959	.963	.963	.963	.967	.963

Note: Forming taps are very effective in ductile materials up to 250 BHN, such as aluminum, copper, low carbon steel, leaded steel, and some stainless steel. They must be power driven at speeds up to 150% to 200% of the speeds for cutting taps. Because forming taps reduce the drilled hole diameter by displacing material from the major diameter toward the minor diameter, a tap drill larger than that which is required for a cutting tap must be used. The tap drill sizes on this Table are for use with forming taps only, and should never be used for cutting taps.

Table 19. Forming Tap Drill Sizes for Class 3B Fit Unified Threads. *(Source, New England Tap Corp.)*

Tap			75% Thread		65% Thread		55% Thread	
Size	UNC	UNF	Theor. Size	Near Drill	Theor. Size	Near Drill	Theor. Size	Near Drill
0		80	.0540	1.35mm	.0546	#54	.0552	1.4mm
1	64		.0654	1.65mm	.0661	1.7mm	.0669	1.7mm
1		72	.0663	1.7mm	.0669	#51	.0676	#51
2	56		.0773	5/64	.0780	5/64	.0789	2mm
2		64	.0784	#47	.0791	2.05mm	.0799	2.05mm
3	48		.0886	2.25mm	.0895	#43	.0905	2.3mm
3		56	.0902	2.3mm	.0910	2.3mm	.0919	2.35mm
4	40		.0995	#39	.1005	#38	.1018	2.6mm
4		48	.1017	2.6mm	.1026	2.6mm	.1036	#37
5	40		.1125	#33	.1136	2.9mm	.1148	2.9mm
5		44	.1137	2.9mm	.1145	2.9mm	.1157	#32
6	32		.1222	3.1mm	.1235	3.1mm	.1251	1/8
6		40	.1255	1/8	.1265	3.2mm	.1278	3.25mm
8	32		.1481	3.8mm	.1494	#25	.1510	#24
8		36	.1500	#25	.1512	#24	.1526	#24
10	24		.1686	4.3mm	.1704	#18	.1724	#17
10		32	.1741	4.4mm	.1755	4.4mm	.1771	#16
12	24		.1941	5mm	.1959	#9	.1980	#8
12		28	.1978	#8	.1993	#8	.2011	#7
1/4	20		.2242	5.7mm	.2263	5.75mm	.2288	5.8mm
1/4		28	.2322	5.9mm	.2337	A	.2356	15/64
5/16	18		.2837	7.2mm	.2861	7.3mm	.2888	7.3mm
5/16		24	.2910	7.4mm	.2928	M	.2949	M
3/8	16		.3425	8.7mm	.3452	8.8mm	.3483	S
3/8		24	.3535	9mm	.3553	9mm	.3575	T
7/16	14		.3999	X	.4032	Y	.4068	13/32

(Continued)

Table 19. *(Continued)* **Forming Tap Drill Sizes for Class 3B Fit Unified Threads.** *(Source, New England Tap Corp.)*

Tap			75% Thread		65% Thread		55% Thread	
Size	UNC	UNF	Theor. Size	Near Drill	Theor. Size	Near Drill	Theor. Size	Near Drill
7/16		20	.4116	Z	.4138	10.5mm	.4163	10.5mm
1/2	13		.4596	15/32	.4629	15/32	.4668	15/32
1/2		20	.4740	12mm	.4762	12mm	.4788	31/64
9/16	12		.5185	33/64	.5222	17/32	.5265	17/32
9/16		18	.5335	17/32	.5360	17/32	.5388	17/32
5/8	11		.5771	37/64	.5810	37/64	.5857	37/64
5/8		18	.5961	19/32	.5985	19/32	.6014	19/32
3/4	10		.6970	45/64	.7014	45/64	.7065	45/64
3/4		16	.7174	23/32	.7201	23/32	.7234	23/32
7/8	9		.818	13/16	.823	.823	.829	.823
7/8		14	.839	21.25mm	.843	27/32	.845	27/32
1	8		.935	15/16	.942	15/16	.948	15/16
1		12	.959	.963	.963	.963	.967	.963

Note: Forming taps are very effective in ductile materials up to 250 BHN, such as aluminum, copper, low carbon steel, leaded steel, and some stainless steel. They must be power driven at speeds up to 150% to 200% of the speeds for cutting taps. Because forming taps reduce the drilled hole diameter by displacing material from the major diameter toward the minor diameter, a tap drill larger than that which is required for a cutting tap must be used. The tap drill sizes on this Table are for use with forming taps only, and should never be used for cutting taps.

Table 20. Forming Tap Drill Sizes for Class 2 Fit Unified Threads. *(Source, New England Tap Corp.)*

Tap			75% Thread		65% Thread		55% Thread	
Size	UNC	UNF	Theor. Size	Near Drill	Theor. Size	Near Drill	Theor. Size	Near Drill
0		80	.0540	1.35mm	.0546	#54	.0552	1.4mm
1	64		.0654	1.65mm	.0661	1.7mm	.0669	1.7mm
1		72	.0662	1.7mm	.0668	#51	.0675	#51
2	56		.0772	5/64	.0779	5/64	.0788	2mm
2		64	.0784	#47	.0791	2.05mm	.0799	2.05mm
3	48		.0886	2.25mm	.0895	#43	.0905	2.3mm
3		56	.0901	2.3mm	.0909	2.3mm	.0918	2.35mm
4	40		.0995	#39	.1005	#38	.1018	2.6mm
4		48	.1016	2.6mm	.1025	2.6mm	.1035	#37
5	40		.1124	#33	.1135	2.9mm	.1147	2.9mm
5		44	.1137	2.9mm	.1145	2.9mm	.1157	#32
6	32		.1222	3.1mm	.1235	3.1mm	.1251	1/8
6		40	.1254	1/8	.1264	3.2mm	.1277	3.25mm
8	32		.1481	3.8mm	.1494	#25	.1510	#24
8		36	.1499	#25	.1511	#24	.1525	#24
10	24		.1687	4.3mm	.1705	#18	.1725	#17
10		32	.1740	4.4mm	.1754	4.4mm	.1770	#16

(Continued)

Table 20. *(Continued)* **Forming Tap Drill Sizes for Class 2 Fit Unified Threads.** *(Source, New England Tap Corp.)*

Tap			75% Thread		65% Thread		55% Thread	
Size	UNC	UNF	Theor. Size	Near Drill	Theor. Size	Near Drill	Theor. Size	Near Drill
12	24		.1941	5mm	.1959	#9	.1980	#8
12		28	.1978	#8	.1993	#8	.2011	#7
1/4	20		.2242	5.7mm	.2263	5.75mm	.2288	5.8mm
1/4		28	.2317	5.9mm	.2332	A	.2351	15/64
5/16	18		.2838	7.2mm	.2862	7.3mm	.2889	7.3mm
5/16		24	.2909	7.4mm	.2927	M	.2948	M
3/8	16		.3426	8.7mm	.3453	8.8mm	.3484	S
3/8		24	.3533	9mm	.3551	9mm	.3573	T
7/16	14		.4001	X	.4034	Y	.4070	13/32
7/16		20	.4113	Z	.4135	10.5mm	.4160	10.5mm
1/2	13		.4598	15/32	.4631	15/32	.4670	15/32
1/2		20	.4737	12mm	.4759	12mm	.4785	.3464
9/16	12		.5187	33/64	.5224	17/32	.5267	17/32
9/16		18	.5334	17/32	.5359	17/32	.5387	17/32
5/8	11		.5774	37/64	.5813	37/64	.5860	37/64
5/8		18	.5959	19/32	.5983	19/32	.6012	19/32
3/4	10		.6973	45/64	.7017	45/64	.7068	45/64
3/4		16	.7172	23/32	.7199	23/32	.7232	23/32
7/8	9		.818	13/16	.823	.823	.829	.823
7/8		14	.839	21.25mm	.843	27/32	.845	27/32
1	8		.935	15/16	.942	15/16	.948	15/16
1		12	.959	.963	.963	.963	.967	24.5mm

Note: Forming taps are very effective in ductile materials up to 250 BHN, such as aluminum, copper, low carbon steel, leaded steel, and some stainless steel. They must be power driven at speeds up to 150% to 200% of the speeds for cutting taps. Because forming taps reduce the drilled hole diameter by displacing material from the major diameter toward the minor diameter, a tap drill larger than that which is required for a cutting tap must be used. The tap drill sizes on this Table are for use with forming taps only, and should never be used for cutting taps.

Table 21. Forming Tap Drill Sizes for Metric Screw Threads. (Source, New England Tap Corp.)

Tap Size	Tap Drill Size	Decimal Equiv. of Tap Drill (Inches)	Prob. % of Thread Engmt.	Tap Size	Tap Drill Size	Decimal Equiv. of Tap Drill (Inches)	Prob. % of Thread Engmt.	Tap Size	Tap Drill Size	Decimal Equiv. of Tap Drill (Inches)	Prob. % of Thread Engmt.
M2×.35	1.39mm	.0547	72	M7×1	6.45mm	.2539	72	M16×1.5	15.1mm	.5945	79
	1.41mm	.0555	64		6.5mm	.2559	65		15.2mm	.5984	69
	1.43mm	.0563	55		6.55mm	.2579	58		15.3mm	.6024	60
M2×.4	1.76mm	.0693	74	M8×1.25	7.3mm	.2874	75	M16×2	14.85mm	.5846	78
	50	.0700	67		L	.2900	67		15.0mm	.5906	67
	1.81mm	.0713	55		7.45mm	.2933	57		19/32	.5938	61
M3×.45	2.24mm	.0882	71	M10×1.25	9.3mm	.3661	74	M18×1.5	17.2mm	.6772	69
	43	.0890	65		U	.3680	69		17.3mm	.6811	59
	2.29mm	.0902	55		9.45mm	.3720	56				
M3×.5	2.7mm	.1063	75	M10×1.5	9.15mm	.3602	77	M18×2.5	21/32	.6563	73
	2.75mm	.1083	61		9.25mm	.3642	67		16.8mm	.6614	65
					9.35mm	.3681	57		16.9mm	.6654	59
M4×.6	3.15mm	.1240	75	M12×1.25	11.3mm	.4449	73	M20×1.5	19.1mm	.7520	78
	3.18mm	.1252	67		11.35mm	.4469	67		19.2mm	.7559	68
	3.22mm	.1268	57		11.4mm	.4488	61		19.3mm	.7598	58
M4×.7	3.6mm	.1417	74	M12×1.75	11.0mm	.4331	78	M20×2.5	18.6mm	.7323	76
	3.65mm	.1437	64		7/16	.4375	68		18.75mm	.7382	67
	3.68mm	.1449	57		11.25mm	.4429	57		18.9mm	.7441	58
M5×.75	4.06mm	.1598	77	M14×1.5	13.2mm	.5197	70	M24×1.5	23.2mm	.9134	66
	4.1mm	.1614	69		13.25mm	.5217	65		23.25mm	.9154	61
	4.15mm	.1634	59		13.3mm	.5236	60				
M5×.8	4.55mm	.1791	73	M14×2	12.9mm	.5079	75	M24×3	22.4mm	.8819	73
	4.6mm	.1811	64		13.0mm	.5118	67		22.5mm	.8858	68
	4.65mm	.1831	55		33/64	.5156	60		57/64	.8906	62
M6×1	5.45mm	.2146	73								
	5.5mm	.2165	66								
	7/32	.2188	57								

The percent of thread engagement in this table is based upon the probable hole size the drill will cut. The actual hole size may vary as a result of the condition of the drill, machine and material being drilled. The actual percent of thread engagement may be determined by pin gaging the hole.

Note: Forming taps are very effective in ductile materials up to 250 BHN, such as aluminum, copper, low carbon steel, leaded steel, and some stainless steel. They must be power driven at speeds up to 150% to 200% of the speeds for cutting taps. Because forming taps reduce the drilled hole diameter by displacing material from the major diameter toward the minor diameter, a tap drill larger than that which is required for a cutting tap must be used. The tap drill sizes on this Table are for use with forming taps only, and should never be used for cutting taps.

Table 22. Forming Tap Recommendations for Classes 2, 2B, and 3B Unified Machine Screw and Fractional Size Screw Threads. *(Source, Fastcut Tool Corp.)*

Size	Threads Per Inch		Recommended Forming Tap for Class of Thread			Pitch Diameter Limits			
	UNC	UNF	Class 2	Class 2B	Class 3B	Min. All Classes (Basic)	Max. Class 2	Max. Class 2B	Max. Class 3B
0		80	G H2	B H3	G H2	.0519	.0536	.0542	.0536
1	64		G H2	G H3	G H2	.0629	.0648	.0655	.0648
1		72	G H2	G H3	G H2	.0640	.0658	.0665	.0659
2	56		G H2	G H3	G H2	.0744	.0764	.0772	.0765
2		64	G H2	G H3	G H2	.0759	.0778	.0786	.0779
3	48		G H2	G H3	G H2	.0855	.0877	.0885	.0877
3		56	G H2	G H3	G H2	.0874	.0894	.0902	.0895
4	40		G H3	G H5	G H3	.0958	.0982	.0991	.0982
4		48	G H3	G H5	G H3	.0985	.1007	.1016	.1008
5	40		G H3	G H5	G H3	.1088	.1112	.1121	.1113
5		44	G H3	G H5	G H3	.1102	.1125	.1134	.1126
6	32		G H3	G H5	G H3	.1177	.1204	.1214	.1204
6		40	G H3	G H5	G H3	.1218	.1242	.1252	.1243
8	32		G H3	G H5	G H3	.1437	.1464	.1475	.1465
8		36	G H3	G H5	G H3	.1460	.1485	.1496	.1487
10	24		G H4	G H6	G H4	.1629	.1662	.1672	.1661
10		32	G H4	G H6	G H4	.1697	.1724	.1736	.1726
12	24		G H4	G H6	G H4	.1889	.1922	.1933	.1922
12		28	G H4	G H6	G H4	.1928	.1959	.1970	.1959
1/4	20		G H4	G H6	G H4	.2175	.2211	.2223	.2211
1/4		28	G H4	G H6	G H4	.2268	.2299	.2311	.2300
5/16	18		G H5	G H7	G H5	.2764	.2805	.2817	.2803
5/16		24	G H5	G H7	G H5	.2854	.2887	.2902	.2890
3/8	16		G H5	G H7	G H5	.3344	.3389	.3401	.3387
3/8		24	G H5	G H7	G H5	.3479	.3512	.3528	.3516
7/16	14		G H5	G H8	G H5	.3911	.3960	.3972	.3957
7/16		20	G H5	G H8	G H5	.4050	.4086	.4104	.4091
1/2	13		G H5	G H8	G H5	.4500	.4552	.4565	.4548
1/2		20	G H5	G H8	G H5	.4675	.4711	.4731	.4717
9/16	12		G H7	G H10	G H7	.5084	.5140	.5152	.5135
9/16		18	G H7	G H10	G H7	.5264	.5305	.5323	.5308
5/8	11		G H7	G H10	G H7	.5660	.5719	.5732	.5714
5/8		18	G H7	G H10	G H7	.5889	.5930	.5949	.5934
3/4	10		G H7	G H10	G H7	.6850	.6914	.6927	.6907
3/4		16	G H7	G H10	G H7	.7094	.7139	.7159	.7143
7/8	9		G H9	G H12	G H9	.8028	.8098	.8110	.8089
7/8		14	G H9	G H12	G H9	.8286	.8335	.8356	.8339

(Continued)

Table 22. *(Continued)* **Forming Tap Recommendations for Classes 2, 2B, and 3B Unified Machine Screw and Fractional Size Screw Threads.** *(Source, Fastcut Tool Corp.)*

Size	Threads Per Inch		Recommended Forming Tap For Class of Thread			Pitch Diameter Limits			
	UNC	UNF	Class 2	Class 2B	Class 3B	Min. All Classes (Basic)	Max. Class 2	Max. Class 2B	Max. Class 3B
1	8		G H9	G H12	G H9	.9188	.9264	.9276	.9254
1		12	G H9	G H12	G H9	.9459	.9515	.9535	.9516

The above recommended taps normally produce the Class of Thread indicated in average materials when used with reasonable care. However, if the tap specified does not give a satisfactory gage fit in the work, a choice of some other limit tap will be necessary. All of the H-Limits shown will produce a class 2B fit.

Percentage of thread in large holes

Even though most types of taps are readily available only up to 1 $1/2$" or 36 mm in diameter, larger sizes are available through tap manufacturers. Sizes up to 3–16 NS are readily obtainable in taper, plug, and bottoming configurations for use on large holes. Holes in excess of this diameter are usually threaded by chasing or by turning. Larger holes are usually bored, and **Table 23** provides recommended bore sizes and resulting percentage of thread engagement for Classes 1B, 2B, and 3B UNC and UNF Unified threads sized between 1 $1/4$–7 and 6–16.

Interference-fit threads

Recommended tap drill sizes for Class 5 interference-fit threads can be found in Table 10 on page 67.

Miniature taps

Miniature taps require special attention during setup. Tap and target hole must be carefully aligned to prevent breakage. Once in the hole, the taps should be rotated forward only a fraction of a turn, then backed off to break and clear the chip. 75% of thread engagement should be considered the maximum allowable for these taps, and that should only be attempted when producing short threaded holes in soft materials. **Table 24** provides recommended tap drill sizes for three lengths of thread engagement for cut threads, and a suggested drill diameter when using forming taps. Dimensions of commonly available miniature plug and forming taps are given in **Table 25**.

Speeds and feeds for tapping

For additional information about speeds and feeds for forming taps, see the above section titled "Calculating hole sizes for desired percentage of thread for forming taps." As with all machining operations, the rigidity of the machine and workpiece is essential to assure repeatable performance in tapping. Since tapping, unlike drilling, is subject to error on two different planes, it is especially important to follow prescribed parameters. In tapping as in drilling, if the machine and workpiece are not absolutely rigid, hole size will be affected. However, in tapping, excessive force on the cutting tool can elongate and distort the pitch diameter of the thread. Hole depth also influences tapping speeds, and a recommended reduction of speed of 5% should be made for every 100% increase in hole depth. For example, if a $1/4$" deep hole being tapped at 40 sfm is increased in depth to 1", it is recommended that the speed should be decreased by 20% to 32 sfm. Achieving a desired percentage of thread engagement also affects speed. Obviously, with a given amount of power, tapping for 60% of thread can be done at higher speed than tapping for 80% of thread. These and other variables make tapping a challenging operation.

Table 39 provides tap speed ranges for various materials. The lower range should be adhered to when tapping deeper holes or when tapping for high percentage of thread engagement, and the higher ranges are to be used when tapping shallow holes with low percentage of thread requirements. Care should be exercised when exiting from holes. Often, the tap may be reversed at up to twice the cutting speed, but high speeds may result in the tap recutting as it is exited. This not only distorts the tapped thread, it also dramatically shortens the tap's life. The selection of the proper coolant is important for tool life and surface finish. RPM for achieving desired cutting speeds in sfm for most Unified and Metric taps can be found in **Tables 40** and **41**. Recommended coolants for a variety of workpiece materials are shown in **Table 42**, **Table 43** is a guide for selecting coatings that act as hardening treatments for high speed steel taps. The selection of the proper coating can greatly enhance tap performance in different workpiece materials. Different specialized coatings are offered by individual manufacturers, so it is advisable to consult with your supplier to determine which coatings are available.   (Text continued on p. 278)

Table 23. Recommended Bores for 1B, 2B, and 3B Unified Threads, 1 1/4" - 6".
(Source, Landis Threading Systems.)

Size and TPI	UNC and UNF Classes 1B, 2B				UNC and UNF Class 3B			
	Min. Hole Size	% of Thread[a]	Max. Hole Size[b]	% of Thread[a]	Min. Hole Size	% of Thread[a]	Max. Hole Size[b]	% of Thread[a]
1 1/4-7	1.095	83.5	1.123	68.4	1.0950	83.5	1.1125	74.1
1 1/4-8	1.115	83.1	1.140	67.7	1.1150	83.1	1.1297	74.1
1 1/4-12	1.160	83.1	1.178	66.5	1.1600	83.1	1.1698	74.1
1 1/4-16	1.182	83.8	1.196	66.5	1.1820	83.8	1.1908	72.9
1 5/16-12	1.222	83.6	1.240	67.0	1.2220	83.6	1.2323	74.1
1 5/16-16	1.245	83.1	1.259	65.9	1.2450	83.1	1.2533	72.9
1 5/16-18	1.252	83.8	1.265	65.8	1.2520	83.8	1.2605	72.1
1 3/8-6	1.195	83.1	1.225	69.3	1.1950	83.1	1.2146	74.1
1 3/8-8	1.240	83.1	1.265	67.7	1.2400	83.1	1.2547	74.1
1 3/8-12	1.285	83.1	1.303	66.5	1.2850	83.1	1.2948	74.1
1 3/8-16	1.307	83.8	1.321	66.5	1.3070	83.8	1.3158	72.9
1 7/16-12	1.347	83.6	1.365	67.0	1.3470	83.6	1.3573	74.1
1 7/16-16	1.370	83.1	1.384	65.9	1.3070	83.8	1.3158	72.9
1 1/2-6	1.320	83.1	1.350	69.3	1.3200	83.1	1.3396	74.1
1 1/2-8	1.365	83.1	1.390	67.7	1.3650	83.1	1.3797	74.1
1 1/2-12	1.410	83.1	1.428	66.5	1.4100	83.1	1.4198	74.1
1 1/2-16	1.432	83.8	1.446	66.5	1.4320	83.8	1.4408	72.9
1 9/16-16	1.495	83.1	1.509	65.9	1.4950	83.1	1.5033	72.9
1 5/8-8	1.490	83.1	1.515	67.7	1.4900	83.1	1.5047	74.1
1 5/8-12	1.535	83.1	1.553	66.5	1.5350	83.1	1.5448	74.1
1 5/8-16	1.557	83.8	1.571	66.5	1.5570	83.8	1.5658	72.9
1 11/16-16	1.620	83.1	1.634	65.9	1.6200	83.1	1.6283	72.9
1 3/4-5	1.534	83.1	1.568	70.1	1.5340	83.1	1.5575	74.1

(Continued)

[a]Based on values as rounded off in preceding column.
[b]Based on length of engagement equal to the nominal diameter.

Table 23. *(Continued)* **Recommended Bores for 1B, 2B, and 3B Unified Threads, 1 ¹/₄" - 6".**
(Source, Landis Threading Systems.)

Size and TPI	UNC and UNF Classes 1B, 2B				UNC and UNF Class 3B			
	Min. Hole Size	% of Thread[a]	Max. Hole Size[b]	% of Thread[a]	Min. Hole Size	% of Thread[a]	Max. Hole Size[b]	% of Thread[a]
1 3/4-8	1.615	83.1	1.640	67.7	1.6150	83.1	1.6297	74.1
1 3/4-12	1.660	83.1	1.678	66.5	1.6600	83.1	1.6698	74.1
1 3/4-16	1.682	83.8	1.696	66.5	1.6820	83.8	1.6908	72.9
1 13/16-16	1.745	83.1	1.759	65.9	1.7450	83.1	1.7533	72.9
1 7/8-8	1.740	83.1	1.765	67.7	1.7400	83.1	1.7547	74.1
1 7/8-12	1.785	83.1	1.803	66.5	1.7850	83.1	1.7948	74.1
1 15/16-16	1.870	83.1	1.884	65.9	1.8700	83.1	1.8783	72.9
2-4 1/2	1.759	83.5	1.795	71.0	1.7590	83.5	1.7861	74.1
2-8	1.865	83.1	1.890	67.7	1.8650	83.1	1.8797	74.1
2-12	1.910	83.1	1.928	66.5	1.9100	83.1	1.9198	74.1
2-16	1.932	83.8	1.946	66.5	1.9320	83.8	1.9408	72.9
2 1/16-16	1.995	83.1	2.009	65.9	1.9950	83.1	2.0033	72.9
2 1/8-8	1.990	83.1	2.015	67.7	1.9900	83.1	2.0047	74.1
2 1/8-12	2.035	83.1	2.053	66.5	2.0350	83.1	2.0448	74.1
2 1/8-16	2.057	83.8	2.071	66.5	2.0570	83.8	2.0658	72.9
2 3/16-16	2.120	83.1	2.134	65.9	2.1200	83.1	2.1283	72.9
2 1/4-4 1/2	2.009	83.5	2.045	71.0	2.0090	83.5	2.0361	74.1
2 1/4-8	2.115	83.1	2.140	67.7	2.1150	83.1	2.1297	74.1
2 1/4-12	2.160	83.1	2.178	66.5	2.1600	83.1	2.1698	74.1
2 1/4-16	2.182	83.8	2.196	66.5	2.1820	83.8	2.1908	72.9
2 5/16-16	2.245	83.1	2.259	65.9	2.2450	83.1	2.2533	72.9
2 3/8-12	2.285	83.1	2.303	66.5	2.2850	83.1	2.2948	74.1
2 3/8-16	2.307	83.8	2.321	66.5	2.3070	83.8	2.3158	72.9
2 7/16-16	2.370	83.1	2.384	65.9	2.3700	83.1	2.3783	72.9
2 1/2-4	2.229	83.4	2.267	71.7	2.2290	83.4	2.2594	74.1
2 1/2-8	2.365	83.1	2.390	67.7	2.3650	83.1	2.3797	74.1
2 1/2-12	2.410	83.1	2.428	66.5	2.4100	83.1	2.4198	74.1
2 1/2-16	2.432	83.8	2.446	66.5	2.4320	83.8	2.4408	72.9
2 5/8-12	2.535	83.1	2.553	66.5	2.5350	83.1	2.5448	74.1
2 5/8-16	2.557	83.8	2.571	66.5	2.5570	83.8	2.5658	72.9
2 3/4-4	2.479	83.4	2.517	71.7	2.4790	83.4	2.5094	74.1
2 3/4-8	2.615	83.1	2.640	67.7	2.6150	83.1	2.6297	74.1
2 3/4-12	2.660	83.1	2.678	66.5	2.6600	83.1	2.6698	74.1
2 3/4-16	2.682	83.8	2.696	66.5	2.6820	83.8	2.6908	72.9
2 7/8-12	2.785	83.1	2.803	66.5	2.7850	83.1	2.7948	74.1

[a]Based on values as rounded off in preceding column.
[b]Based on length of engagement equal to the nominal diameter.

(Continued)

Table 23. *(Continued)* **Recommended Bores for 1B, 2B, and 3B Unified Threads, 1 1/4" - 6".**
(Source, Landis Threading Systems.)

Size and TPI	UNC and UNF Classes 1B, 2B				UNC and UNF Class 3B			
	Min. Hole Size	% of Thread[a]	Max. Hole Size[b]	% of Thread[a]	Min. Hole Size	% of Thread[a]	Max. Hole Size[b]	% of Thread[a]
2 7/8-16	2.807	83.8	2.821	66.5	2.8070	83.8	2.8158	72.9
3-8	2.865	83.1	2.890	67.7	2.8650	83.1	2.8797	74.1
3-12	2.910	83.1	2.928	66.5	2.9100	83.1	2.9198	74.1
3-16	2.932	83.8	2.946	66.5	2.9320	83.8	2.9408	72.9
3 1/8-12	3.035	83.1	3.053	66.5	3.0350	83.1	3.0448	74.1
3 1/8-16	3.057	83.8	3.071	66.5	3.0570	83.8	3.0658	72.9
3 1/4-4	2.979	83.4	3.017	71.7	2.9790	83.4	3.0094	74.1
3 1/4-8	3.115	83.1	3.140	67.7	3.1150	83.1	3.1297	74.1
3 1/4-12	3.160	83.1	3.178	66.5	3.1600	83.1	3.1698	74.1
3 1/4-16	3.182	83.8	3.196	66.5	3.1820	83.8	3.1908	72.9
3 3/8-12	3.285	83.1	3.303	66.5	3.285	83.1	3.303	66.5
3 3/8-16	3.307	83.8	3.321	66.5	3.307	83.8	3.321	66.5
3 1/2-4	3.229	83.4	3.267	71.7	3.229	83.4	3.267	71.7
3 1/2-8	3.365	83.1	3.390	67.7	3.365	83.1	3.390	67.7
3 1/2-12	3.410	83.1	3.428	66.5	3.410	83.1	3.428	66.5
3 1/2-16	3.432	83.8	3.446	66.5	3.432	83.8	3.446	66.5
3 5/8-12	3.535	83.1	3.553	66.5	3.535	83.1	3.553	66.5
3 5/8-16	3.557	83.8	3.571	66.5	3.557	83.8	3.571	66.5
3 3/4-4	3.479	83.4	3.517	71.7	3.479	83.4	3.517	71.7
3 3/4-8	3.615	83.1	3.640	67.7	3.615	83.1	3.640	67.7
3 3/4-12	3.660	83.1	3.678	66.5	3.660	83.1	3.678	66.5
3 3/4-16	3.682	83.8	3.696	66.5	3.682	83.8	3.696	66.5
3 7/8-12	3.785	83.1	3.803	66.5	3.785	83.1	3.803	66.5
3 7/8-16	3.807	83.8	3.821	66.5	3.807	83.8	3.821	66.5
4-4	3.729	83.4	3.767	71.7	3.729	83.4	3.767	71.7
4-8	3.865	83.1	3.890	67.7	3.865	83.1	3.890	67.7
4-12	3.910	83.1	3.928	66.5	3.910	83.1	3.928	66.5
4-16	3.932	83.8	3.946	66.5	3.932	83.8	3.946	66.5
4 1/4-8	4.115	83.1	4.140	67.7	4.115	83.1	4.140	67.7
4 1/4-12	4.160	83.1	4.178	66.5	4.160	83.1	4.178	66.5
4 1/4-16	4.182	83.8	4.196	66.5	4.182	83.8	4.196	66.5
4 1/2-8	4.365	83.1	4.390	67.7	4.365	83.1	4.390	67.7
4 1/2-12	4.410	83.1	4.428	66.5	4.410	83.1	4.428	66.5
4 1/2-16	4.432	83.8	4.446	66.5	4.432	83.8	4.446	66.5
4 3/4-8	4.615	83.1	4.640	67.7	4.615	83.1	4.640	67.7

(Continued)

[a]Based on values as rounded off in preceding column.
[b]Based on length of engagement equal to the nominal diameter.

Table 23. *(Continued)* **Recommended Bores for 1B, 2B, and 3B Unified Threads, 1 $^1/_4$" - 6".** *(Source, Landis Threading Systems.)*

Size and TPI	UNC and UNF Classes 1B, 2B				UNC and UNF Class 3B			
	Min. Hole Size	% of Thread[a]	Max. Hole Size[b]	% of Thread[a]	Min. Hole Size	% of Thread[a]	Max. Hole Size[b]	% of Thread[a]
4 3/4-12	4.660	83.1	4.678	66.5	4.660	83.1	4.678	66.5
4 3/4-16	4.682	83.8	4.696	66.5	4.682	83.8	4.696	66.5
5-8	4.865	83.1	4.890	67.7	4.865	83.1	4.890	67.7
5-12	4.910	83.1	4.928	66.5	4.910	83.1	4.928	66.5
5-16	4.932	83.8	4.946	66.5	4.932	83.8	4.946	66.5
5 1/4-8	5.115	83.1	5.140	67.7	5.115	83.1	5.140	67.7
5 1/4-12	5.160	83.1	5.178	66.5	5.160	83.1	5.178	66.5
5 1/4-16	5.182	83.8	5.196	66.5	5.182	83.8	5.196	66.5
5 1/2-8	5.365	83.1	5.390	67.7	5.365	83.1	5.390	67.7
5 1/2-12	5.410	83.1	5.428	66.5	5.410	83.1	5.428	66.5
5 1/2-16	5.432	83.8	5.446	66.5	5.432	83.8	5.446	66.5
5 3/4-8	5.615	83.1	5.640	67.7	5.615	83.1	5.640	67.7
5 3/4-12	5.660	83.1	5.678	66.5	5.660	83.1	5.678	66.5
5 3/4-16	5.682	83.8	5.696	66.5	5.682	83.8	5.696	66.5
6-8	5.865	83.1	5.890	67.7	5.865	83.1	5.890	67.7
6-12	5.901	83.1	5.928	66.5	5.901	83.1	5.928	66.5
6-16	5.932	83.8	5.946	66.5	5.932	83.8	5.946	66.5

[a]Based on values as rounded off in preceding column.
[b]Based on length of engagement equal to the nominal diameter.

Table 24. Tap Drill Sizes for UNM Miniature Screw Threads. *(Source, Miniature Thread Specialists [MTS].)*

Thread Size	For Thread Cutting Taps						For Forming Taps
	To 2/3D		2/3 to 1.5D		1.5 to 3XD		
	Min.	Max.	Min.	Max.	Min.	Max.	
1.40UNM	0.0439	0.0455	0.0450	0.0471	0.0460	0.0481	0.0498
1.20UNM	0.0379	0.0393	0.0388	0.0406	0.0397	0.0415	0.0438
1.10UNM	0.0340	0.0354	0.0349	0.0367	0.0358	0.0376	0.0397
1.00UNM	0.0300	0.0314	0.0309	0.0327	0.0319	0.0337	0.0358
0.90UNM	0.0270	0.0283	0.0279	0.0295	0.0287	0.0304	0.0318
0.80UNM	0.0241	0.0252	0.0248	0.0263	0.0256	0.0270	0.0283
0.70UNM	0.0211	0.0221	0.0217	0.0231	0.0224	0.0237	0.0248
0.60UNM	0.0181	0.0190	0.0187	0.0198	0.0193	0.0204	0.0212
0.55UNM	0.0170	0.0178	0.0176	0.0186	0.0181	0.0191	0.0195
0.50UNM	0.0150	0.0158	0.0156	0.0166	0.0161	0.0171	
0.45UNM	0.0141	0.0147	0.0145	0.0154	0.0149	0.0158	
0.40UNM	0.0121	0.0127	0.0125	0.0134	0.0130	0.0138	

Tap drill sizes in inches.   *(Continued)*

Table 24. *(Continued)* **Tap Drill Sizes for UNM Miniature Screw Threads.**
(Source, Miniature Thread Specialists [MTS].)

Thread Size	For Thread Cutting Taps						For Forming Taps
	To 2/3D		2/3 to 1.5D		1.5 to 3XD		
	Min.	Max.	Min.	Max.	Min.	Max.	
0.35UNM	0.0105	0.0111	0.0109	0.0117	0.0113	0.0121	
0.30UNM	0.0089	0.0095	0.0093	0.0100	0.0096	0.0104	
00-90	0.0358	0.0380	0.0374	0.0395	0.0385	0.0405	0.0426
00-96	0.0371	0.0385	0.0381	0.0400	0.0390	0.0409	0.0429
00-112	0.0386	0.0398	0.0394	0.0410	0.0402	0.0419	0.0435
000-120	0.0261	0.0273	0.0270	0.0280	0.0277	0.0293	0.0306
0000-160	0.0151	0.0160	0.0157	0.0170	0.0164	0.0176	0.0184

Tap drill sizes in inches.

Table 25. Dimensions of Miniature Taps. *(Source, Miniature Thread Specialists [MTS].)*

Size	Three Flute Plug Taps			Forming Taps		
	Overall Length	Shank Dia.	Thread Length	Overall Length	Shank Dia.	Thread Length
1.40 UNM	30.0	2.0	8.0	25.0	2.0	5.5
1.20 UNM	25.0	1.5	7.0	25.0	1.5	4.5
1.10 UNM	25.0	1.5	7.0	25.0	1.5	4.5
1.00 UNM	25.0	1.5	6.0	25.0	1.5	4.5
.90 UNM	25.0	1.5	5.0	25.0	1.5	3.0
.80 UNM	25.0	1.5	4.0	25.0	1.5	2.5
.70 UNM	25.0	1.5	4.0	25.0	1.5	2.5
.60 UNM	22.0	1.0	3.0	22.0	1.0	1.5
.50 UNM	22.0	1.0	3.0	Not available in Forming Taps		
.45 UNM	22.0	1.0	2.0			
.40 UNM	22.0	1.0	2.0			
.35 UNM	22.0	1.0	2.0			
.30 UNM	22.0	1.0	2.0			
00-90	25.0	1.5	6.0	25.0	1.5	4.5
00-96	25.0	1.5	6.0	25.0	1.5	4.5
00-112	25.0	1.5	6.0	25.0	1.5	4.5
00-120	25.0	1.5	5.0	25.0	1.5	3.0
000-160	25.0	1.5	4.0	22.0	1.0	1.5

Tap dimensions in millimeters.

Table 26. Tap Drill Sizes for Screw Thread Insert Taps. *(Source, Fastcut Tool Corp.)*

Tap Size	—Aluminum—				—Steel, Plastic, Magnesium—			
	Tap Drill Size	Decimal Equiv. of Tap Drill	Minor Dia. Limits (After Tapping)		Tap Drill Size	Decimal Equiv. of Tap Drill	Minor Dia. Limits (After Tapping)	
			Min.	Max.			Min.	Max.
4-40	#31	.1200	.116	.121	#31	.1200	.119	.124
5-40	#30	.1285	.128	.133	#29	.1360	.131	.136
6-32	#25	.1495	.144	.150	#25	.1495	.148	.154
6-40	#26	.1470	.144	.149	#25	.1495	.148	.153
8-32	#17	.1730	.170	.176	#16	.1770	.174	.180
10-24	13/64	.2031	.199	.205	#5	.2055	.203	.209
10-32	#7	.2010	.196	.202	13/64	.2031	.200	.206
12-24	#2	.2210	.221	.227	#1	.2280	.225	.231
1/4-20	17/64	.2656	.261	.267	17/64	.2656	.265	.271
1/4-28	G	.2610	.257	.264	17/64	.2656	.261	.268
5/16-18	Q	.3320	.328	.334	Q	.3320	.331	.337
5/16-24	21/64	.3281	.323	.330	Q	.3320	.327	.334
3/8-16	X	.3970	.390	.398	X	.3970	.396	.402
3/8-24	25/64	.3906	.385	.392	25/64	.3906	.389	.396
7/16-14	29/64	.4531	.453	.463	15/32	.4687	.461	.471
7/16-20	29/64	.4531	.450	.458	29/64	.4531	.453	.461
1/2-13	33/64	.5156	.515	.525	17/32	.5312	.523	.533

All dimensions in inches.

Table 27. Pulley Tap Dimensions. *(Source, Hanson-Whitney Co.)*

Thread Size	Thread Length	Ground Length	Overall Length	Shank Dia.	Neck Length	Square Length	Square Size
1/4-20 UNC	1	1 1/2	6, 8	.255	3/8	5/16	.191
5/16-18 UNC	1 1/8	1 9/16	6, 8	.318	3/8	3/8	.238
3/8-16 UNC	1 1/4	1 5/8	6, 8, 10	.381	3/8	7/16	.286
7/16-14 UNC	1 7/16	1 11/16	6, 8	.444	7/16	1/2	.333
1/2-13 UNC	1 21/32	1 11/16	6, 8, 10, 12	.507	1/2	9/16	.380
5/8-11 UNC	1 13/16	2	6, 8, 10, 12	.633	5/8	11/16	.475
3/4-10 UNC	2	2 1/4	10, 12	.759	3/4	3/4	.569

All taps to G H3 limit. Neck length may vary with manufacturer. All dimensions in inches.

Table 28. Nut Tap Dimensions. *(Source, Hanson-Whitney Co.)*

Thread Size	Thread Length	Overall Length	Shank Dia.	Square Length	Square Size
1/4-20 UNC	1 5/8	5	.185	5/16	.139
5/16-18 UNC	1 13/16	5 1/2	.240	5/8	.180
3/8-16 UNC	2	6	.294	11/16	.220
7/16-14 UNC	2 3/8	6 1/2	.235	3/4	.259
1/2-13 UNC	2 1/2	7	.400	7/8	.300

All taps to G H3 limits. All dimensions in inches.

Table 29. Limits and Tolerances for Taper Pipe Taps (NPT, ANPT, and NPTF Forms).
(Source, OSG Tap and Die, Inc.)

Thread Limits					
Nom. Size	TPI	Gage Measurement		Taper per Foot	
		Projection*	Tolerance ±	Min.	Max.
1/16	27	.312	1/16	23/32	25/32
1/8	27	.312	1/16	23/32	25/32
1/4	18	.459	1/16	23/32	25/32
3/8	18	.454	1/16	23/32	25/32
1/2	14	.579	1/16	23/32	25/32
3/4	14	.565	1/16	23/32	25/32
1	11 1/2	.678	3/32	23/32	25/32
1 1/4	11 1/2	.686	3/32	23/32	25/32
1 1/2	11 1/2	.699	3/32	23/32	25/32
2	11 1/2	.667	3/32	23/32	25/32
2 1/2	8	.925	3/32	47/64	25/32
3	8	.925	3/32	47/64	25/32
3 1/2	8	.938	1/8	47/64	25/32
4	8	.950	1/8	47/64	25/32
Lead Tolerance					
Maximum allowable lead deviation within any two threads not more than one inch apart is ±.0005".					
Angle Tolerance					
Threads per Inch		Deviation in Half Angle			
8		±25°			
11 1/2 to 27 Inclusive		±30°			
Width of Flats at Tap Crest and Root					
TPI	Element	NPT		NPTF	
		Min.	Max.	Min.	Max.
27	Major Dia.	.0014	.0041	.0040	.0055
	Minor Dia.	-	.0041	-	.0040
18	Major Dia.	.0021	.0057	.0050	.0065
	Minor Dia.	-	.0057	-	.0050

* Projection is distance the small end of the tap protrudes through a taper thread ring gage.   *(Continued)*
All dimensions in inches.

Table 29. *(Continued)* **Limits and Tolerances for Taper Pipe Taps (NPT, ANPT, and NPTF Forms).** *(Source, OSG Tap and Die, Inc.)*

Width of Flats at Tap Crest and Root						
TPI	Element	NPT		NPTF		
		Min.	Max.	Min.	Max.	
14	Major Dia.	.0027	.0064	.0050	.0065	
	Minor Dia.	-	.0064	-	.0050	
11 1/2	Major Dia.	.0033	.0073	.0060	.0083	
	Minor Dia.	-	.0073	-	.0060	
8	Major Dia.	.0048	.0090	.0080	.0103	
	Minor Dia.	-	.0090	-	.0030	

Formulas for NPT Pipe Form					
Min. Major Dia. = Measured Pitch Dia. + A below.			Max. Major Dia. = Measured Pitch Dia. + B below.		
Min. Minor Dia. = Measured Pitch Dia. - B below.			Max. Minor Dia. = Measured Pitch Dia. - C below.		
TPI	A	B	C	D	E
27	.0267	.0296	.0257	.0234	.0251
18	.0408	.0444	.0401	.0377	.0395
14	.0535	.0571	.0525	.0515	.0533
11 1/2	.0658	.0696	.0647	.0614	.0649
8	.0966	.1000	.0946	-	-

Formulas for Dryseal NPTF Pipe Form					
Min. Major Dia. = Measured Pitch Dia. + D above.			Max. Major Dia. = Measured Pitch Dia. + E above.		
Min. Minor Dia. = Maximum or smaller.			Max. Minor Dia. = Measured Pitch Dia. - E above.		

Table 30. Reaming Data and Tap Drill Sizes for NPT, NPTF, and SAE Short Pipe Threads. *(Source, Fastcut Tool Corp.)*

Size	Projection NPT/NPTF		Projection SAE-SHORT		Ream Dia. Large End	Gage Width L₁	Reamed Length L₁+L₃	Tap Drill for Use With Reaming	Tap Drill for Use Without Reaming	Forming Tap Drill for Use Without Reaming	Min. Hole Depth for Standard Pipe Tap
	Min.	Max.	Min.	Max.							
1/16-27	.250	.375	.222	.259	.2515	.1600	.2711	15/64	C	I	9/16
1/8-27	.250	.375	.222	.259	.3340	.1615	.2726	21/64	Q	9.25mm	19/3
1/4-18	.397	.521	.333	.389	.4472	.2278	.3945	27/64	7/16	12.1mm	13/16
3/8-18	.392	.516	.333	.389	.5826	.240	.4067	9/16	9/16	5/8	13/16
1/2-14	.517	.641	.429	.500	.7213	.320	.5343	11/16	45/64	19.3mm	1 1/32
3/4-14	.503	.627	.429	.500	.9317	.339	.5533	57/64	29/32	31/32	1 1/32
1-11-1/2	.584	.772			1.1691	.400	.6609	1-1/8	1-9/64		1 1/4
1-1/4-11-1/2	.592	.780			1.5138	.420	.6809	1-15/32	1-31/64		1 9/32
1-1/2-11-1/2	.606	.792			1.7528	.420	.6809	1-45/64	1-23/32		1 5/16
2-11-1/2	.574	.760			2.2267	.436	.6969	2-3/16	2-3/16		1 9/32

All dimensions in inches.

Table 31. Recommended Bores for American National (NPT) and British Standard (BSTP) Taper Pipe Threads.*
(Source, Landis Threading Systems.)

NPT*			BSTP*		
Nominal Pipe Size	Thread Pitch	Bore Size	Nominal Pipe Size	Thread Pitch	Bore Size
1/8	27	11/32	1/8	28	5/16
1/4	18	7/16	1/4	19	27/64
3/8	18	19/32	3/8	19	9/16
1/2	14	23/32	1/2	14	11/16
3/4	14	15/16	3/4	14	29/32
1	11 1/2	1 5/32	1	11	1 1/8
1 1/4	11 1/2	1 1/2	1 1/4	11	1 15/32
1 1/2	11 1/2	1 23/32	1 1/2	11	1 25/32
2	11 1/2	2 3/16	2	11	2 5/32
2 1/2	8	2 5/8	2 1/2	11	2 25/32
3	8	3 1/4	3	11	3 9/32
3 1/2	8	3 3/4	3 1/2	11	3 3/4
4	8	4 1/4	4	11	4 1/4
5	8	5 5/16	5	11	5 1/4
6	8	6 3/8	6	11	6 1/4

*The above bore dimensions will result in a full thread. When using a collapsible tap, it is recommended that the hole be taper bored for best tool results. When boring, leave no more than .010" material per side below the minor diameter of the pipe thread to be tapped. All dimensions in inches.

Table 32. Limits and Tolerances for Straight Pipe Taps (NPS, NPSC, and NPSM Forms).
(Source, OSG Tap and Die, Inc.)

Thread Limits							
Nom. Size	TPI	Major Diameter			Pitch Diameter		
		Plug at Gaging Notch	Min. G	Max. H	Plug at Gaging Notch E	Min. K	Max. L
1/8	27	.3983	.4022	.4032	.3736	.3746	.3751
1/4	18	.5286	.5347	.5357	.4916	.4933	.4938
3/8	18	.6640	.8701	.6711	.6270	.6287	.6292
1/2	14	.8260	.8347	.8357	.7784	.7806	.7811
3/4	14	1.0364	1.0477	1.0457	.9889	.9906	.9916
1	11 1/2	1.2966	1.3062	1.3077	1.2386	1.2402	1.2412
1 1/4	11 1/2	1.6413	1.6507	1.6522	1.5384	1.5847	1.5862
1 1/2	11 1/2	1.8803	1.8897	1.8912	1.8223	1.8237	1.8252
2	11 1/2	2.3542	2.3639	2.3654	2.2963	2.2979	2.2294
2 1/2	8	2.8454	2.8604	2.8619	2.7622	2.7640	2.7660
3	8	3.4718	3.4868	3.4883	3.3885	3.3904	3.3924
3 1/2	8	3.9721	3.9872	3.9887	3.8888	3.8908	3.8928
4	8	4.4704	4.4855	4.4870	4.3871	4.3891	4.3911

(Continued)

Table 32. *(Continued)* Limits and Tolerances for Straight Pipe Taps (NPS, NPSC, and NPSM Forms). *(Source, OSG Tap and Die, Inc.)*

Lead Tolerance	
Maximum allowable lead deviation within any two threads not more than one inch apart is ±.0005".	
Angle Tolerance	
Threads per Inch	**Deviation in Half Angle**
8	± 25'
11 1/2 to 27 Inclusive	± 30'
Minor Diameter	
Minor diameter is equal to actual measured pitch diameter minus the value below.	
TPI	**Value**
27	.0257
18	.0401
14	.0525
11 1/2	.0647
8	.0946

All dimensions in inches. Taps made to these specifications are marked NPS and used for NPS, NPSC, and NPSM straight pipe threads.

Table 33. Limits and Tolerances for Straight Dryseal Pipe Taps (NPSF Form). *(Source, OSG Tap and Die, Inc.)*

		Thread Limits					
Nom. Size	**TPI**	**Major Diameter**			**Pitch Diameter**		**Minor Dia. Flat Max.**
		Plug at Gaging Notch E	**Min. G**	**Max. H**	**Min. K**	**Max. L**	
1/16	27	.2812	.3008	.3018	.2772	.2777	.004
1/8	27	.3736	.3932	.3942	.3696	.3701	.004
1/4	18	.4916	.5239	.5249	.4859	.4864	.005
3/8	18	.6270	.6593	.6603	.6213	.6218	.005
1/2	14	.7784	.8230	.8240	.7712	.7717	.005
3/4	14	.9889	1.0335	1.0345	.9817	.9822	.005
1	11 1/2	1.2386	1.2933	1.2943	1.2295	1.2305	.006
Lead Tolerance							
Maximum allowable lead deviation within any two threads not more than one inch apart is ±.0005".							
Angle Tolerance							
Threads per Inch				**Deviation in Half Angle**			
11 1/2 to 27 Inclusive				± 30'			
Minor Diameter							
Minor diameter is equal to actual measured pitch diameter minus the value below.							
TPI				**Value**			
27				.0251			
18				.0395			
14				.0533			
11 1/2				.0649			

All dimensions in inches. Taps made to these specifications are marked NPSF.

Table 34. Standard Pipe Tap Dimensions and Tolerances, Straight and Taper, Ground Thread.
(Source, Fastcut Tool Corp.)

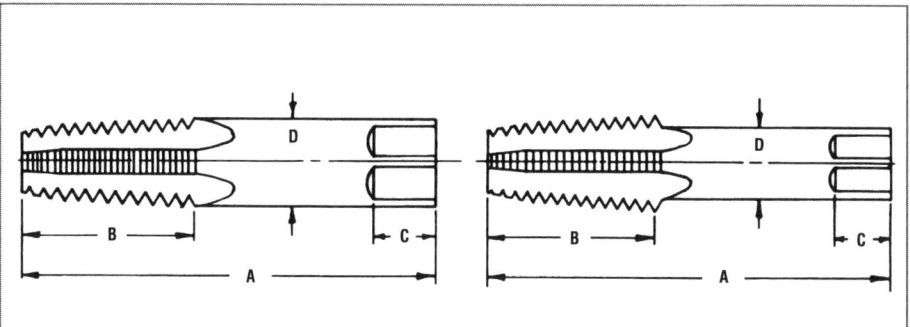

Dimensions					
Nominal Sizes Inches	Overall Length A	Length of Thread B	Length of Square C	Diameter of Shank D	Size of Square E
1/16	2-1/8	11/16	3/8	.3125	.234
1/8*	2-1/8	3/4	3/8	.3125	.234
1/8	2-1/8	3/4	3/8	.4375	.328
1/4	2-7/16	1-1/16	7/16	.5625	.421
3/8	2-9/16	1-1/16	1/2	.7000	.531
1/2	3-1/8	1-3/8	5/8	.6875	.515
3/4	3-1/4	1-3/8	11/16	.9063	.679
1	3-3/4	1-3/4	13/16	1.1250	.843
1-1/4	4	1-3/4	15/16	1.3125	.984
1-1/2	4-1/4	1-3/4	1	1.5000	1.125
2	4-1/2	1-3/4	1-1/8	1.8750	1.406
2-1/2	5-1/2	2-9/16	1-1/4	2.2500	1.687
3	6	2-5/8	1-3/8	2.6250	1.968
3-1/2	6-1/2	2-11/16	1-1/2	2.8125	2.108
4	6-3/4	2-3/4	1-5/8	3.0000	2.250

Tolerances			
Element	Range	Direction	Tolerance
Length Overall - A	1/16" to 3/4" incl.	Plus or Minus	1/32"
	1" to 4" incl.	Plus or Minus	1/16"
Length of Thread - B	1/16" to 3/4" incl.	Plus or Minus	1/16"
	1" to 1-1/4" incl.	Plus or Minus	3/32"
	1-1/2" to 4" incl.	Plus or Minus	1/8"
Length of Square - C	1/16" to 3/4" incl.	Plus or Minus	1/32"
	1" to 4" incl.	Plus or Minus	1/16"
Dia. Of Shank - D	1/16" to 1/8" incl.	Minus	.0015"
	1/4" to 1/2" incl.	Minus	.0020"
	3/4" to 1" incl.	Minus	.0020"
	1-1/4" to 4" incl.	Minus	.0030"
Size of Square - E	1/16" to 1/8" incl.	Minus	.0040"
	1/4" to 3/4" incl.	Minus	.0060"
	1" to 4" incl.	Minus	.0080"

*Small shank. All dimensions in inches.

Table 35. Tap Thread Limits and Tolerances, Unified Threads. See Notes Below. *(Source, OSG Tap and Die, Inc.)*

Threads Per Inch	A 0.130P	B	C			D			
			To 5/8" Inclusive	Over 5/8" to 2 1/2" Incl.	Over 2 1/2"	To 1" Incl.	Over 1" to 1 1/2" Incl.	Over 1 1/2" to 2 1/2" Incl.	Over 2 1/2"
80	.0016	.0011	.0005	.0010	.0015	.0005	.0010	.0010	.0015
72	.0018	.0012	.0005	.0010	.0015	.0005	.0010	.0010	.0015
64	.0020	.0014	.0005	.0010	.0015	.0005	.0010	.0010	.0015
56	.0023	.0016	.0005	.0010	.0015	.0005	.0010	.0010	.0015
48	.0027	.0018	.0005	.0010	.0015	.0005	.0010	.0010	.0015
44	.0030	.0017	.0005	.0010	.0015	.0005	.0010	.0010	.0015
40	.0032	.0019	.0005	.0010	.0015	.0005	.0010	.0010	.0015
36	.0036	.0021	.0005	.0010	.0015	.0005	.0010	.0010	.0015
32	.0041	.0020	.0010	.0010	.0015	.0005	.0010	.0010	.0015
28	.0046	.0023	.0010	.0010	.0015	.0005	.0010	.0010	.0015
24	.0054	.0027	.0010	.0010	.0015	.0005	.0010	.0015	.0015
20	.0065	.0032	.0010	.0010	.0015	.0005	.0010	.0015	.0015
18	.0072	.0036	.0010	.0010	.0015	.0005	.0010	.0015	.0015
16	.0081	.0041	.0010	.0010	.0015	.0005	.0010	.0015	.0020
14	.0093	.0046	.0010	.0015	.0015	.0005	.0010	.0015	.0020
13	.0100	.0050	.0010	.0015	.0015	.0005	.0010	.0015	.0020
12	.0108	.0054	.0010	.0015	.0015	.0005	.0010	.0015	.0020
11	.0118	.0059	.0010	.0015	.0020	.0005	.0010	.0015	.0020
10	.0130	.0065	-	.0015	.0020	.0005	.0010	.0015	.0020
9	.0144	.0072	-	.0015	.0020	.0005	.0010	.0015	.0020
8	.0162	.0081	-	.0015	.0020	.0005	.0010	.0015	.0020
7	.0186	.0093	-	.0015	.0020	.0010	.0010	.0020	.0025
6	.0217	.0108	-	.0015	.0020	.0010	.0010	.0020	.0025
5 1/2	.0236	.0118	-	.0015	.0020	.0010	.0015	.0020	.0025
5	.0260	.0130	-	.0015	.0020	.0010	.0015	.0020	.0025
4 1/2	.0289	.0144	-	.0015	.0020	.0010	.0015	.0020	.0025
4	.0325	.0162	-	.0015	.0020	.0010	.0015	.0020	.0025

Formulas for Unified Inch Screw Threads

Max. Major Dia. = Basic + A	Min. Major Dia. = Max. Major Dia. - B
Max. Pitch Dia. = Min. Pitch Dia. + D	Min. Pitch Dia. = Basic Pitch Dia. + C

Lead Tolerance

Maximum allowable lead deviation within any two threads not more than one inch apart is ± .0005"

Angle Tolerance

Threads per Inch	Deviation in Half Angle
4 to 5 1/2 Inclusive	± 20'
6 to 9 Inclusive	± 25'
10 to 80 Inclusive	± 30'

A (Constant) = 0.130P for all Pitches. **B** = Major Diameter Tolerance as follows: 1) For 48 through 80 tpi, B = 0.087P. 2) For 36 through 47 tpi, B = 0.076P. 3) For 4 through 35 tpi, B = 0.065P. **C** = Amount over Basic for Minimum Pitch Diameter. **D** = Pitch Diameter Tolerance. *Notes:* 1) When the tap major diameter must be determined from a specified pitch diameter, the maximum major diameter equals the minimum specified pitch diameter minus **C**, plus 0.64952P, plus **A**. 2) Dimensions and constants on this table apply only to ground thread taps having a thread lead angle not in excess of 5°, unless otherwise specified.

Table 36. Tap Thread Limits and Tolerances, Metric Threads, in Inches. See Notes Below. (Source, OSG Tap and Die, Inc.)

| Pitch | | Basic Thread Height in Inches | W | X | Tap Limits for Metric Threads, W, X, Y, and Z. Values are in Inches | | | | | | | |
| mm | Inch Equiv. | | | | Y | | | | Z | | | |
					M1.6 to M6.3 Incl.	Over M6.3 to M25 Incl.	Over M25 to M90 Incl.	Over M90	M1.6 to M6.3 Incl.	Over M6.3 to M25 Incl.	Over M25 to M90 Incl.	Over M90
0.3	.011811	.007671	.0009	.0010	.0015	.0015	.0020	.0020	.0006	.0006	.0008	.0008
0.35	.013779	.008950	.0011	.0010	.0015	.0015	.0020	.0020	.0006	.0006	.0008	.0008
0.4	.015748	.010229	.0013	.0010	.0015	.0020	.0020	.0020	.0006	.0006	.0008	.0010
0.45	.017716	.011507	.0014	.0010	.0015	.0020	.0020	.0020	.0006	.0008	.0008	.0010
0.5	.019685	.012786	.0016	.0010	.0015	.0020	.0020	.0025	.0006	.0008	.0010	.0010
0.6	.23622	.015343	.0019	.0010	.0020	.0020	.0025	.0025	.0008	.0008	.0010	.0010
0.7	.027559	.017900	.0022	.0016	.0020	.0020	.0025	.0025	.0008	.0008	.0010	.0010
0.75	.029527	.019178	.0024	.0016	.0020	.0025	.0025	.0030	.0008	.0010	.0010	.0012
0.8	.031496	.020457	.0025	.0016	.0020	.0025	.0025	.0030	.0008	.0010	.0010	.0012
0.9	.035433	.023014	.0028	.0016	.0020	.0025	.0025	.0030	.0008	.0010	.0010	.0012
1	.039370	.025572	.0032	.0016	.0025	.0025	.0030	.0030	.0010	.0010	.0012	.0012
1.25	.049212	.031964	.0039	.0025	.0025	.0025	.0030	.0035	.0010	.0012	.0012	.0012
1.5	.059055	.038357	.0047	.0025	.0025	.0025	.0030	.0035	.0010	.0012	.0012	.0016
1.75	.068897	.044750	.0055	.0025	-	.0030	.0035	.0040	-	.0012	.0016	.0016
2	.078740	.051143	.0063	.0025	-	.0035	.0035	.0040	-	.0016	.0016	.0016
2.5	.098425	.063929	.0079	.0025	-	.0035	.0040	.0045	-	.0016	.0016	.0020
3	.118110	.076715	.0095	.0039	-	.0040	.0040	.0050	-	.0016	.0020	.0020
3.5	.137795	.089501	.0110	.0039	-	.0040	.0045	.0050	-	.0016	.0020	.0020
4	.157480	.102286	.0126	.0039	-	.0040	.0045	.0055	-	.0020	.0020	.0025
4.5	.177165	.115072	.0142	.0039	-	-	.0050	.0055	-	.0020	.0020	.0025
5	.196850	.127858	.0158	.0039	-	-	.0050	.0060	-	-	.0025	.0025

(Continued)

Table 36. *(Continued)* Tap Thread Limits and Tolerances, Metric Threads, in Inches. See Notes Below. *(Source, OSG Tap and Die, Inc.)*

Pitch		Basic Thread Height in Inches	W	X	Tap Limits for Metric Threads, W, X, Y, and Z. Values are in Inches							
mm	Inch Equiv.				Y				Z			
					M1.6 to M6.3 Incl.	Over M6.3 to M25 Incl.	Over M25 to M90 Incl.	Over M90	M1.6 to M6.3 Incl.	Over M6.3 to M25 Incl.	Over M25 to M90 Incl.	Over M90
5.5	.216535	.140644	.0173	.0039	-	-	.0055	.0060	-	-	.0025	.0025
6	.236220	.153430	.0189	.0039	-	-	.0055	.0060	-	-	.0025	.0025

Formulas for Metric Screw Threads

Min. Major Dia. = Basic + W

Max. Major Dia. = Min. Major Dia. + X

Max. Pitch Dia. = Basic Pitch Dia. + Y

Min. Pitch Dia. = Max. Pitch Dia. - Z

Lead Tolerance

Maximum allowable lead deviation within any two threads not more than 25.4 mm apart is ± 0.013 mm

Angle Tolerance

Pitch in millimeters	Deviation in Half Angle
Over 0.25 to 2.5 Inclusive	± 30'
Over 2.5 to 4.0 Inclusive	± 25'
Over 4.0 to 6.0 Inclusive	± 20'

W = Constant to add to Basic Major Diameter (0.080/P). X = Major Diameter Tolerance. Y = Amount over Basic for Maximum Pitch Diameter. Z = Pitch Diameter Tolerance.

Notes: 1) When the tap major diameter must be determined from a specified tap pitch diameter, the minimum major diameter equals the maximum specified tap pitch diameter minus Y, plus the basic single height of the thread, plus W. 2) Dimensions and constants on this table apply only to 60° metric threads with a P/8 flat at the major diameter of the basic thread form.

Table 37. Tap Thread Limits and Tolerances, Metric Threads, in Millimeters. See Notes Below. *(Source, OSG Tap and Die, Inc.)*

Pitch	Basic Thread Height	W	X	Y				Z			
				M1.6 to M6.3 Incl.	Over M6.3 to M25 Incl.	Over M25 to M90 Incl.	Over M90	M1.6 to M6.3 Incl.	Over M6.3 to M25 Incl.	Over M25 to M90 Incl.	Over M90
0.3	.194856	.024	.025	.039	.039	.052	.052	.015	.015	.020	.020
0.35	.227332	.028	.025	.039	.039	.052	.052	.015	.015	.020	.020
0.4	.259808	.032	.025	.039	.052	.052	.052	.015	.015	.020	.025
0.45	.292284	.036	.025	.039	.052	.052	.052	.015	.020	.020	.025
0.5	.324760	.040	.025	.039	.052	.052	.065	.015	.020	.025	.025
0.6	.389712	.048	.025	.052	.052	.065	.065	.020	.020	.025	.025
0.7	.454664	.056	.041	.052	.052	.065	.065	.020	.020	.025	.025
0.75	.487140	.060	.041	.052	.065	.065	.065	.020	.025	.025	.031
0.8	.519616	.064	.041	.052	.065	.065	.078	.020	.025	.025	.031
0.9	.584568	.072	.041	.052	.065	.065	.078	.020	.025	.025	.031
1	.649520	.080	.041	.065	.065	.078	.078	.025	.025	.031	.031
1.25	.811900	.100	.064	.065	.065	.078	.091	.025	.031	.031	.041
1.5	.974280	.120	.064	.065	.078	.078	.091	.025	.031	.031	.041
1.75	1.13666	.140	.064	-	.078	.091	.104	-	.031	.041	.041
2	1.29904	.160	.064	-	.091	.091	.104	-	.041	.041	.041
2.5	1.62380	.200	.064	-	.091	.104	.117	-	.041	.041	.052
3	1.94856	.240	.100	-	.104	.104	.130	-	.041	.052	.052
3.5	2.27332	.280	.100	-	.104	.117	.130	-	.041	.052	.052
4	2.59808	.320	.100	-	.104	.117	.143	-	.052	.052	.064
4.5	2.92284	.360	.100	-	-	.130	.143	-	.052	.052	.064
5	3.24760	.400	.100	-	-	.130	.156	-	-	.064	.064

Tap Limits for Metric Threads, W, X, Y, and Z. Values are in Millimeters

(Continued)

Table 37. *(Continued)* **Tap Thread Limits and Tolerances, Metric Threads, in Millimeters.** See Notes Below. *(Source, OSG Tap and Die, Inc.)*

Tap Limits for Metric Threads, W, X, Y, and Z. Values are in Millimeters

Pitch	Basic Thread Height	W	X	Y				Z			
				M1.6 to M6.3 Incl.	Over M6.3 to M25 Incl.	Over M25 to M90 Incl.	Over M90	M1.6 to M6.3 Incl.	Over M6.3 to M25 Incl.	Over M25 to M90 Incl.	Over M90
5.5	3.57236	.440	.100	-	-	.143	.156	-	-	.064	.064
6	3.89712	.480	.100	-	-	.143	.156	-	-	.064	.064

Formulas for Metric Screw Threads

Min. Major Dia. = Basic + W   Max. Major Dia. = Min. Major Dia. + X

Max. Pitch Dia. = Basic Pitch Dia. + Y   Min. Pitch Dia. = Max. Pitch Dia. - Z

Lead Tolerance

Maximum allowable lead deviation within any two threads not more than 25.4 mm apart is ± 0.013 mm

Angle Tolerance

Pitch in millimeters	Deviation in Half Angle
Over 0.25 to 2.5 Inclusive	± 30'
Over 2.5 to 4.0 Inclusive	± 25'
Over 4.0 to 6.0 Inclusive	± 20'

W = Constant to add to Basic Major Diameter (0.080P). **X** = Major Diameter Tolerance. **Y** = Amount over Basic for Maximum Pitch Diameter. **Z** = Pitch Diameter Tolerance.

Notes: 1) When the tap major diameter must be determined from a specified tap pitch diameter, the minimum major diameter equals the maximum specified tap pitch diameter minus **Y**, plus the basic single height of the thread, plus **W**. 2) Dimensions and constants on this table apply only to 60° metric threads with a P/8 flat at the major diameter of the basic thread form.

Table 38. Cut Thread Tap Limits for Unified Threads. *(Source, Federal Standard GGG-T-60A.)*

Gage No. or Dia.	Threads per Inch			Major Diameter Limits			Pitch Diameter Limits		
	UNC	UNF	UNS	Basic	Min.	Max.	Basic	Min.	Max.
0	-	80	-	.0600	.0609	.0624	.0519	.0521	.0531
1	64	-	-	.0730	.0740	.0755	.0629	.0631	.0641
1	-	72	-	.0730	.0740	.0755	.0640	.0642	.0652
2	56	-	-	.0860	.0872	.0887	.0744	.0746	.0756
2	-	64	-	.0860	.0870	.0885	.0759	.0761	.0771
3	48	-	-	.0990	.1003	.1018	.0855	.0857	.0867
3	-	56	-	.0990	.1002	.1017	.0874	.0876	.0886
4	-	-	32	.1120	.1142	.1162	.0917	.0922	.0937
4	-	-	36	.1120	.1137	.1157	.0940	.0942	.0957
4	40	-	-	.1120	.1136	.1156	.0958	.0960	.0975
4	-	48	-	.1120	.1133	.1153	.0985	.0987	.1002
5	40	-	-	.1250	.1266	.1286	.1088	.1090	.1105
5	-	44	-	.1250	.1264	.1284	.1102	.1104	.1119
6	32	-	-	.1380	.1402	.1422	.1177	.1182	.1197
6	-	-	36	.1380	.1397	.1417	.1200	.1202	.1217
6	-	40	-	.1380	.1396	.1416	.1218	.1220	.1235
8	32	-	-	.1640	.1662	.1682	.1437	.1442	.1457
8	-	36	-	.1640	.1657	.1677	.1460	.1462	.1477
8	-	-	40	.1640	.1656	.1676	.1478	.1480	.1495
10	24	-	-	.1900	.1928	.1948	.1629	.1634	.1649
10	-	-	30	.1900	.1923	.1943	.1684	.1689	.1704
10	-	32	-	.1900	.1922	.1942	.1697	.1702	.1717
12	24	-	-	.2160	.2188	.2208	.1889	.1894	.1909
12	-	28	-	.2160	.2184	.2204	.1928	.1933	.1948
12	-	-	32	.2160	.2182	.2202	.1957	.1962	.1977
14	-	-	20	.2420	.2452	.2477	.2095	.2100	.2120
14	-	-	24	.2420	.2448	.2473	.2149	.2154	.2174
1/16	-	-	64	.0625	.0635	.0650	.0524	.0526	.0536
3/32	-	-	48	.0938	.0951	.0966	.0803	.0805	.0815
1/8	-	-	40	.1250	.1266	.1286	.1088	.1090	.1105
5/32	-	-	32	.1563	.1585	.1605	.1360	.1365	.1380
5/32	-	-	36	.1563	.1580	.1600	.1382	.1384	.1399
3/16	-	-	24	.1875	.1903	.1923	.1604	.1609	.1624
3/16	-	-	32	.1875	.1897	.1917	.1672	.1677	.1692
7/32	-	-	24	.2188	.2216	.2236	.1917	.1922	.1937
7/32	-	-	32	.2188	.2210	.2230	.1985	.1990	.2005
1/4	20	-	-	.2500	.2532	.2557	.2175	.2180	.2200
1/4	-	-	24	.2500	.2528	.2553	.2229	.2234	.2254

All dimensions in inches.

(Continued)

Table 38. *(Continued)* **Cut Thread Tap Limits for Unified Threads.** *(Source, Federal Standard GGG-T-60A.)*

Gage No. or Dia.	Threads per Inch			Major Diameter Limits			Pitch Diameter Limits		
	UNC	UNF	UNS	Basic	Min.	Max.	Basic	Min.	Max.
1/4	-	28	-	.2500	.2524	.2549	.2268	.2273	.2288
1/4	-	-	32	.2500	.2522	.2547	.2297	.2302	.2317
5/16	18	-	-	.3125	.3160	.3185	.2764	.2769	.2789
5/16	-	24	-	.3125	.3153	.3178	.2854	.2859	.2874
5/16	-	-	32	.3125	.3147	.3172	.2922	.2927	.2942
3/8	16	-	-	.3750	.3789	.3814	.3344	.3349	.3369
3/8	-	24	-	.3750	.3778	.3803	.3479	.3484	.3499
7/16	14	-	-	.4375	.4419	.4449	.3911	.3916	.3941
7/16	-	20	-	.4375	.4407	.4437	.4050	.4055	.4075
1/2	13	-	-	.5000	.5047	.5077	.4500	.4505	.4530
1/2	-	20	-	.5000	.5032	.5062	.4675	.4680	.4700
9/16	12	-	-	.5625	.5675	.5705	.5084	.5089	.5114
9/16	-	18	-	.5625	.5660	.5690	.5264	.5269	.5289
5/8	11	-	-	.6250	.6304	.6334	.5660	.5665	.5690
5/8	-	-	12	.6250	.6300	.6330	.5709	.5714	.5739
5/8	-	18	-	.6250	.6285	.6315	.5889	.5894	.5914
11/16	-	-	11	.6875	.6929	.6969	.6285	.6290	.6320
11/16	-	-	16	.6875	.6914	.6954	.6469	.6474	.6499
3/4	10	-	-	.7500	.7559	.7599	.6850	.6855	.6885
3/4	-	16	-	.7500	.7539	.7579	.7094	.7099	.7124
7/8	-	-	-	.8750	.8820	.8860	.8028	.8038	.8068
7/8	14	-	-	.8750	.8799	.8839	.8286	.8296	.8321
1	8	-	-	1.0000	1.0078	1.0118	.9188	.9198	.9228
1	-	12	-	1.0000	1.0055	1.0095	.9459	.9469	.9499
1	-	-	14	1.0000	1.0049	1.0089	.9536	.9546	.9571
1-1/8	7	-	-	1.1250	1.1337	1.1382	1.0322	1.0332	1.0367
1-1/8	-	12	-	1.1250	1.1305	1.1350	1.0709	1.0719	1.0749
1-1/4	7	-	-	1.2500	1.2587	1.2632	1.1572	1.1582	1.1617
1-1/4	-	12	-	1.2500	1.2555	1.2600	1.1959	1.1969	1.1999
1-3/8	6	-	-	1.3750	1.3850	1.3895	1.2667	1.2677	1.2712
1-3/8	-	12	-	1.3750	1.3805	1.3850	1.3209	1.3219	1.3249
1-1/2	6	-	-	1.5000	1.5100	1.5145	1.3917	1.3927	1.3962
1-1/2	-	12	-	1.5000	1.5055	1.5100	1.4459	1.4469	1.4499
1-3/4	5	-	-	1.7500	1.7602	1.7657	1.6201	1.6216	1.6256
2	4-1/2	-	-	2.0000	2.0111	2.0166	1.8557	1.8572	1.8612

All dimensions in inches.

Table 39. Recommended Speeds for Tapping Selected Workpiece Materials. *(Source, Weldon Co.)*

Material	Type	Hardness		Speed	
		Rc	BHN	SFM	SMM
Steel	Low Carbon Free Machining	<15	<180	20-65	6-19
	Medium to High Low Alloyed	<23	<240	20-50	6-15
	Castings and Forgings Heat Treatable Alloys	>24 <=38	>250 <=350	13-26	4-7
	Alloyed Tool Mold	>38 <=44	>350 <=420	6-13	2-4
Stainless Steel	Free Machining	<23	<240	5-15	1.5-4.5
	Heat and Corrosion Resistant	>24	<250	3-13	1-4
	Castings Precipitation Hardening	>38 <=44	>350 <=420	3-10	1-3
Cast Iron	Gray		<=220	20-65	6-19
	Nodular Chilled Meehanite Ductile		>=250	10-40	3-12
Aluminum	Pure Alloys			50-130	15-39
	Alloy Castings			30-65	9-19
Nickel Alloys	718 625 Hastelloy Monel Waspaloy Invar Incoloy	<=38	<=350	<13	<4
	718 Inconel A286	>38 <=44	>350 <=420	<13	<4
Titanium		<=38	<=350	<13	<4
Copper				25-50	7-15
Brass				30-65	9-19
Bronze		<44	<420	6-50	2-15
Zinc				25-50	7-15
Magnesium				30-65	9-19
Plastics	Thermoplastics			30-130	0-39
	Thermosetting Reinforced			15-40	4.5-12
Powder Metal		<44	<420	20-50	6-15
Graphite				15-35	4.5-10
Ceramics	Machinable Glasses			15-35	4.5-10
Special	Bronze Berrilium Copper			<13	<4
	Tungsten Ferro-Tic			5-15	1.5-4.5
	CPM10V			5-15	1.5-4.5

Table 40. Cutting Speeds: SFM Converted to RPM for Machine Screw, Fractional, and Pipe Tap Sizes. *(Source, Weldon Tool Co.)*

UNC/UNF Taps	NPT/NPTF Taps	Surface Feet per Minute																	
		Revolutions per Minute																	
		5'	10'	15'	20'	25'	30'	40'	50'	60'	70'	80'	90'	100'	110'	120'	130'	140'	150'
0		318	637	955	1273	1592	1910	2546	3183	3820	4456	5093	5729	6366	7003	7639	8276	8913	9549
1		273	546	819	1046	1308	1570	2093	2617	3140	3663	4186	4710	5233	5756	6279	6805	7326	7849
2		212	424	637	888	1110	1333	1777	2221	2665	3109	3554	3999	4442	4886	5330	5774	6218	6662
3		191	382	573	772	964	1157	1543	1929	2315	2701	3086	3472	3858	4244	4629	5015	5401	5787
4		174	347	521	682	853	1023	1364	1705	2046	2387	2728	3069	3411	3751	4092	4434	4775	5116
5		147	294	441	611	764	917	1222	1528	1833	2139	2445	2750	3056	3361	3667	3973	4278	4584
6		136	273	409	553	691	829	1106	1382	1659	1935	2212	2488	2766	3042	3318	3595	3871	4148
8		119	239	358	466	583	699	932	1165	1398	1631	1864	2097	2330	2563	2796	3029	3262	3495
10		101	201	302	402	502	603	804	1005	1205	1406	1607	1808	2009	2210	2411	2612	2813	3014
12		87	174	260	354	442	531	707	884	1061	1238	1415	1592	1769	1945	2122	2300	2476	2653
1/4		76	153	229	306	382	458	611	764	917	1070	1222	1375	1528	1681	1833	1986	2139	2292
5/16		62	123	185	245	306	367	489	611	733	856	978	1100	1222	1345	1467	1589	1711	1833
3/8		50	101	151	204	255	305	407	509	611	713	815	917	1019	1120	1222	1324	1426	1528
7/16	1/8	43	87	130	175	219	262	349	437	524	611	698	786	873	960	1048	1135	1222	1310
1/2	-	38	76	115	153	191	229	305	382	458	535	611	688	764	840	917	993	1070	1146
9/16	1/4	34	68	102	137	172	206	274	342	410	478	547	616	683	752	820	888	952	1020
5/8	-	32	64	96	122	153	183	244	306	367	428	489	550	611	672	733	794	856	917
11/16	3/8	28	55	83	111	138	167	222	278	333	389	444	500	556	611	667	722	778	833
3/4	-	25	51	76	102	128	153	203	255	305	357	407	458	509	560	611	662	713	764
7/8	1/2	22	43	65	87	109	131	175	218	262	306	350	392	437	480	524	568	611	655
1	-	19	38	57	76	96	115	153	191	230	268	305	344	382	420	458	497	535	573
1-1/8	3/4	17	34	51	68	84	102	136	170	204	238	272	306	340	373	407	441	475	509

(Continued)

Table 40. (Continued) Cutting Speeds: SFM Converted to RPM for Machine Screw, Fractional, and Pipe Tap Sizes. (Source, Weldon Tool Co.)

UNC/UNF Taps	NPT/NPTF Taps	Surface Feet per Minute																	
		5'	10'	15'	20'	25'	30'	40'	50'	60'	70'	80'	90'	100'	110'	120'	130'	140'	150'
		Revolutions per Minute																	
1-1/4	-	15	31	46	61	76	92	122	153	183	214	244	275	305	336	367	397	428	458
1-3/8	1	14	28	42	56	69	83	111	139	167	194	222	250	278	306	333	361	389	417
1-1/2	-	13	25	38	51	63	76	102	127	153	178	204	229	255	280	305	331	356	382
1-5/8		12	23	35	47	59	71	94	118	141	165	188	212	235	259	282	306	329	353
1-3/4		11	22	33	44	55	65	87	109	131	153	175	196	218	240	262	284	306	327
1-7/8		10	20	30	41	51	61	81	102	122	143	163	183	204	224	244	265	285	306
2		9	19	29	38	48	57	76	96	115	134	153	172	191	210	229	248	267	287

Table 41. Cutting Speeds: SFM Converted to RPM for Metric Tap Sizes. *(Source, Weldon Tool Co.)*

Tap Size	Surface Feet per Minute																	
	5'	10'	15'	20'	25'	30'	40'	50'	60'	70'	80'	90'	100'	110'	120'	130'	140'	150'
	Revolutions per Minute																	
M 1.0	490	979	1469	1959	2449	2938	3918	4897	5877	6856	7836	8815	9795	10774	11754	12733	13713	14692
M 2.0	242	484	725	967	1209	1451	1934	2418	2901	3385	3868	4352	4835	5319	5803	6286	6770	7253
M 3.0	162	324	486	647	809	971	1295	1619	1942	2266	2590	2914	3237	3561	3885	4208	4532	4856
M 3.5	138	277	415	554	692	830	1107	1384	1661	1938	2214	2491	2768	3045	3322	3599	3875	4152
M 4.0	122	243	365	487	608	730	973	1217	1460	1703	1946	2190	2433	2676	2920	3163	3406	3650
M 5.0	97	194	291	388	485	582	776	970	1163	1357	1551	1745	1939	2133	2327	2521	2715	2909
M 6.0	81	162	243	324	405	486	647	809	971	1133	1295	1457	1619	1781	1942	2104	2266	2428
M 7.0	69	138	208	277	346	415	554	692	830	969	1107	1246	1384	1522	1661	1799	1938	2076
M 8.0	61	121	182	243	303	364	485	606	728	849	970	1091	1213	1334	1455	1577	1698	1819
M 10.0	48	97	145	194	242	291	388	485	582	679	776	873	970	1067	1163	1260	1357	1454
M 12.0	40	81	121	162	202	243	324	405	486	567	647	728	809	890	971	1052	1133	1214
M 14.0	35	69	104	139	173	208	277	347	416	485	555	624	693	763	832	901	971	1040
M 16.0	30	61	91	121	152	182	243	303	364	424	485	546	606	667	728	788	849	910
M 18.0	27	54	81	108	135	162	216	269	323	377	431	485	539	593	647	700	754	808
M 20.0	24	49	73	97	121	146	194	243	291	340	388	437	485	534	582	631	680	728
M 22.0	22	44	66	88	110	132	176	221	265	309	353	397	441	485	529	573	617	661
M 24.0	20	40	61	81	101	121	162	202	243	283	323	364	404	445	485	526	566	606
M 27.0	18	36	54	72	90	108	144	180	216	252	287	323	359	395	431	467	503	539
M 30.0	16	32	49	65	81	97	129	162	194	226	259	291	323	356	388	420	453	485

Table 42. Material/Coolant Tapping Recommendations. *(Source, Landis Threading Systems.)*

Material	Recommended Coolant No. See Coolant Schedule
Aluminum	4
Bakelite	2
Brass-All	4
Plastic	2
Copper	4
Fiber	2
Iron	
Cast	3
Malleable	1-6
Wrought	1-6
Magnesium	4-6
Monel Metal	5-7
Nickel	5-7
Rubber	2
Steel-All	6

Coolant Schedule
1. Soluble Oil
2. Dry
3. Dry or Soluble Oil
4. Light Mineral Oil
5. Mineral Lard (20% lard)
6. Sulphur Base
7. Sulphurized Oil

Table 43. Surface Treatments and Coatings for High-Speed Steel Taps. *(Source, Fastcut Tool Corp.)*

Description	Characteristics	Application
Nitride Approx. Hardness 1200 HV Rc 72	Consists of a thin, hardened case 0.0005" to 0.002" deep on the surface of the tool to resist abrasion and reduce galling.	Can be used in most abrasive materials, both ferrous and nonferrous. Not recommended where chipping may be a problem.
Double Nitride Approx. Hardness 1400 HV Rc 74	Consists of a higher hardened case on the surface of the tool to resist abrasion and reduce galling. Prone to brittleness and chipping.	Can be used on nonmetallic, highly abrasive materials such as Bakelite, plastics, hard rubber, and fibers.
Steam Oxide Approx. Hardness No change from Base Material	Consists of a layer of ferrous oxide on the surface of the tool which has good lubricant retaining properties. Improves toughness by relieving grinding stresses.	Can be used in low carbon, stainless, and free machining steels. Not recommended for use in soft, nonferrous materials where it may cause galling.
Nitride and Oxide Approx. Hardness 1200 HV Rc 72	A combination of two treatments which produces the favorable characteristics of both resistance to abrasion and galling.	Can be used in iron and cast iron, stainless, and high tensile steels. Not recommended for use in nonferrous materials where it may cause galling.
Chrome Plate Cr, Hard Chromium Approx. Hardness 1200 HV Rc 72	Consists of a very thin layer of hard chromium on the surface of the tool which reduces friction and prevents galling.	Can be used on most ferrous, nonferrous and nonmetallic materials. While unlikely, it may cause galling in high chromium stainless steels.

(Continued)

Table 43. *(Continued)* **Surface Treatments and Coatings for High Speed Steel Taps.** *(Source, Fastcut Tool Corp.)*

Description	Characteristics	Application
Titanium Nitride TiN, PVD Process Approx. Hardness 2400 HV *Rc 86	Consists of a very hard coating on the surface of the tool which has outstanding wear resistance, reduces friction, and prevents galling.	Can be used on most ferrous, nonferrous and nonmetallic materials. While unlikely, it may cause galling in titanium and titanium alloys.
Titanium Carbonitride TiCN, PVD Process Approx. Hardness 3000 HV *Rc 94	Consists of an extremely hard coating on the surface of the tool which has outstanding wear resistance, reduces friction, and prevents galling.	Can be used on most ferrous, nonferrous and abrasive materials. Very effective at higher speeds. While unlikely, it may cause galling in titanium and titanium alloys.
Chromium Carbide CrC, PVD Process Approx. Hardness 1850 HV Rc 80	Consists of a very hard coating on the surface of the tool which has excellent wear resistance, reduces friction, and prevents galling.	Can be used on titanium, titanium alloys, exotic materials, and die cast aluminum. Very effective at higher speeds and in many tapping applications. Under certain conditions it may cause galling in wrought aluminum.
Chromium Nitride CrN, PVD Process Approx. Hardness 1750 HV Rc 79	Consists of a very hard coating on the surface of the tool which has excellent wear resistance, reduces friction, and prevents galling.	Can be used on titanium, titanium alloys, nickel-base alloys, and copper alloys. Very effective at higher speeds and in many tapping applications. Under certain conditions it may cause galling in wrought aluminum.
Titanium Aluminum Nitride TiAlN, PVD Process Approx. Hardness 2600 HV *Rc 89	Consists of an extremely hard coating on the surface of the tool which has outstanding wear resistance, reduces friction, and prevents galling. Forms an aluminum oxide layer at high speeds and elevated temperatures.	Can be used on titanium, titanium alloys, nickel-base alloys, stainless steel and cast iron. Very effective at higher speeds and in some tapping applications. Not recommended for wrought aluminum, copper, and brass.

*Theoretical values for approximate comparison to the Vickers Hardness values.

Note: While most surface treatments and coatings have antigalling properties, they may cause galling in materials composed of or containing identical base elements. Also, steam oxide and some coatings may cause galling in soft materials such as aluminum.

Solving tapped hole size problems

Tapped holes (internal threads) are inspected with *GO Plug* and *NOT GO Plug* Thread Gages (NOT GO gages are sometimes called NO GO gages). See *Figure 3*. To determine that a thread is acceptable for production, *GO* thread gages must be able to freely enter and pass through the length of the thread, thereby verifying all thread components except the minor diameter. *NOT GO* gages, on the other hand, must fail to pass through (or in some instances enter) the length of the thread. Therefore, a GO/NOT GO check is a pass/fail test only. The size of the part being inspected is never measured.

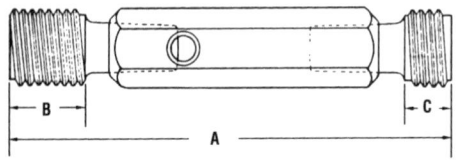

Figure 3. Typical thread plug gage. "A" is overall length, "B" indicates the GO member, and "C" shows the NOT GO member.

GO Plug gages check the extent of the tolerance, as applied to a specific screw thread, in the direction of the limit of maximum material and therefore represent the minimum limit of internal threads. The ideal thread will be a threaded counterpart of the *GO Plug* gage, made to precisely the same material limits. *Thread Snap* gages and *Indicating* gages have two significant advantages over *GO Plug* gages: 1) threads are inspected by applying the gage contacts within the internal thread at various discrete positions, and 2) threads can be inspected for roundness and taper my making additional checks along the thread length.

NOT GO Plug gages inspect the extent of tolerance in the direction of minimum material, which is the high limit of the internal thread. Separate gages are required to check pitch and major and minor diameter at minimum material. Thorough checking of the thread calls for a *NOT GO Thread Snap* gage. Two methods can be used for checking minimum material: functional diameter gaging or single element gaging.

Functional Diameter Gaging uses the NOT GO Plug gage when proper functioning of the thread assembly only requires control of the functional diameter of the thread at the minimum material limit. The gage should not engage for more than three threads. A NOT GO Thread Snap and Indicating gages are also applied at several points along the thread to assure compliance.

Single Element Gaging provides an economical control over the thread variables of lead, uniformity of helix, taper, roundness, and surface condition. Since Indicating gages give size value on production threads, their use alone provides reliable direction for adjusting any production irregularities.

Problems with thread and hole dimensions are indicated when GO gages fail to pass through the thread length, and when NOT GO gages do pass through, either partially or over the thread's full length. The following cases indicate trouble with tapped hole sizes. *Figure 4* shows a typical thread design and the limits controlled by GO and NOT GO gages.

NOT GO gage enters full length. Obviously, this is an indication of an oversize hole, and the pitch diameter of the tap should be inspected to determine how much it is cutting oversize. If the amount is excessive, consider regrinding the tap to provide more hook or less width of land. This will require experimentation, and a more expeditious solution might to apply a different style of tap with a smaller pitch diameter to control the problem.

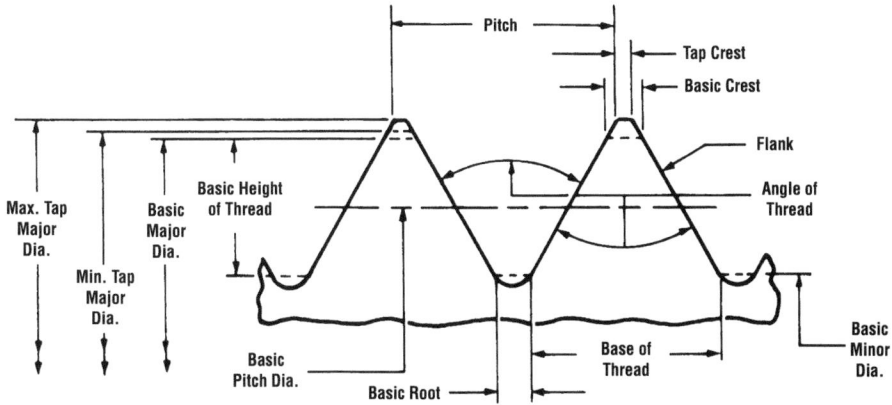

Figure 4. Thread dimensions controlled by thread plug gages. (Source, Kennametal Inc.)

Incorrect hook angle can also cause the tap to push, rather than cut, the thread, which will cause problems with the surface finish of the hole. If the hole is bell-mouthed, the likely cause is the alignment of the spindle and fixture. Spindle alignment can be especially troublesome when using a tapping head in a drill press. Another cause of tapping oversize holes is excessive feed pressure being applied to the tap. A lead screw can be employed or the pressure should be adjusted.

NOT GO gage partially enters the hole. In most cases, this indicates that the tap holder is not concentric with the spindle.

GO gage only partially enters the hole. This is an indication that the thread is tapered or the lead is out of specification. Taper normally indicates that the tap and the hole are not properly aligned. An error in the lead is caused by a defect in the tap and can only be corrected by replacement.

GO gage will not enter hole. Clearly, the first thing to check is the size of the tap and its pitch diameter. Check the tap's and the gage's markings to determine that the correct tap and gage is being used. The wall thickness of the workpiece may be causing the hole to distort. Try tapping the hole with roughing and finishing taps or a tap with more flutes. Another possibility is that the fixture is distorting the part. In this case, either reduce the clamping pressure or try using spiral point taps to remedy the problem.

Gages are available in several materials that can provide extra wear resistance. Those made of chrome plate and carbide are more resistant to wear than tool steel gages. GO gages, especially, are subject to wear through normal use and should be calibrated frequently. Calibration should be done at 68°F (20° C), and calibration schedules should be based on frequency of use, part abrasiveness, tolerance, and applicable quality procedures. With plug gages, GO is normally a plus tolerance and NOT GO is a minus tolerance. Gagemaker's tolerances for different classes of gages (XX, X, Y, Z, and ZZ are symbols for standard gage classes) are given in **Table 44**. See **Table 45** for GO and NOT GO Plug gage diameters for Class 2B and Class 3B Unified threads. **Table 46** gives GO and NOT GO Plug gage diameters for Class 6H Metric threads in both millimeters and inches.

Table 44. Gagemaker's Tolerance Chart for GO/NOT GO Plug Gages. *(Source, Vermont Gage Co.)*

Range	Class				
	XX	X	Y	Z	ZZ
Inch (Dimensions in Inches)					
.0009" to .8250"	.000020	.000040	.000070	.000100	.000200
.8251" to 1.5100"	.000030	.000060	.000090	.000120	.000240
1.5101" to 2.5100"	.000040	.000080	.000120	.000160	.000320
2.5101" to 4.5100"	.000050	.000100	.000150	.000200	.000400
4.5101" to 6.5100"	.000065	.000130	.000190	.000250	.000500
6.5101" to 9.0100"	.000080	.000160	.000240	.000320	.000640
9.0101" to 12.2600"	.000100	.000200	.000300	.000400	.000800
Metric (Dimensions in Millimeters)					
1.00mm to 21.00mm	.0005	.0010	.0018	.0030	.0050
21.01mm to 38.00mm	.0008	.0015	.0023	.0030	.0060
38.01mm to 64.00mm	.0010	.0020	.0030	.0040	.0080
64.01mm to 115.00mm	.0013	.0025	.0038	.0050	.0100
115.01mm to 165.00mm	.0017	.0033	.0048	.0060	.0130
165.01mm to 230.00mm	.0020	.0041	.0061	.0080	.0160
230.01mm to 300.00mm	.0025	.0051	.0076	.0100	.0200

Table 45. GO/NOT GO Thread Plug Gage Pitch Diameters for Class 2B and Class 3B Unified Threads.
(Source, Kennametal Inc.)

Nominal Size	GO Basic All Classes All Series	Unified Pitch Diameters				
		NOT GO		X Tol. GO = + NOT GO = -	X Lead Tol.	X ± Half Angle Tol.
		Class 2B	Class 3B			
#0-80 UNF	.0519	.0542	.0536	.0002	.0002	0°30'
#1-64 UNC	.0629	.0655	.0648	.0002	.0002	0°30'
#1-72 UNF	.0640	.0665	.0659	.0002	.0002	0°30'
#2-56 UNC	.0744	.0772	.0765	.0002	.0002	0°30'
#2-64 UNF	.0759	.0786	.0779	.0002	.0002	0°30'
#3-48 UNC	.0855	.0885	.0877	.0002	.0002	0°30'
#3-56 UNF	.0874	.0902	.0895	.0002	.0002	0°30'
#4-40 UNC	.0958	.0991	.0982	.0002	.0002	0°20'
#4-48 UNF	.0985	.1016	.1008	.0002	.0002	0°30'
#5-40 UNC	.1088	.1121	.1113	.0002	.0002	0°20'
#5-44 UNF	.1102	.1134	.1126	.0002	.0002	0°20'
#6-32 UNC	.1177	.1214	.1204	.0003	.0003	0°15'
#6-40 UNF	.1218	.1252	.1243	.0002	.0002	0°20'
#8-32 UNC	.1437	.1475	.1465	.0003	.0003	0°15'
#8-36 UNF	.1460	.1496	.1487	.0002	.0002	0°20'
#10-24 UNC	.1629	.1672	.1661	.0003	.0003	0°15'
#10-32 UNF	.1697	.1736	.1726	.0003	.0003	0°15'
#12-24 UNC	.1889	.1933	.1922	.0003	.0003	0°15'
#12-28 UNF	.1928	.1970	.1959	.0003	.0003	0°15'
#12-32 UNEF	.1957	.1998	.1988	.0003	.0003	0°15'
1/4-20 UNC	.2175	.2224	.2211	.0003	.0003	0°15'
1/4-28 UNF	.2268	.2311	.2300	.0003	.0003	0°15'
1/4-32 UNEF	.2297	.2339	.2328	.0003	.0003	0°15'
5/16-18 UNC	.2764	.2817	.2803	.0003	.0003	0°10'
5/16-24 UNF	.2854	.2902	.2890	.0003	.0003	0°15'
5/16-32 UNEF	.2922	.2964	.2953	.0003	.0003	0°15'
3/8-16 UNC	.3344	.3401	.3387	.0003	.0003	0°10'
3/8-24 UNF	.3479	.3528	.3516	.0003	.0003	0°15'
3/8-32 UNEF	.3547	.3591	.3580	.0003	.0003	0°15'
7/16-14 UNC	.3911	.3972	.3957	.0003	.0003	0°10'
7/16-20 UNF	.4050	.4104	.4091	.0003	.0003	0°15'
7/16-28 UNEF	.4143	.4189	.4178	.0003	.0003	0°15'
1/2-13 UNC	.4500	.4565	.4548	.0003	.0003	0°10'
1/2-20 UNF	.4675	.4731	.4717	.0003	.0003	0°15'
1/2-28 UNEF	.4768	.4816	.4804	.0003	.0003	0°15'
9/16 UNC	.5084	.5152	.5135	.0003	.0003	0°10'
9/16 UNF	.5264	.5323	.5308	.0003	.0003	0°10'
9/16-24 UNEF	.5354	.5405	.5392	.0003	.0003	0°15'
5/8-11 UNC	.5660	.5732	.5714	.0003	.0003	0°10'
5/8-18 UNF	.5889	.5949	.5934	.0003	.0003	0°10'
5/8-24 UNEF	.5979	.6031	.6018	.0003	.0003	0°15'
11/16-24 UNEF	.6604	.6656	.6643	.0003	.0003	0°15'
3/4-10 UNC	.6850	.6927	.6907	.0003	.0003	0°10'
3/4-16 UNF	.7094	.7159	.7143	.0003	.0003	0°10'
3/4-20 UNEF	.7175	.7232	.7218	.0003	.0003	0°15'
13/16-20 UNEF	.7800	.7857	.7843	.0003	.0003	0°15'

(Continued)

Table 45. *(Continued)* **GO/NOT GO Thread Plug Gage Pitch Diameters for Class 2B and Class 3B Unified Threads.** *(Source, Kennametal Inc.)*

Nominal Size	GO Basic All Classes All Series	Unified Pitch Diameters				
		NOT GO		X Tol. GO = + NOT GO = -	X Lead Tol.	X ± Half Angle Tol.
		Class 2B	Class 3B			
7/8-9 UNC	.8028	.8110	.8089	.0003	.0003	0°10'
7/8-14 UNF	.8286	.8356	.8339	.0003	.0003	0°10'
7/8-20 UNEF	.8425	.8482	.8468	.0003	.0003	0°15'
15/16-20 UNEF	.9050	.9109	.9094	.0003	.0003	0°15'
1"-8 UNC	.9188	.9276	.9254	.0004	.0004	0°05'
1"-12 UNF	.9459	.9535	.9516	.0003	.0003	0°10'
1"-14 UNS	.9536	.9609	.9590	.0003	.0003	0°10'
1"-20 UNEF	.9675	.9734	.9719	.0003	.0003	0°15'
1 1/16-18 UNEF	1.0315	1.0264	1.0310	.0003	.0003	0°10'
1 1/8-7 UNC	1.0322	1.0416	1.0393	.0004	.0004	0°05'
1 1/8-12 UNF	1.0709	1.0787	1.0768	.0003	.0003	0°10'
1 1/8-18 UNEF	1.0889	1.0951	1.0935	.0003	.0003	0°10'
1 3/16-18 UNEF	1.1514	1.1577	1.1561	.0003	.0003	0°10'
1 1/4-7 UNC	1.1572	1.1668	1.1644	.0004	.0004	0°05'
1 1/4-12 UNF	1.1959	1.2039	1.2019	.0003	.0003	0°10'
1 1/4-18 UNEF	1.2139	1.2202	1.2186	.0003	.0003	0°10'
1 5/16-18 UNEF	1.2764	1.2827	1.2811	.0003	.0003	0°10'
1 3/8-6 UNC	1.2667	1.2771	1.2745	.0004	.0004	0°05'
1 3/8-12 UNF	1.3209	1.3291	1.3270	.0003	.0003	0°10'
1 3/8-18 UNEF	1.3389	1.3452	1.3436	.0003	.0003	0°10'
1 7/16-18 UNEF	1.4014	1.4079	1.4062	.0003	.0003	0°10'
1 1/2-6 UNC	1.3917	1.4022	1.3996	.0004	.0004	0°05'
1 1/2-12 UNF	1.4459	1.4542	1.4522	.0003	.0003	0°10'
1 1/2-18 UNEF	1.4639	1.4704	1.4687	.0003	.0003	0°10'

Table 46. GO/NOT GO Thread Plug Gage Pitch Diameters for Class 6H Metric Threads. *(Source, Vermont Gage Co.)*

Size	Pitch	Millimeters		Inch Equivalent	
		GO	NOT GO	GO	NOT GO
M1.6	.35	1.373	1.458	.05406	.05740
M1.8	.35	1.573	1.658	.06193	.06528
M2.0	.40	1.740	1.830	.06850	.07205
M2.2	.45	1.908	2.003	.07512	.07886
M2.5	.45	2.208	2.303	.08693	.09067
M3.0	.50	2.675	2.775	.10531	.10925
M3.5	.60	3.110	3.222	.12244	.12685
M4.0	.70	3.545	3.663	.13957	.14421
M4.5	.75	4.013	4.131	.15799	.16264
M5.0	.80	4.480	4.605	.17638	.18136
M5.5	.75	5.013	5.131	.19736	.20201
M6.0	1.00	5.350	5.500	.21063	.21654
M7.0	1.00	6.350	6.500	.25000	.25591

(Continued)

Table 46. *(Continued)* GO/NOT GO Thread Plug Gage Pitch Diameters for Class 6H Metric Threads. *(Source, Vermont Gage Co.)*

Size	Pitch	Millimeters		Inch Equivalent	
		GO	NOT GO	GO	NOT GO
M8.0	1.25	7.188	7.348	.28299	.28929
M9.0	1.25	8.188	8.348	.32236	.32866
M10.0	1.50	9.026	9.206	.35535	.36244
	1.25	9.188	9.348	.36173	.36803
	.75	9.513	9.645	.37453	.37972
M11.0	1.50	10.026	10.206	.39472	.40181
M12.0	1.75	10.863	11.063	.42768	.43555
	1.50	11.026	11.216	.43409	.44157
	1.25	11.188	11.368	.44047	.44756
M14.0	2.00	12.701	12.913	.50004	.50838
	1.50	13.026	13.216	.51283	.52031
M15.0	1.50	14.026	14.216	.55220	.55968
	1.00	14.350	14.510	.56496	.57126
M16.0	2.00	14.701	14.913	.57878	.58712
	1.50	15.026	15.216	.59157	.59905
M17.0	1.50	16.026	16.216	.63094	.63842
	1.00	16.350	16.510	.64370	.65000
M18.0	2.50	16.376	16.600	.64472	.65354
	1.50	17.026	17.216	.67031	.67779
M20.0	2.50	18.376	18.600	.72346	.73228
	1.50	19.026	19.216	.74905	.75653
	1.00	19.350	19.510	.76181	.76811
M22.0	2.50	20.376	20.600	.80220	.81102
	1.50	21.026	21.216	.82779	.83557
M24.0	3.00	22.051	22.316	.86815	.87858
	2.00	22.701	22.925	.89374	.90256
M25.0	2.00	23.701	23.925	.93311	.94193
	1.50	24.026	24.226	.94590	.95378
M26.0	1.50	25.026	25.226	.98527	.99315
	1.00	25.350	25.520	.99803	1.00472
M27.0	3.00	25.051	25.316	.98626	.99669
	2.00	25.701	25.925	1.01185	1.02067
M28.0	2.00	26.701	26.925	1.05122	1.06004
	1.50	27.026	27.226	1.06401	1.07189
	1.00	27.350	27.520	1.07677	1.08346
M30.0	3.50	27.727	28.007	1.09161	1.10264
	2.00	28.701	28.925	1.12996	1.13878
	1.50	29.026	29.226	1.14276	1.15063
M32.0	2.00	30.701	30.925	1.20870	1.21752
	1.50	31.026	31.226	1.22149	1.22937
M33.0	3.50	30.727	31.007	1.20972	1.22075
	2.00	31.701	31.925	1.24807	1.25689
M35.0	1.50	34.026	34.226	1.33960	1.34748
M36.0	4.00	33.402	33.702	1.31504	1.32685
	3.00	34.051	34.316	1.34059	1.35102
	2.00	34.701	34.925	1.36618	1.37500

(Continued)

Table 46. *(Continued)* **GO/NOT GO Thread Plug Gage Pitch Diameters for Class 6H Metric Threads.** *(Source, Vermont Gage Co.)*

Size	Pitch	Millimeters		Inch Equivalent	
		GO	NOT GO	GO	NOT GO
M38.0	1.50	37.026	37.226	1.45771	1.46559
M39.0	4.00	36.402	36.702	1.43320	1.44500
	3.00	37.051	37.316	1.45870	1.46910
	2.00	37.701	37.925	1.48429	1.49311

Thread Rolling

Manufacturing male threads

Generally speaking, threads are manufactured either by rolling or cutting. In fact, over 90% of commercially available bolts up to 2 inches (51 millimeters) were manufactured by rolling. In addition, it is an economically efficient process. Since rolled threads are cold formed, no material is removed from the blank, resulting in significant material savings: for example, to cut a $^1/_2$ – 13 UNC bolt requires almost 20% more material than rolling the same bolt. The information in this handbook pertaining to thread rolling was provided by Landis Threading Systems, who have been producing threads by this method for almost 100 years.

Rolling versus cutting in thread production

Rolling is a chipless cold forming process capable of producing threads at high threading speeds with good tool life. Not all threads should be rolled. The following application considerations should be taken into account before beginning any rolling operation.

1. In the majority of instances, materials that are well suited to rolling do not cut well, and those well suited to cutting do not roll well.
2. In order to be suitable for rolling, the material should have an elongation factor of at least 12%. This is the characteristic that allows material to be plastically and permanently deformed. Rollability of materials will be covered in more detail later in this section.
3. The design or end use of the workpiece may dictate the use of a certain material. Since cast iron, for example, does not flow, it is not a rollable material and its use would require cutting the threads.
4. Since thread roll dies will accept only a specific volume of rolled material, using an oversize blank will result in excessive material flow that overfills the dies and can result in die breakage. Therefore, the blank diameter must be tightly controlled within specified limits. Recommended blank diameters for classes of threads are given later in this section.

Where finish is a prime consideration, rolling is superior to cutting. Some workpiece prints specify a finish on the thread, and rolling can be relied on to produce finishes 32 μin. (.813 μm) or less. Cutting, on the other hand, rarely produces threads better than 63 μin. (1.6 μm). Strength is another factor to consider. Because rolling does not cut across the grain structure, and also due to the high compressive forces created by the process, rolled threads are as much as 20% stronger than cut threads. Cutting shears the grain at every thread form, resulting in weakened threads (see *Figure 5a*).

When threading from the end of a workpiece, it is often necessary to roll close to a shoulder, and in such instances cutting is the preferred option because the cutting edge

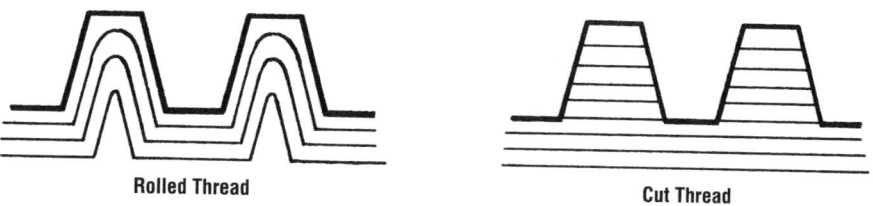

Rolled Thread   **Cut Thread**

Figure 5a. Cross section of rolled thread and cut thread. (Source, Landis Threading Systems.)

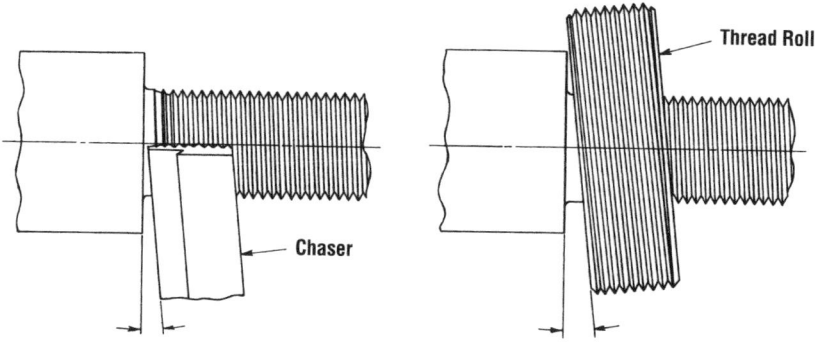

Figure 5b. Difference in proximity to shoulder in cutting (left) and rolling (right).
(Source, Landis Threading Systems.)

lies at the helix angle on the center line of the work and tangentially stops at that point. The thread roll also lies at the helix angle, and contacts the workpiece on the center line; but instead of stopping at the point of contact, it must continue full circle. In doing so it is often unable to produce a thread close enough to the shoulder because the curve of the roll beyond the point of contact is beyond the center line and will strike the shoulder, as shown in *Figure 5b*.

On certain nonferrous metals (zinc and aluminum die castings, some aluminum and brass bars), thread cutting can be performed at relatively high speeds. However, on steels (especially the higher alloys), it is necessary to stay within the 5 to 60 sfm (1.5 to 19 smm) range. By comparison, a rolled thread can usually be initially cut at 100 sfm (30 smm) and in some instances taken as high as 400 sfm (120 smm). At higher speeds, caution should be exercised because the axial travel is very rapid, and short threads or threading going up to the shoulder can be difficult to control. Where the rolling head is being used on an automatic screw machine, and the threading operation does not govern the cycle time, using high speeds and having the head idle for some other operation is counterproductive. A speed of 100 sfm (30 smm) is usually sufficient for most operations.

When very deep, coarse, or multiple threads are required, cutting is preferred over rolling. It is possible when using cutting heads to make more than one pass without damage to the tool. This distributes the chip load over more cutting edges, thus improving both finish and tool life. This is not good practice in thread rolling because any tendency to workharden on the first pass will require the roll to work much harder on the second. Multiple threads introduce the problem of long throats and higher pressures than single starts. In cutting, the second pass can be adjusted so that it removes just enough metal to smooth up the thread if desired. On multiple start threads, especially Acme and trapezoidal, where the helix angle is high and the rear flank on the chaser has a negative rake, the throat teeth can be ground individually to provide positive rake for better cutting action. With a rolling head, the rolls cannot be turned to an angle great enough to accommodate many types of multiple threads.

Threading behind the shoulder can be done using a straddle type thread rolling attachment. This tool generally operates from the cross slide of an automatic screw machine simultaneous with other operations, and it is capable of rolling on the collet side of a shoulder even when the shoulder is larger than the thread. Attachments permit rolling very close to a shoulder or into relatively small relief. They can roll either straight or tapered threads, and have all the advantageous characteristics of other roll threading procedures.

Although thread cutting is acceptable for long runs, one of its advantages is that it requires less downtime than rolling in changeover for short runs. Also, there is a significant difference in the initial cost of a thread chaser compared to the greater cost of a roll. Many times, the workpiece for cut threads does not require the accuracy that a rolling blank does because the chaser can be made with a throat chamfer below the root of the new thread that acts as a hollow mill to remove excess metal.

Where it isn't necessary to have the full thread diameter on the workpiece back of the thread, such as with bent bolts, it is less expensive, from a blank material standpoint, to roll the threads. For instance, if rolling stock is purchased to blank diameter, there is approximately about a 20% saving in weight realized for $1/2$–13 UNC, and more than 15% for 1–8 UNC over the bar necessary for a cut thread.

Steel stock containing small amounts of lead is used, especially on automatic screw machines, for increasing machinability by as much as 25%. These steels do not work well for roll threading because the lead inclusions have a tendency to be squeezed out of the parent material as flakes, and will both foul the coolant and do serious damage to the surface finish. The same result occurs when rolling high sulfur steels.

Rollability of materials

A material's hardness and its percent of elongation factor indicate whether it can be plastically and permanently deformed. Generally, 32 R_c is the hardest material that is practical to roll. The elongation factor should be at least 12%. Other factors that influence a material's rollability are its yield point, microstructure, and the degree and speed at which it workhardens. Also, its modulus of elasticity, the nonmetallic content (in steel), and the workpiece diameter and pitch can be factors.

By its nature, thread rolling generates a great deal of pressure and heat, and coolant should be applied at the point of roll contact with the workpiece. Applying the coolant to the point of contact is especially important with revolving type heads which tend to sling the coolant away from the work. Cutting oil offers superior lubrication and works well for certain applications, but soluble oils are superior in carrying heat from the operation. See the "Lubricoolants" heading below for lubricant/coolant recommendations.

Basic open-hearth steels in the soft state are rollable providing carbon content does not exceed 1.5%. Sulfurized steels are rollable depending on the severity of cold working to be done and the percentage of sulfur content. While sulfur additives enhance the machinability of a cutting operation, they are not desirable for cold forming. Therefore, sulfur content should not exceed .13%, as higher rates cause extremely hard sulfur inclusions that require higher rolling pressures. In addition to shortening die life, inclusions resist cold forming, cause flaking, and can result in roll breakage.

Soft and malleable leaded steels are thought to have good rollability. However, while desirable for enhancing machinability, it is not conclusive that additives are compatible with rolling. Lead inclusions are softer than the base material, and lead inclusions result in intermittent and varying die loading that contributes to poor die life. Therefore, lead content should be no more than 0.1%.

Stainless steels are prone to workhardening. Most of the 400 Series steels are considered best for rolling, with 410 yielding the best results. Because of their extreme workhardening characteristics, the 300-Series should be avoided if possible even though their chemical analysis suggests that they would be appropriate.

To what degree workhardening will occur depends on the analysis of the material, its initial hardness, the severity of deformation, and the number of cold working cycles imposed on the workpiece. The greatest amount of deformation with the greatest hardness occurs in the root area. The least occurs on the flanks. Since the grain structure flows along the thread contour, the deeper and finer grain in the root reflects the high degree of hardness in that area. Workhardening resulting from rolling is superficial, and rolled materials retain core hardness 0.030" to 0.040" (0.76 mm to 1.02 mm) beneath the surface. Where a material has higher workhardening tendencies, hardening can extend as far as 0.150" (3.81 mm) below the surface. Die life will proportionally increase/decrease in direct ratio to any increase in workhardening. See **Tables 47** through **49**, which list the rollability factor and die life longevity for various materials.

Table 47. Rollability of Materials: Carbon and Alloy Steels. *(Source, Landis Threading Systems.)*

Material Designation	Thread Finish	Proportional Die Life				Remarks
		Soft	R_c 15-24	R_c 25-32	R_c 33+	
AISI 1008-1095	E	H	H-M	M	L	Excellent rollability
AISI 1108-1151 AISI 1211-1215 AISI B1111-B1113	G F F	H M H	H-M	M		These are free machining steels with high sulfur content. The highest sulfur materials (1110, 1144, and 1200) should be avoided when possible.
AISI 1330-1345 AISI 4012-4047	E E	H H-M	M M	M-L M-L	L L	These are medium alloy steels such as manganese, molybdenum, chrome, and nickel. Workhardening of material requires higher pressures, and some reduction in roll life over the 1000-1200 series will be experienced.
AISI 4118-4161	E	H-M	M	M-L	L	
AISI 4320-4340 AISI 4419 AISI 4615-4626 AISI 4718-4720 AISI 4815-4820	E	H-M	M	M-L	L	
AISI 5015-5060 AISI 5115-5160 AISI E51100-E52100	E	H-M	M	M-L	L	
AISI 6118-6150	E	H-M	M	M-L	L	
AISI 8615-8655 AISI 8720-8740 AISI 8822 AISI 9255-9260	E	H-M	M	M-L	L	
Stainless 302-304 309-317	E		M-L			These are nonhardenable austenitic steels containing higher quantities of nickel and chromium. High workhardening occurs with higher percent alloys. Also nonmagnetic. Recommend Carpenter #10 wherever possible. Material does not seam.
Stainless 305, 321, 347, 348 Carpenter & 12-20CB	E		M	L		
Stainless Carpenter #10	E		H-M	M-L		

(Continued)

Table 47. *(Continued)* **Rollability of Materials: Carbon and Alloy Steels.** *(Source, Landis Threading Systems.)*

Material Designation	Thread Finish	Proportional Die Life				Remarks
		Soft	R$_c$ 15-24	R$_c$ 25-32	R$_c$ 33+	
Stainless 329-430F-446	E		M-L	L		Nonhardenable ferritic chromium stainless but magnetic. Lower work hardening but higher pressures required due to carbon.
Stainless 430-443	E		M	L		
Stainless 414-420F-440F	E		M-L	L		Hardenable martensitic chromium steels, magnetic. Most suited for rolling of the stainless grades, low workhardening.
Stainless 410-431-440C-501-502	E		H-M	L		
High Speed T1, M1, M2	E		M-L	L		Not generally recommended, however can be rolled under proper conditions.
Nitralloy 135-230	E		M	M-L	L	Not rollable after nitriding.

Letter designations for finish: E-excellent, G-good, F-fair, P-poor.
Letter designations for die life: H-high, M-medium, L-low.
Elongation factor: generally acceptable results can be achieved when percent elongation equals twelve (12) or more.

Table 48. **Rollability of Materials: Wrought Copper and Copper Alloys.** *(Source, Landis Threading Systems.)*

Material Designation		Alloy Name	Max. Hardness	Finish	Die Life	Remarks
SAE No.	ASTM No.					
CA102	B124 #12	Oxygen Free Copper	R$_F$ 40	E	H	More than 90% copper.
CA110	B124 #12	Electrolytic Copper (ETP)	R$_F$ 40	E	H	Excellent rollability.
CA122	B124 #12	Phosphorus Deoxidized (DHP)	R$_F$ 45	E	H	
CA210	B36 #1	Gilding 95%	R$_B$40	E	H	Copper-zinc alloys basically good for rolling except when zinc exceeds 30%. This tends to produce poor finish as indicated in CA270-CA280.
CA220	B36 #2	Commercial Bronze 90%	R$_B$42	E	H	
CA230	B36 #3	Red Brass	R$_B$55	E	H	
CA240	B36 #4	Low Brass 80%	R$_B$55	E	H	
CA260	B36 #6	Cartridge Brass 70%	R$_B$60	E	H	
CA270	B36 #8	Yellow Brass	R$_B$55	F	H	
CA280	B135 #5	Muntz Metal	R$_B$78	P	M	
CA314	B140-B	Leaded Comm. Bronze	R$_B$65	P	M-H	Copper zinc alloys with lead added for improved machining characteristics. Poor to fair for rolling. Higher lead produces poorer thread finish and is not recommended for rolling.
CA335	B121 #2	Low Lead Brass	R$_B$60	G	M-H	
CA340	B121 #3	Medium Leaded Brass	R$_B$60	G	M-H	
CA342	B121 #5	High Leaded Brass	R$_B$55	P	M	
CA345		Thread Rolling Brass	R$_B$75	G	M-H	
CA353	B121 #5	High Leaded Brass 62%	R$_B$55	P	M	
CA356	B121 #6	Extra High Leaded Brass	R$_B$55	P	L-M	
CA360	B16	Free Cutting Brass	R$_B$70	P	L	
CA365 to CA368	B171	Leaded Muntz Metal	R$_B$70	P	M	
CA370	B135 #6	Free Cutting Muntz Metal	R$_B$70	P	L	
CA377	B124 #2	Forging Brass	R$_B$65	P	L	
CA385		Architectural Bronze	R$_B$65	P	L	
CA443 to CA445	B171	Inhibited Admiralty	R$_B$75	E	H	Copper zinc alloy with 1.0% tin excellent rollability.

(Continued)

Table 48. *(Continued)* **Rollability of Materials: Wrought Copper and Copper Alloys.**
(Source, Landis Threading Systems.)

Material Designation		Alloy Name	Max. Hardness	Finish	Die Life	Remarks
SAE No.	**ASTM No.**					
CA464 to CA467 CA485	B124 #3 B21-C	Naval Brass Leaded Naval Brass	R_B75 R_B80	P-F P	M L	Copper zinc alloy with lead and tin not conducive to good rolling characteristics. Alternate material should be used.
CA502 CA510 CA521 CA524 CA544	B105 B139-A B139-C B139-D B139-B2	Phosphor Bronze E Phosphor Bronze A Phosphor Bronze C Phosphor Bronze D Free Cutting Phosphor Bronze	R_B50 R_B65 R_B70 R_B70 R_B70	G G P-F P P	H H M-H M L	Copper tin alloy generally good for rolling but increasing tin content reduces rollability. CA544 contains some lead & zinc thereby reducing its rollability.
CA606 CA614 CA617 CA360 CA639	B150-3 B150-1 B150-2	Aluminum Bronze Aluminum Bronze D Aluminum Bronze Aluminum Bronze Aluminum Silicon Bronze	R_B70 R_B70 R_B70 R_B70 R_B75	G G P P G	M M L-M L L	Copper aluminum alloy fair to good rolling characteristics. Increased quantities of silicon nickel, introduce workhardening and reduce rollability.
CA651 CA655	B98-B B98-A	Low Silicon Bronze B High Silicon Bronze A	R_B70 R_B75	E FG	M M	Copper with silicon as basic alloy. Average rollability.
CA675	B138-A	Manganese Bronze A	R_B70	P	M	High zinc alloy, alternate material should be used.
CA706 CA715	B111 B111	Copper Nickel-10% Copper Nickel-30%	R_B70 R_P70	G G	M-H M	High nickel alloy reduces rollability proportionally.
CA745 CA752 CA754 CA757 CA770	B151-E B151-A B151-D B151-B	Nickel Silver 65-10 Nickel Silver 65-18 Nickel Silver 65-15 Nickel Silver 65-12 Nickel Silver 55-18	R_B70 R_B70 R_B70 R_P70 R_B70	E G-E E E G	H H H H M	Copper with zinc and nickel as alloy, rollability good to excellent. As alloy increases, rollability decreases.

Copper Casting Alloys, Annealed

Copper casting alloys, in the annealed condition are, for the most part, rated as poor rollability and poor die life. The copper alloys with basic quantities of tin, zinc, or silicon rate slightly better in die life with poor to fair finish. It is recommended that these materials be avoided where possible, and should only be considered for low production quantities.

Letter designations for finish: E-excellent, G-good, F-fair, P-poor.
Letter designations for die life: H-high, M-medium, L-low.
Elongation factor: generally acceptable results can be achieved when percent elongation equals twelve (12) or more.

Table 49. Rollability of Materials: Wrought Aluminum and Aluminum Alloys. *(Source, Landis Threading Systems.)*

SAE No.	ASTM No.	Condition	Max. Hardness	% Elongation	Finish	Die Life	Remarks
1100-0	990A	Annealed	R_B23	45	E	H	99% aluminum
1100-H12	990A	1/4 Hard	R_B28	25	E	H	recommended for rolling.
110-H14	990A	1/2 Hard	R_B32	20	G-E	H	Workhardens very slowly,
1100-H16	990A	3/4 Hard	R_B38	17	G	H	cannot be heat treated.
1100-H18	990A	Full Hard	R_B44	15	F-G	M	Major alloy is silicon.
2011-T3	CB60A	Heat treated and cold worked	R_B95	15	FG	M-H	Lower quality finish is a result of lead and
2011-T6C	B60A	Heat treated and aged	R_B97	17	F	M-H	Bismuth alloys,
2011-T8	CB60A	Heat treated, cold worked & aged	R_B100	12	P-F	M-H	not generally recommended for rolling.
2014-0	CS41A	Annealed	R_B45	18	G	M-H	Copper, silicon,
2014-T4	CS41A	Heat treated and aged	R_B105	20	G-E	M-H	manganese major alloys,
2014-T6	CS41A	Heat treated and aged	R_B135	13	F	M-H	higher strength requires greater roll pressure.
2017-0	CM41A	Annealed	R_B45	22	E	H	Good rollability. Most
2017-T4	CM41A	Heat treated and aged	R_B105	22	E	H	commonly used for
2024-0	CG42A	Annealed	R_B47	22	E	H	rolling.
2024-T3	CG42A	Heat treated, cold worked	R_B120	18	E	H	
2024-T4	CG42A	Heat treated and aged	R_B120	18	E	H	
2117-T4	CG30A	Heat treated and aged	R_B70	27	E	H	
3003-0	MIA	Annealed	R_B28	40	E	H	99% aluminum,
3003-H12	MIA	1/4 Hard	R_B35	20	G-E	H	recommended for rolling.
3003-H14	MIA	1/2 Hard	R_B40	16	G	H	Work hardens very
3003-H16	MIA	3/4 Hard	R_B47	14	F	M	slowly, cannot be heat
3003-H18	MIA	Full Hard	R_B55	10	P-F	L-M	treated. Major alloy is manganese.
5052-0	CR20A	Annealed	R_B47	30	E	H	Fair to good rollability in
5052-H32	CR20A	1/4 Hard	R_B60	18	G	M	the lower hardness
5052-H34	CR20A	1/2 Hard	R_B68	14	F	M	condition, major alloy
5052-H36	CR20A	3/4 Hard	R_B73	10	P-F	L-M	manganese with
5053-H38	CR20A	Full Hard	R_B77	8	P	L	chromium.
5056-0	GM50A	Annealed	R_B65	35	E	H	Major alloy magnesium.
5056-H18	GM50A	Strain Hardened	R_B105	10	P	L-M	Recommend rolling in
5056-H38	GM50A	Strain Hardened and Stabilized	R_B100	15	P-F	L-M	annealed condition only.
6061-0	GS11A	Annealed	R_B30	30	E	H	Good to excellent
6061-T4	GS11A	Heat treated and aged	R_B25	65	G-E	H	rollability in conditions.
6061-T6	GS11A	Heat treated and aged	R_B17	95	G	H	
7075-0	ZG62A	Annealed	R_B60	16	F	H	Generally not
7075-T6	ZG62A	Heat treated and aged	R_B150	11	P	M	recommended for rolling.

Letter designations for finish: E-excellent, G-good, F-fair, P-poor.
Letter designations for die life: H-high, M-medium, L-low.
Elongation factor: generally acceptable results can be achieved when percent elongation equals twelve (12) or more.

Preparing and controlling blank diameters

The principles of rolling techniques will vary somewhat depending upon whether a rolling head, attachment, or machine is being used. However, the basic principles apply to all types of equipment. One requirement is the need to maintain the blank diameter within specific limits. Failure to do so can result in premature wear or early failure of the dies. Rolling dies can be compared to a vessel that holds liquid: they can only hold an amount of flowed metal equal to (but not more than) their displacement. While a vessel can overflow, rolling dies cannot. Since the roll life is much greater than that of a cutting tool, extra effort spent monitoring and controlling the blank to prevent breakage or premature wear is more than offset by the performance maintained. Unlike the cutting process which machines excess material away from the workpiece, rolling requires that the blank contain only enough material to fill the die cavity and form the thread. When too much material is present, the die cavity is overfilled. Conversely, too little material will result in an incorrectly formed thread.

An oversize blank exerts pressure against the dies, forcing them outward and damaging the die or producing an oversize thread. Oversize threads cannot be corrected by sizing the head smaller as that remedy would aggravate the condition and produce even greater pressure on the dies. The blank diameters given in **Tables 50** through **55** are recommended starting points. Depending on the material and its metallurgical makeup, the final blank diameter must finally be established by actual rolling. Although not always visible to the naked eye, rolled threads have a seam that is the result of material being flowed from the root area up to and folded in at the crest. Depending on the coarseness of the pitch and how tightly the thread is being rolled, the seam will be more visible on some threads than on others. Better die life will be realized by starting with the minimal acceptable blank diameter (which advantageously allows for growth) and accepting a larger seam. Seams have no bearing on thread strength, but they are often objected to on the basis of appearance. If appearance is a factor, threads can be rolled tighter (die life will be shortened) or the thread can be rolled with an oversize addendum which can be ground away to remove the seam.

With balanced threads such as the UN 60° included thread angle, the volume of the thread above the pitch diameter will approximate that below. The volume below is that which is flowed and displaced upward to form the addendum. Thus, it is apparent that, with a balanced form, the prepared blank's outside diameter will approximate the thread's pitch diameter. Starting at the minimum recommended blank diameter allows a certain amount of blank growth due to tool wear and allows the longest possible running time before the maximum allowable blank diameter is reached. Therefore, in practice, the blank diameter should be less than the maximum thread pitch diameter, and tolerances on the blank diameter should be as small as practical. When rolling shorter length threads, it may be necessary to increase the blank diameter slightly to compensate for endwise stretching.

Die thread form chipping can be minimized by beveling the end of the blank as shown in *Figure 6*. The small bevel diameter at the blank end should be equal to the thread minor diameter minus 0.10" (0.255 mm). Rolling will force the end threads outward and the finished bevel will be approximately 45°. Blanks should be as round and as straight as possible as rolling will not correct these inaccuracies. Avoid variations in diameter along the blank length. These cause uneven pressure distribution that can overload the rolls and result in premature failure. Maintain the smoothest finish possible on the blank, as it will influence the final thread smoothness. Finally, be aware that blank diameter may vary according to material hardness.

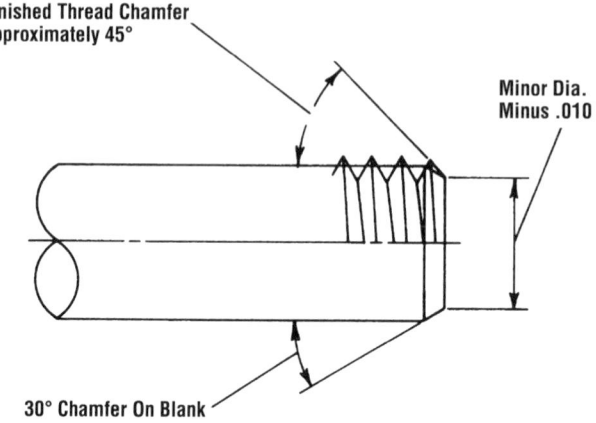

Figure 6: Proper beveling of the end of a blank. (Source, Landis Threading Systems.)

Lubricoolant recommendations for thread rolling

The pressure developed between the workpiece and the dies of a thread rolling machine can theoretically approach 500,000 psi (3,450 MPa) depending on the type of thread and workpiece hardness. During the rolling operation, the thread surfaces of the dies and workpiece are running together at a surface speed differential, because the largest surface of each die is in contact with and running on the smallest diameter of the thread. Accordingly, slippage under extreme pressure occurs between these two surfaces at diameters larger and smaller than the resolved, effective, rotating diameter which—for purposes of consideration—can be regarded as the pitch diameter. While this contributes to the heat generated by the rolling operation, the greatest amount of total heat is generated by the material being flowed or deformed to produce the thread.

Depending on the efficiency of the thread rolling machine, most of the input energy is converted to heat that must be dissipated by the dies and the lubricant, which acts as a coolant (hence, lubricoolant). The remaining heat in the workpiece is dissipated into the environment. It is desirable to remove as much heat as possible, as quickly as possible. Thus the lubricant with the best wetting and extreme pressure properties should be used.

Water, with a specific heat of 1.000, will absorb heat better than any other readily available substance. The specific heat of most oils ranges between .350 and .550, so water will clearly *absorb* twice as much heat as most oils. For this reason, it is desirable to use a water emulsion for thread rolling. Since the die spindle bearings in some thread rolling machines rely on the lubricant for cooling, the properties of the lubricoolant must have adequate bearing lubrication properties. However, cutting oils and emulsions should not contain active sulfur or chlorine, as both have been shown to have an erosive effect on thread rolling dies.

The following guidelines should be followed when selecting a lubricoolant.

1. The lubricant should be of the water emulsion or water miscible type for maximum heat dissipation.

2. Generally, a mix of 9 parts of water to 1 part concentrate works best. The mix may vary somewhat, but no combination should contain less than 4 parts of water to 1 part concentrate.

(Text continued on p. 310)

Table 50. Recommended Steel Blank Diameters for Parallel Rolled UNF and UNC Class 2 Threads. *(Source, Landis Threading Systems.)*

| Size | | Steel | | | | | | | | | | | |
| O.D. | Pitch | 10-50 C Soft | | 30-50 C Soft | | 30-50 C or Alloy 15-25 R_c | | 30-50 C or Alloy 26-32 R_c | | Stainless/Chrome Nickel 300 Series | | Stainless/Chrome 400 Series* | |
		Min.	Max.	Min.	Max.	Min.	Max.	Min.	Max.	Min.	Max.	Min.	Max.
#0	80	0.0504	0.0498	0.0507	0.0501	0.0509	0.0503	0.0511	0.0505	0.0513	0.0507	0.0515	0.0509
#1	72	0.0623	0.0617	0.0627	0.0621	0.0629	0.0622	0.0631	0.0625	0.0633	0.0626	0.0635	0.0628
#1	64	0.0612	0.0605	0.0616	0.0609	0.0618	0.0611	0.0620	0.0613	0.0622	0.0615	0.0624	0.0617
#2	64	0.0742	0.0735	0.0746	0.0739	0.0748	0.0741	0.0750	0.0743	0.0752	0.0745	0.0754	0.0747
#2	56	0.0726	0.0719	0.0730	0.0723	0.0732	0.0725	0.0734	0.0728	0.0737	0.0730	0.0739	0.0732
#3	56	0.0855	0.0847	0.0859	0.0852	0.0861	0.0854	0.0863	0.0856	0.0866	0.0858	0.0868	0.0860
#3	48	0.0835	0.0827	0.0840	0.0832	0.0842	0.0834	0.0844	0.0837	0.0846	0.0839	0.0849	0.0841
#4	48	0.0964	0.0956	0.0969	0.0961	0.0971	0.0964	0.0974	0.0966	0.0976	0.0968	0.0979	0.0971
#4	40	0.0936	0.0928	0.0941	0.0933	0.0943	0.09335	0.0946	0.0938	0.0948	0.0940	0.0951	0.0943
#5	44	0.1081	0.1073	0.1086	0.1078	0.1088	0.1080	0.1091	0.1082	0.1093	0.1085	0.1096	0.1088
#5	40	0.1065	0.1057	0.1070	0.1062	0.1073	0.1064	0.1075	0.1067	0.1078	0.1070	0.1081	0.1072
#6	40	0.1195	0.1187	0.1200	0.1192	0.1203	0.1194	0.1205	0.1197	0.1208	0.1200	0.1211	0.1202
#6	32	0.1153	0.1144	0.1159	0.1149	0.1161	0.1152	0.1164	0.1155	0.1167	0.1158	0.1170	0.1161
#8	36	0.1436	0.1427	0.1442	0.1432	0.1444	0.1435	0.1447	0.1438	0.1450	0.1441	0.1453	0.1444
#8	32	0.1412	0.1402	0.1417	0.1408	0.1420	0.1411	0.1423	0.1414	0.1426	0.1416	0.1429	0.1419
#10	32	0.1671	0.1661	0.1677	0.1667	0.1680	0.1670	0.1683	0.1673	0.1686	0.1676	0.1689	0.1679
#10	24	0.1591	0.1589	0.1608	0.1596	0.1611	0.1599	0.1614	0.1602	0.1618	0.1606	0.1621	0.1609
#12	28	0.1900	0.1889	0.1907	0.1896	0.1910	0.1899	0.1913	0.1902	0.1916	0.1905	0.1920	0.1908
#12	24	0.1861	0.1848	0.1868	0.1855	0.1871	0.1859	0.1874	0.1862	0.1878	0.1865	0.1881	0.1869
1/4	20	0.2144	0.2131	0.2151	0.2138	0.2155	0.2142	0.2159	0.2146	0.2166	0.2153	0.2159	0.2146
1/4	28	0.2241	0.2228	0.2248	0.2235	0.2251	0.2238	0.2255	0.2242	0.2261	0.2248	0.2255	0.2242

(Continued)

Table 50. *(Continued)* Recommended Steel Blank Diameters for Parallel Rolled UNF and UNC Class 2 Threads. *(Source, Landis Threading Systems.)*

Size		Steel											
		10-50 C Soft		30-50 C Soft		30-50 C or Alloy 15-25 R$_c$		30-50 C or Alloy 26-32 R$_c$		Stainless/Chrome Nickel 300 Series		Stainless/Chrome 400 Series*	
O.D.	Pitch	Min.	Max.	Min.	Max.	Min.	Max.	Min.	Max.	Min.	Max.	Min.	Max.
5/16	18	0.2729	0.2716	0.2737	0.2724	0.2741	0.2728	0.2745	0.2732	0.2753	0.2740	0.2745	0.2732
5/16	24	0.2823	0.2810	0.2830	0.2817	0.2834	0.2821	0.2837	0.2824	0.2845	0.2832	0.2837	0.2824
3/8	16	0.3306	0.3291	0.3315	0.3300	0.3320	0.3305	0.3324	0.3309	0.3333	0.3318	0.3324	0.3309
3/8	24	0.3448	0.3434	0.3455	0.3441	0.3459	0.3445	0.3463	0.3449	0.3471	0.3457	0.3463	0.3449
7/16	14	0.3871	0.3855	0.3880	0.3864	0.3885	0.3869	0.3890	0.3874	0.3899	0.3883	0.3890	0.3874
7/16	20	0.4012	0.3999	0.4021	0.4008	0.4025	0.4012	0.4029	0.4016	0.4037	0.4024	0.4029	0.4016
1/2	13	0.4458	0.4440	0.4468	0.4450	0.4473	0.4455	0.4478	0.4460	0.4488	0.4470	0.4478	0.4460
1/2	20	0.4637	0.4623	0.4646	0.4632	0.4650	0.4636	0.4655	0.4641	0.4663	0.4649	0.4655	0.4641
9/16	12	0.5039	0.5021	0.5050	0.5032	0.5055	0.5037	0.5060	0.5042	0.5070	0.5052	0.5060	0.5042
9/16	18	0.5225	0.5210	0.5234	0.5219	0.5238	0.5223	0.5243	0.5228	0.5252	0.5237	0.5243	0.5228
5/8	11	0.5614	0.5595	0.5625	0.5606	0.5630	0.5611	0.5636	0.5617	0.5647	0.5628	0.5636	0.5617
5/8	18	0.5850	0.5833	0.5859	0.5842	0.5864	0.5847	0.5869	0.5852	0.5878	0.5861	0.5869	0.5852
3/4	10	0.6799	0.6779	0.6811	0.6791	0.6817	0.6797	0.6823	0.6803	0.6834	0.6814	0.6823	0.6803
3/4	16	0.7052	0.7034	0.7062	0.7044	0.7067	0.7049	0.7072	0.7054	0.7082	0.7064	0.7072	0.7054
7/8	9	0.7972	0.7952	0.7985	0.7965	0.7991	0.7971	0.7998	0.7978	0.8010	0.7990	0.7998	0.7978
7/8	14	0.8240	0.8221	0.8251	0.8232	0.8257	0.8238	0.8262	0.8243	0.8273	0.8259	0.8262	0.8243
1	8	0.9131	0.9107	0.9144	0.9120	0.9151	0.9127	0.9158	0.9134	0.9172	0.9148	0.9158	0.9134
1	12	0.9408	0.9388	0.9420	0.9400	0.9426	0.9406	0.9432	0.9412	0.9443	0.9423	0.9432	0.9412
1-1/8	7	1.0262	1.0235	1.0276	1.0250	1.0283	1.0257	1.0290	1.0264	1.0304	1.0278	1.0290	1.0264
1-1/8	12	1.0657	1.0637	1.0669	1.0649	1.0675	1.0655	1.0681	1.0661	1.0693	1.0673	1.0681	1.0661
1-1/4	7	1.1510	1.1483	1.1525	1.1498	1.1533	1.1506	1.1540	1.1513	1.1555	1.1528	1.1540	1.1513

(Continued)

Table 50. *(Continued)* **Recommended Steel Blank Diameters for Parallel Rolled UNF and UNC Class 2 Threads.** *(Source, Landis Threading Systems.)*

| Size | | 10-50 C Soft | | 30-50 C Soft | | 30-50 C or Alloy 15-25 R_c | | 30-50 C or Alloy 26-32R_c | | Stainless/Chrome Nickel 300 Series | | Stainless/Chrome 400 Series* | |
O.D.	Pitch	Min.	Max.	Min.	Max.	Min.	Max.	Min.	Max.	Min.	Max.	Min.	Max.
1-1/4	12	1.1906	1.1885	1.1919	1.1898	1.1925	1.1904	1.1931	1.1910	1.1943	1.1922	1.1931	1.1910
1-3/8	6	1.2601	1.2571	1.2617	1.2587	1.2625	1.2595	1.2633	1.2603	1.2649	1.2619	1.2633	1.2603
1-3/8	12	1.3153	1.3133	1.3166	1.3146	1.3171	1.3152	1.3178	1.3158	1.3191	1.3171	1.13178	1.3158
1-1/2	6	1.3850	1.3820	1.3866	1.3836	1.3874	1.3844	1.3882	1.3852	1.3898	1.3868	1.3882	1.3851
1-1/2	12	1.4402	1.4382	1.4415	1.4395	1.4422	1.4402	1.4428	1.4408	1.4441	1.4421	1.4428	1.4408
1-3/4	5	1.6124	1.6094	1.6142	1.6112	1.6150	1.6120	1.6159	1.6129	1.6177	1.6147	1.6159	1.6129
2	4-1/2	1.8474	1.8442	1.8493	1.8461	1.8503	1.8471	1.8512	1.8480	1.8531	1.8499	1.8512	1.8480

Note: These dimensions are for set-up reference. Diameters must be finally established by actual rolling.
*Only certain grades rollable.

Table 51. Recommended Brass, Bronze, and Aluminum Blank Diameters for Parallel Rolled UNF and UNC Class 2 Threads. *(Source, Landis Threading Systems.)*

Size		Brass & Bronze		Aluminum Alloy			
				Soft		Hard	
O.D.	Pitch	Max.	Min.	Max.	Min.	Max.	Min.
#0	80	0.0507	0.0501	0.0509	0.0503	0.0507	0.0501
#1	72	0.0627	0.06321	0.0629	0.0622	0.0627	0.0621
#1	64	0.0616	0.0609	0.0618	0.0611	0.0616	0.0609
#2	64	0.0746	0.0739	0.0748	0.0741	0.0746	0.0739
#2	56	0.0730	0.0723	0.0732	0.0725	0.0730	0.0723
#3	56	0.0859	0.0852	0.0861	0.0854	0.0859	0.0852
#3	48	0.0840	0.0832	0.0842	0.0834	0.0840	0.0832
#4	48	0.0969	0.0961	0.0971	0.0964	0.0969	0.0961
#4	40	0.0941	0.0938	0.0943	0.0935	0.0941	0.0933
#5	44	0.1086	0.1078	0.1088	0.1080	0.1086	0.1078
#5	40	0.1070	0.1062	0.1073	0.1064	0.1070	0.1062
#6	40	0.1200	0.1192	0.1203	0.1194	0.1200	0.1192
#6	32	0.1159	0.1149	0.1161	0.1152	0.1159	0.1149
#8	36	0.1442	0.1432	0.1444	0.1435	0.1442	0.1432
#8	32	0.1417	0.1408	0.1420	0.1411	0.1417	0.1408
#10	32	0.1677	0.1667	0.1680	0.1670	0.1677	0.1667
#10	24	0.1608	0.1596	0.1611	0.1599	0.1608	0.1596
#12	28	0.1907	0.1896	0.1910	0.1899	0.1907	0.1896
#12	24	0.1868	0.1855	0.1871	0.1859	0.1868	0.1855
1/4	20	0.2151	0.2138	0.2155	0.2142	0.2151	0.2138
1/4	28	0.2248	0.2235	0.2251	0.2238	0.2248	0.2235
5/16	18	0.2737	0.2724	0.2741	0.2728	0.2737	0.2724
5/16	24	0.2830	0.2817	0.2834	0.2821	0.2830	0.2817
3/8	16	0.3315	0.3300	0.3320	0.3305	0.3315	0.3300
3/8	24	0.3455	0.3441	0.3459	0.3445	0.3455	0.3441
7/16	14	0.3880	0.3864	0.3885	0.3869	0.3880	0.3864
7/16	20	0.4021	0.4008	0.4025	0.4012	0.4021	0.4008
1/2	13	0.4468	0.4450	0.4473	0.4455	0.4468	0.4450
1/2	20	0.4646	0.4632	0.4650	0.4636	0.4646	0.4632
9/16	12	0.5050	0.5032	0.5055	0.5037	0.5050	0.5032
9/16	18	0.5234	0.5219	0.5238	0.5223	0.5234	0.5219
5/8	11	0.5625	0.5606	0.5630	0.5611	0.5625	0.5606
5/8	18	0.5859	0.5842	0.5864	0.5847	0.5859	0.5842
3/4	10	0.6811	0.6791	0.6817	0.6797	0.6811	0.6791
3/4	16	0.7062	0.7044	0.7067	0.7049	0.7062	0.7044
7/8	9	0.7985	0.7965	0.7991	0.7971	0.7985	0.7965
7/8	14	0.8251	0.8232	0.8257	0.8238	0.8251	0.8232

(Continued)

Table 51. *(Continued)* **Recommended Brass, Bronze, and Aluminum Blank Diameters for Parallel Rolled UNF and UNC Class 2 Threads.** *(Source, Landis Threading Systems.)*

Size		Brass & Bronze		Aluminum Alloy			
				Soft		Hard	
O.D.	Pitch	Max.	Min.	Max.	Min.	Max.	Min.
1	8	0.9144	0.9120	0.9151	0.9127	0.9144	0.9120
1	12	0.9420	0.9400	0.9426	0.9406	0.9420	0.9400
1-1/8	7	1.0276	1.0250	1.0283	1.0257	1.0276	1.0250
1-1/8	12	1.0669	1.0649	1.0675	1.0655	1.0669	1.0649
1-1/4	7	1.1525	1.1498	1.1533	1.1506	1.1525	1.1498
1-1/4	12	1.1919	1.1898	1.1925	1.1904	1.1919	1.1898
1-3/8	6	1.2617	1.2587	1.2625	1.2595	1.2617	1.2587
1-3/8	12	1.3166	1.3146	1.3172	1.3152	1.3166	1.3146
1-1/2	6	1.3866	1.3836	1.3874	1.3844	1.3866	1.3836
1-1/2	12	1.4415	1.4395	1.4422	1.4402	1.4415	1.4395
1-3/4	5	1.6142	1.6112	1.6150	1.6120	1.6142	1.6112
2	4-1/2	1.8442	1.8461	1.8503	1.8471	1.8442	1.8461

Note: These dimensions are for setup reference. Diameters must be finally established by actual rolling.

Table 52. Recommended Steel Blank Diameters for Parallel Rolled UNF and UNC Class 3 Threads. (Source, Landis Threading Systems.)

| Size | | Steel | | | | | | | | Stainless/Chrome Nickel 300 Series | | Stainless/Chrome 400 Series* | |
| | | 10-50 C Soft | | 30-50 C Soft | | 30-50 C or Alloy 15-25 Rc | | 30-50 C or Alloy 26-32Rc | | | | | |
O.D.	Pitch	Min.	Max.	Min.	Max.	Min.	Max.	Min.	Max.	Min.	Max.	Min.	Max.
#0	80	0.0512	0.0507	0.0514	0.0510	0.0515	0.0511	0.0517	0.0513	0.0519	0.0515	0.0517	0.0513
#1	72	0.0632	0.0627	0.0635	0.0630	0.0637	0.0632	0.0638	0.0633	0.0641	0.0636	0.0638	0.0633
#1	64	0.0621	0.0616	0.0624	0.0619	0.0625	0.0620	0.0627	0.0622	0.0630	0.0625	0.0627	0.0622
#2	64	0.0751	0.0746	0.0754	0.0749	0.0755	0.0750	0.0757	0.0752	0.0760	0.0755	0.0757	0.0752
#2	56	0.0735	0.0730	0.0738	0.0733	0.0739	0.0734	0.0741	0.0736	0.0744	0.0739	0.0741	0.0736
#3	56	0.0865	0.0860	0.0868	0.0863	0.0869	0.0864	0.0871	0.0866	0.0874	0.0869	0.0871	0.0866
#3	48	0.0845	0.0840	0.0848	0.0843	0.0850	0.0845	0.0852	0.0847	0.0855	0.0850	0.0852	0.0847
#4	48	0.0975	0.0969	0.0978	0.0972	0.0980	0.0974	0.0982	0.0976	0.0986	0.0980	0.0982	0.0976
#4	40	0.0947	0.0941	0.0950	0.0945	0.0953	0.0947	0.0955	0.0949	0.0958	0.0952	0.0955	0.0949
#5	44	0.1091	0.1085	0.1095	0.1088	0.1097	0.1091	0.1099	0.1093	0.1102	0.1096	0.1099	0.1093
#5	40	0.1077	0.1071	0.1081	0.1075	0.1083	0.1077	0.1085	0.1079	0.1088	0.1082	0.1085	0.1079
#6	40	0.1207	0.1200	0.1211	0.1204	0.1213	0.1206	0.1215	0.1208	0.1219	0.1212	0.1215	0.1208
#6	32	0.1164	0.1157	0.1169	0.1162	0.1171	0.1164	0.1174	0.1167	0.1178	0.1171	0.1174	0.1167
#8	36	0.1448	0.1441	0.1452	0.1445	0.1454	0.1447	0.1457	0.1450	0.1461	0.1454	0.1457	0.1450
#8	32	0.1424	0.1417	0.1429	0.1422	0.1431	0.1424	0.1433	0.1426	0.1437	0.1430	0.1433	0.1426
#10	32	0.1683	0.1676	0.1688	0.1681	0.1690	0.1683	0.1693	0.1686	0.1697	0.1690	0.1693	0.1686
#10	24	0.1615	0.1607	0.1620	0.1612	0.1622	0.1614	0.1625	0.1617	0.1630	0.1622	0.1625	0.1617
#12	28	0.1913	0.1906	0.1918	0.1911	0.1921	0.1914	0.1923	0.1916	0.1928	0.1921	0.1923	0.1916
#12	24	0.1874	0.1866	0.1879	0.1871	0.1881	0.1873	0.1884	0.1876	0.1889	0.1881	0.1884	0.1876
1/4	20	0.2159	0.2150	0.2164	0.2155	0.2167	0.2158	0.2170	0.2161	0.2176	0.2167	0.2170	0.2161
1/4	28	0.2254	0.2246	0.2259	0.2251	0.2261	0.2253	0.2264	0.2256	0.2269	0.2261	0.2264	0.2256

(Continued)

Table 52. (Continued) Recommended Steel Blank Diameters for Parallel Rolled UNF and UNC Class 3 Threads. (Source, Landis Threading Systems.)

Size		Steel												
		10-50 C Soft		30-50 C Soft		30-50 C or Alloy 15-25 R$_c$		30-50 C or Alloy 26-32R$_c$		Stainless/Chrome Nickel 300 Series		Stainless/Chrome 400 Series*	
O.D.	Pitch	Min.	Max.	Min.	Max.	Min.	Max.	Min.	Max.	Min.	Max.	Min.	Max.
5/16	18	0.2724	0.2737	0.2753	0.2743	0.2756	0.2746	0.2759	0.2749	0.2765	0.2755	0.2759	0.2749
5/16	24	0.2839	0.2830	0.2845	0.2836	0.2847	0.2838	0.2850	0.2841	0.2855	0.2846	0.2850	0.2841
3/8	16	0.3326	0.3314	0.3333	0.3321	0.3336	0.3324	0.3340	0.3328	0.3346	0.3334	0.3340	.3328
3/8	24	0.3463	0.3453	0.3469	0.3459	0.3472	0.3462	0.3475	0.3465	0.3480	0.3470	0.3475	0.3465
7/16	14	0.3893	0.3880	0.3900	0.3887	0.3903	0.3890	0.3907	0.3894	0.3914	0.3901	0.3907	0.3894
7/16	20	0.4032	0.4022	0.4038	0.4028	0.4041	0.4031	0.4044	0.4034	0.4051	0.4041	0.4044	0.4034
1/2	13	0.4480	0.4467	0.4487	0.4474	0.4491	0.4478	0.4495	0.4482	0.4502	0.4489	0.4495	0.4482
1/2	20	0.4657	0.4646	0.4664	0.4653	0.4654	0.4643	0.4670	0.4659	0.4676	0.4665	0.4670	0.4659
9/16	12	0.5062	0.5049	0.5070	0.5057	0.5070	0.5057	0.5078	0.5065	0.5085	0.5072	0.5078	0.5065
9/16	18	0.5246	0.5233	0.5253	0.5240	0.5253	0.5240	0.5260	0.5247	0.5267	0.5254	0.5260	0.5247
5/8	11	0.5636	0.5623	0.5644	0.5631	0.5644	0.5631	0.5653	0.5640	0.5661	0.5648	0.5653	.5640
5/8	18	0.5871	0.5858	0.5878	0.5865	0.5878	0.5865	0.5885	0.5872	0.5892	0.5879	0.5885	0.5872
3/4	10	0.6825	0.6810	0.6834	0.6819	0.6834	0.6819	0.6843	0.6828	0.6852	0.6837	0.6843	0.6828
3/4	16	0.7073	0.7060	0.7080	0.7067	0.7080	0.7067	0.7088	0.7075	0.7096	0.7083	0.7088	0.7075
7/8	9	0.8002	0.7986	0.8010	0.7994	0.8011	0.7995	0.8021	0.8005	0.8030	0.8014	0.8021	0.8005
7/8	14	0.8262	0.8249	0.8270	0.8257	0.8270	0.8257	0.8279	0.8266	0.8287	0.8274	0.8279	0.8266
1	8	0.9160	0.9142	0.9170	0.9151	0.9170	0.9152	0.9180	0.9162	0.9191	0.9173	0.9180	0.9162
1	12	0.9434	0.9419	0.9443	0.9428	0.9428	0.9415	0.9452	0.9437	0.9461	0.9446	0.9452	0.9437
1-1/8	7	1.0292	1.0273	1.0303	1.0284	1.0303	1.0284	1.0314	1.0295	1.0325	1.0306	1.0314	1.0295
1-1/8	12	1.0684	1.0669	1.0693	1.0678	1.0693	1.0678	1.0693	1.0678	1.0711	1.0696	1.0693	1.0678
1-1/4	7	1.1542	1.1523	1.1553	1.1534	1.1553	1.1534	1.1553	1.1534	1.1575	1.1556	1.1553	1.1534

(Continued)

Table 52. (*Continued*) Recommended Steel Blank Diameters for Parallel Rolled UNF and UNC Class 3 Threads. (*Source, Landis Threading Systems.*)

Size		Steel											
O.D.	Pitch	10-50 C Soft		30-50 C Soft		30-50 C or Alloy 15-25 R$_c$		30-50 C or Alloy 26-32R$_c$		Stainless/Chrome Nickel 300 Series		Stainless/Chrome 400 Series*	
		Min.	Max.	Min.	Max.	Min.	Max.	Min.	Max.	Min.	Max.	Min.	Max.
1-1/4	12	1.1933	1.1918	1.1942	1.1927	1.1942	1.1927	1.1951	1.1936	1.1960	1.1945	1.1951	1.1936
1-3/8	6	1.2633	1.2613	1.2645	1.2625	1.2651	1.2631	1.2657	1.2637	1.2665	1.2649	1.2657	1.2637
1-3/8	12	1.3182	1.3167	1.3191	1.3176	1.3196	1.3181	1.3200	1.3185	1.3210	1.3195	1.3200	1.3185
1-1/2	6	1.3883	1.3862	1.3894	1.3874	1.3900	1.3880	1.3906	1.3886	1.3918	1.3898	1.3906	1.3886
1-1/2	12	1.4433	1.4416	1.4442	1.4425	1.4447	1.4430	1.4451	1.4434	1.4461	1.4444	1.4451	1.4434
1-3/4	5	1.6165	1.6141	1.6178	1.6154	1.6185	1.6161	1.6191	1.6167	1.6205	1.6181	1.6191	1.6167
2	4-1/2	1.8518	1.8493	1.8532	1.8507	1.8539	1.8514	1.8546	1.8521	1.8561	1.8536	1.8546	1.8521

Note: These dimensions are for setup reference. Diameters must be finally established by actual rolling.

Table 53. Recommended Brass, Bronze, and Aluminum Blank Diameters for Parallel Rolled UNF and UNC Class 3 Threads. *(Source, Landis Threading Systems.)*

Size		Brass & Bronze		Aluminum Alloy			
				Soft		Hard	
O.D.	Pitch	Max.	Min.	Max.	Min.	Max.	Min.
#0	80	0.0514	0.0510	0.0515	0.0511	0.0514	0.0510
#1	72	0.0635	0.0630	0.0637	0.0632	0.0635	0.0630
#1	64	0.0624	0.0619	0.0625	0.0620	0.0624	0.0619
#2	64	0.0754	0.0749	0.0755	0.0750	0.0754	0.0749
#2	56	0.0738	0.0733	0.0739	0.0734	0.0738	0.0733
#3	56	0.0868	0.0863	0.0869	0.0864	0.0868	0.0863
#3	48	0.0848	0.0843	0.0850	0.0845	0.0848	0.0843
#4	48	0.0978	0.0972	0.0980	0.0974	0.0978	0.0972
#4	40	0.0950	0.0945	0.0953	0.0947	0.0950	0.0945
#5	44	0.1095	0.1088	0.1097	0.1091	0.1095	0.1088
#5	40	0.1081	0.1075	0.1083	0.1077	0.1081	0.1075
#6	40	0.1211	0.1204	0.1213	0.1206	0.1211	0.1204
#6	32	0.1169	0.1162	0.1171	0.1164	0.1169	0.1162
#8	36	0.1452	0.1445	0.1454	0.1447	0.1452	0.1445
#8	32	0.1429	0.1422	0.1431	0.1424	0.1429	0.1422
#10	32	0.1688	0.1681	0.1690	0.1683	0.1688	0.1681
#10	24	0.1620	0.1612	0.1622	0.1614	0.1620	0.1612
#12	28	0.1918	0.1911	0.1921	0.1914	0.1918	0.1911
#12	24	0.1879	0.1871	0.1881	0.1873	0.1879	0.1871
1/4	20	0.2164	0.2155	0.2167	0.2158	0.2164	0.2155
1/4	28	0.2259	0.2251	0.2261	0.2253	0.2259	0.2251
5/16	18	0.2753	0.2743	0.2756	0.2746	0.2753	0.2743
5/16	24	0.2845	0.2836	0.2847	0.2838	0.2845	0.2836
3/8	16	0.3346	0.3334	0.3340	0.3328	0.3333	0.3321
3/8	24	0.3469	0.3459	0.3472	0.3462	0.3469	0.3459
7/16	14	0.3900	0.3887	0.3903	0.3890	0.3900	0.3887
7/16	20	0.4038	0.4028	0.4041	0.4031	0.4038	0.4028
1/2	13	0.4487	0.4474	0.4491	0.4478	0.4487	0.4474
1/2	20	0.4664	0.4653	0.4654	0.4643	0.4664	0.4653
9/16	12	0.5070	0.5057	0.5070	0.5057	0.5070	0.5057
9/16	18	0.5253	0.5240	0.5253	0.5240	0.5253	0.5240
5/8	11	0.5644	0.5631	0.5644	0.5631	0.5644	0.5631
5/8	18	0.5878	0.5865	0.5878	0.5865	0.5878	0.5865
3/4	10	0.6834	0.6819	0.6834	0.6819	0.6834	0.6819
3/4	16	0.7080	0.7067	0.7080	0.7067	0.7087	0.7067
7/8	9	0.8010	0.7994	0.8011	0.7995	0.8010	0.7994
7/8	14	0.8270	0.8257	0.8270	0.8257	0.8270	0.8257

(Continued)

Table 53. *(Continued)* **Recommended Brass, Bronze, and Aluminum Blank Diameters for Parallel Rolled UNF and UNC Class 3 Threads.** *(Source, Landis Threading Systems.)*

Size		Brass & Bronze		Aluminum Alloy			
				Soft		Hard	
O.D.	Pitch	Max.	Min.	Max.	Min.	Max.	Min.
1	8	0.9170	0.9152	0.9170	0.9152	0.9170	0.9152
1	12	0.9443	0.9428	0.9428	0.9415	0.9443	0.9428
1-1/8	7	1.0303	1.0284	1.0303	1.0284	1.0303	1.0284
1-1/8	12	1.0693	1.0678	1.0693	1.0678	1.0693	1.0678
1-1/4	7	1.1553	1.1539	1.1553	1.1534	1.1553	1.1534
1-1/4	12	1.1942	1.1927	1.1942	1.1927	1.1942	1.1927
1-3/8	6	1.2645	1.2625	1.2651	1.2631	1.2645	1.2625
1-3/8	12	1.3191	1.3176	1.3196	1.3161	1.3191	1.3176
1-1/2	6	1.3894	1.3874	1.3900	1.3880	1.3894	1.3874
1-1/2	12	1.4442	1.4425	1.4447	1.4430	1.4442	1.4425
1-3/4	5	1.6178	1.6154	1.6185	1.6161	1.6178	1.6154
2	4-1/2	1.8532	1.8507	1.8539	1.8514	1.8532	1.8507

Note: These dimensions are for setup reference. Diameters must be finally established by actual rolling.

Table 54. Recommended Steel Blank Diameters for Straight, Rolled Metric Threads. *(Source, Landis Threading Systems.)*

Series Designation		Dia. in mm & inches	Steel											
			10-50 C Soft		30-50 C Soft		30-50 C or Alloy 15-25 R_c		30-50 C or Alloy 26-32R_c		Stainless/Chrome Nickel 300 Series		Stainless/Chrome 400 Series*	
Size	Pitch		Min.	Max.	Min.	Max.	Min.	Max.	Min.	Max.	Min.	Max.	Min.	Max.
3	35	mm	2.756	2.743	2.764	2.751	2.769	2.756	2.771	2.758	2.779	2.766	2.771	2.758
		inch	0.1085	0.1080	0.1088	0.1083	0.1090	0.1085	0.1091	0.1086	0.1094	0.1089	0.1091	0.1086
3	.5	mm	2.652	2.639	2.659	2.647	2.664	2.652	2.667	2.654	2.675	2.662	2.667	2.654
		inch	0.1044	0.1039	0.1047	0.1042	0.1049	0.1044	0.1050	0.1045	0.1053	0.1048	0.1050	0.1045
3.5	.35	mm	3.256	3.244	3.261	3.249	3.266	3.254	3.269	3.256	3.277	3.264	3.269	3.256
		inch	0.1282	0.1277	0.1284	0.1279	0.1286	0.1281	0.1287	0.1282	0.1290	0.1285	0.1287	0.1282
3.5	6	mm	3.084	3.071	3.094	3.081	3.096	3.084	3.101	3.089	3.109	3.096	3.101	3.089
		inch	0.1214	0.1209	0.1218	0.1213	0.1219	0.1214	0.1221	0.1216	0.1224	0.1219	0.1221	0.1216
4	.5	mm	3.653	3.640	3.660	3.647	3.663	3.650	3.668	3.655	3.675	3.663	3.668	3.655
		inch	0.1438	0.1433	0.1441	0.1436	0.1442	0.1437	0.1444	0.1439	0.1447	0.1442	0.1444	0.1439
4	.7	mm	3.520	3.505	3.528	3.513	3.533	3.518	3.538	3.523	3.548	3.533	3.538	3.523
		inch	0.1386	0.1380	0.1389	0.1383	0.1391	0.1385	0.1393	0.1387	0.1397	0.1391	0.1393	0.1387
4.5	.5	mm	4.153	4.140	4.161	4.148	4.163	4.150	4.168	4.155	4.176	4.163	4.168	4.155
		inch	0.1635	0.1630	0.1638	0.1633	0.1639	0.1634	0.1641	0.1636	0.1644	0.1639	0.1641	0.1636
5	.5	mm	4.651	4.638	4.661	4.648	4.663	4.651	4.666	4.653	4.676	4.663	4.666	4.653
		inch	0.1831	0.1826	0.1835	0.1830	0.1836	0.1831	0.1837	0.1832	0.1841	0.1836	0.1837	0.1832
5	.8	mm	4.455	4.437	4.463	4.448	4.468	4.453	4.473	4.458	4.483	4.465	4.473	4.463
		inch	0.1754	0.1747	0.1757	0.1751	0.1759	0.1753	0.1761	0.1755	0.1765	0.1758	0.1761	0.1755
6	.75	mm	5.484	5.469	5.494	5.479	5.499	5.484	5.504	5.489	5.512	5.496	5.504	5.489
		inch	0.2159	0.2153	0.2163	0.2157	0.2165	0.2159	0.2167	0.2161	0.2170	0.2164	0.2167	0.2161
6	1.0	mm	5.314	5.293	5.326	5.306	5.334	5.314	5.339	5.319	5.352	5.331	5.339	5.319
		inch	0.2092	0.2084	0.2097	0.2089	0.2100	0.2092	0.2102	0.2094	0.2107	0.2099	0.2102	0.2094
7	.75	mm	6.485	6.469	6.495	6.480	6.500	6.485	6.505	6.490	6.513	6.497	6.505	6.490
		inch	0.2553	0.2547	0.2557	0.2551	0.2559	0.2553	0.2561	0.2555	0.2564	0.2558	0.2561	0.2561

(Continued)

Table 54. *(Continued)* **Recommended Steel Blank Diameters for Straight, Rolled Metric Threads.** *(Source, Landis Threading Systems.)*

Series Designation		Dia. in mm & inches	Steel											
			10-50 C Soft		30-50 C Soft		30-50 C or Alloy 15-25 Rc		30-50 C or Alloy 26-32Rc		Stainless/Chrome Nickel 300 Series		Stainless/Chrome 400 Series*	
Size	Pitch		Min.	Max.	Min.	Max.	Min.	Max.	Min.	Max.	Min.	Max.	Min.	Max.
7	1.0	mm	6.314	6.294	6.327	6.307	6.335	6.314	6.340	6.320	6.353	6.332	6.340	6.320
		inch	0.2486	0.2478	0.2491	0.2483	0.2494	0.2486	0.2496	0.2488	0.2501	0.2493	0.2496	0.2488
8	1.0	mm	7.315	7.295	7.328	7.308	7.336	7.315	7.341	7.320	7.353	7.333	7.341	7.320
		inch	0.2880	0.2872	0.2885	0.2877	0.2888	0.2880	0.2890	0.2882	0.2895	0.2887	0.2890	0.2882
8	1.25	mm	7.150	7.130	7.163	7.142	7.168	7.148	7.176	7.155	7.188	7.168	7.176	7.155
		inch	0.2815	0.2807	0.2820	0.2812	0.2822	0.2814	0.2825	0.2817	0.2830	0.2822	0.2825	0.2817
9	1.0	mm	8.313	8.293	8.326	8.306	8.334	8.313	8.339	8.319	8.352	8.331	8.339	8.319
		inch	0.3273	0.3265	0.3278	0.3270	0.3281	0.3273	0.3283	0.3275	0.3288	0.3280	0.3283	0.3275
9	1.25	mm	8.151	8.131	8.164	8.143	8.169	8.148	8.176	8.156	8.189	8.169	8.176	8.156
		inch	0.3209	0.3201	0.3214	0.3206	0.3216	0.3208	0.3219	0.3211	0.3224	0.3216	0.3219	0.3211
10	1.0	mm	9.314	9.294	9.327	9.307	9.335	9.314	9.340	9.319	9.352	9.332	9.340	9.319
		inch	0.3667	0.3659	0.3672	0.3664	0.3675	0.3667	0.3677	0.3669	0.3682	0.3674	0.3677	0.3669
10	1.5	mm	8.984	8.956	8.999	8.971	9.007	8.979	9.014	8.989	9.030	9.004	9.014	8.989
		inch	0.3537	0.3526	0.3543	0.3532	0.3546	0.3535	0.3549	0.3539	0.3555	0.3545	0.3549	0.3539
11	1.5	mm	9.982	9.954	9.997	9.970	10.005	9.977	10.012	9.987	10.027	10.003	10.012	9.987
		inch	0.3930	0.3919	0.3936	0.3925	0.3939	0.3928	0.3942	0.3932	0.3948	0.3938	0.3942	0.3932
12	1.5	mm	10.983	10.955	10.998	10.970	11.006	10.978	11.013	10.988	11.029	11.003	11.013	10.988
		inch	0.4324	0.4313	0.4330	0.4319	0.4333	0.4322	0.4336	0.4326	0.4342	0.4332	0.4336	0.4326
12	1.75	mm	10.818	10.785	10.836	10.803	10.843	10.810	10.853	10.820	10.871	10.838	10.853	10.820
		inch	0.4259	0.4246	0.4266	0.4253	0.4269	0.4256	0.4273	0.4260	0.4280	0.4267	0.4273	0.4260
14	1.5	mm	12.982	12.954	12.997	12.969	13.005	12.977	13.012	12.987	13.028	13.002	13.012	12.987
		inch	0.5111	0.5100	0.5117	0.5106	0.5120	0.5109	0.5123	0.5113	0.5129	0.5119	0.5123	0.5113
14	2	mm	12.647	12.614	12.664	12.631	12.675	12.642	12.685	12.652	12.705	12.672	12.685	12.652
		inch	0.4979	0.4966	0.4986	0.4973	0.4990	0.4977	0.4994	0.4981	0.5002	0.4989	0.4994	0.4981

(Continued)

Table 54. (Continued) Recommended Steel Blank Diameters for Straight, Rolled Metric Threads. (Source, Landis Threading Systems.).

Series Designation		Dia. in mm & inches	Steel											
			10-50 C Soft		30-50 C Soft		30-50 C or Alloy 15-25 R_c		30-50 C or Alloy 26-32 R_c		Stainless/Chrome Nickel 300 Series		Stainless/Chrome 400 Series*	
Size	Pitch		Min.	Max.	Min.	Max.	Min.	Max.	Min.	Max.	Min.	Max.	Min.	Max.
16	1.5	mm	14.983	14.956	14.999	14.971	15.006	14.978	15.014	14.989	15.029	15.004	15.014	14.989
		inch	0.5899	0.5888	0.5905	0.5894	0.5908	0.5897	0.5911	0.5901	0.5917	0.5907	0.5911	0.5901
16	2	mm	14.648	14.615	14.666	14.633	14.676	14.643	14.686	14.653	14.707	14.674	14.686	14.653
		inch	0.5767	0.5754	0.5774	0.5761	0.5778	0.5765	0.5782	0.5769	0.5790	0.5777	0.5782	0.5769
18	1.5	mm	16.982	16.955	16.998	16.970	17.005	16.977	17.013	16.988	17.028	17.003	17.013	16.988
		inch	0.6686	0.6675	0.6692	0.6681	0.6695	0.6684	0.6698	0.6688	0.6704	0.6694	0.6698	0.6688
18	2.5	mm	16.312	16.274	16.335	16.297	16.345	16.307	16.358	16.320	16.380	16.342	16.358	16.320
		inch	0.6422	0.6407	0.6431	0.6416	0.6435	0.6420	0.6440	0.6425	0.6449	0.6434	0.6440	0.6425
20	1.5	mm	18.984	18.956	18.999	18.971	19.007	18.979	19.014	18.989	19.030	19.004	19.014	18.989
		inch	0.7474	0.7463	0.7480	0.7469	0.7483	0.7472	0.7486	0.7476	0.7492	0.7482	0.7486	0.7476
20	2.5	mm	18.313	18.275	18.336	18.298	18.346	18.308	18.359	18.321	18.382	18.344	18.359	18.321
		inch	0.7210	0.7195	0.7219	0.7204	0.7223	0.7208	0.7228	0.7213	0.7237	0.7222	0.7228	0.7213
22	1.5	mm	20.983	20.955	20.998	20.970	21.006	20.978	21.013	20.988	21.029	21.003	21.013	20.988
		inch	0.8261	0.8250	0.8267	0.8256	0.8270	0.8259	0.8273	0.8263	0.8279	0.8269	0.8273	0.8263
22	2.5	mm	20.312	20.274	20.335	20.297	20.345	20.307	20.358	20.320	20.381	20.343	20.358	20.320
		inch	0.7997	0.7982	0.8006	0.7991	0.8010	0.7995	0.8015	0.8000	0.8024	0.8009	0.8015	0.8000
24	2	mm	22.647	22.614	22.664	22.631	22.675	22.642	22.685	22.652	22.705	22.672	22.685	22.652
		inch	0.8916	0.8903	0.8923	0.8910	0.8927	0.8927	0.8931	0.8918	0.8939	0.8926	0.8931	0.8918
24	3	mm	21.979	21.930	22.004	21.958	22.017	21.971	22.029	21.984	22.055	22.007	22.029	21.984
		inch	0.8653	0.8634	0.8663	0.8645	0.8668	0.8650	0.8673	0.8655	0.8683	0.8664	0.8673	0.8655
27	2	mm	25.646	25.613	25.664	15.631	25.674	25.641	25.684	25.651	25.705	25.672	25.684	25.651
		inch	1.0097	1.0084	1.0104	1.0091	1.0108	1.0095	1.0112	1.0099	1.012	1.0107	1.0112	1.0099
27	3	mm	24.981	24.933	25.006	24.961	25.019	24.973	25.032	24.986	25.060	25.011	25.032	24.986
		inch	0.9835	0.9816	0.9845	0.9827	0.9850	0.9832	0.9855	0.9837	0.9866	0.9847	0.9855	0.9837

(Continued)

Table 54. (Continued) Recommended Steel Blank Diameters for Straight, Rolled Metric Threads. (Source, Landis Threading Systems.)

Series Designation		Dia. in mm & inches	Steel											
Size	Pitch		10-50 C Soft		30-50 C Soft		30-50 C or Alloy 15-25 Rc		30-50 C or Alloy 26-32Rc		Stainless/Chrome Nickel 300 Series		Stainless/Chrome 400 Series*	
			Min.	Max.	Min.	Max.	Min.	Max.	Min.	Max.	Min.	Max.	Min.	Max.
30	2	mm	28.646	28.613	28.664	2.8631	28.674	28.641	2.8684	28.651	28.705	28.672	28.684	2.8651
		inch	1.1278	1.1265	1.1285	1.1272	1.1289	1.1276	1.1293	1.1280	1.1301	1.1288	1.1293	1.1280
30	3.5	mm	27.643	27.595	27.763	27.623	27.689	27.638	27.704	27.653	27.732	27.683	27.704	27.653
		inch	1.0883	1.0864	1.0895	1.0875	1.0901	1.0881	1.0907	1.0887	1.0918	1.0899	1.0907	1.0887
33	2	mm	31.648	31.615	31.666	31.633	31.676	31.643	31.687	31.653	31.707	31.674	31.687	31.653
		inch	1.2460	1.2447	1.2467	1.2454	1.2471	1.2458	1.2475	1.2462	1.2483	1.2470	1.2475	1.2462
33	3.5	mm	30.643	30.594	30.673	30.622	30.688	30.637	30.704	30.653	30.731	30.683	30.704	30.653
		inch	1.2064	1.2045	1.2076	1.2056	1.2082	1.2062	1.2088	1.2068	1.2099	1.2080	1.2088	1.2068
36	3	mm	33.973	33.925	34.001	33.953	34.016	33.967	34.029	33.981	34.057	34.009	34.029	33.981
		inch	1.3375	1.3356	1.3367	1.3392	1.3373	1.3397	1.3378	1.3408	1.3389	1.3397	1.3378	1.3386
36	4	mm	33.316	33.265	33.346	33.295	33.361	33.311	33.377	33.326	33.407	33.356	33.377	33.326
		inch	1.3116	1.3096	1.3128	1.3108	1.3134	1.3114	1.3140	1.3120	1.3152	1.3132	1.3140	1.3120
39	3	mm	36.973	36.925	37.001	36.953	37.015	36.967	37.029	36.981	37.057	37.009	37.029	36.981
		inch	1.4556	1.4537	1.4567	1.4548	1.4573	1.4554	1.4578	1.4559	1.4589	1.4569	1.4578	1.4559
39	4	mm	36.316	36.265	36.346	36.295	36.361	36.311	36.377	36.326	36.407	36.356	36.377	36.326
		inch	1.4297	1.4277	1.4309	1.4289	1.4315	1.4295	1.4321	1.4301	1.4333	1.4313	1.4321	1.4301
42	3	mm	39.971	39.923	39.999	39.951	40.014	39.966	40.028	39.980	40.056	40.008	40.028	39.980
		inch	1.5848	1.5829	1.5860	1.5841	1.5865	1.5846	1.5591	1.5572	1.5882	1.5863	1.5591	1.5572
42	4.5	mm	38.987	38.928	39.020	38.961	39.036	38.978	39.053	38.994	39.086	39.027	39.053	38.994
		inch	1.5348	1.5325	1.5361	1.5338	1.5368	1.5345	1.5374	1.5351	1.5387	1.5364	1.5374	1.5351
45	3	mm	42.971	42.923	42.999	42.951	43.014	42.966	43.028	42.980	43.056	43.008	43.028	42.980
		inch	1.7029	1.7010	1.7041	1.7022	1.7046	1.7027	1.6772	1.6753	1.7063	1.7044	1.6772	1.6753
45	4.5	mm	41.987	41.928	42.020	41.961	42.036	41.978	42.053	41.994	42.086	42.027	42.053	41.994
		inch	1.6529	1.6506	1.6542	1.6519	1.6549	1.6526	1.6555	1.6532	1.6568	1.6545	1.6555	1.6532

(Continued)

Table 54. (Continued) Recommended Steel Blank Diameters for Straight, Rolled Metric Threads. (Source, Landis Threading Systems.)

Series Designation		Dia. in mm & inches	Steel												
			10-50 C Soft		30-50 C Soft		30-50 C or Alloy 15-25 R_c		30-50 C or Alloy 26-32 R_c		Stainless/Chrome Nickel 300 Series		Stainless/Chrome 400 Series*	
Size	Pitch		Min.	Max.	Min.	Max.	Min.	Max.	Min.	Max.	Min.	Max.	Min.	Max.
48	3	mm	45.971	45.921	46.000	45.950	46.015	45.964	46.029	45.979	46.058	46.008	46.029	45.979
		inch	1.8099	1.8079	1.8111	1.8091	1.8117	1.8097	1.8123	1.8103	1.8135	1.8115	1.8123	1.8103
48	5	mm	44.655	44.592	44.691	44.628	44.709	44.645	44.727	44.663	44.762	44.699	44.727	44.663
		inch	1.7581	1.7556	1.7595	1.7570	1.7601	1.7577	1.7609	1.7584	1.7623	1.7598	1.7609	1.7584
52	3	mm	49.969	49.918	49.999	49.948	50.013	49.963	50.028	49.977	50.058	50.007	50.028	49.977
		inch	1.9673	1.9653	1.9685	1.9665	1.9691	1.9671	1.9697	1.9677	1.9709	1.9689	1.9697	1.9677
52	5	mm	48.643	48.592	48.678	48.628	48.696	48.645	48.714	48.663	48.749	48.699	48.714	48.663
		inch	1.9151	1.9131	1.9165	1.9145	1.9172	1.9152	1.9179	1.9159	1.9193	1.9173	1.9179	1.9159

Note: These dimensions are for setup reference. Diameters must be finally established by actual rolling.
*Only certain grades rollable.

Table 55. Recommended Brass, Bronze, and Aluminum Blank Diameters for Straight, Rolled Metric Threads.
(Source, Landis Threading Systems.)

Series Designation		Dia. in mm & inches	Brass & Bronze		Aluminum Alloy			
					Soft		Hard	
Size	Pitch		Max.	Min.	Max.	Min.	Max.	Min.
3	.35	mm	2.764	2.751	2.769	2.756	2.764	2.751
		inch	0.1088	0.1083	0.1090	0.1085	0.1088	0.1083
3	.5	mm	2.659	2.647	2.664	2.652	2.659	2.647
		inch	0.1047	0.1042	0.1049	0.1044	0.1047	0.1042
3.5	.35	mm	3.261	3.249	3.266	3.254	3.261	3.249
		inch	0.1284	0.1279	0.1286	0.1281	0.1284	0.1279
3.5	6	mm	3.094	3.081	3.096	3.084	3.094	3.081
		inch	0.1218	0.1213	0.1219	0.1214	0.1218	0.1213
4	.5	mm	3.660	3.647	3.663	3.650	3.660	3.647
		inch	0.1441	0.1436	0.1442	0.1437	0.1441	0.1436
4	.7	mm	3.528	3.513	3.533	3.518	3.528	3.513
		inch	0.1389	0.1383	0.1391	0.1385	0.1389	0.1383
4.5	.5	mm	4.161	4.148	4.163	4.150	4.161	4.148
		inch	0.1638	0.1633	0.1639	0.1634	0.1638	0.1633
5	.5	mm	4.661	4.648	4.663	4.651	4.661	4.648
		inch	0.1835	0.1830	0.1836	0.1831	0.1835	0.1830
5	.8	mm	4.448	4.468	4.453	4.463	4.448	4.448
		inch	0.1757	0.1751	0.1759	0.1753	0.1757	0.1751
6	.75	mm	5.494	5.479	5.499	5.484	5.494	5.479
		inch	0.2163	0.2157	0.2165	0.2159	0.2163	0.2157
6	1.0	mm	5.326	5.306	5.334	5.314	5.326	5.306
		inch	0.2097	0.2089	0.2100	0.2092	0.2097	0.2089
7	.75	mm	6.495	6.480	6.500	6.485	6.495	6.480
		inch	0.2557	0.2551	0.2559	0.2553	0.2557	0.2551
7	1.0	mm	6.327	6.307	6.335	6.314	6.327	6.307
		inch	0.2491	0.2483	0.2494	0.2486	0.2491	0.2483
8	1.0	mm	7.328	7.308	7.336	7.315	7.328	7.308
		inch	0.2885	0.2877	0.2888	0.2880	0.2885	0.2877
8	1.25	mm	7.163	7.142	7.168	7.148	7.163	7.142
		inch	0.2820	0.2812	0.2822	0.2814	0.2820	0.2812
9	1.0	mm	8.326	8.306	8.334	8.313	8.326	8.306
		inch	0.3278	0.3270	0.3281	0.3273	0.3278	0.3270
9	1.25	mm	8.164	8.143	8.169	8.148	8.164	8.143
		inch	0.3214	0.3206	0.3216	0.3208	0.3214	0.3206
10	1.0	mm	9.327	9.307	9.335	9.314	9.327	9.307
		inch	0.3672	0.3664	0.3675	0.3667	0.3672	0.3664
10	1.5	mm	8.999	8.971	9.007	8.979	8.999	8.971
		inch	0.3543	0.3532	0.3546	0.3535	0.3543	0.3532
11	1.5	mm	9.997	9.970	10.005	9.977	9.997	9.970
		inch	0.3936	0.3925	0.3939	0.3928	0.3936	0.3925
12	1.5	mm	10.998	10.970	11.006	10.978	10.998	10.970
		inch	0.4330	0.4319	0.4333	0.4322	0.4330	0.4319
12	1.75	mm	10.836	10.803	10.843	10.810	10.836	10.803
		inch	0.4266	0.4253	0.4269	0.4256	0.4266	0.4253

(Continued)

Table 55. *(Continued)* **Recommended Brass, Bronze, and Aluminum Blank Diameters for Straight, Rolled Metric Threads.** *(Source, Landis Threading Systems.)*

Series Designation		Dia. in mm & inches	Brass & Bronze		Aluminum Alloy			
					Soft		Hard	
Size	Pitch		Max.	Min.	Max.	Min.	Max.	Min.
14	1.5	mm	12.997	12.969	13.005	12.977	12.997	12.969
		inch	0.5117	0.5106	0.5120	0.5109	0.5117	0.5106
14	2	mm	12.664	12.631	12.657	12.642	12.664	12.631
		inch	0.4986	0.4973	0.4990	0.4977	0.4986	0.4973
16	1.5	mm	14.999	14.971	15.006	14.978	14.999	14.971
		inch	0.5905	0.5894	0.5908	0.5897	0.5905	0.5894
16	2	mm	14.666	14.633	14.676	14.643	14.666	14.633
		inch	0.5774	0.5761	0.5778	0.5765	0.5774	0.5761
18	1.5	mm	16.998	16.970	17.005	16.977	16.998	16.970
		inch	0.6692	0.6681	0.6695	0.6684	0.6692	0.6681
18	2.5	mm	16.335	16.297	16.345	16.307	16.335	16.297
		inch	0.6431	0.6416	0.6435	0.6420	0.6431	0.6416
20	1.5	mm	18.999	18.971	19.007	18.979	18.999	18.971
		inch	0.7480	0.7469	0.7483	0.7472	0.7480	0.7469
20	2.5	mm	18.336	18.298	18.346	18.308	18.336	18.298
		inch	0.7219	0.7204	0.7223	0.7208	0.7219	0.7204
22	1.5	mm	20.998	20.970	21.006	20.978	20.998	20.970
		inch	0.8267	0.8256	0.8270	0.8259	0.8267	0.8256
22	2.5	mm	20.335	20.297	20.345	20.307	20.335	20.297
		inch	0.8006	0.7991	0.8010	0.7995	0.8006	0.7991
24	2	mm	22.664	22.631	22.675	22.642	22.664	22.631
		inch	0.8923	0.8910	0.8927	0.8914	0.8923	0.8910
24	3	mm	22.004	21.958	22.017	21.971	22.004	21.958
		inch	0.8663	0.8645	0.8668	0.8650	0.8663	0.8645
27	2	mm	25.664	25.631	25.674	25.641	25.664	25.631
		inch	1.0104	1.0091	1.0208	1.0095	1.0104	1.0091
27	3	mm	25.006	24.961	25.109	24.973	25.006	24.961
		inch	0.9845	0.9827	0.9850	0.9832	0.9845	0.9827
30	2	mm	28.664	28.631	28.674	28.641	28.664	28.631
		inch	1.1285	1.1272	1.1289	1.1276	1.1285	1.1272
30	3.5	mm	27.673	27.623	27.689	27.638	27.673	27.623
		inch	1.0895	1.0875	1.0901	1.0881	1.0895	1.0875
33	2	mm	31.666	31.633	31.676	31.643	31.666	31.633
		inch	1.2467	1.2454	1.2471	1.2458	1.2467	1.2454
33	3.5	mm	30.673	30.622	30.688	30.637	30.673	30.622
		inch	1.2076	1.2056	1.2082	1.2062	1.2076	1.2056
36	3	mm	34.001	33.953	34.015	33.967	34.001	33.953
		inch	1.3367	1.3392	1.3373	1.3386	1.3386	1.3367
36	4	mm	33.346	33.295	33.361	33.311	33.346	33.295
		inch	1.3128	1.3108	1.3134	1.3114	1.3128	1.3108
39	3	mm	37.001	36.953	37.015	36.967	37.001	36.953
		inch	1.4567	1.4548	1.4573	1.4554	1.4567	1.4548
39	4	mm	36.346	36.295	36.361	36.311	36.346	36.295
		inch	1.4309	1.4289	1.4315	1.4295	1.4309	1.4289

(Continued)

Table 55. *(Continued)* **Recommended Brass, Bronze, and Aluminum Blank Diameters for Straight, Rolled Metric Threads.** *(Source, Landis Threading Systems.)*

Series Designation		Dia. in mm & inches	Brass & Bronze		Aluminum Alloy			
					Soft		Hard	
Size	Pitch		Max.	Min.	Max.	Min.	Max.	Min.
42	3	mm	39.999	39.951	40.014	39.966	39.999	39.951
		inch	1.5860	1.5841	1.5865	1.5846	1.5860	1.5841
42	4.5	mm	39.020	38.961	39.036	38.978	39.020	38.961
		inch	1.5361	1.5338	1.5368	1.5345	1.5361	1.5338
45	3	mm	42.999	42.951	43.014	42.966	42.999	42.951
		inch	1.7041	1.7022	1.7046	1.7027	1.7041	1.7022
45	4.5	mm	42.020	41.961	42.036	41.978	42.020	41.961
		inch	1.6542	1.6519	1.6549	1.6526	1.6542	1.6519
48	3	mm	46.000	45.950	46.015	45.964	46.000	45.950
		inch	1.8111	1.8091	1.8117	1.8097	1.8111	1.8091
48	5	mm	44.691	44.628	44.709	44.645	44.691	44.628
		inch	1.7595	1.7570	1.7602	1.7577	1.7595	1.7570
52	3	mm	49.999	49.948	50.013	49.963	49.999	49.948
		inch	1.9685	1.9665	1.9691	1.9671	1.9685	1.9665
52	5	mm	48.678	48.628	48.696	48.645	48.678	48.628
		inch	1.9165	1.9145	1.9172	1.9152	1.9165	1.9145

Note: These dimensions are for setup reference. Diameters must be finally established by actual rolling.

3. The concentrate, when mixed with water, must form a thorough, stable mixture with no clotting or livering.
4. The mix should not foster the growth of fungi, mold, or bacteria, nor shall it cause dermatological effects.
5. The lubricoolant should not contain sulfur, chlorine, or other elements that may have a corrosive effect on iron, steel, copper, brass, aluminum, or other materials used for rolling.
6. The mix must remain stable in the presence of lime, aluminum-stearate, or other drawing compounds that may coat cold-drawn steel.
7. It should provide maximum rust preventative properties and have no tendency to remove paint.

Cutting speeds for thread rolling

Recommended rolling sfm recommendations range from 75 sfm (23 smm) to 250 sfm (75 smm), depending on the workpiece material, thread specification, and available machine power. With softer, more ductile material that is readily cold worked, rates of 250 sfm or even higher can be achieved. 100 sfm (30 smm) is considered a good reference point, and most threads are rolled in the 100 – 125 sfm (30 – 38 smm) range. **Table 56** gives recommended rolling head RPM based on a speed of 100 sfm (30 smm).

Maintenance and troubleshooting in thread rolling operations

Die Life Expectancy. As blank hardness increases, the difficulty in rolling and the degree of die life decreases at a proportionally faster rate due to the corresponding reduction in the material elongation (ductility) factor. **Table 57**, based on rolls for thread rolling machines, serves to illustrate the effect of hardness on die life. These

Table 56. RPM Chart for Thread Rolling Heads and Threading Machines. *(Source, Landis Threading Systems.)*

Thread Size	RPM*	Thread Size	RPM*	Thread Size	RPM*	Thread Size	RPM*
1/4–28	1680	1/2–20	820	7/8–14	460	1 3/8–12	289
5/16–18	1380	9/16–12	750	1–8	420	1 1/2–6	274
5/16–24	1340	9/16–18	730	1–12	400	1 1/2–12	264
3/8–16	1140	5/8–11	680	1 1/8–7	370	1 3/4–5	235
3/8–24	1100	5/8–18	650	1 1/8–12	360	1 3/4–12	225
7/16–14	975	3/4–10	560	1 1/4–7	330	2–4 1/2	206
7/16–20	940	3/4–16	540	1 1/4–12	320	2–12	196

* Based on speed of 100 surface feet per minute.

Table 57. Effect of Workpiece Hardness on Die Life. *(Source, Landis Threading Systems.)*

Workpiece Hardness	Yield (PSI)	Yield (MPa)	% Elongation	Pressure Factor	Average Die Life*
200 BHN (15 RC)	65,000	448	22.5	1.0	1,000,000
250 BHN (25 RC)	95,000	655	19.0	1.6	500,000
300 BHN (32 RC)	135,000	930	16.0	2.3	100,000
350 BHN (38 RC)	162,000	1117	14.0	2.8	10,000

* Average die life measured in number of pieces produced.

values are not absolute, but do serve as an indicator of the critical effect that only a few points increase of R_C has on die life.

Even though hardness and the elongation factor are regarded as indices of rollability, they are not absolute deciding factors. Certain high-temperature alloys may be comparatively soft, have high elongation and low yield considerations, yet develop a yield point that makes cold work impossible. Rolling affects the workhardening characteristics of certain materials to such a degree that the yield is increased to such a degree that permanent deformation cannot be accomplished. An example of a material with these characteristics is Haynes Alloy #25.

Head Maintenance. Thread rolling heads are generally equipped with multiple bushing type bearings. When extreme pressures are involved, such as when producing coarse pitches or threading heat treated or stainless steels, these bushings tend to mushroom or elongate endwise. Either of these conditions will sometimes cause the bearings to bind, causing the rolls to skip or stall. When this occurs, the bearings can be honed or filed to relieve the condition. An alternative is to substitute single bronze or carbide bushings for multiple bushing bearings. See **Table 58** for specific information to correct threading problems.

Prestart Checklist. The following checks should be made before beginning any rolling operation.

1. Is the blank diameter within specified tolerances? Start with the recommended minimum tolerance to assure maximum running time before tool wear results in the blank "growing" over the maximum allowable blank outside diameter.
2. Does the blank have the recommended chamfer angle?
3. Has the head been properly aligned with the work?
4. Are the rolls installed in the proper rotation? Install in 1,2,3 clockwise sequence

Table 58.Troubleshooting Chart for Thread Rolling. *(Source, Landis Threading Systems.)*

Problem	Cause	Correction
Excessive truncation on thread crest and oversize pitch diameter.	Material too hard or prone to workhardening, causing the head to deflect.	Use softer material or material less prone to workharden.
Taper at leave-off end of full threads.	Using improper helix angle bushings or rolls too low at rear end.	Check helix angle bushings. Confirm that helix angle is correct for threat being produced.
Highlights or marks on crests and flanks of threads where rolls leave work.	Overrolling. Thread is being piled and pushed ahead of the rolls.	If o.d. of blank is correct, size head to increase pitch diameter. If blank o.d. is oversize, reduce blank to recommended tolerances.
Enlarged and burnished radius in root, normally discernable with a comparator and thread chart.	Rolls not laying close enough to the helix angle of the thread.	Use correct helix angle bushings.
Variations in outside diameter of thread.	Variable blank diameter or variable hardness.	Confirm blank diameter is within recommended tolerances. Hold heat treatment to closer limits and assure uniformity.
Taper on diameter at beginning of the thread.	Axial flow of surface material toward end of work, leaving insufficient material to finish the thread.	Change to less ductile material or, if there is an unthreaded section in front of the thread, cut off the thread after rolling.
Taper on pitch diameter at beginning of thread. Usually on the first two or three threads.	Head deflecting due to material hardness.	Use softer material or use rolls with a longer throat angle.
Cupped end on work (first thread forced over the end of the work).	Material is flowing over an insufficient chamfer on the work.	Start with a chamfer angle of 30° from the centerline of the work. Some materials may require a chamfer angle of 15° to 12°.
Material flakes and adheres to thread flanks.	Material is either aluminum or steel that is leaded or high in sulfur content. Also caused by overrolling.	Change to a more suitable material. Decrease blank diameter.
Excessive truncation with correct pitch diameter.	More ductile materials may extrude excessively. Rough finish on the blank.	Increase blank diameter. Use blank with fine finish.
Rolls break and separate into washer-type segments several threads thick.	Overrolling or rolls laying at a higher-than-standard helix angle.	Reduce blank diameter. Use correct helix angle bushings for the application.
Premature thread crest breakdown and chipping.	Excessively hard material, incorrect helix angle, or incorrect starting lead.	Reduce workpiece hardness. Use suitable helix angle bushings. Use correct starting lead rate.
Head overheats.	Insufficient coolant supply or improper coolant application.	Increase coolant volume. Apply coolant to the workpiece at the point of roll contact, not to the rolls from the head periphery.
Thread eccentric to workpiece.	Improper alignment.	Check head and machine for misalignment.
Broken roll shafts.	Excessively hard material, striking the shoulder, or overrolling.	Use softer material. Reset head tripping point. Use recommended blank diameter.

for right-hand threading, the opposite for left-hand. For left-hand threading, install left-hand helix angle bushings.

5. Deliberately oversize the head for the first piece rolled. Then size down on subsequent pieces until the desired thread size is obtained. This will prevent overrolling and possible damage to the rolls.

Die Threading (Chasing)

Thread chasers are self-opening die heads and collapsing taps that are used to cut uniform internal or external helical threads. Die heads are capable of reproducing most thread forms including tapered pipe threads. A notable advantage of chasing threads is that, upon completion of cutting the thread form, the die head self-opens and retracts from the workpiece, thereby eliminating the need to be backed off the length of the thread. Die heads can be used on automatic screw machines, universal threading machines, and some varieties of lathes, and they may be either rotating or nonrotating.

Chaser types

External thread die chasers are fitted to either rotating or non-rotating heads, and are available in three styles: tangent, circular, or radial. *Tangential chasers* are flat and threaded on one side. They provide long life, free cutting action, and natural clearance. These chasers may either have straight "threads" ground across its length, or have the helix angle incorporated into the cutting edges. When setting straight cut chasers, refer to the helix angle Tables on pages 1360-1394. As shown in *Figure 7*, straight thread tangential chasers have two angles that are important to cutting operations. The *lead angle* is the angle created by the end of the chaser and the stamped edge. It varies with the helix angle at which the chaser is used, the type of die head, and whether or not a leadscrew feed is used. When cutting UNC (as well as metric and Whitworth coarse) standard diameter and

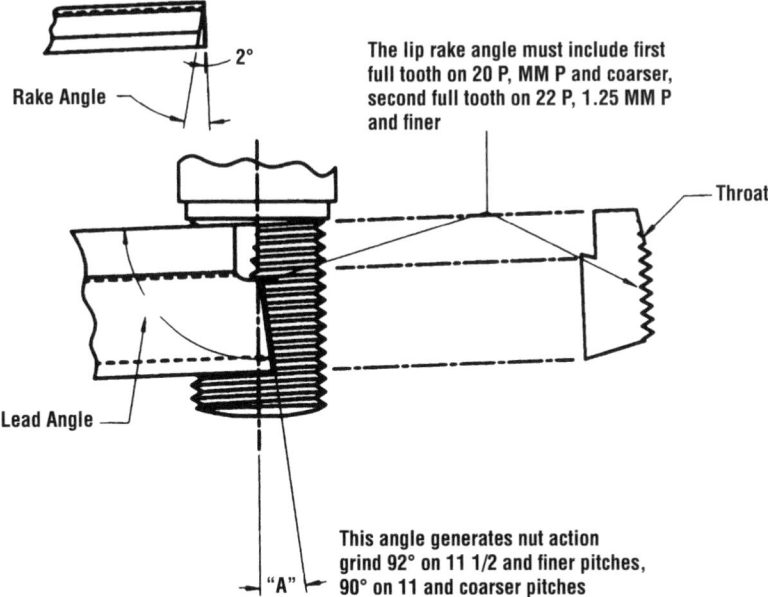

Figure 7. Rake and lead angles on tangential chasers "Lip rake" angle shown.
(Source, Landis Threading Systems.)

Figure 8. Adjustable die head for tangential cutters. (Source, Greenfield Industries, Inc.)

pitch combinations, and UNF (as well as metric and BSF) standard threads of 11 pitch or coarser, the lead angle should be ground to 90°. For UNF and other fine threads 12 pitch and finer, the suggested lead angle is 92°. The *rake angle* establishes the cutting edge of the chaser. It varies with the machinability of the workpiece, and **Table 59** provides recommended rake angles for selected materials for tangential cutters. A third angle, the *throat angle* is set and does not require regrinding. It is predetermined to provide maximum chaser life and is important in determining thread finish quality. *Figure 8* shows a solid adjustable die head with four tangential cutters locked into place. This versatile head is available for tapping machines, Goss and De Leeuw machines, radial drills, drill presses, and boring mills. Chamfer angle is discussed below.

    Although tangential cutters are generally considered to provide the highest accuracy of all thread chasers, *circular thread cutters* have several significant features. First, they can

Table 59. Rake Angles for Tangential Thread Chasing Cutters. *(Source, Landis Threading Systems.)*

Material	Rake Angle	Material	Rake Angle
Cast iron	15°	Aluminum alloy bars and shapes	25°
Wrought/gray cast iron Malleable/ductile iron	18°	Copper	28°
		Bronze or brass, cast	5° Neg.–0°
Low carbon steel free machining (B1112, C1117, etc.), and low carbon steel, non-free machining (C1010, C1018, etc.)	22°	Brass, forged or ruled, except free cutting	22°
		Brass, free cutting bars, forgings; bronze, rolled or forged	10°
Alloy steel (SAE 2000 to 6000 series, etc.), 160–200 BHN	25°	Manganese bronze	0°–10°
		Silicon bronze (Everdur)	22°
Alloy steel (SAE 2000 to 6000 series, etc.), 200–300 BHN	18°–22°	Aluminum bronze (Ampco)	18°–22°
		Naval bronze	0°–10°
Stainless steel	25°	Plastics and fibers	0°–35° Neg.
Aluminum bars and castings Aluminum alloy castings	10°	Bakelite	0°–10°
		Lucite	0°–15° Neg.

Note: All rake angles on this table are positive unless otherwise noted.

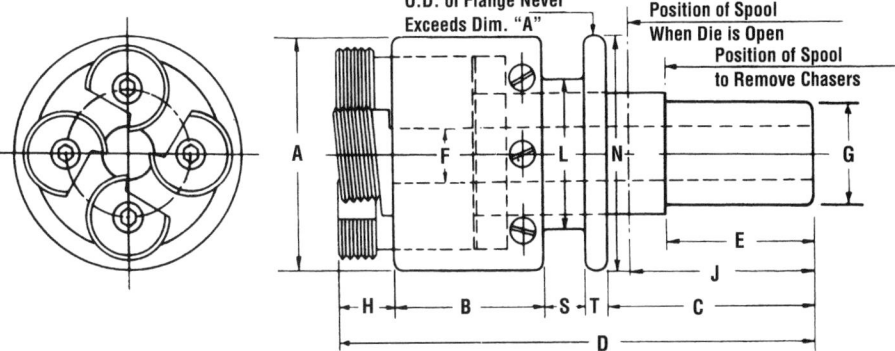

Figure 9. Revolving type, circular thread chaser head. (Source, Cleveland Twist Drill.)

be preset on chaser blocks with a micrometer, and then mounted directly on the machine tool, thereby eliminating setup time. Since they are circular, with a single flute to aid in chip removal, their cutting face can be ground through a full 270°, giving them long life and reducing replacement costs to a minimum. Their comparatively large bodies contribute to heat dissipation. Finally, since their thread grooves are annular (no lead is cut into the threading grooves), the helix angle is contained in the chaser holding block that mounts the cutter on the head. A front and profile view of a revolving style head with circular dies is shown in *Figure 9*. This head can cut straight, taper, and right- or left-hand threads. The pitch diameter is adjustable by turning opposing screws on the periphery, and when fitted with special dies, it can be used for end forming or hollow milling. It is available in a range of head diameters as small as 1.937 inches (49.20 mm) and as large as 18 inches (457.20 mm). Rake angles for circular cutters are, in most cases, the same as for tangential chasers. Chamfer angles for circular thread cutters are discussed below.

Radial thread chasers are available in several styles, with variations available from different manufacturers. In general, *milled radial chasers*, because their hardness—even though they are milled prior to heat treatment—is not equal to ground chasers, are used where misalignment and subsequent chipping due to worn machinery and/or inexperienced operators are expected. The helix angle is milled into the cutter, and the threads are resharpened on the chamfer. *Ground thread radial chasers* are held to tighter tolerances than miller chasers, and hold a much closer tolerance (within class 3 thread limits). Both of these chasers are made in sizes to fit most die heads and collapsing taps. *Insert chasers* allow interchangeability between die head and cutter. A single insert style cutter can be used in up to 100 different heads, and their small size permits economical storage—up to 3500 sets can be stored in one cubic foot. *Figure 10* provides nomenclature for insert chasers. Many of the terms and dimensions of insert chasers apply to all radial thread chasers. Rake angles for radial cutters are, in most cases, one-half that of tangential chasers. Chamfer angles for circular thread cutters are discussed below.

Chamfer angle

The chamfer angle (or throat length) on a chaser is specified to allow the chaser to thread close to the shoulder of a bolt head. Since chip size decreases, and chaser life and stress on the cutting edge increase with smaller chamfer angles, a minimal angle should be used unless threading as close to the shoulder as possible is required. A short chamfer (larger angle) is required to provide a thread close to the head, and good tool life and thread quality. **Tables 60** and **61** provide the amount of undercut for chamfer angles

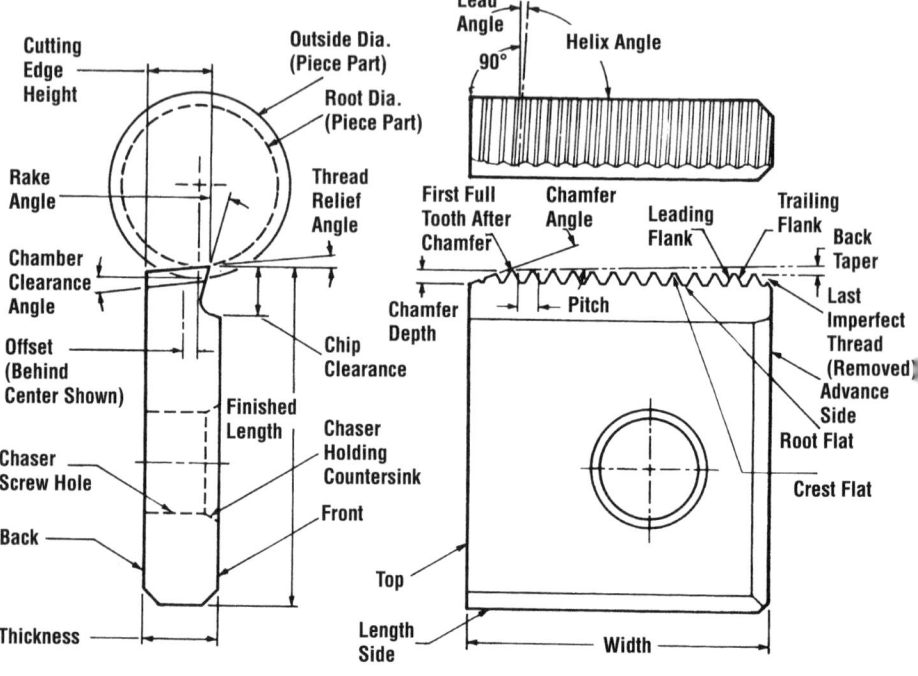

Figure 10. Insert chaser nomenclature. (Source, Cleveland Twist Drill.)

from 10° to 45°. However, it should be remembered that these distances do not allow for clearance for the die head's forward travel during opening at the completion of cutting. The following distances should be added to the undercuts given in the tables in order to assure that no damage is done from the chasers contacting the workpiece when the die head opens.

For 32–14 pitch chasers, add $^1/_{32}$" (.03125 inch/.7938 millimeter).

For 13–8 pitch chasers, add $^3/_{64}$" (.046875 inch/1.1906 millimeters).

For 7–4 pitch chasers, add $^1/_{16}$" (.0625 inch/1.5875 millimeters).

For .5–1.75 mm pitch chasers, add .8 millimeter.

For 2.0–3.0 mm pitch chasers, add 1.2 millimeters.

For 3.5–6.0 mm pitch chasers, add 1.5 millimeters.

Workpiece design should always allow for the widest relief possible, so that a chaser with the longest possible throat can be used. This will allow the chip to be divided over a longer cutting edge, which will enhance chaser life and produce a superior thread finish.

Cutting fluids for thread chasing

As in all machining operations, efficiency and quality can be maximized with the proper tools and fluids. The following cutting fluids are suggested by Cleveland Twist Drill for chasing operations.

Steel, Free Machining. B1112, B1113, C1117, C1118, and leaded high sulfur grades. Use noncorrosive sulfurized animal fatty oil; viscosity 125/175 SUS at 100° F; or water emulsifiable oil.

Carbon Steel. Carbon range of .10 to .40. Use chemically combined sulfurized

Table 60. Chamfer Length for Thread Chasers—Unified Threads. *(Source, Cleveland Twist Drill.)*

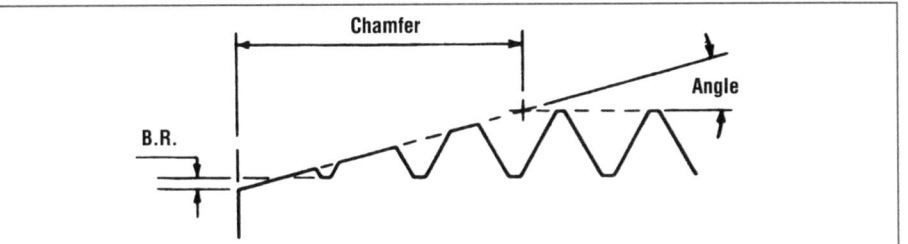

Threads Per Inch	Chamfer Angle							
	10°	15°	20°	25°	30°	35°	40°	45°
	.010" Below Root					.005" Below Root		
5	.797	.520	.383	.299	.241	.192	.160	.135
5-1/2	.727	.476	.350	.274	.221	.175	.146	.123
6	.674	.440	.324	.253	.204	.161	.134	.113
7	.585	.383	.281	.219	.177	.139	.116	.097
8	.519	.342	.252	.197	.159	.124	.103	.087
9	.467	.305	.224	.175	.141	.110	.091	.077
10	.426	.278	.205	.160	.129	.099	.083	.072
-	.008" Below Root					.003" Below Root		
11	.382	.252	.186	.145	.117	.089	.075	.062
11-1/2	.368	.240	.177	.138	.112	.085	.071	.059
12	.354	.233	.172	.134	.108	.082	.069	.058
13	.330	.216	.159	.124	.100	.076	.063	.053
14	.310	.203	.150	.117	.094	.071	.059	.049
16	.277	.180	.132	.103	.083	.062	.051	.043
-	.005" Below Root					.002" Below Root		
18	.234	.155	.114	.088	.072	.055	.046	.038
19	.222	.146	.107	.084	.068	.052	.043	.036
20	.213	.140	.103	.081	.065	.050	.041	.035
22	.196	.129	.095	.074	.060	.045	.038	.032
24	.183	.118	.087	.067	.055	.041	.034	.029
-	.002" Below Root					.001" Below Root		
26	.154	.101	.074	.058	.047	.037	.031	.026
27	.149	.097	.071	.056	.045	.036	.030	.025
28	.142	.094	.069	.054	.044	.035	.029	.024
30	.135	.088	.065	.050	.041	.032	.027	.023
32	.127	.085	.062	.048	.039	.031	.026	.021
36	.114	.075	.055	.043	.035	.027	.023	.019
40	.104	.067	.052	.039	.032	.025	.021	.017
44	.095	.063	.046	.036	.029	.023	.019	.016
48	.088	.058	.042	.033	.026	.021	.017	.014
56	.077	.051	.037	.029	.024	.018	.015	.013

All dimensions in inches.

Table 61. Chamfer Length for Thread Chasers—Metric Threads. *(Source, Cleveland Twist Drill.)*

Metric Pitch	Chamfer Angle							
	10°	15°	20°	25°	30°	35°	40°	45°
	.25 Millimeter Below Root					.13 Millimeter Below Root		
6.00	23.06	15.19	11.18	8.71	7.04	5.64	4.70	3.94
5.50	21.26	14.00	10.31	8.05	6.50	5.18	4.32	3.63
5.00	19.46	12.80	9.42	7.37	5.94	4.72	3.94	3.30
4.50	17.65	11.63	8.56	6.68	5.38	4.27	3.56	3.00
4.00	15.85	10.44	7.70	5.99	4.85	3.81	3.18	2.67
3.50	14.05	9.25	6.81	5.31	4.29	3.35	2.79	2.36
3.00	12.25	8.08	5.94	4.62	3.73	2.90	2.41	2.03
2.50	10.46	6.88	5.05	3.96	3.20	2.44	2.06	1.73
-	.20 Millimeter Below Root					.08 Millimeter Below Root		
2.00	8.36	5.51	4.04	3.15	2.57	1.93	1.60	1.35
1.75	7.47	4.90	3.61	2.82	2.29	1.70	1.42	1.19
1.50	6.55	4.32	3.18	2.49	2.01	1.47	1.21	1.04
-	.13 Millimeter Below Root					.05 Millimeter Below Root		
1.25	5.23	3.43	2.54	1.98	1.60	1.22	1.02	.84
-	.05 Millimeter Below Root					.025 Millimeter Below Root		
1.00	3.89	2.57	1.88	1.47	1.19	.94	.79	.66
.80	3.18	2.08	1.55	1.19	.97	.76	.64	.53
.75	3.00	1.98	1.45	1.14	.91	.71	.61	.51
.70	2.82	1.85	1.37	1.07	.86	.66	.56	.48
.60	2.43	1.63	1.19	.91	.74	.58	.48	.41
.50	2.08	1.37	1.02	.79	.64	.48	.41	.36
.45	1.91	1.24	.91	.71	.58	.46	.38	.30
.40	1.73	1.14	.84	.66	.53	.41	.33	.28
.35	1.55	1.02	.76	.58	.48	.36	.30	.25
.30	1.37	.89	.66	.51	.41	.30	.25	.23

All dimensions in millimeters.

oil—active sulfur 1.5% min., plus animal fatty oil, 4% max.; viscosity 150 to 200 SUS at 100° F, or water emulsifiable oil.

Carbon Steel. Carbon range of .45 and higher. Use sulfur-chlorinated animal fatty oil—compounded in petroleum vehicle; viscosity 250 SUS at 100° F.

Alloy Steel. Carbon range of .10 to .35. Same as for carbon steel, .10 to .40 carbon.

Alloy Steel. Carbon range of .40 and higher. Use 10% to 30% sulfurized fatty oil, or sulfur-chlorinated fatty oil compounded in petroleum vehicle; viscosity 250 SUS at 100° F.

Stainless Steel, Type 300 series. Same as carbon steel, .10 to .40 carbon, or water emsulsifiable oil.

Stainless Steel, Type 400 series. Same as alloy steel, .40 carbon and higher, or water emsulsifiable oil.

Cast Iron, Malleable. Use water emulsifiable oil. For superior finish, use sulfurized animal fatty oil, active sulfur 0.5% to 1% in finish viscosity petroleum vehicle of 60 SUS at 100° F.

Cast Iron, Wrought. Same as carbon steel, .10 to .40 carbon.

Aluminum, Pure Alloy. Use petroleum oil containing 1%–2% animal fatty oil, viscosity of finished oil 40–60 SUS at 100° F, or use water emulsifiable oil.

Bronze, Regular Commercial. Use petroleum oil containing 15%–25% animal fatty oil in compound with viscosity 150–200 SUS at 100° F, or water emulsifiable oil.

Bronze, Silicon, Phosphor, or Aluminum. Use petroleum oil containing in compound sulfurized animal fatty oil (active sulfur 5% max.) or water emulsifiable oil.

Copper. Same as commercial bronze.

Brass, Free Cutting. Use water emulsifiable oil, or paraffin neutral oil, viscosity range 75–100 SUS at 100° F.

Brass, Forging. Same as silicon bronze.

Monel. Same as silicon bronze, but clean immediately with petroleum solvent.

Magnesium. Same petroleum oil as pure alloy aluminum. Do not use water. Tools should always be flooded.

Fiber. Dry.

Rubber. Use water emulsifiable oil, plus the addition of sodium carbonate in the amount of 25% total emulsion.

Power requirements for thread chasing

Calculating the power required for a thread chasing operation can be useful in evaluating machining methods, tools, tool life, and production rates. The method provided below is a simple way to estimate the horsepower needed to produce a given pitch (tpi) on a given workpiece material. Due to the variables involved, it is not 100% accurate, and the rating is for horsepower required at the tool. Due to machine inefficiency, the actual power required at the motor may be 10% to 20% higher than the values provided by the following formula. Therefore, an appropriate allowance, based on machine age and condition, should be made with regard to age and condition of the machine.

$$P.P.V. \times M.P.F. \times sfm \times T.S. = HP$$

where: P.P.V. = Pitch Power Value from **Table 62**

M.P.F. = Material Power Factor from **Table 63**

sfm = surface feet per minute

T.S. = Tool Sharpness factor.

Table 62 provides power requirements based on threading free machining steels B1111 and 1213. The P.P.V. value (in the second column) has already been factored into the values given in the Table, and is given for reference when machining other materials. While the values given in the Table are based on the Material Power Factor rating for free machining steel (1.0), they can be used for other materials by multiplying the chart value by the M.P.F. of the material being cut. For example, the M.P.F. factor for stainless steel (from **Table 63**) is 2.40. Thus the power required to cut 16 tpi in stainless steel at 30 sfm can be calculated as follows.

$$2.40 \times .405 = .972 \text{ HP at the tool,}$$

where 2.40 is the M.P.F. for stainless steel, and .405 is the HP needed to cut 16 tpi into free machining steel (from **Table 62**).

The Tool Sharpness (T.S.) is factored as 1.0 for a sharp tool; 1.5 for a medium sharp tool; and 2.0 for a dull tool.

Example. Find the power at the tool required to machine 14 tpi onto alloy steel (BHN 220) with a dull tool at 25 sfm.

Table 62. Power Requirements for Thread Chasing (See Notes). *(Source, Landis Threading Systems.)*

SFM		10	15	20	25	30	35	40	45	50
TPI	P.P.V.	Horsepower								
4	.120	1.800	2.700	3.600	4.500	5.400	6.300	7.200	8.100	9.000
4.5	.0951	1.427	2.1398	2.853	3.5663	4.2795	4.9928	5.706	6.4193	7.1325
5	.077	1.155	1.7325	2.310	2.8875	3.465	4.0425	4.620	5.1975	5.775
6	.0532	.7984	1.1976	1.5968	1.996	2.3952	2.7944	3.1936	3.5928	3.992
7	.0393	.589	.8835	1.178	1.4725	1.767	2.0615	2.356	2.6505	2.945
8	.030	.450	.675	.900	1.125	1.350	1.575	1.800	2.025	2.250
9	.0248	.3716	.5574	.7432	.929	1.1148	1.3006	1.4864	1.6722	1.858
10	.020	.300	.450	.600	.750	.900	1.050	1.200	1.350	1.500
11	.016	.240	.360	.480	.600	.720	.840	.960	1.080	1.200
11.5	.0155	.232	.349	.465	.581	.698	.814	.930	1.046	1.163
12	.015	.225	.3375	.450	.5625	.675	.7875	.900	1.0125	1.125
13	.012	.180	.270	.360	.450	.540	.630	.720	.810	.900
14	.0103	.1545	.2318	.309	.3863	.4635	.5408	.618	.6953	.7725
16	.009	.135	.2025	.270	.3375	.405	.4725	.540	.6075	.675
18	.007	.105	.1575	.210	.2625	.315	.3675	.420	.4725	.525
20	.006	.090	.135	.180	.225	.270	.315	.360	.405	.450
24	.004	.060	.090	.120	.150	.180	.210	.240	.270	.300
28	.003	.045	.0675	.090	.1125	.135	.1575	.180	.2025	.225
32	.002	.030	.045	.060	.075	.090	.105	.120	.135	.150
36	.0016	.0238	.0357	.0476	.0595	.0714	.0833	.0952	.1071	.119
40	.0013	.0192	.0288	.0384	.048	.0576	.0672	.0768	.0864	.0960
44	.0010	.0154	.0231	.0308	.0385	.0462	.0539	.0616	.0693	.077
48	.0009	.0131	.0197	.0262	.0328	.0393	.0459	.0524	.0590	.0655
56	.0007	.0100	.015	.020	.025	.030	.035	.040	.045	.050
64	.0005	.0077	.0116	.0154	.0193	.0231	.0270	.0308	.0347	.0385

Notes: For speeds in excess of 50 sfm, add the totals of existing columns. For example, to find the horsepower required to chase a 12 tpi thread at 75 sfm, add the totals from the columns for 50 sfm (1.125 HP) and 25 sfm (.5625 HP): the power required is 1.6875. To convert horsepower to kilowatts (kW), multiply HP by 0.7456999. To convert surface feet per minute (sfm) to surface meters per minute (smm), multiply sfm by 0.3048.

The values in this Table are for threading B1111 and 1213 free machining steel with a Material Power Factor (M.P.F.) of 1.0 with medium sharp tools having a Tool Sharpness (T.S.) rating of 1.5 (medium-sharp tool). P.P.V. stands for Pitch Power Factor. See text for full explanation of terms and derivations of factors used for calculating power requirements for thread chasing operations.

$$.0103 \times 2.10 \times 25 \times 2.0 = 1.0815 \text{ HP}$$

where: 0.0103 is the P.P.V. for 14 tpi (**Table 62**)
2.10 is the M.P.F. for alloy steel (BHN 220) from **Table 63**
25 is the speed in surface feet per minute
2.0 is the T.S. for a dull tool.

Cutting speeds for thread chasing

As in any cutting operation, tool rigidity, workpiece material, required tolerances, and desired finish are all instrumental in setting cutting speeds. In addition, tool balance can be critical at higher speeds when using thread chasing heads. Recommended surface speeds for high speed steel chasers cutting unified or metric threads are given in **Table 64**.

Collapsing taps

Collapsing taps are widely used for threading larger holes, especially for pipe threads. These taps quickly thread large diameter holes, and then collapse so that withdrawal does not require backing out by reverse threading. Since they are adjustable, a single tapping head can thread a large range of pipe sizes. They can be used for cutting standard 60° threads as well as taper pipe threads. These tools can be highly versatile, and *Figure 11* shows a collapsing tap fitted with rose reamer blades, facing blades, and threading chasers. This tool will ream a hole to tap drill size, face the workpiece, tap the hole, and collapse and rapidly extract from the workpiece. Collapsing taps are available in either stationary or revolving designs.

Table 63. Material Power Factors for Thread Chasing (See Notes). *(Source, Landis Threading Systems.)*

Material	Hardness BHN	Power Factor (M.P.F.)	Material	Hardness BHN	Power Factor (M.P.F.)
Aluminum	-	.28	Free Cutting Steel (111-1213)	140	1.00
Brass	-	.50			
Bronze	-	.50	Alloy Steel (1330-8642)	175	1.40
Cast Iron	-	.90		190	1.50
Copper	-	.62		200	1.60
Magnesium	-	.70		203	1.70
Malleable Iron	-	1.10		205	1.80
Stainless Steel	-	2.40		210	1.90
Titanium	-	2.15		215	2.00
Zinc	-	.50		220	2.10
	90	1.50		230	2.20
	110	1.60		240	2.30
	140	1.70		250	2.40
Carbon Steel (1008-1095)	170	1.80		330	3.20
	190	1.90		390	3.80
	200	2.00		470	4.50
	250	2.10	Structural Steel (A-36)	160	1.40

Notes: Hardness is expressed in Brinell hardness (BHN). M.P.F is Material Power Factor, with free cutting steel having a reference value M.P.F. of 1.0. See text for explanation of how to use M.P.F. in calculating power required for thread chasing operations.

Table 64. Recommended Surface Speeds for Thread Chasers (See Notes). (Source, Cleveland Twist Drill.)

Material	Threads per Inch								
	Surface Feet per Minute								
	3-4 1/2	5-6	7-8	9-11 1/2	12-15	16-19	20-24	25-32	33-120
Carbon Steels-Plain Carbon									
1052, 1064*, 1065*, 1066*, 1069*, 1070*, 1071*, 1074*, 1075*, 1080*, 1084*, 1085*, 1086*, 1090*, 1095*	5	6	7	9	11	13	14	15	16
1036, 1038, 1039, 1040, 1041, 1042, 1043, 1045, 1046, 1049, 1051, 1052, 1053, 1054*, 1055*, 1059*, 1060*, 1061*	8	11	14	17	20	22	24	25	27
1008, 1010, 1011, 1012, 1013, 1015, 1017, 1020, 1024, 1025, 1027, 1029, 1030, 1031, 1033, 1034, 1035, 1037, 1045*, 1050*	12	18	23	28	32	36	38	40	43
1016, 1018, 1019, 1021, 1022, 1023, 1026	17	26	35	43	49	54	58	61	66
Carbon Steels-Free Machining									
1137, 1140, 1141, 1145, 1146, 1151	12	18	23	28	32	36	38	40	43
1108, 1109, 1115, 1120, 1125, 1126, 1132, 1138, 1139, 1141*, 1144, 1145*, 1151*	17	26	35	43	49	54	58	61	66
1116, 1117, 1118, 1119, 1121	23	37	49	61	70	79	85	91	100
1112, 1113, 1119, 1212, 1213, 12L14	30	48	66	83	98	112	112	112	112
Magnesium Steels (1.60-1.90% Mn)									
1130*, 1335, 1340	8	11	14	17	20	22	24	25	27
Nickel Steel (3.50-5.00% Ni)									
2330	8	11	14	17	20	22	24	25	27
Nickel Chromium Steels (1.25-3.50% Ni, 0.60-1.50% Cr)									
3140	5	6	7	9	11	13	14	15	16
3310*	8	11	14	17	20	22	24	25	27
3115, 3140*	12	18	23	28	32	36	38	40	43

(Continued)

Table 64. *(Continued)* Recommended Surface Speeds for Thread Chasers (See Notes). *(Source, Cleveland Twist Drill.)*

Material	Threads per Inch								
	Surface Feet per Minute								
	3-4 1/2	5-6	7-8	9-11 1/2	12-15	16-19	20-24	25-32	33-120
Molybdenum Steels									
4817*, 4820*	5	6	7	9	11	13	14	15	16
4063*, 4145*, 4147*, 4150*, 4320*, 4337*, 4340*, 4718, 4720, 4815*	8	11	14	17	20	22	24	25	27
4027, 4037*, 4042*, 4047*, 4130*, 4135*, 4137*, 4140*, 4142*, 4422, 4427, 4520, 4615, 4620, 4621	12	18	23	28	32	36	38	40	43
4023, 4024, 4118	17	26	35	43	49	54	58	61	66
Chromium Steels									
50100*, 51100*, 52100*	5	6	7	9	11	13	14	15	16
5150, 5155*, 5160*, 51B60*	8	11	14	17	20	22	24	25	27
50B40*, 50B44*, 5046*, 50B46*, 50B50*, 5132*, 5135*, 5140*, 5145*, 5147*	12	18	23	28	32	36	38	40	43
5015, 5115, 5120	17	26	35	43	49	54	58	61	66
Chromium Vanadium Steels									
6120, 6150*	8	11	14	17	20	22	24	25	27
6118	12	18	23	28	32	36	38	40	43
Nickel Chromium Molybdenum Steels									
8625, 8627, 8645*, 86B45*, 8650*, 8655*, 8660*, 8822, 9310*, 9840*	8	11	14	17	20	22	24	25	27
8115, 81B45*, 8615, 8617, 8620, 8622, 8630*, 8637*, 8640*, 8642*, 8720, 8735, 8740*, 8742*, 94B30	12	18	23	28	32	36	38	40	43
Silicon Manganese Steels									
9262*	5	6	7	9	11	13	14	15	16
9255*, 9260*, 9840*	8	11	14	17	20	22	24	25	27

(Continued)

Table 64. *(Continued)* **Recommended Surface Speeds for Thread Chasers** (See Notes). *(Source, Cleveland Twist Drill.)*

Material	Threads per Inch								
	3-4 1/2	5-6	7-8	9-11 1/2	12-15	16-19	20-24	25-32	33-120
	Surface Feet per Minute								
Nickel Chromium Molybdenum Manganese Steels									
94B15, 94B17, 94B40*	12	18	23	28	32	36	38	40	43
Stainless Steels									
200 Series, 302, 304, 321, 440A	5	6	7	9	11	13	14	15	16
303, 330, 410, 430	8	11	14	17	20	22	24	25	27
416, 430F	12	18	23	28	32	36	38	40	43
Aluminum Alloys									
1100	30	48	66	83	98	112	112	112	112
6061, Die Castings, Sand Castings	38	61	84	108	129	151	165	182	213
2011, 2017, 2024, 6262, 7075, 7178	57	96	136	186	237	300	332	370	418
Copper Alloys									
Red Brass, Manganese Bronze, Naval Brass, Electrolitic Tough Pitch Copper, Commercial Bronze, Phosphor Bronze, Al Bronze	17	26	35	43	49	54	58	61	66
Aluminum Silicon Bronze, Leaded Phosphor Bronze	23	37	49	61	70	79	85	91	100
Leaded Commercial Bronze, Forging Brass	30	48	66	83	98	112	112	112	112
Free Cutting Copper	38	61	84	108	129	151	165	182	213
Free Cutting Brass	47	77	106	140	170	208	232	259	302
Iron									
Wrought	8	11	14	17	20	22	24	25	27
Cast	17	26	35	43	49	54	58	61	66
Malleable	30	48	66	83	98	112	112	112	112
Magnesium	57	96	136	186	237	300	332	370	418

(Continued)

Table 64. *(Continued)* **Recommended Surface Speeds for Thread Chasers** (See Notes). *(Source, Cleveland Twist Drill.)*

Material	Threads per Inch								
	3-4 1/2	5-6	7-8	9-11 1/2	12-15	16-19	20-24	25-32	33-120
	Surface Feet per Minute								
Nickel Alloys									
Inconel	5	6	7	9	11	13	14	15	16
Monel and others	8	11	14	17	20	22	24	25	27
Plastics (Machinable)	47	77	106	140	170	208	232	259	302
Rubber (Machinable)	57	96	136	186	237	300	332	370	418
Zinc (Die Castings)	30	48	66	83	98	112	112	112	112

* Indicates steel has been annealed.

Notes: Reduce speeds by 25% when cutting NPT and NPTF threads. To convert surface feet per minute (sfm) values into surface meters per minute (smm), divide by 0.3048.

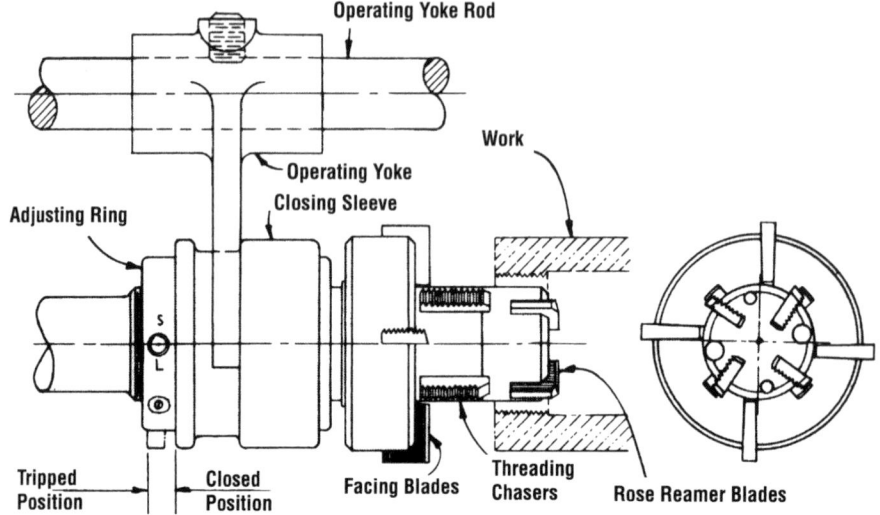

Figure 11. Multipurpose collapsing tap. (Source, Greenfield Industries Inc.)

Alignment is crucial and must be properly maintained between the tap and the hole being tapped. Misalignment will cause the tap to feed into the work at an angle that will spoil the threads and possibly damage the chasers. Rapid wearing or chipping on the tops of the chaser teeth is frequently caused by misalignment. It is advisable to use a lead screw or other positive feed to assure that the tap is fed into the work at the proper lead. If a positive feed is not used, the lead of the thread becomes largely dependent upon the operator.

The use of a proper lubricant assures the production of smooth and well formed threads. For tapping all grades of steel, including steel and wrought iron pipe, a good grade of sulfur base cutting oil is recommended. Cast and malleable iron can be tapped successfully with soluble oil as a lubricant. Cast brass, fiber, and hard rubber should be tapped dry, and kerosene works well for tapping aluminum.

Thread Grinding

While not widespread, thread grinding offers high precision and excellent surface finish. The most common use of thread grinding is for producing threads on taps and thread gages, and it is also often used to produce threads lead or adjusting screws used in measuring instruments. The process is also highly useful for threading materials that have a hardness in excess of R_C 36, and for materials with a hardness below R_C 17 that are difficult to machine cleanly with traditional threading processes. This makes threading of heat treated workpieces practical, thereby eliminating the possibility of distortion in a screw or bolt that is heat treated after threading. One of the primary benefits of grinding is that it is a gradual removal process, and stress cracks—that can cause failure under extreme pressure—are sometimes induced into sharp thread roots by traditional processes, but are eliminated by precision grinding. Ideally, ground threads should have a root radius of at least 0.003 in. (0.076 mm), and 0.010 in. (0.254 mm) is preferred. **Table 65** provides recommended grinding wheel specifications for internal and external threads. **Table 66** gives grain size requirements for grinding root widths from 0.002 to 0.010 in. (0.05 to 0.254 mm), and **Table 67** shows the coarsest grain sizes permissible for pitches ranging from 4 tpi to 80 tpi (0.3 to 6 mm pitch). The Norton Co. recommends the following cutting speeds for grinding: 9,000 to 10.500 sfm (2,750 to 3,200 smm) for resinoid wheels and 7,500 to 9,500 sfm (2,285 to 2,900 smm) for vitrified wheels.

Parts as small as 0.015 in. to as large as 16 in. (0.381 to 406 mm) can be ground with threads holding a 0.0002 in. (0.005 mm) pitch diameter tolerance limit, a lead tolerance of 0.0005 in./in. (0.0127 mm/mm), and a surface finish of 4 rms. Either center (cylindrical) or centerless grinding may be used, but centerless grinding provides increased efficiency because it allows the complete thread to be ground in one pass. Center grinding normally requires five to six passes when grinding from the solid. Centerless grinding uses

Table 65. Recommended Grinding Wheel Specifications for Thread Grinding. *(Source, Norton Co.)*

Application	Thread Type	TPI	Precision Quality	Commercial Quality
		External Grinding		
Precision Screws	Fine Pitch	32-40 (from solid)	32A220-M9VG	-
		Finer than 40 (from solid)	A320-09B	-
Screws and Studs (Heat treated alloy steel)	American National Form Threads	8-12 (from solid)	-	23A100-R9BH
		14-20 (from solid)	-	23A120-R9BH
		24 and finer (from solid)	-	23A150-S9BH or
			-	23A150-T9BH or
			-	23A180-T9BH
Taps (High speed steel)	Acme Threads	8-18 (from solid)	-	23A100-R9BH
		20-26 (from solid)	-	23A150-S9BH
		27 and finer (from solid)	-	23A150-T9BH or
			-	23A180-T9BH
	American National Form Threads	4-12 (from solid)	32A100-K8VG	23A100-R9BH
		14-20 (from solid)	32A120-L9VG	23A120-R9BH
		24-36 (from solid)	32A180-N9VG	23A180-T9BH or
			-	A220-U9BH
	American Standard Pipe Threads	8-18 (from solid)	-	23A100-R9BH
		20-26 (from solid)	-	23A150-S9BH
		27 and finer (from solid)	-	23A150-T9BH or
			-	23A180-T9BH
Whitworth Form Threads	Whitworth Form Threads	8-16	32A120-K9VG	-
		18-24	32A150-L9VG	-
Worms and Lead Screws	Acme and Worm Threads	4-6 (precut)	32A80-J8VG	-
		8-12 (precut)	32A100-K8VG	-
		12-16 (from solid)	32A120-K9VG	23A100-R9BH
		Internal Grinding		
Tool and Alloy Steel Hardened	Acme Threads	2-8 (precut)	38A80-J8VG	-
		10-20 (precut)	38A120-K9VG	-
	American National Form Threads	6-16 (precut)	38A120-L9VG	23A100-V9BH
		20-24 (precut)	38A220-M9VG	23A150-V9BH
		6-20 (from solid)	-	23A100-V9BH
		24 and finer (from solid)	-	23A150-V9BH

(Continued)

Table 65. *(Continued)* **Recommended Grinding Wheel Specifications for Thread Grinding.** *(Source, Norton Co.)*

Application	Thread Type	TPI	Precision Quality	Commercial Quality
		Multirib Grinding		
		Wheel Size ➤	Up to 14" Dia.	15" Dia. and Over
Hardened Steel	Centerless	5-10	-	32A120-Q9VG
	Cylindrical		38A120-N9VG	32A120-N9VG
	Surfacing		38A120-K9VG	32A120-K9VG
	Centerless	11-14	38A150-N9VG	38A150-Q10VG
	Cylindrical		38A150-N9VG	32A150-N9VG
	Surfacing		38A150-K9VG	32A150-K9VG
	Centerless	16-24	38A2203-N9VG	38A220-Q10VG
	Cylindrical		38A2203-N9VG	38A220-N9VG
	Surfacing		38A2203-K9VG	32A220-K9VG
	Centerless	24-32	38A320-N10VG	A320-Q11VG
	Cylindrical		38A320-N10VG	A320-N10VG
	Surfacing		38A320-K9VG	A320-K9VG

Table 66. Grain Size Requirements for Given Thread Roots. *(Source, Norton Co.)*

Root Width in./mm	Grain Size	
	Vitrified Wheel	Resinoid Wheel
0.002 / 0.05	–	180
0.003 / 0.076	220	180
0.004 / 0.10	180	150
0.005 / 0.127	150	120
0.006 / 0.15	120	120
0.007 / 0.178	120	100
0.008 / 0.20	100	100
0.009 / 0.229	100	90
0.010 / 0.254	90	90

Note: Even though resinoid wheels remove stock faster than vitrified wheels, they are considerably less rigid than vitrified wheels and care must be taken to assure that they do not deflect during cutting operations.

Table 67. Grain Size Requirements for Given Thread Pitches (Vitrified Wheels). (See Notes.)

Pitch tpi (mm)	Grain Size	Pitch tpi (mm)	Grain Size
4 (6)	70	16-20 (1.25-1.5)	120 / 220*
4 1/2–8 (3–5.5)	80	24-28 (.8-1)	150
5-10 (2.5-5)	120*	24-32 (.8-1)	320*
9-12 (2-2.5)	90	32-36 (.7-.75)	180
11-14 (1.75-2)	150*	40-64 (.4-.6)	220
13-14 (1.75-2)	100	72-80 (.3-.35)	240

Notes: *For multirib wheels. All other sizes are for single rib wheels. Metric pitch conversions are equivalent. All grain sizes are the coarsest allowable for pitch of thread.

multirib wheels and offers speed and economy. For example, production rates of 60 to 70 $1/4$–20 hardened socket set-screws can be ground from solid in less than a minute, with the grinding wheel having a useful life of up to 8 hours (up to 560 set-screws) before redressing.

Single-Rib Wheel. Single rib wheels have the thread profile on their cutting edge and must traverse, while inclined to the thread's helix angle, the length of the thread. Lead screw control and some other form of automatic infeed of the grinding machine is required, and multiple passes are normally required to finish a thread, with about two-thirds of the thread depth being removed on the first pass or passes, and the final finish cut removing the remaining one-third of thread. The exceptions are fine threads 12 tpi and finer that can often be cut in one pass. The maximum infeed for a single-rib wheel should be 0.040 in. (0.10 mm); 0.020 in. (0.05 mm) is commonly used for roughing, and 0.010 in. (0.025 mm) or less is used for achieving high accuracy. Grinding threads with a single-rib wheel is analogous to single-point threading on a lathe. It is the most common method of thread grinding and, with care, can be the most accurate.

Multirib Wheel. When production rate takes precedence over ultimate accuracy, multirib wheels are used. In instances where the thread is longer than the width of the wheel, traverse grinding is required, but it should be noted that the leading ribs on the wheel will experience accelerated wear and will eventually cut oversize threads while the trailing ribs have retained their accuracy. When using traverse grinding with multirib wheels, the thread pitch should not be in excess of one-eighth of the wheel's width, and the root width should be at least 0.007 in. (0.178 mm) wide due to the difficulty inherent in dressing the wheel for smaller root widths. In instances where the wheel width exceeds the thread length by 1.5 thread pitches, plunge grinding can be used to produce the complete thread in less than $1^1/_2$ revolutions of the workpiece, making this method the most productive type of thread grinding. In plunge grinding, the wheel is introduced to the work gradually over about one-half revolution and then cuts the complete thread and withdraws in just over a full revolution. In addition to conventional multirib wheels that contain several identical thread profiles, other varieties are the skip-rib (or alternate-rib) and three-rib

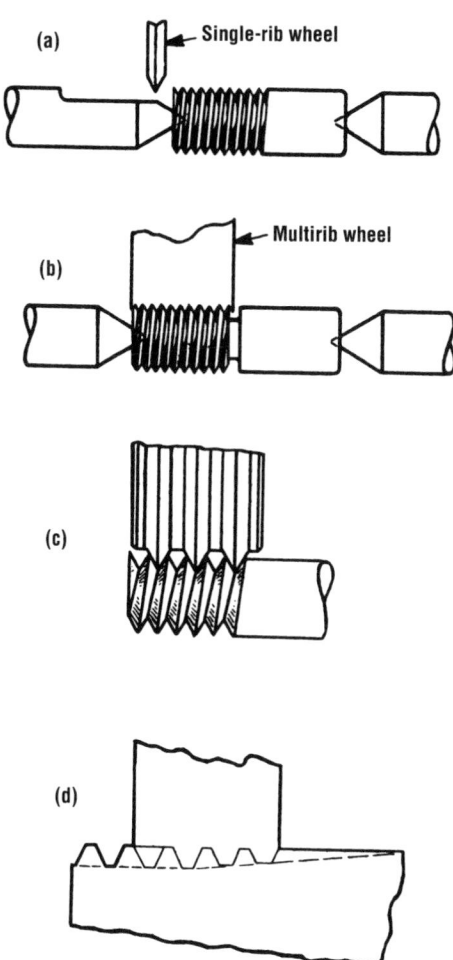

Figure 12. (a) Single-rib wheel. (b) Multirib wheel. (c) Skip-rib wheel. (d) Three-rib wheel with flattening rib.

wheels. The thread spacing on the skip-rib wheel is equal to twice the thread pitch, and approximately $2\frac{1}{2}$ workpiece revolutions are required to complete a thread. Due to the thread spacing, cutting fluids have better access to the cutting area, allowing for improved finish and making it easier to cut fine pitch threads. The three-rib design is named for its three individual purpose cutting ribs: a roughing rib that cuts to $\frac{2}{3}$ of the total thread depth, an intermediate rib that removes further material down to within approximately 0.005 in. (0.127 mm) of the total thread depth, and a finishing rib that completes the cut. An optional "flattening rib" can be used to finish grind the thread crests. Grinding wheel profiles are shown in *Figure 12*.

Thread Milling

Traditional thread milling is performed with either single- or multiform cutters. Single-form cutters approach the workpiece at an angle equal to the helix angle of the thread being cut, and advance and revolve around the workpiece at a preset rate to cut one thread at a time as the workpiece also rotates in the opposite direction of the cutter. While this method is capable of producing quality threads, it has grown out of favor due to the necessarily long setup and production time. It remains, however, a viable method for cutting large diameter workpieces or in instances where the thread length is too long for a multiform cutter. Using a multiform cutter reduces setup and production time. The complete thread form is contained on the tool which travels parallel to the workpiece which it circulates 1.1 revolutions before exiting. The most obvious limitation of the multiform cutter is that the length of the thread being cut cannot exceed the length of the cutter. Also, because the cutters are arbor or shank mounted, there are unavoidable size limitations when cutting internal threads. Most thread milling operations are now performed with full profile thread milling inserts that allow fast cutting and, as shown in **Table 68**, have been miniaturized to cut internal threads in bores as small as 0.374" (9.5 mm). **Table 69** provides troubleshooting solutions for CNC thread milling, and **Table 70** shows common tolerance classes for thread cutting inserts. The thread milling system that follows has been developed by Kennametal Inc. who produce a full range of milling holders and inserts.

When milling with inserts, the following points should be considered.

1) Inserts are designed to mill the full thread depth in one revolution.
2) When machining difficult materials, it may be desirable to make two passes. A 60% thread depth on the first pass, and a 40% thread depth on the second pass is recommended.
3) Thread relief grooves in blind holes are not necessary.
4) Thread milling large parts requires considerably less horsepower than other threading methods.
5) Thread milling produces short chips compared to stringy chips of other threading methods.
6) One holder is suitable for many different thread pitches.
7) PVD coated inserts provide maximum tool life for a wide variety of materials.

CNC thread milling

In order to perform the following thread milling operations with full profile inserts, a milling machine with three-axis control, capable of helical interpolation, is required. Helical interpolation is a CNC function producing tool movement along a helical path. Helical travel combines circular movement in one plane with a simultaneous linear motion in a plane perpendicular to the first. For example, the path from point A to point B [*Figure 13(a)*] on the envelope of the cylinder combines a circular movement in the X- and Y-plane with a linear movement in the Z-direction.

Table 68. Minimum Bore Diameters for Thread Milling with Thread Milling Inserts (UN-ISO-BSW). (Source, Kennametal Inc.)

tpi ▶	48	32	24	20	16	12	10	8
pitch (mm) ▶	0.5	0.75	1.0	1.25	1.5	2.0	2.5	3.0
Cutter Dia.	Minimum Bore Diameter inch/mm							
.35 / 8.9	.374 / 9.5	.394 / 10	.421 / 10.7	.449 / 11.4	-	-	-	-
.45 / 11.4	.472 / 12	.492 / 12.5	.520 / 13.2	.547 / 13.9	.571 / 14.5	-	-	-
.49 / 12.4	.512 / 13	.531 / 13.5	.559 / 14.2	.587 / 14.9	.610 / 15.5	-	-	-
.61 / 15.5	.630 / 16	.650 / 16.5	.667 / 16.9	.705 / 17.9	.728 / 18.5	.768 / 19.5	-	-
.67 / 17	.693 / 17.6	.717 / 18.2	.748 / 19	.772 / 19.6	.787 / 20	.827 / 21	-	-
.75 / 19	.776 / 19.7	.803 / 20.4	.827 / 21	.850 / 21.6	.866 / 22	.906 / 23	-	-
.79 / 19.3	.815 / 20.7	.843 / 21.4	.866 / 22	.890 / 22.6	.906 / 23	.945 / 24	-	-
.87 / 22.1	.893 / 22.7	.921 / 23.4	.945 / 24	.969 / 24.6	.984 / 25	1.024 / 26	-	-
1.02 / 25.9	1.051 / 26.7	1.079 / 27.4	1.102 / 28	1.130 / 28.7	1.154 / 29.4	1.193 / 30.3	-	-
1.18 / 30	1.209 / 29.7	1.236 / 31.4	1.260 / 32	1.291 / 32.8	1.319 / 33.5	1.362 / 34.6	1.441 / 36.6	1.535 / 39
1.46 / 37.1	1.496 / 38	1.520 / 38.6	1.555 / 39.5	1.591 / 40.4	1.614 / 41	1.654 / 42	1.732 / 44	1.830 / 47.2
1.65 / 42	1.701 / 43.2	1.724 / 43.8	1.772 / 45	1.811 / 46	1.831 / 46.5	1.866 / 47.4	1.929 / 49	2.047 / 52

tpi ▶	7	6	-	5	-	4.5	-	4
pitch (mm) ▶	3.5	4.0	4.5	5.0	5.5	-	6.0	-
Cutter Dia.	Minimum Bore Diameter inch/mm							
1.18 / 30	1.654 / 42	1.772 / 45	1.890 / 48	-	-	-	-	-
1.46 / 37.1	1.929 / 49	2.047 / 52	2.185 / 55.5	-	-	-	-	-
1.65 / 42	2.146 / 54.5	2.268 / 57.6	2.401 / 56.9	-	-	-	-	-
1.38 / 35 (UN)	-	1.969 / 75.4	-	1.843 / 46.8	-	1.756 / 44.6	-	2.228 / 55.6
1.38 / 35 (ISO)	-	1.969 / 75.4	2.102 / 53.4	1.673 / 42.5	1.969 / 50	-	2.264 / 57.5	-
1.38 / 35 (BSW)	-	1.961 / 75.2	-	1.831 / 46.5	-	1.866 / 47.4	-	-

All diameters given as inch / millimeter. Millimeters are expressed to the nearest tenth.

Table 69. Troubleshooting Thread Milling Operations. *(Source, Kennametal Inc.)*

Problem	Possible Cause	Solution
Excessive insert flank wear	Cutting speed too high	Reduce cutting speed.
	Chip is too thin	Increase feed rate.
	Insufficient coolant	Increase coolant quality/pressure.
Chipping of cutting edge	Chip is too thick	Reduce feed rate. Use the tangential arc method of entering. Increase rpm.
	Vibration	Check rigidity.
Material buildup on cutting edge	Cutting speed too low	Increase cutting speed.
	Chip thickness too small	Increase feed rate.
Chatter / vibration	Feed rate too high	Reduce feed rate.
	Profile is too deep (coarse threads)	Execute two passes, each with increased cutting depth. Execute two passes, each cutting only half the thread length.
	Thread length is too long	Execute two passes, each cutting only half the thread length.
Insufficient thread accuracy	Tool deflection	Reduce feed rate. Execute a "zero" cut.

Table 70. Insert Tolerance Classes. *(Source, Kennametal Inc.)*

Thread Designations	Standard Designations	Tolerance Class
UN	ANSI B1.174	2A/2B
UNJ	MIL-S-8879A	3A/3B
ISO	R262 (DIN 13)	6g/6H
NPT	USAS B2.1:1968	Standard NPT
NPTF	ANSI B1.30.3-1976	Standard
BSW	B.S. 84:1956, DIN 259, ISO 228/1:1982	Medium Class A
BSPT	B.S. 21:1985	Standard BSPT
ACME	ANSI B1/5:1988	3G
PG	DIN 40430	Standard
TR	DIN 103	7e/7H

On most CNC systems, this function can be executed in two different ways:

GO2: helical interpolation in a *clockwise* direction, or

GO3: helical interpolation in a *counterclockwise* direction.

The thread milling operation shown in *Figure 13(b)* consists of a circular rotation of the tool about its own axis together with an orbiting motion along the bore or workpiece circumference. During one such orbit, the tool will move vertically one pitch length. These movements, combined with the insert geometry, create the required thread form.

Approaching the workpiece

There are three acceptable ways to approach the workpiece with the tool to initiate the

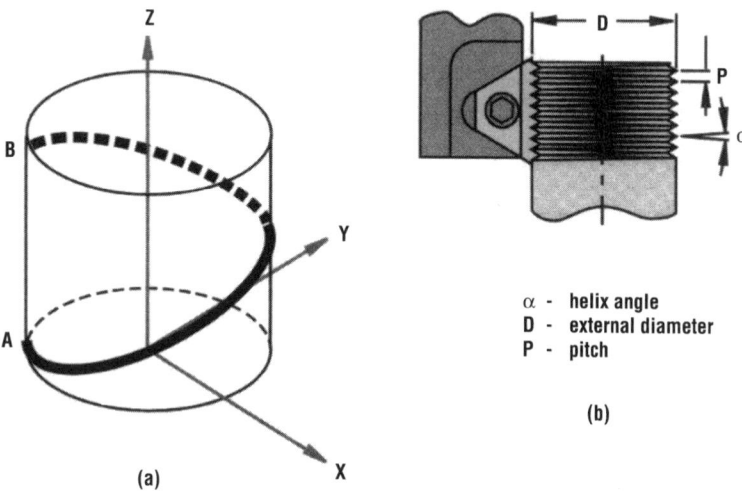

α - helix angle
D - external diameter
P - pitch

Figure 13. (a) Helical interpolation. (b) Thread milling. (Source, Kennametal Inc.)

thread: 1) along a tangential arc; 2) radially; and 3) along a tangential straight line. In all instances, climb milling is preferred.

Tangential Approach (Arc). With this method, shown in *Figure 14*, the tool enters and exits the workpiece smoothly. No marks are left on the workpiece, and there is no vibration, even with harder materials. Although the arc method requires slightly more complex programming than the radial approach discussed below, this method is recommended for machining the highest quality threads.

Radial Approach. Although this is the simplest method, there are two characteristics that should be noted about this approach. 1) A very small vertical mark that is of no significance to the thread itself may be left at the entry (and exit) point. 2) When using this method with very hard materials, there may be a tendency for the tool to vibrate as it approaches full cutting depth. When using this approach, radial feed during the entry to full profile depth should be only $1/3$ of the subsequent circular feed. *Figure 15* shows the radial approach.

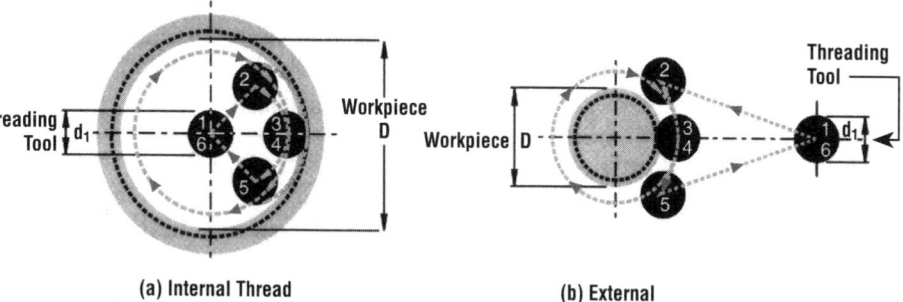

(a) Internal Thread   (b) External

Figure 14. Tangential arc approach. (a) Internal thread. (b) External thread. (Source, Kennametal Inc.)
Points 1–2 show rapid approach path. 2–3 shows tool entry along tangential arc, with simultaneous feed along the Z-axis. 3–4 show the helical movement during one full orbit (360°). 4–5 show the tool exit path along the tangential arc, with continuing feed along the Z-axis. 5–6 show the rapid return path.

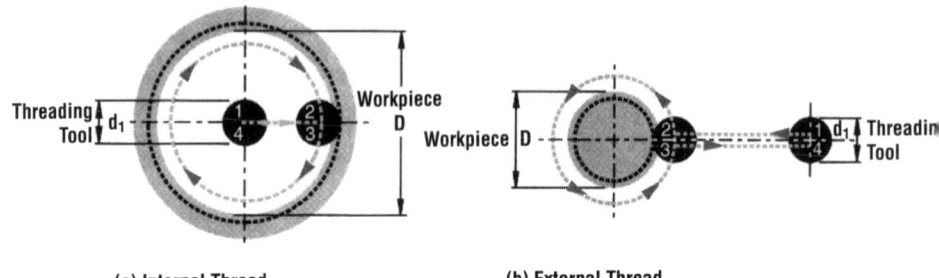

(a) Internal Thread   **(b) External Thread**

Figure 15. Radial approach. (a) Internal thread. (b) External thread. (Source, Kennametal Inc.)
Points 1–2 show the radial entry. 2–3 show the helical movement during one full orbit (360°).
3–4 show the return path.

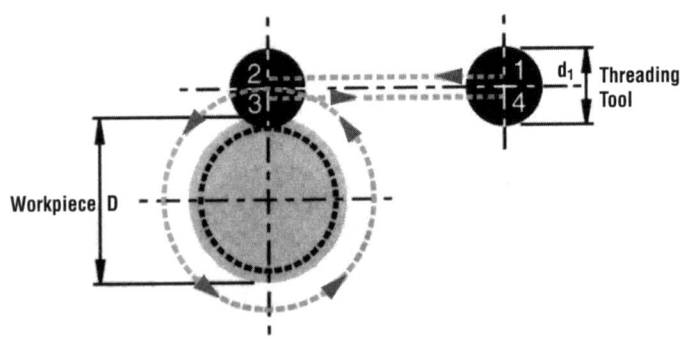

Figure 16. Tangential line approach. (Source, Kennametal Inc.)
Points 1–2 show the radial entry with simultaneous feed along the Z-axis. 2–3 show the helical movement during
one full orbit (360°). 3–4 shows the radial exit path.

Tangential Approach (Line). This method, shown in *Figure 16*, is very simple, yet shares the advantages of the tangential arc method. However, it is applicable only to external threads.

Preparing a thread milling operation

The following description uses inch system units. However, the formulas for feed rate in metric units are given in the milling section of this book.

Calculating Feed Rate at the Cutting Edge. Before beginning the operation, the feed rates at both the cutting edge and the tool centerline must be calculated. The feed rate at the cutting edge can be determined with the following equation:

$$F_1 = \text{ipt} \times \text{nt} \times \text{rpm}$$

where F_1 = tool feed rate at the cutting edge
ipt = inch/tooth (feed rate)
nt = number of effective inserts in cutter
rpm = rotational speed (spindle rpm).

The rotational speed (rpm) is calculated with the following equation:

$$\text{rpm} = (12 \times \text{sfm}) \div (\pi \times d_1)$$

where   sfm = cutting speed, surface feet per minute
d_1 = cutter diameter over insert.
π = 3.1416.

Calculating Feed Rate at the Tool Centerline. On most CNC machines, the feed rate required for programming is at the centerline of the tool. When dealing with linear tool movement, the feed rate at the cutting edge and centerline are identical, but, with circular tool movement, this is not the case. The following equations define the relationship between feed rates at the cutting edge and at the tool centerline.

For *internal threads.* (*Figure 17*)

$$F_2 = (F_1 \times [D - d_1]) \div D$$

and for *external threads.* (*Figure 18*)

$$F_2 = (F_1 \% [d_1 + D]) \div D$$

where F_2 = centerline feed rate (in./min)
       D = minor diameter, *internal thread*
       D = major diameter, *external thread*
       d_1 = cutter diameter over insert.

Calculating CNC program parameters

The following recommendations provided by Kennametal Inc. are a sure method for programming thread milling operations. For mass production, more complex methods exist. The equations in this section can be converted to metric units by substituting metric units (millimeters) for inch units.

Internal Threads (Climb Milling). The simplest way to perform an internal thread milling operation follows five programming steps. (See *Figure 19.*)

1) Rapidly move the tool to point "A." This point leaves a safe distance or clearance (C_L) between the cutter and the workpiece (usually about 0.02").

2) Move the cutter along the tool path (dashed line) from point "A" to point "B" so that within 90° (about the workpiece's centerline) it reaches maximum cutting depth, moving

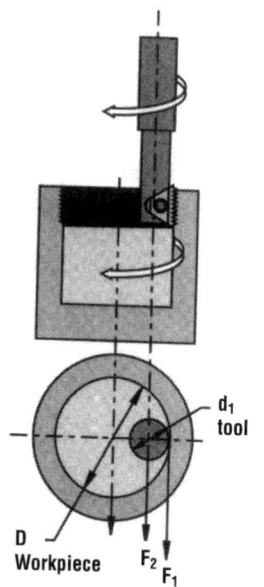
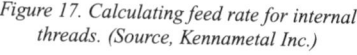

Figure 17. Calculating feed rate for internal threads. (Source, Kennametal Inc.)

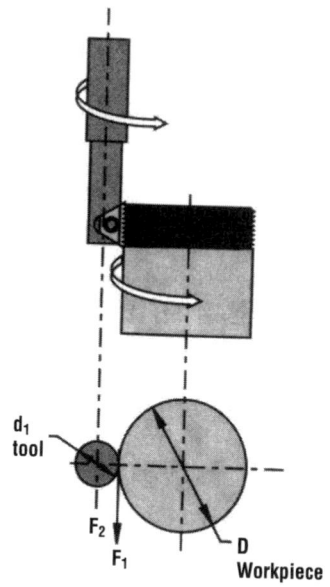

Figure 18. Calculating feed rate for external threads. (Source, Kennametal Inc.)

simultaneously $^1/_4$ pitch length in the Z-axis direction. This path along the entrance angle (β) is defined as radius R_e about a point "C" on the X-axis.

3) Machine the thread by completing one orbit around the circumference (starting and ending at point "B"), moving simultaneously one pitch length in the Z-axis direction.

4) Upon returning to point "B," exit the workpiece to point "D" along the tool path (dashed line), moving simultaneously $^1/_4$ pitch length in the Z-axis direction.

5) Rapidly return to point "O."

To program the path described above, six parameters much be calculated.

1) The entrance radius, R_e. The right triangle OAC enables R_e to be simply solved (see *Figure 20*).

$$OA = R_i - C_L$$
$$CA = R_e$$
$$OC = R_o - R_e$$
$$OA^2 + OC^2 = AC^2 \text{ (Pythagoras' Law)}$$

Replacing the actual values, $(R_i - C_L)^2 + (R_o - R_e)^2 = R_e{}^2$.

Therefore

$$R_e = ([R_i - C_L]^2 + R_o{}^2) \div 2 \times R_o$$

where $R_i = D \div 2$ (D = minor diameter, internal thread)
$R_o = D_o \div 2$ (D_o = nominal diameter, internal thread)
C_L = clearance between cutter and workpiece.

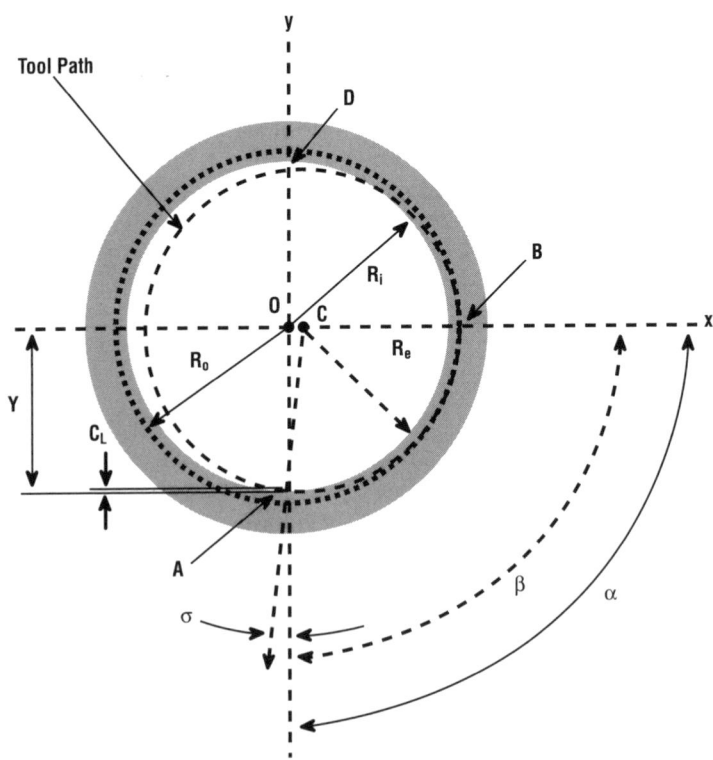

Figure 19. Machining internal threads with climb milling. (Source, Kennametal Inc.)

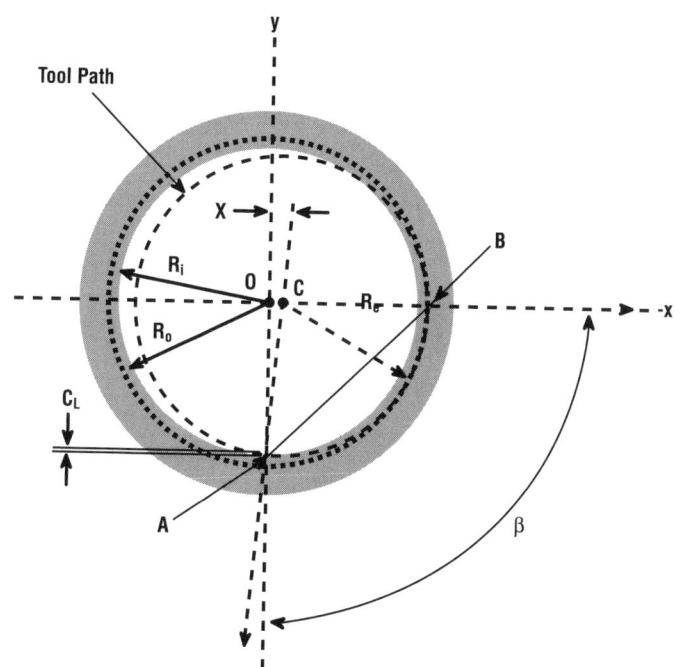

Figure 20. Solving for R_e. (Source, Kennametal Inc.)

2) The entrance angle about point "C," β (see *Figure 20*). β can be easily found using the same right triangle.

$$\beta = 90° + \sigma$$
$$\sin \sigma = OC \div CA = (R_o - R_e) \div R_e$$
$$\sigma = \text{arc sin} ([R_o - R_e] \div R_e).$$

Therefore,

$$\beta = 90° + \text{arc sin} ([R_o - R_e] \div R_e).$$

3) The vertical movement between points "A" and "B," Z_α.

$$Z_\alpha = P \times (\alpha \div 360°) = P \div 4 \text{ (since } \alpha = 90°)$$

where: P = pitch, in inches.

4) The values at the start of the entry approach, "X" and "Y" (see *Figure 19*).

$$X = 0$$
$$Y = - R_i + C_L.$$

Note: in conventional milling, $Y = R_i - C_L$.

5) The Z-axis location at the start of the entry approach, Z.

$$Z = - (L + Z_\alpha)$$

where: L = thread length.

Note: in conventional milling, $Z = -L + P + Z_\alpha$.

6) The starting point, O.

$$X_a = 0$$
$$Y_a = 0.$$

External Threads (Climb Milling). External thread milling operations are similar to the internal application. It should be noted that, for climb milling, the orbit is clockwise, and the vertical displacement is in the positive direction. The simplest way to perform an external thread milling operation follows five programming steps. (See *Figure 21*).

1) Rapidly move the cutter to point "A." This point leaves a safe distance or clearance (C_L) between the cutter and the workpiece (usually about 0.02").

2) Move the cutter along the tool path (dashed line) to point "B" so that within 90° (about the workpiece's center) it reaches maximum cutting depth, moving simultaneously $1/4$ pitch length in the Z-axis direction. In this trajectory, the entrance angle (β) is defined by a radius R_e about a point "C" on the X-axis.

3) Machine the thread by completing one orbit around the circumference, moving simultaneously one pitch length in the Z-axis direction.

4) Upon returning to point "B," exit the workpiece along the tool path (dashed line) to point "D," moving simultaneously $1/4$ pitch length in the Z-axis direction.

5) Rapidly return to point "O."

To program the path described above, seven parameters much be calculated.

1) The entry radius, R_e. The right triangle OAC enables R_e to be simply solved (see *Figure 21*).

$$OA = R_o + C_L$$
$$CA = R_e$$
$$OC = R_e - R_i$$
$$OA^2 + OC^2 = AC^2 \text{ (Pythagoras' Law)}$$

Replacing the actual values, $(R_o + C_L)^2 + (R_e - R_i)^2 = R_e^2$.

Therefore,

$$R_e = ([R_o + C_L]^2 + R_i^2) \div 2 \times R_i$$

where: $R_i = D \div 2$ (D = minor diameter, external thread)
$R_o = D_o \div 2$ (D_o = nominal diameter, external thread)
C_L = clearance between cutter and workpiece.

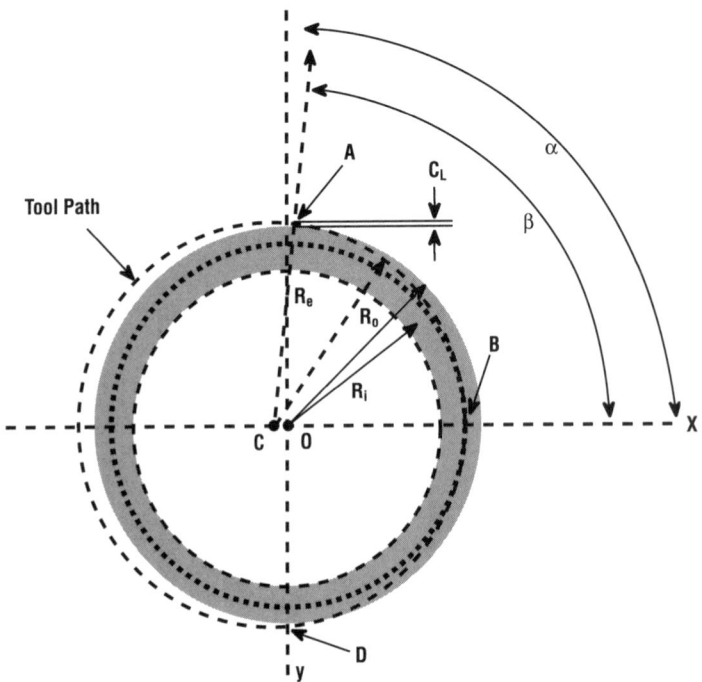

Figure 21. Machining external threads with climb milling. (Source, Kennametal Inc.)

2) The entrance angle about point "C," β (see *Figure 21*). β can be easily found using the same triangle.

$\sin \beta = AO \div AC = (R_o + C_L) \div R_e$

$\beta = \text{arc sin } (R_o + C_L \div R_e)$.

3) The vertical movement between points "A" and "B," Z_α.

$Z_\alpha = P \times (\alpha \div 360°) = P \div 4$ (since $\alpha = 90°$)

where P = pitch, in inches.

4) The values at the start of the entry approach, "X" and "Y" (see *Figure 21*).

$X = 0$

$Y = - R_o + C_L$.

Note: in conventional milling, $Y = - R_o - C_L$.

5) The Z-axis location at the start of the entry approach, Z.

$Z = L - P - Z_\alpha$

where L = thread length.

6) The tool's initiation point (prior to compensation).

$X_a = 0$

$Y_a = (R_o + C_L) + R_t$

where R_t = tool radius $(d_1 \div 2)$.

Note: in conventional milling, $Y_a = - (R_o + C_L) - R_t$.

7) X_{CNC} and Y_{CNC} defined.

$X_{CNC} = R_i$

$Y_{CNC} = R_o + C_L$.

Step-by-step thread milling example

Machining parameters.

Thread: internal right hand $1^1/_4 \times 16$ UN-2B-RH(21)

Material: AISI 4140 (300 BHN)

Thread diameters: $D = 1.182"$ (minimum bore diameter)

$D_o = 1.25"$ (nominal diameter)

Thread length: 0.50 inch

Thread milling method: climb milling.

For the best thread quality, the cutter with the largest d_1 (cutter diameter) possible should be used. As can be seen in **Table 68**, any cutter with a diameter of 1.02" or less can be used for this operation. Even though a smaller cutter will perform the operation in less time, it will provide less tool rigidity and should be used with caution on tough materials. Given the material being used for this example, a cutter diameter of 0.79" will be selected. Next, an insert with the proper pitch and thread length must be mated to the cutter, and a feed rate must be specified based on the insert grade. For this example, a cutting speed of 500 sfm and a feed per tooth of 0.004" will be used.

1) Calculate the feed rate. Begin by finding the rpm.

$\text{rpm} = (12 \times \text{sfm}) \div (\pi \times d_1) = (12 \times 500) \div (3.1416 \times .79) = 2418 \text{ rpm}$.

Next, calculate F_1, the feed rate at the insert cutting edge.

$F_1 = \text{ipt} \times \text{nt} \times \text{rpm} = 0.004 \times 1 \times 2418 = 9.67 \text{ in./min}$.

Finally, calculate F_2 the feed rate at the cutter centerline.

$F_2 = (F_1 \times [D - d_1]) \div D = (9.67 \times [1.182 - 0.79]) \div 1.182 = 3.207 \text{ in./min}$.

2) Calculate the radius of the tangential arc, R_e.

$R_e = ([R_i - C_L]^2 + R_o^2) \div 2 \times R_o = ([0.591 - 0.02]^2 + 0.625^2) \div 2 \times 0.625$

$= 0.573333 \text{ in}$.

3) Calculate the angle β.

$$\beta = 90° + \text{arc sin} ([R_o - R_e] \div R_e)$$
$$\beta = 90° + \text{arc sin} ([0.625 - 0.573333] \div 0.573333$$
$$\beta = 90° + 5.17° = 95.17° = 95° \ 10'.$$

4) Calculate the movement, Z_α, along the Z-axis during the entry approach from point "A" to point "B."

$$Z_\alpha = P \times (\alpha \div 360°) = P \div 4 \text{ (since } \alpha = 90°) = 0.0625 \div 4 = 0.0156 \text{ in.}$$

5) Calculate the "X" and "Y" values at the start of the entry approach.

$$X = 0$$
$$Y = - R_i + C_L = -0.591 + 0.02 = -0.571 \text{ in.}$$

6) Define the Z-axis location at the start of the entry approach.

$$Z = -(L + Z_\alpha) = -(0.50 + 0.0156) = -0.5156 \text{ in.}$$

7) Define the starting point.

$$X_a = 0$$
$$Y_a = 0.$$

The CNC program (for Fanuc 11M) would be written as follows.

```
%
N10G90G00G57X0.000Y0.000
N20G43H10Z0.M3S2417
N30G91G00X0.Y0.Z-0.5156
N40G41D60X0.000Y-0.5710Z0.
N50G03X0.6250Y0.5710Z0.0156R0.5733F3.206
N60G03X0.Y0.Z0.0625I-0.625J0.
N70G03X-0.625Y0.5710Z0.0156R0.5733
N80G00G40X0.Y-0.5710Z0.
N90G49G57G00Z8.0M5
N100M30
%
```

Methods of thread milling

Climb milling results in lower cutting forces, better chip development, higher thread surface quality, and longer insert life. For these reasons, it is recommended whenever possible. However, in some cases of hardened materials or when milling certain difficult-to-machine exotic materials, conventional milling may be preferred. *Figure 22 a,b,c,d* illustrates four methods of external thread milling, both conventional and climb milling; and *Figure 23 a,b,c,d* shows how the same operations can be performed when milling internal threads.

Single Point Threading

In the earliest days of threadmaking, threads were often cut on engine, bench, and toolroom lathes, but today single point threading is usually restricted to CNC machines. While manual lathes are still used for very limited production threading jobs, the setup operation is very complicated and will not be detailed in this book. Suffice to say that the setup requires selecting a correct change gear for the required pitch and, of course, applying a brazed tool with the proper relief angle.

CNC, toolholder, and insert technology, when combined, have made single point threading much less troublesome than in the past. Careful selection of the toolholder, insert, and machining data is essential if the operation is to achieve optimum performance. Unlike conventional turning, only a limited number of machining parameters can be modified in threading. These include the thread form, which dictates the insert's physical size and shape. Surface speed, number of passes, depth of each pass, and infeed angle are all critical to successful single point threading.

Most standard 60° thread forms (UN and ISO) allow for root and crest truncation, so the selection of a full profile insert (as opposed to a full thread height insert) will increase

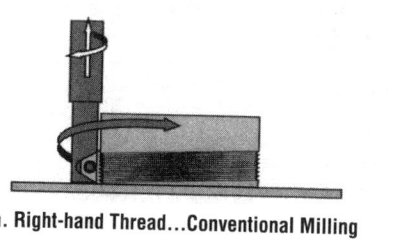

a. Right-hand Thread...Conventional Milling

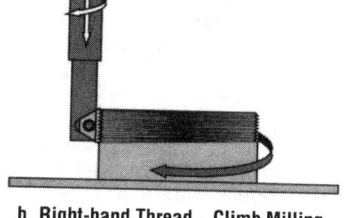

b. Right-hand Thread...Climb Milling

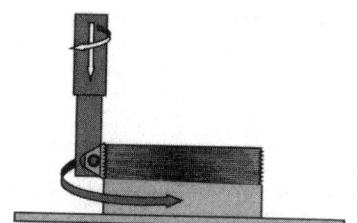

c. Left-hand Thread...Conventional Milling

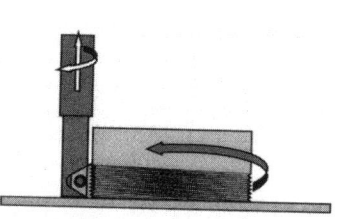

d. Left-hand Thread...Climb Milling

Figure 22. Methods of external thread milling. (Source, Kennametal Inc.)

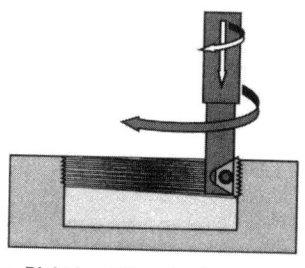

a. Right-hand Thread...Conventional Milling

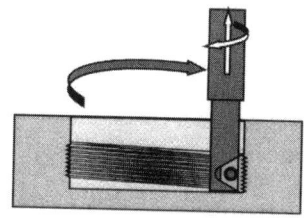

b. Right-hand Thread...Climb Milling

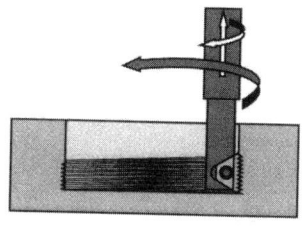

c. Left-hand Thread...Conventional Milling

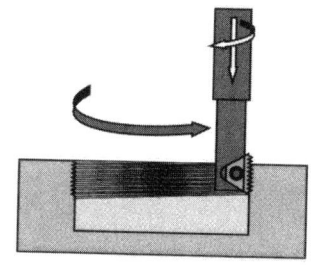

d. Left-hand Thread...Climb Milling

Figure 23. Methods of internal thread milling. (Source, Kennametal Inc.)

edge strength and reduce the required number of passes. The main advantage of a full profile insert over a partial profile insert is that, in addition to producing an essentially burr free thread, less material is removed with a stronger tool and the resulting thread is stronger. *Figure 24* illustrates a partial and full 8UN thread. It can be seen that a partial profile insert would have to enter the workpiece a depth of 0.031" deeper than the full profile insert in order to achieve the same pitch diameter. There are various types of inserts for single point threading, as shown in *Figure 25*.

Infeed angle

In single point threading, the weakest part of the insert removes the greatest amount of material. There are four standard infeed angles, and each has specific advantages and disadvantages that should be considered before choosing the best method for a given operation. *Radial* angle infeed, as shown in *Figure 26*, appears to be a logical choice and is the common selection of inexperienced machinists. The obvious advantage is that it cuts on both sides of the thread form, placing all of the cutting edge in the cut—thereby minimizing the possibility of edge chipping. Unfortunately, this method concentrates frictional heat on the point of the insert, dramatically decreasing tool life. Additional disadvantages are that it forms a channel chip, which can be difficult to handle; burrs will be maximized; and there is an increased tendency to chatter because the entire cutting edge is engaged in the thread. *Flank* angle infeed feeds into the workpiece at the exact angle of the thread (see *Figure 27*), but for best results the angle should be slightly reduced. Since the leading edge of the tool is cutting, the chip flows out of the thread form area, reducing the possibility of burrs on the trailing edge of the tool. For best results when using this approach, the infeed angle should be 3° to 5° smaller than the angle of the thread (25° to 27° for UN and ISO threads) in order to avoid poor surface finish, chipping, or excessive flank wear from rubbing the insert's trailing edge. Problems to expect when using the flank angle are torn or poor surface finish, especially when cutting soft, gummy materials like low carbon steels, aluminum, and stainless steels. Also, the trailing edge of the insert may drag or rub, leading to a tendency to chip. *Modified flank* angle infeed (*Figure 28*) is similar to radial angle infeed—the tool cuts both sides of the thread form, offering some protection from chipping. And, even though a channel chip is produced, the uneven chip thickness assists in chip removal. *Alternating flank* angle infeed is difficult when using conventional machinery (*Figure 29*). Although some machine tools will require special programming techniques to use this method, it offers significantly increased tool life because both edges of the tool are used equally.

Laydown style inserts

Laydown triangular inserts (see *Figure 30*) are popular for single point threading, especially for use in small diameter bores. However, unlike other insert types, these inserts

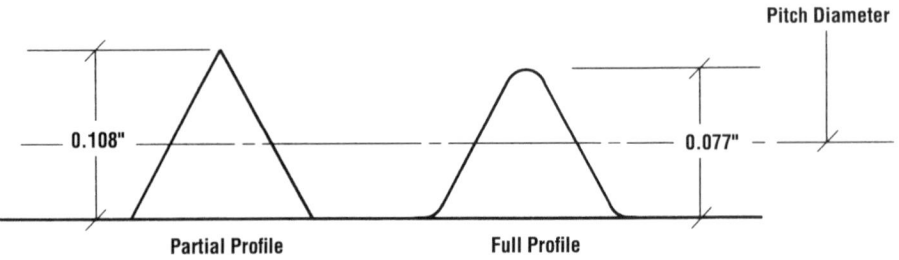

Figure 24. Partial- and full-profile for 8UN thread form. (Source, Duramet.)

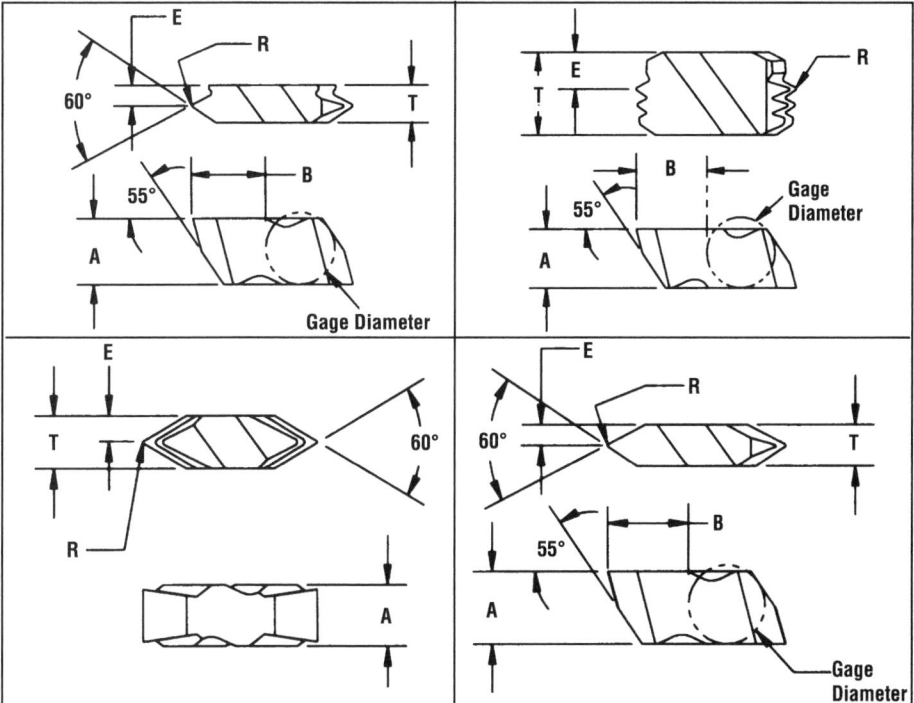

Figure 25. Single point thread cutting inserts. (Source, Kennametal Inc.)

require shims for thread helix angles. When setting up a thread cutting operation with laydown inserts, the first step is to determine the direction of the cut. If machining toward the chuck (when cutting a right hand thread with a right hand tool, or when cutting a left-hand thread with a left-hand tool), the standard helix method is used. If machining away from the chuck (when cutting a right-hand thread with a left-hand tool, or when cutting a left-hand thread with a right-hand tool), the reverse helix method is used. The part's shape and stability, and the flow of chips, are fundamental considerations when choosing to machine toward or away from the chuck.

Once the method of cutting has been established, the angle of the tool in the holder is adjusted, with shims or "anvils," to match the helix angle of the thread being cut. The correct helix angle is especially important when producing multistart and small-diameter threads in order to allow for adequate side clearance for the insert. Almost all toolholders, as shown in *Figures 31* and *32*, are supplied with a 1 $^{1}/_{2}°$ lead angle, and shims are available in sizes to match helix angles from 4.5° to –1.5°. The helix angle of a UN or ISO 60° thread (and other threads) is a component of the pitch (or lead) and the effective pitch diameter of the thread (see *Figure 33*), and can be derived from the equation:

$$\tan \lambda = l \div (\pi \times d)$$

where: λ = helix angle
l = lead (1/tpi)
d = pitch diameter of thread.
Note: for multistart threads, l = 1/tpi × number of starts.

Helix angles for UN, Metric, and various Acme threads are provided in **Tables 75** through **80** at the end of this section.

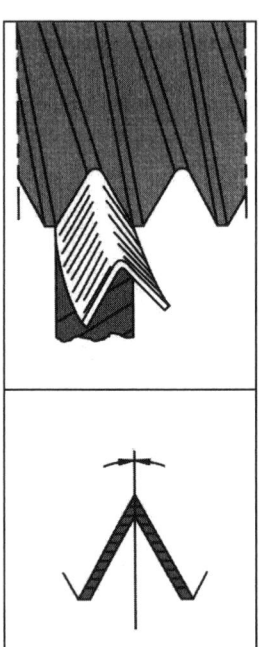

Figure 26. Radial angle infeed. (Source, Kennametal Inc.)

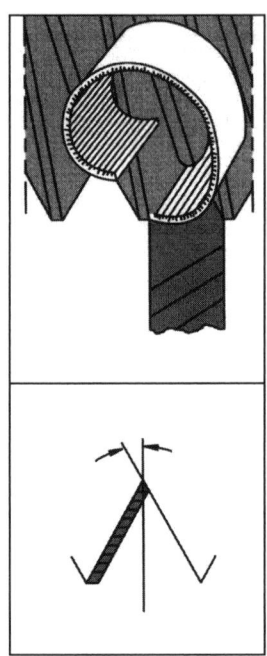

Figure 27. Flank angle infeed. (Source, Kennametal Inc.)

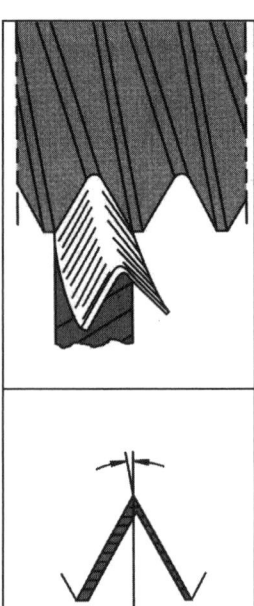

Figure 28. Modified flank angle infeed. (Source, Kennametal Inc.)

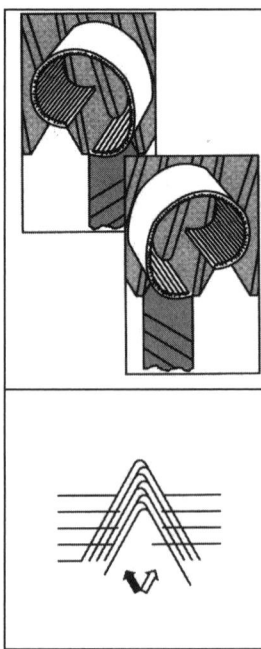

Figure 29. Alternating flank angle infeed. (Source, Kennametal Inc.)

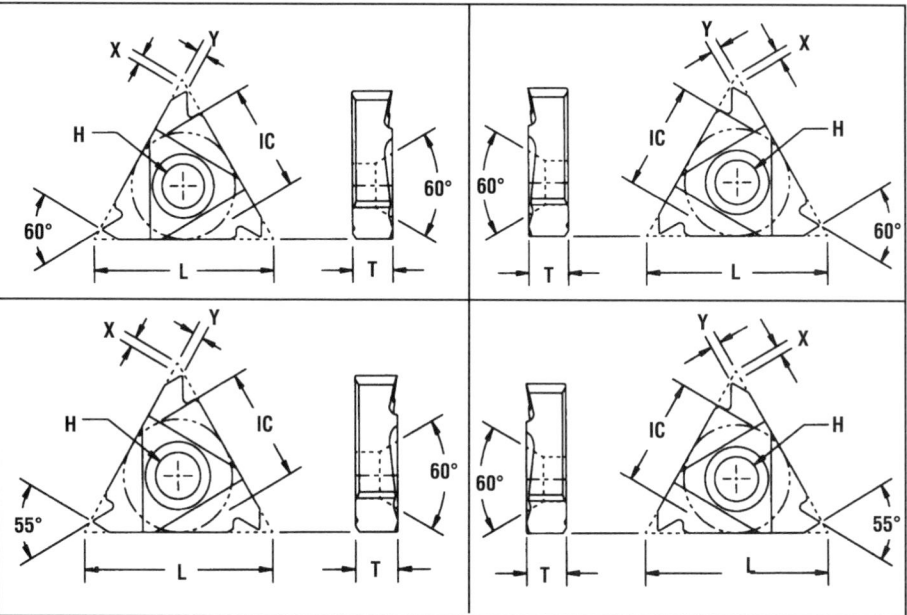

Figure 30. Laydown style inserts.

Infeed values

Single point threading requires making several passes, each with varying depths of cut, over the workpiece. Because a larger portion of the segment of the cutting edge of the insert is engaged in each successive pass, each pass should be made at a shallower depth than the preceding pass. Most CNC lathes have canned programs for threading operations, but **Table 71** and **Table 72** offer guidelines for externally and internally threading steel workpieces of <300 BHN. The depths of cut given are for full profile inserts. These depths are derived from a formula that assumes that the first pass will remove a recommended percentage of the final depth, and that each succeeding pass will remove that percentage of the total depth times the square root of the number of the pass, minus the initial depth of cut.

Example: Find the number of passes and recommended depths of cut for each pass for an 8 pitch UN external thread.

From the table, it can be seen that the total thread depth will be 0.0789" and the recommended depth of cut for the first pass is 0.0197", which is approximately 25% of the total thread depth. To determine the depth of cut for the second pass, multiply the depth of the first by the square root of 2 (for the second pass), and subtract the initial depth of cut from the total. For the third pass, multiply the depth of cut of the first pass by the square root of 3 (for the third pass), and subtract the initial depth of cut from the total. Continue until the accumulated cuts remove the total thread depth.

Second pass: $\quad 0.0197 \times \sqrt{2} = 0.0278$
$\qquad\qquad 0.0278 - 0.0197 = 0.0082$ (d.o.c. for the second pass).

Third pass: $\quad 0.0197 \times \sqrt{3} = 0.0341$
$\qquad\qquad 0.0341 - 0.0197 = 0.0063$ (d.o.c. for the third pass).

Fourth pass: $\quad 0.0197 \times \sqrt{4} = 0.0394$
$\qquad\qquad 0.0394 - 0.0197 = 0.0053$ (d.o.c. for the fourth pass).

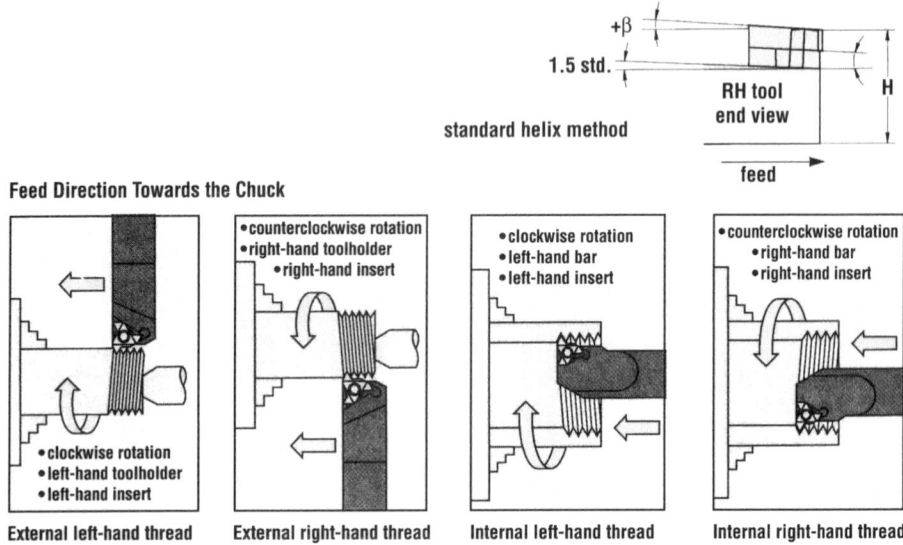

Figure 31. Standard helix toolholder and cutting diagrams for laydown inserts. (Source, Kennametal Inc.)

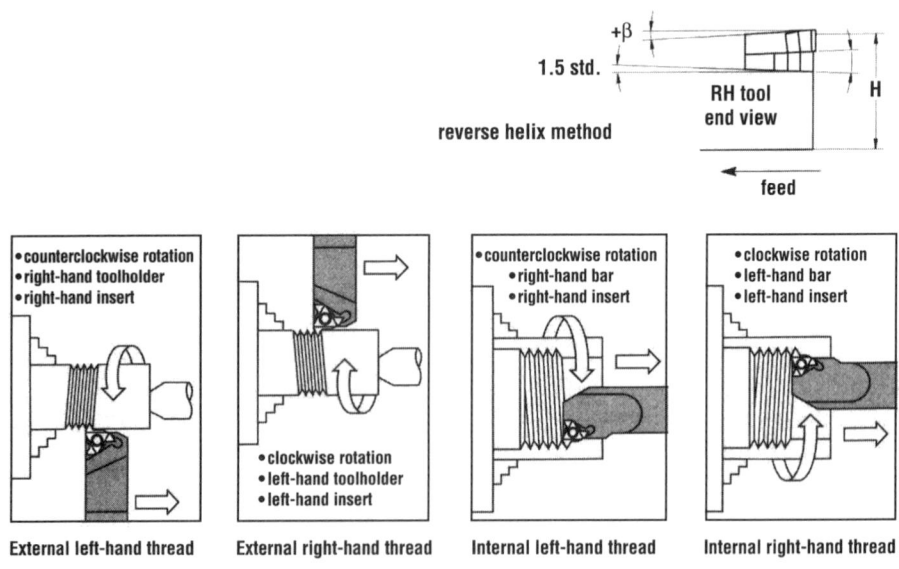

Figure 32. Reverse helix toolholder and cutting diagrams for laydown style inserts. (Source, Kennametal Inc.)

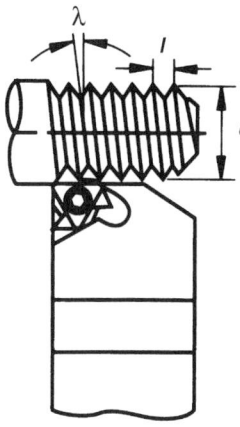

Figure 33. Values for determining helix angle of a thread. λ = helix angle; l = lead; and d = pitch diameter of thread. (Source, Kennametal Inc.)

If, when using this method, the final pass will remove excessive material, adjust the final pass so that the depth of cut exactly equals the amount needed to reach the total thread depth. In the case of the 8UN external thread on the table, the final depth of cut was reduced so that all sixteen passes resulted in the removal of precisely 0.0789" of material.

Surface speeds

The ideal surface speed will combine maximum tool life with required surface finish. In addition, in single point threading, side velocities must not exceed the capabilities of the machine, or thread quality will be adversely affected. Small diameter workpieces can also present problems as suggested surface feet per minute or surface meters per minute values may exceed practical tool parameters. To check whether maximum cutting speed recommendations exceed the maximum travel speed of the tool (as allowed by the machine in inches per minute or millimeters per minute), use the following equation.

Inch units: $\text{sfm (max)} = (D_3 \times \pi \times \text{tpi}) \times (\text{ipm (max)} \div 12)$
Metric units: $\text{smm (max)} = (D_3 \times \pi \times 1/\text{pitch}) \times (\text{mmpm (max)} \div 1000)$
where D_3 = part diameter in inches or millimeters.

The suggested speeds chart in **Table 73** provides speeds in sfm and smm for a variety of insert materials identified by their ISO grade classification. For an explanation of this system, see the Inserts section of this book.

Troubleshooting single point threading operations

Due to the large numbers of parameters that apply to single point threading, troubleshooting unsatisfactory results can be confusing. The solutions outlined in **Table 74** can help pinpoint solutions to a wide selection of problems.

Table 71. Infeed Values for Externally Threading Steel Workpieces (<300 BHN). (Source, Kennametal Inc.)

tpi (UN)	4	5	6	7	8	9	10	11	12	13
Thread Depth	.1578	.1262	.1052	.0902	.0789	.0701	.0631	.0574	.0526	.0485
1st pass	.0353	.0298	.0248	.0213	.0197	.0175	.0169	.0157	.0152	.0142
2nd pass	.0146	.0122	.0105	.0088	.0082	.0073	.0070	.0066	.0064	.0057
3rd pass	.0113	.0094	.0078	.0077	.0063	.0056	.0053	.0048	.0048	.0044
4th pass	.0095	.0079	.0067	.0059	.0053	.0047	.0045	.0041	.0042	.0037
5th pass	.0084	.0070	.0058	.0050	.0047	.0042	.0039	.0036	.0036	.0033
6th pass	.0076	.0063	.0052	.0045	.0043	.0037	.0036	.0031	.0032	.0030
7th pass	.0070	.0058	.0048	.0041	.0039	.0034	.0031	.0028	.0029	.0027
8th pass	.0065	.0054	.0045	.0038	.0036	.0032	.0030	.0026	.0027	.0025
9th pass	.0061	.0051	.0042	.0036	.0034	.0030	.0029	.0025	.0026	.0024
10th pass	.0057	.0048	.0040	.0034	.0032	.0028	.0028	.0024	.0025	.0023
11th pass	.0054	.0045	.0038	.0032	.0031	.0027	.0027	.0023	.0023	.0022
12th pass	.0052	.0043	.0036	.0031	.0029	.0026	.0026	.0022	.0022	.0021
13th pass	.0049	.0042	.0035	.0030	.0027	.0025	.0025	.0021	-	-
14th pass	.0048	.0041	.0034	.0029	.0026	.0024	.0024	.0020	-	-
15th pass	.0046	.0040	.0033	.0028	.0025	.0023	-	-	-	-
16th pass	.0044	.0039	.0032	.0027	.0025	.0022	-	-	-	-
17th pass	.0043	.0038	.0031	.0026	-	-	-	-	-	-
18th pass	.0042	.0037	.0030	.0025	-	-	-	-	-	-
19th pass	.0041	-	-	-	-	-	-	-	-	-
20th pass	.0039	-	-	-	-	-	-	-	-	-
tpi (UN)	14	16	18	20	24	28	32	36	40	44
Thread Depth	.0451	.0394	.0350	.0315	.0263	.0225	.0197	.0175	.0157	.0143
1st pass	.0136	.0125	.0124	.0119	.0118	.0112	.0098	.0087	.0078	.0073

(Continued)

Table 71. *(Continued)* **Infeed Values for Externally Threading Steel Workpieces (<300 BHN).** *(Source, Kennametal Inc.)*

tpi (UN)	14	16	18	20	24	28	32	36	40	44
Thread Depth	.0451	.0394	.0350	.0315	.0263	.0225	.0197	.0175	.0157	.0143
2nd pass	.0059	.0054	.0053	.0049	.0048	.0046	.0042	.0036	.0032	.0028
3rd pass	.0043	.0039	.0039	.0039	.0039	.0036	.0031	.0028	.0024	.0022
4th pass	.0036	.0034	.0033	.0032	.0031	.0031	.0026	.0024	.0020	.0020
5th pass	.0032	.0029	.0029	.0028	.0027	-	-	-	-	-
6th pass	.0029	.0026	.0026	.0025	-	-	-	-	-	-
7th pass	.0026	.0024	.0024	.0023	-	-	-	-	-	-
8th pass	.0024	.0022	.0022	-	-	-	-	-	-	-
9th pass	.0023	.0021	-	-	-	-	-	-	-	-
10th pass	.0022	.0020	-	-	-	-	-	-	-	-
11th pass	.0021	-	-	-	-	-	-	-	-	-

Note: These are nominal thread depths, in inches, for full profile inserts. When using partial profile inserts, reduce the initial depth of cut and increase the number of passes. When threading workhardening materials (e.g., stainless steel), the infeed should never be less than 0.003″ .

Table 72. Infeed Values for Internally Threading Steel Workpieces (<300 BHN). *(Source, Kennametal Inc.)*

tpi (UN)	4	5	6	7	8	9	10	11	12	13
Thread Depth	**.1353**	**.1082**	**.0902**	**.0773**	**.0676**	**.0601**	**.0541**	**.0492**	**.0451**	**.0416**
1st pass	.0303	.0255	.0213	.0183	.0169	.0150	.0145	.0132	.0131	.0120
2nd pass	.0125	.0105	.0090	.0076	.0073	.0062	.0064	.0055	.0054	.0050
3rd pass	.0096	.0083	.0069	.0058	.0053	.0047	.0046	.0044	.0041	.0038
4th pass	.0081	.0068	.0057	.0049	.0047	.0040	.0038	.0035	.0035	.0032
5th pass	.0071	.0060	.0050	.0043	.0041	.0035	.0034	.0031	.0031	.0028
6th pass	.0064	.0054	.0045	.0039	.0036	.0032	.0031	.0028	.0028	.0025
7th pass	.0059	.0050	.0041	.0036	.0033	.0029	.0028	.0026	.0026	.0023
8th pass	.0055	.0046	.0038	.0033	.0030	.0027	.0026	.0024	.0024	.0022
9th pass	.0052	.0043	.0036	.0031	.0028	.0025	.0024	.0022	.0022	.0021
10th pass	.0049	.0041	.0034	.0029	.0027	.0024	.0023	.0021	.0021	.0020
11th pass	.0046	.0039	.0032	.0028	.0026	.0023	.0022	.0020	.0020	.0019
12th pass	.0044	.0037	.0031	.0027	.0025	.0022	.0021	.0019	.0019	.0018
13th pass	.0042	.0036	.0030	.0026	.0024	.0021	.0020	.0018	-	-
14th pass	.0041	.0035	.0029	.0025	.0023	.0020	.0019	.0017	-	-
15th pass	.0040	.0034	.0028	.0024	.0022	.0019	-	-	-	-
16th pass	.0039	.0033	.0027	.0023	.0021	.0019	-	-	-	-
17th pass	.0038	.0032	.0026	.0022	-	-	-	-	-	-
18th pass	.0037	.0031	.0025	.0021	-	-	-	-	-	-
19th pass	.0036	-	-	-	-	-	-	-	-	-
20th pass	.0035	-	-	-	-	-	-	-	-	-
tpi (UN)	**14**	**16**	**18**	**20**	**24**	**28**	**32**	**36**	**40**	**44**
Thread Depth	**.0386**	**.0338**	**.0300**	**.0270**	**.0225**	**.0193**	**.0169**	**.0150**	**.0135**	**.0123**
1st pass	.0117	.0107	.0106	.0120	.0101	.0096	.0084	.0075	.0067	.0061

(Continued)

Table 72. *(Continued)* **Infeed Values for Internally Threading Steel Workpieces (<300 BHN).** *(Source, Kennametal Inc.)*

tpi (UN)	14	16	18	20	24	28	32	36	40	44
Thread Depth	.0386	.0338	.0300	.0270	.0225	.0193	.0169	.0150	.0135	.0123
2nd pass	.0048	.0043	.0044	.0042	.0042	.0039	.0035	.0031	.0029	.0025
3rd pass	.0037	.0034	.0033	.0032	.0032	.0033	.0027	.0023	.0021	.0019
4th pass	.0031	.0028	.0028	.0027	.0027	.0025	.0023	.0021	.0018	.0018
5th pass	.0027	.0025	.0025	.0024	.0023	-	-	-	-	-
6th pass	.0025	.0029	.0023	.0022	-	-	-	-	-	-
7th pass	.0023	.0021	.0021	.0021	-	-	-	-	-	-
8th pass	.0021	.0020	.0029	-	-	-	-	-	-	-
9th pass	.0020	.0019	-	-	-	-	-	-	-	-
10th pass	.0019	.0018	-	-	-	-	-	-	-	-
11th pass	.0018	-	-	-	-	-	-	-	-	-

Note: These are nominal thread depths, in inches, for full profile inserts. When using partial profile inserts, reduce the initial depth of cut and increase the number of passes. When threading workhardening materials (e.g., stainless steel), the infeed should never be less than 0.003".

Table 73. Recommended Grades and Speeds for Threading Various Workpiece Materials. *(Source, Kennametal Inc.)*

Workpiece Material		UCC* M20R†	PVD Coated P20R M15-25 K05-15	PVD Coated P30-45 M40R K25-35	PVD Coated M5-25 K05-15	M40R K25-35	CVD Coated P10-20 M25-35	CVD Coated P25-45 M35>	CVD Coated P5-15 M10-25 K10	Cermet P10 M20	PCD M5-15 K01
Free-machining Carbon Steel	sfm	-	300-700	150-500	300-550	350-600	450-900	500-800	500-1000	500-1000	-
	smm	-	91-213	46-152	91-168	107-183	137-274	152-244	152-305	152-305	-
Plain Carbon Steel	sfm	-	250-600	150-400	250-500	300-500	300-750	400-700	400-800	450-900	-
	smm		76-183	46-122	76-152	91-152	91-229	122-213	122-244	137-274	-
Alloy Steels 190-330 BHN	sfm	-	300-500	150-350	250-500	250-450	400-600	350-600	350-700	400-800	-
	smm	-	91-152	46-107	76-152	76-137	122-183	107-183	107-213	122-244	-
Alloy Steels 330-450 BHN	sfm	-	-	-	200-400	-	-	-	-	100-500	-
	smm	-	-	-	61-122	-	-	-	-	31-152	-
Gray Cast Iron 190-330 BHN	sfm	200-300	-	150-300	200-600	200-500	-	-	300-800	-	-
	smm	61-91	-	46-91	61-183	61-152	-	-	91-244	-	-
Gray Cast Iron 330-450 BHN	sfm	150-250	-	50-200	150-450	-	-	-	-	-	-
	smm	46-76	-	15-61	46-137	-	-	-	-	-	-
Alloy/Ductile Irons	sfm	150-250	300-500	100-400	150-500	250-500	-	300-600	300-750	400-800	-
	smm	46-76	91-152	31-122	46-152	76-152	-	91-183	91-229	122-244	-
Austenitic Stainless Steel*	sfm	200-350	-	150-350	200-500	300-500	-	400-650	-	500-850	-
	smm	61-107	-	46-107	61-152	91-152	-	122-198	-	152-259	-
Martensitic/ Ferritic Stainless	sfm	-	300-500	100-300	150-400	250-500	-	300-500	-	600-800	-
	smm	-	91-152	31-91	46-122	76-152	-	91-152	-	183-244	-
Hi-Temp Alloys** 200-260 BHN	sfm	80-120	-	50-150	80-300	-	-	-	-	600-800	-
	smm	24-35	-	15-46	-	-	-	-	-	183-344	-

Recommended Surface Speeds - sfm and smm

(Continued)

Table 73. *(Continued)* **Recommended Grades and Speeds for Threading Various Workpiece Materials.** *(Source, Kennametal Inc.)*

Workpiece Material		UCC*	PVD Coated				CVD Coated			Cermet	PCD
		M20R†	P20R M 15-25 K 05-15	P 30-45 M40R K 25-35	M 5-25 K 05-15	M 40R K 25-35	P10-20 M25-35	P25-45 M35>	P5-15 M10-25 K10	P10 M20	M5-15 K01
Hi-Temp Alloys** 260-450 BHN	sfm	80-100	-	30-80	100-200	-	-	-	-	-	-
	smm	24-31	-	9-24	31-61	-	-	-	-	-	-
Titanium Alloys (Ti 6Al-4V)	sfm	110-180	-	-	110-250	-	-	-	-	-	-
	smm	34-55	-	-	34-76	-	-	-	-	-	-
Free-Machining Aluminum Alloys	sfm	400-800	-	-	400-1000	-	-	-	-	700-4000	Up to 4000
	smm	122-244	-	-	122-305	-	-	-	-	213-1220	Up to 1220
High-Silicon Aluminum Alloys	sfm	-	-	-	-	-	-	-	-	-	Up to 3000
	smm	-	-	-	-	-	-	-	-	-	Up to 915
Copper/Zinc/Brass	sfm	250-600	-	150-500	250-800	400-800	-	-	-	700-4000	Up to 3000
	smm	76-183	-	46-152	76-244	122-244	-	-	-	213-1220	Up to 915
Non-Metallics	sfm	400-1500	-	150-800	400-1500	-	-	-	-	700-4000	Up to 4500
	smm	122-457	-	46-244	122-457	-	-	-	-	213-1220	Up to 1220

Recommended Surface Speeds - sfm and smm

*Uncoated carbide. ** Example: Inconel 718. † ISO grades are expressed as the insert appears on the grade chart. "R" as in 20R indicates that the grade occupies that range (from 20 to 30, for example) on the chart. A rating of 25-35 indicates that the grade spans the center of the 20 range to the center of the 30 range on the chart. A rating of 30-40 indicates that the grade fills the entire 30 and 40 grids on the chart. The symbol ">" indicates that the grade rating begins at the position given and continues to the end of the chart section.

Table 74. Troubleshooting Single Point Threading Operations. *(Source, Kennametal Inc.)*

Quick Solution Guide

Problem → / Solution →	Increase sfm/smm	Reduce sfm/smm	Increase chip load	Decrease chip load where failure occurs	Use tougher carbide grade	Use harder carbide grade	Apply coolant	Use coated carbide	Change infeed angle	Check for insert movement and reseat	Reduce tool overhang	Reselect shim	Reselect d.o.c.	Adjust center height	Begin cutting threads 0.5 inch before workpiece
Chatter	•			•						•	•			•	
Burr on crest	•														
Short tool life		•				•		•							
Chipped leading edge			•	•	•										
Chipped trailing edge			•	•	•										
Broken nose (first pass)	•								•						
Broken nose (after first pass)			•	•	•				•			•	•	•	
Build-up on cutting edge	•		•				•	•							
Premature topping												•			
Splitting threads															•

Detailed Solution Guide

Problem	Cause	Possible Solution
Thread with torn finish	• Burrs	• Use positive rake or full profile insert. • Increase coolant concentration.
	• Torn finish	• Alter infeed • Use PVD grade insert. • Use positive rake insert. • Increase speed.

(Continued)

Table 74. *(Continued)* **Troubleshooting Single Point Threading Operations.** *(Source, Kennametal Inc.)*

Problem	Cause	Possible Solution
		Detailed Solution Guide
Thread with torn finish	• Steps	• Check machine Z-axis travel. • Avoid cutting chips. • Check insert form. • Check for correct shim (laydown system).
Chatter	• Poor rigidity	• Minimize tool overhang • Check for workpiece deflection
	• Incorrect speed	• Adjust speed.
	• Wrong edge prep	• Adjust hone level.
	• Insert movement	• Check insert and clamp.
	• Improper infeed	• Use modified feed angle.
	• Off centerline	• Verify that tool cutting position is at workpiece centerline.
Built-up edge	• Speed too low	• Increase speed.
	• Insufficient coolant	• Increase coolant concentration and/or flow
	• Wrong edge prep	• Adjust hone size.
	• Chip load	• Adjust infeed angle. • Adjust depth of cut per pass.
Deformation of insert	• Speed too high	• Reduce speed.
	• Incorrect grade	• Use grade with a higher hot hardness.
	• Improper infeed angle	• Alter infeed angle.
Chipping of insert	• Light chip load	• Adjust chip load. • Increase or decrease number of passes.
	• Wrong grade	• Use tougher grade.
	• Improper infeed	• Alter infeed to modified flank

(Continued)

Table 74. *(Continued)* **Troubleshooting Single Point Threading Operations.** *(Source, Kennametal Inc.)*

Problem	Detailed Solution Guide	
	Cause	**Possible Solution**
Chipping of insert	• Incorrect speed	• Increase speed if chipping on trailing edge. • Decrease speed if chipping on leading edge
	• Wrong edge prep	• Increase hone size.
	• Poor rigidity	• Minimize tool overhang. • Check for insert movement/check clamp. • Check for possible part deflection.
Broken nose on insert	• Heavy chip load	• Decrease chip load
	• Small nose radius	• Use larger nose radius if allowable.
	• Wrong grade	• Use tougher grade.
	• Wrong edge prep	• Increase hone size.
Flank wear on insert	• Wrong grade	• Use a more wear resistant grade.
	• Insufficient coolant	• Increase coolant flow.
	• Off centerline	• Check centerline height of the tool. (The smaller the diameter, the more critical the need for centerline accuracy.)

Table 75. Helix Angles for UN Threads #0 Through 1 5/16 Inch. (*Source, Landis Threading Systems.*)

Nom. Dia.	Threads per Inch													
	80	72	64	56	48	44	40	36	32	28	24	20	18	16
#0	4°23'	-	-	-	-	-	-	-	-	-	-	-	-	-
#1	3°31'	3°57'	4°31'	-	-	-	-	-	-	-	-	-	-	-
#2	2°55'	3°17'	3°45'	4°22'	-	-	-	-	-	-	-	-	-	-
#3	2°30'	2°49'	3°12'	3°43'	4°26'	-	-	-	-	-	-	-	-	-
#4	2°12'	2°28'	2°48'	3°14'	3°51'	4°15'	4°45'	-	-	-	-	-	-	-
#5	1°56'	2°11'	2°29'	2°52'	3°24'	3°45'	4°11'	4°43'	5°26'	6°22'	7°43'	-	-	-
#6	1°45'	1°58'	2°14'	2°34'	3°3'	3°22'	3°44'	4°13'	4°50'	5°39'	6°49'	-	-	-
#8	1°28'	1°38'	1°51'	2°8'	2°31'	2°46'	3°5'	3°28'	3°58'	4°37'	5°32'	-	-	-
#10	1°15'	1°24'	1°35'	1°49'	2°9'	2°22'	2°37'	2°57'	3°21'	3°54'	4°39'	-	-	-
#12	1°6'	1°13'	1°23'	1°35'	1°52'	2°4'	2°17'	2°33'	2°54'	3°22'	4°1'	-	-	-
3/16	-	-	1°36'	1°51'	2°11'	2°24'	2°40'	2°59'	3°24'	3°57'	4°44'	5°52'	6°40'	7°43'
1/4	-	-	1°11'	1°22'	1°36'	1°46'	1°57'	2°11'	2°29'	2°52'	3°24'	4°11'	4°44'	5°25'
5/16	-	-	0°57'	1°5'	1°16'	1°24'	1°32'	1°43'	1°57'	2°15'	2°40'	3°15'	3°40'	4°11'
3/8	-	-	-	0°54'	1°3'	1°9'	1°16'	1°25'	1°36'	1°51'	2°11'	2°40'	2°59'	3°24'
7/16	-	-	-	0°46'	0°54'	0°59'	1°5'	1°12'	1°22'	1°34'	1°51'	2°15'	2°31'	2°52'
1/2	-	-	-	0°40'	0°47'	0°51'	0°57'	1°3'	1°11'	1°22'	1°36'	1°57'	2°11'	2°29'
9/16	-	-	-	0°35'	0°42'	0°45'	0°50'	0°56'	1°3'	1°12'	1°25'	1°43'	1°55'	2°11'
5/8	-	-	-	0°32'	0°37'	0°41'	0°45'	0°50'	0°57'	1°5'	1°16'	1°32'	1°43'	1°57'
11/16	-	-	-	-	-	-	0°41'	0°45'	0°51'	0°59'	1°9'	1°24'	1°33'	1°46'
3/4	-	-	-	-	-	-	0°37'	0°42'	0°47'	0°54'	1°3'	1°16'	1°25'	1°36'
13/16	-	-	-	-	-	-	0°34'	0°38'	0°43'	0°50'	0°58'	1°10'	1°18'	1°29'
7/8	-	-	-	-	-	-	0°32'	0°35'	0°40'	0°46'	0°54'	1°5'	1°12'	1°22'
15/16	-	-	-	-	-	-	0°30'	0°33'	0°37'	0°43'	0°50'	1°0'	1°7'	1°16'

(Continued)

Table 75. *(Continued)* **Helix Angles for UN Threads #0 Through 1 $\frac{5}{16}$ Inch.** *(Source, Landis Threading Systems.)*

Nom. Dia.	Threads per Inch													
	80	72	64	56	48	44	40	36	32	28	24	20	18	16
1	-	-	-	-	-	-	0°28'	0°31'	0°35'	0°40'	0°47'	0°57'	1°3'	1°11'
1 1/16	-	-	-	-	-	-	-	0°29'	0°33'	0°38'	0°44'	0°53'	0°59'	1°7'
1 1/8	-	-	-	-	-	-	-	-	0°31'	0°35'	0°42'	0°50'	0°56'	1°3'
1 3/16	-	-	-	-	-	-	-	-	0°29'	0°34'	0°39'	0°47'	0°53'	1°0'
1 1/4	-	-	-	-	-	-	-	-	0°28'	0°32'	0°37'	0°45'	0°50'	0°57'
1 5/16	-	-	-	-	-	-	-	-	0°26'	0°30'	0°35'	0°43'	0°48'	0°54'

Nom. Dia.	Threads per Inch													
	14	13	12	11	10	9	8	7	6	5.5	5	4.5	4	3.5
5/16	4°53'	5°20'	5°52'	-	-	-	-	-	-	-	-	-	-	-
3/8	3°57'	4°18'	4°44'	5°14'	-	-	-	-	-	-	-	-	-	-
7/16	3°20'	3°37'	3°57'	4°22'	4°53'	-	-	-	-	-	-	-	-	-
1/2	2°52'	3°7'	3°24'	3°45'	4°11'	4°43'	5°26'	-	-	-	-	-	-	-
9/16	2°31'	2°44'	2°59'	3°17'	3°40'	4°8'	4°44'	-	-	-	-	-	-	-
5/8	2°15'	2°26'	2°40'	2°56'	3°15'	3°40'	4°11'	4°53'	-	-	-	-	-	-
11/16	2°1'	2°12'	2°24'	2°38'	2°56'	3°17'	3°45'	4°22'	5°14'	-	-	-	-	-
3/4	1°51'	2°0'	2°11'	2°24'	2°40'	2°57'	3°24'	3°57'	4°44'	-	-	-	-	-
13/16	1°42'	1°50'	2°0'	2°12'	2°26'	2°44'	3°7'	3°37'	4°16'	4°46'	5°20'	-	-	-
7/8	1°34'	1°42'	1°51'	2°2'	2°15'	2°31'	2°52'	3°20'	3°57'	4°22'	4°53'	-	-	-
15/16	1°28'	1°35'	1°43'	1°53'	2°5'	2°20'	2°40'	3°5'	3°40'	4°2'	4°30'	-	-	-
1	1°22'	1°29'	1°36'	1°46'	1°57'	2°11'	2°29'	2°52'	3°24'	3°45'	4°11'	4°44'	5°26'	-
1 1/16	1°17'	1°23'	1°30'	1°39'	1°50'	2°3'	2°19'	2°41'	3°11'	3°30'	3°54'	4°24'	5°3'	-
1 1/8	1°12'	1°18'	1°25'	1°33'	1°43'	1°55'	2°11'	2°31'	2°59'	3°17'	3°40'	4°8'	4°44'	5°32'
1 3/16	1°8'	1°14'	1°20'	1°28'	1°37'	1°49'	2°4'	2°23'	2°49'	3°6'	3°27'	3°53'	4°26'	5°11'

(Continued)

Table 75. *(Continued)* Helix Angles for UN Threads #0 Through 1 5/16 Inch. *(Source, Landis Threading Systems.)*

Nom. Dia.	Threads per Inch													
	3.5	4	4.5	5	5.5	6	7	8	9	10	11	12	13	14
1 1/4	4°53'	4°11'	3°40'	3°15'	2°56'	2°40'	2°15'	1°57'	1°43'	1°32'	1°24'	1°16'	1°10'	1°5'
1 5/16	4°37'	3°58'	3°28'	3°5'	2°46'	2°31'	2°8'	1°51'	1°38'	1°28'	1°19'	1°12'	1°7'	1°2'

Nom. Dia.	Threads per Inch													
	7	8	9	10	11	12	13	14	16	18	20	24	28	32
1 3/8	2°2'	1°46'	1°33'	1°24'	1°16'	1°9'	1°4'	0°59'	0°51'	0°45'	0°41'	0°34'	0°29'	0°25'
1 7/16	1°56'	1°41'	1°29'	1°20'	1°12'	1°6'	1°1'	0°56'	0°49'	0°43'	0°39'	0°32'	0°28'	0°24'
1 1/2	1°51'	1°36'	1°25'	1°16'	1°9'	1°3'	0°58'	0°54'	0°47'	0°42'	0°37'	0°31'	0°26'	0°23'
1 9/16	1°46'	1°32'	1°22'	1°13'	1°6'	1°0'	0°56'	0°51'	0°45'	0°40'	0°36'	0°30'	0°25'	0°22'
1 5/8	1°42'	1°29'	1°18'	1°10'	1°4'	0°58'	0°53'	0°49'	0°43'	0°38'	0°34'	0°29'	0°24'	0°21'
1 11/16	1°38'	1°25'	1°15'	1°7'	1°1'	0°56'	0°51'	0°48'	0°42'	0°37'	0°33'	0°27'	0°23'	0°21'
1 3/4	1°34'	1°22'	1°12'	1°5'	0°59'	0°54'	0°49'	0°46'	0°40'	0°35'	0°32'	0°26'	0°23'	0°20'
1 13/16	1°31'	1°19'	1°10'	1°3'	0°57'	0°52'	0°48'	0°44'	0°39'	0°34'	0°31'	0°26'	0°22'	0°19'
1 7/8	1°28'	1°16'	1°7'	1°0'	0°55'	0°50'	0°46'	0°43'	0°37'	0°33'	0°30'	0°25'	0°21'	0°18'
1 15/16	1°25'	1°14'	1°5'	0°58'	0°53'	0°48'	0°45'	0°41'	0°36'	0°32'	0°29'	0°24'	0°20'	0°18'
2	1°22'	1°11'	1°3'	0°57'	0°51'	0°47'	0°43'	0°40'	0°35'	0°31'	0°28'	0°23'	0°20'	0°17'
2 1/8	1°17'	1°7'	0°59'	0°53'	0°48'	0°44'	0°41'	0°37'	0°33'	0°29'	0°26'	0°22'	0°19'	0°16'
2 1/4	1°12'	1°3'	0°56'	0°50'	0°45'	0°42'	0°38'	0°35'	0°31'	0°27'	0°25'	0°21'	0°17'	0°15'
2 3/8	1°8'	1°0'	0°53'	0°47'	0°43'	0°39'	0°36'	0°33'	0°29'	0°26'	0°23'	0°19'	0°17'	0°15'
2 1/2	1°5'	0°56'	0°50'	0°45'	0°41'	0°37'	0°34'	0°32'	0°28'	0°25'	0°22'	0°18'	0°16'	0°14'
2 5/8	1°2'	0°54'	0°48'	0°43'	0°39'	0°35'	0°33'	0°30'	0°26'	0°23'	0°21'	0°18'	0°15'	0°13'
2 3/4	0°59'	0°51'	0°45'	0°41'	0°37'	0°34'	0°31'	0°29'	0°25'	0°22'	0°20'	0°17'	0°14'	0°13'
2 7/8	0°56'	0°49'	0°43'	0°39'	0°35'	0°32'	0°30'	0°28'	0°24'	0°21'	0°19'	0°16'	0°14'	0°12'
3	0°54'	0°47'	0°42'	0°37'	0°34'	0°31'	0°29'	0°26'	0°23'	0°21'	0°18'	0°15'	0°13'	0°11'

(Continued)

Table 75. (*Continued*) Helix Angles for UN Threads #0 Through 1 5/16 Inch. (*Source, Landis Threading Systems.*)

Nom. Dia.	Threads per Inch													
	7	8	9	10	11	12	13	14	16	18	20	24	28	32
3 1/8	0°52'	0°45'	0°40'	0°36'	0°32'	0°30'	0°27'	0°25'	0°22'	0°20'	0°18'	0°15'	0°13'	0°11'
3 1/4	0°50'	0°43'	0°38'	0°34'	0°31'	0°29'	0°26'	0°24'	0°21'	0°19'	0°17'	0°14'	0°12'	0°11'
3 3/8	0°48'	0°42'	0°37'	0°33'	0°30'	0°27'	0°25'	0°24'	0°21'	0°18'	0°16'	0°14'	0°12'	0°10'
3 1/2	0°46'	0°40'	0°35'	0°32'	0°29'	0°26'	0°24'	0°23'	0°20'	0°18'	0°16'	0°13'	0°11'	0°10'
3 5/8	0°44'	0°39'	0°34'	0°31'	0°28'	0°26'	0°24'	0°22'	0°19'	0°17'	0°15'	0°13'	0°11'	0°9'
3 3/4	0°43'	0°37'	0°33'	0°30'	0°27'	0°25'	0°23'	0°21'	0°18'	0°16'	0°15'	0°12'	0°10'	0°9'
3 7/8	0°41'	0°36'	0°32'	0°29'	0°26'	0°24'	0°22'	0°20'	0°18'	0°16'	0°14'	0°12'	0°10'	0°9'
4	0°40'	0°35'	0°31'	0°28'	0°25'	0°22'	0°21'	0°20'	0°17'	0°15'	0°13'	0°11'	0°10'	0°9'
4 1/8	0°39'	0°34'	0°30'	0°27'	0°24'	0°22'	0°21'	0°19'	0°17'	0°15'	0°13'	0°11'	0°10'	0°8'
4 1/4	0°38'	0°33'	0°29'	0°26'	0°24'	0°22'	0°20'	0°18'	0°16'	0°14'	0°13'	0°11'	0°9'	0°8'
4 1/2	0°35'	0°31'	0°27'	0°25'	0°22'	0°21'	0°19'	0°18'	0°15'	0°14'	0°12'	0°10'	0°9'	0°8'

Nom. Dia.	Threads per Inch					
	3.5	4	4.5	5	5.5	6
2	2°52'	2°29'	2°11'	1°57'	1°46'	1°36'
2 1/8	2°41'	2°19'	2°3'	1°50'	1°39'	1°30'
2 1/4	2°31'	2°11'	1°55'	1°43'	1°33'	1°25'
2 3/8	2°23'	2°4'	1°49'	1°37'	1°28'	1°20'
2 1/2	2°15'	1°57'	1°43'	1°32'	1°24'	1°16'
2 5/8	2°8'	1°51'	1°38'	1°28'	1°19'	1°12'
2 3/4	2°2'	1°46'	1°33'	1°24'	1°15'	1°9'
2 7/8	1°56'	1°41'	1°29'	1°20'	1°12'	1°6'
3	1°51'	1°36'	1°25'	1°16'	1°9'	1°3'
3 1/8	1°46'	1°32'	1°22'	1°13'	1°6'	1°0'

Nom. Dia.	Threads per Inch					
	3.5	4	4.5	5	5.5	6
1 3/8	4°22'	3°45'	3°17'	2°56'	2°38'	2°24'
1 7/16	4°9'	3°34'	3°8'	2°47'	2°31'	2°17'
1 1/2	3°57'	3°24'	2°59'	2°40'	2°24'	2°11'
1 9/16	3°47'	3°15'	2°51'	2°33'	2°18'	2°5'
1 5/8	3°37'	3°7'	2°44'	2°26'	2°12'	2°0'
1 11/16	3°28'	2°59'	2°37'	2°20'	2°7'	1°55'
1 3/4	3°20'	2°52'	2°31'	2°15'	2°2'	1°51'
1 13/16	3°12'	2°46'	2°26'	2°10'	1°57'	1°47'
1 7/8	3°5'	2°40'	2°20'	2°5'	1°53'	1°43'
1 15/16	2°58'	2°34'	2°15'	2°1'	1°49'	1°40'

(*Continued*)

Table 75. *(Continued)* **Helix Angles for UN Threads #0 Through 1 $5/16$ Inch.** *(Source, Landis Threading Systems.)*

Nom. Dia.	Threads per Inch					
	6	5.5	5	4.5	4	3.5
3 1/4	0°58'	1°4'	1°10'	1°18'	1°29'	1°42'
3 3/8	0°56'	1°1'	1°7'	1°15'	1°25'	1°38'
3 1/2	0°54'	0°59'	1°5'	1°12'	1°22'	1°34'
3 5/8	0°52'	0°57'	1°3'	1°10'	1°19'	1°31'
3 3/4	0°50'	0°55'	1°0'	1°7'	1°16'	1°28'
3 7/8	0°48'	0°53'	0°58'	1°5'	1°14'	1°25'
4	0°47'	0°51'	0°57'	1°3'	1°11'	1°22'
4 1/8	0°45'	0°50'	0°55'	1°1'	1°9'	1°19'
4 1/4	0°44'	0°48'	0°53'	0°59'	1°7'	1°17'
4 1/2	0°41'	0°45'	0°50'	0°56'	1°3'	1°12'

Table 76. Helix Angles for ISO and Metric Threads 1 mm Through 24 mm. *(Source, Landis Threading Systems.)*

Nom. Dia.	Pitch in Millimeters													
	0.25	0.3	0.35	0.4	0.45	0.5	0.6	0.7	0.75	0.8	1	1.25	1.5	1.75
1	5°26'	6°46'	8°12'	-	-	-	-	-	-	-	-	-	-	-
1.1	4°51'	6°1'	7°17'	-	-	-	-	-	-	-	-	-	-	-
1.2	4°23'	5°26'	6°32'	6°22'	-	-	-	-	-	-	-	-	-	-
1.4	3°41'	4°32'	5°26'	6°22'	7°22'	-	-	-	-	-	-	-	-	-
1.6	3°10'	3°53'	4°38'	5°26'	6°15'	-	-	-	-	-	-	-	-	-
1.7	2°58'	3°38'	4°20'	5°3'	5°49'	6°36'	-	-	-	-	-	-	-	-
1.8	2°47'	3°24'	4°3'	4°44'	5°26'	6°9'	-	-	-	-	-	-	-	-
2	2°29'	3°2'	3°36'	4°11'	4°48'	5°26'	6°46'	-	-	-	-	-	-	-
2.2	2°14'	2°44'	3°14'	3°45'	4°18'	4°51'	6°1'	-	-	-	-	-	-	-
2.3	2°8'	2°36'	3°5'	3°34'	4°5'	4°36'	5°43'	-	-	-	-	-	-	-
2.5	1°57'	2°22'	2°48'	3°15'	3°43'	4°11'	5°10'	6°13'	6°46'	-	-	-	-	-
2.6	1°52'	2°16'	2°41'	3°7'	3°33'	4°0'	4°56'	5°56'	6°27'	-	-	-	-	-
3	1°36'	1°57'	2°18'	2°40'	3°2'	3°24'	4°11'	5°0'	5°26'	5°52'	-	-	-	-
3.5	1°22'	1°39'	1°57'	2°15'	2°33'	2°52'	3°31'	4°11'	4°32'	4°53'	-	-	-	-
4	-	-	-	1°57'	2°13'	2°29'	3°2'	3°35'	3°53'	4°11'	5°26'	7°7'	8°58'	11°1'
4.5	-	-	-	1°43'	1°57'	2°11'	2°40'	3°9'	3°24'	3°40'	4°44'	6°9'	7°43'	9°24'
5	-	-	-	1°32'	1°45'	1°57'	2°22'	2°48'	3°2'	3°15'	4°11'	5°26'	6°46'	8°12'
5.5	-	-	-	1°24'	1°35'	1°46'	2°8'	2°32'	2°44'	2°56'	3°45'	4°51'	6°1'	7°17'
6	-	-	-	-	1°26'	1°36'	1°57'	2°18'	2°29'	2°40'	3°24'	4°23'	5°26'	6°32'
7	-	-	-	-	1°13'	1°22'	1°39'	1°57'	2°6'	2°15'	2°52'	3°41'	4°32'	5°26'
8	-	-	-	-	1°4'	1°11'	1°26'	1°41'	1°49'	1°57'	2°29'	3°10'	3°53'	4°38'
9	-	-	-	-	0°57'	1°3'	1°16'	1°30'	1°36'	1°43'	2°11'	2°47'	3°24'	4°3'
10	-	-	-	-	0°51'	0°57'	1°8'	1°20'	1°26'	1°32'	1°57'	2°29'	3°2'	3°36'

(Continued)

Table 76. *(Continued)* Helix Angles for ISO and Metric Threads 1 mm Through 24 mm. *(Source, Landis Threading Systems.)*

Nom. Dia.	Pitch in Millimeters													
	0.25	0.3	0.35	0.4	0.45	0.5	0.6	0.7	0.75	0.8	1	1.25	1.5	1.75
11	-	-	-	-	-	0°51'	1°2'	1°13'	1°18'	1°24'	1°46'	2°14'	2°44'	3°14'
12	-	-	-	-	0°46'	-	0°57'	1°6'	1°11'	1°16'	1°36'	2°2'	2°29'	2°56'
14	-	-	-	-	-	-	0°48'	0°57'	1°1'	1°5'	1°22'	1°44'	2°6'	2°29'
15	-	-	-	-	-	-	0°45'	0°53'	0°57'	1°0'	1°16'	1°36'	1°57'	2°18'
16	-	-	-	-	-	-	0°42'	0°49'	0°53'	0°57'	1°11'	1°30'	1°49'	2°9'
17	-	-	-	-	-	-	0°40'	0°46'	0°50'	0°53'	1°7'	1°24'	1°42'	2°1'
18	-	-	-	-	-	-	0°37'	0°44'	0°47'	0°50'	1°3'	1°20'	1°36'	1°54'
20	-	-	-	-	-	-	0°33'	0°39'	0°42'	0°45'	0°57'	1°11'	1°26'	1°42'
22	-	-	-	-	-	-	0°30'	0°36'	0°38'	0°41'	0°51'	1°5'	1°18'	1°32'
24	-	-	-	-	-	-	0°28'	0°33'	0°35'	0°37'	0°47'	0°59'	1°11'	1°24'

Nom. Dia.	Pitch in Millimeters										
	2	2.5	3	3.5	4	4.5	5	5.5	6	6.5	7
6	7°43'	-	-	-	-	-	-	-	-	-	-
7	6°22'	8°25'	-	-	-	-	-	-	-	-	-
8	5°26'	7°7'	-	-	-	-	-	-	-	-	-
9	4°44'	6°9'	-	-	-	-	-	-	-	-	-
10	4°11'	5°26'	6°46'	8°12'	-	-	-	-	-	-	-
11	3°45'	4°51'	6°1'	7°16'	8°37'	-	-	-	-	-	-
12	3°24'	4°23'	5°26'	6°32'	7°43'	8°58'	-	-	-	-	-
14	2°52'	3°41'	4°32'	5°26'	6°22'	7°22'	8°25'	-	-	-	-
15	2°40'	3°24'	4°11'	5°0'	5°52'	6°46'	7°43'	-	-	-	-
16	2°29'	3°10'	3°53'	4°38'	5°26'	6°15'	7°7'	8°1'	-	-	-
17	2°19'	2°58'	3°38'	4°20'	5°3'	5°49'	6°36'	7°26'	-	-	-

(Continued)

Table 76. (Continued) Helix Angles for ISO and Metric Threads 1 mm Through 24 mm. (Source, Landis Threading Systems.)

Nom. Dia.						Pitch in Millimeters						
	2	2.5	3	3.5	4	4.5	5	5.5	6	6.5	7	
18	2°11'	2°47'	3°24'	4°3'	4°44'	5°26'	6°9'	6°55'	7°43'	-	-	
20	1°57'	2°29'	3°2'	3°36'	4°11'	4°48'	5°26'	6°5'	6°46'	7°28'	-	
22	1°46'	2°14'	2°44'	3°14'	3°45'	4°18'	4°51'	5°26'	6°1'	6°38'	-	
24	1°36'	2°2'	2°29'	2°56'	3°24'	3°53'	4°23'	4°54'	5°26'	5°58'	6°32'	

Nom. Dia.							Pitch in Millimeters							
	0.6	0.7	0.75	0.8	1	1.25	1.5	1.75	2	2.5	3	4	4.5	5
25	0°27'	0°31'	0°33'	0°36'	0°45'	0°57'	1°8'	1°20'	1°32'	1°57'	2°22'	3°15'	3°43'	4°11'
27	0°25'	0°29'	0°31'	0°33'	0°42'	0°52'	1°3'	1°14'	1°25'	1°48'	2°11'	2°59'	3°24'	3°50'
28	0°24'	0°28'	0°30'	0°32'	0°40'	0°50'	1°1'	1°11'	1°22'	1°44'	2°6'	2°52'	3°16'	3°41'
30	0°22'	0°26'	0°28'	0°30'	0°37'	0°47'	0°57'	1°6'	1°16'	1°36'	1°57'	2°40'	3°2'	3°24'
32	0°21'	0°24'	0°26'	0°28'	0°35'	0°44'	0°53'	1°2'	1°11'	1°30'	1°49'	2°29'	2°49'	3°10'
33	0°20'	0°24'	0°25'	0°27'	0°34'	0°42'	0°51'	1°0'	1°9'	1°27'	1°46'	2°24'	2°44'	3°4'
35	0°19'	0°22'	0°24'	0°25'	0°32'	0°40'	0°48'	0°57'	1°5'	1°22'	1°39'	2°15'	2°33'	2°52'
36	0°18'	0°22'	0°23'	0°25'	0°31'	0°39'	0°47'	0°55'	1°3'	1°20'	1°36'	2°11'	2°29'	2°47'
39	0°17'	0°20'	0°21'	0°23'	0°29'	0°36'	0°43'	0°51'	0°58'	1°13'	1°29'	2°0'	2°16'	2°33'
40	0°17'	0°19'	0°21'	0°22'	0°28'	0°35'	0°42'	0°49'	0°57'	1°11'	1°26'	1°57'	2°13'	2°29'
42	0°16'	0°18'	0°20'	0°20'	0°26'	0°33'	0°40'	0°47'	0°54'	1°8'	1°22'	1°57'	2°6'	2°21'
45	0°15'	0°17'	0°18'	0°20'	0°25'	0°31'	0°37'	0°44'	0°50'	1°3'	1°16'	1°43'	1°57'	2°11'
48	0°14'	0°16'	0°17'	0°18'	0°23'	0°29'	0°35'	0°41'	0°47'	0°59'	1°11'	1°36'	1°49'	2°2'
50	0°13'	0°15'	0°17'	0°18'	0°22'	0°28'	0°33'	0°39'	0°45'	0°57'	1°8'	1°32'	1°45'	1°57'
52	0°13'	0°15'	0°16'	0°17'	0°21'	0°27'	0°32'	0°38'	0°43'	0°54'	1°6'	1°29'	1°40'	1°52'
55	-	-	0°15'	0°16'	0°20'	0°25'	0°30'	0°36'	0°41'	0°51'	1°2'	1°24'	1°35'	1°46'
56	-	-	0°15'	0°16'	0°20'	0°25'	0°30'	0°35'	0°40'	0°50'	1°1'	1°22'	1°33'	1°44'

(Continued)

Table 76. *(Continued)* Helix Angles for ISO and Metric Threads 1 mm Through 24 mm. *(Source, Landis Threading Systems.)*

Nom. Dia.	Pitch in Millimeters													
	0.6	0.7	0.75	0.8	1	1.25	1.5	1.75	2	2.5	3	4	4.5	5
58	-	-	-	-	0°19'	0°24'	0°29'	0°34'	0°39'	0°49'	0°59'	1°19'	1°29'	1°40'
60	-	-	-	-	0°19'	0°23'	0°28'	0°33'	0°37'	0°47'	0°57'	1°16'	1°26'	1°36'
62	-	-	-	-	0°18'	0°22'	0°27'	0°31'	0°36'	0°45'	0°55'	1°14'	1°23'	1°33'
64	-	-	-	-	0°17'	0°22'	0°26'	0°30'	0°35'	0°44'	0°53'	1°11'	1°21'	1°30'
65	-	-	-	-	0°17'	0°21'	0°26'	0°30'	0°34'	0°43'	0°52'	1°10'	1°19'	1°29'
68	-	-	-	-	0°16'	0°20'	0°24'	0°29'	0°33'	0°41'	0°50'	1°7'	1°16'	1°24'
70	-	-	-	-	0°16'	0°20'	0°24'	0°28'	0°32'	0°40'	0°48'	1°5'	1°13'	1°22'
72	-	-	-	-	0°15'	0°19'	0°23'	0°27'	0°31'	0°39'	0°47'	1°3'	1°11'	1°20'
75	-	-	-	-	0°15'	0°18'	0°22'	0°26'	0°30'	0°37'	0°45'	1°0'	1°8'	1°16'
76	-	-	-	-	0°15'	0°18'	0°22'	0°26'	0°29'	0°37'	0°44'	1°0'	1°6'	1°15'
80	-	-	-	-	0°14'	0°17'	0°21'	0°24'	0°28'	0°35'	0°42'	0°57'	1°4'	1°11'
84	-	-	-	-	-	-	0°20'	0°23'	0°26'	0°33'	0°40'	0°54'	1°1'	1°8'
85	-	-	-	-	-	-	0°20'	0°23'	0°26'	0°33'	0°40'	0°53'	1°0'	1°7'
88	-	-	-	-	-	-	0°19'	0°22'	0°25'	0°32'	0°38'	0°51'	0°58'	1°5'
90	-	-	-	-	-	-	0°18'	0°22'	0°25'	0°31'	0°37'	0°50'	0°57'	1°3'

Nom. Dia.	Pitch in Millimeters					
	5.5	6	6.5	7	7.5	8
25	4°40'	5°10'	5°41'	6°13'	-	-
27	4°16'	4°43'	5°11'	5°40'	6°9'	-
28	4°6'	4°32'	4°58'	5°26'	5°54'	-
30	3°47'	4°11'	4°35'	5°0'	5°26'	5°52'
32	3°31'	3°53'	4°16'	4°38'	5°2'	5°26'
33	3°24'	3°45'	4°7'	4°29'	4°51'	5°14'
35	3°11'	3°31'	3°51'	4°11'	4°32'	4°53'
36	3°5'	3°24'	3°44'	4°3'	4°23'	4°43'
39	2°50'	3°7'	3°24'	3°42'	4°0'	4°19'
40	2°45'	3°2'	3°19'	3°36'	3°53'	4°11'
42	2°36'	2°52'	3°8'	3°24'	3°41'	3°58'
45	2°25'	2°40'	2°54'	3°9'	3°24'	3°40'

(Continued)

Table 76. *(Continued)* Helix Angles for ISO and Metric Threads 1 mm Through 24 mm. *(Source, Landis Threading Systems.)*

Nom. Dia.	Pitch in Millimeters					
	5.5	6	6.5	7	7.5	8
48	2°15'	2°29'	2°42'	2°56'	3°10'	3°24'
50	2°10'	2°22'	2°35'	2°48'	3°2'	3°15'
52	2°4'	2°17'	2°29'	2°41'	2°54'	3°7'
55	1°57'	2°8'	2°20'	2°32'	2°44'	2°56'
56	1°55'	2°6'	2°17'	2°29'	2°40'	2°52'
58	1°51'	2°1'	2°12'	2°23'	2°34'	2°46'
60	1°47'	1°57'	2°7'	2°18'	2°29'	2°40'
62	1°43'	1°53'	2°3'	2°13'	2°24'	2°34'
64	1°40'	1°49'	1°59'	2°9'	2°19'	2°29'
65	1°38'	1°47'	1°57'	2°7'	2°16'	2°26'

Nom. Dia.	Pitch in Millimeters					
	5.5	6	6.5	7	7.5	8
68	1°33'	1°42'	1°51'	2°1'	2°10'	2°19'
70	1°31'	1°39'	1°48'	1°57'	2°6'	2°15'
72	1°28'	1°36'	1°45'	1°54'	2°2'	2°11'
75	1°24'	1°32'	1°40'	1°49'	1°57'	2°5'
76	1°23'	1°31'	1°39'	1°47'	1°55'	2°4'
80	1°19'	1°26'	1°34'	1°41'	1°49'	1°57'
84	1°15'	1°22'	1°29'	1°36'	1°44'	1°51'
85	1°14'	1°21'	1°28'	1°35'	1°42'	1°50'
88	1°11'	1°18'	1°25'	1°32'	1°39'	1°46'
90	1°10'	1°16'	1°23'	1°30'	1°36'	1°43'

Table 77. Helix Angles for General Purpose Acme Threads 3/16 Through 1 1/2 Inch. (*Source, Landis Threading Systems.*)

Dia.	Threads per Inch											
	28	26	24	22	20	18	16	14	13	12	11	10
3/16	4°2'	4°25'	4°51'	5°22'	5°57'	6°44'	7°45'	-	-	-	-	-
1/4	2°56'	3°11'	3°28'	3°49'	4°14'	4°47'	5°28'	6°22'	6°56'	7°26'	8°28'	9°31'
5/16	2°18'	2°29'	2°42'	2°58'	3°17'	3°41'	4°12'	4°53'	5°18'	5°42'	6°17'	7°11'
3/8	1°53'	2°2'	2°12'	2°25'	2°41'	2°59'	3°25'	3°57'	3°54'	4°41'	5°10'	5°43'
7/16	1°35'	1°43'	1°49'	2°3'	2°16'	2°31'	2°53'	3°19'	3°36'	3°56'	4°20'	4°49'
1/2	1°20'	1°30'	1°37'	1°46'	1°58'	2°12'	2°29'	2°52'	3°6'	3°23'	3°43'	4°9'
9/16	1°13'	1°18'	1°26'	1°34'	1°43'	1°56'	2°11'	2°31'	2°43'	2°59'	3°16'	3°39'
5/8	1°6'	1°11'	1°16'	1°24'	1°32'	1°43'	1°57'	2°15'	2°27'	2°39'	2°54'	3°12'
11/16	1°3'	1°4'	1°10'	1°16'	1°24'	1°34'	1°46'	2°3'	2°12'	2°23'	2°37'	2°54'
3/4	54'	59'	1°3'	1°10'	1°16'	1°25'	1°36'	1°51'	2°0'	2°11'	2°23'	2°39'
13/16	50'	54'	58'	1°4'	1°10'	1°18'	1°28'	1°43'	1°50'	2°0'	2°11'	2°25'
7/8	46'	50'	54'	59'	1°5'	1°12'	1°21'	1°34'	1°42'	1°51'	2°1'	2°14'
15/16	43'	46'	50'	55'	1°1'	1°7'	1°15'	1°28'	1°34'	1°43'	1°53'	2°5'
1	40'	43'	47'	51'	57'	1°3'	1°11'	1°22'	1°28'	1°36'	1°45'	1°56'
1 1/16	38'	40'	44'	48'	53'	1°0'	1°7'	1°18'	1°23'	1°30'	1°39'	1°49'
1 1/8	36'	38'	41'	45'	50'	56'	1°3'	1°13'	1°18'	1°25'	1°33'	1°45'
1 3/16	34'	36'	39'	43'	48'	53'	59'	1°9'	1°14'	1°19'	1°28'	1°38'
1 1/4	32'	34'	38'	41'	45'	50'	56'	1°5'	1°10'	1°16'	1°23'	1°32'
1 5/16	30'	32'	36'	39'	43'	48'	54'	1°2'	1°6'	1°12'	1°19'	1°27'
1 3/8	29'	31'	34'	37'	41'	46'	51'	59'	1°3'	1°9'	1°15'	1°23'
1 7/16	-	30'	32'	36'	39'	44'	49'	57'	1°0'	1°6'	1°12'	1°19'
1 1/2	-	29'	31'	34'	37'	43'	47'	54'	58'	1°3'	1°9'	1°16'

(*Continued*)

Table 77. *(Continued)* **Helix Angles for General Purpose Acme Threads** $3/16$ **Through 1** $1/2$ **Inch.** *(Source, Landis Threading Systems.)*

Dia.	Threads per Inch											
	9	8	7	6	5.5	5	4.5	4	3.5	3	2.5	2
1/4	10°52'	-	-	-	-	-	-	-	-	-	-	-
5/16	8°9'	9°25'	-	-	-	-	-	-	-	-	-	-
3/8	6°31'	7°30'	8°48'	10°40'	-	-	-	-	-	-	-	-
7/16	5°26'	6°14'	7°16'	8°45'	-	-	-	-	-	-	-	-
1/2	4°39'	5°19'	6°12'	7°25'	8°15'	-	-	-	-	-	-	-
9/16	4°5'	4°38'	5°24'	6°32'	7°9'	-	-	-	-	-	-	-
5/8	3°37'	4°7'	4°47'	5°42'	6°12'	7°2'	-	-	-	-	-	-
11/16	3°15'	3°41'	4°18'	5°6'	5°38'	6°17'	7°7'	-	-	-	-	-
3/4	2°57'	3°26'	3°54'	4°37'	5°5'	5°41'	6°27'	7°22'	8°40'	10°29'	-	-
13/16	2°43'	3°5'	3°33'	4°13'	4°40'	5°10'	5°51'	6°42'	7°50'	9°26'	-	-
7/8	2°29'	2°50'	3°17'	3°53'	4°17'	4°45'	5°23'	6°8'	7°13'	8°38'	10°49'	-
15/16	2°19'	2°37'	3°3'	3°36'	3°57'	4°24'	4°58'	5°40'	6°37'	7°56'	9°57'	-
1	2°10'	2°27'	2°53'	3°21'	3°41'	4°5'	4°37'	5°15'	6°8'	7°14'	9°10'	-
1 1/16	2°2'	2°18'	2°39'	3°8'	3°26'	3°49'	4°19'	4°54'	5°43'	6°49'	8°30'	-
1 1/8	1°55'	2°10'	2°29'	2°57'	3°13'	3°35'	4°2'	4°36'	5°21'	6°23'	7°56'	10°24'
1 3/16	1°48'	2°4'	2°22'	2°42'	3°3'	3°23'	3°49'	4°20'	5°2'	5°59'	7°25'	9°45'
1 1/4	1°42'	1°56'	2°14'	2°37'	2°53'	3°12'	3°36'	4°5'	4°45'	5°39'	6°59'	9°8'
1 5/16	1°37'	1°50'	2°7'	2°29'	2°44'	3°3'	3°24'	3°53'	4°29'	5°21'	6°34'	8°37'
1 3/8	1°33'	1°45'	2°1'	2°22'	2°36'	2°53'	3°14'	3°40'	4°15'	5°4'	6°13'	8°8'
1 7/16	1°29'	1°40'	1°56'	2°16'	2°29'	2°44'	3°5'	3°30'	4°3'	4°50'	5°55'	7°43'
1 1/2	1°25'	1°35'	1°50'	2°10'	2°22'	2°38'	2°57'	3°20'	3°51'	4°35'	5°38'	7°18'

(Continued)

Based on General Purpose Acme Basic Dimensions.

Table 77. *(Continued)* Helix Angles for General Purpose Acme Threads 1 9/16 Through 4 Inch. *(Source, Landis Threading Systems.)*

Dia.	Threads per Inch											
	28	26	24	22	20	18	16	14	13	12	11	10
1 9/16	-	-	-	32'	36'	40'	45'	52'	56'	1°0'	1°6'	1°13'
1 5/8	-	-	-	31'	35'	38'	43'	50'	54'	58'	1°3'	1°10'
1 3/4	-	-	-	-	32'	35'	40'	46'	49'	54'	59'	1°5'
1 7/8	-	-	-	-	30'	33'	37'	43'	45'	50'	55'	1°0'
2	-	-	-	-	28'	31'	35'	40'	43'	47'	51'	56'
2 1/8	-	-	-	-	-	-	-	-	40'	45'	48'	53'
2 1/4	-	-	-	-	-	-	-	-	38'	42'	45'	50'
2 3/8	-	-	-	-	-	-	-	-	36'	40'	43'	47'
2 1/2	-	-	-	-	-	-	-	-	35'	37'	41'	45'
2 5/8	-	-	-	-	-	-	-	-	-	35'	39'	43'
2 3/4	-	-	-	-	-	-	-	-	-	34'	37'	41'
2 7/8	-	-	-	-	-	-	-	-	-	32'	35'	39'
3	-	-	-	-	-	-	-	-	-	31'	34'	37'
3 1/8	-	-	-	-	-	-	-	-	-	-	33'	36'
3 1/4	-	-	-	-	-	-	-	-	-	-	31'	34'
3 3/8	-	-	-	-	-	-	-	-	-	-	30'	33'
3 1/2	-	-	-	-	-	-	-	-	-	-	29'	32'
3 5/8	-	-	-	-	-	-	-	-	-	-	-	31'
3 3/4	-	-	-	-	-	-	-	-	-	-	-	30'
3 7/8	-	-	-	-	-	-	-	-	-	-	-	29'
4	-	-	-	-	-	-	-	-	-	-	-	28'

(Continued)

Table 77. *(Continued)* Helix Angles for General Purpose Acme Threads 1 9/16 Through 4 Inch. *(Source, Landis Threading Systems.)*

Dia.	9	8	7	6	5.5	5	4.5	4	3.5	3	2.5	2
1 9/16	1°21'	1°32'	1°46'	2°4'	2°16'	2°32'	2°49'	3°12'	3°42'	4°32'	5°24'	6°57'
1 5/8	1°18'	1°28'	1°41'	1°59'	2°10'	2°25'	2°42'	3°4'	3°32'	4°11'	5°9'	6°37'
1 3/4	1°12'	1°22'	1°33'	1°50'	2°0'	2°13'	2°30'	2°49'	3°15'	3°51'	4°43'	6°5'
1 7/8	1°8'	1°16'	1°27'	1°42'	1°52'	2°4'	2°19'	2°37'	3°1'	3°34'	4°21'	5°37'
2	1°3'	1°11'	1°21'	1°36'	1°44'	1°55'	2°10'	2°27'	2°48'	3°20'	4°4'	5°13'
2 1/8	59'	1°7'	1°17'	1°30'	1°38'	1°49'	2°2'	2°17'	2°38'	3°6'	3°48'	4°54'
2 1/4	56'	1°3'	1°13'	1°24'	1°32'	1°43'	1°55'	2°11'	2°29'	2°56'	3°33'	4°34'
2 3/8	53'	1°0'	1°9'	1°20'	1°28'	1°37'	1°48'	2°2'	2°21'	2°46'	3°21'	4°18'
2 1/2	50'	57'	1°5'	1°16'	1°23'	1°32'	1°43'	1°56'	2°13'	2°37'	3°10'	4°3'
2 5/8	47'	54'	1°2'	1°13'	1°19'	1°27'	1°37'	1°50'	2°7'	2°28'	3°1'	3°51'
2 3/4	45'	51'	59'	1°9'	1°15'	1°23'	1°33'	1°45'	2°1'	2°20'	2°52'	3°38'
2 7/8	43'	49'	56'	1°6'	1°12'	1°19'	1°29'	1°40'	1°55'	2°15'	2°44'	3°29'
3	41'	47'	53'	1°3'	1°9'	1°16'	1°25'	1°36'	1°50'	2°9'	2°37'	3°19'
3 1/8	40'	45'	52'	1°0'	1°6'	1°13'	1°21'	1°32'	1°46'	2°3'	2°30'	3°11'
3 1/4	38'	43'	50'	58'	1°3'	1°10'	1°18'	1°28'	1°41'	1°57'	2°24'	3°3'
3 3/8	37'	42'	48'	56'	1°1'	1°7'	1°15'	1°24'	1°37'	1°53'	2°19'	2°56'
3 1/2	36'	40'	46'	54'	59'	1°4'	1°12'	1°21'	1°33'	1°49'	2°13'	2°48'
3 5/8	35'	38'	45'	52'	56'	1°2'	1°9'	1°18'	1°30'	1°44'	2°8'	2°42'
3 3/4	33'	37'	43'	50'	54'	1°0'	1°7'	1°16'	1°27'	1°39'	2°3'	2°36'
3 7/8	32'	36'	42'	49'	53'	59'	1°5'	1°13'	1°24'	1°37'	1°59'	2°31'
4	30'	35'	41'	47'	51'	57'	1°3'	1°11'	1°21'	1°34'	1°55'	2°26'

Based on General Purpose Acme Basic Dimensions.

Table 78. Helix Angles for General Purpose Double Start Acme Threads 3/16 Through 1 1/2 Inch. (Source, Landis Threading Systems.)

Dia.	Threads per Inch										
	28	20	18	16	14	12	11.5	11	10	9	8
3/16	7°51'	11°25'	12°52'	14°44'	17°51'						
1/4	5°43'	8°14'	9°14'	10°32'	12°15'	15°38'	15°22'	16°35'	18°32'	21°0'	18°21'
5/16	4°29'	6°26'	7°12'	8°12'	9°30'	11°33'	11°50'	12°25'	14°9'	15°59'	14°45'
3/8	3°41'	5°16'	5°54'	6°42'	7°45'	9°18'	9°37'	10°15'	11°20'	12°52'	12°19'
7/16	3°8'	4°28'	4°59'	5°40'	6°32'	7°50'	8°6'	8°37'	9°34'	10°46'	10°33'
1/2	2°44'	3°52'	4°20'	4°54'	5°39'	6°45'	6°59'	7°25'	8°16'	9°14'	9°13'
9/16	2°25'	3°25'	3°49'	4°19'	4°59'	5°57'	6°9'	6°31'	7°16'	8°6'	8°11'
5/8	2°10'	3°4'	3°25'	3°52'	4°27'	5°17'	5°29'	6°9'	6°23'	7°12'	7°22'
11/16	1°57'	2°46'	3°5'	3°30'	4°1'	4°46'	4°57'	5°13'	5°47'	6°29'	6°42'
3/4	1°47'	2°32'	2°49'	3°11'	3°40'	4°22'	4°31'	4°45'	5°17'	5°53'	6°9'
13/16	1°39'	2°20'	2°36'	2°56'	3°22'	4°0'	4°9'	4°22'	4°50'	5°25'	5°39'
7/8	1°32'	2°9'	2°24'	2°43'	3°7'	3°42'	3°50'	4°2'	4°28'	4°57'	5°14'
15/16	1°25'	2°0'	2°14'	2°32'	2°54'	3°26'	3°34'	3°46'	4°10'	4°37'	4°53'
1	1°20'	1°53'	2°6'	2°22'	2°43'	3°12'	3°20'	3°30'	3°52'	4°20'	4°36'
1 1/16	-	1°46'	1°58'	2°13'	2°33'	3°0'	3°7'	3°18'	3°38'	4°4'	4°20'
1 1/8	-	1°40'	1°51'	2°5'	2°24'	2°50'	2°57'	3°6'	3°30'	3°49'	4°8'
1 3/16	-	1°34'	1°45'	1°59'	2°16'	2°40'	2°47'	2°56'	3°16'	3°36'	3°52'
1 1/4	-	1°30'	1°40'	1°53'	2°9'	2°32'	2°38'	2°46'	3°4'	3°24'	3°40'
1 5/16	-	1°25'	1°35'	1°47'	2°3'	2°24'	2°30'	2°38'	2°54'	3°14'	3°30'
1 3/8	-	1°21'	1°30'	1°42'	1°57'	2°18'	2°23'	2°30'	2°46'	3°6'	3°20'
1 7/16	-	1°18'	1°26'	1°38'	1°52'	2°12'	2°17'	2°24'	2°38'	2°58'	3°10'
1 1/2	-	1°14'	1°23'	1°33'	1°47'	2°6'	2°11'	2°18'	2°32'	2°50'	

(Continued)

Based on General Purpose Acme Basic Dimensions.

Table 78. *(Continued)* **Helix Angles for General Purpose Double Start Acme Threads** $3/16$ **Through** $1\,1/2$ **Inch.** *(Source, Landis Threading Systems.)*

Dia.	Threads per Inch									
	2	2.5	3	3.5	4	4.5	5	5.5	6	7
3/8	-	-	-	-	-	-	-	-	20°39'	17°12'
7/16	-	-	-	-	-	-	-	-	17°7'	14°18'
1/2	-	-	-	-	-	-	-	16°7'	14°36'	12°15'
9/16	-	-	-	-	-	-	-	14°5'	12°54'	10°43'
5/8	-	-	-	-	-	-	-	12°16'	11°17'	9°30'
11/16	-	-	-	-	-	-	-	11°10'	10°8'	8°33'
3/4	-	-	20°34'	16°57'	14°30'	12°41'	11°16'	10°5'	9°11'	7°46'
13/16	-	-	18°23'	15°23'	13°13'	11°34'	10°15'	9°16'	8°23'	7°4'
7/8	-	20°55'	17°2'	14°13'	12°8'	10°38'	9°34'	8°31'	7°43'	6°32'
15/16	-	19°20'	15°34'	13°4'	11°12'	9°50'	8°45'	7°52'	7°10'	6°5'
1	-	17°53'	14°16'	12°8'	10°25'	9°9'	8°9'	7°20'	6°41'	5°45'
1 1/16	-	16°38'	13°27'	11°19'	9°44'	8°33'	7°37'	6°50'	6°15'	5°17'
1 1/8	20°9'	15°35'	12°37'	10°37'	9°8'	8°1'	7°9'	6°25'	5°53'	4°57'
1 3/16	18°58'	14°36'	11°50'	9°59'	8°37'	7°33'	6°44'	6°5'	5°23'	4°43'
1 1/4	17°49'	13°46'	11°11'	9°26'	8°7'	7°9'	6°22'	5°45'	5°13'	4°28'
1 5/16	16°52'	12°58'	10°37'	8°55'	7°42'	6°46'	6°5'	5°27'	4°57'	4°14'
1 3/8	15°57'	12°17'	10°3'	8°27'	7°19'	6°26'	5°45'	5°11'	4°43'	4°2'
1 7/16	15°10'	11°43'	9°37'	8°4'	6°59'	6°8'	5°27'	4°57'	4°32'	3°52'
1 1/2	14°22'	11°9'	9°10'	7°40'	6°39'	5°51'	5°15'	4°43'	4°19'	3°40'

(Continued)

Based on General Purpose Acme Basic Dimensions.

Table 78. *(Continued)* **Helix Angles for General Purpose Double Start Acme Threads 1 9/16 Through 4 Inch.** *(Source, Landis Threading Systems.)*

Dia.	Threads per Inch									
	20	18	16	14	12	11.5	11	10	9	8
1 9/16	1°11'	1°19'	1°30'	1°43'	2°0'	2°6'	2°12'	2°26'	2°42'	3°3'
1 5/8	1°9'	1°16'	1°26'	1°39'	1°56'	2°1'	2°6'	2°20'	2°36'	2°56'
1 3/4	1°4'	1°11'	1°20'	1°31'	1°48'	1°52'	1°58'	2°10'	2°24'	2°44'
1 7/8	59'	1°6'	1°14'	1°25'	1°40'	1°44'	1°50'	2°0'	2°16'	2°32'
2	55'	1°2'	1°10'	1°20'	1°34'	1°37'	1°42'	1°52'	2°6'	2°22'
2 1/8	-	-	-	-	1°30'	1°32'	1°36'	1°46'	1°58'	2°14'
2 1/4	-	-	-	-	1°24'	1°26'	1°30'	1°40'	1°52'	2°6'
2 3/8	-	-	-	-	-	1°22'	1°26'	1°34'	1°46'	2°0'
2 1/2	-	-	-	-	-	1°18'	1°22'	1°30'	1°40'	1°54'
2 5/8	-	-	-	-	-	1°14'	1°18'	1°26'	1°34'	1°48'
2 3/4	-	-	-	-	-	1°10'	1°14'	1°22'	1°30'	1°42'
2 7/8	-	-	-	-	-	1°7'	1°10'	1°18'	1°26'	1°38'
3	-	-	-	-	-	1°4'	1°8'	1°14'	1°22'	1°34'
3 1/8	-	-	-	-	-	1°2'	1°6'	1°12'	1°20'	1°30'
3 1/4	-	-	-	-	-	59'	1°2'	1°8'	1°18'	1°26'
3 3/8	-	-	-	-	-	57'	1°0'	1°6'	1°14'	1°24'
3 1/2	-	-	-	-	-	55'	58'	1°4'	1°4'	1°20'
3 5/8	-	-	-	-	-	-	-	-	1°2'	1°16'
3 3/4	-	-	-	-	-	-	-	-	-	1°14'
3 7/8	-	-	-	-	-	-	-	-	-	1°12'
4	-	-	-	-	-	-	-	-	-	1°10'

Based on General Purpose Acme Basic Dimensions.

(Continued)

Table 78. *(Continued)* Helix Angles for General Purpose Double Start Acme Threads 1 9/16 Through 4 Inch. *(Source, Landis Threading Systems.)*

Dia.	Threads per Inch									
	2	2.5	3	3.5	4	4.5	5	5.5	6	7
1 9/16	13°42'	10°42'	8°45'	7°22'	6°23'	5°36'	5°3'	4°32'	4°8'	3°32'
1 5/8	13°4'	10°13'	8°19'	7°3'	6°6'	5°22'	4°50'	4°20'	3°58'	3°21'
1 3/4	12°2'	9°22'	7°40'	6°29'	5°37'	4°58'	4°25'	4°0'	3°40'	3°6'
1 7/8	11°8'	8°39'	7°6'	6°1'	5°14'	4°37'	4°8'	3°44'	3°24'	2°54'
2	10°21'	8°6'	6°39'	5°35'	4°48'	4°18'	3°50'	3°28'	3°12'	2°42'
2 1/8	9°44'	7°34'	6°11'	5°15'	4°33'	4°2'	3°34'	3°16'	3°0'	2°34'
2 1/4	9°5'	7°4'	5°51'	4°57'	4°22'	3°48'	3°26'	3°4'	2°48'	2°26'
2 3/8	8°33'	6°40'	5°31'	4°41'	4°4'	3°35'	3°14'	2°56'	2°40'	2°18'
2 1/2	8°4'	6°19'	5°12'	4°26'	3°52'	3°24'	3°4'	2°46'	2°32'	2°10'
2 5/8	7°40'	6°1'	4°56'	4°14'	3°40'	3°14'	2°55'	2°38'	2°26'	2°4'
2 3/4	7°14'	5°43'	4°40'	4°2'	3°30'	3°5'	2°46'	2°30'	2°18'	1°58'
2 7/8	6°57'	5°27'	4°30'	3°50'	3°20'	2°56'	2°39'	2°24'	2°12'	1°52'
3	6°37'	5°13'	4°18'	3°40'	3°12'	2°49'	2°32'	2°18'	2°6'	1°46'
3 1/8	6°21'	5°0'	4°6'	3°32'	3°4'	2°42'	2°26'	2°12'	2°0'	1°44'
3 1/4	6°5'	4°48'	3°54'	3°22'	2°56'	2°35'	2°20'	2°6'	1°56'	1°40'
3 3/8	5°51'	4°38'	3°46'	3°14'	2°48'	2°29'	2°14'	2°2'	1°52'	1°36'
3 1/2	5°35'	4°26'	3°38'	3°6'	2°42'	2°24'	2°8'	1°58'	1°48'	1°32'
3 5/8	5°23'	4°16'	3°29'	3°0'	2°36'	2°19'	2°4'	1°52'	1°44'	1°30'
3 3/4	5°11'	4°6'	3°18'	2°54'	2°32'	2°14'	2°0'	1°48'	1°40'	1°26'
3 7/8	5°2'	3°58'	3°14'	2°48'	2°26'	2°9'	1°57'	1°46'	1°38'	1°24'
4	4°51'	3°50'	3°8'	2°42'	2°22'	2°5'	1°54'	1°42'	1°34'	1°22'

Based on General Purpose Acme Basic Dimensions.

Table 79. Helix Angles for Standard Stub Acme Threads $3/16$ Through $1 1/2$ Inch. (Source, Landis Threading Systems.)

Dia.	Threads per Inch											
	28	26	24	22	20	18	16	14	12	11	10	9
3/16	3°47'	4°6'	4°28'	4°54'	5°26'	6°5'	6°56'	-	-	-	-	-
1/4	2°47'	3°0'	3°16'	3°35'	3°58'	4°26'	5°2'	5°48'	6°52'	7°45'	8°37'	9°43'
5/16	2°12'	2°22'	2°34'	2°49'	3°7'	3°29'	3°57'	4°33'	5°22'	6°0'	6°40'	7°29'
3/8	1°49'	1°57'	2°7'	2°19'	2°34'	2°52'	3°15'	3°44'	4°24'	4°54'	5°26'	6°5'
7/16	1°33'	1°40'	1°49'	1°59'	2°11'	2°26'	2°45'	3°10'	3°43'	4°8'	4°35'	5°8'
1/2	1°21'	1°27'	1°34'	1°43'	1°54'	2°7'	2°24'	2°45'	3°14'	3°35'	3°58'	4°26'
9/16	1°11'	1°17'	1°24'	1°31'	1°41'	1°52'	2°7'	2°26'	2°51'	3°9'	3°29'	3°54'
5/8	1°4'	1°9'	1°15'	1°22'	1°30'	1°41'	1°54'	2°11'	2°33'	2°49'	3°7'	3°29'
11/16	0°58'	1°3'	1°8'	1°14'	1°22'	1°31'	1°43'	1°58'	2°19'	2°33'	2°49'	3°9'
3/4	0°53'	0°57'	1°2'	1°8'	1°15'	1°23'	1°34'	1°48'	2°7'	2°20'	2°34'	2°52'
13/16	0°49'	0°53'	0°57'	1°3'	1°9'	1°17'	1°27'	1°39'	1°56'	2°8'	2°22'	2°38'
7/8	0°45'	0°49'	0°53'	0°58'	1°4'	1°11'	1°20'	1°32'	1°48'	1°59'	2°11'	2°26'
15/16	0°42'	0°46'	0°50'	0°54'	1°0'	1°6'	1°15'	1°26'	1°40'	1°50'	2°2'	2°16'
1	0°40'	0°43'	0°46'	0°51'	0°56'	1°2'	1°10'	1°20'	1°34'	1°43'	1°54'	2°7'
1 1/16	0°37'	0°40'	0°44'	0°48'	0°52'	0°58'	1°6'	1°15'	1°28'	1°37'	1°47'	1°59'
1 1/8	0°35'	0°38'	0°41'	0°45'	0°50'	0°55'	1°2'	1°11'	1°23'	1°31'	1°41'	1°52'
1 3/16	0°33'	0°36'	0°39'	0°43'	0°47'	0°52'	0°59'	1°7'	1°19'	1°26'	1°35'	1°46'
1 1/4	0°32'	0°34'	0°37'	0°40'	0°44'	0°49'	0°56'	1°4'	1°15'	1°22'	1°30'	1°41'
1 5/16	0°30'	0°32'	0°35'	0°38'	0°42'	0°47'	0°53'	1°1'	1°11'	1°18'	1°26'	1°36'
1 3/8	0°29'	0°31'	0°34'	0°37'	0°40'	0°45'	0°51'	0°58'	1°8'	1°14'	1°22'	1°31'
1 7/16	-	0°30'	0°32'	0°35'	0°39'	0°43'	0°48'	0°55'	1°5'	1°11'	1°18'	1°27'
1 1/2	-	0°28'	0°31'	0°34'	0°37'	0°41'	0°46'	0°53'	1°2'	1°8'	1°15'	1°23'

Based on General Purpose Acme Basic Dimensions.

(Continued)

Table 79. (*Continued*) Helix Angles for Standard Stub Acme Threads $^3/_{16}$ Through 1 $^1/_2$ Inch. (*Source, Landis Threading Systems.*)

Dia.	Threads per Inch										
	8	7	6	5.5	5	4.5	4	3.5	3	2.5	2
3/8	6°56'	8°2'	9°34'	-	-	-	-	-	-	-	-
7/16	5°50'	6°45'	8°0'	-	-	-	-	-	-	-	-
1/2	5°2'	5°48'	6°53'	7°34'	-	-	-	-	-	-	-
9/16	4°25'	5°6'	6°2'	6°38'	-	-	-	-	-	-	-
5/8	3°56'	4°33'	5°22'	5°54'	6°33'	-	-	-	-	-	-
11/16	3°33'	4°6'	4°50'	5°18'	5°53'	6°36'	-	-	-	-	-
3/4	3°15'	3°44'	4°24'	4°50'	5°21'	6°0'	6°50'	7°55'	9°25'	-	-
13/16	2°59'	3°26'	4°2'	4°26'	4°54'	5°29'	6°15'	7°14'	8°35'	-	-
7/8	2°45'	3°10'	3°43'	4°5'	4°31'	5°4'	5°45'	6°39'	7°54'	9°42'	-
15/16	2°34'	2°57'	3°28'	3°48'	4°12'	4°42'	5°20'	6°10'	7°18'	8°58'	-
1	2°24'	2°45'	3°14'	3°32'	3°55'	4°23'	4°58'	5°45'	6°48'	8°20'	-
1 1/16	2°15'	2°35'	3°2'	3°19'	3°40'	4°6'	4°39'	5°23'	6°21'	7°47'	-
1 1/8	2°7'	2°26'	2°51'	3°7'	3°27'	3°52'	4°23'	5°3'	5°58'	7°18'	9°22'
1 3/16	2°0'	2°18'	2°42'	2°57'	3°16'	3°39'	4°8'	4°46'	5°37'	6°52'	8°48'
1 1/4	1°54'	2°11'	2°33'	2°48'	3°5'	3°27'	3°55'	4°30'	5°19'	6°29'	8°19'
1 5/16	1°48'	2°4'	2°26'	2°39'	2°56'	3°17'	3°43'	4°17'	5°3'	6°9'	7°52'
1 3/8	1°43'	1°58'	2°19'	2°32'	2°48'	3°7'	3°32'	4°4'	4°48'	5°50'	7°28'
1 7/16	1°38'	1°53'	2°12'	2°25'	2°40'	2°59'	3°22'	3°53'	4°34'	5°34'	7°6'
1 1/2	1°34'	1°48'	2°7'	2°19'	2°33'	2°51'	3°13'	3°42'	4°22'	5°19'	6°47'

(*Continued*)

Based on General Purpose Acme Basic Dimensions.

Table 79. *(Continued)* Helix Angles for Standard Stub Acme Threads 1 9/16 Through 4 Inch. *(Source, Landis Threading Systems.)*

Dia.	Threads per Inch										
	26	24	22	20	18	16	14	12	11	10	9
1 9/16	-	-	0°32'	0°35'	0°39'	0°44'	0°51'	0°59'	1°5'	1°12'	1°20'
1 5/8	-	-	0°31'	0°34'	0°38'	0°43'	0°49'	0°57'	1°3'	1°9'	1°17'
1 3/4	-	-	-	0°32'	0°35'	0°40'	0°45'	0°53'	0°58'	1°4'	1°11'
1 7/8	-	-	-	0°29'	0°33'	0°37'	0°42'	0°49'	0°54'	1°0'	1°6'
2	-	-	-	0°28'	0°31'	0°35'	0°40'	0°46'	0°51'	0°56'	1°2'
2 1/8	-	-	-	-	-	-	-	0°44'	0°48'	0°52'	0°58'
2 1/4	-	-	-	-	-	-	-	0°41'	0°45'	0°50'	0°55'
2 3/8	-	-	-	-	-	-	-	0°39'	0°43'	0°47'	0°52'
2 1/2	-	-	-	-	-	-	-	0°37'	0°40'	0°44'	0°49'
2 5/8	-	-	-	-	-	-	-	0°35'	0°38'	0°42'	0°47'
2 3/4	-	-	-	-	-	-	-	0°34'	0°37'	0°40'	0°45'
2 7/8	-	-	-	-	-	-	-	0°32'	0°35'	0°39'	0°43'
3	-	-	-	-	-	-	-	0°31'	0°34'	0°37'	0°41'
3 1/8	-	-	-	-	-	-	-	-	0°32'	0°35'	0°39'
3 1/4	-	-	-	-	-	-	-	-	0°31'	0°34'	0°38'
3 3/8	-	-	-	-	-	-	-	-	0°30'	0°33'	0°36'
3 1/2	-	-	-	-	-	-	-	-	0°29'	0°32'	0°35'
3 5/8	-	-	-	-	-	-	-	-	-	0°31'	0°34'
3 3/4	-	-	-	-	-	-	-	-	-	0°29'	0°33'
3 7/8	-	-	-	-	-	-	-	-	-	0°29'	0°32'
4	-	-	-	-	-	-	-	-	-	0°28'	0°31'

(Continued)

Based on General Purpose Acme Basic Dimensions.

Table 79. (Continued) Helix Angles for Standard Stub Acme Threads 1 9/16 Through 4 Inch. (Source, Landis Threading Systems.)

Dia.	Threads per Inch										
	8	7	6	5.5	5	4.5	4	3.5	3	2.5	2
1 9/16	1°30'	1°44'	2°1'	2°13'	2°27'	2°44'	3°5'	3°33'	4°11'	5°5'	6°29'
1 5/8	1°27'	1°39'	1°56'	2°8'	2°21'	2°37'	2°58'	3°24'	4°0'	4°52'	6°12'
1 3/4	1°20'	1°32'	1°48'	1°58'	2°10'	2°25'	2°44'	3°9'	3°42'	4°30'	5°43'
1 7/8	1°15'	1°26'	1°40'	1°50'	2°1'	2°15'	2°33'	2°56'	3°26'	4°10'	5°18'
2	1°10'	1°20'	1°34'	1°43'	1°53'	2°6'	2°23'	2°44'	3°13'	3°54'	4°57'
2 1/8	1°6'	1°15'	1°28'	1°37'	1°47'	1°59'	2°14'	2°34'	3°1'	3°39'	4°38'
2 1/4	1°2'	1°11'	1°23'	1°31'	1°40'	1°52'	2°6'	2°25'	2°50'	3°26'	4°21'
2 3/8	0°59'	1°7'	1°19'	1°26'	1°35'	1°46'	1°59'	2°17'	2°41'	3°15'	4°7'
2 1/2	0°56'	1°4'	1°15'	1°22'	1°30'	1°40'	1°53'	2°10'	2°33'	3°5'	3°54'
2 5/8	0°53'	1°1'	1°11'	1°18'	1°26'	1°35'	1°48'	2°4'	2°25'	2°55'	3°42'
2 3/4	0°51'	0°58'	1°8'	1°14'	1°22'	1°31'	1°43'	1°58'	2°18'	2°47'	3°31'
2 7/8	0°48'	0°55'	1°5'	1°11'	1°18'	1°27'	1°38'	1°53'	2°12'	2°39'	3°21'
3	0°46'	0°53'	1°2'	1°8'	1°15'	1°23'	1°34'	1°48'	2°6'	2°33'	3°13'
3 1/8	0°44'	0°51'	1°0'	1°5'	1°12'	1°20'	1°30'	1°43'	2°1'	2°26'	3°4'
3 1/4	0°43'	0°49'	0°57'	1°3'	1°9'	1°17'	1°27'	1°39'	1°56'	2°20'	2°57'
3 3/8	0°41'	0°47'	0°55'	1°1'	1°6'	1°14'	1°23'	1°35'	1°52'	2°15'	2°50'
3 1/2	0°40'	0°45'	0°53'	0°58'	1°4'	1°11'	1°20'	1°32'	1°48'	2°10'	2°44'
3 5/8	0°38'	0°44'	0°51'	0°56'	1°2'	1°9'	1°17'	1°29'	1°44'	2°5'	2°38'
3 3/4	0°37'	0°42'	0°49'	0°54'	1°0'	1°6'	1°15'	1°26'	1°40'	2°1'	2°32'
3 7/8	0°36'	0°41'	0°48'	0°52'	0°58'	1°4'	1°12'	1°23'	1°37'	1°57'	2°27'
4	0°35'	0°40'	0°46'	0°51'	0°56'	1°2'	1°10'	1°20'	1°34'	1°53'	2°22'

Based on General Purpose Acme Basic Dimensions.

Table 80. Helix Angles for Double Start Standard Stub Acme Threads 3/16 Through 1 1/2 Inch. *(Source, Landis Threading Systems.)*

Dia.	Threads per Inch										
	28	20	18	16	14	13	12	11.5	11	10	9
3/16	7°20'	10°28'	11°42'	13°16'	15°19'	-	-	-	-	-	-
1/4	5°26'	7°43'	8°38'	9°46'	11°15'	12°11'	13°16'	13°54'	14°34'	16°8'	18°5'
5/16	4°19'	6°7'	6°49'	7°43'	8°53'	9°37'	10°28'	10°57'	11°29'	12°42'	14°13'
3/8	3°35'	5°4'	5°39'	6°23'	7°20'	7°56'	8°37'	9°1'	9°27'	10°28'	11°42'
7/16	3°3'	4°19'	4°49'	5°26'	6°15'	6°45'	7°20'	7°40'	8°2'	8°53'	9°56'
1/2	2°40'	3°46'	4°11'	4°44'	5°26'	5°52'	6°23'	6°40'	6°59'	7°43'	8°37'
9/16	2°22'	3°20'	3°43'	4°11'	4°49'	5°12'	5°39'	5°54'	6°11'	6°49'	7°37'
5/8	2°8'	3°0'	3°20'	3°46'	4°19'	4°39'	5°4'	5°17'	5°32'	6°7'	6°49'
11/16	1°56'	2°43'	3°1'	3°25'	3°55'	4°13'	4°35'	4°47'	5°1'	5°32'	6°11'
3/4	1°46'	2°29'	2°46'	3°7'	3°35'	3°52'	4°12'	4°23'	4°35'	5°4'	5°39'
13/16	1°38'	2°17'	2°33'	2°53'	3°18'	3°33'	3°52'	4°2'	4°13'	4°39'	5°12'
7/8	1°31'	2°8'	2°22'	2°40'	3°3'	3°18'	3°35'	3°44'	3°55'	4°19'	4°49'
15/16	1°25'	1°59'	2°12'	2°29'	2°51'	3°4'	3°20'	3°29'	3°39'	4°1'	4°29'
1	1°19'	1°51'	2°4'	2°20'	2°40'	2°53'	3°7'	3°16'	3°25'	3°46'	4°11'
1 1/16	-	1°45'	1°57'	2°11'	2°30'	2°42'	2°56'	3°4'	3°12'	3°32'	3°56'
1 1/8	-	1°39'	1°50'	2°4'	2°22'	2°33'	2°46'	2°53'	3°1'	3°30'	3°43'
1 3/16	-	1°34'	1°44'	1°57'	2°14'	2°25'	2°37'	2°44'	2°52'	3°9'	3°31'
1 1/4	-	1°29'	1°39'	1°51'	2°8'	2°17'	2°29'	2°36'	2°43'	3°0'	3°20'
1 5/16	-	1°25'	1°34'	1°46'	2°1'	2°11'	2°22'	2°28'	2°35'	2°51'	3°10'
1 3/8	-	1°21'	1°30'	1°41'	1°56'	2°5'	2°15'	2°21'	2°28'	2°43'	3°2'
1 7/16	-	1°17'	1°26'	1°37'	1°51'	1°59'	2°9'	2°15'	2°21'	2°36'	2°53'
1 1/2	-	1°14'	1°22'	1°33'	1°46'	1°54'	2°4'	2°9'	2°15'	2°29'	2°46'

Based on General Purpose Acme Basic Dimensions.

(Continued)

Table 80. *(Continued)* Helix Angles for Double Start Standard Stub Acme Threads 3/16 Through 1 1/2 Inch. *(Source, Landis Threading Systems.)*

Dia.	Threads per Inch										
	8	7	6	5.5	5	4.5	4	3.5	3	2.5	2
5/16	16°9'	-	-	-	-	-	-	-	-	-	-
3/8	13°16'	15°19'	18°5'	-	-	-	-	-	-	-	-
7/16	11°15'	12°59'	15°19'	-	-	-	-	-	-	-	-
1/2	9°46'	11°15'	13°16'	14°34'	-	-	-	-	-	-	-
9/16	8°37'	9°56'	11°42'	12°51'	-	-	-	-	-	-	-
5/8	7°43'	8°53'	10°28'	11°29'	-	-	-	-	-	-	-
11/16	6°59'	8°2'	9°27'	10°22'	-	-	-	-	-	-	-
3/4	6°23'	7°20'	8°37'	9°27'	10°28'	11°42'	13°16'	15°19'	18°5'	-	-
13/16	5°52'	6°45'	7°56'	8°41'	9°37'	10°45'	12°11'	14°3'	16°36'	-	-
7/8	5°26'	6°14'	7°20'	8°2'	8°53'	9°56'	11°15'	12°59'	15°19'	18°39'	-
15/16	5°4'	5°49'	6°49'	7°28'	8°16'	9°14'	10°28'	12°4'	14°13'	17°18'	-
1	4°44'	5°26'	6°23'	6°59'	7°43'	8°37'	9°46'	11°15'	13°16'	16°9'	-
1 1/16	4°27'	5°6'	5°59'	6°33'	7°15'	8°5'	9°10'	10°33'	12°26'	15°7'	-
1 1/8	4°11'	4°49'	5°39'	6°11'	6°49'	7°37'	8°37'	9°56'	11°42'	14°13'	18°5'
1 3/16	3°58'	4°33'	5°20'	5°50'	6°27'	7°12'	8°9'	9°23'	11°3'	13°25'	17°4'
1 1/4	3°46'	4°19'	5°4'	5°32'	6°7'	6°49'	7°43'	8°53'	10°28'	12°42'	16°9'
1 5/16	3°35'	4°6'	4°49'	5°16'	5°49'	6°29'	7°20'	8°26'	9°56'	12°4'	15°19'
1 3/8	3°25'	3°55'	4°35'	5°1'	5°32'	6°11'	6°58'	8°2'	9°27'	11°29'	14°34'
1 7/16	3°16'	3°44'	4°23'	4°47'	5°17'	5°54'	6°40'	7°40'	9°1'	10°57'	13°54'
1 1/2	3°7'	3°35'	4°11'	4°35'	5°4'	5°39'	6°23'	7°20'	8°37'	10°28'	13°16'

(Continued)

Based on General Purpose Acme Basic Dimensions.

Table 80. *(Continued)* **Helix Angles for Double Start Standard Stub Acme Threads 1 9/16 Through 4 Inch.** *(Source, Landis Threading Systems.)*

Dia.	Threads per Inch										
	8	9	10	11	11.5	12	13	14	16	18	20
1 9/16	3°0'	2°39'	2°23'	2°10'	2°4'	1°59'	1°50'	1°42'	1°29'	1°19'	1°11'
1 5/8	2°53'	2°33'	2°17'	2°5'	1°59'	1°54'	1°45'	1°38'	1°25'	1°16'	1°8'
1 3/4	2°40'	2°22'	2°8'	1°56'	1°51'	1°46'	1°38'	1°31'	1°19'	1°10'	1°3'
1 7/8	2°29'	2°12'	1°59'	1°48'	1°43'	1°39'	1°31'	1°25'	1°14'	1°6'	0°59'
2	2°20'	2°4'	1°51'	1°41'	1°37'	1°33'	1°25'	1°19'	1°9'	1°2'	0°55'
2 1/8	2°11'	1°57'	1°45'	1°35'	1°31'	1°27'	1°20'	-	-	-	-
2 1/4	2°4'	1°50'	1°39'	1°30'	1°26'	1°22'	1°16'	-	-	-	-
2 3/8	1°57'	1°44'	1°34'	1°25'	1°21'	1°18'	1°12'	-	-	-	-
2 1/2	1°51'	1°39'	1°29'	1°21'	1°17'	1°14'	1°8'	-	-	-	-
2 5/8	1°46'	1°34'	1°25'	1°17'	1°14'	1°10'	-	-	-	-	-
2 3/4	1°41'	1°30'	1°21'	1°13'	1°10'	1°7'	-	-	-	-	-
2 7/8	1°37'	1°26'	1°17'	1°10'	1°7'	1°4'	-	-	-	-	-
3	1°33'	1°22'	1°14'	1°7'	1°4'	1°2'	-	-	-	-	-
3 1/8	1°29'	1°19'	1°11'	1°5'	1°2'	-	-	-	-	-	-
3 1/4	1°25'	1°16'	1°8'	1°2'	0°59'	-	-	-	-	-	-
3 3/8	1°22'	1°13'	1°6'	1°0'	0°57'	-	-	-	-	-	-
3 1/2	1°19'	1°10'	1°3'	0°58'	0°55'	-	-	-	-	-	-
3 5/8	1°17'	1°8'	-	-	-	-	-	-	-	-	-
3 3/4	1°14'	1°4'	-	-	-	-	-	-	-	-	-
3 7/8	1°12'	-	-	-	-	-	-	-	-	-	-
4	1°9'	-	-	-	-	-	-	-	-	-	-

Based on General Purpose Acme Basic Dimensions.

(Continued)

Table 80. (*Continued*) **Helix Angles for Double Start Standard Stub Acme Threads 1 ⁹/₁₆ Through 4 Inch.** (*Source, Landis Threading Systems.*)

Dia.	Threads per Inch									
	7	6	5.5	5	4.5	4	3.5	3	2.5	2
1 9/16	3°26'	4°4'	4°24'	4°51'	5°24'	6°7'	7°2'	8°16'	10°1'	12°42'
1 5/8	3°18'	3°52'	4°13'	4°39'	5°12'	5°52'	6°45'	7°56'	9°37'	12°11'
1 3/4	3°3'	3°35'	3°55'	4°19'	4°49'	5°26'	6°15'	7°20'	8°53'	11°15'
1 7/8	2°51'	3°20'	3°39'	4°1'	4°29'	5°4'	5°49'	6°49'	8°16'	10°28'
2	2°40'	3°7'	3°25'	3°46'	4°11'	4°44'	5°26'	6°23'	7°43'	9°46'
2 1/8	2°30'	2°56'	3°12'	3°32'	3°56'	4°27'	5°6'	5°59'	7°15'	9°10'
2 1/4	2°22'	2°46'	3°1'	3°20'	3°43'	4°11'	4°49'	5°39'	6°49'	8°37'
2 3/8	2°14'	2°37'	2°52'	3°9'	3°31'	3°58'	4°33'	5°20'	6°27'	8°9'
2 1/2	2°8'	2°29'	2°43'	3°0'	3°20'	3°46'	4°19'	5°4'	6°7'	7°43'
2 5/8	2°1'	2°22'	2°35'	2°51'	3°10'	3°35'	4°6'	4°49'	5°49'	7°20'
2 3/4	1°56'	2°15'	2°28'	2°43'	3°1'	3°25'	3°55'	4°35'	5°32'	6°59'
2 7/8	1°51'	2°9'	2°21'	2°36'	2°53'	3°16'	3°44'	4°23'	5°17'	6°40'
3	1°46'	2°4'	2°15'	2°29'	2°46'	3°7'	3°35'	4°11'	5°4'	6°23'
3 1/8	1°42'	1°59'	2°10'	2°23'	2°39'	3°0'	3°26'	4°1'	4°51'	6°7'
3 1/4	1°38'	1°54'	2°5'	2°17'	2°33'	2°53'	3°18'	3°52'	4°39'	5°52'
3 3/8	1°34'	1°50'	2°0'	2°12'	2°27'	2°46'	3°10'	3°43'	4°29'	5°39'
3 1/2	1°31'	1°46'	1°56'	2°8'	2°22'	2°40'	3°3'	3°35'	4°19'	5°26'
3 5/8	1°28'	1°42'	1°52'	2°3'	2°17'	2°35'	2°57'	3°27'	4°10'	5°14'
3 3/4	1°25'	1°39'	1°48'	1°59'	2°12'	2°29'	2°51'	3°20'	4°1'	5°4'
3 7/8	1°22'	1°36'	1°44'	1°55'	2°8'	2°24'	2°45'	3°13'	3°53'	4°53'
4	1°19'	1°33'	1°41'	1°51'	2°4'	2°20'	2°40'	3°7'	3°46'	4°44'

Based on General Purpose Acme Basic Dimensions.

Threaded Fasteners

This section primarily contains dimensions and specifications for threaded fasteners. Explanations of standard fastener grades, plus the ratings and head and nut markings for each grade, were provided in the "Bolted Joints" section. Also, because washers are used in conjunction with threaded fasteners, they are included in this section so that they are easily accessible for review when considering bolt/screw/washer combinations. As anyone who has specified a fastener for a specific application is aware, there are vast numbers of new, innovative styles of threaded fasteners for almost any need. It is impossible to chronicle each of these ingenious devices in this book, as they have neither been standardized nor used in sufficient numbers to make them universally available. With this in mind, creative designers should, by consulting manufacturer's catalogs, be able to find a nearly ideal threaded fastener for almost any requirement.

After this introduction containing brief descriptions of the general types of threaded fasteners, Tables follow with dimensions for the most widely used English/U.S. Customary unit fasteners: Hexagon head and other bolt forms; hex nuts; socket head screws; machine, tapping, and other screws. Tables providing Metric measurement threaded fasteners are then grouped similarly. It is hoped that this arrangement will minimize confusion and make locating specific fasteners more efficient and less time consuming. For full information on specific fasteners, the following standards should be consulted.

English/U.S. Customary Unit Threaded Fasteners and Washers
ANSI/ASME B18.2.1. American National Standard Square Bolts, Hex Bolts, and Hex Screws.
ANSI/ASME B18.2.2. American National Standard Hex Nuts.
ANSI/ASME B18.3. American National Standard Socket Head Screws.
ANSI/ASME B18.5. American National Standard Round Head, Countersunk Head, and Special Head Bolts.
ANSI/ASME B18.6.2. American National Standard Slotted Head Cap Screws and Slotted and Square Head Set Screws.
ANSI/ASME B18.6.3. American National Standard Machine Screws and Machine Screw Nuts.
ANSI/ASME B18.6.4. American National Standard Self-Tapping and Metallic Drive Screws.
ANSI B18.17. American National Standard Wing Nuts and Screws, and Thumb Screws.
ANSI/ASME B18.21.1. American National Standard Lock Washers.
ANSI/ASME B18.22.1. American National Standard Plain Washers.

Metric Unit Fasteners and Washers*
ANSI B18.2.3.1M. Metric Hexagon Cap Screws.
ISO/DIN 4017 and DIN 933. Hexagon Head Screw, product grades A and B.
ISO/DIN 4018 and DIN 558. Hexagon Head Screw, product grade C.
ISO 8676 and DIN 961. Hexagon Head Screw with Fine Thread.
ANSI B18.2.3.4M. Metric Hexagon Flange Screws.
ISO 8100/04 and DIN 6921. Hexagon Flange Bolt.
ANSI B18.2.3.5M. Metric Hexagon Bolts.
ISO/DIN 4014 and DIN 931. Hexagon Head Bolt, product grades A and B.
ISO/DIN 4016 and DIN 601. Hexagon Head Bolt, product grade C.
ISO 8765 and DIN 960. Hexagon Head Bolt with Fine Thread.
ANSI B18.2.3.7M. Metric Heavy Hexagon Structural Bolts.

DIN 9614. Hexagon Bolt for High Tensile Structural Bolting.
ANSI B18.2.3-8M. Metric Hex Lag Screws.
ANSI B18.3.1M. Metric Socket Head Cap Screws.
ISO 4762 and DIN 912. Socket Head Cap Screw.
ANSI B18.2.4.1M and B18.2.4.2M. Metric Hexagon Nuts.
ISO/DIN 4032 and DIN 934. Hexagon Nut, style 1, product grades A and B.
ISO/DIN 4034 and DIN 555. Hexagon Nut, style 1, product grade C.
ANSI B18.2.4.3M. Metric Slotted Hexagon Nuts.
DIN 935. Hexagon Castle Nut.
DIN 979. Hexagon Thin Castle Nut.
ANSI B18.2.4.4M. Metric Hexagon Flange Nuts.
ISO 4161 and DIN 6923. Hexagon Flange Nut.
ANSI B19.2.4.5M and 2.4.6M. Metric Hex Jam Nuts and Heavy Hexagon Nuts.
ANSI B18.3.3M. Metric Socket Head Shoulder Screws.
ISO 7379 and DIN 9841. Socket Head Shoulder Screws.
ANSI B18.3.4M. Metric Socket Button Head Cap Screws.
ANSI B18.3.5M. Metric Socket Flat Head Cap Screws.
DIN 7991. Hexagon Socket Countersunk Head Screw.
ANSI B18.3.6M. Metric Socket Set Screws.
ISO 4026, 4027, 4028, 4029, and DIN 913, 914, 915, 915. Socket Set Screws.
ANSI B18.5.2.2M. Metric Round Head Square Neck Bolts.
ANSI B18.18.3M-1998. Metric Prevailing Torque Hex Flange Nuts.
ISO 7043 and DIN 6926. Prevailing Torque Hex Nut with Nylon Insert.
ISO 7044 and DIN 6927. Prevailing Torque Hex Nut–All Metal.
ANSI B18.22M. Metric Plain Washers.
ISO 7089/90 and DIN-125-T1. Hex Head Bolt and Nut Washers up to Hardness 250 HV.
ISO 7089/90 and DIN-125-T2. Hex Head Bolt and Nut Washers from Hardness 300 HV.
ISO 7092 and DIN 433-T-1. Cheese Head Bolt and Nut Washers up to Hardness 250 HV.
ISO 7092 and DIN 433-T-2. Cheese Head Bolt and Nut Washers from Hardness 300 HV.
ISO 1207 and DIN 84. Slotted Cheese Head Screws.
ISO/DIN 1479. Hexagon Head Tapping Screws.
ISO/DIN 1481. Slotted Pan Head Tapping Screws.
ISO/DIN 1482. Slotted Countersunk Flat Head Tapping Screws.
ISO/DIN 1483. Slotted Raised Countersunk Oval Head Tapping Screws.
ISO/DIN 1580 and DIN 85. Slotted Pan Head Screws.
ISO/DIN 7045. Cross Recessed Pan Head Screws.
ISO/DIN 7046. Cross Recessed Countersunk Flat Head Screws.
ISO/DIN 7047. Cross Recessed Raised Countersunk Head Screws.
ISO/DIN 7049 and DIN 7981. Cross Recessed Pan Head Tapping Screws.
ISO/DIN 7050 and DIN 7982. Cross Recessed Countersunk Flat Head Tapping Screws.
ISO/DIN 7051 and DIN 7983. Cross Recessed Raised Countersunk Oval Head Tapping Screws.

* Grade A and B Metric fasteners are minimum property class 8.8 for bolts, and minimum property class 8 for nuts. Bolts adhering to these Grades, in diameters of 5 mm or higher, have their property class stamped on the head of the bolt. Nuts adhering to

these grades in nominal sizes of 5 mm or higher have their property class on one side flat (preferred), or on the clock face. Grade C (commercial) Metric bolts conform to property class 4.6 or better, and nuts of this Grade conform to property class 4 or better.

The Metric threaded fastener Tables reference appropriate ISO/DIN and DIN standard numbers. They should be consulted for additional standard reference numbers. The values in these tables are generally nominal values based on the relevant standards, but they, as with all the Tables in this section, should not be taken as absolute measurements. Appropriate standards should be consulted when dimensional references are critical.

ISO tolerances for Metric threaded fasteners are provided in **Table 75**, and ISO tolerances for socket screws are given in **Table 124**.

Bolts and screws

Even though individual bolts and screws may be interchangeable, a bolt is intended to be used with a nut. A screw, on the other hand, mates with preexisting threads, or, in some configurations such as self-tapping screws, a screw can make its own threads. Screws and bolts are both externally threaded fasteners that come in a variety of head/shoulder/shank/thread combinations, and are used to join separate elements, or to fasten something into place. A stud is used for similar applications, but does not have a head that can be used to torque it into place.

As seen in the following material provided by Unbrako/SPS Technologies, Inc. fabrication methods can impact the quality of a fastener. Producing threads on bolts and screws is covered in detail elsewhere in this book, but bolt and screw heads are produced by entirely different operations. There are two general methods of making bolt heads—forging and machining. The economy and grain flow resulting from forging make it the preferred method.

The temperature of forging can vary from room temperature to 2000° F (1094° C). By far, the greatest number of parts are cold upset on forging machines known as headers, or boltmakers. For materials lacking sufficient formability for cold forging, hot forging must be used. Hot forging is also used for bolts too large for cold upsetting due to machine capacity. In fact, the largest cold forming machines can make bolts up to 1 $^1/_2$ inches in diameter. For large quantities of bolt heads, hot forging is, unfortunately, more expensive than cold forging.

Some materials, such as stainless steel, are warm forged at temperatures up to 1,000° F (538° C). This heating provides two benefits. First, lower forging pressures are necessary due to the reduced yield strength of the material, and, second, work hardening rates are reduced.

Machining is used for very large diameters or small production runs. The disadvantage of machining is that the process cuts the metal grain flow, thus creating planes of weakness at the critical head-to-shank fillet area. This can result in reduced tension performance resulting from fracture planes.

The head-to-shank transition (fillet) represents a sizable change in cross section at a critical area of bolt performance. It is important that this notch effect be minimized. A generous radius in the fillet reduces the notch effect, but a compromise is necessary because too much radius will reduce the load bearing area under the head. Composite fillets, such as elliptical fillets, maximize curvature on the shank side of the fillet, and minimize it on the head side—thereby reducing loss of bearing area on the load bearing surface. The head-shank fillet must not be restricted or bound by the bolt hole. A sufficient chamfer or radius on the edge of the hole will prevent interference that could seriously reduce fatigue life. Also, if the bolt should seat on an unchamfered edge, there might be serious loss of preload if the edge breaks under load.

Bolt forms

Hexagon head bolts are available in two basic configurations: standard hex bolts, and heavy hex bolts/heavy hex structural bolts. The threads may be Class 2A Unified Coarse, Fine, or 8 thread series. Heavy hex bolts have head dimensions that are approximately $1/8$ inch wider than standard hex bolts throughout the range, thereby usefully increasing the bearing surface area. The material specification for standard and heavy Grade A bolts is ASTM A307. Heavy hex structural bolts have generally the same dimensions as heavy hex bolts, but the thread length is shorter (0.25 inch shorter on $1/2$ inch nominal size bolts, and 1 inch shorter on $1 \, 1/2$ inch nominal size bolts). Structural bolts are made to ASTM A325 and ASTM A490 specifications, and the head should be stamped with the appropriate specification. Since hex bolts and hex head cap screws essentially share dimensions other than head height (the head on screws is not as high as on bolts), many suppliers will substitute one for the other in smaller diameters up to $3/4$ inch. Therefore, if head height is critical, orders should specify bolt or screw. Also, different material standards (SAE J429, ASTM A449, or ASTM A354) are specified for screws.

Many round head and flat head bolts are in wide use, and the most common varieties are shown in the Tables.

Metric standard hex bolts and standard hex cap screws are almost dimensionally identical, but the screw is normally specified with a washer face while the bearing surface is often flat on the bolt.

Dimensions of English/U.S. Customary unit bolts are given in **Tables 1** through **15**, and dimensions for Metric bolts are given in **Tables 76** through **85**.

Nuts

The most common shapes for nuts are square and hex. Square nuts are not normally made to exacting tolerances and their use is generally restricted to lighter-duty applications. Hex nuts come in many configurations, in standard and heavy versions, and in thinner sizes known as jam nuts. Jam nuts are used as locking nuts. Normally, a jam nut is tensioned to seating torque on a screw or stud, and then held stationary while a full nut is torqued to full preload on top of the jam nut. This results in the two nuts bearing in opposite directions and "jamming" the threads to serve as a locking nut. Standard hex and jam nuts have slightly smaller dimensions than heavy hex and heavy jam nuts: up to $9/16$, the standard nut is $1/16$ inch narrower across flats; above $9/16$, standard nuts are $1/8$ inch narrower across flats. In addition, jam nuts are always thinner than hex nuts of the same nominal size. For standard nuts and jam nuts, the difference in thickness ranges from $1/16$ inch to $7/16$ inch, and for heavy nuts and heavy jam nuts the difference in thickness ranges from $2/16$ inch to $1 \, 3/4$ inch. Hex nuts are available in many other varieties including locking nuts with nylon inserts, flanged nuts, and castle (or slotted) nuts. Hex nuts for high temperature, high pressure service should be specified as follows. For moderate temperatures and pressures, specify ASTM A194 Grade 2 nuts. For quenched nuts suitable for high temperature and high pressure conditions, specify ASTM A194 Grade 2H nuts. For severe conditions, specify ASTM A194 Grades 4 or 7 for heat treated carbon molybdenum and chromium molybdenum nuts.

Hex nuts in English/U.S. Customary units are available in a wider variety of sizes than metric hex nuts, but metric versions of most hex shaped nuts are available including jam nuts, nylon insert locking nuts, flanged nuts, and castle nuts.

Dimensions of English/U.S. Customary unit nuts are given in **Tables 16** through **31**, and dimensions for Metric nuts are given in **Tables 86** through **95**.

Washers
Due to the fact that washers are manufactured in almost every conceivable width, internal diameter, and thickness, only standardized washers are included in this book. Companies such as WCL (West Coast Lockwasher) and ASM (Accurate Screw Machine) publish extensive lists of available sizes and materials and should be contacted for dimensions of flat, curved, Belleville, and wave form washers. See the section in this book on "Bolted Joints" for more information about these washer forms.

Dimensions of selected English/U.S. Customary unit washers are given in **Tables 32** (which provides comparative dimensions of both inch unit and Metric unit washers), **33, 34,** and **35,** and dimensions for Metric washers are given in **Tables 96** through **107.**

Common screws, and machine screw head and thread configurations
Hexagon head cap screws are the most common screws in use. They are briefly compared to hexagon head bolts in the section on bolts (above). There are many basic and standardized head configurations for machine screws. In addition, there are probably twice as many different "drive" systems (Phillips head, slotted head, Torx, etc.), and each of the more popular head styles can most likely be obtained with any of the drive systems. With the exception of slotted, cross recessed or Phillips, and socket head screws, most drive systems have not been standardized and their intended use can be very specialized. Therefore, before specifying a screw or drive system, it is often a good idea to shop around for the one best suited to a particular application. With the exception of the hex head screw, all machine screws have round heads when viewed from above.

Binding Head. This design is only available with a flat bearing surface. In appearance, it resembles the fillister head, but is wider and lower. Due to their low head height, binding head screws are available only with a slotted drive.

Cheese Head. This design is more popular with Metric screws in Europe than it is in the U.S. It resembles a tall fillister head, but has a flat or recessed slotted top surface.

Fillister Head. The fillister head is easily identified by its relatively high sides and its rounded top surface.

Flat Countersunk Head. There are two established designs for this head: one with a head angle of approximately 82° (usually preferred), and a less common head with an angle of approximately 100° (the 100° thread is, however, used exclusively in the miniature series). When installed correctly, this head design sits flush with the surrounding surface. A "flat countersunk trim head" provides either a head one size larger (large trim head) or two sizes smaller (small trim head) than a conventionally sized flat countersunk head. Trim heads are available only with cross-recessed drives.

Hex Head. Identified by six flat sides, a flat bearing surface, and a flat or slightly indented top surface, this design offers the choice of fastening or releasing either by wrenching or by a slotted driver.

Hex Washer Head. This is a refined version of the hex head that incorporates a round, flat bearing surface that extends beyond the raised head of the screw. It may be either wrenched or tightened/removed with a slotted driver.

Oval Countersunk Head. This head surface is rounded and the head angle is approximately 82°. An "oval countersunk trim head" provides either a head one size larger (large trim head) or two sizes smaller (small trim head) than a conventionally sized oval countersunk head. Trim heads are available only with cross-recessed drives.

Pan Head. The top surface of a pan head screw is flat if the screw is slotted, but rounded if it is cross-recessed. The pan head is generally preferred over the round head screw.

Round Head. This is the only semi-elliptical machine screw head. It has been to some degree replaced by the pan head, which provides a slot or recess of almost uniform depth across the head, and protrudes less from the bearing surface.

Truss Head. The truss head is the second widest head available (after the binding head). It is low and rounded on top, allowing for very little depth between the bearing surface and slot or recess bottom. Therefore, it is weak by design and should only be used with low driving torques.

Except for the smallest sizes, machine screws are available with UNC and UNF Class 2A threads, or UNRC and UNRF series threads. Thread lengths are:

Size	Thread Length
No. 5 and smaller, nominal length three diameters or less	Threads extend to at least one pitch of bearing surface
Nominal lengths greater than three diameters, to 1 $^1/_8$ inches	Threads extend to at least two pitches of bearing surface
No. 6 and larger, nominal length three diameters or less	Threads extend to at least one pitch of bearing surface
Nominal lengths greater than three diameters, to 2 inches	Threads extend to at least two pitches of bearing surface
Over 2 inches	1 $^1/_2$ inches, minimum.

Ironically, there are machine screw nuts as well. Although the use of a nut transforms the screw into a "bolt," hex nuts are available in standard plus small pattern sizes, and square nuts are available as well.

The Metric dimensions for machine screw lengths are based on a minimal screw length, which is equal to three diameters, and governs the unthreaded portion of the screw beneath the head. For the minimal lengths, this unthreaded portion is equal to approximately 20% of the diameter. For mid-range sizes (up to 30 mm for M × 2 and M × 2.5, and up to 50 mm for all other sizes), the length of the unthreaded area is approximately one-third of the diameter. For longer screws, the full form thread length is a minimum of 25 mm on M × 2 and M × 2.5 diameters, and 30 mm on larger diameters.

Hex head cap screws and machine screws are covered in **Tables 36** through **47**, and instrument (miniature) screws are covered in **Tables 48** through **50** (English/U.S. Customary units). Metric hex head and machine screw dimensions are given in **Tables 108** through **116**.

Self-tapping and thread cutting screws

For many years, the Type A point was the most widely specified thread forming tapping screw style. Its thread was very coarse (25% fewer threads per inch in the $^3/_8$ or #24 nominal diameter than the "AB" thread), and generally it fell out of favor and was relegated to the "not recommended" category. The AB thread has generally replaced the A, and it shares general dimensions with the DIN 7940 and ISO 1478 Type ST tapping screw thread. The Type AB thread combines the 45° gimlet point of the Type A, and the pitches of the Type B. (The ISO designation for this point is also "AB," but the DIN nomenclature is Type B.) The Type AB can be used for most applications, and recommended core hole sizes are provided in the Tables for tapping screws.

The Type B thread forming tapping screw is used for sheet metal, nonferrous castings, plywood, and plastics. It is essentially a flat end Type AB screw, and the Metric version is covered by DIN 7940 and ISO 1478. (The ISO standard refers to the flat end as a Type B point, while the DIN standard calls it a Type BZ.) The Type BT, also known as the Type 25, shares the threads and point of the Type B, but has a cutting groove on the point. It is commonly used in plastics.

Thread Type D (also known as Type 1), Type F, and Type T (also known as Type 23) are thread cutting screws, but they have finer threads approximately equivalent to those found on machine screws. They also feature blunt points and tapered entering threads with chip cavities. These screws are commonly used in sheet metal, aluminum, zinc, and lead die castings, cast iron, and plastics.

The Type U has a smooth, round, unslotted head. It is inserted with pressure and is not intended to be removed once in position.

Tapping and thread forming screws use many of the same head styles covered above in the discussion of machine screws, and most commonly with slotted or Phillips recessed drives.

Recommended hole sizes for English/U.S. Customary unit tapping and thread forming screws are given in **Tables 59** through **62**.

Tapping and thread cutting screws are covered in **Tables 55** through **58** (English/U.S. Customary units), and **Tables 117** through **120** (Metric screws).

Sems

Sems are screws with captive washers. While most commonly available as slotted or Phillips pan head screws available in size from No. 2 to $^1/_2$" nominal diameter, they may also have round, flat, oval, fillister, truss, binding, socket, hex, or hex washer heads. Thread styles can be specified to meet specific needs, as sems can be obtained off the shelf with machine screw threads, or the following tapping threads: Type AB, Type B, Type D/Type1, Type F, and Type T/Type 23. Lock washer styles include internal tooth, external tooth, square cone, helical, and conical, and plain flat washer versions are also available. New versions of sems-type captive washer screws are constantly being introduced, so several suppliers should be contacted if an unusual thread/washer combination is desirable for a specific application.

Typically, sems are available in thread diameters of 0.0470" (size 00) to $^1/_2$", and Metric diameters of M1 through M12. Lengths range from $^1/_8$ to 3 inches, and 3 to 75 millimeters.

Socket screws

Socket screws were developed for applications with limited space, and their cylindrical head and internal wrenching features allow their use in locations where externally wrenched fasteners would be impractical. As originally designed, consistent relationships were not maintained among the nominal shank diameter, head diameter, and socket size throughout the range of sizes. However, in 1960, industry standards were issued that provided consistent relationships throughout the size range. Screws made to this standardized design are commonly referred to as "1960 Series" socket head cap screws. The older design, termed the "1936 Series" after the year of their design, are no longer in production. The designations "1936" and "1960" have caused some confusion because the numbers are sometimes thought to refer to the materials used in the screw's construction. The terms refer only to the dates of acceptance for dimensional relationships, and do not in any way define the mechanical properties of a screw.

The International Organization for Standardization standard ISO 898 contains a system of property classes that perform a function similar to the Society of Automobile Engineers SAE J429 grade designations. These property classes differ in material, and have up to 100% difference in strength. The two most pertinent classes, for socket head screws, are 10.9 and 12.9. The numerals used in property class designations refer to the nominal ultimate tensile strength and nominal percent yield strength. For example, a Property Class 10.9 fastener has 1,000 MPa nominal ultimate strength, and a yield of 90 percent of ultimate. The strength of Property Class 10.9 allows the use of plain carbon steels, while Grade 8 calls for only alloy steels.

The U.S. industry material standard for Metric, alloy steel socket head cap screws, ASTM A574M, specifies one strength level, which is equivalent to ISO Property Class 12.9 and is similar to the strength level for inch unit socket heads. As noted, the U.S. standard specifies only one strength level for all socket head cap screws, metric socket screws are manufactured around the world to various standards, many of which are not equal to ASTM A547M. Therefore, metric socket screws may look similar but may be Property Class 8.8, Property Class 10.9, or Property Class 12.9. U.S. standards call for all socket head cap screws to meet ASTM A574M, which is equal to Property Class 12.9 (1,220 MPa [177,000 PSI]) ultimate tensile strength.

Socket head set screws have less bearing area than hex head screws of the same nominal diameter, and a hardened and ground washer should be used with socket head set screws when the bolt is stressed in tension. Otherwise, there is the risk that the head will sink into the washer and relieve the residual stress within the screw.

Low head socket hex screws. These screws are normally used in parts too thin for standard height socket cap screws, and in applications with limited clearance.

Socket head shoulder screws. Shoulder screws have an undercut portion between the thread and shoulder, allowing for a close fit. They are used in many punch and die operations, such as the location and retention of stripper plates, and act as a guide in blanking and forming presses. Other applications include bearing pins for swing arms, links, and levers; shafts for cam rolls and other rotating parts, pivots, and stud bolts. They are sometimes called "stripper bolts," resulting from their use with stripper plates and springs.

Flat head and button head socket screws. These screws were designed to be used for moderate fastening applications such as machining guards, hinges, covers, etc. They are not recommended for use in critical high strength applications where socket head cap screws should be used.

Set screw points. These points are common to both socket and square head set screws.

Knurled cup. Designed for quick and permanent location of gears, collars, pulleys, knobs, or shafts. This style resists loosening even during severe vibration.

Plain cup. Use against hardened shafts, in zinc, die castings, and other soft materials where high tightening torques are impractical.

Flat. Designed for applications where parts must be frequently reset. Flat socket screws cause minimal damage to the surface they bear against. They can be used with hardened shafts, and as adjusting screws. Preferred for thin walled thicknesses and on soft plugs.

Oval. Like the flat point, oval points are popular for applications where frequent adjustment without deforming the part is required. Also for seating against an angular surface.

Cone. For permanent location of parts. Deep penetration gives highest axial and holding power. In material over R_C 15, point should be spotted to half its length to develop shear strength across point. Used for pivots and fine adjustment.

Dog/half dog. Used for permanent location of one part to another. Point is spotted in hole drilled in shaft or against flat (milled). Often used to replace dowel pins. Works well against hardened members or hollow tubing.

Torsional and axial holding power of English/U.S. Customary unit cup point socket set screws can be found in **Table 71**.

Dimensions for English/U.S. Customary socket screws are given in **Tables 63** through

71. Metric socket screws are covered in **Tables 121** through **130**. Grip and body lengths for English/U.S. Customary unit socket head cap screws can be found in **Table 64**, and hole and counterbore sizes for these screws are given in **Table 65**. The same information for Metric socket head cap screws is provided in **Tables 122** and **123**.

Retained nuts and speed nuts

These are essentially "clip on" internally threaded fasteners that offer several benefits for assembly operations. Since they are, essentially, floating nuts, they do not require drilling in the fixture being attached, and they do not require drilling or tapping. This system of floating alignment does not require special tools, skills, or other equipment. Retained nuts actually use a threaded square nut mounted on a retaining clip, while speed nuts provide a hole that accommodates a tapping nut or sem. Dimensions for Type-J and Type-U clip on fasteners in both English/U.S. Customary units and Metric units are given in **Tables 25–28**.

Studs

There are three basic configurations of studs in wide use. The most common is the continuous thread stud, which is available in two types. Type 1 continuous thread studs are measured from end to end, and threads are UNRC-2A. Type 2 continuous thread studs are measured from first thread to first thread, and the points are flat and chamfered. Type 2 studs are made to standards specified in ANSI B16.5, and threads are UNRC-2A for sizes one inch and under, and 8UNR-2A for longer sizes (for high temperature/high pressure service).

Tap-end studs have different types of thread on each end, and an unthreaded section (the "body") in the center. The end with the shorter threaded section is the tap-end, with Class NC5 interference fit threads or Class UNRC-3A threads. The tap-end has a chamfered point and is designed to be threaded into a tapped hole. The opposite end (with the longer thread) has UNRC-2A threads and either a chamfered or round point, and receives the nut. Length is measured overall. Type 1 tap-end studs are unthreaded blanks. Type 2 tap-end studs may have either an undersize body (rolled threads) or a full size body (cut threads). The body is finished to either a minimum Class 2A pitch diameter, or the maximum basic diameter of the nut-end thread. Type 3 tap-end studs have body tolerances equal to the major diameter of Class 2A threads, and Type 4 tap-end studs have a body milled to tolerances specified by the user.

Double-end studs are designed to receive nuts on both ends. The body separates Class-2A threaded sections of equal length, and the points are chamfered or rounded. Length is measured overall.

Table 1. Dimensions of Standard Hexagon Head Bolts.

Bolt Size		E Body Dia. (Max.)	F Width Across Flats			G Width Across Corners		H Head Height			R Fillet Radius		L_T Thread Length	
Nom.	Basic		Basic	Max.	Min.	Max.	Min.	Basic	Max.	Min.	Max.	Min.	<6 in. (Basic)	>6 in. (Basic)
1/4	0.2500	0.260	7/16	0.438	0.425	0.505	0.498	11/64	0.188	0.150	0.03	0.01	0.750	1.000
5/16	0.3125	0.324	1/2	0.500	0.484	0.577	0.552	7/32	0.235	0.195	0.03	0.01	0.875	1.125
3/8	0.3750	0.388	9/16	0.562	0.544	0.650	0.620	1/4	0.268	0.226	0.03	0.01	1.000	1.250
7/16	0.4375	0.452	5/8	0.625	0.603	0.722	0.687	19/64	0.316	0.272	0.03	0.01	1.125	1.375
1/2	0.5000	0.515	3/4	0.750	0.725	0.866	0.826	11/32	0.364	0.302	0.03	0.01	1.250	1.500
5/8	0.6250	0.642	15/16	0.938	0.906	1.083	1.033	27/64	0.444	0.378	0.06	0.02	1.500	1.750
3/4	0.7500	0.768	1 1/8	1.125	1.088	1.299	1.240	1/2	0.524	0.455	0.06	0.02	1.750	2.000
7/8	0.8750	0.895	1 5/16	1.312	1.269	1.516	1.447	37/64	0.604	0.531	0.06	0.02	2.000	2.250
1	1.0000	1.022	1 1/2	1.500	1.450	1.732	1.653	43/64	0.700	0.591	0.09	0.03	2.250	2.500
1 1/8	1.1250	1.149	1 11/16	1.688	1.631	1.949	1.859	3/4	0.780	0.658	0.09	0.03	2.500	2.750
1 1/4	1.2500	1.277	1 7/8	1.875	1.812	2.165	2.066	27/32	0.876	0.749	0.09	0.03	2.750	3.000
1 3/8	1.3750	1.404	2 1/16	2.062	1.994	2.382	2.273	29/32	0.940	0.810	0.09	0.03	3.000	3.250

(Continued)

All dimensions in inches. Source: Extracted from ANSI B18.2.1-1972, published by the American Society of Mechanical Engineers.

Table 1. *(Continued)* **Dimensions of Standard Hexagon Head Bolts.**

Bolt Size		E Body Dia. (Max.)	F Width Across Flats			G Width Across Corners		H Head Height			R Fillet Radius		LT Thread Length	
Nom.	Basic		Basic	Max.	Min.	Max.	Min.	Basic	Max.	Min.	Max.	Min.	<6 in. (Basic)	>6 in. (Basic)
1 1/2	1.5000	1.531	2 1/4	2.250	2.175	2.598	2.480	1	1.036	0.902	0.09	0.03	3.250	3.500
1 3/4	1.7500	1.785	2 5/8	2.625	2.538	3.031	2.893	1 5/32	1.196	1.054	0.12	0.04	3.750	4.000
2	2.0000	2.039	3	3.000	2.900	3.464	3.306	1 11/32	1.388	1.175	0.12	0.04	4.250	4.500
2 1/4	2.2500	2.305	3 3/8	3.375	3.262	3.897	3.719	1 1/2	1.548	1.327	0.19	0.06	4.750	5.000
2 1/2	2.5000	2.559	3 3/4	3.750	3.625	4.330	4.133	1 21/32	1.708	1.479	0.19	0.06	5.250	5.500
2 3/4	2.7500	2.827	4 1/8	4.125	3.988	4.763	4.546	1 13/16	1.869	1.632	0.19	0.06	5.750	6.000
3	3.0000	3.081	4 1/2	4.500	4.350	5.196	4.959	2	2.060	1.815	0.19	0.06	6.250	6.500
3 1/4	3.2500	3.335	4 7/8	4.875	4.712	5.629	5.372	2 3/16	2.251	1.936	0.19	0.06	6.750	7.000
3 1/2	3.5000	3.589	5 1/4	5.250	5.075	6.062	5.786	2 5/16	2.380	2.057	0.19	0.06	7.250	7.500
3 3/4	3.7500	3.858	5 5/8	5.625	5.437	6.495	6.198	2 1/2	2.572	2.241	0.19	0.06	7.750	8.000
4	4.0000	4.111	6	6.000	5.800	6.928	6.612	2 11/16	2.764	2.424	0.19	0.06	8.250	8.500

All dimensions in inches. Source: Extracted from ANSI B18.2.1-1972, published by the American Society of Mechanical Engineers.

Table 2. Dimensions of Heavy Hexagon Head Structural Bolts.

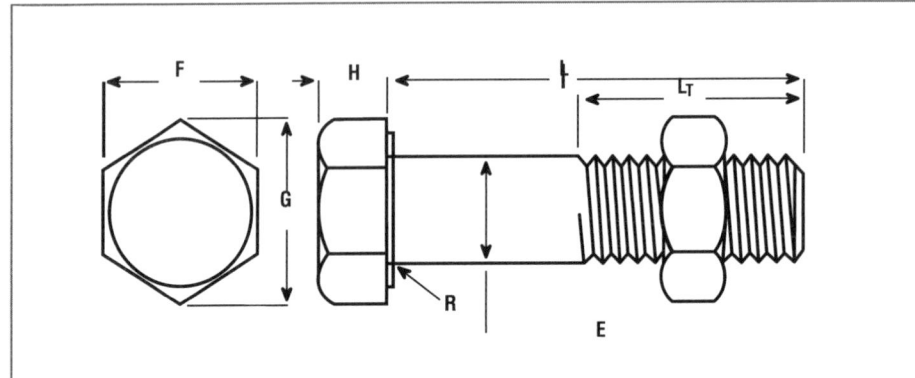

Bolt Size		E Body Dia.		F Width Across Flats			G Width Across Corners	
Nom.	Basic	Max.	Min.	Basic	Max.	Min.	Max.	Min.
1/2	0.5000	0.515	0.482	7/8	0.875	0.850	1.010	0.969
5/8	0.6250	0.642	0.605	1 1/16	1.062	1.031	1.227	1.175
3/4	0.7500	0.768	0.729	1 1/4	1.250	1.212	1.443	1.383
7/8	0.8750	0.895	0.852	1 7/16	1.438	1.394	1.660	1.589
1	1.0000	1.022	0.976	1 5/8	1.625	1.575	1.876	1.796
1 1/8	1.1250	1.149	1.098	1 13/16	1.812	1.756	2.093	2.002
1 1/4	1.2500	1.277	1.223	2	2.000	1.938	2.309	2.209
1 3/8	1.3750	1.404	1.345	2 3/16	2.188	2.119	2.526	2.416
1 1/2	1.5000	1.531	1.470	2 3/8	2.375	2.300	2.742	2.622

Bolt Size		H Height			R Fillet Radius		L_T Thread Length	V Transition Thread Length	Runout of Bearing Surface, *FIR*
Nom.	Basic	Basic	Max.	Min.	Max.	Min.	Basic	Max.	Max.
1/2	0.5000	5/16	0.323	0.302	0.031	0.009	1.00	0.19	0.016
5/8	0.6250	25/64	0.403	0.378	0.062	0.021	1.25	0.22	0.019
3/4	0.7500	15/32	0.483	0.455	0.062	0.021	1.38	0.25	0.022
7/8	0.8750	35/64	0.563	0.531	0.062	0.031	1.50	0.28	0.025
1	1.0000	39/64	0.627	0.591	0.093	0.062	1.75	0.31	0.028
1 1/8	1.1250	11/16	0.718	0.658	0.093	0.062	2.00	0.34	0.032
1 1/4	1.2500	25/32	0.813	0.749	0.093	0.062	2.00	0.38	0.035
1 3/8	1.3750	27/32	0.878	0.810	0.093	0.062	2.25	0.44	0.038
1 1/2	1.5000	15/16	0.974	0.902	0.093	0.062	2.25	0.44	0.041

All dimensions in inches. Source: Extracted from ANSI B18.2.1-1972, published by the American Society of Mechanical Engineers.

Table 3. Dimensions of Square Head Bolts.

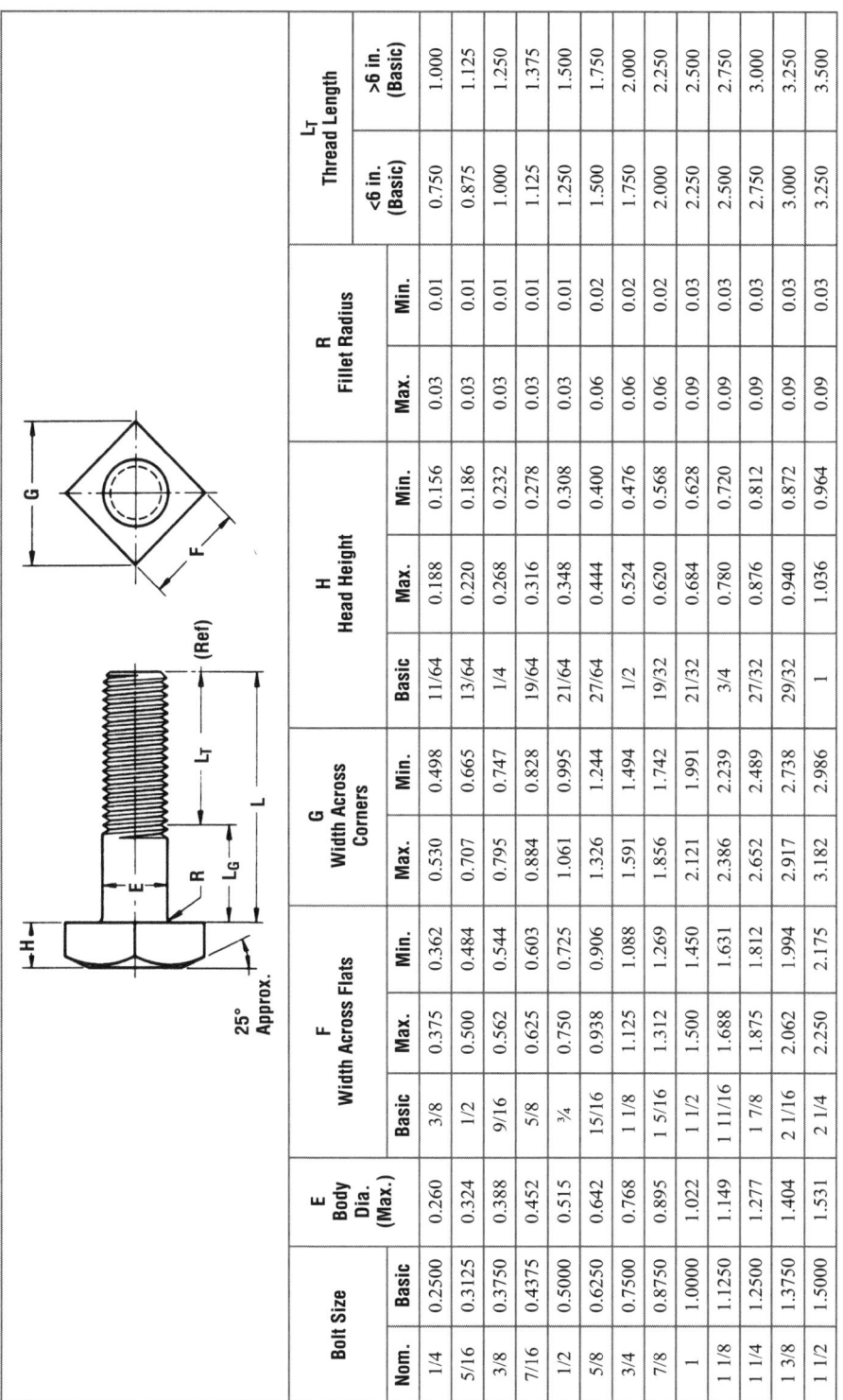

Bolt Size		E Body Dia. (Max.)	F Width Across Flats			G Width Across Corners		H Head Height			R Fillet Radius		L_T Thread Length	
Nom.	Basic		Basic	Max.	Min.	Max.	Min.	Basic	Max.	Min.	Max.	Min.	<6 in. (Basic)	>6 in. (Basic)
1/4	0.2500	0.260	3/8	0.375	0.362	0.530	0.498	11/64	0.188	0.156	0.03	0.01	0.750	1.000
5/16	0.3125	0.324	1/2	0.500	0.484	0.707	0.665	13/64	0.220	0.186	0.03	0.01	0.875	1.125
3/8	0.3750	0.388	9/16	0.562	0.544	0.795	0.747	1/4	0.268	0.232	0.03	0.01	1.000	1.250
7/16	0.4375	0.452	5/8	0.625	0.603	0.884	0.828	19/64	0.316	0.278	0.03	0.01	1.125	1.375
1/2	0.5000	0.515	3/4	0.750	0.725	1.061	0.995	21/64	0.348	0.308	0.03	0.01	1.250	1.500
5/8	0.6250	0.642	15/16	0.938	0.906	1.326	1.244	27/64	0.444	0.400	0.06	0.02	1.500	1.750
3/4	0.7500	0.768	1 1/8	1.125	1.088	1.591	1.494	1/2	0.524	0.476	0.06	0.02	1.750	2.000
7/8	0.8750	0.895	1 5/16	1.312	1.269	1.856	1.742	19/32	0.620	0.568	0.06	0.02	2.000	2.250
1	1.0000	1.022	1 1/2	1.500	1.450	2.121	1.991	21/32	0.684	0.628	0.09	0.03	2.250	2.500
1 1/8	1.1250	1.149	1 11/16	1.688	1.631	2.386	2.239	3/4	0.780	0.720	0.09	0.03	2.500	2.750
1 1/4	1.2500	1.277	1 7/8	1.875	1.812	2.652	2.489	27/32	0.876	0.812	0.09	0.03	2.750	3.000
1 3/8	1.3750	1.404	2 1/16	2.062	1.994	2.917	2.738	29/32	0.940	0.872	0.09	0.03	3.000	3.250
1 1/2	1.5000	1.531	2 1/4	2.250	2.175	3.182	2.986	1	1.036	0.964	0.09	0.03	3.250	3.500

All dimensions in inches. Source: Extracted from ANSI B18.2.1-1972, published by the American Society of Mechanical Engineers.

Table 4. Dimensions of 12-Point Flange Screws. (See Notes.)

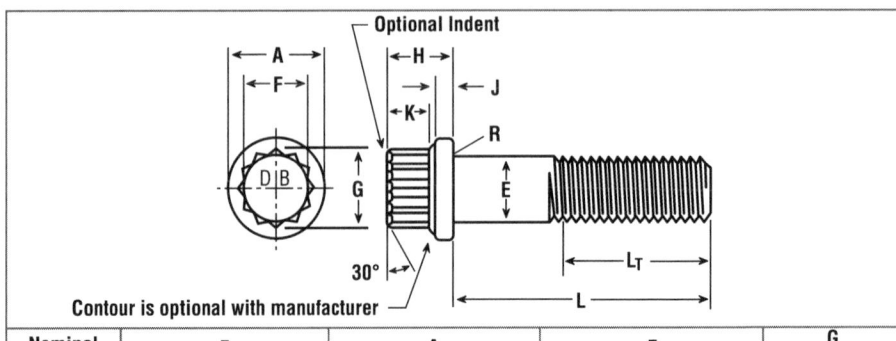

Nominal Dia. and Socket Wrench Size	E Body Dia.		A Flange Dia.		F Width Across Flats		G Width Across Corners
	Max.	Min.	Max.	Min.	Max.	Min.	Min.
1/4	0.2500	0.2435	0.375	0.365	0.252	0.244	0.278
5/16	0.3125	0.3053	0.469	0.457	0.315	0.306	0.348
3/8	0.3750	0.3678	0.562	0.550	0.377	0.368	0.420
7/16	0.4375	0.4294	0.656	0.642	0.438	0.429	0.489
1/2	0.5000	0.4919	0.750	0.735	0.502	0.493	0.562
5/8	0.6250	0.6163	0.938	0.921	0.627	0.618	0.705
3/4	0.7500	0.7406	1.125	1.107	0.752	0.743	0.847
7/8	0.8750	0.8647	1.312	1.293	0.870	0.866	0.987
1	1.0000	0.9886	1.500	1.479	1.003	0.991	1.130
1 1/8	1.1250	1.1086	1.688	1.665	1.128	1.115	1.271
1 1/4	1.2500	1.2336	1.875	1.852	1.253	1.240	1.414
1 1/2	1.5000	1.4818	2.250	2.224	1.503	1.489	1.697

Nominal Dia. and Socket Wrench Size	H Head Height	J Flange Thickness	K Wrenching Height	R Radius		L_T Thread Length
	Max.	Min.	Min.	Max.	Min.	Min.
1/4	0.250	0.058	0.15	0.014	0.009	1.000
5/16	0.312	0.074	0.18	0.017	0.012	1.125
3/8	0.375	0.095	0.21	0.020	0.015	1.250
7/16	0.438	0.109	0.26	0.023	0.018	1.375
1/2	0.500	0.129	0.29	0.026	0.020	1.500
5/8	0.625	0.166	0.36	0.032	0.024	1.750
3/4	0.750	0.200	0.44	0.039	0.030	2.000
7/8	0.875	0.234	0.51	0.044	0.034	2.250
1	1.000	0.268	0.60	0.050	0.040	2.500
1 1/8	1.125	0.310	0.66	0.055	0.045	2.750
1 1/4	1.250	0.350	0.73	0.060	0.050	3.000
1 1/2	1.500	0.433	0.87	0.070	0.060	3.500

Notes: 12-point flange screws and bolts are made to numerous dimensions as specified by Military (MS) and Aerospace (NAS) Standards. Material specifications will vary. The above screw is typical and produced from alloy steel with a hardness of RC 37-43, tensile strength of 170,000 PSI (min), yield strength of 153,000 PSI (min), and proof load of 140,000 PSI (min). Threads are UNRC/UNRF Class 3A for sizes through one inch, and UNRC/UNRF Class 2A for sizes 1 $^1/_8$ to 1 $^1/_2$ inch. Source: Darling Bolt Company.

Table 5. Dimensions of Round-Head Bolts.

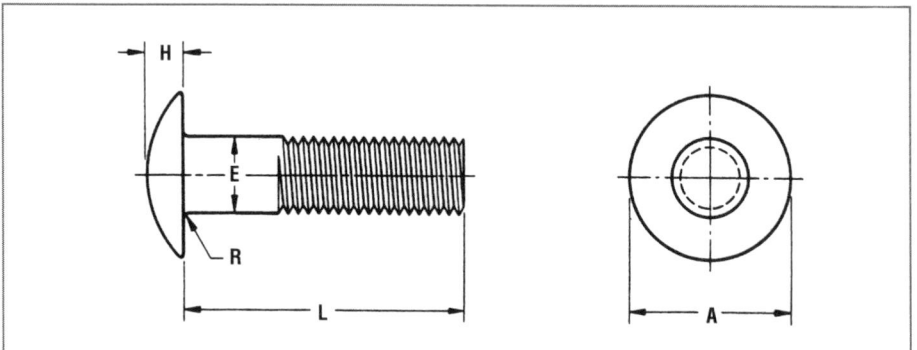

Bolt Size		E Body Dia.		A Head Dia.		H Head Height		R Fillet Radius
Nom.	Basic	Max.	Min.	Max.	Min.	Max.	Min.	Max.
No. 10	0.1900	0.199	0.182	0.469	0.438	0.114	0.094	0.031
1/4	0.2500	0.260	0.237	0.594	0.563	0.145	0.125	0.031
5/16	0.3125	0.324	0.298	0.719	0.688	0.176	0.156	0.031
3/8	0.3750	0.388	0.360	0.844	0.782	0.208	0.188	0.031
7/16	0.4375	0.452	0.421	0.969	0.907	0.239	0.219	0.031
1/2	0.5000	0.515	0.483	1.094	1.032	0.270	0.250	0.031
5/8	0.6250	0.642	0.605	0.344	1.219	0.344	0.313	0.062
3/4	0.7500	0.768	0.729	1.594	1.469	0.406	0.375	0.062

All dimensions in inches. Source: Extracted from ANSI B18.5-1971, published by the American Society of Mechanical Engineers.

Table 6. Dimensions of Round-Head Square-Neck Bolts.

Bolt Size		E Body Dia.		A Head Dia.		H Head Height		O Square Width		P Square Depth		Q Corner Radius	R Fillet Radius
Nom.	Basic	Max.	Min.	Max.	Min.	Max.	Min.	Max.	Min.	Max.	Min.	Max.	Max.
No. 10	0.1900	0.199	0.182	0.469	0.438	0.114	0.094	0.199	0.185	0.125	0.094	0.031	0.031
1/4	0.2500	0.260	0.237	0.594	0.563	0.145	0.125	0.260	0.245	0.156	0.125	0.031	0.031
5/16	0.3125	0.324	0.298	0.719	0.688	0.176	0.156	0.324	0.307	0.187	0.156	0.031	0.031
3/8	0.3750	0.388	0.360	0.844	0.782	0.208	0.188	0.388	0.368	0.219	0.188	0.047	0.031
7/16	0.4375	0.452	0.421	0.969	0.907	0.239	0.219	0.452	0.431	0.250	0.219	0.047	0.031
1/2	0.5000	0.515	0.483	1.094	1.032	0.270	0.250	0.515	0.492	0.281	0.250	0.047	0.031
5/8	0.6250	0.642	0.605	1.344	1.219	0.344	0.313	0.642	0.616	0.344	0.313	0.078	0.062
3/4	0.7500	0.768	0.729	1.594	1.469	0.406	0.375	0.768	0.741	0.406	0.375	0.078	0.062
7/8	0.8750	0.895	0.852	1.844	1.719	0.469	0.438	0.895	0.865	0.469	0.438	0.094	0.062
1	1.0000	1.022	0.976	2.094	1.969	0.531	0.500	1.022	0.990	0.531	0.500	0.094	0.062

All dimensions in inches. Source: Extracted from ANSI B18.5-1971, published by the American Society of Mechanical Engineers.

Table 7. Dimensions of Round-Head Fin-Neck Bolts.

Bolt Size		E Body Dia.		A Head Dia.		H Head Height		M Fin Thickness		N Distance Across Fins		U Fin Depth		R Fillet Radius
Nom.	Basic	Max.	Min.	Max.	Min.	Max.	Min.	Max.	Min.	Max.	Min.	Max.	Min.	Max.
No. 10	0.1900	0.199	0.182	0.469	0.438	0.114	0.094	0.098	0.078	0.395	0.375	0.088	0.078	0.031
1/4	0.2500	0.260	0.237	0.594	0.563	0.145	0.125	0.114	0.094	0.458	0.438	0.104	0.094	0.031
5/16	0.3125	0.324	0.298	0.719	0.688	0.176	0.156	0.145	0.125	0.551	0.531	0.135	0.125	0.031
3/8	0.3750	0.388	0.360	0.844	0.782	0.208	0.188	0.161	0.141	0.645	0.625	0.151	0.141	0.031
7/16	0.4375	0.452	0.421	0.969	0.907	0.239	0.219	0.192	0.172	0.739	0.719	0.182	0.172	0.031
1/2	0.5000	0.515	0.483	1.094	1.032	0.270	0.250	0.208	0.188	0.833	0.813	0.198	0.188	0.031

All dimensions in inches. Source: Extracted from ANSI B18.5-1971, published by the American Society of Mechanical Engineers.

Table 8. Dimensions of Round-Head Ribbed-Neck Bolts.

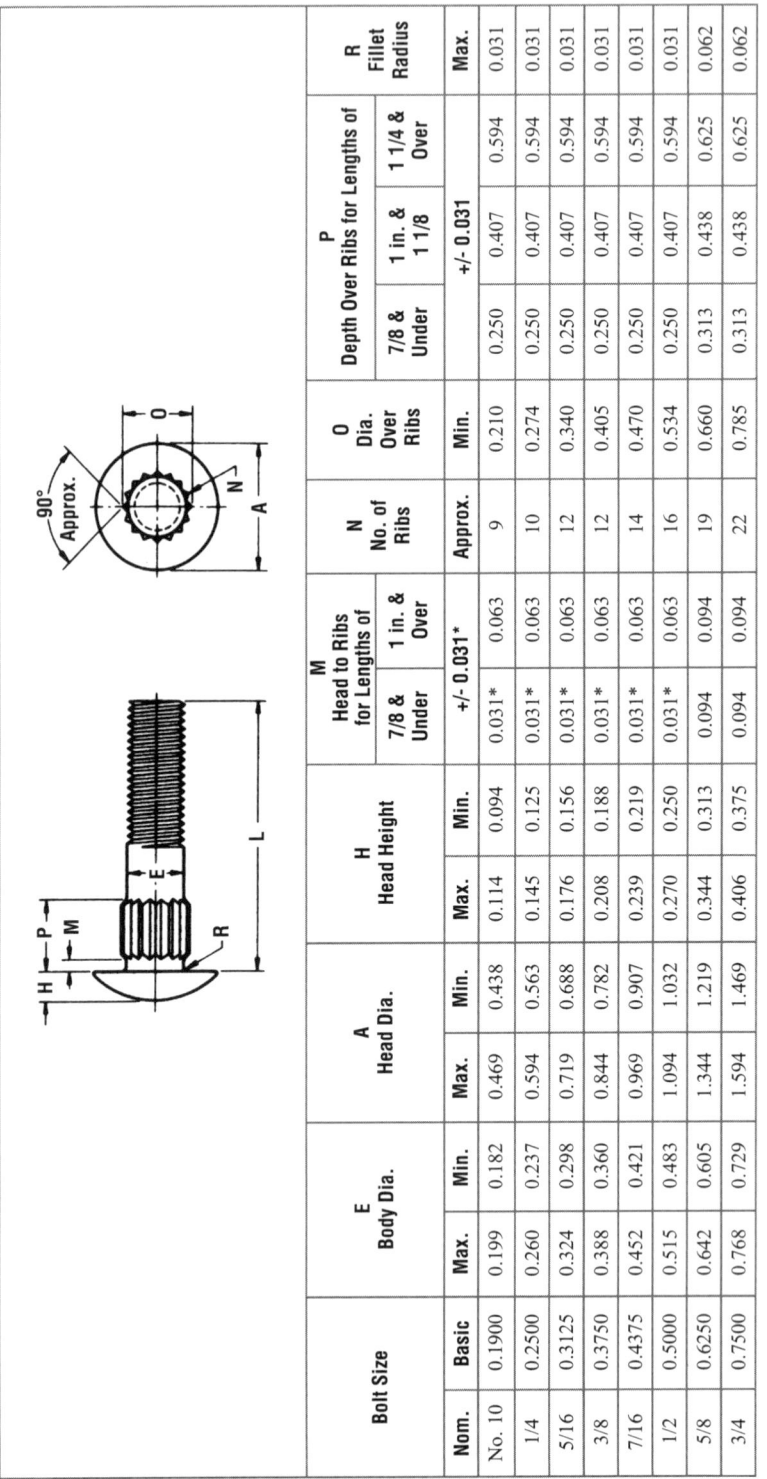

Bolt Size		E Body Dia.		A Head Dia.		H Head Height		M Head to Ribs for Lengths of		N No. of Ribs	O Dia. Over Ribs	P Depth Over Ribs for Lengths of			R Fillet Radius
Nom.	Basic	Max.	Min.	Max.	Min.	Max.	Min.	7/8 & Under	1 in. & Over	Approx.	Min.	7/8 & Under	1 in. & 1 1/8	1 1/4 & Over	Max.
								+/- 0.031*					+/- 0.031		
No. 10	0.1900	0.199	0.182	0.469	0.438	0.114	0.094	0.031*	0.063	9	0.210		0.407	0.594	0.031
1/4	0.2500	0.260	0.237	0.594	0.563	0.145	0.125	0.031*	0.063	10	0.274	0.250	0.407	0.594	0.031
5/16	0.3125	0.324	0.298	0.719	0.688	0.176	0.156	0.031*	0.063	12	0.340	0.250	0.407	0.594	0.031
3/8	0.3750	0.388	0.360	0.844	0.782	0.208	0.188	0.031*	0.063	12	0.405	0.250	0.407	0.594	0.031
7/16	0.4375	0.452	0.421	0.969	0.907	0.239	0.219	0.031*	0.063	14	0.470	0.250	0.407	0.594	0.031
1/2	0.5000	0.515	0.483	1.094	1.032	0.270	0.250	0.031*	0.063	16	0.534	0.250	0.407	0.594	0.031
5/8	0.6250	0.642	0.605	1.344	1.219	0.344	0.313	0.094	0.094	19	0.660	0.313	0.438	0.625	0.062
3/4	0.7500	0.768	0.729	1.594	1.469	0.406	0.375	0.094	0.094	22	0.785	0.313	0.438	0.625	0.062

* Tolerance on the No. 10 through 1/2 sizes for nominal lengths 7/8 in. and under shall be +0.031 and -0.000.
All dimensions in inches. Source: Extracted from ANSI B18.5-1971, published by the American Society of Mechanical Engineers.

Table 9. Dimensions of Step Bolts.

Bolt Size		E Body Dia.		A Head Dia.		H Head Height		O Square Width		P Square Depth		Q Corner Radius	R Fillet Radius
Nom.	Basic	Max.	Min.	Max.	Min.	Max.	Min.	Max.	Min.	Max.	Min.	Max.	Max.
No. 10	0.1900	0.199	0.182	0.656	0.625	0.114	0.094	0.199	0.185	0.125	0.094	0.031	0.031
1/4	0.2500	0.260	0.237	0.844	0.813	0.145	0.125	0.260	0.245	0.156	0.125	0.031	0.031
5/16	0.3125	0.324	0.298	1.031	1.000	0.176	0.156	0.324	0.307	0.187	0.156	0.031	0.031
3/8	0.3750	0.388	0.360	1.219	1.188	0.208	0.188	0.388	0.268	0.219	0.188	0.031	0.031
7/16	0.4375	0.452	0.421	1.406	1.375	0.239	0.219	0.452	0.431	0.250	0.219	0.047	0.031
1/2	0.5000	0.515	0.483	1.594	1.563	0.270	0.250	0.515	0.492	0.281	0.250	0.047	0.031

All dimensions in inches. Source: Fastbolt Corp.

Table 10. Dimensions of Countersunk Head Square-Neck Plow Bolts.

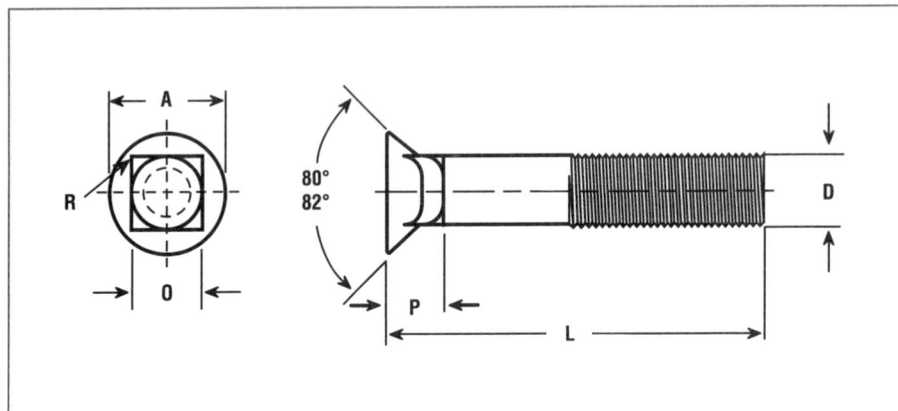

Bolt Size		A Head Dia.			K Feed Thickness	P Depth of Square and Head		O Width of Square		R Radius on Corners
		Max.	Min. Sharp	Abs. Min. With Flat						
Nom.	Basic				Max.	Max.	Min.	Max.	Min.	Max.
5/16	0.3125	0.605	0.578	0.563	0.025	0.269	0.243	0.325	0.313	1/32
3/8	0.3750	0.708	0.671	0.656	0.031	0.312	0.281	0.387	0.375	3/64
7/16	0.4375	0.826	0.781	0.766	0.036	0.364	0.328	0.450	0.438	3/64
1/2	0.2500	0.945	0.890	0.875	0.042	0.417	0.375	0.515	0.500	3/64
*9/16	0.5625	1.045	1.000	0.969	0.045	0.461	0.416	0.578	0.563	5/64
5/8	0.6250	1.147	1.094	1.063	0.050	0.506	0.456	0.640	0.625	5/64
3/4	0.7500	1.303	1.250	1.219	0.050	0.541	0.491	0.765	0.750	5/64
7/8	0.8750	1.512	1.469	1.406	0.063	0.626	0.563	0.906	0.875	3/32
1	1.0000	1.700	1.656	1.594	0.063	0.690	0.627	1.031	1.000	3/32
Repair †										
5/16	0.3125	0.556	0.531	0.516	0.020	0.232	0.212	0.325	0.313	1/32
3/8	0.3750	0.659	0.624	0.609	0.025	0.272	0.247	0.387	0.375	3/64
7/16	0.4375	0.779	0.734	0.719	0.030	0.324	0.294	0.450	0.438	3/64
1/2	0.2500	0.898	0.843	0.828	0.035	0.375	0.340	0.515	0.500	3/64
*9/16	0.5625	1.002	0.953	0.922	0.040	0.423	0.383	0.578	0.563	5/64
5/8	0.6250	1.096	1.047	1.016	0.040	0.458	0.418	0.640	0.625	5/64
3/4	0.7500	1.252	1.203	1.172	0.040	0.493	0.453	0.765	0.750	5/64
7/8	0.8750	1.465	1.422	1.359	0.050	0.573	0.523	0.906	0.875	3/32
1	1.0000	1.653	1.609	1.547	0.050	0.637	0.587	1.031	1.000	3/32

† The letter "R" shall be shown on top of the repair head, to distinguish it from the regular head bolt.
* This size is not recommended.
If the method of manufacture permits, it is recommended that the same radius be maintained on each of all four corners of the square.
All dimensions in inches. Source: Extracted from ANSI B18.9-1958, published by the American Society of Mechanical Engineers.

Table 11. Dimensions of Countersunk Bolts and Slotted Countersunk Bolts.

Slotted Head

Bolt Size		E Body Dia.		A Head Dia.			F⁴ Flat on Min. Dia. Head	H Head Height		J¹ Slot Width		T¹ Slot Depth	
Nom.	Basic	Max.	Min.	Max. Edge Sharp²	Max. Edge Sharp³	Min. Edge Rounded or Flat	Max.	Max.	Min.	Max.	Min.	Max.	Min.
1/4	0.2500	0.260	0.237	0.493	0.477	0.445	0.018	0.150	0.131	0.075	0.064	0.068	0.045
5/16	0.3125	0.324	0.298	0.618	0.598	0.558	0.023	0.189	0.164	0.084	0.072	0.086	0.057
3/8	0.3750	0.388	0.360	0.740	0.715	0.668	0.027	0.225	0.196	0.094	0.081	0.103	0.068
7/16	0.4375	0.452	0.421	0.803	0.778	0.726	0.030	0.226	0.196	0.094	0.081	0.103	0.068
1/2	0.5000	0.515	0.483	0.935	0.905	0.845	0.035	0.269	0.233	0.106	0.091	0.103	0.068
5/8	0.6250	0.642	0.605	1.169	1.132	1.066	0.038	0.336	0.292	0.133	0.116	0.137	0.091
3/4	0.7500	0.768	0.729	1.402	1.357	1.285	0.041	0.403	0.349	0.149	0.131	0.171	0.115
7/8	0.8750	0.895	0.852	1.637	1.584	1.511	0.042	0.470	0.408	0.167	0.147	0.206	0.138
1	1.0000	1.022	0.976	1.869	1.810	1.735	0.043	0.537	0.466	0.188	0.166	0.240	0.162

(Continued)

Table 11. (Continued) Dimensions of Countersunk Bolts and Slotted Countersunk Bolts.

Bolt Size		E Body Dia.		A Head Dia.			F[4] Flat on Min. Dia. Head	H Head Height		J[1] Slot Width		T[1] Slot Depth	
Nom.	Basic	Max.	Min.	Max. Edge Sharp[2]	Max. Edge Sharp[3]	Min. Edge Rounded or Flat	Max.	Max.	Min.	Max.	Min.	Max.	Min.
1 1/8	1.1250	1.149	1.098	2.104	2.037	1.962	0.043	0.604	0.525	0.196	0.178	0.257	0.173
1 1/4	1.2500	1.277	1.223	2.337	2.262	2.187	0.043	0.671	0.582	0.211	0.193	0.291	0.197
1 3/8	1.3750	1.404	1.345	2.571	2.489	2.414	0.043	0.738	0.641	0.226	0.208	0.326	0.220
1 1/2	1.5000	1.531	1.470	2.804	2.715	2.640	0.043	0.805	0.698	0.258	0.240	0.360	0.244

[1] Head shall be unslotted, unless otherwise specified. Slot dimensions are same as Slotted Flat Countersunk Head Cap Screws in American National Standard, ANSI B18.6.2.
[2] Maximum head height calculated on maximum sharp head diameter, basic bolt diameter and 78° head angle.
[3] Minimum head height calculated on minimum sharp head diameter, basic bolt diameter and 80° head angle.
[4] Flat on minimum diameter head calculated on minimum sharp and absolute minimum head diameters and 82° head angle.
All dimensions in inches. Source: Extracted from ANSI B18.5-1971, published by the American Society of Mechanical Engineers.

Table 12. Dimensions of 114-Degree Countersunk Neck Bolts.

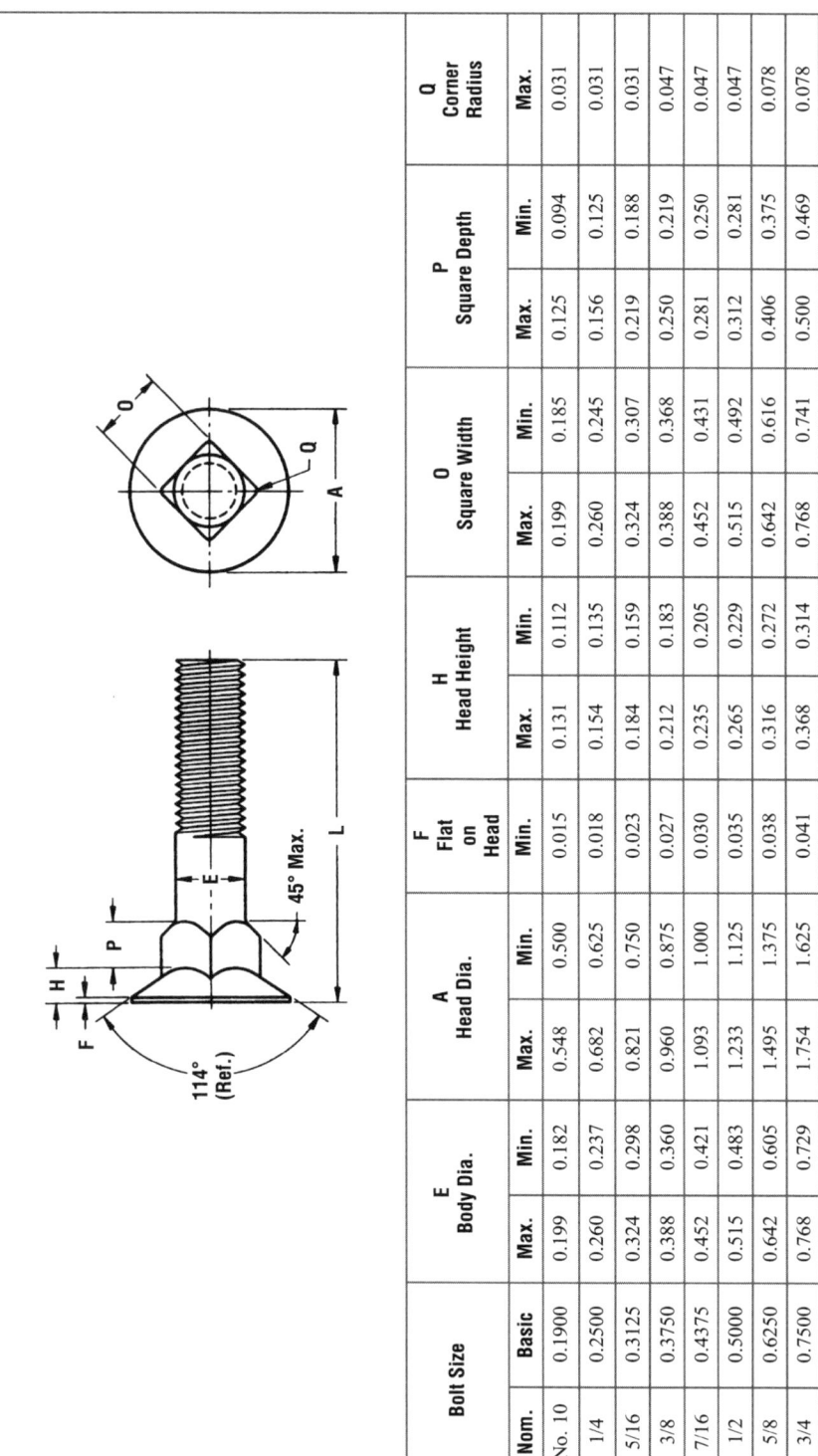

Bolt Size		E Body Dia.		A Head Dia.		F Flat on Head	H Head Height		O Square Width		P Square Depth		Q Corner Radius
Nom.	Basic	Max.	Min.	Max.	Min.	Min.	Max.	Min.	Max.	Min.	Max.	Min.	Max.
No. 10	0.1900	0.199	0.182	0.548	0.500	0.015	0.131	0.112	0.199	0.185	0.125	0.094	0.031
1/4	0.2500	0.260	0.237	0.682	0.625	0.018	0.154	0.135	0.260	0.245	0.156	0.125	0.031
5/16	0.3125	0.324	0.298	0.821	0.750	0.023	0.184	0.159	0.324	0.307	0.219	0.188	0.031
3/8	0.3750	0.388	0.360	0.960	0.875	0.027	0.212	0.183	0.388	0.368	0.250	0.219	0.047
7/16	0.4375	0.452	0.421	1.093	1.000	0.030	0.235	0.205	0.452	0.431	0.281	0.250	0.047
1/2	0.5000	0.515	0.483	1.233	1.125	0.035	0.265	0.229	0.515	0.492	0.312	0.281	0.047
5/8	0.6250	0.642	0.605	1.495	1.375	0.038	0.316	0.272	0.642	0.616	0.406	0.375	0.078
3/4	0.7500	0.768	0.729	1.754	1.625	0.041	0.368	0.314	0.768	0.741	0.500	0.469	0.078

All dimensions in inches. Source: Extracted from ANSI B18.5-1971, published by the American Society of Mechanical Engineers.

Table 13. Dimensions of Flat Countersunk Head Elevator Bolts.

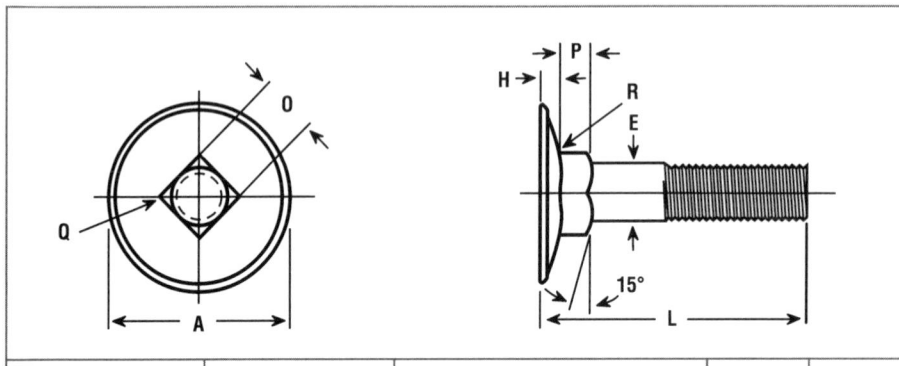

Bolt Size		E Body Dia.		A Head Dia.			C Head Angle, Degrees (Ref.)	F Flat on Min. Dia. Head
				Edge Sharp	Edge Sharp	Edge Flat		
Nom.	Basic	Max.	Min.	Max.	Min.	Min.		
No. 10	0.1900	0.199	0.182	0.790	0.750	0.740	9	0.025
1/4	0.2500	0.260	0.237	1.008	0.969	0.938	9	0.035
5/16	0.3125	0.324	0.298	1.227	1.188	1.157	9	0.035
3/8	0.3750	0.388	0.360	1.352	1.312	1.272	11	0.040
7/16	0.4375	0.452	0.421	1.477	1.438	1.397	13	0.040
1/2	0.5000	0.515	0.483	1.602	1.562	1.522	12	0.040

Bolt Size		H Head Height		O Square Width		P Square Depth		Q Corner Radius	R Fillet Radius
Nom.	Basic	Max.	Min.	Max.	Min.	Max.	Min.	Max.	Max.
No. 10	0.1900	0.082	0.062	0.210	0.185	0.125	0.094	0.031	0.031
1/4	0.2500	0.098	0.078	0.280	0.245	0.219	0.188	0.031	0.031
5/16	0.3125	0.114	0.094	0.342	0.307	0.250	0.219	0.031	0.031
3/8	0.3750	0.145	0.125	0.405	0.368	0.250	0.219	0.047	0.031
7/16	0.4375	0.176	0.156	0.468	0.431	0.281	0.250	0.047	0.031
1/2	0.5000	0.176	0.156	0.530	0.492	0.281	0.250	0.047	0.031

All dimensions in inches. Source: Fastbolt Corp.

Table 14. Type 1 Regular-Pattern Eyebolt.

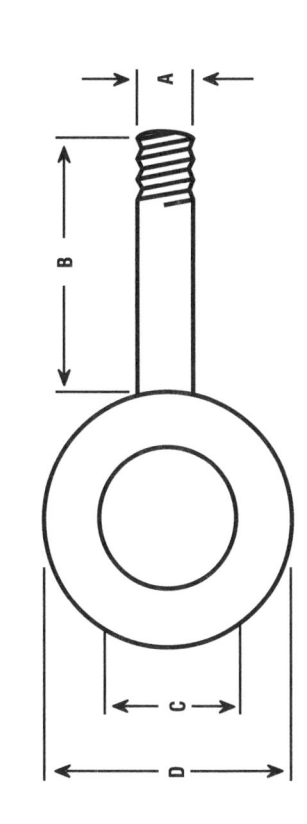

Nominal Size	A Shank Dia.	B Shank Length	C Eye ID	D Nominal Eye OD	Eye Sect. Dia.	Overall Length	Min. Length Full Thread	Thread Size UNC - 2A	Safe Working Load, lb. at 0° [1]	Safe Working Load, lb. at 45° [1]	Safe Working Load, lb. at 90° [1]
1/4* 0.25	0.25 0.28	1.00 1.06	0.69 0.81	1.19	0.19 0.25	2.06 2.38	0.75	1/4 - 20 or .250 - 20	400	80	60
5/16* 0.31	0.31 0.34	1.12 1.19	0.81 0.94	1.44	0.25 0.31	2.44 2.75	0.81	5/16 - 18 or .3125 - 18	800	160	120
3/8* 0.38	0.38 0.41	1.25 1.38	0.94 1.06	1.69	0.31 0.38	2.81 3.19	0.88	3/8 - 16 or .375 - 16	1400	280	210
7/16 0.44	0.44 0.47	1.38 1.50	1.00 1.12	1.81	0.34 0.41	3.06 3.44	1.00	7/16 - 14 or .4375 - 14	2000	400	300
1/2* 0.50	0.50 0.53	1.50 1.62	1.12 1.25	2.12	0.44 0.50	3.50 3.88	1.12	1/2 - 13 or .500 - 13	2600	520	390

(Continued)

Table 14. (Continued) Type 1 Regular-Pattern Eyebolt.

Nominal Size	A Shank Dia.	B Shank Length	C Eye ID	D Nominal Eye OD	Eye Sect. Dia.	Overall Length	Min. Length Full Thread	Thread Size UNC - 2A	Safe Working Load, lb. at 0° [1]	Safe Working Load, lb. at 45° [1]	Safe Working Load, lb. at 90° [1]
9/16	0.56 0.59	1.62 1.75	1.19 1.31	2.25	0.47 0.53	3.75 4.12	1.25	9/16 - 12 or .5625 - 12	3000	600	450
5/8*	0.62 0.66	1.75 1.88	1.31 1.44	2.56	0.56 0.62	4.19 4.56	1.38	5/8 - 11 or .625 - 11	4000	800	600
3/4*	0.75 0.78	2.00 2.12	1.44 1.56	2.81	0.62 0.69	4.69 5.06	1.62	3/4 - 10 or .750 - 10	6000	1200	900
7/8*	0.88 0.91	2.25 2.38	1.56 1.69	3.19	0.75 0.81	5.31 5.69	1.81	7/8 - 9 or .875 - 9	6600	1300	1000
1*	1.00 1.06	2.50 2.62	1.69 1.81	3.56	0.88 0.94	5.94 6.31	2.06	1 - 8 or 1.000-8	8000	1600	1200
1 1/8	1.12 1.19	2.75 2.88	1.94 2.06	4.06	1.00 1.06	6.69 7.06	2.31	1 1/8 - 7 or 1.125 - 7	10000	2000	1500
1 1/4*	1.25 1.34	3.00 3.12	2.12 2.25	4.44	1.09 1.16	7.31 7.69	2.50	1 1/4 - 7 or 1.250 - 7	15000	3000	2250
1 1/2*	1.50 1.59	3.50 3.62	2.44 2.56	5.19	1.31 1.38	8.56 8.94	3.00	1 1/2 - 6 or 1.500 - 6	18000	3600	2700
1 3/4*	1.75 1.84	3.75 3.88	2.75 3.00	6.00	1.50 1.62	9.50 10.12	3.19	1 3/4 - 5 or 1.750-5	22000	4400	3300
2*	2.00 2.09	4.00 4.25	3.06 3.44	6.88	1.75 1.88	10.56 11.44	3.38	2 - 4 1/2 or 2.000 - 4.50	26000	5200	3900
2 1/2*	2.50 2.62	5.00 5.25	3.81 4.19	8.50	2.19 2.31	13.19 14.06	4.25	2 1/2 - 4 or 2.500 - 4	32000	6400	4800

* Preferred. [1] These safe working loads are based on the following percentages of minimum proof loads shown in ASTM A 489; 0°—66.6 percent; 45°—13.3 percent, 90°—10.0 percent. All dimensions in inches. Source: Extracted from ANSI B18.15-1969, published by the American Society of Mechanical Engineers.

Table 15. Type 2 Shoulder-Pattern Eyebolt.

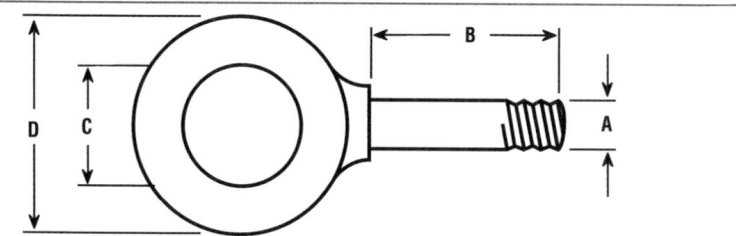

Nominal Size		A Shank Dia.	B Shank Length	C Eye ID	D Nominal Eye OD	Eye Sect. Dia.	Overall Length	Min. Length Full Thread
1/4*	0.25	0.25-0.28	1.00-1.06	0.69-0.81	1.19	0.19-0.25	2.22-2.53	0.75
5/16*	0.31	0.31-0.34	1.12-1.19	0.81-0.94	1.44	0.25-0.31	2.66-2.97	0.81
3/8*	0.38	0.38-0.41	1.25-1.38	0.94-1.06	1.69	0.31-0.38	3.09-3.47	0.88
7/16	0.44	0.44-0.47	1.38-1.50	1.00-1.12	1.81	0.34-0.41	3.41-3.78	1.00
1/2*	0.50	0.50-0.53	1.50-1.62	1.12-1.25	2.12	0.44-0.50	3.81-4.19	1.12
9/16	0.56	0.56-0.59	1.62-1.75	1.19-1.31	2.25	0.47-0.53	4.19-4.56	1.25
5/8*	0.62	0.62-0.66	1.75-1.88	1.31-1.44	2.56	0.56-0.62	4.56-4.94	1.38
3/4*	0.75	0.75-0.78	2.00-2.12	1.44-1.56	2.81	0.62-0.69	5.06-5.50	1.62
7/8*	0.88	0.88-0.91	2.25-2.38	1.56-1.69	3.19	0.75-0.81	5.75-6.19	1.81
1*	1.00	1.00-1.06	2.50-2.62	1.69-1.81	3.56	0.88-0.94	6.44-6.88	2.06
1 1/8	1.12	1.12-1.19	2.75-2.88	1.94-2.06	4.06	1.00-1.06	7.31-7.75	2.31
1 1/4*	1.25	1.25-1.34	3.00-3.12	2.12-2.25	4.44	1.09-1.16	8.00-8.44	2.50
1 1/2*	1.50	1.50-1.59	3.50-3.62	2.44-2.56	5.19	1.31-1.38	9.22-9.72	3.00
1 3/4	1.75	1.75-1.84	3.75-3.88	2.75-3.00	6.00	1.50-1.62	10.50-11.12	3.19
2*	2.00	2.00-2.09	4.00-4.25	3.06-3.44	6.88	1.75-1.88	11.53-12.22	3.38

Nominal Size		Thread Size UNC-2A	C_L Eye to Shoulder	Shoulder Dia.	Shoulder Height	Radius Under Shoulder	Safe Working Load, lb. at [1]		
							0°	45°	90°
1/4*	0.25	1/4 - 20 or .250 - 20	0.69 0.81	0.50 0.56	0.12 0.19	0.015 0.025	400	100	80
5/16*	0.31	5/16 - 18 or .3125 - 18	0.88 1.00	0.56 0.32	0.12 0.19	0.015 0.025	800	200	160
3/8*	0.38	3/8 - 16 or .375 - 16	1.06 1.19	0.62 0.69	0.12 0.19	0.015 0.025	1400	350	280
7/16	0.44	7/16 - 14 or .4375 - 14	1.19 1.31	0.75 0.81	0.19 0.25	0.015 0.025	2000	500	400
1/2*	0.50	1/2 - 13 or .500 - 13	1.31 1.44	0.88 0.94	0.19 0.25	0.015 0.025	2600	650	520
9/16	0.56	9/16 - 12 or .5625 - 12	1.50 1.62	0.94 1.00	0.22 0.28	0.020 0.045	3000	750	600
5/8*	0.62	5/8 - 11 or .625 - 11	1.59 1.72	1.00 1.06	0.25 0.31	0.020 0.045	4000	1000	800

(Continued)

Table 15. *(Continued)* Type 2 Shoulder-Pattern Eyebolt.

Nominal Size		Thread Size UNC-2A	C_L Eye to Shoulder	Shoulder Dia.	Shoulder Height	Radius Under Shoulder	Safe Working Load, lb. at [1]		
							0°	45°	90°
3/4*	0.75	3/4 - 10 or .750 - 10	1.72 1.91	1.12 1.25	0.25 0.31	0.020 0.045	6000	1500	1200
7/8*	0.88	7/8 - 9 or .875 - 9	2.03 2.22	1.31 1.44	0.31 0.38	0.040 0.065	6600	1670	1330
1*	1.00	1 - 8 or 1.000 - 8	2.22 2.41	1.50 1.62	0.38 0.44	0.060 0.095	8000	2000	1600
1 1/8	1.12	1 1/8 - 7 or 1.125 - 7	2.59 2.78	1.69 1.81	0.44 0.50	0.060 0.095	10000	2500	2000
1 1/4*	1.25	1 1/4 - 7 or 1.250 - 7	2.84 3.03	1.88 2.00	0.50 0.56	0.060 0.095	15000	3750	3000
1 1/2*	1.50	1 1/2 - 6 or 1.500 - 6	3.19 3.44	2.12 2.25	0.50 0.62	0.060 0.095	18000	4500	3600
1 3/4	1.75	1 3/4 - 5 or 1.750 - 5	3.88 4.12	2.50 2.62	0.50 0.62	0.060 0.095	22000	5500	4400
2*	2.00	2 - 4 1/2 or 2.000 - 4.50	4.25 4.50	2.88 3.00	0.62 0.75	0.060 0.095	26000	6500	5200

* Preferred. [1] Applies if shoulder is properly seated. Otherwise safe working load of Type 1 applies. These safe working loads are based on the following percentages of minimum proof loads shown in ASTM A 489; 0°—66.6 percent, 45°—16.7 percent, 90°—13.3 percent. All dimensions in inches. Source: Extracted from ANSI B18.15-1969, published by the American Society of Mechanical Engineers.

Table 16. Dimensions of Hex Nuts and Hex Jam Nuts.

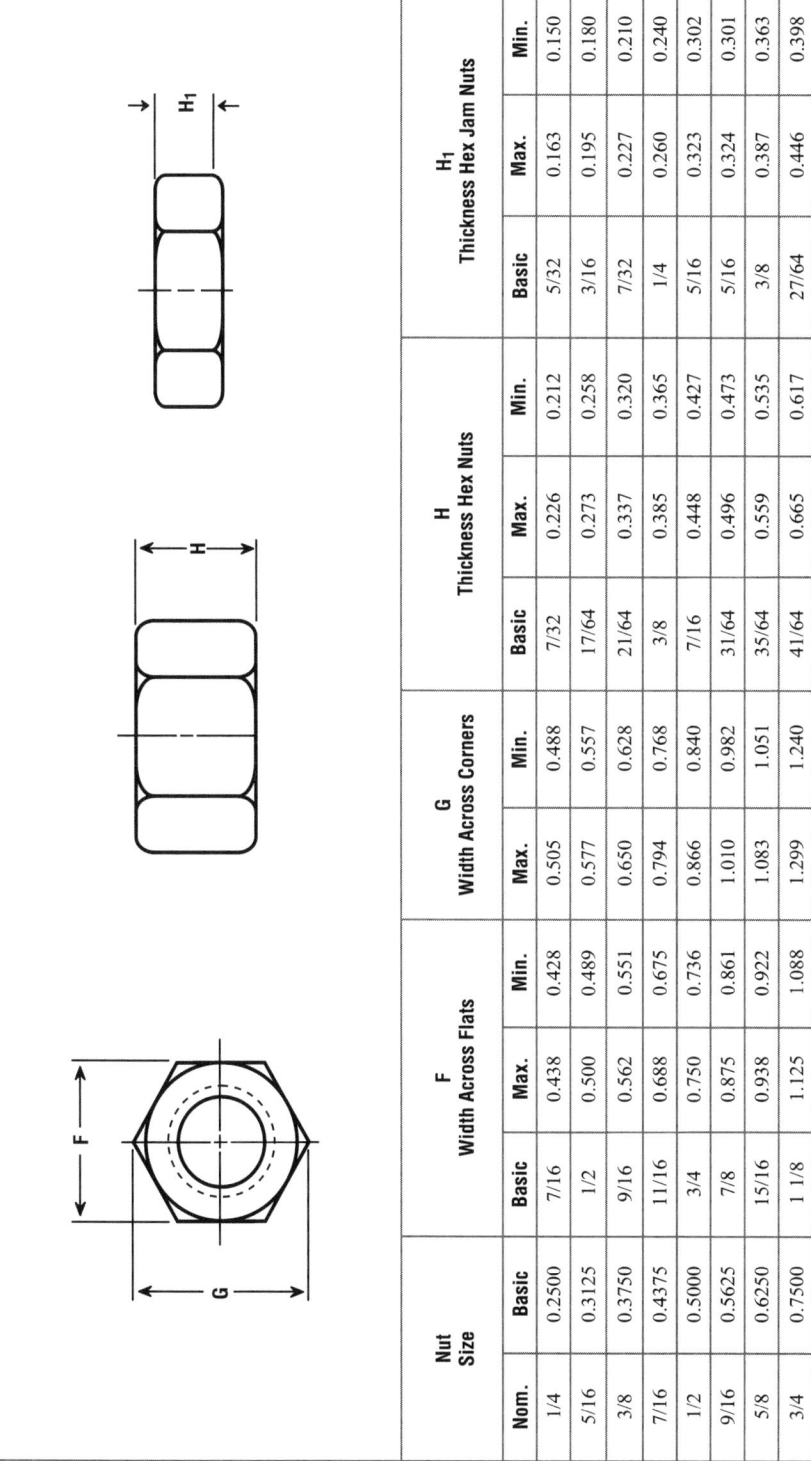

Nut Size		F Width Across Flats			G Width Across Corners		H Thickness Hex Nuts			H₁ Thickness Hex Jam Nuts		
Nom.	Basic	Basic	Max.	Min.	Max.	Min.	Basic	Max.	Min.	Basic	Max.	Min.
1/4	0.2500	7/16	0.438	0.428	0.505	0.488	7/32	0.226	0.212	5/32	0.163	0.150
5/16	0.3125	1/2	0.500	0.489	0.577	0.557	17/64	0.273	0.258	3/16	0.195	0.180
3/8	0.3750	9/16	0.562	0.551	0.650	0.628	21/64	0.337	0.320	7/32	0.227	0.210
7/16	0.4375	11/16	0.688	0.675	0.794	0.768	3/8	0.385	0.365	1/4	0.260	0.240
1/2	0.5000	3/4	0.750	0.736	0.866	0.840	7/16	0.448	0.427	5/16	0.323	0.302
9/16	0.5625	7/8	0.875	0.861	1.010	0.982	31/64	0.496	0.473	5/16	0.324	0.301
5/8	0.6250	15/16	0.938	0.922	1.083	1.051	35/64	0.559	0.535	3/8	0.387	0.363
3/4	0.7500	1 1/8	1.125	1.088	1.299	1.240	41/64	0.665	0.617	27/64	0.446	0.398

(Continued)

Table 16. *(Continued)* Dimensions of Hex Nuts and Hex Jam Nuts.

Nut Size			F Width Across Flats			G Width Across Corners			H Thickness Hex Nuts			H₁ Thickness Hex Jam Nuts		
Nom.	Basic		Basic	Max.	Min.		Max.	Min.	Basic	Max.	Min.	Basic	Max.	Min.
7/8	0.8750		1 5/16	1.312	1.269		1.516	1.447	3/4	0.776	0.724	31/64	0.510	0.458
1	1.0000		1 1/2	1.500	1.450		1.732	1.653	55/64	0.887	0.831	35/64	0.575	0.519
1 1/8	1.1250		1 11/16	1.688	1.631		1.949	1.859	31/32	0.999	0.939	39/64	0.639	0.579
1 1/4	1.2500		1 7/8	1.875	1.812		2.165	2.066	1 1/16	1.094	1.030	23/32	0.751	0.687
1 3/8	1.3750		2 1/16	2.062	1.994		2.382	2.273	1 11/64	1.206	1.138	25/32	0.815	0.747
1 1/2	1.5000		2 1/4	2.250	2.175		2.598	2.480	1 9/32	1.317	1.245	27/32	0.880	0.808

All dimensions in inches. Source: Fastbolt Corp.

Table 17. Dimensions of Hex Nuts and Hex Jam Nuts.

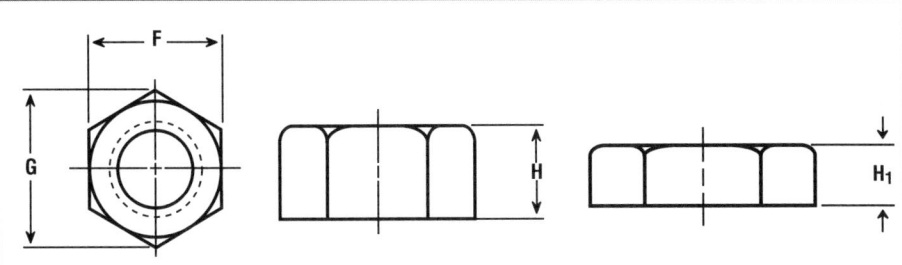

Nut Size		F Width Across Flats			G Width Across Corners		H Thickness Hex Nuts		
Nom.	Basic	Basic	Max.	Min.	Max.	Min.	Basic	Max.	Min.
1/4	0.2500	7/16	0.438	0.428	0.505	0.488	7/32	0.226	0.212
5/16	0.3125	1/2	0.500	0.489	0.577	0.557	17/64	0.273	0.258
3/8	0.3750	9/16	0.562	0.551	0.650	0.628	21/64	0.337	0.320
7/16	0.4375	11/16	0.688	0.675	0.794	0.768	3/8	0.385	0.365
1/2	0.5000	3/4	0.750	0.736	0.866	0.840	7/16	0.448	0.427
9/16	0.5625	7/8	0.875	0.861	1.010	0.982	31/64	0.496	0.473
5/8	0.6250	15/16	0.938	0.922	1.083	1.051	35/64	0.559	0.535
3/4	0.7500	1 1/8	1.125	1.088	1.299	1.240	41/64	0.665	0.617
7/8	0.8750	1 5/16	1.312	1.269	1.516	1.447	3/4	0.776	0.724
1	1.0000	1 1/2	1.500	1.450	1.732	1.653	55/64	0.887	0.831
1 1/8	1.1250	1 11/16	1.688	1.631	1.949	1.859	31/32	0.999	0.939
1 1/4	1.2500	1 7/8	1.875	1.812	2.165	2.066	1 1/16	1.094	1.030
1 3/8	1.3750	2 1/16	2.062	1.994	2.382	2.273	1 11/64	0.206	1.138
1 1/2	1.5000	2 1/4	2.250	2.175	2.598	2.480	1 9/32	1.317	1.245

Nut Size		H_1 Thickness Hex Jam Nuts			Hex Nuts, Specified Proof Load		Runout of Bearing Face, *FIR* Max
					Up to 150,000 psi	150,000 psi & Greater	Jam Nuts, All Strength Levels
Nom.	Basic	Basic	Max.	Min.			
1/4	0.2500	5/32	0.163	0.150	0.015	0.010	0.015
5/16	0.3125	3/16	0.195	0.180	0.016	0.011	0.016
3/8	0.3750	7/32	0.227	0.210	0.017	0.012	0.017
7/16	0.4375	1/4	0.260	0.240	0.018	0.013	0.018
1/2	0.5000	5/16	0.323	0.302	0.019	0.014	0.019
9/16	0.5625	5/16	0.324	0.301	0.020	0.015	0.020
5/8	0.6250	3/8	0.387	0.363	0.021	0.016	0.021
3/4	0.7500	27/64	0.446	0.398	0.023	0.018	0.023
7/8	0.8750	31/64	0.510	0.458	0.025	0.020	0.025

(Continued)

Table 17. *(Continued)* **Dimensions of Hex Nuts and Hex Jam Nuts.**

Nut Size		H₁ Thickness Hex Jam Nuts			Hex Nuts, Specified Proof Load		Runout of Bearing Face, *FIR* Max
					Up to 150,000 psi	150,000 psi & Greater	Jam Nuts, All Strength Levels
Nom.	Basic	Basic	Max.	Min.			
1	1.0000	35/64	0.575	0.519	0.027	0.022	0.027
1 1/8	1.1250	39/64	0.639	0.579	0.030	0.025	0.030
1 1/4	1.2500	23/32	0.751	0.687	0.033	0.028	0.033
1 3/8	1.3750	25/32	0.815	0.747	0.036	0.031	0.036
1 1/2	1.5000	27/32	0.880	0.808	0.039	0.034	0.039

All dimensions in inches. Source: ANSI B18.2.2-1972, as published by the American Society of Mechanical Engineers.

Table 18. Dimensions of Thick Pattern and Thin Pattern Nylon Insert Lock Nuts.

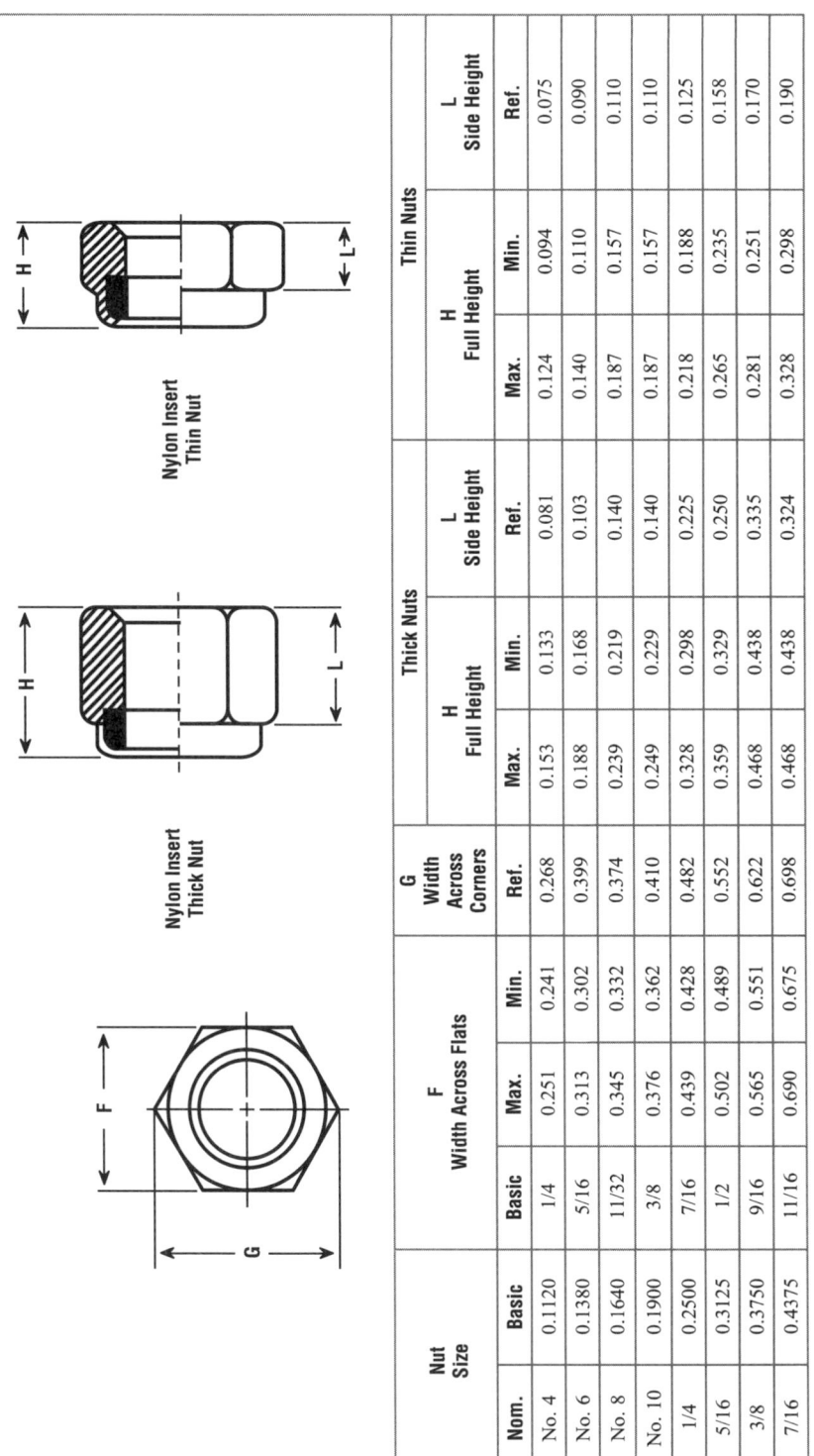

Nylon Insert Thin Nut

Nylon Insert Thick Nut

Nut Size		F Width Across Flats			G Width Across Corners	Thick Nuts				Thin Nuts			
						H Full Height		L Side Height		H Full Height		L Side Height	
Nom.	Basic	Basic	Max.	Min.	Ref.	Max.	Min.	Ref.		Max.	Min.	Ref.	
No. 4	0.1120	1/4	0.251	0.241	0.268	0.153	0.133	0.081		0.124	0.094	0.075	
No. 6	0.1380	5/16	0.313	0.302	0.399	0.188	0.168	0.103		0.140	0.110	0.090	
No. 8	0.1640	11/32	0.345	0.332	0.374	0.239	0.219	0.140		0.187	0.157	0.110	
No. 10	0.1900	3/8	0.376	0.362	0.410	0.249	0.229	0.140		0.187	0.157	0.110	
1/4	0.2500	7/16	0.439	0.428	0.482	0.328	0.298	0.225		0.218	0.188	0.125	
5/16	0.3125	1/2	0.502	0.489	0.552	0.359	0.329	0.250		0.265	0.235	0.158	
3/8	0.3750	9/16	0.565	0.551	0.622	0.468	0.438	0.335		0.281	0.251	0.170	
7/16	0.4375	11/16	0.690	0.675	0.698	0.468	0.438	0.324		0.328	0.298	0.190	

(Continued)

Table 18. (Continued) Dimensions of Thick Pattern and Thin Pattern Nylon Insert Lock Nuts.

Nut Size		F Width Across Flats			G Width Across Corners	Thick Nuts				Thin Nuts			
						H Full Height		L Side Height		H Full Height			L Side Height
Nom.	Basic	Basic	Max.	Min.	Ref.	Max.	Min.	Ref.	Max.	Min.	Ref.		
1/2	0.5000	3/4	0.752	0.736	0.837	0.609	0.579	0.464	0.328	0.298	0.225		
9/16	0.5625	7/8	0.877	0.861	0.978	0.656	0.626	0.469	-	-	-		
5/8	0.6250	15/16	0.940	0.922	1.051	0.765	0.735	0.593	0.406	0.376	0.264		
3/4	0.7500	1 1/8	1.127	1.088	1.191	0.890	0.860	0.742	0.421	0.391	0.288		
7/8	0.8750	1 5/16	1.315	1.269	1.430	0.999	0.969	0.790	-	-	-		
1	1.0000	1 1/2	1.502	1.450	1.615	1.078	1.061	0.825	-	-	-		
1 1/8	1.1250	1 11/16	1.690	1.631	1.826	1.203	1.141	0.930	-	-	-		
1 1/4	1.2500	1 7/8	1.877	1.812	2.038	1.422	1.360	1.125	-	-	-		
1 1/2	1.5000	2 1/4	2.252	2.175	2.444	1.640	1.578	1.313	-	-	-		

All dimensions in inches.

Table 19. Dimensions of Hex Slotted Nuts.

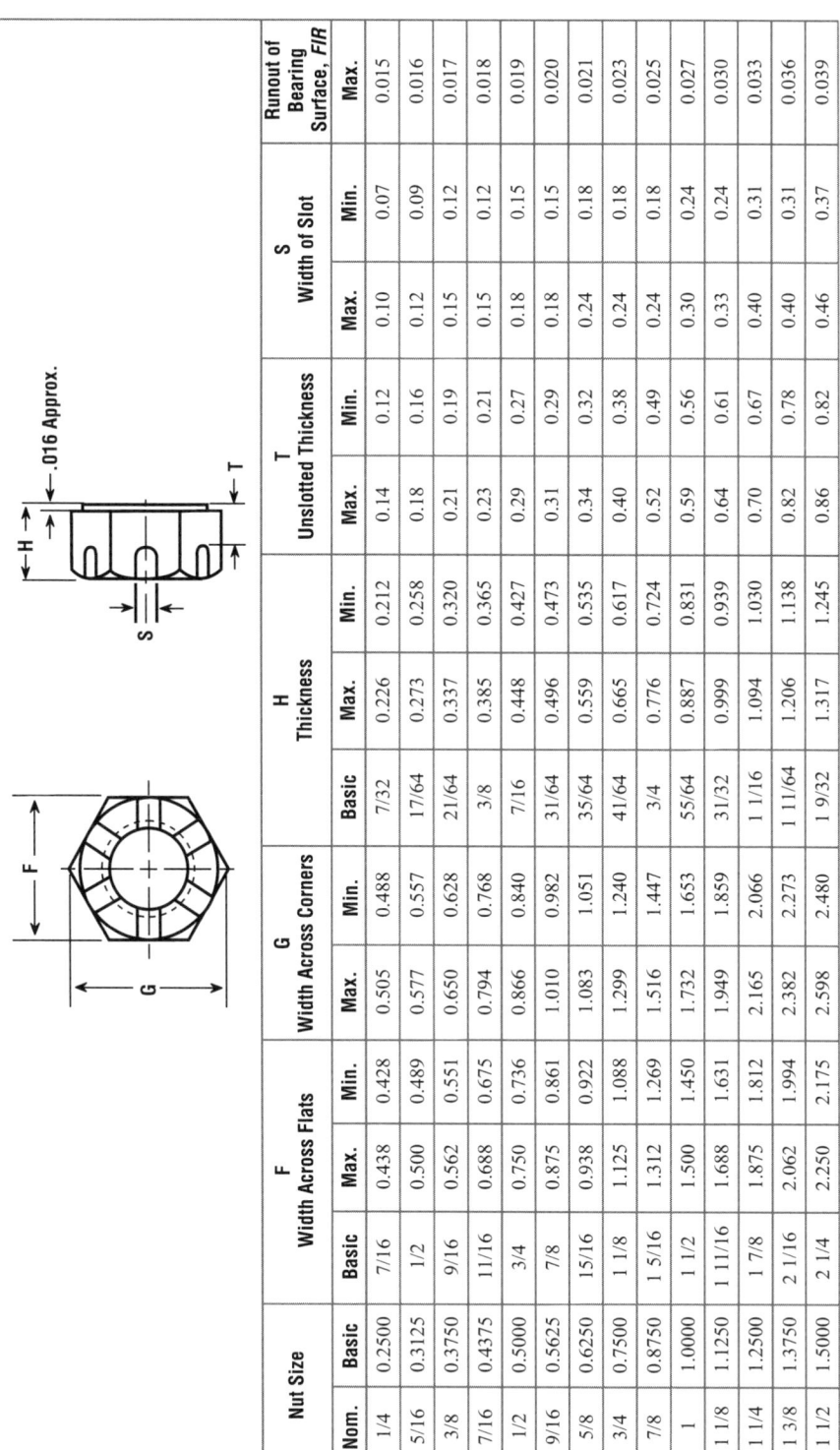

Nut Size			F Width Across Flats			G Width Across Corners			H Thickness			T Unslotted Thickness		S Width of Slot		Runout of Bearing Surface, F/R
Nom.	Basic	Basic	Max.	Min.	Max.	Min.	Basic	Max.	Min.	Max.	Min.	Max.	Min.	Max.	Min.	Max.
1/4	0.2500	7/16	0.438	0.428	0.505	0.488	7/32	0.226	0.212	0.14	0.12	0.10	0.07	0.015
5/16	0.3125	1/2	0.500	0.489	0.577	0.557	17/64	0.273	0.258	0.18	0.16	0.12	0.09	0.016
3/8	0.3750	9/16	0.562	0.551	0.650	0.628	21/64	0.337	0.320	0.21	0.19	0.15	0.12	0.017
7/16	0.4375	11/16	0.688	0.675	0.794	0.768	3/8	0.385	0.365	0.23	0.21	0.15	0.12	0.018
1/2	0.5000	3/4	0.750	0.736	0.866	0.840	7/16	0.448	0.427	0.29	0.27	0.18	0.15	0.019
9/16	0.5625	7/8	0.875	0.861	1.010	0.982	31/64	0.496	0.473	0.31	0.29	0.18	0.15	0.020
5/8	0.6250	15/16	0.938	0.922	1.083	1.051	35/64	0.559	0.535	0.34	0.32	0.24	0.18	0.021
3/4	0.7500	1 1/8	1.125	1.088	1.299	1.240	41/64	0.665	0.617	0.40	0.38	0.24	0.18	0.023
7/8	0.8750	1 5/16	1.312	1.269	1.516	1.447	3/4	0.776	0.724	0.52	0.49	0.24	0.18	0.025
1	1.0000	1 1/2	1.500	1.450	1.732	1.653	55/64	0.887	0.831	0.59	0.56	0.30	0.24	0.027
1 1/8	1.1250	1 11/16	1.688	1.631	1.949	1.859	31/32	0.999	0.939	0.64	0.61	0.33	0.24	0.030
1 1/4	1.2500	1 7/8	1.875	1.812	2.165	2.066	1 1/16	1.094	1.030	0.70	0.67	0.40	0.31	0.033
1 3/8	1.3750	2 1/16	2.062	1.994	2.382	2.273	1 11/64	1.206	1.138	0.82	0.78	0.40	0.31	0.036
1 1/2	1.5000	2 1/4	2.250	2.175	2.598	2.480	1 9/32	1.317	1.245	0.86	0.82	0.46	0.37	0.039

All dimensions in inches. Source: Extracted from ANSI B18.2.2-1972, published by the American Society of Mechanical Engineers.

Table 20. Dimensions of Hex Thick Slotted Nuts.

Nut Size		F Width Across Flats			G Width Across Corners		H Thickness			T Unslotted Thickness		S Width of Slot		Runout of Bearing Surface, F/R
Nom.	Basic	Basic	Max.	Min.	Max.	Min.	Basic	Max.	Min.	Max.	Min.	Max.	Min.	Max.
1/4	0.2500	7/16	0.438	0.428	0.505	0.488	9/32	0.288	0.274	0.20	0.18	0.10	0.07	0.015
5/16	0.3125	1/2	0.500	0.489	0.577	0.557	21/64	0.336	0.320	0.24	0.22	0.12	0.09	0.016
3/8	0.3750	9/16	0.562	0.551	0.650	0.628	13/32	0.415	0.398	0.29	0.27	0.15	0.12	0.017
7/16	0.4375	11/16	0.688	0.675	0.794	0.768	29/64	0.463	0.444	0.31	0.29	0.15	0.12	0.018
1/2	0.5000	3/4	0.750	0.736	0.866	0.840	9/16	0.573	0.552	0.42	0.40	0.18	0.15	0.019
9/16	0.5625	7/8	0.875	0.861	1.010	0.982	39/64	0.621	0.598	0.43	0.41	0.18	0.15	0.020
5/8	0.6250	15/16	0.938	0.922	1.083	1.051	23/32	0.731	0.706	0.51	0.49	0.24	0.18	0.021
3/4	0.7500	1 1/8	1.125	1.088	1.299	1.240	13/16	0.827	0.798	0.57	0.55	0.24	0.18	0.023
7/8	0.8750	1 5/16	1.312	1.269	1.516	1.447	29/32	0.922	0.890	0.67	0.64	0.24	0.18	0.025
1	1.0000	1 1/2	1.500	1.450	1.732	1.653	1	1.018	0.982	0.73	0.70	0.30	0.24	0.027
1 1/8	1.1250	1 11/16	1.688	1.631	1.949	1.859	1 5/32	1.176	1.136	0.83	0.80	0.33	0.24	0.030
1 1/4	1.2500	1 7/8	1.875	1.812	2.165	2.066	1 1/4	1.272	1.228	0.89	0.86	0.40	0.31	0.033
1 3/8	1.3750	2 1/16	2.062	1.994	2.382	2.273	1 3/8	1.399	1.351	1.02	0.98	0.40	0.31	0.036
1 1/2	1.5000	2 1/4	2.250	2.175	2.598	2.480	1 1/2	1.526	1.474	1.08	1.04	0.46	0.37	0.039

All dimensions in inches. Source: Extracted from ANSI B18.2.2-1972, published by the American Society of Mechanical Engineers.

Table 21. Dimensions of Hex Castle Nuts.

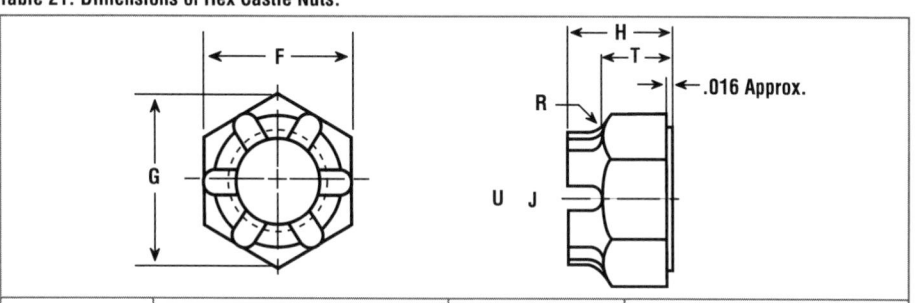

Nut Size		F Width Across Flats			G Width Across Corners		H Thickness		
Nom.	Basic	Basic	Max.	Min.	Max.	Min.	Basic	Max.	Min.
1/4	0.2500	7/16	0.438	0.428	0.505	0.488	9/32	0.228	0.274
5/16	0.3125	1/2	0.500	0.489	0.577	0.557	21/64	0.336	0.320
3/8	0.3750	9/16	0.562	0.551	0.650	0.628	13/32	0.415	0.398
7/16	0.4375	11/16	0.688	0.675	0.794	0.768	29/64	0.463	0.444
1/2	0.5000	3/4	0.750	0.736	0.866	0.840	9/16	0.573	0.552
9/16	0.5625	7/8	0.875	0.861	1.010	0.982	39/64	0.621	0.598
5/8	0.6250	15/16	0.938	0.922	1.083	1.051	23/32	0.731	0.706
3/4	0.7500	1 1/8	1.125	1.088	1.299	1.240	13/16	0.827	0.798
7/8	0.8750	1 5/16	1.312	1.269	1.516	1.447	29/32	0.922	0.890
1	1.0000	1 1/2	1.500	1.450	1.732	1.653	1	1.018	0.982
1 1/8	1.1250	1 11/16	1.688	1.631	1.949	1.859	1 5/32	1.176	1.136
1 1/4	1.2500	1 7/8	1.875	1.812	2.165	2.066	1 1/4	1.272	1.228
1 3/8	1.3750	2 1/16	2.062	1.994	2.382	2.273	1 3/8	1.399	1.351
1 1/2	1.5000	2 1/4	2.250	2.175	2.598	2.480	1 1/2	1.526	1.474

Nut Size		T Unslotted Thickness & Height of Flats			J Width of Slot		R Radius of Fillet	U Dia. of Cylindrical Part	Runout of Bearing Surface, FIR
Nom.	Basic	Nom.	Max.	Min.	Max.	Min.	+/- 0.010	Min.	Max.
1/4	0.2500	3/16	0.20	0.18	0.10	0.07	0.094	0.371	0.015
5/16	0.3125	15/64	0.24	0.22	0.12	0.09	0.094	0.425	0.016
3/8	0.3750	9/32	0.29	0.27	0.15	0.12	0.094	0.478	0.017
7/16	0.4375	19/64	0.31	0.29	0.15	0.12	0.094	0.582	0.018
1/2	0.5000	13/32	0.42	0.40	0.18	0.15	0.125	0.637	0.019
9/16	0.5625	27/64	0.43	0.41	0.18	0.15	0.156	0.744	0.020
5/8	0.6250	1/2	0.51	0.49	0.24	0.18	0.156	0.797	0.021
3/4	0.7500	9/16	0.57	0.55	0.24	0.18	0.188	0.941	0.023
7/8	0.8750	21/32	0.67	0.64	0.24	0.18	0.188	1.097	0.025
1	1.0000	23/32	0.73	0.70	0.30	0.24	0.188	0.254	0.027

(Continued)

Table 21. *(Continued)* Dimensions of Hex Castle Nuts.

Nut Size		T Unslotted Thickness & Height of Flats			J Width of Slot		R Radius of Fillet	U Dia. of Cylindrical Part	Runout of Bearing Surface, *FIR*
Nom.	Basic	Nom.	Max.	Min.	Max.	Min.	+/- 0.010	Min.	Max.
1 1/8	1.1250	13/16	0.83	0.80	0.33	0.24	0.250	1.411	0.030
1 1/4	1.2500	7/8	0.89	0.86	0.40	0.31	0.250	1.570	0.033
1 3/8	1.3750	1	1.02	0.98	0.40	0.31	0.250	1.726	0.036
1 1/2	1.5000	1 1/16	1.08	1.04	0.46	0.37	0.250	1.881	0.039

All dimensions in inches. Source: Extracted from ANSI B18.2.2-1972, published by the American Society of Mechanical Engineers.

Table 22. Dimensions of Serrated Flange Lock Nuts.

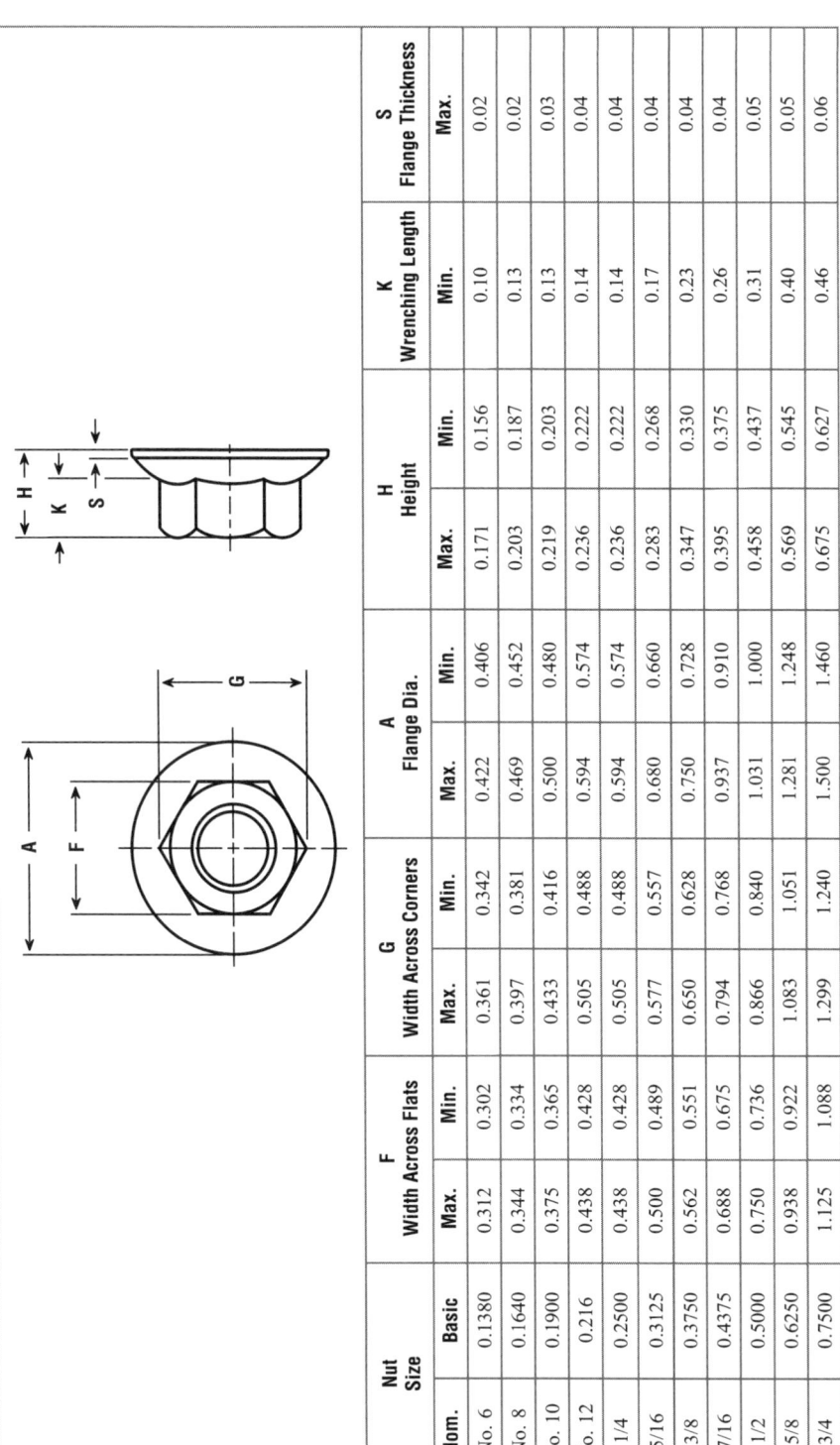

Nut Size		F Width Across Flats		G Width Across Corners		A Flange Dia.		H Height		K Wrenching Length	S Flange Thickness
Nom.	Basic	Max.	Min.	Max.	Min.	Max.	Min.	Max.	Min.	Min.	Max.
No. 6	0.1380	0.312	0.302	0.361	0.342	0.422	0.406	0.171	0.156	0.10	0.02
No. 8	0.1640	0.344	0.334	0.397	0.381	0.469	0.452	0.203	0.187	0.13	0.02
No. 10	0.1900	0.375	0.365	0.433	0.416	0.500	0.480	0.219	0.203	0.13	0.03
No. 12	0.216	0.438	0.428	0.505	0.488	0.594	0.574	0.236	0.222	0.14	0.04
1/4	0.2500	0.438	0.428	0.505	0.488	0.594	0.574	0.236	0.222	0.14	0.04
5/16	0.3125	0.500	0.489	0.577	0.557	0.680	0.660	0.283	0.268	0.17	0.04
3/8	0.3750	0.562	0.551	0.650	0.628	0.750	0.728	0.347	0.330	0.23	0.04
7/16	0.4375	0.688	0.675	0.794	0.768	0.937	0.910	0.395	0.375	0.26	0.04
1/2	0.5000	0.750	0.736	0.866	0.840	1.031	1.000	0.458	0.437	0.31	0.05
5/8	0.6250	0.938	0.922	1.083	1.051	1.281	1.248	0.569	0.545	0.40	0.05
3/4	0.7500	1.125	1.088	1.299	1.240	1.500	1.460	0.675	0.627	0.46	0.06

All dimensions in inches.

Table 23. Dimensions of Square Nuts.

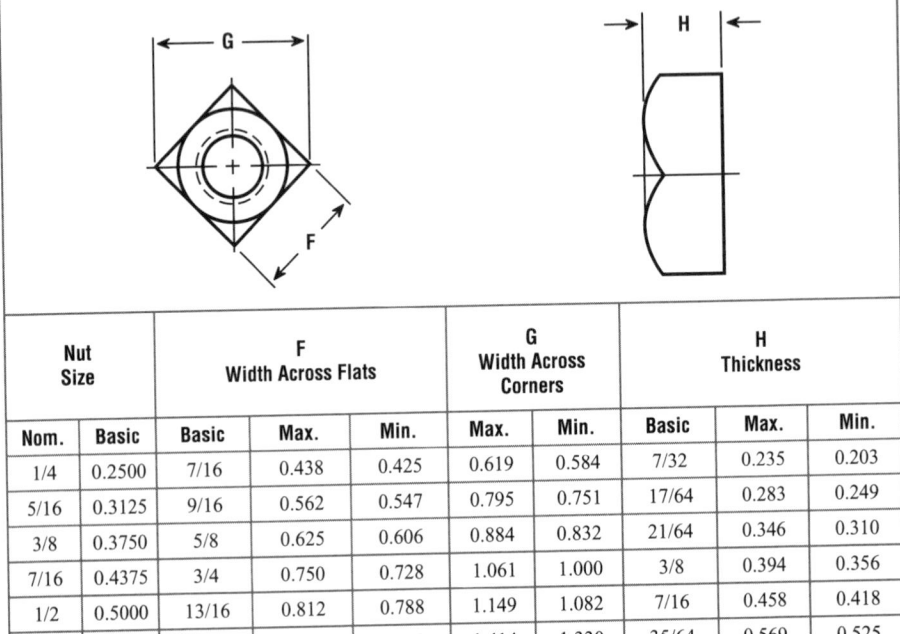

Nut Size		F Width Across Flats			G Width Across Corners		H Thickness		
Nom.	Basic	Basic	Max.	Min.	Max.	Min.	Basic	Max.	Min.
1/4	0.2500	7/16	0.438	0.425	0.619	0.584	7/32	0.235	0.203
5/16	0.3125	9/16	0.562	0.547	0.795	0.751	17/64	0.283	0.249
3/8	0.3750	5/8	0.625	0.606	0.884	0.832	21/64	0.346	0.310
7/16	0.4375	3/4	0.750	0.728	1.061	1.000	3/8	0.394	0.356
1/2	0.5000	13/16	0.812	0.788	1.149	1.082	7/16	0.458	0.418
5/8	0.6250	1	1.000	0.969	1.414	1.330	35/64	0.569	0.525
3/4	0.7500	1 1/8	1.125	1.088	1.591	1.494	21/32	0.680	0.632
7/8	0.8750	1 5/16	1.312	1.269	1.856	1.742	49/64	0.792	0.740
1	1.0000	1 1/2	1.500	1.450	2.121	1.991	7/8	0.903	0.847
1 1/8	1.1250	1 11/16	1.688	1.631	2.386	2.239	1	1.030	0.970
1 1/4	1.2500	1 7/8	1.875	1.812	2.652	2.489	1 3/32	1.126	1.062
1 3/8	1.3750	2 1/16	2.062	1.994	2.917	2.738	1 13/64	1.237	1.169
1 1/2	1.5000	2 1/4	2.250	2.175	3.182	2.986	1 5/16	1.348	1.276

All dimensions in inches. Source: Extracted from ANSI B18.2.2-1972, published by the American Society of Mechanical Engineers.

Table 24. Dimensions of Square and Hex Machine Screw Nuts. (See Notes.)

Nut Size		F Width Across Flats			Width Across Corners				H Thickness		
					G Square Nut		G₁ Hex Nut				
Nom.	Basic	Basic	Max.	Min.	Max.	Min.	Max.	Min.	Basic	Max.	Min.
2	0.0860	3/16	0.188	0.180	0.265	0.247	0.217	0.205	1/16	0.066	0.057
3	0.0990	3/16	0.188	0.180	0.265	0.247	0.217	0.205	1/16	0.066	0.057
4	0.1120	1/4	0.250	0.241	0.354	0.331	0.289	0.275	3/32	0.098	0.087
5	0.1250	5/16	0.312	0.302	0.442	0.415	0.361	0.344	7/64	0.114	0.102
6	0.1380	5/16	0.312	0.302	0.442	0.415	0.361	0.344	7/64	0.114	0.102
8	0.1640	11/32	0.344	0.332	0.486	0.456	0.397	0.378	1/8	0.130	0.117
10	0.1900	3/8	0.375	0.362	0.530	0.497	0.433	0.413	1/8	0.130	0.117
12	0.2160	7/16	0.438	0.423	0.619	0.581	0.505	0.482	5/32	0.161	0.148
1/4	0.2500	7/16	0.438	0.423	0.619	0.581	0.505	0.482	3/16	0.193	0.178
5/16	0.3125	9/16	0.562	0.545	0.795	0.748	0.650	0.621	7/32	0.225	0.208
3/8	0.3750	5/8	0.625	0.607	0.884	0.883	0.722	0.692	1/4	0.257	0.239

Notes: Small pattern nuts are available in the following sizes: 4-40 ($^3/_{16}$" across flats, $^1/_{16}$" thick); 6-32 ($^1/_4$" across flats, $^3/_{32}$" thick); 8-32 ($^1/_4$" across flats, $^3/_{32}$" thick); 8-32 ($^5/_{16}$" across flats, $^7/_{64}$" thick); 10-32 ($^5/_{16}$" across flats, $^7/_{64}$" thick). All dimensions in inches.

Table 25. Dimensions of J-Type Nut Retainers. (See Notes.)

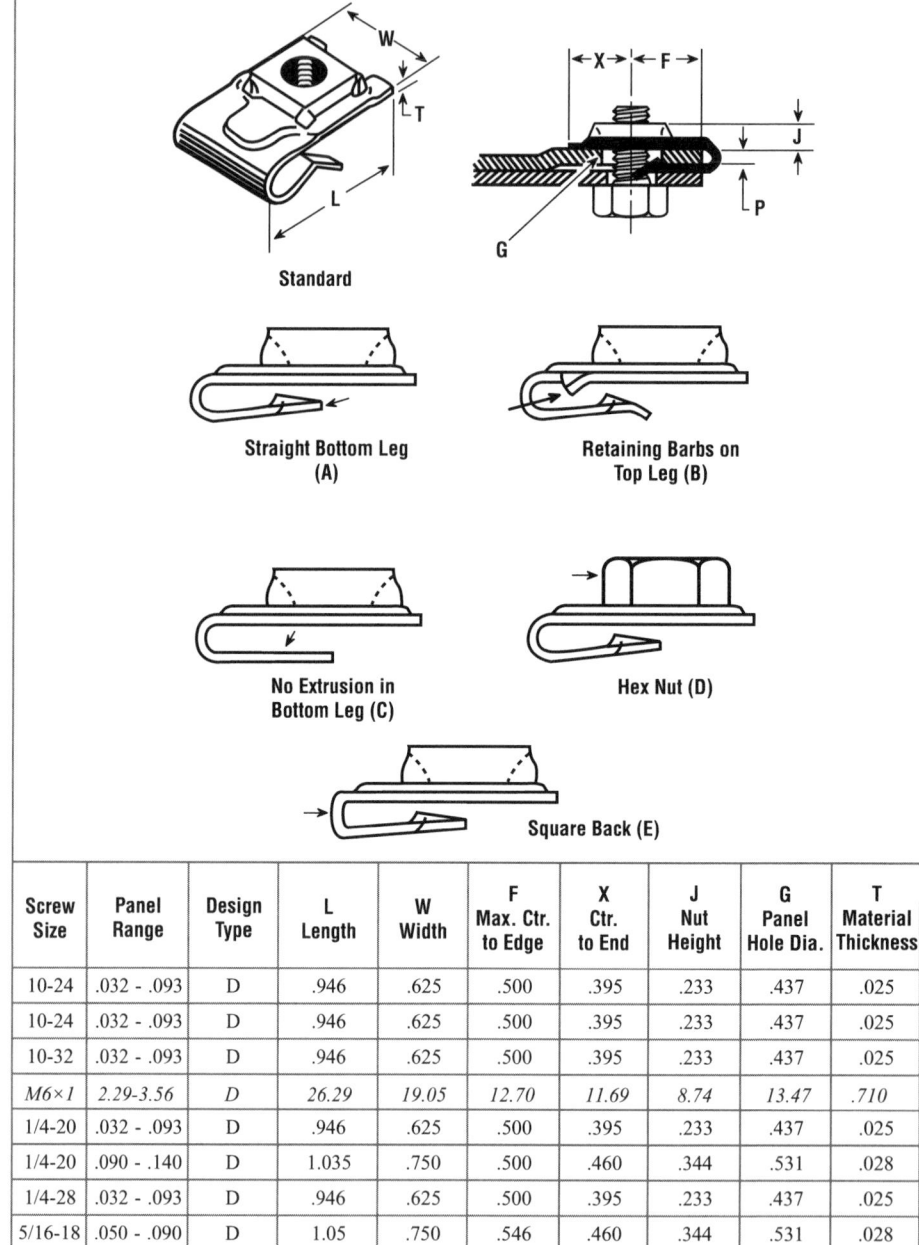

Standard

Straight Bottom Leg (A)

Retaining Barbs on Top Leg (B)

No Extrusion in Bottom Leg (C)

Hex Nut (D)

Square Back (E)

Screw Size	Panel Range	Design Type	L Length	W Width	F Max. Ctr. to Edge	X Ctr. to End	J Nut Height	G Panel Hole Dia.	T Material Thickness
10-24	.032 - .093	D	.946	.625	.500	.395	.233	.437	.025
10-24	.032 - .093	D	.946	.625	.500	.395	.233	.437	.025
10-32	.032 - .093	D	.946	.625	.500	.395	.233	.437	.025
M6×1	*2.29-3.56*	*D*	*26.29*	*19.05*	*12.70*	*11.69*	*8.74*	*13.47*	*.710*
1/4-20	.032 - .093	D	.946	.625	.500	.395	.233	.437	.025
1/4-20	.090 - .140	D	1.035	.750	.500	.460	.344	.531	.028
1/4-28	.032 - .093	D	.946	.625	.500	.395	.233	.437	.025
5/16-18	.050 - .090	D	1.05	.750	.546	.460	.344	.531	.028
5/16-18	.130-.140	ACDE	.820	.687	.400	.385	.330	.343	.031
3/8-16	.090-.140	D	1.035	.750	.500	.460	.344	.531	.028
3/8-16	.093-.203	Std.	1.48	.750	.750	.440	.230	.437	.028

Notes: Dimensions given are for Tinnerman brand fasteners. Dimensions for inch screw sizes are in inches, and dimensions for metric sizes (shown in italics) are in millimeters. Source: Tinnerman Fasteners.

Table 26. Dimensions of U-Type Nut Retainers. (See Notes.)

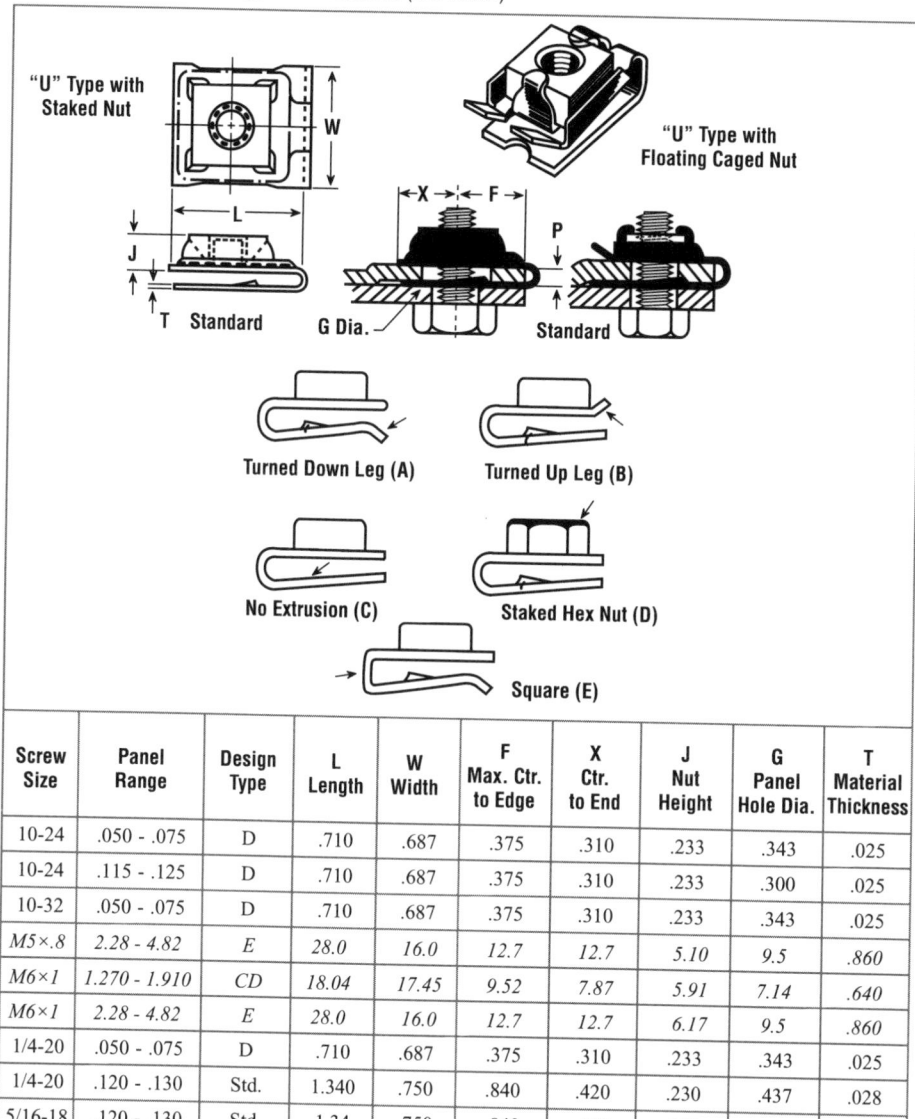

Screw Size	Panel Range	Design Type	L Length	W Width	F Max. Ctr. to Edge	X Ctr. to End	J Nut Height	G Panel Hole Dia.	T Material Thickness
10-24	.050 - .075	D	.710	.687	.375	.310	.233	.343	.025
10-24	.115 - .125	D	.710	.687	.375	.310	.233	.300	.025
10-32	.050 - .075	D	.710	.687	.375	.310	.233	.343	.025
M5×.8	2.28 - 4.82	E	28.0	16.0	12.7	12.7	5.10	9.5	.860
M6×1	1.270 - 1.910	CD	18.04	17.45	9.52	7.87	5.91	7.14	.640
M6×1	2.28 - 4.82	E	28.0	16.0	12.7	12.7	6.17	9.5	.860
1/4-20	.050 - .075	D	.710	.687	.375	.310	.233	.343	.025
1/4-20	.120 - .130	Std.	1.340	.750	.840	.420	.230	.437	.028
5/16-18	.120 - .130	Std.	1.34	.750	.840	.420	.230	.437	.028

Notes: Dimensions given are for Tinnerman brand fasteners. Dimensions for inch screw sizes are in inches, and dimensions for metric sizes (shown in italics) are in millimeters. Source: Tinnerman Fasteners.

Table 27. Dimensions of J-Type Speed Nuts. (See Notes.)

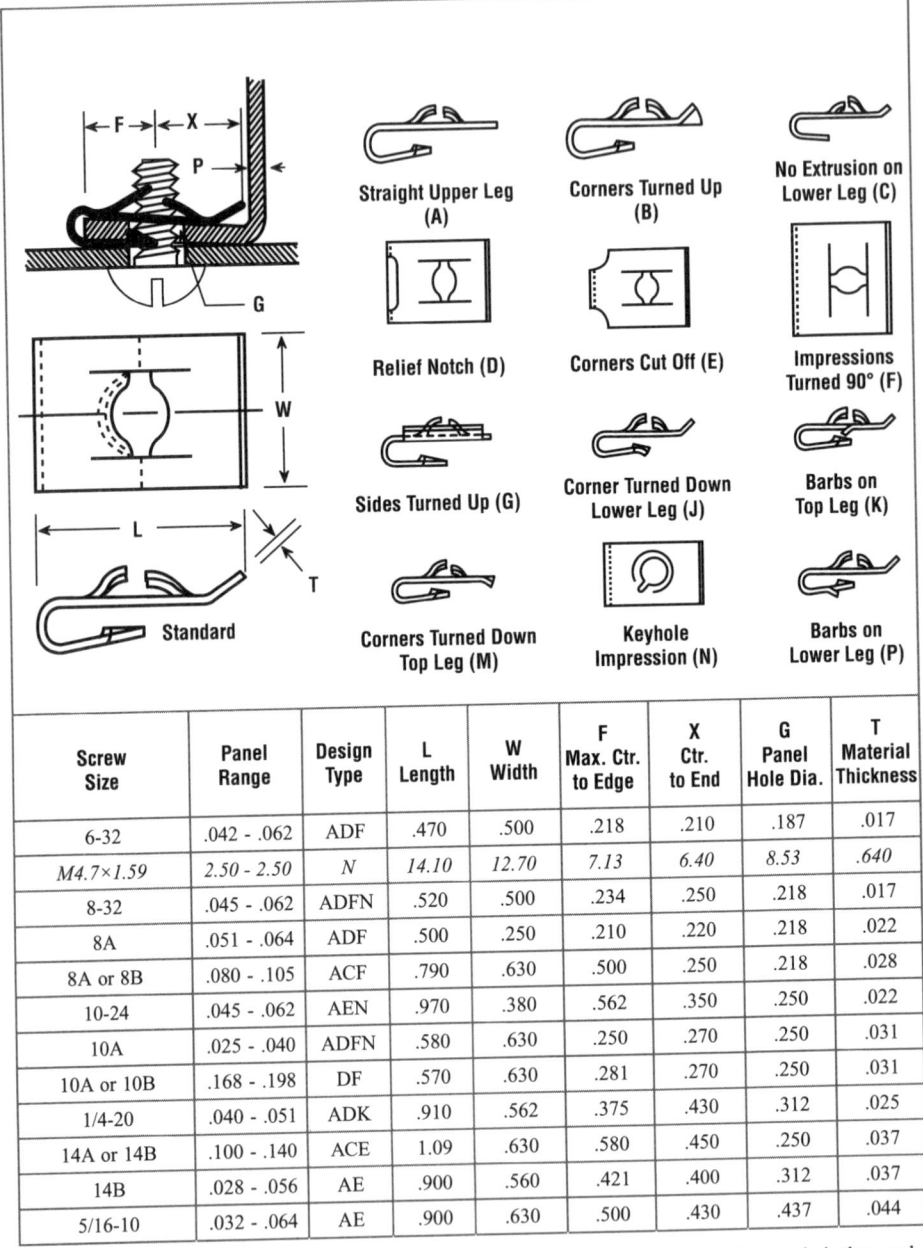

Screw Size	Panel Range	Design Type	L Length	W Width	F Max. Ctr. to Edge	X Ctr. to End	G Panel Hole Dia.	T Material Thickness
6-32	.042 - .062	ADF	.470	.500	.218	.210	.187	.017
M4.7×1.59	*2.50 - 2.50*	*N*	*14.10*	*12.70*	*7.13*	*6.40*	*8.53*	*.640*
8-32	.045 - .062	ADFN	.520	.500	.234	.250	.218	.017
8A	.051 - .064	ADF	.500	.250	.210	.220	.218	.022
8A or 8B	.080 - .105	ACF	.790	.630	.500	.250	.218	.028
10-24	.045 - .062	AEN	.970	.380	.562	.350	.250	.022
10A	.025 - .040	ADFN	.580	.630	.250	.270	.250	.031
10A or 10B	.168 - .198	DF	.570	.630	.281	.270	.250	.031
1/4-20	.040 - .051	ADK	.910	.562	.375	.430	.312	.025
14A or 14B	.100 - .140	ACE	1.09	.630	.580	.450	.250	.037
14B	.028 - .056	AE	.900	.560	.421	.400	.312	.037
5/16-10	.032 - .064	AE	.900	.630	.500	.430	.437	.044

Notes: Dimensions given are for Tinnerman brand fasteners. Dimensions for inch screw sizes are in inches, and dimensions for metric sizes (shown in italics) are in millimeters. Source: Tinnerman Fasteners.

Table 28. Dimensions of U-Type Speed Nuts. (See Notes.)

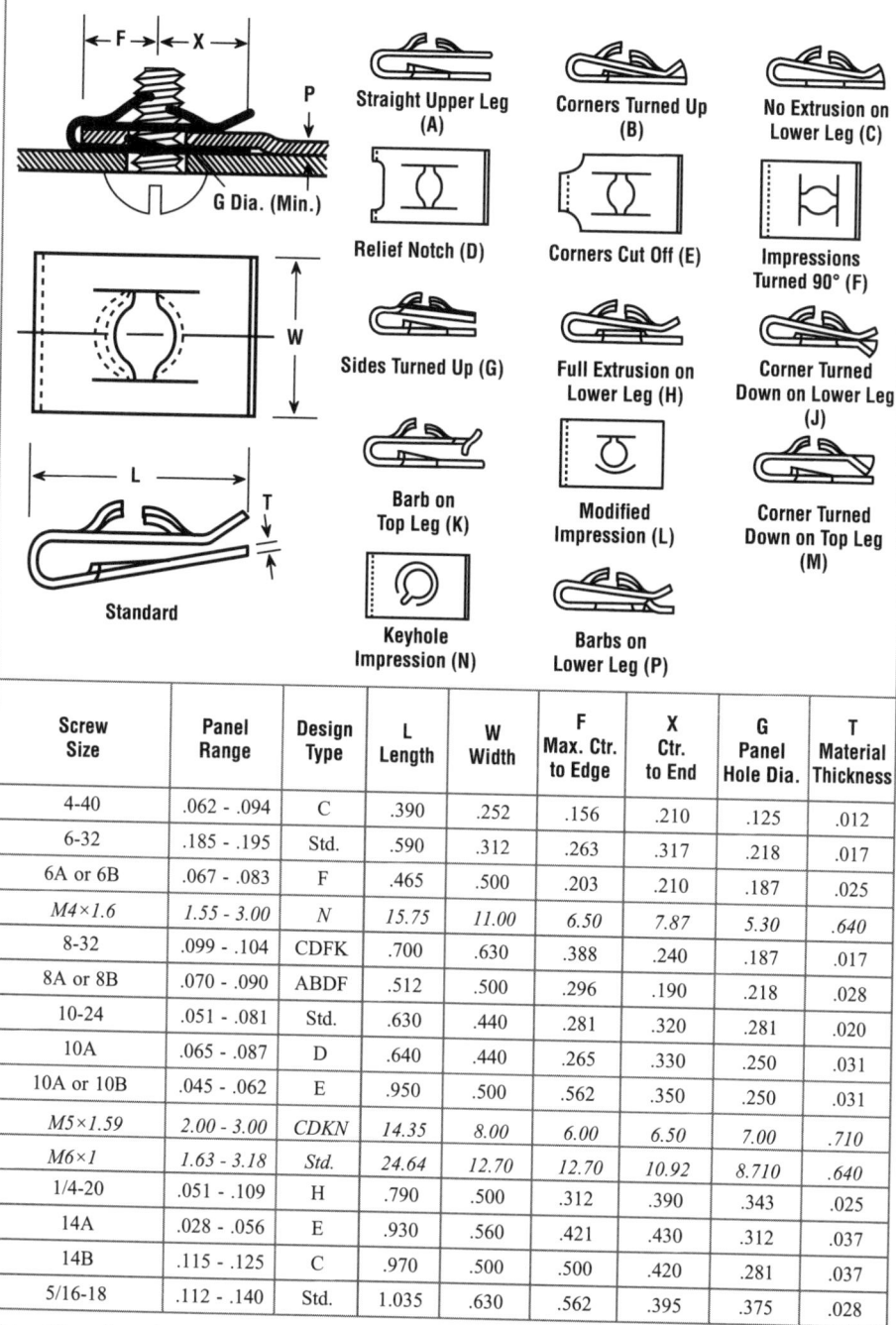

Screw Size	Panel Range	Design Type	L Length	W Width	F Max. Ctr. to Edge	X Ctr. to End	G Panel Hole Dia.	T Material Thickness
4-40	.062 - .094	C	.390	.252	.156	.210	.125	.012
6-32	.185 - .195	Std.	.590	.312	.263	.317	.218	.017
6A or 6B	.067 - .083	F	.465	.500	.203	.210	.187	.025
M4×1.6	*1.55 - 3.00*	*N*	*15.75*	*11.00*	*6.50*	*7.87*	*5.30*	*.640*
8-32	.099 - .104	CDFK	.700	.630	.388	.240	.187	.017
8A or 8B	.070 - .090	ABDF	.512	.500	.296	.190	.218	.028
10-24	.051 - .081	Std.	.630	.440	.281	.320	.281	.020
10A	.065 - .087	D	.640	.440	.265	.330	.250	.031
10A or 10B	.045 - .062	E	.950	.500	.562	.350	.250	.031
M5×1.59	*2.00 - 3.00*	*CDKN*	*14.35*	*8.00*	*6.00*	*6.50*	*7.00*	*.710*
M6×1	*1.63 - 3.18*	*Std.*	*24.64*	*12.70*	*12.70*	*10.92*	*8.710*	*.640*
1/4-20	.051 - .109	H	.790	.500	.312	.390	.343	.025
14A	.028 - .056	E	.930	.560	.421	.430	.312	.037
14B	.115 - .125	C	.970	.500	.500	.420	.281	.037
5/16-18	.112 - .140	Std.	1.035	.630	.562	.395	.375	.028

Notes: Dimensions given are for Tinnerman brand fasteners. Dimensions for inch screw sizes are in inches, and dimensions for metric sizes (shown in italics) are in millimeters. Source: Tinnerman Fasteners.

Table 29. Dimensions of Type A Wing Nuts.

Nut Size		Threads per inch	Series	Nut Blank Size (Ref.)	A Wing Spread		B Wing Height		C Wing Thickness		D Between Wings		E Boss Diameter		G Boss Height	
Nom.	Basic				Max.	Min.	Max.	Min.	Max.	Min.	Max.	Min.	Max.	Min.	Max.	Min.
No. 3	0.0990	48 & 56	Heavy	AA	0.72	0.59	0.41	0.28	0.11	0.07	0.21	0.17	0.33	0.29	0.14	0.10
No. 4	0.1120	40 & 38	Heavy	AA	0.72	0.59	0.41	0.28	0.11	0.07	0.21	0.17	0.33	0.29	0.14	0.10
No. 5	0.1250	40 & 44	Light	AA	.072	0.59	0.41	0.28	0.11	0.07	0.21	0.17	0.33	0.29	0.14	0.10
			Heavy	A	0.91	0.78	0.47	0.34	0.14	0.10	0.27	0.22	0.43	0.39	0.18	0.14
No. 6	0.1380	32 & 40	Light	AA	0.72	0.59	0.41	0.28	0.11	0.07	0.21	0.17	0.33	0.29	0.14	0.10
			Heavy	A	0.91	0.78	0.47	0.34	0.14	0.10	0.27	0.22	0.43	0.39	0.18	0.14
No. 8	0.1640	32 & 36	Light	A	0.91	0.78	0.47	0.34	0.14	0.10	0.27	0.22	0.43	0.39	0.18	0.14
			Heavy	B	1.10	0.97	0.57	0.43	0.18	0.14	0.33	0.26	0.50	0.45	0.22	0.17
No.10	0.1900	24 & 32	Light	A	0.91	0.78	0.47	0.34	0.14	0.10	0.27	0.22	0.43	0.39	0.18	0.14
			Heavy	B	1.10	0.97	0.57	0.43	0.18	0.14	0.33	0.26	0.50	0.45	0.22	0.17
No.12	0.2160	24 & 28	Light	B	1.10	0.97	0.57	0.43	0.18	0.14	0.33	0.26	0.50	0.45	0.22	0.17
			Heavy	C	1.25	1.12	0.66	0.53	0.21	0.17	0.39	0.32	0.58	0.51	0.25	0.20

(Continued)

Table 29. (Continued) Dimensions of Type A Wing Nuts.

Nom.	Basic	Threads per inch	Series	Nut Blank Size (Ref.)	A Wing Spread Max.	A Wing Spread Min.	B Wing Height Max.	B Wing Height Min.	C Wing Thickness Max.	C Wing Thickness Min.	D Between Wings Max.	D Between Wings Min.	E Boss Diameter Max.	E Boss Diameter Min.	G Boss Height Max.	G Boss Height Min.
1/4	0.2500	20 & 28	Light	B	1.10	0.97	0.57	0.43	0.18	0.14	0.39	0.26	0.50	0.45	0.22	0.17
			Regular	C	1.25	1.12	0.66	0.53	0.21	0.17	0.39	0.32	0.58	0.51	0.25	0.20
			Heavy	D	1.44	1.31	0.79	0.65	0.24	0.20	0.48	0.42	0.70	0.64	0.30	0.26
5/16	0.3125	18 & 24	Light	C	1.25	1.12	0.66	0.53	0.21	0.17	0.39	0.32	0.58	0.51	0.25	0.20
			Regular	D	1.44	1.31	0.79	0.65	0.24	0.20	0.48	0.42	0.70	0.64	0.30	0.26
			Heavy	E	1.94	1.81	1.00	0.87	0.33	0.26	0.65	0.54	0.93	0.86	0.39	0.35
3/8	0.3750	16 & 24	Light	D	1.44	1.31	0.79	0.65	0.24	0.20	0.48	0.42	0.70	0.64	0.30	0.26
			Regular	E	1.94	1.81	1.00	0.87	0.33	0.26	0.65	0.54	0.93	0.86	0.39	0.35
7/16	0.4375	14 & 20	Light	E	1.94	1.81	1.00	0.87	0.33	0.26	0.65	0.54	0.93	0.86	0.39	0.35
			Heavy	F	2.76	2.62	1.44	1.31	0.40	0.34	0.90	0.80	1.19	1.13	0.55	0.51
1/2	0.5000	13 & 20	Light	E	1.94	1.81	1.00	0.87	0.33	0.26	0.65	0.54	0.93	0.86	0.39	0.35
			Heavy	F	2.76	2.62	1.44	1.31	0.40	0.34	0.90	0.80	1.19	1.13	0.55	0.51
9/16	0.5825	12 & 18	Heavy	F	2.76	2.62	1.44	1.31	0.40	0.34	0.90	0.80	1.19	1.13	0.55	0.51
5/8	0.6250	11 & 18	Heavy	F	2.76	2.62	1.44	1.31	0.40	0.34	0.90	0.80	1.19	1.13	0.55	0.51
3/4	0.7500	10 & 16	Heavy	F	2.76	2.62	1.44	1.31	0.40	0.34	0.90	0.80	1.19	1.13	0.55	0.51

All dimensions in inches. Source: Extracted from ANSI B18.17-1968, published by the American Society of Mechanical Engineers.

Table 30. Dimensions of Type D, Style 2 Wing Nuts.

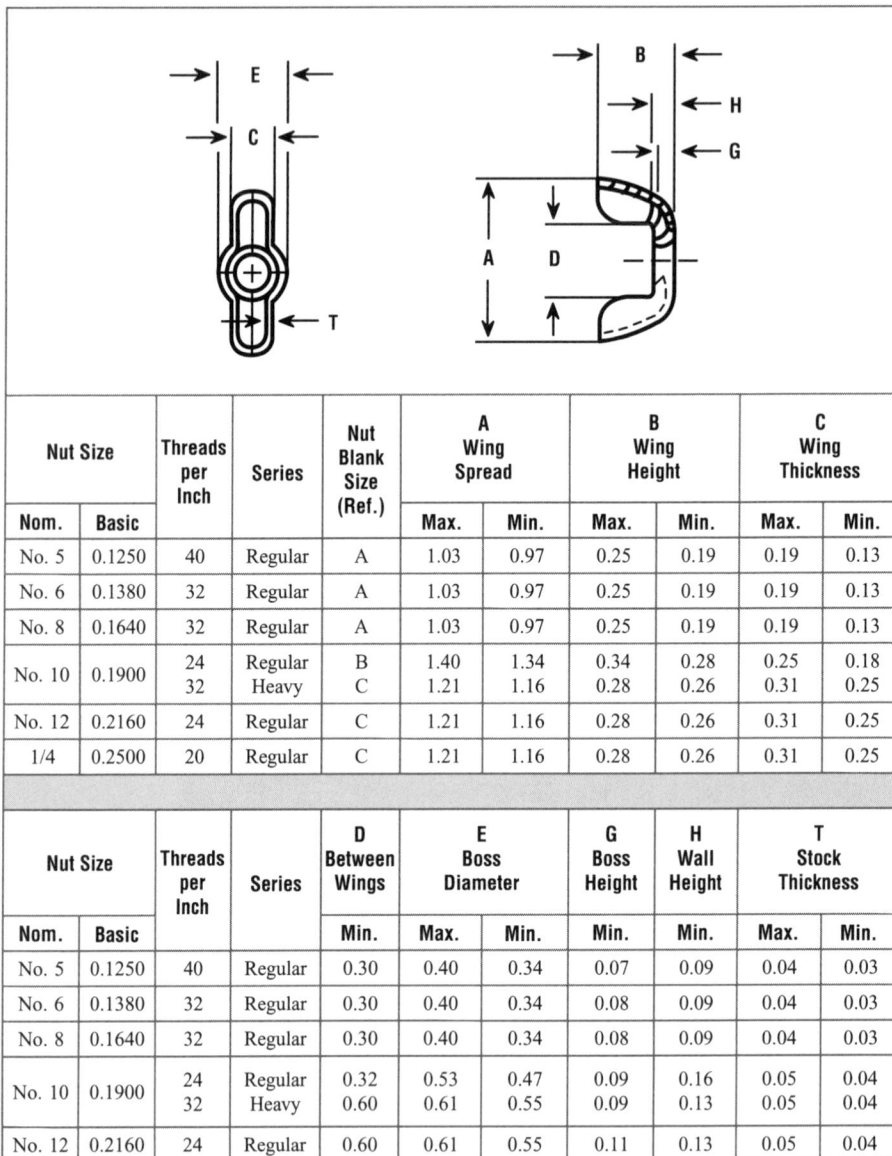

Nut Size		Threads per Inch	Series	Nut Blank Size (Ref.)	A Wing Spread		B Wing Height		C Wing Thickness	
Nom.	Basic				Max.	Min.	Max.	Min.	Max.	Min.
No. 5	0.1250	40	Regular	A	1.03	0.97	0.25	0.19	0.19	0.13
No. 6	0.1380	32	Regular	A	1.03	0.97	0.25	0.19	0.19	0.13
No. 8	0.1640	32	Regular	A	1.03	0.97	0.25	0.19	0.19	0.13
No. 10	0.1900	24	Regular	B	1.40	1.34	0.34	0.28	0.25	0.18
		32	Heavy	C	1.21	1.16	0.28	0.26	0.31	0.25
No. 12	0.2160	24	Regular	C	1.21	1.16	0.28	0.26	0.31	0.25
1/4	0.2500	20	Regular	C	1.21	1.16	0.28	0.26	0.31	0.25

Nut Size		Threads per Inch	Series	D Between Wings	E Boss Diameter		G Boss Height	H Wall Height	T Stock Thickness	
Nom.	Basic			Min.	Max.	Min.	Min.	Min.	Max.	Min.
No. 5	0.1250	40	Regular	0.30	0.40	0.34	0.07	0.09	0.04	0.03
No. 6	0.1380	32	Regular	0.30	0.40	0.34	0.08	0.09	0.04	0.03
No. 8	0.1640	32	Regular	0.30	0.40	0.34	0.08	0.09	0.04	0.03
No. 10	0.1900	24	Regular	0.32	0.53	0.47	0.09	0.16	0.05	0.04
		32	Heavy	0.60	0.61	0.55	0.09	0.13	0.05	0.04
No. 12	0.2160	24	Regular	0.60	0.61	0.55	0.11	0.13	0.05	0.04
1/4	0.2500	20	Regular	0.60	0.61	0.55	0.11	0.13	0.05	0.04

All dimensions in inches. Source: Extracted from ANSI B18.17-1968, published by the American Society of Mechanical Engineers.

Table 31. Dimensions of Type D, Style 3 Wing Nuts.

Nominal Size or Basic Major Dia. of Thread	Threads per Inch	Series	Nut Blank Size (Ref.)	A Wing Spread		B Wing Height		C Wing Thickness	
				Max.	Min.	Max.	Min.	Max.	Min.
10 - 0.1900	24	Light	A	1.31	1.25	0.48	0.42	0.29	0.23
	32	Regular	C	1.40	1.34	0.53	0.47	0.25	0.19
12 - 0.2160	24	Regular	B	1.28	1.22	0.40	0.34	0.23	0.17
1/4 - 0.2500	20	Light	B	1.28	1.22	0.40	0.34	0.23	0.17
		Regular	E	1.78	1.72	0.66	0.60	0.31	0.25
		Heavy	D	1.47	1.40	0.50	0.44	0.37	0.31
5/16 - 0.3125	18	Regular	E	1.78	1.72	0.66	0.60	0.31	0.25
		Heavy	D	1.47	1.40	0.50	0.44	0.37	0.31

Nominal Size or Basic Major Dia. of Thread	Threads per Inch	Series	Nut Blank Size (Ref.)	D Between Wings	E Boss Diameter		G Boss Height	H Wall Height	T Stock Thickness	
				Min.	Max.	Min.	Min.	Min.	Max.	Min.
10 - 0.1900	24	Light	A	0.47	0.65	0.59	0.08	0.12	0.04	0.03
	32	Regular	C	0.50	0.75	0.69	0.08	0.14	0.04	0.03
12 - 0.2160	24	Regular	B	0.59	0.73	0.67	0.11	0.12	0.04	0.03
1/4 - 0.2500	20	Light	B	0.59	0.73	0.67	0.11	0.12	0.04	0.03
		Regular	E	0.70	1.03	0.97	0.14	0.17	0.06	0.04
		Heavy	D	0.66	1.03	0.97	0.14	0.14	0.08	0.06
5/16 - 0.3125	18	Regular	E	0.70	1.03	0.97	0.14	0.17	0.06	0.04
		Heavy	D	0.66	1.03	0.97	0.14	0.14	0.08	0.06

All dimensions in inches. Source: Extracted from ANSI B18.17-1968, published by the American Society of Mechanical Engineers.

Table 32. Dimensions of Selected Metallic Flat Washers. (See Notes.)

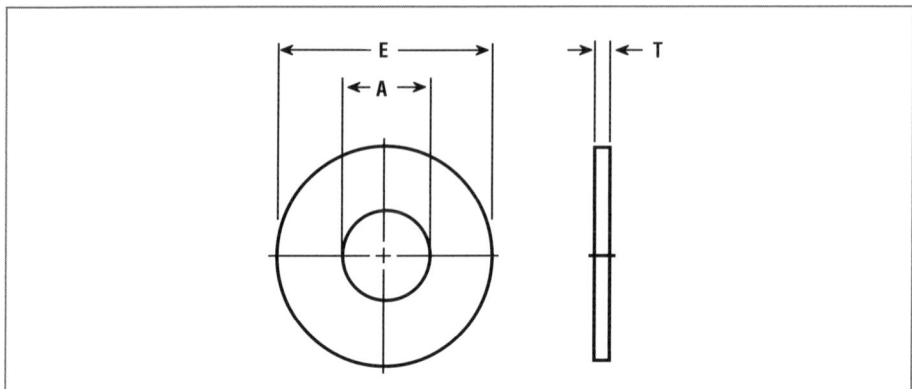

Size inches	Nominal Dimensions			Reference (see notes)	Size inches	Nominal Dimensions			Reference (see notes)
	A	E	Thickness			A	E	Thickness	
000	.040	.090	.010	-	M2.5	.106	.197	.020	DIN 433
M1	.043	.099	.012	DIN 433	M2.5	.106	.256	.020	DIN 125 A
M1	.043	.126	.012	DIN 125A	M2.5	.106	.315	.032	DIN 9021B
M1.2	.051	.118	.012	DIN 433	3	.109	.219	.025	ANS B-N
M1.2	.051	.149	.012	DIN 125A	3	.109	.312	.032	ANS B-R
00	.055	.156	.010	-	3	.109	.406	.040	ANS B-W
0	.065	.099	.016	NAS 620	M3	.119	.236	.032	DIN 1440
0	.065	.125	.025	MS1496	M2.5	.121	.231	.030	ANSI-N
M1.6	.067	.140	.012	DIN 433	M2.5	.121	.308	.030	ANSI-R
M1.6	.067	.158	.012	DIN 125A	M2.5	.121	.387	.030	ANSI-W
0	.068	.125	.025	ANS B-N	4	.125	.250	.022	MS15795
0	.068	.188	.025	ANS B-R	4	.125	.250	.032	ANS B-N
0	.068	.250	.025	ANS B-W	4	.125	.312	.032	ANS A-P
M1.8	.080	.152	.024	ANSI-N	4	.125	.375	.040	ANS B-R
M1.8	.080	.191	.024	ANSI-R	4	.125	.438	.040	ANS B-W
M1.8	.080	.230	.030	ANSI-W	M3	.126	.236	.020	DIN 433
1	.084	.156	.025	ANS B-N	M3	.126	.276	.020	DIN 125A
1	.084	.219	.025	ANS B-R	M3	.126	.355	.030	DIN 9021B
1	.084	.281	.032	ANS B-W	M3	.126	.355	.040	DIN 7349
M2	.086	.177	.012	DIN 433	M3	.126	.269	.030	ANSI-N
M2	.086	.197	.020	DIN 125A	M3	.141	.387	.040	ANSI-R
2	.094	.188	.025	ANS B-N	M3	.141	.464	.047	ANSI-W
2	.094	.250	.020	ANS A-P	5	.141	.281	.032	ANS B-N
2	.094	.250	.032	ANS B-R	5	.141	.406	.040	ANS B-R
2	.094	.344	.035	ANS B-W	5	.141	.500	.040	ANS B-W
M2	.101	.191	.030	ANS B-N	6	.156	.312	.032	ANS B-N
M2	.101	.230	.030	ANS B-R	6	.156	.375	.049	ANS A-P
M2	.101	.308	.030	ANS B-R	6	.156	.438	.040	ANS B-R

(Continued)

Table 32. *(Continued)* Dimensions of Selected Metallic Flat Washers. (See Notes.)

Size inches	Nominal Dimensions			Reference (see notes)	Size inches	Nominal Dimensions			Reference (see notes)
	A	E	Thickness			A	E	Thickness	
6	.156	.562	.040	ANS B-W	3/8	.406	.812	.065	ANS A-P
M4	.158	.315	.032	DIN 1440	3/8	.406	1.250	.100	ANS B-W
M3.5	.161	.348	.040	ANSI-N	M10	.414	.709	.063	DIN 433
M3.5	.161	.583	.058	ANSI-W	M10	.414	.827	.079	DIN 125A
M4	.169	.315	.020	DIN 433	M10	.433	1.093	.095	ANSI-N
M4	.169	.355	.032	DIN 125A	M10	.433	1.524	.118	ANSI-W
M4	.169	.473	.040	DIN 9021B	3/8	.438	1.000	.083	ANS A-P
8	.188	.375	.040	ANS B-N	7/16	.469	.922	.062	ANS A-P
8	.188	.375	.049	ANS A-P	7/16	.469	1.125	.063	ANS B-R
8	.188	.625	.063	ANS B-W	7/16	.469	1.469	.100	ANS B-W
10	.203	.406	.040	ANS B-N	7/16	.500	1.625	.083	ANS A
10	.203	.469	.049	ANS A	M12	.512	.788	.079	DIN 433
10	.203	.734	.063	ANS B-W	M12	.512	.946	.099	DIN 125A
M5	.209	.374	.040	DIN 433	M12	.529	.994	.095	ANSI-N
M5	.209	.394	.040	DIN 125A	M12	.529	1.721	.118	ANSI-W
M5	.209	.591	.063	DIN 9021B	1/2	.531	1.062	.095	ANS A-P
10	.219	.500	.049	ANS A-P	1/2	.531	1.250	.100	ANS B-R
M5	.224	.425	.047	ANSI-N	1/2	.531	1.750	.100	ANS B-W
M5	.224	.778	.077	ANSI-W	1/2	.562	1.250	.109	ANS A-P
12	.234	.438	.040	ANS B-N	9/16	.594	1.125	.063	ANS B-N
12	.234	.875	.063	ANS B-W	9/16	.594	1.156	.095	ANS A-P
12	.250	.562	.049	ANS A-P	9/16	.594	2.000	.100	ANS B-W
12	.250	.562	.065	ANS A-P	9/16	.625	1.469	.109	ANS A-P
M6	.253	.493	.062	DIN 125A	M16	.630	1.103	.118	DIN 1440
M6	.253	.670	.118	DIN 7349	5/8	.656	1.312	.095	ANS A-P
M6	.260	.493	.062	DIN 126	5/8	.656	1.750	.100	ANS B-R
1/4	.266	.625	.049	ANS A	5/8	.656	2.250	.160	ANS B-W
1/4	.281	.500	.063	ANS B-N	M16	.670	1.064	.079	DIN 433
1/4	.281	.625	.062	ANS A-P	M16	.670	1.182	.118	DIN 125A
1/4	.281	1.000	.063	ANS B-W	5/8	.688	1.750	.134	ANS A-P
1/4	.312	.734	.062	ANS A-P	M16	.709	1.182	.118	DIN 126
1/4	.312	.875	.062	ANS A	3/4	.812	1.469	.134	ANS A-P
M8	.331	.591	.062	DIN 433	3/4	.812	2.000	.100	ANS B-R
M8	.331	.670	.062	DIN 125A	3/4	.812	2.000	.148	ANS A-P
5/16	.344	.625	.063	ANS B-N	3/4	.812	2.500	.160	ANS B-W
5/16	.344	1.125	.063	ANS B-W	M20	.827	1.300	.099	DIN 433
5/16	.375	.734	.065	ANS A	M20	.867	1.458	.118	DIN 125
3/8	.390	1.125	.062	ANS A	M20	.865	1.154	.138	ANSI
3/8	.344	.625	.063	ANS B-N	M22	.906	1.537	.118	DIN 125A

(Continued)

Table 32. *(Continued)* **Dimensions of Selected Metallic Flat Washers.** (See Notes.)

Size inches	Nominal Dimensions			Reference (see notes)	Size inches	Nominal Dimensions			Reference (see notes)
	A	E	Thickness			A	E	Thickness	
7/8	.938	1.750	.134	ANS A-P	1 1/2	1.625	3.500	.181	ANS A-P
7/8	.938	2.000	.165	ANS A	1 5/8	1.750	3.000	.160	ANS B-N
7/8	.938	2.250	.165	ANS A-P	1 5/8	1.750	3.750	.180	ANS A-P
7/8	.938	2.750	.160	ANS B-W	1 5/8	1.750	4.250	.266	ANS B-W
1"	1.062	1.750	.100	ANS B-N	M42	1.773	3.073	.276	DIN 126
1"	1.062	2.000	.134	ANS A-P	1 3/4	1.875	3.250	.160	ANS B-N
1"	1.062	2.250	.165	ANS A	1 3/4	1.875	4.000	.180	ANS A-P
1"	1.062	2.500	.165	ANS A-P	1 3/4	1.875	4.500	.250	ANS B-W
1 1/16	1.078	1.812	.032	AN 960	1 3/4	1.890	2.625	.090	AN 960
1 1/8	1.188	2.000	.100	ANS B-N	1 7/8	2.000	3.500	.250	ANS B-N
1 1/8	1.188	2.750	.160	ANS B-R	1 7/8	2.000	4.250	.180	ANS A-P
1 1/8	1.188	3.325	.160	ANS B-W	1 7/8	2.000	4.750	.250	ANS B-W
M30	1.221	2.206	.158	DIN 125A	2"	2.125	3.750	.250	ANS B-N
1 1/8	1.250	2.840	.165	ANS A	2"	2.125	4.500	.180	ANS A-P
M30	1.300	2.206	.158	DIN 126	2"	2.125	5.000	.250	ANS B-W
1 1/4	1.312	2.250	.160	ANS B-N	2 1/4	2.265	3.000	.032	AN 960
1 1/4	1.312	3.500	.250	ANS B-W	2 1/4	2.375	4.750	.220	ANS A-P
1 5/16	1.375	2.500	.165	ANS A-P	2 1/2	2.515	3.250	.090	AN 960
1 5/16	1.375	3.000	.165	ANS A-P	2 1/2	2.625	5.000	.238	ANS A-P
1 3/8	1.500	3.250	.181	ANS A-P	2 3/4	2.875	5.250	.259	ANS A-P
1 1/2	1.515	1.250	.090	NAS 1149	2 7/8	2.968	3.385	.010	SHIM
M36	1.537	2.600	.058	DIN 126	3"	3.125	5.500	.284	ANS A-P
1 1/2	1.562	2.750	.160	ANS B-N	3"	3.155	3.500	.025	-
1 1/2	1.562	4.000	.250	ANS B-W					

All dimensions in inches. See text for information about available sizes, materials, and tolerances for flat washers. This table was compiled from information published by West Coast Lockwasher (WCL) Company and referenced standards.

Notes: Washers to DIN standards conform to German national standards. Full specifications for metric washers to ANSI specifications can be found in ANSI B18.22M. American National Standard washers are designated above as ANS, with suffixes A and B identifying the two standard series; and N designates narrow, R is regular, and W is wide. P indicates preferred series. For complete specifications for inch unit national standard washers, consult ANSI B 18.22.1. Air Force-Navy (AN) 960 washers are steel washers made to military specifications. This standard is scheduled to be replaced by NAS 1149. Multipurpose military washers conform to Mil-Std 15795. Mil-Std 15496 washers are made from stainless steel to reduced diameters.

Table 33. Dimensions of Regular (Medium) and High Collar Helical Spring Lock Washers.

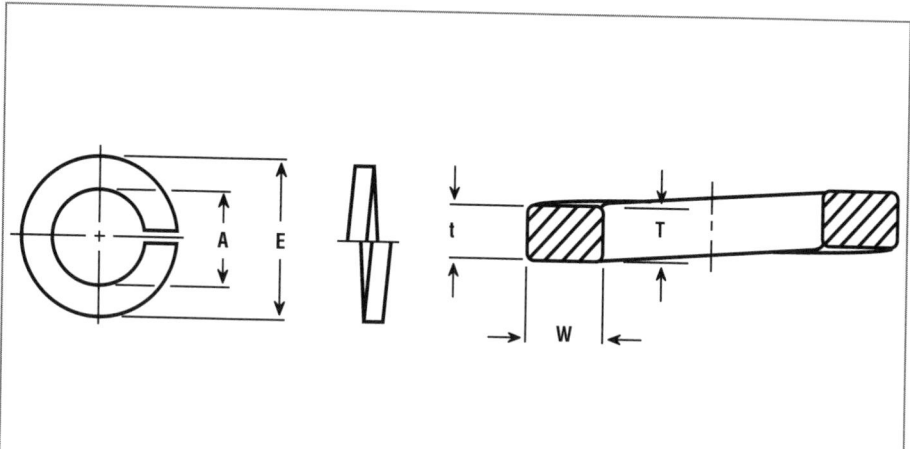

Nom. Washer Size	A Inside Dia.		Regular Lock Washer			High Collar Lock Washer		
			E Outside Dia.	T Mean Thickness[1]	W Section Width	E Outside Dia.	T Mean Thickness[1]	W Section Width
	Max.	Min.	Max.	Min.	Min.	Max.	Min.	Min.
#2	0.094	0.088	0.172	0.020	0.035	-	-	-
#3	0.107	0.101	0.195	0.025	0.040	-	-	-
#4	0.120	0.114	0.209	0.025	0.040	0.173	0.022	0.022
#5	0.133	0.127	0.236	0.031	0.047	0.202	0.030	0.030
#6	0.148	0.141	0.250	0.031	0.047	0.216	0.030	0.030
#8	0.174	0.167	0.293	0.040	0.055	0.267	0.047	0.042
#10	0.200	0.193	0.334	0.047	0.062	0.294	0.047	0.042
#12	0.227	0.220	0.377	0.056	0.070	-	-	-
1/4	0.262	0.254	0.489	0.062	0.109	0.365	0.078	0.047
5/16	0.326	0.317	0.586	0.078	0.125	0.460	0.093	0.062
3/8	0.390	0.380	0.683	0.094	0.141	0.553	0.125	0.076
7/16	0.455	0.443	0.779	0.109	0.156	0.647	0.140	0.090
1/2	0.518	0.506	0.873	0.125	0.171	0.737	0.172	0.103
9/16	0.582	0.570	0.971	0.141	0.188	-	-	-
5/8	0.650	0.635	1.079	0.156	0.203	0.923	0.203	0.125
11/16	0.713	0.698	1.176	0.172	0.219	-	-	-
3/4	0.775	0.760	1.271	0.188	0.234	1.111	0.218	0.154
13/16	0.843	0.824	1.367	0.203	0.250	-	-	-
7/8	0.905	0.887	1.464	0.219	0.266	1.296	0.234	0.182
15/16	0.970	0.950	1.560	0.234	0.281	-	-	-

(Continued)

Table 33. *(Continued)* Dimensions of Regular (Medium) and High Collar Helical Spring Lock Washers.

Nom. Washer Size	A Inside Dia.		Regular Lock Washer			High Collar Lock Washer		
			E Outside Dia.	T Mean Thickness[1]	W Section Width	E Outside Dia.	T Mean Thickness[1]	W Section Width
	Max.	Min.	Max.	Min.	Min.	Max.	Min.	Min.
1	1.042	1.017	1.661	0.250	0.297	1.483	0.250	0.208
1 1/16	1.107	1.080	1.756	0.266	0.312	-	-	-
1 1/8	1.172	1.144	1.853	0.281	0.328	1.669	0.313	0.236
1 3/16	1.237	1.208	1.950	0.297	0.344	-	-	-
1 1/4	1.302	1.271	2.045	0.312	0.359	1.799	0.313	0.236
1 5/16	1.366	1.334	2.141	0.328	0.375	-	-	-
1 3/8	1.432	1.398	2.239	0.344	0.391	2.041	0.375	0.292
1 7/16	1.497	1.462	2.334	0.359	0.406	-	-	-
1 1/2	1.561	1.525	2.430	0.375	0.422	2.170	0.375	0.292
1 3/4	1.811	1.775	-	-	-	2.602	0.469	0.383
2	2.061	2.025	-	-	-	2.852	0.469	0.383
2 1/4	2.311	2.275	-	-	-	3.352	0.508	0.508
2 1/2	2.561	2.525	-	-	-	3.602	0.508	0.508
2 3/4	2.811	2.775	-	-	-	4.102	0.633	0.633
3	3.061	3.025	-	-	-	4.352	0.633	0.633

Notes: [1] Mean thickness is inside diameter thickness (T) plus outside diameter thickness (t), divided by two. All dimensions in inches.

Table 34. Dimensions of External Tooth Lock Washers.

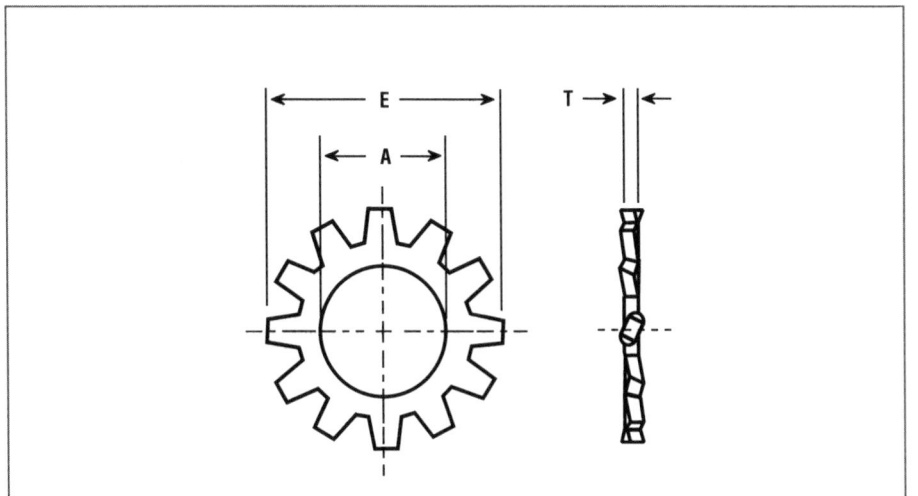

Nom. Washer Size	A Inside Dia.		E Outside Dia.		T Thickness
	Max.	Min.	Max.	Min.	
#4	0.123	0.116	0.255	0.245	0.015
#6	0.150	0.142	0.317	0.306	0.020
#8	0.176	0.168	0.381	0.370	0.020
#10	0.204	0.195	0.406	0.395	0.022
1/4	0.267	0.256	0.506	0.494	0.025
5/16	0.326	0.320	0.601	0.588	0.029
3/8	0.398	0.384	0.695	0.680	0.035
7/16	0.463	0.448	0.760	0.740	0.035
1/2	0.529	0.512	0.898	0.882	0.039
9/16	0.594	0.576	0.985	0.965	0.039
5/8	0.659	0.640	1.070	1.045	0.045
3/4	0.790	0.768	1.260	1.240	0.050
7/8	0.927	0.897	1.140	1.380	0.055
1	1.060	1.025	1.620	1.590	0.062
1 1/4	1.142	1.128	1.827	1.797	0.062

Source: MS35335.

Table 35. Dimensions of Internal Tooth Lock Washers.

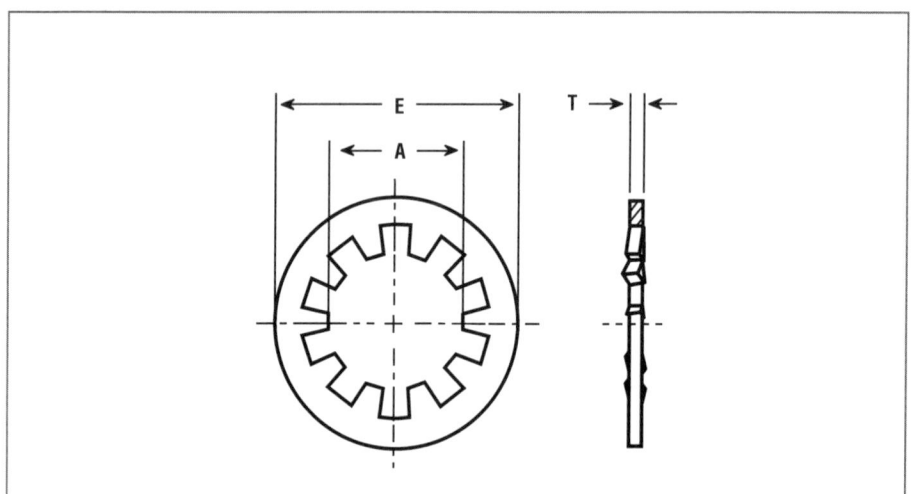

Nom. Washer Size	A Inside Dia.		E Outside Dia.		T Thickness
	Max.	Min.	Max.	Min.	
#2	0.095	0.089	0.185	0.175	0.013
#4	0.123	0.116	0.265	0.255	0.015
#6	0.150	0.142	0.288	0.278	0.017
#8	0.176	0.168	0.336	0.325	0.020
#10	0.204	0.195	0.381	0.370	0.022
1/4	0.267	0.256	0.478	0.466	0.025
5/16	0.332	0.320	0.607	0.594	0.029
3/8	0.398	0.384	0.692	0.678	0.035
7/16	0.463	0.448	0.789	0.774	0.035
7/16	0.480	0.472	0.607	0.593	0.017
1/2	0.529	0.512	0.883	0.867	0.039
9/16	0.572	0.565	0.692	0.678	0.017
5/8	0.659	0.640	1.071	1.053	0.045
3/4	0.795	0.769	1.245	1.220	0.050
3/4	0.785	0.775	1.077	1.047	0.022
7/8	0.918	0.894	1.386	1.364	0.055
1	1.043	1.019	1.637	1.613	0.062
1 1/4	1.280	1.260	1.950	1.921	0.062

Source: MS35333.

Table 36. Dimensions of Hexagon Head Cap Screws (Finished Hex Bolts).

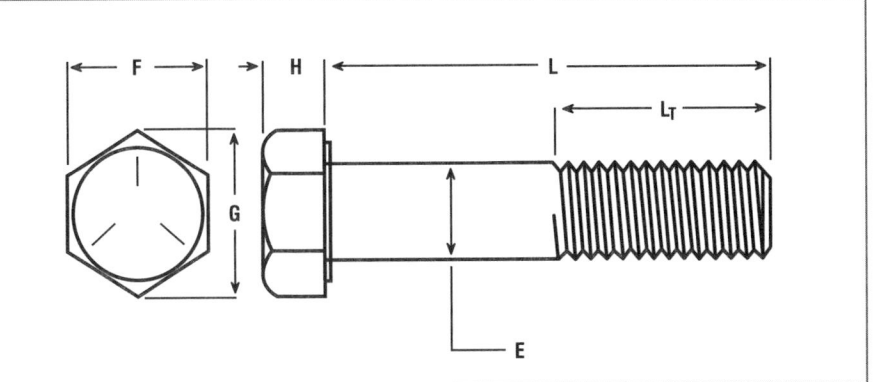

Screw/Bolt Size		E Body Dia.		F Width Across Flats			G Width Across Corners		Wrenching Height
Nom.	Basic	Max.	Min.	Basic	Max.	Min.	Max.	Min.	Min.
1/4	0.2500	0.2500	0.2450	7/16	0.438	0.428	0.505	0.488	0.106
5/16	0.3125	0.3125	0.3065	1/2	0.500	0.489	0.577	0.557	0.140
3/8	0.3750	0.3750	0.3690	9/16	0.562	0.551	0.650	0.628	0.160
7/16	0.4375	0.4375	0.4305	5/8	0.625	0.612	0.722	0.698	0.195
1/2	0.5000	0.5000	0.4930	3/4	0.750	0.736	0.866	0.840	0.215
9/16	0.5625	0.5625	0.5545	13/16	0.812	0.798	0.938	0.910	0.250
5/8	0.6250	0.6250	0.6170	15/16	0.938	0.922	1.083	1.051	0.269
3/4	0.7500	0.7500	0.7410	1-1/8	1.125	1.100	1.299	1.254	0.324
7/8	0.8750	0.8750	0.8660	1-5/16	1.312	1.285	1.516	1.465	0.378
1	1.0000	1.0000	0.9900	1-1/2	1.500	1.469	1.732	1.675	0.416
1-1/8	1.1250	1.1250	1.1140	1-11/16	1.688	1.631	1.949	1.859	0.461
1-1/4	1.2500	1.2500	1.2390	1-7/8	1.875	1.812	2.165	2.066	0.530
1-3/8	1.3750	1.3750	1.3630	2-1/16	2.062	1.994	2.382	2.273	0.569
1-1/2	1.5000	1.5000	1.4880	2-1/4	2.250	2.175	2.598	2.480	0.640
1-3/4	1.7500	1.7500	1.7380	2-5/8	2.625	2.538	3.031	2.893	0.748
2	2.0000	2.0000	1.9880	3	3.000	2.900	3.464	3.306	0.825
2-1/4	2.2500	2.2500	2.2380	3-3/8	3.375	3.262	3.897	3.719	0.933
2-1/2	2.5000	2.5000	2.4880	3-3/4	3.750	3.625	4.330	4.133	1.042
2-3/4	2.7500	2.7500	2.7380	4-1/8	4.125	3.988	4.763	4.546	1.151
3	3.0000	3.0000	2.9880	4-1/2	4.500	4.350	5.196	4.959	1.290

(Continued)

Table 36. *(Continued)* Dimensions of Hexagon Head Cap Screws (Finished Hex Bolts).

Screw/Bolt Size		H Height			L$_T$ Thread Length for Screw Lengths		Transition Thread Lengths for Screw Lengths		Runout of Bearing Surface FIR
					6 in. and shorter	Over 6 in.	6 in. and shorter	Over 6 in.	
Nom.	Basic	Basic	Max.	Min.	Basic	Basic	Max.	Max.	Max.
1/4	0.2500	5/32	0.163	0.150	0.750	1.000	0.400	0.650	0.010
5/16	0.3125	13/64	0.211	0.195	0.875	1.125	0.417	0.667	0.011
3/8	0.3750	15/64	0.243	0.226	1.000	1.250	0.438	0.688	0.012
7/16	0.4375	9/32	0.291	0.272	1.125	1.375	0.464	0.714	0.013
1/2	0.5000	5/16	0.323	0.302	1.250	1.500	0.481	0.731	0.014
9/16	0.5625	23/64	0.371	0.348	1.375	1.625	0.750	0.750	0.015
5/8	0.6250	25/64	0.403	0.378	1.500	1.750	0.773	0.773	0.017
3/4	0.7500	15/32	0.483	0.455	1.750	2.000	0.800	0.800	0.020
7/8	0.8750	35/64	0.563	0.531	2.000	2.250	0.833	0.833	0.023
1	1.0000	39/64	0.627	0.591	2.250	2.500	0.875	0.875	0.026
1-1/8	1.1250	11/16	0.718	0.658	2.500	2.750	0.929	0.929	0.029
1-1/4	1.2500	25/32	0.813	0.749	2.750	3.000	0.929	0.929	0.033
1-3/8	1.3750	27/32	0.878	0.810	3.000	3.250	1.000	1.000	0.036
1-1/2	1.5000	15/16	0.974	0.902	3.250	3.500	1.000	1.000	0.039
1-3/4	1.7500	1-3/32	1.134	1.054	3.750	4.000	1.100	1.100	0.046
2	2.0000	1-7/32	1.263	1.175	4.250	4.500	1.167	1.167	0.052
2-1/4	2.2500	1-3/8	1.423	1.327	4.750	5.000	1.167	1.167	0.059
2-1/2	2.5000	1-17/32	1.583	1.479	5.250	5.500	1.250	1.250	0.065
2-3/4	2.7500	1-11/16	1.744	1.632	5.750	6.000	1.250	1.250	0.072
3	3.0000	1-7/8	1.935	1.815	6.250	6.500	1.250	1.250	0.079

All dimensions in inches. Source: Extracted from ANSI B18.2.1-1972, published by the American Society of Mechanical Engineers.

Table 37. Dimensions of Slotted Fillister Head Cap Screws.

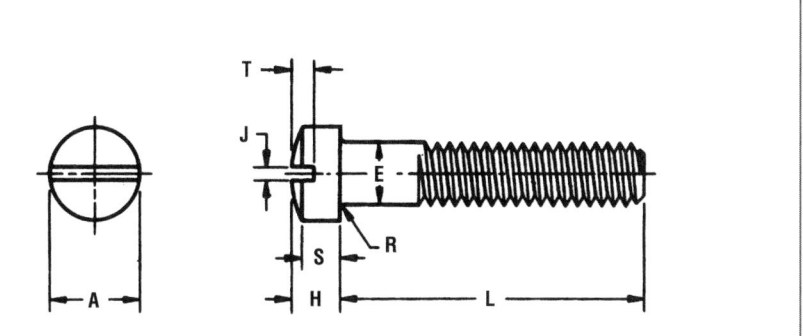

Screw/Bolt Size		E Body Dia.		A Head Dia.		S Head Side Height		H Total Head Height	
Nom.	Basic	Max.	Min.	Max.	Min.	Max.	Min.	Max.	Min.
1/4	0.2500	0.2500	0.2450	0.375	0.363	0.172	0.157	0.216	0.194
5/16	0.3125	0.3125	0.3070	0.437	0.424	0.203	0.186	0.253	0.230
3/8	0.3750	0.3750	0.3690	0.562	0.547	0.250	0.229	0.314	0.284
7/16	0.4375	0.4375	0.4310	0.625	0.608	0.297	0.274	0.368	0.336
1/2	0.5000	0.5000	0.4930	0.750	0.731	0.328	0.301	0.413	0.376
9/16	0.5625	0.5625	0.5550	0.812	0.792	0.375	0.346	0.467	0.427
5/8	0.6250	0.6250	0.6170	0.875	0.853	0.422	0.391	0.521	0.478
3/4	0.7500	0.7500	0.7420	1.000	0.976	0.500	0.466	0.612	0.566
7/8	0.8750	0.8750	0.8660	1.125	1.098	0.594	0.556	0.720	0.668
1	1.0000	1.0000	0.9900	1.312	1.282	0.656	0.612	0.803	0.743

Screw/Bolt Size		J Slot Width		T Slot Depth		R Fillet Radius	
Nom.	Basic	Max.	Min.	Max.	Min.	Max.	Min.
1/4	0.2500	0.075	0.064	0.097	0.077	0.031	0.016
5/16	0.3125	0.084	0.072	0.115	0.090	0.031	0.016
3/8	0.3750	0.094	0.081	0.142	0.112	0.031	0.016
7/16	0.4375	0.094	0.081	0.168	0.133	0.047	0.016
1/2	0.5000	0.106	0.091	0.193	0.153	0.047	0.016
9/16	0.5625	0.118	0.102	0.213	0.168	0.047	0.016
5/8	0.6250	0.133	0.116	0.239	0.189	0.062	0.031
3/4	0.7500	0.149	0.131	0.283	0.223	0.062	0.031
7/8	0.8750	0.167	0.147	0.334	0.264	0.062	0.031
1	1.0000	0.188	0.166	0.371	0.291	0.062	0.031

Note: A slight rounding on the edges at the periphery of the head is permissible if the diameter of the bearing circle is equal to no less than 90% of the specified minimum head diameter. All dimensions in inches. Source: Extracted from ANSI B16.2-1972, published by the American Society of Mechanical Engineers.

Table 38. Dimensions of Square Head and Hexagon Head Lag Screws.

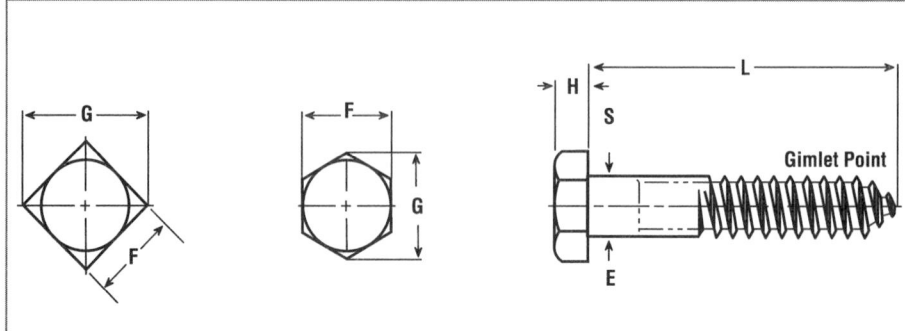

Screw Size		E Body or Shoulder Dia.		F Width Across Flats			S Shoulder Length	Radius of Fillet	
Nom.	Inch	Max.	Min.	Basic	Max.	Min.	Min.	Max.	Min.
No. 10	0.1900	0.199	0.178	9/32	0.281	0.271	0.094	0.003	0.001
1/4*	0.2500	0.260	0.237	3/8	0.375	0.362	0.094	0.003	0.001
1/4 †	0.2500	0.260	0.237	7/16	0.438	0.425	0.094	0.003	0.001
5/16	0.3125	0.324	0.298	1/2	0.500	0.484	0.125	0.003	0.001
3/8	0.3750	0.388	0.360	9/16	0.562	0.544	0.125	0.003	0.001
7/16	0.4375	0.452	0.421	5/8	0.625	0.603	0.156	0.003	0.001
1/2	0.5000	0.515	0.482	3/4	0.750	0.725	0.156	0.003	0.001
5/8	0.6250	0.642	0.605	15/16	0.938	0.906	0.312	0.006	0.002
3/4	0.7500	0.768	0.729	1 1/8	1.125	1.088	0.375	0.006	0.002
7/8	0.8750	0.895	0.852	1 5/16	1.312	1.269	0.375	0.006	0.002
1	1.0000	1.022	0.976	1 1/2	1.500	1.450	0.625	0.009	0.003
1 1/8	1.1250	1.149	1.098	1 11/16	1.688	1.631	0.625	0.009	0.003
1 1/4	1.2500	1.277	1.223	1 7/8	1.875	1.812	0.625	0.009	0.003

Screw Size		H Height of Square Head			G Width Across Square Corners		H Height of Hex Head			G Width Across Hex Corners	
Nom.	Inch	Basic	Max.	Min.	Max.	Min.	Basic	Max.	Min.	Max.	Min.
No. 10	0.1900	1/8	0.140	0.110	0.398	0.372	1/8	0.140	0.110	0.323	0.309
1/4	0.2500	11/64	0.188	0.156	0.530	0.498	11/64	0.188	0.150	0.505	0.484
5/16	0.3125	13/64	0.220	0.186	0.707	0.665	7/32	0.235	0.195	0.577	0.552
3/8	0.3750	1/4	0.268	0.232	0.795	0.747	1/4	0.268	0.226	0.650	0.620
7/16	0.4375	19/64	0.316	0.278	0.884	0.828	19/64	0.316	0.272	0.722	0.687
1/2	0.5000	11/32	0.348	0.308	1.061	0.995	11/32	0.364	0.302	0.866	0.826
5/8	0.6250	27/64	0.444	0.400	0.326	1.244	27/64	0.444	0.378	1.083	1.033
3/4	0.7500	1/2	0.524	0.476	1.591	1.494	1/2	0.524	0.455	1.299	1.240
7/8	0.8750	37/64	0.620	0.568	1.856	1.742	37/64	0.604	0.531	1.516	1.447
1	1.0000	43/64	0.684	0.628	2.121	1.991	43/64	0.700	0.591	1.732	1.653
1 1/8	1.1250	3/4	0.780	0.720	2.386	2.239	3/4	0.780	0.658	1.949	1.859
1 1/4	1.2500	27/32	0.876	0.812	2.652	2.489	27/32	0.876	0.749	2.165	2.066

* Square head lag screw. † Hex head lag screw. All dimensions in inches.
Source: Extracted from ANSI B18.2-1-1972, published by the American Society of Mechanical Engineers.

Table 39. Dimensions of Slotted Binding Head Machine Screws.

Screw Size	A Head Dia.		H Head Height		F Oval Height		J Slot Width		T Slot Width		U¹ Undercut Dia.		X¹ Undercut Depth	
	Max.	Min.	Max.	Min.	Max.	Min.	Max.	Min.	Max.	Min.	Max.	Min.	Max.	Min.
0000	0.046	0.040	0.014	0.009	0.006	0.003	0.008	0.004	0.009	0.005	-	-	-	-
000	0.073	0.067	0.021	0.015	0.008	0.005	0.012	0.006	0.013	0.009	-	-	-	-
00	0.098	0.090	0.028	0.023	0.011	0.007	0.017	0.010	0.018	0.012	-	-	-	-
0	0.126	0.119	0.032	0.026	0.012	0.008	0.023	0.016	0.018	0.009	0.098	0.086	0.007	0.002
1	0.153	0.145	0.041	0.035	0.015	0.011	0.026	0.019	0.024	0.014	0.120	0.105	0.008	0.003
2	0.181	0.171	0.050	0.043	0.018	0.013	0.031	0.023	0.030	0.020	0.141	0.124	0.010	0.005
3	0.208	0.197	0.059	0.052	0.022	0.016	0.035	0.027	0.036	0.025	0.162	0.143	0.011	0.006
4	0.235	0.223	0.068	0.061	0.025	0.018	0.039	0.031	0.042	0.030	0.184	0.161	0.012	0.007
5	0.263	0.249	0.078	0.069	0.029	0.021	0.043	0.035	0.048	0.035	0.205	0.180	0.014	0.009
6	0.290	0.275	0.087	0.078	0.032	0.024	0.048	0.039	0.053	0.040	0.226	0.199	0.015	0.010
8	0.344	0.326	0.105	0.095	0.039	0.029	0.054	0.045	0.065	0.050	0.269	0.236	0.017	0.012

(Continued)

Table 39. *(Continued)* Dimensions of Slotted Binding Head Machine Screws.

Screw Size	A Head Dia.		H Head Height		F Oval Height		J Slot Width		T Slot Width		U[1] Undercut Dia.		X[1] Undercut Depth	
	Max.	Min.	Max.	Min.	Max.	Min.	Max.	Min.	Max.	Min.	Max.	Min.	Max.	Min.
10	0.399	0.378	0.123	0.112	0.045	0.034	0.060	0.050	0.077	0.060	0.312	0.274	0.020	0.015
12	0.454	0.430	0.141	0.130	0.052	0.039	0.067	0.056	0.089	0.070	0.354	0.311	0.023	0.018
1/4	0.525	0.498	0.165	0.152	0.061	0.046	0.075	0.064	0.105	0.084	0.410	0.360	0.026	0.021
5/16	0.656	0.622	0.209	0.194	0.077	0.059	0.084	0.072	0.134	0.108	0.513	0.450	0.032	0.027
3/8	0.788	0.746	0.253	0.235	0.094	0.071	0.094	0.081	0.163	0.132	0.615	0.540	0.039	0.034

Note: [1] Unless otherwise specified, screws will not be undercut. All dimensions in inches. Source: Fastbolt Corp.

Table 40. Dimensions of Regular Hex Head and Hex Washer Head Machine Screws, Tapping Screws, and Self-Drilling Screws.

Screw Size	F Width Across Flats		G Width Across Corners	H Head Height		D Washer Dia.		U Washer Thickness		J Slot Width		T Slot Depth	
	Max.	Min.	Min.	Max.	Min.	Max.	Min.	Max.	Min.	Max.	Min.	Max.	Min.
4	0.187	0.181	0.202	0.060	0.049	0.243	0.225	0.019	0.011	0.039	0.031	0.042	0.025
6	0.250	0.244	0.272	0.093	0.080	0.328	0.302	0.025	0.015	0.048	0.039	0.053	0.033
8	0.250	0.244	0.272	0.110	0.096	0.348	0.322	0.031	0.019	0.054	0.045	0.074	0.052
10	0.312	0.305	0.340	0.120	0.105	0.414	0.384	0.031	0.019	0.060	0.050	0.080	0.057
12	0.312	0.305	0.340	0.155	0.139	0.432	0.398	0.039	0.022	0.067	0.056	0.103	0.077
14	0.375	0.367	0.409	0.190	0.172	0.520	0.480	0.050	0.030	0.075	0.064	0.111	0.083
1/4	0.375	0.367	0.409	0.190	0.172	0.520	0.480	0.050	0.030	0.075	0.064	0.111	0.083
5/16	0.500	0.489	0.545	0.230	0.208	0.676	0.624	0.055	0.035	0.084	0.072	0.134	0.100

All dimensions in inches. Source: Fastbolt Corp.

Table 41. Dimensions of Slotted and Phillips Fillister Head Machine Screws.

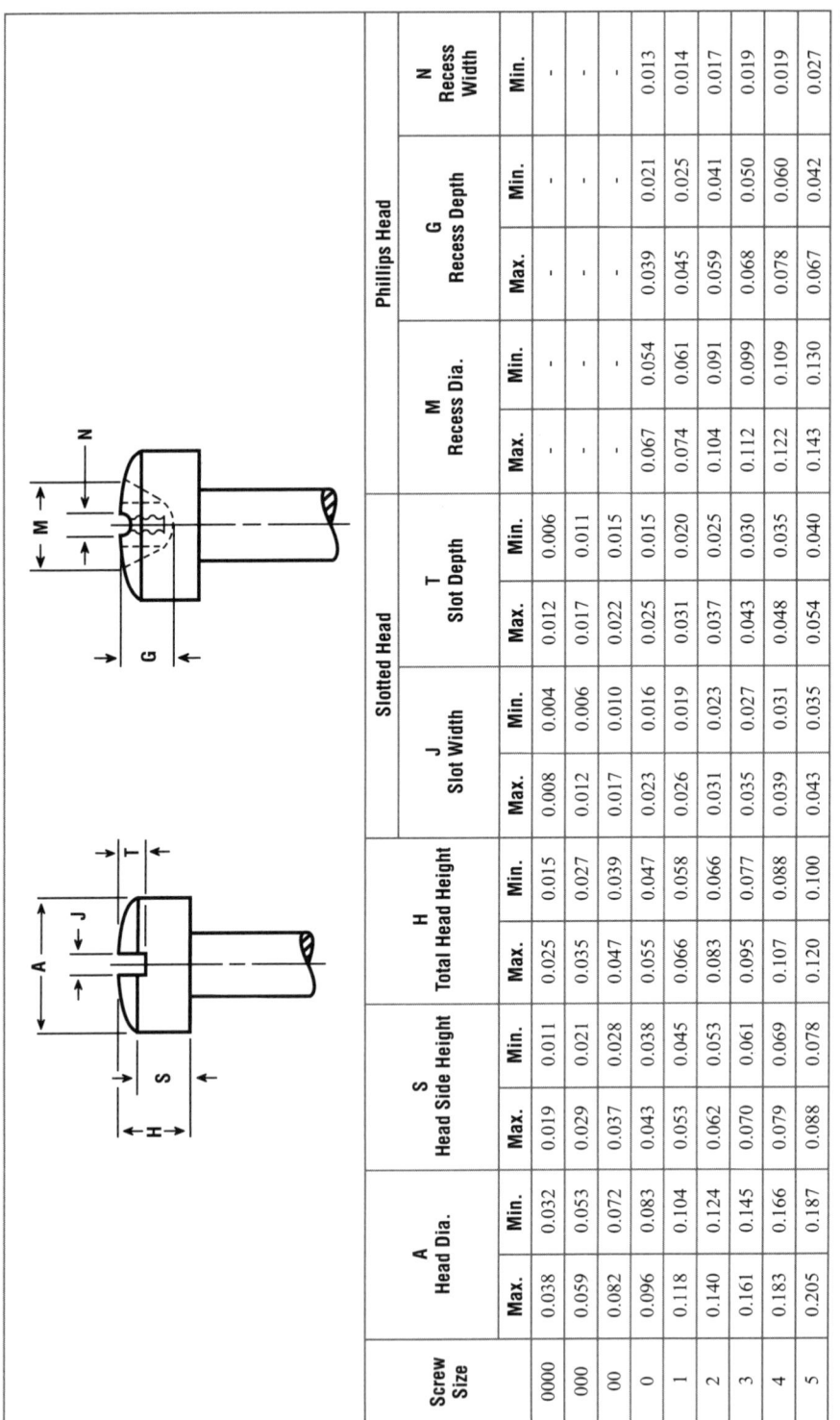

Screw Size	A Head Dia.		S Head Side Height		H Total Head Height		Slotted Head J Slot Width		T Slot Depth		Phillips Head M Recess Dia.		G Recess Depth		N Recess Width
	Max.	Min.	Max.	Min.	Max.	Min.	Max.	Min.	Max.	Min.	Max.	Min.	Max.	Min.	Min.
0000	0.038	0.032	0.019	0.011	0.025	0.015	0.008	0.004	0.012	0.006	-	-	-	-	-
000	0.059	0.053	0.029	0.021	0.035	0.027	0.012	0.006	0.017	0.011	-	-	-	-	-
00	0.082	0.072	0.037	0.028	0.047	0.039	0.017	0.010	0.022	0.015	-	-	-	-	-
0	0.096	0.083	0.043	0.038	0.055	0.047	0.023	0.016	0.025	0.015	0.067	0.054	0.039	0.021	0.013
1	0.118	0.104	0.053	0.045	0.066	0.058	0.026	0.019	0.031	0.020	0.074	0.061	0.045	0.025	0.014
2	0.140	0.124	0.062	0.053	0.083	0.066	0.031	0.023	0.037	0.025	0.104	0.091	0.059	0.041	0.017
3	0.161	0.145	0.070	0.061	0.095	0.077	0.035	0.027	0.043	0.030	0.112	0.099	0.068	0.050	0.019
4	0.183	0.166	0.079	0.069	0.107	0.088	0.039	0.031	0.048	0.035	0.122	0.109	0.078	0.060	0.019
5	0.205	0.187	0.088	0.078	0.120	0.100	0.043	0.035	0.054	0.040	0.143	0.130	0.067	0.042	0.027

(Continued)

Table 41. *(Continued)* Dimensions of Slotted and Phillips Fillister Head Machine Screws.

Screw Size	A Head Dia.		S Head Side Height		H Total Head Height		Slotted Head				Phillips Head					
							J Slot Width		T Slot Depth		M Recess Dia.		G Recess Depth		N Recess Width	
	Max.	Min.	Max.	Min.	Max.	Min.	Max.	Min.	Max.	Min.	Max.	Min.	Max.	Min.	Min.	
6	0.226	0.208	0.096	0.086	0.132	0.111	0.048	0.039	0.060	0.045	0.166	0.153	0.091	0.066	0.028	
8	0.270	0.250	0.113	0.102	0.156	0.133	0.054	0.045	0.071	0.054	0.182	0.169	0.108	0.082	0.030	
10	0.313	0.292	0.130	0.118	0.180	0.156	0.060	0.050	0.083	0.064	0.199	0.186	0.124	0.100	0.031	
12	0.357	0.334	0.148	0.134	0.205	0.178	0.067	0.056	0.094	0.074	0.259	0.246	0.141	0.115	0.034	
1/4	0.414	0.389	0.170	0.155	0.237	0.207	0.075	0.064	0.109	0.087	0.281	0.268	0.161	0.135	0.036	
5/16	0.518	0.490	0.211	0.194	0.295	0.262	0.084	0.072	0.137	0.110	0.322	0.309	0.203	0.177	0.042	
3/8	0.622	0.590	0.253	0.233	0.355	0.315	0.094	0.081	0.164	0.133	0.389	0.376	0.233	0.210	0.065	
7/16	0.625	0.589	0.265	0.242	0.368	0.321	0.094	0.081	0.170	0.135	0.413	0.400	0.259	0.234	0.068	
1/2	0.750	0.710	0.297	0.273	0.412	0.362	0.106	0.091	0.190	0.151	0.435	0.422	0.280	0.255	0.071	
9/16	0.812	0.768	0.336	0.308	0.466	0.410	0.118	0.102	0.214	0.172	0.470	0.442	0.312	0.288	0.076	
5/8	0.875	0.827	0.375	0.345	0.521	0.461	0.133	0.116	0.240	0.193	0.587	0.564	0.343	0.314	0.081	
3/4	1.000	0.945	0.441	0.406	0.612	0.542	0.149	0.131	0.281	0.226	0.633	0.610	0.382	0.355	0.086	

All dimensions in inches. Source: Fastbolt Corp.

Table 42. Dimensions of Slotted and Phillips Flat Countersunk Head Machine Screws, Tapping Screws, and Wood Screws.

Screw Size	A Head Dia.		H Head Height	Slotted Head				Phillips Head					
				J Slot Width		T Slot Depth		M Recess Dia.		Recess Depth		N Recess Width	
	Max.¹	Min.²	Ref.	Max.	Min.	Max.	Min.	Max.	Min.	Max.	Min.		Min.
0000	0.043	0.037	0.011	0.008	0.004	0.007	0.003	-	-	-	-		-
000	0.064	0.058	0.016	0.011	0.007	0.009	0.005	-	-	-	-		-
00	0.092	0.076	0.028	0.017	0.010	0.014	0.009	-	-	-	-		-
0	0.119	0.099	0.035	0.023	0.016	0.015	0.010	0.069	0.056	0.043	0.027		0.014
1	0.146	0.123	0.043	0.026	0.019	0.019	0.012	0.077	0.064	0.051	0.035		0.015
2	0.172	0.147	0.051	0.031	0.023	0.023	0.015	0.102	0.089	0.063	0.047		0.017
3	0.199	0.171	0.059	0.035	0.027	0.027	0.017	0.107	0.094	0.068	0.052		0.018
4	0.225	0.195	0.067	0.039	0.031	0.030	0.020	0.128	0.115	0.089	0.073		0.018
5	0.252	0.220	0.075	0.043	0.035	0.034	0.022	0.154	0.141	0.086	0.063		0.027
6	0.279	0.244	0.083	0.048	0.039	0.038	0.024	0.174	0.161	0.106	0.083		0.029

(Continued)

Table 42. *(Continued)* Dimensions of Slotted and Phillips Flat Countersunk Head Machine Screws, Tapping Screws, and Wood Screws.

Screw Size	A Head Dia.		H Head Height	Slotted Head				Phillips Head				
				J Slot Width		T Slot Depth		M Recess Dia.		Recess Depth		N Recess Width
	Max.[1]	Min.[2]	Ref.	Max.	Min.	Max.	Min.	Max.	Min.	Max.	Min.	Min.
8	0.332	0.292	0.100	0.054	0.045	0.045	0.029	0.189	0.176	0.121	0.098	0.030
10	0.385	0.340	0.116	0.060	0.050	0.053	0.034	0.204	0.191	0.136	0.113	0.032
12	0.438	0.389	0.132	0.067	0.056	0.060	0.039	0.268	0.255	0.156	0.133	0.035
1/4	0.507	0.452	0.153	0.075	0.064	0.070	0.046	0.283	0.270	0.171	0.148	0.036
5/16	0.635	0.568	0.191	0.084	0.072	0.088	0.058	0.365	0.352	0.216	0.194	0.061
3/8	0.762	0.685	0.230	0.094	0.081	0.106	0.070	0.393	0.380	0.245	0.223	0.065
7/16	0.812	0.723	0.223	0.094	0.081	0.103	0.066	0.409	0.396	0.261	0.239	0.068
1/2	0.875	0.775	0.223	0.106	0.091	0.103	0.065	0.424	0.411	0.276	0.254	0.069
9/16	1.000	0.889	0.260	0.118	0.102	0.120	0.077	0.454	0.431	0.300	0.278	0.073
5/8	1.125	1.002	0.298	0.133	0.116	0.137	0.088	0.576	0.553	0.342	0.316	0.079
3/4	1.375	1.230	0.372	0.149	0.131	0.171	0.111	0.640	0.617	0.406	0.380	0.087

Notes: [1] Maximum head diameter is to top of sharp edge. [2] Minimum head diameter is to rounded edge or flat. All dimensions in inches. Source: Fastbolt Corp.

Table 43. Dimensions of Slotted and Phillips Undercut Flat Countersunk Head Machine Screws.

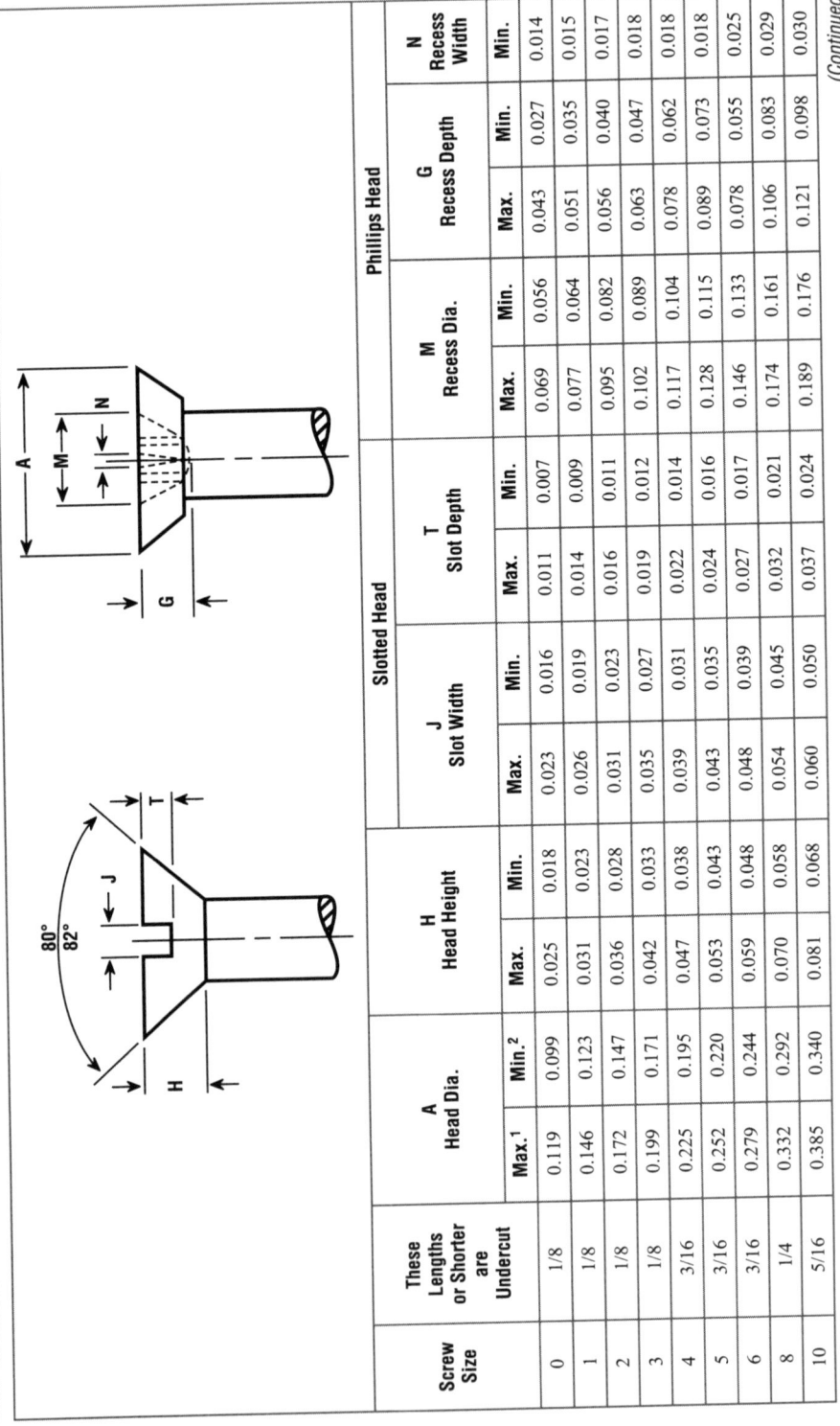

Screw Size	These Lengths or Shorter are Undercut	A Head Dia. Max.[1]	A Head Dia. Min.[2]	H Head Height Max.	H Head Height Min.	Slotted Head J Slot Width Max.	Slotted Head J Slot Width Min.	Slotted Head T Slot Depth Max.	Slotted Head T Slot Depth Min.	Phillips Head M Recess Dia. Max.	Phillips Head M Recess Dia. Min.	Phillips Head G Recess Depth Max.	Phillips Head G Recess Depth Min.	Phillips Head N Recess Width Min.
0	1/8	0.119	0.099	0.025	0.018	0.023	0.016	0.011	0.007	0.069	0.056	0.043	0.027	0.014
1	1/8	0.146	0.123	0.031	0.023	0.026	0.019	0.014	0.009	0.077	0.064	0.051	0.035	0.015
2	1/8	0.172	0.147	0.036	0.028	0.031	0.023	0.016	0.011	0.095	0.082	0.056	0.040	0.017
3	1/8	0.199	0.171	0.042	0.033	0.035	0.027	0.019	0.012	0.102	0.089	0.063	0.047	0.018
4	3/16	0.225	0.195	0.047	0.038	0.039	0.031	0.022	0.014	0.117	0.104	0.078	0.062	0.018
5	3/16	0.252	0.220	0.053	0.043	0.043	0.035	0.024	0.016	0.128	0.115	0.089	0.073	0.018
6	3/16	0.279	0.244	0.059	0.048	0.048	0.039	0.027	0.017	0.146	0.133	0.078	0.055	0.025
8	1/4	0.332	0.292	0.070	0.058	0.054	0.045	0.032	0.021	0.174	0.161	0.106	0.083	0.029
10	5/16	0.385	0.340	0.081	0.068	0.060	0.050	0.037	0.024	0.189	0.176	0.121	0.098	0.030

(Continued)

Table 43. *(Continued)* Dimensions of Slotted and Phillips Undercut Flat Countersunk Head Machine Screws.

Screw Size	These Lengths or Shorter are Undercut	A Head Dia.		H Head Height		Slotted Head				Phillips Head				
						J Slot Width		T Slot Depth		M Recess Dia.		G Recess Depth		N Recess Width
		Max.¹	Min.²	Max.	Min.	Max.	Min.	Max.	Min.	Max.	Min.	Max.	Min.	Min.
12	3/8	0.438	0.389	0.092	0.078	0.067	0.056	0.043	0.028	0.233	0.220	0.121	0.098	0.030
1/4	7/16	0.507	0.452	0.107	0.092	0.075	0.064	0.050	0.032	0.250	0.237	0.136	0.113	0.032
5/16	1/2	0.635	0.568	0.134	0.116	0.084	0.072	0.062	0.041	0.317	0.304	0.168	0.146	0.053
3/8	9/16	0.762	0.685	0.161	0.140	0.094	0.081	0.075	0.049	0.365	0.352	0.216	0.194	0.061
7/16	5/8	0.812	0.723	0.156	0.133	0.094	0.081	0.072	0.045	0.393	0.380	0.245	0.223	0.065
1/2	3/4	0.875	0.775	0.156	0.130	0.106	0.091	0.072	0.046	0.409	0.396	0.261	0.242	0.068

Notes: [1] Maximum head diameter is to top of sharp edge. [2] Minimum head diameter is to rounded edge or flat. All dimensions in inches. Source: Fastbolt Corp.

Table 44. Dimensions of Slotted and Phillips Undercut Oval Head Countersunk Machine Screws, Tapping Screws, and Wood Screws.

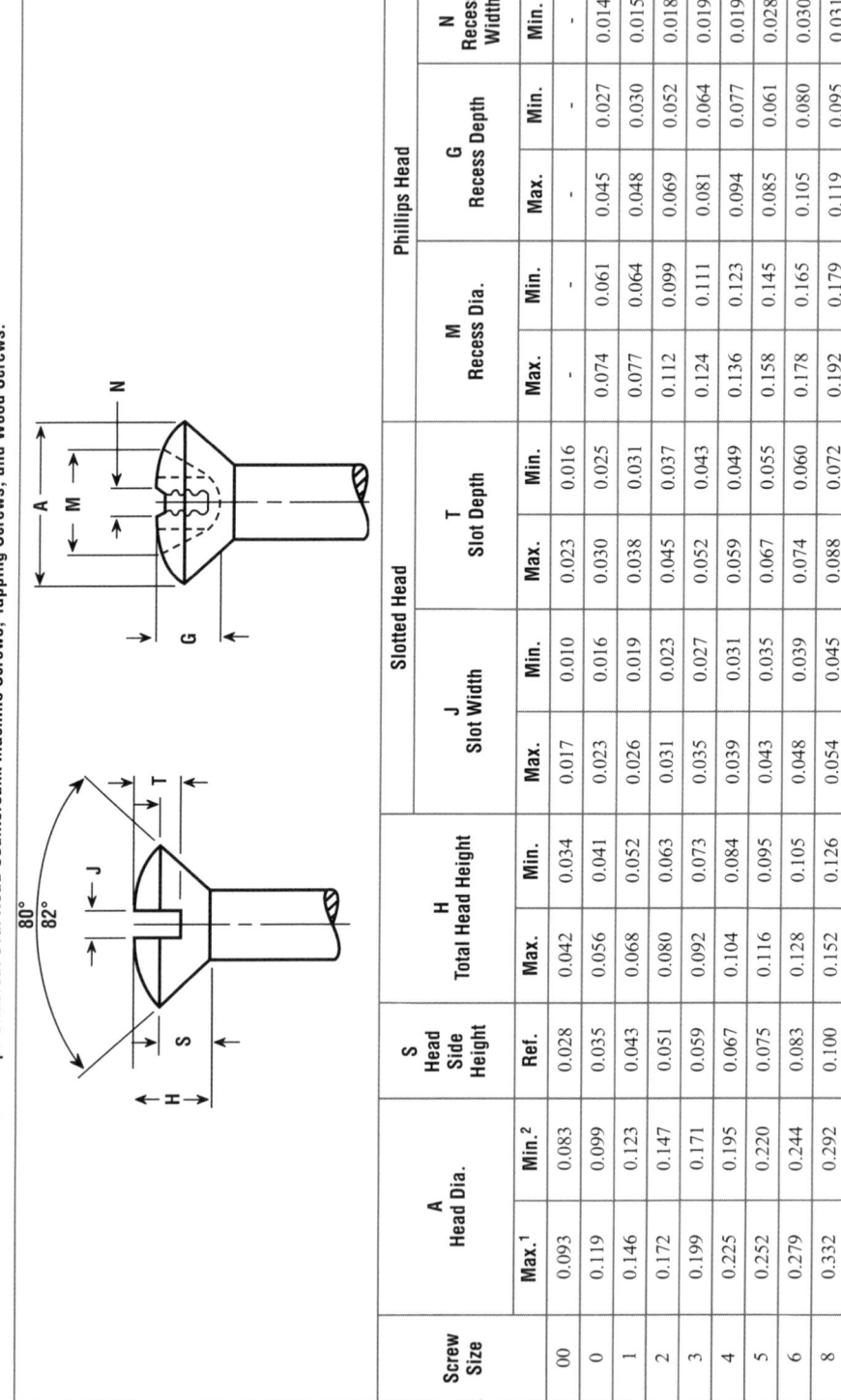

Screw Size	A Head Dia. Max.[1]	A Head Dia. Min.[2]	S Head Side Height Ref.	H Total Head Height Max.	H Total Head Height Min.	Slotted Head J Slot Width Max.	Slotted Head J Slot Width Min.	Slotted Head T Slot Depth Max.	Slotted Head T Slot Depth Min.	Phillips Head M Recess Dia. Max.	Phillips Head M Recess Dia. Min.	Phillips Head G Recess Depth Max.	Phillips Head G Recess Depth Min.	Phillips Head N Recess Width Min.
00	0.093	0.083	0.028	0.042	0.034	0.017	0.010	0.023	0.016	-	-	-	-	-
0	0.119	0.099	0.035	0.056	0.041	0.023	0.016	0.030	0.025	0.074	0.061	0.045	0.027	0.014
1	0.146	0.123	0.043	0.068	0.052	0.026	0.019	0.038	0.031	0.077	0.064	0.048	0.030	0.015
2	0.172	0.147	0.051	0.080	0.063	0.031	0.023	0.045	0.037	0.112	0.099	0.069	0.052	0.018
3	0.199	0.171	0.059	0.092	0.073	0.035	0.027	0.052	0.043	0.124	0.111	0.081	0.064	0.019
4	0.225	0.195	0.067	0.104	0.084	0.039	0.031	0.059	0.049	0.136	0.123	0.094	0.077	0.019
5	0.252	0.220	0.075	0.116	0.095	0.043	0.035	0.067	0.055	0.158	0.145	0.085	0.061	0.028
6	0.279	0.244	0.083	0.128	0.105	0.048	0.039	0.074	0.060	0.178	0.165	0.105	0.080	0.030
8	0.332	0.292	0.100	0.152	0.126	0.054	0.045	0.088	0.072	0.192	0.179	0.119	0.095	0.031

80°
82°

(Continued)

Table 44. (Continued) Dimensions of Slotted and Phillips Undercut Oval Head Countersunk Machine Screws, Tapping Screws, and Wood Screws.

Screw Size	A Head Dia.		S Head Side Height	H Total Head Height		Slotted Head				Phillips Head				
						J Slot Width		T Slot Depth		M Recess Dia.		G Recess Depth		N Recess Width
	Max.¹	Min.²	Ref.	Max.	Min.	Max.	Min.	Max.	Min.	Max.	Min.	Max.	Min.	Min.
10	0.385	0.340	0.116	0.176	0.148	0.060	0.050	0.103	0.084	0.209	0.196	0.137	0.113	0.033
12	0.438	0.389	0.132	0.200	0.169	0.067	0.056	0.117	0.096	0.270	0.257	0.152	0.128	0.038
1/4	0.507	0.452	0.153	0.232	0.197	0.075	0.064	0.136	0.112	0.290	0.277	0.173	0.148	0.040
5/16	0.635	0.568	0.191	0.290	0.249	0.084	0.072	0.171	0.141	0.381	0.368	0.226	0.202	0.064
3/8	0.762	0.685	0.230	0.347	0.300	0.094	0.081	0.206	0.170	0.400	0.387	0.245	0.221	0.066
7/16	0.812	0.723	0.223	0.345	0.295	0.094	0.081	0.210	0.174	0.410	0.397	0.257	0.233	0.068
1/2	0.875	0.775	0.223	0.354	0.299	0.106	0.091	0.216	0.176	0.422	0.409	0.269	0.245	0.070
9/16	1.000	0.889	0.260	0.410	0.350	0.118	0.102	0.250	0.207	-	-	-	-	-
5/8	1.125	1.002	0.298	0.467	0.399	0.133	0.116	0.285	0.235	-	-	-	-	-
3/4	1.375	1.230	0.372	0.578	0.497	0.149	0.131	0.353	0.293	-	-	-	-	-

Notes: ¹ Maximum head diameter is to top of sharp edge. ² Minimum head diameter is to rounded edge or flat. All dimensions in inches. Source: Fastbolt Corp.

Table 45. Dimensions of Slotted and Phillips Pan Head Machine Screws, Tapping Screws, and Self-drilling Screws.

| Screw Size | A Head Dia. | | Slotted Head | | | | | | Phillips Head | | | | | | | N Recess Width |
| | Max. | Min. | H Head Height | | J Slot Width | | T Slot Depth | | H₁ Head Height | | M Recess Dia. | | G Recess Depth | | | |
			Max.	Min.	Max.	Min.	Max.	Min.	Max.	Min.	Max.	Min.	Max.	Min.	Min.	Min.
0000	0.042	0.036	0.016	0.010	0.008	0.004	0.008	0.004	-	-	-	-	-	-	-	-
000	0.066	0.060	0.023	0.017	0.012	0.008	0.012	0.008	-	-	-	-	-	-	-	-
00	0.090	0.082	0.025	0.032	0.017	0.010	0.016	0.010	-	-	-	-	-	-	-	-
0	0.116	0.104	0.039	0.031	0.023	0.016	0.022	0.014	0.044	0.036	0.067	0.054	0.039	0.021	0.013
1	0.142	0.130	0.046	0.038	0.026	0.019	0.027	0.018	0.053	0.044	0.074	0.061	0.045	0.025	0.014
2	0.167	0.155	0.053	0.045	0.031	0.023	0.031	0.022	0.062	0.053	0.104	0.091	0.059	0.041	0.017
3	0.193	0.180	0.060	0.051	0.035	0.027	0.036	0.026	0.071	0.062	0.112	0.099	0.068	0.050	0.019
4	0.219	0.205	0.068	0.058	0.039	0.031	0.040	0.030	0.080	0.070	0.122	0.109	0.078	0.060	0.019
5	0.245	0.231	0.075	0.065	0.043	0.035	0.045	0.034	0.089	0.079	0.158	0.145	0.083	0.057	0.028

(Continued)

Table 45. *(Continued)* Dimensions of Slotted and Phillips Pan Head Machine Screws, Tapping Screws, and Self-drilling Screws.

Screw Size	A Head Dia.		Slotted Head						Phillips Head						
			H Head Height		J Slot Width		T Slot Depth		H_1 Head Height		M Recess Dia.		G Recess Depth		N Recess Width
	Max.	Min.	Max.	Min.	Max.	Min.	Max.	Min.	Max.	Min.	Max.	Min.	Max.	Min.	Min.
6	0.270	0.256	0.082	0.072	0.048	0.039	0.050	0.037	0.097	0.087	0.166	0.153	0.091	0.066	0.028
8	0.322	0.306	0.096	0.085	0.054	0.045	0.058	0.045	0.115	0.105	0.182	0.169	0.108	0.082	0.030
10	0.373	0.357	0.110	0.099	0.060	0.050	0.068	0.053	0.133	0.122	0.199	0.186	0.124	0.100	0.031
12	0.425	0.407	0.125	0.112	0.067	0.056	0.077	0.061	0.151	0.139	0.259	0.246	0.141	0.115	0.034
1/4	0.492	0.473	0.144	0.130	0.075	0.064	0.087	0.070	0.175	0.162	0.281	0.268	0.161	0.135	0.036
5/16	0.615	0.594	0.178	0.162	0.084	0.072	0.106	0.085	0.218	0.203	0.350	0.337	0.193	0.169	0.059
3/8	0.740	0.716	0.212	0.195	0.094	0.081	0.124	0.100	0.261	0.244	0.389	0.376	0.233	0.210	0.065
7/16	0.863	0.837	0.247	0.228	0.094	0.081	0.142	0.116	0.305	0.284	0.413	0.400	0.259	0.234	0.068
1/2	0.987	0.958	0.281	0.260	0.106	0.091	0.161	0.131	0.348	0.325	0.435	0.422	0.280	0.255	0.071
9/16	1.041	1.000	0.315	0.293	0.118	0.102	0.179	0.146	0.391	0.366	0.470	0.447	0.312	0.288	0.076
5/8	1.172	1.125	0.350	0.325	0.133	0.116	0.197	0.162	0.434	0.406	0.587	0.564	0.343	0.314	0.081
3/4	1.435	1.375	0.419	0.390	0.149	0.131	0.234	0.192	0.521	0.488	0.633	0.610	0.382	0.355	0.086

All dimensions in inches. Source: Fastbolt Corp.

Table 46. Dimensions of Slotted and Phillips Round Head Machine Screws and Wood Screws.

Screw Size	A Head Diameter		H Head Height		Slotted Head				Phillips Head				N Recess Width
					J Slot Width		T Slot Depth		M Recess Diameter		G Recess Depth		
	Max.	Min.	Max.	Min.	Max.	Min.	Max.	Min.	Max.	Min.	Max.	Min.	Min.
0000	0.041	0.035	0.022	0.016	0.008	0.004	0.017	0.013	-	-	-	-	-
000	0.062	0.056	0.031	0.025	0.012	0.008	0.018	0.012	-	-	-	-	-
00	0.089	0.080	0.045	0.036	0.017	0.010	0.026	0.018	-	-	-	-	-
0	0.113	0.099	0.053	0.043	0.023	0.016	0.039	0.029	0.073	0.060	0.042	0.022	0.014
1	0.138	0.122	0.061	0.051	0.026	0.019	0.044	0.033	0.082	0.069	0.052	0.033	0.015
2	0.162	0.146	0.069	0.059	0.031	0.023	0.048	0.037	0.100	0.087	0.053	0.034	0.017
3	0.187	0.169	0.078	0.067	0.035	0.027	0.053	0.040	0.109	0.096	0.062	0.042	0.018
4	0.211	0.193	0.086	0.075	0.039	0.031	0.058	0.044	0.118	0.105	0.072	0.053	0.019
5	0.236	0.217	0.095	0.083	0.043	0.035	0.063	0.047	0.154	0.141	0.074	0.046	0.027

(Continued)

Table 46. *(Continued)* Dimensions of Slotted and Phillips Round Head Machine Screws and Wood Screws.

Screw Size	A Head Diameter		H Head Height		Slotted Head				Phillips Head				
					J Slot Width		T Slot Depth		M Recess Diameter		G Recess Depth		N Recess Width
	Max.	Min.	Max.	Min.	Max.	Min.	Max.	Min.	Max.	Min.	Max.	Min.	Min.
6	0.260	0.240	0.103	0.091	0.048	0.039	0.068	0.051	0.162	0.149	0.084	0.156	0.027
8	0.309	0.287	0.120	0.107	0.054	0.045	0.077	0.058	0.178	0.165	0.101	0.075	0.030
10	0.359	0.334	0.137	0.123	0.060	0.050	0.087	0.065	0.195	0.182	0.119	0.093	0.031
12	0.408	0.382	0.153	0.139	0.067	0.056	0.096	0.073	0.249	0.236	0.125	0.099	0.032
1/4	0.472	0.443	0.175	0.160	0.075	0.064	0.109	0.082	0.268	0.255	0.147	0.121	0.034
5/16	0.590	0.557	0.216	0.198	0.084	0.072	0.132	0.099	0.308	0.295	0.187	0.161	0.040
3/8	0.708	0.670	0.256	0.237	0.094	0.081	0.155	0.117	0.387	0.374	0.228	0.202	0.064
7/16	0.750	0.707	0.328	0.307	0.094	0.081	0.196	0.148	0.402	0.389	0.241	0.216	0.066
1/2	0.813	0.766	0.355	0.332	0.106	0.091	0.211	0.159	0.416	0.403	0.256	0.231	0.068
9/16	0.938	0.887	0.410	0.385	0.118	0.102	0.242	0.183	0.459	0.436	0.292	0.265	0.075
5/8	1.000	0.944	0.438	0.411	0.133	0.116	0.258	0.195	0.554	0.531	0.318	0.277	0.077
3/4	1.250	1.185	0.547	0.516	0.149	0.131	0.320	0.242	0.654	0.631	0.418	0.379	0.088

All dimensions in inches. Source: Fastbolt Corp.

Table 47. Dimensions of Slotted and Phillips Truss Head Machine Screws and Tapping Screws.

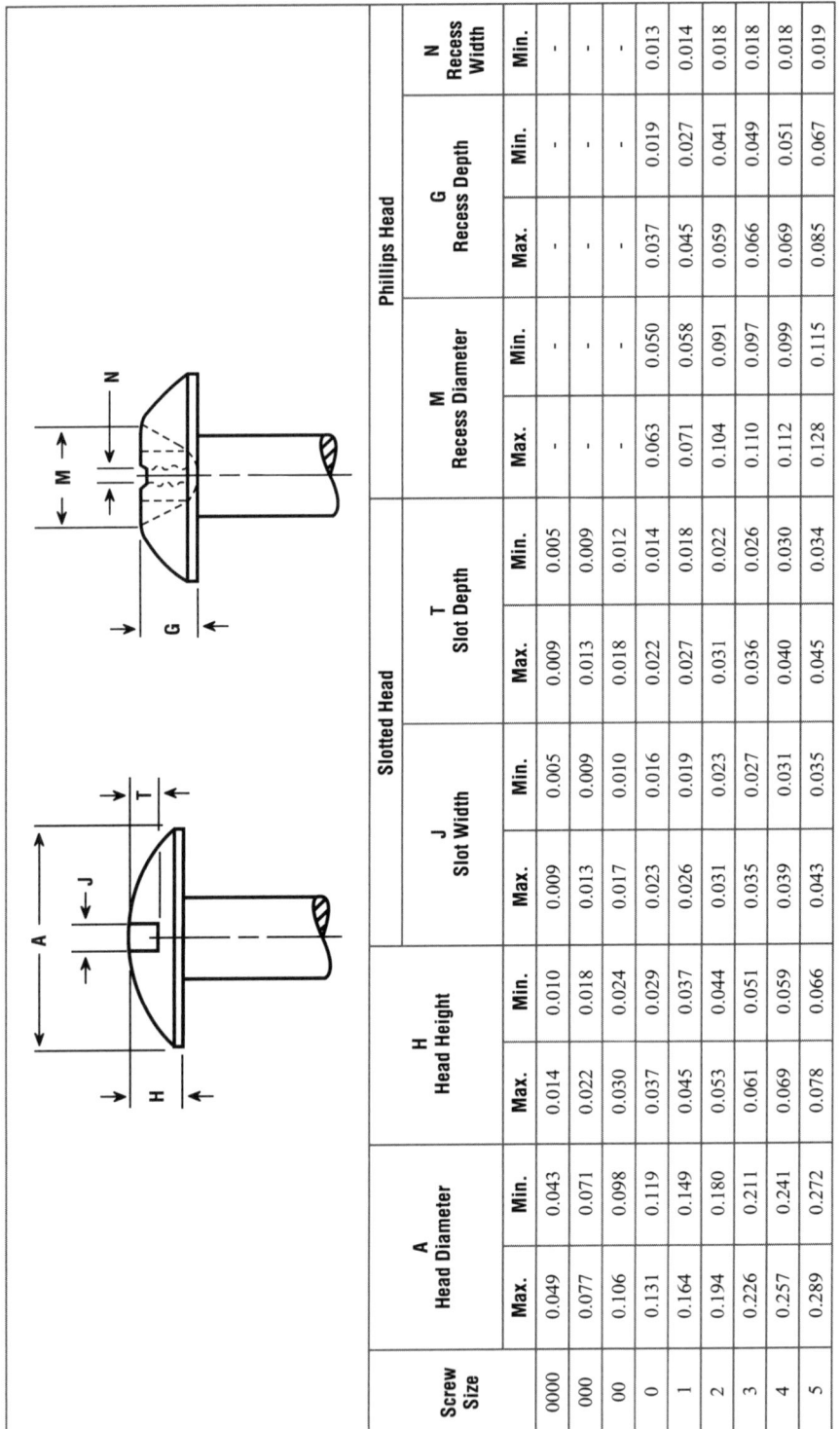

Screw Size	A Head Diameter Max.	A Head Diameter Min.	H Head Height Max.	H Head Height Min.	Slotted Head J Slot Width Max.	Slotted Head J Slot Width Min.	Slotted Head T Slot Depth Max.	Slotted Head T Slot Depth Min.	Phillips Head M Recess Diameter Max.	Phillips Head M Recess Diameter Min.	Phillips Head G Recess Depth Max.	Phillips Head G Recess Depth Min.	Phillips Head N Recess Width Min.
0000	0.049	0.043	0.014	0.010	0.009	0.005	0.009	0.005	-	-	-	-	-
000	0.077	0.071	0.022	0.018	0.013	0.009	0.013	0.009	-	-	-	-	-
00	0.106	0.098	0.030	0.024	0.017	0.010	0.018	0.012	-	-	-	-	-
0	0.131	0.119	0.037	0.029	0.023	0.016	0.022	0.014	0.063	0.050	0.037	0.019	0.013
1	0.164	0.149	0.045	0.037	0.026	0.019	0.027	0.018	0.071	0.058	0.045	0.027	0.014
2	0.194	0.180	0.053	0.044	0.031	0.023	0.031	0.022	0.104	0.091	0.059	0.041	0.018
3	0.226	0.211	0.061	0.051	0.035	0.027	0.036	0.026	0.110	0.097	0.066	0.049	0.018
4	0.257	0.241	0.069	0.059	0.039	0.031	0.040	0.030	0.112	0.099	0.069	0.051	0.018
5	0.289	0.272	0.078	0.066	0.043	0.035	0.045	0.034	0.128	0.115	0.085	0.067	0.019

(Continued)

Table 47. (Continued) Dimensions of Slotted and Phillips Truss Head Machine Screws and Tapping Screws.

Screw Size	A Head Diameter		Slotted Head						Phillips Head				
			H Head Height		J Slot Width		T Slot Depth		M Recess Diameter		G Recess Depth		N Recess Width
	Max.	Min.	Max.	Min.	Max.	Min.	Max.	Min.	Max.	Min.	Max.	Min.	Min.
6	0.321	0.303	0.086	0.074	0.048	0.039	0.050	0.037	0.158	0.145	0.084	0.059	0.027
8	0.384	0.364	0.102	0.088	0.054	0.045	0.058	0.045	0.173	0.160	0.099	0.074	0.029
10	0.448	0.425	0.118	0.103	0.060	0.050	0.068	0.053	0.188	0.175	0.115	0.090	0.030
12	0.511	0.487	0.134	0.118	0.067	0.056	0.077	0.061	0.248	0.235	0.128	0.103	0.032
1/4	0.573	0.546	0.150	0.133	0.075	0.064	0.087	0.070	0.263	0.250	0.143	0.118	0.033
5/16	0.698	0.666	0.183	0.162	0.084	0.072	0.106	0.085	0.352	0.339	0.193	0.168	0.059
3/8	0.823	0.787	0.215	0.191	0.094	0.081	0.124	0.100	0.383	0.370	0.226	0.202	0.063
7/16	0.948	0.907	0.248	0.221	0.094	0.081	0.142	0.116	0.414	0.401	0.257	0.232	0.068
1/2	1.073	1.028	0.280	0.250	0.106	0.091	0.161	0.131	0.444	0.431	0.288	0.263	0.072
9/16	1.198	1.149	0.312	0.279	0.118	0.102	0.179	0.146	0.451	0.428	0.302	0.278	0.074
5/8	1.323	1.269	0.345	0.309	0.133	0.116	0.196	0.162	0.559	0.536	0.322	0.289	0.077
3/4	1.573	1.511	0.410	0.368	0.149	0.131	0.234	0.182	0.620	0.597	0.384	0.352	0.085

All dimensions in inches. Source: Fastbolt Corp.

Table 48. Dimensions of Binding Head Instrument (Miniature) Screws. [1]

Screw Size	Threads Per Inch	Major Diameter Max.	Head Dimensions										
			A Head Diameter		C Chamfer	H Head Height		J Slot Width		T Slot Depth[2]		R Radius
			Max.	Min.	Max.	Max.	Min.	Max.	Min.	Max.	Min.	Max.
.40 UNM	254	0.0157	0.041	0.039	0.002	0.010	0.008	0.006	0.004	0.005	0.003	0.004
.50 UNM	203	0.0197	0.051	0.049	0.003	0.012	0.010	0.008	0.005	0.006	0.004	0.004
.60 UNM	169	0.0236	0.062	0.058	0.004	0.016	0.013	0.010	0.007	0.008	0.006	0.006
.80 UNM	127	0.0315	0.082	0.078	0.005	0.020	0.017	0.012	0.008	0.010	0.007	0.008
1.00 UNM	102	0.0394	0.103	0.097	0.006	0.025	0.021	0.016	0.012	0.012	0.009	0.010
1.20 UNM	102	0.0472	0.124	0.116	0.008	0.032	0.028	0.020	0.015	0.016	0.012	0.012

Notes: [1] For screw lengths four times the maximum major diameter or less, thread length (Lt) shall extend to within two threads of the head bearing surface. Screws of greater length shall have complete threads for a minimum of four maximum major diameters. [2] Slot depth measured from bearing surface. All dimensions except screw size are in inches. Source: MS21271.

Table 49. Dimensions of Fillister Head Instrument (Miniature) Screws. [1]

Screw Size	Threads Per Inch	Major Diameter	Head Dimensions										
			A Head Diameter		C Chamfer	H Head Height		J Slot Width		T Slot Depth[2]		R Radius
		Max.	Max.	Min.	Max.	Max.	Min.	Max.	Min.	Max.	Min.	Max.
.30 UNM	318	0.0118	0.021	0.019	0.002	0.012	0.010	0.004	0.003	0.006	0.004	0.002
.40 UNM	254	0.0157	0.025	0.023	0.002	0.016	0.013	0.005	0.003	0.008	0.006	0.002
.50 UNM	203	0.0197	0.033	0.031	0.003	0.020	0.017	0.006	0.004	0.010	0.007	0.002
.60 UNM	169	0.0236	0.041	0.039	0.004	0.025	0.021	0.008	0.005	0.012	0.008	0.003
.80 UNM	127	0.0315	0.051	0.049	0.005	0.032	0.020	0.010	0.007	0.016	0.012	0.004
1.00 UNM	102	0.0394	0.062	0.058	0.006	0.040	0.035	0.012	0.008	0.020	0.016	0.005
1.20 UNM	102	0.0472	0.082	0.078	0.008	0.050	0.045	0.016	0.012	0.025	0.020	0.006

Notes: [1] For screw lengths four times the maximum major diameter or less, thread length (Lt) shall extend to within two threads of the head bearing surface. Screws of greater length shall have complete threads for a minimum of four maximum major diameters. [2] Slot depth measured from bearing surface. All dimensions except screw size are in inches. Source: MS21268.

Table 50. Dimensions of 100° Flat Head Instrument (Miniature) Screws. [1]

Screw Size	Threads Per Inch	Major Diameter Max.	A Head Diameter Max.	A Head Diameter Min.	Av Dia. at Full Cone[2]	H Head Height Max.	H Head Height Min.	J Slot Width Max.	J Slot Width Min.	T Slot Depth Max.	T Slot Depth Min.	R Radius Max.
.30 UNM	318	0.0118	0.023	0.021	0.0285	0.007	0.005	0.004	0.003	0.004	0.002	0.005
.40 UNM	254	0.0157	0.029	0.027	0.0348	0.008	0.006	0.005	0.003	0.005	0.003	0.006
.50 UNM	203	0.0197	0.037	0.035	0.0459	0.011	0.008	0.006	0.004	0.006	0.004	0.008
.60 UNM	169	0.0236	0.045	0.043	0.0546	0.013	0.010	0.008	0.005	0.008	0.005	0.010
.80 UNM	127	0.0315	0.056	0.054	0.0696	0.016	0.012	0.010	0.007	0.010	0.006	0.012
1.00 UNM	102	0.0394	0.072	0.068	0.0847	0.019	0.015	0.012	0.008	0.012	0.008	0.016
1.20 UNM	102	0.0472	0.092	0.088	0.1068	0.025	0.020	0.016	0.012	0.016	0.010	0.020

Notes: [1] For screw lengths four times the maximum major diameter or less, thread length (Lt) shall extend to within two threads of the head bearing surface. Screws of greater length shall have complete threads for a minimum of four maximum major diameters. [2] "Av" derived from D (Max), H (Max), and mean angle. All dimensions except screw size are in inches. Source: MS21270.

Table 51. Dimensions of Square Head Set Screws.

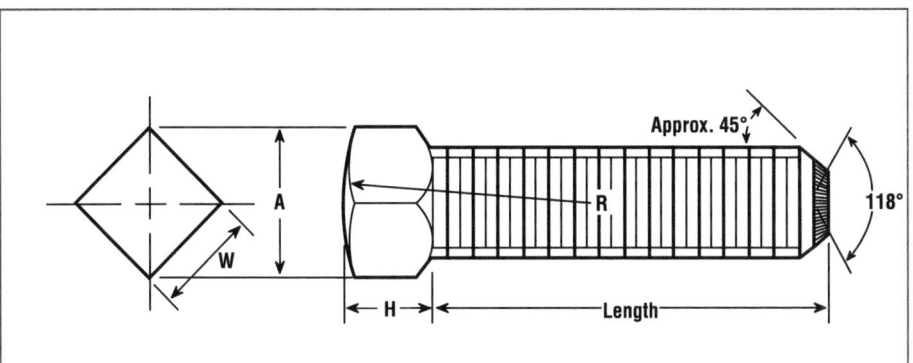

Screw Size		TPI Class 2A	W Width Across Flats		A Width Across Corners		H Height of Head		R Head Radius
Nom.	Basic		Max.	Min.	Max.	Min.	Max.	Min.	Nom.
No. 10	0.1900	24	0.188	0.180	0.265	0.247	0.148	0.134	31/64
1/4	0.2500	20	0.250	0.241	0.354	0.331	0.196	0.178	5/8
5/16	0.3120	18	0.312	0.302	0.442	0.415	0.245	0.224	25/32
3/8	0.3750	16	0.375	0.362	0.530	0.497	0.293	0.270	15/16
7/16	0.4370	14	0.437	0.423	0.619	0.581	0.341	0.315	1-3/32
1/2	0.5000	13	0.500	0.484	0.707	0.665	0.389	0.361	1-1/4
9/16	0.5620	12	0.562	0.545	0.795	0.748	0.437	0.407	1-13/32
5/8	0.6520	11	0.625	0.606	0.884	0.833	0.485	0.452	1-9/16
3/4	0.7500	10	0.750	0.729	1.060	1.001	0.582	0.544	1-7/8
7/8	0.8750	9	0.875	0.852	1.237	1.170	0.678	0.635	2-3/16
1	1.000	8	1.000	0.974	1.414	1.337	0.774	0.726	2-1/2
1-1/8	1.125	7	1.125	1.096	1.591	1.505	0.870	0.817	2-13/16
1-1/4	1.250	7	1.250	1.219	1.768	1.674	0.966	0.908	3-1/8
1-3/8	1.375	6	1.375	1.342	1.945	1.843	1.063	1.000	3-7/16
1-1/2	1.500	6	1.500	1.464	2.121	2.010	1.159	1.091	3-3/4

Length Tolerances for Square Head Set Screws			
Size	Up to 1" incl.	Over 1" to 2" incl.	2" and over
No. 10 thru 5/8"	- 0.03	- 0.06	- 0.09
3/4" and over	- 0.06	- 0.12	- 0.18

All dimensions in inches. Source: Unbrako/SPS Technologies, Inc.

Table 52. Point Dimensions for Square Head Set Screws.

Screw Size Nom.	Screw Size Basic	C Point Dia. Cup and Flat Max.	C Point Dia. Cup and Flat Min.	P Point Dia. Dog and Half Dog Max.	P Point Dia. Dog and Half Dog Min.	Q Point Length Dog Max.	Q Point Length Dog Min.	Q Point Length Half Dog Max.	Q Point Length Half Dog Min.	R Radius Oval Point[1]	Y Cone Point Angle[2]
No. 10	0.1900	0.102	0.088	0.127	0.120	0.095	0.085	0.050	0.040	0.142	1/4
1/4	0.2500	0.132	0.118	0.156	0.149	0.130	0.120	0.068	0.058	0.188	5/16
5/16	0.3125	0.172	0.156	0.203	0.195	0.161	0.151	0.083	0.073	0.234	3/8
3/8	0.3750	0.212	0.194	0.250	0.241	0.193	0.183	0.099	0.089	0.281	7/16
7/16	0.4375	0.252	0.232	0.297	0.287	0.224	0.214	0.114	0.104	0.328	1/2
1/2	0.5000	0.291	0.270	0.344	0.334	0.255	0.245	0.130	0.120	0.375	9/16
9/16	0.5625	0.332	0.309	0.391	0.379	0.287	0.275	0.146	0.134	0.422	5/8

Plain Cup — Approx. 45° — C — 118°

Flat — Approx. 45° — C

Dog and Half Dog — P — Q

Cone — 90°

Oval — R

(Continued)

Table 52. *(Continued)* **Point Dimensions for Square Head Set Screws.**

Screw Size		C Point Dia. Cup and Flat		P Point Dia. Dog and Half Dog		Q Point Length Dog		Q Point Length Half Dog		R Radius Oval Point[1]	Y Cone Point Angle[2]
Nom.	Basic	Max.	Min.	Max.	Min.	Max.	Min.	Max.	Min.		
5/8	0.6250	0.371	0.347	0.469	0.456	0.321	0.305	0.164	0.148	0.469	3/4
3/4	0.7500	0.450	0.425	0.562	0.549	0.383	0.367	0.196	0.180	0.562	7/8
7/8	0.8750	0.530	0.502	0.656	0.642	0.446	0.430	0.227	0.211	0.656	1
1	1.0000	0.609	0.579	0.750	0.734	0.510	0.490	0.260	0.240	0.750	1 1/8
1 1/8	1.1250	0.689	0.655	0.844	0.826	0.572	0.552	0.291	0.271	0.844	1 1/4
1 1/4	1.2500	0.767	0.733	0.938	0.920	0.635	0.615	0.323	0.303	0.938	1 1/2
1 3/8	1.3750	0.848	0.808	1.031	1.011	0.698	0.678	0.354	0.334	1.031	1 5/8
1 1/2	1.5000	0.926	0.886	1.125	1.105	0.760	0.740	0.385	0.365	1.125	1 3/4

Notes: Point angle "X" for flat and cup points is 45° +5°/-0° for screws of nominal lengths equal to or longer than those listed in column Y, and 30° minimum for screws of shorter nominal lengths. The extent of rounding or flat at the apex of the cone point is not to exceed an amount equivalent to 10% of the basic screw diameter. [1] Radius of oval point value in column R is +0.031 / -0.000. [2] Cone point angle is 90° ± 2° for the nominal lengths shown in column Y, and for longer lengths. Cone point angle is 118° ± 2° for lengths shorter than those shown in column Y. All dimensions in inches. Source: Extracted from ANSI B18.6.2-1972, published by the American Society of Mechanical Engineers.

Table 53. Dimensions of Type A, Regular Thumb Screws.

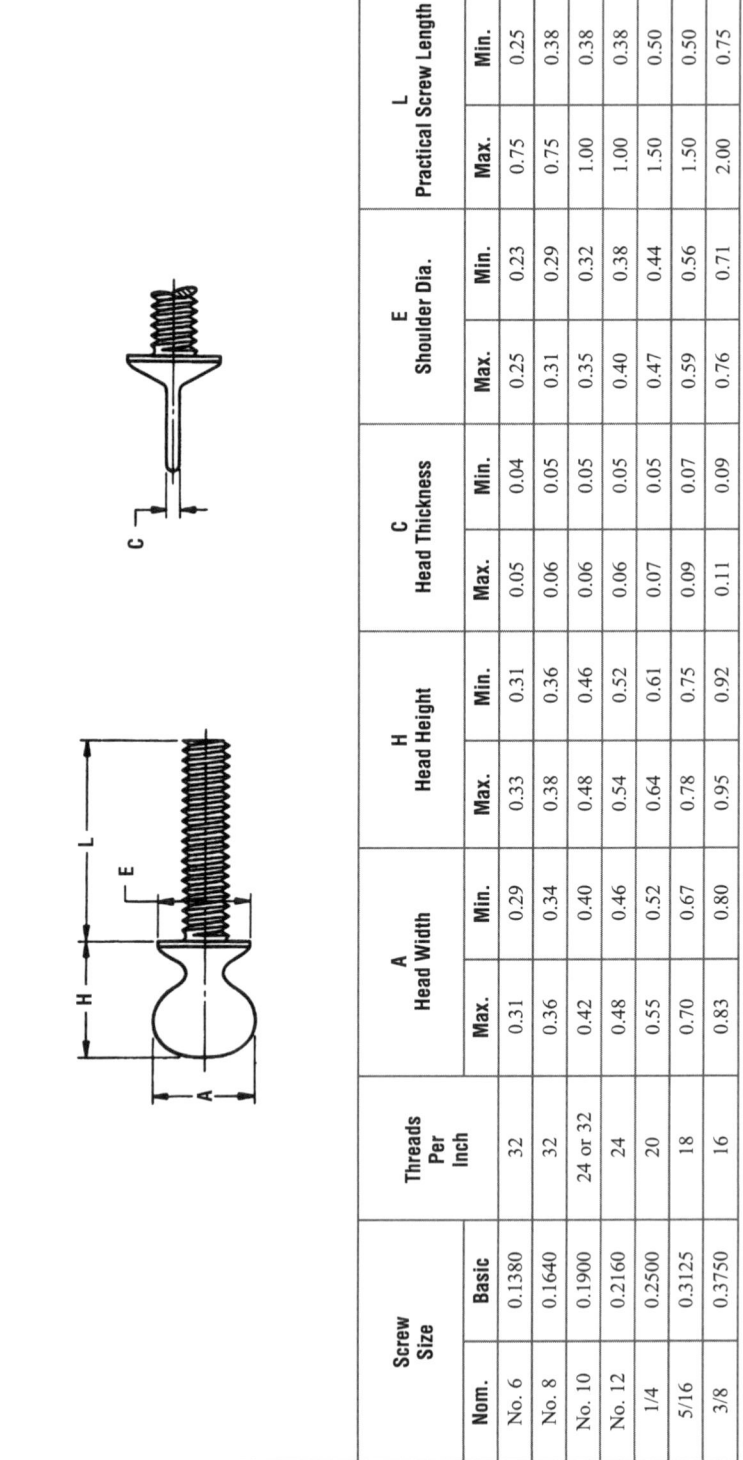

Screw Size		Threads Per Inch	A Head Width		H Head Height		C Head Thickness		E Shoulder Dia.		L Practical Screw Length	
Nom.	Basic		Max.	Min.	Max.	Min.	Max.	Min.	Max.	Min.	Max.	Min.
No. 6	0.1380	32	0.31	0.29	0.33	0.31	0.05	0.04	0.25	0.23	0.75	0.25
No. 8	0.1640	32	0.36	0.34	0.38	0.36	0.06	0.05	0.31	0.29	0.75	0.38
No. 10	0.1900	24 or 32	0.42	0.40	0.48	0.46	0.06	0.05	0.35	0.32	1.00	0.38
No. 12	0.2160	24	0.48	0.46	0.54	0.52	0.06	0.05	0.40	0.38	1.00	0.38
1/4	0.2500	20	0.55	0.52	0.64	0.61	0.07	0.05	0.47	0.44	1.50	0.50
5/16	0.3125	18	0.70	0.67	0.78	0.75	0.09	0.07	0.59	0.56	1.50	0.50
3/8	0.3750	16	0.83	0.80	0.95	0.92	0.11	0.09	0.76	0.71	2.00	0.75

All dimensions in inches. Source: Extracted from ANSI B18.17-1968, published by the American Society of Mechanical Engineers.

Table 54. Dimensions of Type A, Heavy Thumb Screws.

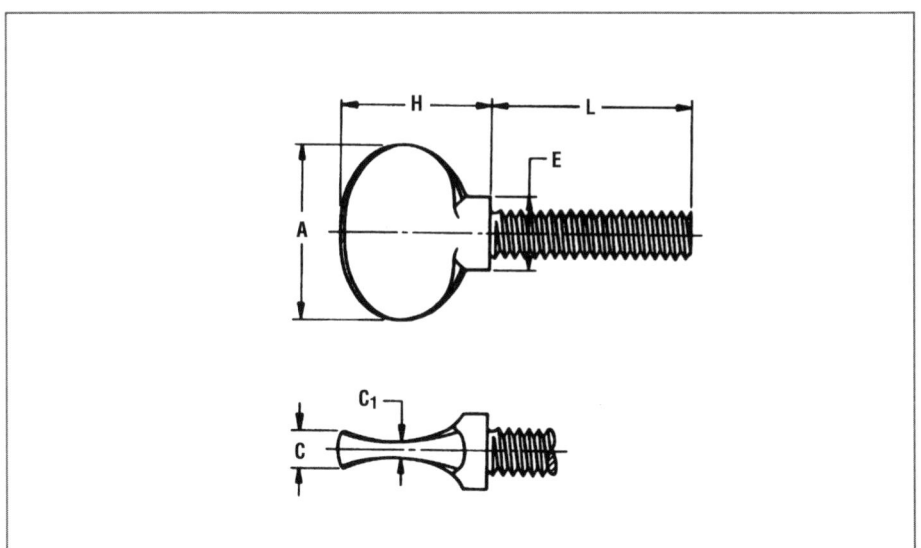

Screw Size		Threads per Inch	A Head Width		H Head Height		C Head Thickness	
Nom.	Basic		Max.	Min.	Max.	Min.	Max.	Min.
No. 10	0.1900	24	0.89	0.83	0.84	0.72	0.18	0.16
1/4	0.2500	20	1.05	0.99	0.94	0.81	0.24	0.22
5/16	0.3125	18	1.21	1.15	1.00	0.88	0.27	0.25
3/8	0.3750	16	1.41	1.34	1.16	1.03	0.30	0.28
7/16	0.4375	14	1.59	1.53	1.22	1.09	0.36	0.34
1/2	0.5000	13	1.81	1.72	1.28	1.16	0.40	0.38

Screw Size		Threads per Inch	C_1 Head Thickness		E Shoulder Dia.		L Practical Screw Length	
Nom.	Basic		Max.	Min.	Max.	Min.	Max.	Min.
No. 10	0.1900	24	0.10	0.08	0.33	0.31	2.00	0.50
1/4	0.2500	20	0.10	0.08	0.40	0.38	3.00	0.50
5/16	0.3125	18	0.11	0.09	0.46	0.44	4.00	0.50
3/8	0.3750	16	0.11	0.09	0.55	0.53	4.00	0.50
7/16	0.4375	14	0.13	0.11	0.71	0.69	2.50	1.00
1/2	0.5000	13	0.14	0.12	0.83	0.81	3.00	1.00

All dimensions in inches. Source: Extracted from ANSI B18.17-1968, published by the American Society of Mechanical Engineers.

Table 55. Dimensions of Type A and Type AB Self-Tapping Screw Threads. (See Notes.)

Type A Threads[1]								
Screw Size		Threads per Inch	Major Dia.		Minor Dia.		Length[2]	
Nom.	Basic		Max.	Min.	Max.	Min.	90° Head	Csk Head[3]
#0	0.0600	40	0.060	0.057	0.042	0.039	1/8	3/16
#1	0.0730	32	0.075	0.072	0.051	0.048	1/8	3/16
#2	0.0860	32	0.088	0.084	0.061	0.056	5/32	3/16
#3	0.0990	28	0.101	0.097	0.076	0.071	3/16	7/32
#4	0.1120	24	0.114	0.110	0.083	0.078	3/16	1/4
#5	1.1250	20	0.130	0.126	0.095	0.090	3/16	1/4
#6	0.1380	18	0.141	0.136	0.102	0.096	1/4	5/16
#7	0.1510	16	0.158	0.152	0.114	0.108	5/16	3/8
#8	0.1640	15	0.168	0.162	0.123	0.116	3/8	7/16
#10	0.1900	12	0.194	0.188	0.133	0.126	3/8	1/2
#12	0.2160	11	0.221	0.215	0.162	0.155	7/16	9/16
#14	0.2420	10	0.254	0.248	0.185	0.178	1/2	5/8
#16	0.2680	10	0.280	0.274	0.197	0.189	9/16	3/4
#18	0.2940	9	0.306	0.300	0.217	0.209	5/8	13/16
#20	0.3200	9	0.333	0.327	0.234	0.226	11/16	13/16
#24	0.3720	9	0.390	0.383	0.291	0.282	3/4	1

Type AB Threads								
Screw Size		Threads per Inch	Major Dia.		Minor Dia.		Minimum Length	
Nom.	Basic		Max.	Min.	Max.	Min.	90° Head	Csk Head[3]
#0	0.0600	48	0.060	0.057	0.036	0.033	3/32	7/64
#1	0.0730	42	0.075	0.072	0.049	0.046	1/8	9/64
#2	0.0860	32	0.088	0.084	0.064	0.060	9/64	11/64
#3	0.0990	28	0.101	0.097	0.075	0.071	11/64	3/16
#4	0.1120	24	0.114	0.110	0.086	0.082	3/16	7/32
#5	0.1250	20	0.130	0.126	0.094	0.090	3/16	1/4
#6	0.1380	20	0.139	0.135	0.104	0.099	7/32	17/64
#7	0.1510	19	0.154	0.149	0.115	0.109	17/64	5/16
#8	0.1640	18	0.166	0.161	0.122	0.116	9/32	21/64
#10	0.1900	16	0.189	0.183	0.141	0.135	21/64	3/8
#12	0.2160	14	0.215	0.209	0.164	0.157	3/8	13/32
1/4	0.2500	14	0.246	0.240	0.192	0.185	13/32	15/32
5/16	0.3125	12	0.315	0.308	0.244	0.236	17/32	19/32

Notes: [1] Although still widely available, the A thread is generally considered obsolete and is no longer recommended. [2] Type A threads in these lengths or shorter should not be used. Substitute AB screws. [3] Csk = Countersunk Head. All dimensions in inches.

Table 56. Dimensions of Type B, Type BF, Type BP, and Type BT/Type 25 Self-Tapping Screw Threads.

Type B

Type 25

Screw Size		Threads per Inch	D Major Dia.		d Minor Dia.		P Point Dia.	
Nom.	Basic		Max.	Min.	Max.	Min.	Max.	Min.
#0	0.0600	48	0.060	0.057	0.036	0.033	0.031	0.027
#1	0.0730	42	0.075	0.072	0.049	0.046	0.044	0.040
#2	0.0860	32	0.088	0.084	0.064	0.060	0.058	0.054
#3	0.0990	28	0.101	0.097	0.075	0.071	0.068	0.063
#4	0.1120	24	0.114	0.110	0.086	0.082	0.079	0.074
#5	0.1250	20	0.130	0.126	0.094	0.090	0.087	0.082
#6	0.1380	20	0.139	0.135	0.104	0.099	0.095	0.089
#7	0.1510	19	0.154	0.149	0.115	0.109	0.105	0.099
#8	0.1640	18	0.166	0.161	0.122	0.116	0.112	0.106
#10	0.1900	16	0.189	0.183	0.141	0.135	0.130	0.123
#12	0.2160	14	0.215	0.209	0.164	0.157	0.152	0.145
1/4	0.2500	14	0.246	0.240	0.192	0.185	0.179	0.171
5/16	0.3125	12	0.315	0.308	0.244	0.236	0.230	0.222

(Continued)

Table 56. *(Continued)* Dimensions of Type B, Type BF, Type BP, and Type BT/Type 25 Self-Tapping Screw Threads.

Screw Size		Threads per Inch	D Major Dia.		d Minor Dia.		P Point Dia.	
Nom.	Basic		Max.	Min.	Max.	Min.	Max.	Min.
3/8	0.3750	12	0.380	0.371	0.309	0.299	0.293	0.285
7/16	0.4375	10	0.440	0.431	0.359	0.349	0.343	0.335
1/2	0.5000	10	0.504	0.495	0.423	0.413	0.407	0.399

Nom. Size	S Point Taper Length				L Type B, Type BF, Type BT				L Type BP	
	Short Screws		Long Screws		Determine Length for Point Taper[1]		Minimum Practical Screw Length		Determine Min. Practical Length[2]	
	Max.	Min.	Max.	Min.	90° Head	Csk Head[3]	90° Head	Csk Head[3]	90° Head	Csk Head[3]
#0	0.042	0.031	0.052	0.042	5/64	1/8	5/64	3/32	5/32	1 3/64
#1	0.048	0.036	0.060	0.048	5/64	5/32	5/64	1/8	11/64	1/4
#2	0.062	0.047	0.078	0.062	7/64	3/16	7/64	5/32	13/64	9/32
#3	0.071	0.054	0.089	0.071	9/64	7/32	9/64	3/16	1/4	21/64
#4	0.083	0.063	0.104	0.083	3/16	1/4	9/64	3/16	5/16	3/8
#5	0.100	0.075	0.125	0.100	3/16	1/4	5/32	3/16	21/64	25/64
#6	0.100	0.075	0.125	0.100	1/4	5/16	11/64	1/4	25/64	15/32
#7	0.105	0.079	0.132	0.105	5/16	3/8	3/16	1/4	15/32	17/32
#8	0.111	0.083	0.139	0.111	5/16	7/16	3/16	1/4	31/64	39/64
#10	0.125	0.094	0.156	0.125	3/8	1/2	15/64	5/16	9/16	11/16
#12	0.143	0.107	0.179	0.143	7/16	9/16	9/32	3/8	21/32	25/32
1/4	0.143	0.107	0.179	0.143	1/2	5/8	9/32	3/8	3/4	7/8
5/16	0.167	0.125	0.208	0.167	1/2	5/8	5/16	7/16	53/64	61/64
3/8	0.167	0.125	0.208	0.137	1/2	5/8	5/16	7/16	29/32	1 1/32

(Continued)

Table 56. *(Continued)* Dimensions of Type B, Type BF, Type BP, and Type BT/Type 25 Self-Tapping Screw Threads.

Nom. Size	S Point Taper Length				L Type B, Type BF, Type BT				L Type BP	
	Short Screws		Long Screws		Determine Length for Point Taper[1]		Minimum Practical Screw Length		Determine Min. Practical Length[2]	
	Max.	Min.	Max.	Min.	90° Head	Csk Head[3]	90° Head	Csk Head[3]	90° Head	Csk Head[3]
7/16	0.200	0.150	0.250	0.200	5/8	3/4	15/32	5/8	1 7/64	1 15/64
1/2	0.200	0.150	0.250	0.200	5/8	3/4	15/32	5/8	1 3/16	1 5/16

Notes: [1] Screws of the stated nominal length or shorter have point taper length specified for short screws. Longer lengths use the point taper length for long screws.
[2] Type BP screws of the stated nominal length or shorter have point taper length specified for short screws. Longer lengths use the point taper length for long screws.
[3] Csk = Countersunk Head. All dimensions in inches.

Table 57. Dimensions of Type D/Type 1, Type F, and Type T/Type 23 Thread Cutting Tapping Screws.

Type 1   Type F   Type 23

Screw Size		TPI¹	D Major Dia.		P Point Dia.		S Point Taper Length				Determine Length for Point Taper²		Min. Practical Screw Length	
Nom.	Basic		Max.	Min.	Max.	Min.	Short Screws		Long Screws		90° Head	Csk Head³	90° Head	Csk Head³
#2	0.0860	56	0.0860	0.0820	0.067	0.061	0.062	0.045	0.080	0.062	9/64	3/16	1/8	5/32
#2	0.0860	64	0.0860	0.0822	0.070	0.064	0.055	0.039	0.070	0.055	1/8	11/64	1/8	5/32
#3	0.0990	48	0.0990	0.0946	0.077	0.070	0.073	0.052	0.094	0.073	11/64	7/32	1/8	3/16
#3	0.0990	56	0.0990	0.0950	0.080	0.074	0.062	0.045	0.080	0.062	9/64	3/16	1/8	3/16
#4	0.1120	40	0.1120	0.1072	0.086	0.077	0.088	0.062	0.112	0.088	13/64	1/4	1/8	3/16
#4	0.1120	48	0.1120	0.1076	0.090	0.083	0.073	0.052	0.094	0.073	11/64	7/32	1/8	3/16
#5	0.1250	40	0.1250	0.1202	0.099	0.090	0.088	0.062	0.112	0.088	13/64	9/32	1/8	3/16
#5	0.1250	44	0.1250	0.1204	0.101	0.093	0.080	0.057	0.102	0.080	3/16	1/4	1/8	3/16
#6	0.1380	32	0.1380	0.1326	0.106	0.095	0.109	0.078	0.141	0.109	1/4	5/16	3/16	1/4
#6	0.1380	40	0.1380	0.1332	0.112	0.103	0.088	0.062	0.112	0.088	13/64	17/64	3/16	1/4

(Continued)

Table 57. *(Continued)* **Dimensions of Type D/Type 1, Type F, and Type T/Type 23 Thread Cutting Tapping Screws.**

Screw Size		TPI¹	D Major Dia.		P Point Dia.		S Point Taper Length				L — Determine Length for Point Taper²		L — Min. Practical Screw Length	
Nom.	Basic		Max.	Min.	Max.	Min.	Short Screws		Long Screws		90° Head	Csk Head³	90° Head	Csk Head³
#8	0.1640	32	0.1640	0.1586	0.132	0.121	0.109	0.078	0.141	0.109	1/4	21/64	3/16	1/4
#8	0.1640	36	0.1640	0.1590	0.135	0.125	0.097	0.069	0.125	0.097	15/64	19/64	3/16	1/4
#10	0.1900	24	0.1900	0.1834	0.147	0.133	0.146	0.104	0.188	0.146	11/32	27/64	15/64	5/16
#10	0.1900	32	0.1900	0.1846	0.158	0.147	0.109	0.178	0.141	0.109	1/4	11/32	15/64	5/16
#12	0.2160	24	0.2160	0.2094	0.173	0.159	0.146	0.104	0.188	0.146	11/32	7/16	17/64	3/8
#12	0.2160	28	0.2160	0.2098	0.179	0.167	0.125	0.089	0.161	0.125	19/64	25/64	17/64	3/8
1/4	0.2500	20	0.2500	0.2428	0.198	0.181	0.175	0.125	0.225	0.175	13/32	33/64	17/64	3/8
1/4	0.2500	28	0.2500	0.2438	0.213	0.201	0.125	0.089	0.161	0.125	19/64	13/32	17/64	3/8
5/16	0.3125	18	0.3125	0.3043	0.255	0.236	0.194	0.139	0.250	0.194	29/64	19/32	9/32	7/16
5/16	0.3125	24	0.3125	0.3059	0.269	0.255	0.146	0.104	0.188	0.146	11/32	15/32	9/32	7/16
3/8	0.3750	16	0.3750	0.3660	0.310	0.289	0.219	0.156	0.281	0.219	1/2	47/64	9/32	7/16
3/8	0.3750	24	0.3750	0.3684	0.332	0.318	0.146	0.104	0.188	0.146	11/32	1/2	9/32	7/16

Notes: ¹ TPI = threads per inch. ² Screws of the stated nominal length or shorter have point taper length specified for short screws. Longer lengths use the point taper length for long screws. ³ Csk = Countersunk Head. All dimensions in inches.

Table 58. Dimensions of Type U Round Head Drive Screws.

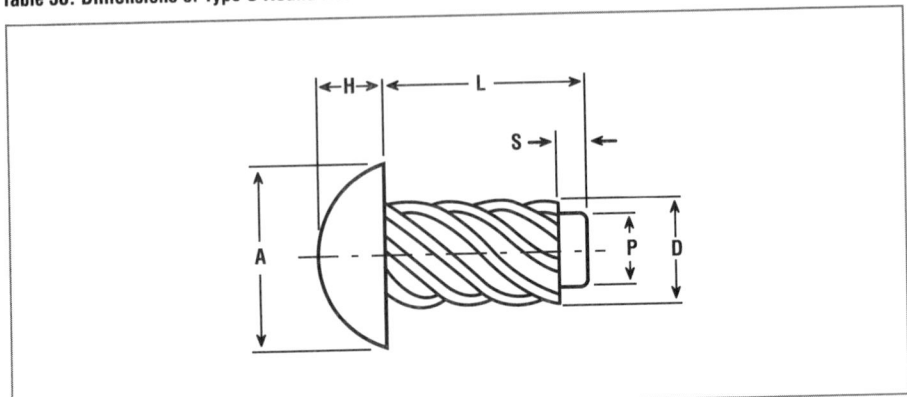

Nom. Screw Size	No of Thread Starts	D Body Dia.		A Head Dia.		H Head Height		P Pilot Dia.		Hole Size	
		Max.	Min.	Max.	Min.	Max.	Min.	Max.	Min.	Drill Size	Hole Dia.
00	6	0.060	0.057	0.099	0.090	0.034	0.026	0.049	0.046	#55	0.052
0	6	0.075	0.072	0.127	0.118	0.049	0.041	0.063	0.060	#51	0.067
2	8	0.100	0.097	0.162	0.146	0.069	0.059	0.083	0.080	#41	0.086
4	7	0.116	0.112	0.211	0.193	0.086	0.075	0.096	0.092	#37	0.104
6	7	0.140	0.136	0.260	0.240	0.103	0.091	0.116	0.112	#31	0.120
7	8	0.154	0.150	0.285	0.264	0.111	0.099	0.126	0.122	#29	0.136
8	8	0.167	0.162	0.309	0.287	0.120	0.107	0.136	0.132	#27	0.144
10	8	0.182	0.177	0.359	0.334	0.137	0.123	0.150	0.146	#20	0.161
12	8	0.212	0.206	0.408	0.382	0.153	0.139	0.177	0.173	#11	0.191
14	9	0.236	0.242	0.457	0.429	0.170	0.155	0.202	0.198	#2	0.221

Nom. Screw Size (L)	1/8	3/16	1/4	5/16	3/8	1/2	5/8	3/4	≥ 1
Pilot Length (S)	0.047	0.047	0.047	0.047	0.062	0.062	0.078	0.078	0.125

All dimensions in millimeters. Source: Fastbolt Corp.

Table 59. Hole Sizes for Types AB, B, and BP Tapping and Thread Forming Screws.

Screw Size	Sheet Metal Thickness	Steel, Stainless Steel, Monel, Brass Sheet Metal			Aluminum Alloy Steel Metal		
		Pierced or Extruded Hole	Drilled or Clean Punched Hole		Pierced or Extruded Hole	Drilled or Clean Punched Hole	
		Hole Dia.	Hole Dia.	Drill Size	Hole Dia.	Hole Dia.	Drill Size
2	.015		.064	52			
	.018		.064	52			
	.024		.067	51		.064	52
	.030		.070	50		.064	52
	.036		.073	49		.064	52
	.048		.073	49		.067	51
	.060		.076	48		.070	50
4	.015	.086	.086	44			
	.018	.086	.086	44			
	.024	.098	.089	43	.086		
	.030	.098	.094	42	.086	.086	44
	.036	.098	.094	42	.086	.086	44
	.048		.096	41	.086	.086	44
	.060		.100	39		.089	43
	.075		.102	38		.089	43
	.105					.094	42
6	.015	.111	.104	37			
	.018	.111	.104	37			
	.024	.111	.106	36	.111		
	.030	.111	.106	36	.111	.104	37
	.036	.111	.110	35	.111	.104	37
	.048		.111	34	.111	.104	37
	.060		.116	32		.106	36
	.075		.120	31		.110	35
	.105		.128	30		.111	34
	.128 to .250					.120	31
8	.018	.136					
	.024	.136	.125	1/8	.136		
	.030	.136	.125	1/8	.136	.116	32
	.036	.136	.125	1/8	.136	.120	31
	.048	.136	.128	30	.136	.128	30
	.060		.136	29		.136	29
	.075		.140	28		.140	28
	.105		.150	25		.147	26
	.125		.150	25		.147	26
	.135		.152	24		.149	25
	.162 to .375					.152	24
10	.018	.157					
	.024	.157	.144	27	.157		
	.030	.157	.144	27	.157		
	.036	.157	.147	26	.157	.144	27
	.048	.157	.152	24	.157	.144	27
	.060		.152	24		.144	27
	.075		.157	22		.147	26
	.105		.161	20		.147	26
	.125		.170	18		.154	23
	.135		.170	18		.154	23
	.164		.173	17		.159	21
	.200 to .375					.166	19

(Continued)

Table 59. *(Continued)* Hole Sizes for Types AB, B, and BP Tapping and Thread Forming Screws.

Screw Size	Sheet Metal Thickness	Steel, Stainless Steel, Monel, Brass Sheet Metal			Aluminum Alloy Steel Metal		
		Pierced or Extruded Hole	Drilled or Clean Punched Hole		Pierced or Extruded Hole	Drilled or Clean Punched Hole	
		Hole Dia.	Hole Dia.	Drill Size	Hole Dia.	Hole Dia.	Drill Size
1/4	.030	.209	.194	10			
	.036	.209	.194	10			
	.048	.209	.194	10			
	.060		.199	8		.199	8
	.075		.204	6		.201	7
	.105		.209	4		.204	6
	.125		.228	1		.209	4
	.135		.228	1		.209	4
	.164		.234	15/64		.213	3
	.187		.234	15/64		.213	3
	.194		.234	15/64		.221	2
	.200 to .375					.228	1

Notes: Hole sizes for metal thicknesses 0.105 inch and above apply to Types B and BP only. All dimensions in inches. Source: MS24632.

Table 60. Approximate Hole Sizes for Types AB, B, and BP Tapping and Thread Forming Screws.

Screw Size	Cast Metals - Aluminum, Magnesium, Zinc, Brass, Bronze[1]			Resin Impregnated Plywood				
	Hole Dia.	Drill Size	Min. Penetration in Blind Hole	Hole Dia.	Drill Size	Min. Material Thickness	Penetration in Blind Hole	
							Min.	Max.
#2	0.078	#47	0.125	0.073	#49	0.125	0.188	0.500
#4	0.104	#37	0.188	0.100	#39	0.188	0.250	0.625
#6	0.128	#30	0.250	0.125	1/8	0.188	0.250	0.625
#8	0.152	#24	0.250	0.144	#27	0.188	0.250	0.750
#10	0.177	#16	0.250	0.173	#17	0.250	0.312	1.000
1/4	0.234	15/64	0.312	0.228	#1	0.312	0.375	1.000

Plastics[1]						
Screw Size	Phenol Formaldehyde			Cellulose Acetate and Nitrate, Acrylic and Styrene Resin		
	Hole Dia.	Drill Size	Min. Penetration in Blind Hole	Hole Dia.	Drill Size	Min. Penetration in Blind Hole
#2	0.078	#47	0.188	0.078	#47	0.188
#4	0.100	#39	0.250	0.094	#42	0.250
#6	0.128	#30	0.250	0.120	#31	0.250
#8	0.150	#25	0.312	0.144	#27	0.312
#10	0.177	#16	0.312	0.170	#18	0.312
1/4	0.214	15/64	0.375	0.221	#2	0.375

Notes: [1] These Tables refer to Type B and Type BP only. All dimensions in inches. Source: MS24632.

Table 61. Approximate Hole Sizes for Type BF and Type BT/Type 25 Self-Tapping Screws.

Screw Size	Die Cast Zinc and Aluminum			Screw Size	Die Cast Zinc and Aluminum		
	Material Thickness	Hole Dia.	Drill Size		Material Thickness	Hole Dia.	Drill Size
#2	0.060	0.073	#49	#8	0.125	0.149	#25
	0.083	0.073	#49		0.140	0.149	#25
	0.109	0.076	#48		0.188	0.149	#25
	0.125	0.076	#48		0.250	0.152	#24
	0.140	0.076	#48		0.312	0.152	#24
#4	0.109	0.098	#40	#10	0.125	0.166	#19
	0.125	0.100	#39		0.140	0.166	#19
	0.140	0.100	#39		0.188	0.166	#19
	0.188	0.100	#39		0.250	0.170	#18
	0.250	0.102	#38		0.312	0.172	11/64
					0.375	0.172	11/64
#6	0.125	0.120	#31	1/4	0.125	0.221	#2
	0.140	0.120	#31		0.140	0.221	#2
	0.188	0.120	#31		0.188	0.221	#2
	0.250	0.125	1/8		0.250	0.228	#1
	0.312	0.125	1/8		0.312	0.228	#1
					0.375	0.228	#1

Plastics[1]							
Screw Size	Phenol Formaldehyde				Cellulose Acetate and Nitrate, Acrylic and Styrene Resin		
	Hole Dia.	Drill Size	Penetration		Hole Dia.	Drill Size	Penetration	
			Min.	Max.			Min.	Max.
#2	0.078	5/64	0.094	0.250	0.076	48	0.094	0.250
#4	0.104	#37	0.125	0.312	0.100	#39	0.125	0.312
#6	0.125	1/8	0.188	0.375	0.120	#31	0.188	0.375
#8	0.147	#26	0.250	0.500	0.144	#27	0.250	0.500
#10	0.170	#18	0.312	0.625	0.166	#19	0.312	0.625
1/4	0.228	#1	0.375	0.750	0.221	#2	0.375	0.750

All dimensions in inches. Source: MS24633.

Table 62. Approximate Hole Sizes for Type D/Type 1, Type F, and Type T/Type 23 "Machine Screw" Thread Cutting Tapping Screws.

Hardest Material Tapped		Screw Size No.	Thickness of Hardest Material Tapped									
			0.015 to 0.018 incl.	0.018 to 0.040 incl.	0.041 to 0.056 incl.	0.057 to 0.071 incl.	0.072 to 0.089 incl.	0.090 to 0.110 incl.	0.103 to 0.115 incl.	0.116 to 0.131 incl.	0.132 to 0.148 incl.	Over 0.148
Steel	Carbon Mild Quarter Hard 55,000 PSI Min. Tensile Strength	4-40	0.086	0.086	0.086	0.093	0.093	-	0.096	0.096	0.096	-
		6-32	0.104	0.104	0.104	0.113	0.113	-	0.113	0.120	0.120	-
		8-32	0.128	0.128	0.128	0.140	0.140	-	0.140	0.144	0.144	-
		10-24	0.149	0.149	0.149	0.154	0.154	-	0.154	0.161	0.161	-
		1/4-20	0.199	0.199	0.199	0.209	0.209	-	0.209	0.213	0.213	-
	Chrome-Molybdenum Normalized [1] 90,000 PSI Min. Tensile Strength	4-40	0.189	0.189	0.189	0.089	0.096	0.096	-	0.098	-	-
		6-32	0.110	0.110	0.110	0.110	0.116	0.120	-	0.120	-	-
		8-32	0.136	0.136	0.136	0.136	0.140	0.144	-	0.144	-	-
		10-24	0.157	0.157	0.157	0.157	0.159	0.161	-	0.161	-	-
		1/4-20	0.209	0.209	0.209	0.209	0.213	0.218	-	0.218	0.218	-
	Corrosion Resistant Half Hard [1] 150,000 PSI Min. Tensile Strength	4-40	0.093	0.093	0.096	-	-	-	-	-	-	-
		6-32	0.110	0.110	0.110	0.116	-	-	-	-	-	-
		8-32	0.136	0.136	0.140	0.140	0.144	-	-	-	-	-
		10-24	-	-	0.159	0.161	0.161	0.166	-	-	-	-
		1/4-20	-	-	0.205	0.205	0.209	0.213	0.218	0.221	-	-

(Continued)

Notes: [1] Lubricate with heavy cutting oil. All dimensions in inches. Source: AND10326.

Table 62. (Continued) Approximate Hole Sizes for Type D/Type 1, Type F, and Type T/Type 23 "Machine Screw" Thread Cutting Tapping Screws.

Hardest Material Tapped	Screw Size No.	Thickness of Hardest Material Tapped									
		0.015 to 0.018 incl.	0.018 to 0.040 incl.	0.041 to 0.056 incl.	0.057 to 0.071 incl.	0.072 to 0.089 incl.	0.090 to 0.110 incl.	0.103 to 0.115 incl.	0.116 to 0.131 incl.	0.132 to 0.148 incl.	Over 0.148
Aluminum Alloy											
Al-Mg-Cr Half Hard 34,000 PSI Min. Tensile Strength	4-40	0.086	0.086	0.086	0.086	0.089	-	-	0.093	-	-
	6-32	0.104	0.104	0.104	0.104	0.106	-	-	0.106	-	-
	8-32	0.128	0.128	0.128	0.128	0.136	-	-	0.140	-	-
	10-24	0.149	0.149	0.149	0.149	0.152	-	-	0.154	-	-
	1/4-20	0.193	0.193	0.199	0.199	0.204	-	-	0.204	-	-
Al-Cu-Mg-Mn Heat Treated 52,000 PSI Min. Tensile Strength	4-40	0.086	0.086	0.086	0.086	0.089	0.089	-	0.093	-	-
	6-32	0.104	0.104	0.104	0.104	0.106	0.106	-	0.106	-	-
	8-32	0.128	0.128	0.128	0.128	0.136	0.140	-	0.140	-	-
	10-24	0.149	0.149	0.149	0.149	0.152	0.152	-	0.154	-	-
	1/4-20	0.199	0.199	0.199	0.199	0.204	0.204	-	0.204	-	-
Al-Cu-Mg (1.5% Mn) Heat Treated 62,000 PSI Min. Tensile Strength	4-40	0.086	0.086	0.086	0.089	0.089	-	-	0.	-	-
	6-32	0.104	0.104	0.104	0.109	0.109	-	-	0.	-	-
	8-32	0.129	0.129	0.0129	0.136	0.136	-	-	0.	-	-
	10-24	0.149	0.149	0.149	0.152	0.152	-	-	0.	-	-
	1/4-20	0.199	0.199	0.199	0.203	0.203	-	-	0.	-	-
Castings Aluminum Alloy or Aluminum	4-40	-	-	-	0.096	0.096	0.096	0.096	0.098	0.098	0.098
	6-32	-	-	-	0.116	0.116	0.116	0.116	0.120	0.120	0.120
	8-32	-	-	-	0.144	0.144	0.144	0.144	0.147	0.147	0.147
	10-24	-	-	-	0.161	0.161	0.161	0.161	0.166	0.166	0.166
	1/4-20	-	-	-	0.221	0.221	0.221	0.221	0.228	0.228	0.228

Notes: [1] Lubricate with heavy cutting oil. All dimensions in inches. Source: AND10326.

Table 63. Dimensions of Socket Head Cap Screws, Hexagon Recess, 1960 Series.

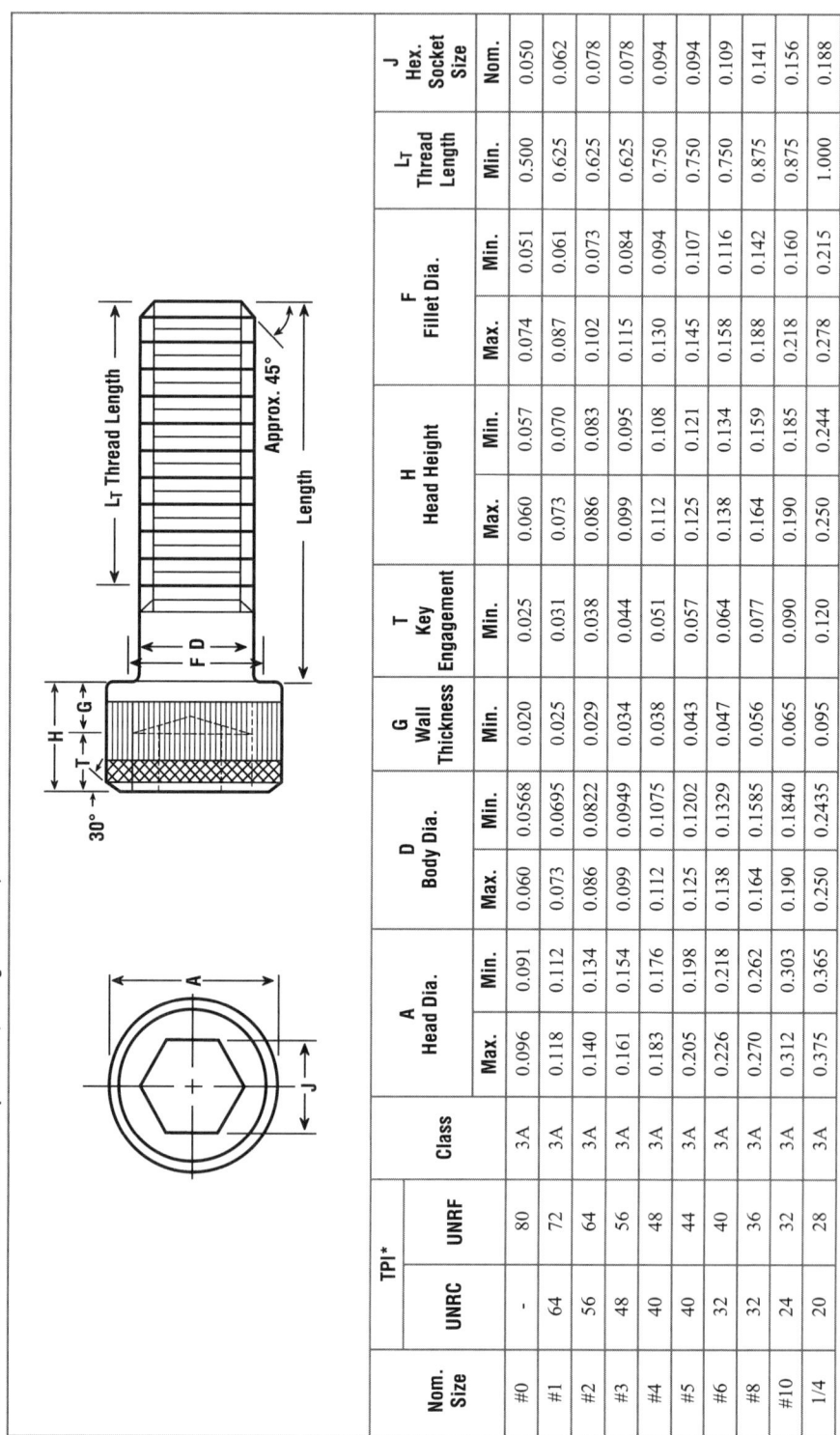

Nom. Size	TPI* UNRC	TPI* UNRF	Class	A Head Dia. Max.	A Head Dia. Min.	D Body Dia. Max.	D Body Dia. Min.	G Wall Thickness Min.	T Key Engagement Min.	H Head Height Max.	H Head Height Min.	F Fillet Dia. Max.	F Fillet Dia. Min.	L_T Thread Length Min.	J Hex. Socket Size Nom.
#0	–	80	3A	0.096	0.091	0.060	0.0568	0.020	0.025	0.060	0.057	0.074	0.051	0.500	0.050
#1	64	72	3A	0.118	0.112	0.073	0.0695	0.025	0.031	0.073	0.070	0.087	0.061	0.625	0.062
#2	56	64	3A	0.140	0.134	0.086	0.0822	0.029	0.038	0.086	0.083	0.102	0.073	0.625	0.078
#3	48	56	3A	0.161	0.154	0.099	0.0949	0.034	0.044	0.099	0.095	0.115	0.084	0.625	0.078
#4	40	48	3A	0.183	0.176	0.112	0.1075	0.038	0.051	0.112	0.108	0.130	0.094	0.750	0.094
#5	40	44	3A	0.205	0.198	0.125	0.1202	0.043	0.057	0.125	0.121	0.145	0.107	0.750	0.094
#6	32	40	3A	0.226	0.218	0.138	0.1329	0.047	0.064	0.138	0.134	0.158	0.116	0.750	0.109
#8	32	36	3A	0.270	0.262	0.164	0.1585	0.056	0.077	0.164	0.159	0.188	0.142	0.875	0.141
#10	24	32	3A	0.312	0.303	0.190	0.1840	0.065	0.090	0.190	0.185	0.218	0.160	0.875	0.156
1/4	20	28	3A	0.375	0.365	0.250	0.2435	0.095	0.120	0.250	0.244	0.278	0.215	1.000	0.188

(Continued)

Table 63. *(Continued)* **Dimensions of Socket Head Cap Screws, Hexagon Recess, 1960 Series.**

| Nom. Size | TPI* | | Class | A Head Dia. | | D Body Dia. | | G Wall Thickness | T Key Engagement | H Head Height | | F Fillet Dia. | | L_T Thread Length | J Hex. Socket Size |
	UNRC	UNRF		Max.	Min.	Max.	Min.	Min.	Min.	Max.	Min.	Max.	Min.	Min.	Nom.
5/16	18	24	3A	0.469	0.457	0.3125	0.3053	0.119	0.151	0.312	0.306	0.347	0.273	1.125	0.250
3/8	16	24	3A	0.562	0.550	0.375	0.3678	0.143	0.182	0.375	0.368	0.415	0.331	1.250	0.312
7/16	14	20	3A	0.656	0.642	0.4375	0.4294	0.166	0.213	0.437	0.430	0.484	0.388	1.375	0.375
1/2	13	20	3A	0.750	0.735	0.500	0.4919	0.190	0.245	0.500	0.492	0.552	0.446	1.500	0.375
9/16	12	18	3A	0.843	0.827	0.5625	0.5538	0.214	0.265	0.562	0.554	0.6185	0.525	1.625	0.438
5/8	11	18	3A	0.938	0.921	0.625	0.6163	0.238	0.307	0.625	0.616	0.689	0.562	1.750	0.500
3/4	10	16	3A	1.125	1.107	0.750	0.7406	0.285	0.370	0.750	0.740	0.828	0.681	2.000	0.625
7/8	9	14	3A	1.312	1.293	0.875	0.8647	0.333	0.432	0.875	0.864	0.963	0.798	2.250	0.750
1	8	12	3A	1.500	1.479	1.000	0.9886	0.380	0.495	1.000	0.988	1.100	0.914	2.500	0.750
1**	-	14*	3A	1.500	1.479	1.000	0.9886	0.380	0.495	1.000	0.988	1.100	0.914	2.500	0.750
1 1/8	7	12	2A	1.688	1.665	1.125	1.1086	0.428	0.557	1.125	1.111	1.235	1.023	2.812	0.875
1 1/4	7	12	2A	1.875	1.852	1.250	1.2336	0.475	0.620	1.250	1.236	1.370	1.148	3.125	0.875
1 3/8	6	12	2A	2.062	2.038	1.375	1.3568	0.523	0.682	1.375	1.360	1.505	1.256	3.437	1.000
1 1/2	6	12	2A	2.250	2.224	1.500	1.4818	0.570	0.745	1.500	1.485	1.640	1.381	3.750	1.000
1 3/4	5	12	2A	2.625	2.597	1.750	1.7295	0.665	0.870	1.750	1.734	1.910	1.609	4.375	1.250
2	4 1/2	12	2A	3.000	2.970	2.000	1.9780	0.760	0.995	2.000	1.983	2.180	1.843	5.000	1.500
2 1/4	4 1/2	12	2A	3.375	3.344	2.250	2.2280	0.855	1.120	2.250	2.232	2.450	2.093	5.625	1.750
2 1/2	4	12	2A	3.750	3.717	2.500	2.4762	0.950	1.245	2.500	2.481	2.720	2.324	6.250	1.750
2 3/4	4	12	2A	4.125	4.090	2.750	2.7262	1.045	1.370	2.750	2.730	2.990	2.574	6.875	2.000
3	4	12	2A	4.500	4.464	3.000	2.9762	1.140	1.495	3.000	2.979	3.260	2.824	7.500	2.250

Notes: * Threads per inch. ** 1 inch size also available in 1-14 UNRS (special) thread form.

(Continued)

Table 63. *(Continued)* **Dimensions of Socket Head Cap Screws, Hexagon Recess, 1960 Series.**

Size	Length tolerances for head cap screws, hexagon recess, 1960 series				
	Up to 1" incl.	Over 1" to 2 1/2" incl.	Over 2 1/2" to 6" incl.	Over 6"
#0 thru 3/8"	- 0.03	- 0.04	- 0.06	- 0.12
7/16" thru 3/4"	- 0.03	- 0.06	- 0.08	- 0.12
7/8" thru 1 1/2"	- 0.05	- 0.10	- 0.14	- 0.20
Over 1 1/2"	-	- 0.18	- 0.20	- 0.24

All dimensions in inches. Source: Unbrako/SPS Technologies, Inc.

Table 64. Grip (L_G) and Body (L_B) Lengths for 1960 Series Socket Head Cap Screws. (See Notes.)

Length	#1		#2		#3		#4		#5		#6	
	L_G	L_B	L_G	L_B	L_G	L_B	L_G	L_B	L_G	L_B	L_G	L_B
7/8	.250	.172	.250	.161	.250	.146						
1	.250	.172	.250	.161	.250	.146	.250	.125	.250	.125		
1 1/4	.625	.547	.625	.536	.625	.521	.250	.125	.250	.125	.500	.344
1 1/2	.875	.797	.875	.786	.875	.771	.750	.625	.750	.625	.500	.344
1 3/4			1.125	1.036	1.125	1.021	.750	.625	.750	.625	1.000	.844
2						1.271	1.250	1.125	1.250	1.125	1.000	.844
2 1/4									1.250	1.125	1.500	1.344
2 1/2									1.750	1.625	1.500	1.344
2 3/4											2.000	1.844

(Continued)

Table 64. *(Continued)* **Grip (L_G) and Body (L_B) Lengths for 1960 Series Socket Head Cap Screws.** (See Notes.)

Length	#8		#10		1/4		5/16		3/8		7/16	
	L_G	L_B	L_G	L_B	L_G	L_B	L_G	L_B	L_G	L_B	L_G	L_B
1 1/4	.375	.219	.375	.167								
1 1/2	.375	.219	.375	.167	.500	.250						
1 3/4	.875	.719	.875	.667	.500	.250	.625	.347	.500	.187	.625	.268
2	.875	.719	.875	.667	1.000	.750	.625	.347	.500	.187	.625	.268
2 1/4	1.375	1.219	1.375	1.167	1.000	.750	1.125	.847	1.000	.687	1.125	.768
2 1/2	1.375	1.219	1.375	1.167	1.500	1.250	1.125	.847	1.000	.687	1.125	.768
2 3/4	1.875	1.719	1.875	1.667	1.500	1.250	1.625	1.187	1.500	1.187	1.625	1.268
3	1.875	1.719	1.875	1.667	2.000	1.750	1.625	1.347	1.500	1.187	1.625	1.268
3 1/4	2.375	2.219	2.375	2.167	2.000	1.750	2.125	1.847	2.000	1.687	2.125	1.768
3 1/2			2.375	2.167	2.500	2.250	2.125	1.847	2.000	1.687	2.125	1.768
3 3/4			2.875	2.667	2.500	2.250	2.625	2.347	2.500	2.187	2.625	2.268
4					3.000	2.750	2.625	2.347	2.500	2.187	2.625	2.268
4 1/4					3.000	2.750	3.125	2.847	3.000	2.687	3.125	2.768
4 1/2					3.500	3.250	3.125	2.847	3.000	2.687	3.125	2.768
4 3/4					3.500	3.250	3.625	3.347	3.500	3.187	3.625	3.268
5					4.000	3.750	3.625	3.347	3.500	3.187	3.625	3.268
5 1/4					4.000	3.750	4.125	3.847	4.000	3.687	4.125	3.768
5 1/2							4.125	3.847	4.000	3.687	4.125	3.768
5 3/4							4.625	4.347	4.500	4.187	4.625	4.268
6							4.625	4.347	4.500	4.187	4.625	4.268
6 1/4							5.125	4.847	5.000	4.687	5.125	4.768
6 1/2									5.000	4.687	5.125	4.768
6 3/4									5.500	5.187		

(Continued)

Table 64. *(Continued)* **Grip (L$_G$) and Body (L$_B$) Lengths for 1960 Series Socket Head Cap Screws.** *(See Notes.)*

Length	#8 L$_G$	#8 L$_B$	#10 L$_G$	#10 L$_B$	1/4 L$_G$	1/4 L$_B$	5/16 L$_G$	5/16 L$_B$	3/8 L$_G$	3/8 L$_B$	7/16 L$_G$	7/16 L$_B$
7									5.500	5.187	5.625	5.268
7 1/4									6.000	5.687	5.625	5.268
7 1/2											6.125	5.768
7 3/4											6.125	5.768
8											6.625	6.268
8 1/2											7.125	6.768
9											7.625	7.268

Length	1/2 L$_G$	1/2 L$_B$	9/16 L$_G$	9/16 L$_B$	5/8 L$_G$	5/8 L$_B$	3/4 L$_G$	3/4 L$_B$	7/8 L$_G$	7/8 L$_B$	1 L$_G$	1 L$_B$
2 1/4	.750	.365	.875	.458	.750	.295						
2 1/2	.750	.365	.875	.458	.750	.295						
2 3/4	.750	.365	.875	.458	.750	.295						
3	1.500	1.115	1.625	1.208	1.500	1.045	1.000	.500	1.000	.444		
3 1/4	1.500	1.115	1.625	1.208	1.500	1.045	1.000	.500	1.000	.444	1.000	.375
3 1/2	1.500	1.115	1.625	1.208	1.500	1.045	1.000	.500	1.000	.444	1.000	.375
3 3/4	2.250	1.865	2.375	1.958	2.250	1.795	1.000	.500	1.000	.444	1.000	.375
4	2.250	1.865	2.375	1.958	2.250	1.795	2.000	1.500	2.000	1.444	1.000	.375
4 1/4	2.250	1.865	2.375	1.958	2.250	1.795	2.000	1.500	2.000	1.444	2.000	1.375
4 1/2	3.000	2.615	3.125	2.708	3.000	2.545	2.000	1.500	2.000	1.444	2.000	1.375
4 3/4	3.000	2.615	3.125	2.708	3.000	2.545	2.000	1.500	2.000	1.444	2.000	1.375
5	3.000	2.615	3.125	2.708	3.000	2.545	3.000	2.500	2.000	1.444	2.000	1.375
5 1/4	3.750	3.365	3.875	3.458	3.750	3.295	3.000	2.500	3.000	2.444	2.000	1.375
5 1/2	3.750	3.365	3.875	3.458	3.750	3.295	3.000	2.500	3.000	2.444	3.000	2.375

(Continued)

Table 64. *(Continued)* Grip (L_G) and Body (L_B) Lengths for 1960 Series Socket Head Cap Screws. (See Notes.)

Length	1/2		9/16		5/8		3/4		7/8		1	
	L_G	L_B	L_G	L_B	L_G	L_B	L_G	L_B	L_G	L_B	L_G	L_B
5 3/4	3.750	3.365	3.875	3.458	3.750	3.295	3.000	2.500	3.000	2.444	3.000	2.375
6	4.500	4.115	3.875	3.458	3.750	3.295	4.000	3.500	3.000	2.444	3.000	2.375
6 1/4	4.500	4.115	4.625	4.208	4.500	4.045	4.000	3.500	4.000	3.444	3.000	2.375
6 1/2	4.500	4.115	4.625	4.208	4.500	4.045	4.000	3.500	4.000	3.444	4.000	3.375
6 3/4	5.250	4.865	4.625	4.208	4.500	4.045	4.000	3.500	4.000	3.444	4.000	3.375
7	5.250	4.865	5.375	4.958	5.250	4.795	5.000	4.500	4.000	3.444	4.000	3.375
7 1/4	5.250	4.865	5.375	4.958	5.250	4.795	5.000	4.500	5.000	4.444	4.000	4.375
7 1/2	6.000	5.615	5.375	4.958	5.250	4.795	5.000	4.500	5.000	4.444	5.000	4.375
7 3/4	6.000	5.615	6.125	5.708	6.000	5.545	5.000	4.500	5.000	4.444	5.000	4.375
8	6.000	5.615	6.125	5.708	6.000	5.545	6.000	5.500	5.000	4.444	5.000	4.375
8 1/2	7.000	6.615	6.875	6.458	6.750	6.295	6.000	5.500	6.000	5.444	5.000	5.375
9	7.000	6.615	6.875	6.458	6.750	6.295	7.000	6.500	6.000	5.444	5.000	5.375
9 1/2	8.000	7.615	7.625	7.208	7.750	7.295	7.000	6.500	7.000	6.444	7.000	6.375
10	8.000	7.615	7.625	7.208	7.750	7.295	8.000	7.500	7.000	6.444	7.000	6.375
11			9.125	8.708	9.250	8.795	9.000	8.500	8.000	7.444	8.000	7.375
12			10.125	9.708	10.250	9.795	10.000	9.000	9.000	8.444	9.000	8.375

Notes: L_G is the maximum grip length and is the distance from the bearing surface to the first complete thread. L_B is the minimum body length and is the length of the unthreaded cylindrical portion of the shank. Note that 3/4", 7/8", and 1 inch diameters are available in one-inch increments in lengths up to 20 inches. For sizes larger than 1 inch, the minimum complete thread length shall be equal to the basic thread length, and the total thread length including imperfect threads shall be basic thread length plus five pitches. Lengths too short to apply this formula shall be threaded to the head. All dimensions in inches. Source: Unbrako/SPS Technologies, Inc.

Table 65. Drill and Counterbore Sizes for 1960 Series Socket Head Cap Screws.

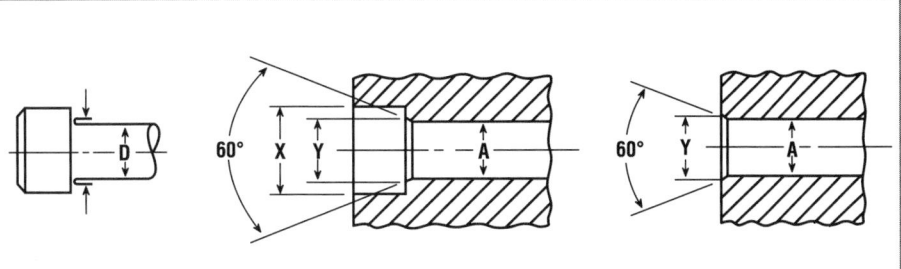

Nom. Size	Basic Screw Dia.	A Drill Size for Hole A				X Counter-bore Dia.	Y Counter-sink Dia.[3]	Hole Dimensions Tap Drill Dia.	
		Close Fit[1]		Normal Fit[2]					
		Nom.	Decimal	Nom.	Decimal			UNRC	UNRF
0	0.0600	#51	0.0670	#49	0.0730	1/8	0.074	-	3/64
1	0.0730	#46	0.0810	#43	0.0890	5/32	0.087	1.5mm	#53
2	0.0860	3/32	0.0937	#36	0.1065	3/16	0.102	#50	#50
3	0.0990	#36	0.1065	#31	0.1200	7/32	0.115	#47	#45
4	0.1120	1/8	0.1250	#29	0.1360	7/32	0.130	#43	#42
5	0.1250	9/64	0.1406	#23	0.1540	1/4	0.145	#38	#38
6	0.1380	#23	0.1540	#18	0.1695	9/32	0.158	#36	#33
8	0.1640	#15	0.1800	#10	0.1935	5/16	0.188	#29	#29
10	0.1900	#5	0.2055	#2	0.2210	3/8	0.218	#25	#21
1/4	0.2500	17/64	0.2656	9/23	0.2812	7/16	0.278	#7	#3
5/16	0.3125	21/64	0.3281	11/32	0.3437	17/32	0.346	F	I
3/8	0.0375	25/64	0.3906	13/32	0.4062	5/8	0.415	5/16	Q
7/16	0.4375	29/64	0.4531	15/32	0.4687	23/32	0.183	U	25/64
1/2	0.5000	33/64	0.5156	17/32	0.5312	13/16	0.552	27/64	29/64
5/8	0.6250	41/64	0.6406	21/32	0.6562	1	0.689	35/64	14.5mm
3/4	0.7500	49/64	0.7656	25/32	0.7812	1 3/16	0.828	21/32	11/16
7/8	0.8750	57/64	0.8906	29/32	0.9062	1 3/8	0.963	49/64	20.5mm
1	1.0000	1 1/64	1.0156	1 1/32	1.0312	1 5/8	1.100	7/8	59/64
1 1/4	1.2500	1 9/32	1.2812	1 5/32	1.3125	2	1.370	1 7/64	1 11/64
1 1/2	1.5000	1 17/32	1.5312	1 9/16	1.5625	2 3/8	1.640	34mm	36mm

Notes: [1] Close fit is limited to holes for those lengths of screws threaded to the head in assemblies in which: A) only one screw is used; or B) two or more screws are used and the mating holes are produced at assembly or by matched and coordinated tooling. [2] Normal fit is intended for: A) screws of relatively long length; or B) assemblies that involve two or more screws and where the mating holes are produced by conventional tolerancing methods. This fit provides the maximum allowable eccentricity of the longest standard screws and for certain deviations in the parts being fastened. [3] It is considered good practice to chamfer or break the edges of holes that are smaller than the screw diameter plus twice the fillet height (D Max. + 2F Max.) in parts in which the hardness approaches, equals, or exceeds the screw hardness. Dimension "C" on the Table provides this value. All dimensions in inches. Source: Unbrako/SPS Technologies. Inc.

Table 66. Dimensions of Low Head Socket Head Cap Screws, Hexagon Recess.

| Nom. Size | TPI* | | Class | A Head Dia. | | D Body Dia. | | T Key Engagement | H Head Height | | Fillet Extension | | J Hex. Socket Size |
	UNRC	UNRF		Max.	Min.	Max.	Min.	Min.	Max.	Min.	Max.	Min.	Nom.
#4	40	48	3A	0.183	0.178	0.1120	0.1075	0.038	0.059	0.053	0.009	0.005	0.0500
#5	40	44	3A	0.205	0.200	0.1250	0.1202	0.044	0.065	0.059	0.010	0.006	0.0625
#6	32	40	3A	0.226	0.221	0.1380	0.1329	0.050	0.072	0.066	0.010	0.006	0.0625
#8	32	36	3A	0.270	0.265	0.1640	0.1585	0.060	0.085	0.079	0.012	0.007	0.0781
#10	24	32	3A	0.312	0.307	0.1900	0.1840	0.072	0.098	0.092	0.014	0.009	0.0938
1/4	20	28	3A	0.375	0.369	0.2500	0.2435	0.094	0.127	0.121	0.014	0.009	0.1250
5/16	18	24	3A	0.437	0.431	0.3125	0.3053	0.110	0.158	0.152	0.017	0.012	0.1562
3/8	16	24	3A	0.562	0.556	0.3750	0.3678	0.115	0.192	0.182	0.020	0.015	0.1875
1/2	13	20	3A	0.750	0.743	0.5000	0.4919	0.151	0.254	0.244	0.026	0.020	0.2500
5/8	11	18	3A	0.875	0.867	0.6250	0.6163	0.225	0.316	0.306	0.032	0.024	0.3125
3/4	10	16	3A	1.000	0.987	0.7500	0.7406	0.247	0.376	0.366	0.039	0.030	0.3750
7/8	9	14	3A	1.125	1.111	0.8750	0.8647	0.304	0.438	0.428	0.044	0.034	0.5000
1	8	12	3A	1.312	1.297	1.0000	0.9886	0.359	0.500	0.489	0.050	0.040	0.5620

* TPI = Threads per inch. All dimensions in inches. Source: SPS Technologies, Inc.

Table 67. Dimensions of Button Head, Hexagon Recess Screws.

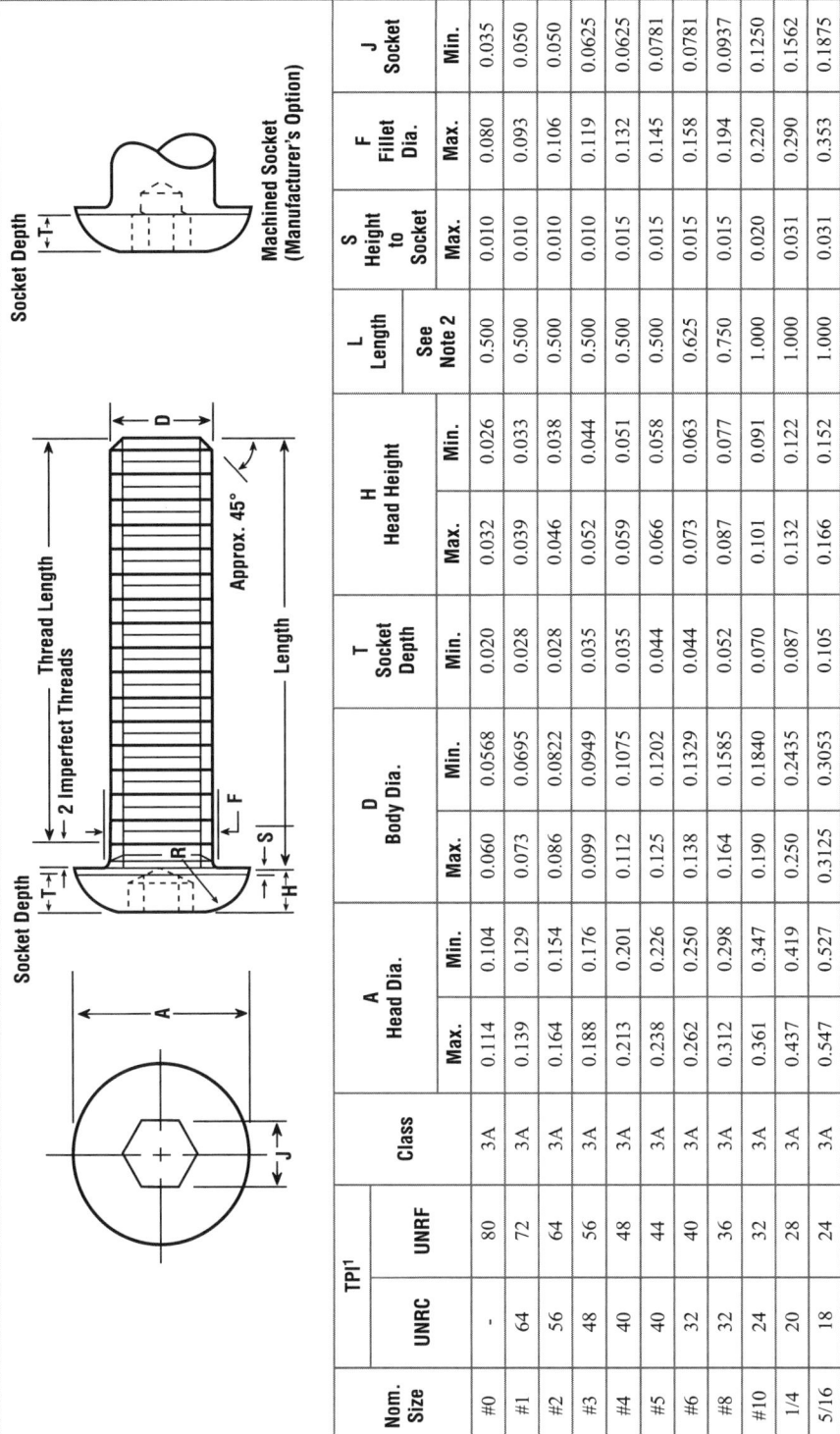

Nom. Size	TPI[1]		Class	A Head Dia.		D Body Dia.		T Socket Depth		H Head Height		L Length	S Height to Socket	F Fillet Dia.	J Socket
	UNRC	UNRF		Max.	Min.	Max.	Min.	Min.		Max.	Min.	See Note 2	Max.	Max.	Min.
#0	-	80	3A	0.114	0.104	0.060	0.0568	0.020		0.032	0.026	0.500	0.010	0.080	0.035
#1	64	72	3A	0.139	0.129	0.073	0.0695	0.028		0.039	0.033	0.500	0.010	0.093	0.050
#2	56	64	3A	0.164	0.154	0.086	0.0822	0.028		0.046	0.038	0.500	0.010	0.106	0.050
#3	48	56	3A	0.188	0.176	0.099	0.0949	0.035		0.052	0.044	0.500	0.010	0.119	0.0625
#4	40	48	3A	0.213	0.201	0.112	0.1075	0.035		0.059	0.051	0.500	0.015	0.132	0.0625
#5	40	44	3A	0.238	0.226	0.125	0.1202	0.044		0.066	0.058	0.500	0.015	0.145	0.0781
#6	32	40	3A	0.262	0.250	0.138	0.1329	0.044		0.073	0.063	0.625	0.015	0.158	0.0781
#8	32	36	3A	0.312	0.298	0.164	0.1585	0.052		0.087	0.077	0.750	0.015	0.194	0.0937
#10	24	32	3A	0.361	0.347	0.190	0.1840	0.070		0.101	0.091	1.000	0.020	0.220	0.1250
1/4	20	28	3A	0.437	0.419	0.250	0.2435	0.087		0.132	0.122	1.000	0.031	0.290	0.1562
5/16	18	24	3A	0.547	0.527	0.3125	0.3053	0.105		0.166	0.152	1.000	0.031	0.353	0.1875

(Continued)

Table 67. *(Continued)* **Dimensions of Button Head, Hexagon Recess Screws.**

| Nom. Size | TPI[1] | | Class | A Head Dia. | | D Body Dia. | | T Socket Depth | H Head Height | | L Length | S Height to Socket | F Fillet Dia. | J Socket |
	UNRC	UNRF		Max.	Min.	Max.	Min.	Min.	Max.	Min.	See Note 2	Max.	Max.	Min.
3/8	16	24	3A	0.656	0.636	0.375	0.3678	0.122	0.199	0.185	1.250	0.031	0.415	0.2187
1/2	13	20	3A	0.875	0.851	0.500	0.4919	0.175	0.265	0.245	2.000	0.046	0.560	0.3125
5/8	11	18	3A	1.000	0.970	0.625	0.6163	0.210	0.331	0.311	2.000	0.062	0.685	0.3750

Notes: [1] TPI = Threads per inch. [2] Screw lengths equal to or shorter than listed in column L will be threaded to head. For screws longer than listed, the minimum useable thread length shall be equal to twice the diameter + 0.500 inch. [3] Imperfect threads not to enter into fillet area.

Length tolerances for button head, hexagon recess screws

Up to 1" incl.	Over 1" to 2" incl.	Over 2"
- 0.03	- 0.04	- 0.06

All dimensions in inches. Source: Unbrako/SPS Technologies, Inc.

Table 68. Dimensions of 82° Flat Countersunk Head, Hexagon Recess Screws.

Nom. Size	TPI* UNRC	TPI* UNRF	Class	A Head Dia. Max.¹	A Head Dia. Min.²	D Body Dia. Max.	D Body Dia. Min.	T Key Engagement Min.	G Gage Dia. Max.	H Head Height Ref.	P Head Protrusion Max.	P Head Protrusion Min.	Z Flat Height Max.	J Hex Socket Size Nom.
#0	–	80	3A	0.138	0.117	0.060	0.0568	0.025	0.078	0.044	0.034	0.029	0.011	0.035
#1	64	72	3A	0.168	0.143	0.073	0.0695	0.031	0.101	0.054	0.038	0.032	0.014	0.050
#2	56	64	3A	0.197	0.168	0.086	0.0822	0.038	0.124	0.064	0.042	0.034	0.015	0.050
#3	48	56	3A	0.226	0.193	0.099	0.0949	0.044	0.148	0.073	0.044	0.035	0.018	0.0625
#4	40	48	3A	0.255	0.218	0.112	0.1075	0.055	0.172	0.083	0.047	0.037	0.021	0.0625
#5	40	44	3A	0.281	0.240	0.125	0.1202	0.061	0.196	0.090	0.048	0.037	0.022	0.0781
#6	32	40	3A	0.307	0.263	0.138	0.1329	0.066	0.220	0.097	0.049	0.037	0.024	0.0781
#8	32	36	3A	0.359	0.311	0.164	0.1585	0.076	0.267	0.112	0.051	0.039	0.026	0.0937
#10	24	32	3A	0.411	0.359	0.190	0.1840	0.087	0.313	0.127	0.054	0.041	0.028	0.1250
1/4	20	28	3A	0.531	0.480	0.250	0.2435	0.111	0.424	0.161	0.059	0.046	0.029	0.1562
5/16	18	24	3A	0.656	0.600	0.3125	0.3053	0.135	0.539	0.198	0.063	0.050	0.031	0.1875

(Continued)

Table 68. *(Continued)* Dimensions of 82° Flat Countersunk Head, Hexagon Recess Screws.

Nom. Size	TPI*		Class	A Head Dia.		D Body Dia.		T Key Engagement	G Gage Dia.	H Head Height	P Head Protrusion		Z Flat Height	J Hex Socket Size
	UNRC	UNRF		Max.[1]	Min.[2]	Max.	Min.	Min.	Max.	Ref.	Max.	Min.	Max.	Nom.
3/8	16	24	3A	0.781	0.720	0.375	0.3678	0.159	0.653	0.234	0.069	0.056	0.033	0.2187
7/16	14	20	3A	0.844	0.781	0.4375	0.4294	0.159	0.690	0.234	0.084	0.071	0.034	0.2500
1/2	13	20	3A	0.937	0.872	0.500	0.4919	0.172	0.739	0.251	0.110	0.096	0.034	0.3125
5/8	11	18	3A	1.188	1.112	0.625	0.6163	0.220	0.962	0.324	0.123	0.108	0.040	0.3750
3/4	10	16	3A	1.438	1.355	0.750	0.7406	0.220	1.186	0.396	0.136	0.121	0.040	0.5000
7/8	9	14	3A	1.688	1.605	0.875	0.8647	0.248	1.411	0.468	0.149	0.134	0.040	0.5625
1	8	12	3A	1.938	1.855	1.000	0.9886	0.297	1.635	0.540	0.162	0.146	0.040	0.6250

Notes: [1] Maximum to theoretical sharp corners. [2] Minimum-Absolute with a flat.

Length tolerances for 82° countersunk head, hexagon recess screws

Size	Up to 1" incl.	Over 1" to 2 1/2" incl.	Over 2 1/2" to 6" incl.	Over 6"
#0 thru 3/8"	- 0.03	- 0.04	- 0.06	- 0.12
7/16" thru 3/4"	- 0.03	- 0.06	- 0.08	- 0.12
7/8" and up	- 0.05	- 0.10	- 0.14	- 0.20

All dimensions in inches. Source: Unbrako/SPS Technologies, Inc.

Table 69. Dimensions for Hexagon Recess Socket Head Shoulder Bolts.

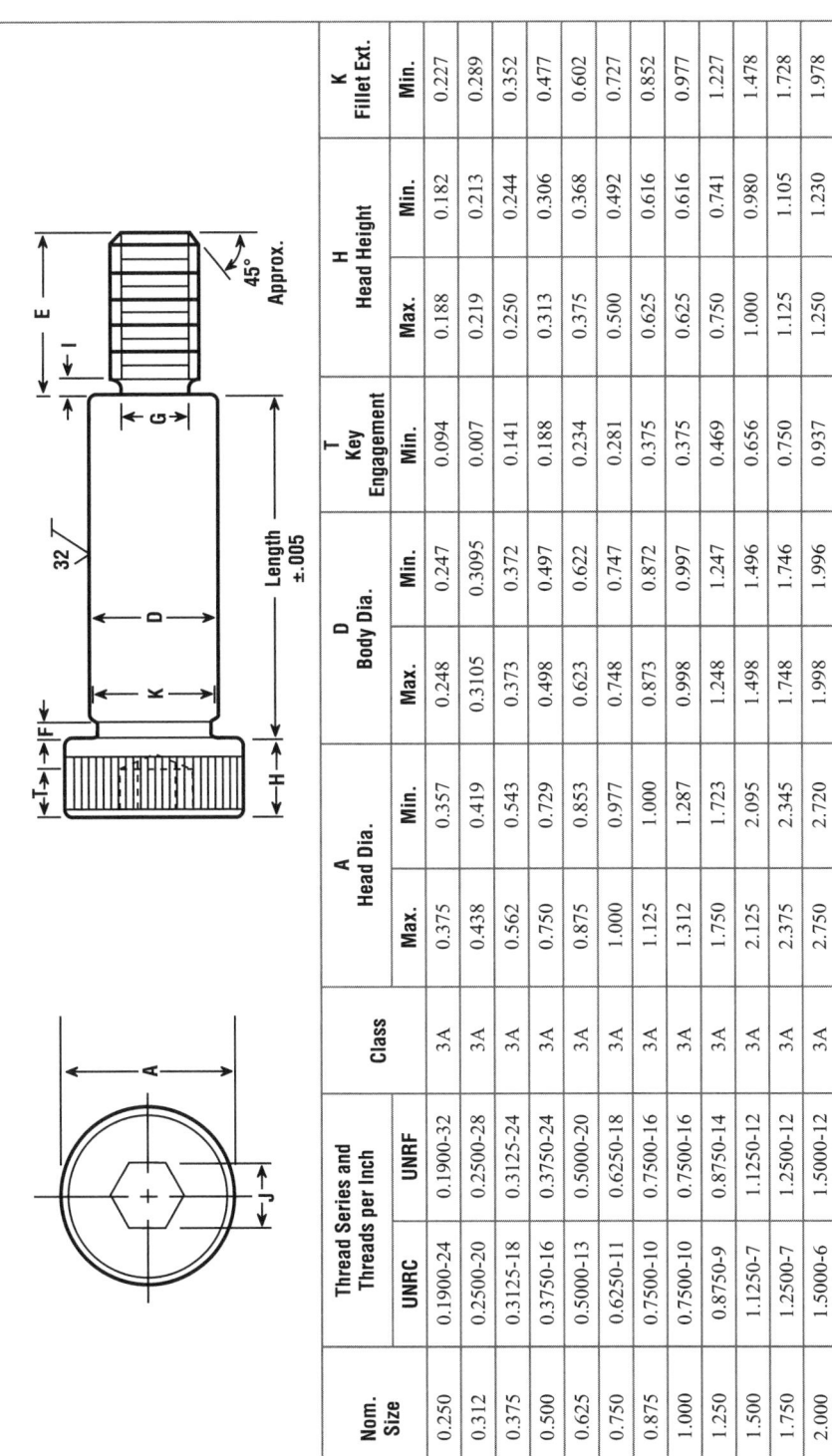

Nom. Size	Thread Series and Threads per Inch		Class	A Head Dia.		D Body Dia.		T Key Engagement	H Head Height		K Fillet Ext.
	UNRC	UNRF		Max.	Min.	Max.	Min.	Min.	Max.	Min.	Min.
0.250	0.1900-24	0.1900-32	3A	0.375	0.357	0.248	0.247	0.094	0.188	0.182	0.227
0.312	0.2500-20	0.2500-28	3A	0.438	0.419	0.3105	0.3095	0.007	0.219	0.213	0.289
0.375	0.3125-18	0.3125-24	3A	0.562	0.543	0.373	0.372	0.141	0.250	0.244	0.352
0.500	0.3750-16	0.3750-24	3A	0.750	0.729	0.498	0.497	0.188	0.313	0.306	0.477
0.625	0.5000-13	0.5000-20	3A	0.875	0.853	0.623	0.622	0.234	0.375	0.368	0.602
0.750	0.6250-11	0.6250-18	3A	1.000	0.977	0.748	0.747	0.281	0.500	0.492	0.727
0.875	0.7500-10	0.7500-16	3A	1.125	1.000	0.873	0.872	0.375	0.625	0.616	0.852
1.000	0.7500-10	0.7500-16	3A	1.312	1.287	0.998	0.997	0.375	0.625	0.616	0.977
1.250	0.8750-9	0.8750-14	3A	1.750	1.723	1.248	1.247	0.469	0.750	0.741	1.227
1.500	1.1250-7	1.1250-12	3A	2.125	2.095	1.498	1.496	0.656	1.000	0.980	1.478
1.750	1.2500-7	1.2500-12	3A	2.375	2.345	1.748	1.746	0.750	1.125	1.105	1.728
2.000	1.5000-6	1.5000-12	3A	2.750	2.720	1.998	1.996	0.937	1.250	1.230	1.978

(Continued)

Table 69. *(Continued)* **Dimensions for Hexagon Recess Socket Head Shoulder Bolts.**

Nom. Size	Thread Series and Threads per Inch		Class	G Thread Neck Dia.				F Shoulder Neck Width	I Thread Neck Width	E Thread & Neck Length	J Hex Socket Size
				UNRC		UNRF					
	UNRC	UNRF		Max.	Min.	Max.	Min.	Max.	Max.	See Note 1	Nom.
0.250	0.1900-24	0.1900-32	3A	0.142	0.133	0.156	0.147	0.093	0.083	0.375	0.125
0.312	0.2500-20	0.2500-28	3A	0.193	0.182	0.210	0.201	0.093	0.100	0.437	0.156
0.375	0.3125-18	0.3125-24	3A	0.249	0.237	0.265	0.256	0.093	0.111	0.500	0.187
0.500	0.3750-16	0.3750-24	3A	0.304	0.291	0.327	0.318	0.093	0.125	0.625	0.250
0.625	0.5000-13	0.5000-20	3A	0.414	0.397	0.443	0.432	0.093	0.154	0.750	0.312
0.750	0.6250-11	0.6250-18	3A	0.521	0.502	0.561	0.549	0.093	0.182	0.875	0.375
0.875	0.7500-10	0.7500-16	3A	0.638	0.616	0.678	0.665	0.093	0.200	1.000	0.500
1.000	0.7500-10	0.7500-16	3A	0.638	0.616	0.678	0.665	0.125	0.200	1.000	0.500
1.250	0.8750-9	0.8750-14	3A	0.750	0.726	0.796	0.778	0.125	0.222	1.125	0.625
1.500	1.1250-7	1.1250-12	3A	0.964	0.934	1.022	1.014	0.125	0.286	1.500	0.875
1.750	1.2500-7	1.2500-12	3A	1.089	1.059	1.147	1.139	0.125	0.286	1.750	1.000
2.000	1.5000-6	1.5000-12	3A	1.307	1.277	1.397	1.389	0.125	0.333	2.000	1.250

Notes: [1] Dimension is +0.000 / -0.020.

* TPI = Threads per inch.

All dimensions in inches. Source: SPS Technologies, Inc.

Table 70. Dimensions for Socket Hexagon Recess Set Screws. (See Notes.)

Plain Cup, Flat, Half Dog, Cone, Knurled Cone, Oval

Nom. Size	Threads per Inch		Class	A Diameter				C Point Dia. Cup and Flat		P Point Dia. Dog and Half Dog	
	UNRC	UNRF		UNRC		UNRF					
				Max.	Min.	Max.	Min.	Max.	Min.	Max.	Min.
-	-	80	3A	-	-	0.0600	0.0568	0.033	0.027	0.040	0.037
#1	64	72	3A	0.0730	0.0692	0.0730	0.0695	0.040	0.033	0.049	0.045
#2	56	64	3A	0.0860	0.0819	0.0860	0.0822	0.047	0.039	0.057	0.053
#3	48	56	3A	0.0990	0.0945	0.0990	0.0949	0.054	0.045	0.066	0.062
#4	40	48	3A	0.1120	0.1069	0.1120	0.1075	0.061	0.051	0.075	0.070
#5	40	44	3A	0.1250	0.1199	0.1250	0.1202	0.067	0.057	0.083	0.078

(Continued)

Table 70. (Continued) Dimensions for Socket Hexagon Recess Set Screws. (See Notes.)

Nom. Size	Threads per Inch		Class	A Diameter				C Point Dia. Cup and Flat		P Point Dia. Dog and Half Dog	
	UNRC	UNRF		UNRC		UNRF					
				Max.	Min.	Max.	Min.	Max.	Min.	Max.	Min.
#6	32	40	3A	0.1380	0.1320	0.1380	0.1329	0.074	0.064	0.092	0.087
#8	32	36	3A	0.1640	0.1580	0.1640	0.1585	0.087	0.076	0.109	0.103
#10	24	32	3A	0.1900	0.1828	0.1900	0.1840	0.102	0.088	0.127	0.120
1/4	20	28	3A	0.2500	0.2419	0.2500	0.2435	0.132	0.118	0.156	0.149
5/16	18	24	3A	0.3125	0.3038	0.3125	0.3053	0.172	0.156	0.203	0.195
3/8	16	24	3A	0.3750	0.3656	0.3750	0.3678	0.212	0.194	0.250	0.241
7/16	14	20	3A	0.4375	0.4272	0.4375	0.4294	0.252	0.232	0.296	0.287
1/2	13	20	3A	0.5000	0.4891	0.5000	0.4919	0.291	0.270	0.343	0.334
9/16	12	18	3A	0.5625	0.5511	0.5625	0.5538	0.332	0.309	0.390	0.379
5/8	11	18	3A	0.6250	0.6129	0.6250	0.6163	0.371	0.347	0.468	0.456
3/4	10	16	3A	0.7500	0.7371	0.7500	0.7406	0.450	0.425	0.562	0.549
7/8	9	14	3A	0.8750	0.8611	0.8750	0.8647	0.530	0.502	0.656	0.642
1	8	12	3A	1.0000	0.9850	1.0000	0.9897	0.609	0.579	0.750	0.734
1 1/8	7	12	3A	1.1250	1.1086	1.1250	1.1136	0.689	0.655	0.843	0.826
1 1/4	7	12	3A	1.2500	1.2336	1.2500	1.2386	0.767	0.733	0.937	0.920
1 3/8	6	12	3A	1.3750	1.3568	1.3750	1.3636	0.848	0.808	1.031	1.011
1 1/2	6	12	3A	1.5000	1.4818	1.5000	1.4886	0.926	0.886	1.125	1.105

(Continued)

Table 70. *(Continued)* **Dimensions for Socket Hexagon Recess Set Screws.** (See Notes.)

Nom. Size	Threads per Inch		Class	Q Point Length Half Dog		Point Length Dog (Not Shown)		R Radius Oval Point	T¹ Socket Depth	J Hex Socket Size
	UNRC	UNRF		Max.	Min.	Max.	Min.	Basic	Min.	Nom.
-	-	80	3A	0.017	0.013	0.033	0.027	0.045	0.035	0.028
#1	64	72	3A	0.021	0.017	0.040	0.034	0.055	0.035	0.035
#2	56	64	3A	0.024	0.020	0.047	0.039	0.064	0.035	0.035
#3	48	56	3A	0.027	0.023	0.054	0.046	0.074	0.060	0.050
#4	40	48	3A	0.030	0.026	0.060	0.052	0.084	0.075	0.050
#5	40	44	3A	0.033	0.027	0.064	0.056	0.094	0.075	0.0625
#6	32	40	3A	0.038	0.032	0.074	0.066	0.104	0.075	0.0625
#8	32	36	3A	0.043	0.037	0.084	0.076	0.123	0.075	0.0781
#10	24	32	3A	0.049	0.041	0.095	0.085	0.142	0.105	0.0937
1/4	20	28	3A	0.0665	0.0585	0.130	0.120	0.188	0.105	0.125
5/16	18	24	3A	0.082	0.074	0.164	0.148	0.234	0.140	0.1562
3/8	16	24	3A	0.0987	0.0887	0.1955	0.1795	0.281	0.140	0.1875
7/16	14	20	3A	0.114	0.104	0.2267	0.2107	0.328	0.190	0.2187
1/2	13	20	3A	0.130	0.120	0.260	0.240	0.375	0.210	0.250
9/16	12	18	3A	0.1456	0.1356	0.291	0.271	0.422	0.265	0.250
5/8	11	18	3A	0.164	0.148	0.3225	0.3025	0.469	0.265	0.3125
3/4	10	16	3A	0.1955	0.1795	0.385	0.365	0.562	0.330	0.375
7/8	9	14	3A	0.2267	0.2107	0.4475	0.4275	0.656	0.450	0.500
1	8	12	3A	0.260	0.240	0.510	0.490	0.750	0.550	0.5625
1 1/8	7	12	3A	0.291	0.271	0.5775	0.5475	0.844	0.650	0.5625
1 1/4	7	12	3A	0.3225	0.3025	0.640	0.610	0.938	0.700	0.625

(Continued)

Table 70. (Continued) Dimensions for Socket Hexagon Recess Set Screws. (See Notes.)

Nom. Size	Threads per Inch		Class	Q Point Length Half Dog		Point Length Dog (Not Shown)		R Radius Oval Point	T¹ Socket Depth	J Hex Socket Size
	UNRC	UNRF		Max.	Min.	Max.	Min.	Basic	Min.	Nom.
1 3/8	6	12	3A	0.3537	0.3337	0.7025	0.6725	1.031	0.700	0.625
1 1/2	6	12	3A	0.385	0.365	0.765	0.735	1.125	0.750	0.750

Notes: When cone cup set screw length equals nominal diameter or less, included angle is 118°; for other lengths the included angle is 90°. When plain cup set screw length equals nominal diameter or less, included angle is 130°; for other lengths the included angle is 118°.
¹ Values shown in column T are for minimum stock length cup point screws. Screws shorter than nominal minimum length shown do not have sockets deep enough to utilize full key capacity which can result in failure of socket, key, or mating threads.

Length tolerance table for socket hexagon recess set screws

Up to 0.630"	Over 0.630" to 2.0"	Over 2.0" to 6.0"	Over 6.0"
± 0.01	± 0.02	± 0.03	± 0.06

All dimensions in inches. Source: Unbrako/SPS Technologies, Inc.

Table 71. Torsional and Axial Holding Power of Cup Point Socket Set Screws. (See Notes.)

Nom. Size	Seating Torque (in.-lb)	Axial Holding Power (Pounds)	Shaft Diameter (shaft hardness R_C 15 to R_C 35)							
			Torsional Holding Power (in.-lb)							
			1/16	3/32	1/8	5/32	3/16	7/32	1/4	5/16
#0	1.0	50	**1.5**	2.3	3.1	3.9	4.7	5.4	6.2	
#1	1.8	65	2.0	**3.0**	4.0	5.0	6.1	7.1	8.1	10.0
#2	1.8	85	2.6	4.0	**5.3**	6.6	8.0	9.3	10.6	13.2
#3	5	120	3.2	5.6	7.5	9.3	11.3	13.0	15.0	18.7
#4	5	160		7.5	10.0	**9.3**	15.0	17.5	20.0	25.0
#5	10	200			12.5	15.6	**15.0**	**21.8**	**25.0**	31.2
#6	10	250				19	23	27	31	**39**
#8	20	385				30	36	42	48	60
#10	36	540					51	59	68	84
1/4	87	1,000							125	156
5/16	165	1,500								234

Nom. Size	Seating Torque (in.-lb)	Axial Holding Power (Pounds)	Shaft Diameter (shaft hardness R_C 15 to R_C 35)							
			Torsional Holding Power (in.-lb)							
			3/8	7/16	1/2	9/16	5/8	3/4	7/8	1
#2	1.8	85	16.0							
#3	5	120	22.5	26.3						
#4	5	160	30.0	35.0	40.0					
#5	10	200	37.5	43.7	50.0	56.2	62			
#6	10	250	47	55	62	70	78	94	109	
#8	20	385	**72**	84	96	108	120	144	168	192
#10	36	540	101	**118**	135	152	169	202	236	270
1/4	87	1,000	187	218	**250**	281	312	357	437	500

(Continued)

Table 71. (Continued) Torsional and Axial Holding Power of Cup Point Socket Set Screws. (See Notes.)

Nom. Size	Seating Torque (in.-lb)	Axial Holding Power (Pounds)	Shaft Diameter (shaft hardness R_C 15 to R_C 35) Torsional Holding Power (in.-lb)							
			3/8	7/16	1/2	9/16	5/8	3/4	7/8	1
5/16	165	1,500	280	327	375	421	468	562	656	750
3/8	290	2,000	375	437	500	562	625	750	875	1000
7/16	430	2,500		545	625	702	780	937	1095	1250
1/2	620	3,000			750	843	937	1125	1310	1500
9/16	620	3,500				985	1090	1310	1530	1750
5/8	1,325	4,000					1250	1500	1750	2000
3/4	2,400	5,000						1875	2190	2500
7/8	3,600	5,600							2620	3000
1	5,000	6,500								3500

Nom. Size	Seating Torque (in.-lb)	Axial Holding Power (Pounds)	Shaft Diameter (shaft hardness R_C 15 to R_C 35) Torsional Holding Power (in.-lb)							
			1 1/4	1 1/2	1 3/4	2	2 1/2	3	3 1/2	4
#10	36	540	338							
1/4	87	1,000	625	750						
5/16	165	1,500	937	1125	1310	1500				
3/8	290	2,000	1250	1500	1750	2000				
7/16	430	2,500	1560	1875	2210	2500	3125			
1/2	620	3,000	1875	2250	2620	3000	3750	4500		
9/16	620	3,500	2190	2620	3030	3500	4370	5250	6120	
5/8	1,325	4,000	2500	3000	3750	4000	5000	6000	7000	8000
3/4	2,400	5,000	3125	3750	4500	5000	6250	7500	8750	10000

(Continued)

Table 71. *(Continued)* **Torsional and Axial Holding Power of Cup Point Socket Set Screws.** (See Notes.)

Nom. Size	Seating Torque (in.-lb)	Axial Holding Power (Pounds)	Shaft Diameter (shaft hardness R_C 15 to R_C 35)							
			1 1/4	1 1/2	1 3/4	2	2 1/2	3	3 1/2	4
			Torsional Holding Power (in.-lb)							
7/8	3,600	5,600	3750	4500	5250	6000	7500	9000	**10500**	12000
1	5,000	6,500	4375	5250	6120	7000	8750	10500	12250	**14000**

Notes: Values are for cup point socket screws seated at recommended installation torques, using Class 3A threads in Class 2A tapped holes. Holding power was defined as the minimum load to produce 0.010 inch relative movement of shaft and collar. Tabulated axial and torsional holding powers are typical strengths and should be used accordingly, with specific safety factors appropriate to the given application and load conditions. Values in bold type indicate recommended set screw sizes on the basis that screw diameter should be roughly one-half shaft diameter. Source: Unbrako/SPS Technologies, Inc.

Table 72. Dimensions of Hexagon Keys.

W Key Size		C Length of Long Arm				B Length of Short Arm	
		Short Series		Long Series			
Max.	Min.	Max.	Min.	Max.	Min.	Max.	Min.
0.0280	0.0275	1.312	1.125	2.688	2.500	0.312	0.125
0.0350	0.0345	1.312	1.125	2.766	2.578	0.438	0.250
0.0500	0.0490	1.750	1.562	2.938	2.750	0.625	0.438
1/16	0.0615	1.844	1.656	3.094	2.906	0.656	0.469
5/64	0.0771	1.969	1.781	3.281	3.094	0.703	0.516
3/32	0.0927	2.094	1.906	3.469	3.281	0.750	0.562
7/64	0.1079	2.219	2.031	3.656	3.469	0.797	0.609
1/8	0.1235	2.344	2.156	3.844	3.656	0.844	0.656
9/64	0.1391	2.469	2.281	4.031	3.844	0.891	0.703
5/32	0.1547	2.594	2.406	4.219	4.031	0.938	0.750
3/16	0.1860	2.844	2.656	4.594	4.406	1.031	0.844
7/32	0.2172	3.094	2.906	4.969	4.781	1.125	0.938
1/4	0.2485	3.344	3.156	5.344	5.156	1.219	1.031
5/16	0.3110	3.844	3.656	6.094	5.906	1.344	1.156
3/8	0.3735	4.344	4.156	6.844	6.656	1.469	1.281
7/16	0.4355	4.844	4.656	7.594	7.406	1.594	1.406
1/2	0.4975	5.344	5.156	8.344	8.156	1.719	1.531
9/16	0.5600	5.844	5.656	9.094	8.906	1.844	1.656
5/8	0.6225	6.344	6.156	9.844	9.656	1.969	1.781
3/4	0.7470	7.344	7.156	11.344	11.156	2.219	2.031
7/8	0.8720	8.344	8.156	12.844	12.656	2.469	2.281
1	0.9970	9.344	9.156	14.344	14.156	2.719	2.531
1-1/4	1.2430	11.500	11.000	-	-	3.250	2.750
1-1/2	1.4930	13.500	13.000	-	-	3.750	3.250
1-3/4	1.7430	15.500	15.000	-	-	4.250	3.750
2	1.9930	17.500	17.000	-	-	4.750	4.250

All dimensions in inches. Source: Unbrako/SPS Technologies, Inc.

Table 73. Dimensions of Spline Sockets.

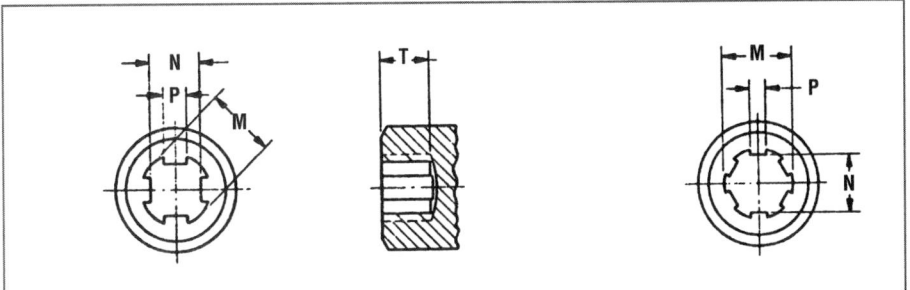

Nom. Socket and Key Size	No. of Teeth	M Socket Major Dia.		N Socket Minor Dia.		P Width of Tooth	
		Max.	Min.	Max.	Min.	Max.	Min.
0.033	4	0.0350	0.0340	0.0260	0.0255	0.0120	0.0115
0.048	6	0.050	0.049	0.041	0.040	0.011	0.010
0.060	6	0.062	0.061	0.051	0.050	0.014	0.013
0.072	6	0.074	0.073	0.064	0.063	0.016	0.015
0.096	6	0.098	0.097	0.082	0.080	0.022	0.021
0.111	6	0.115	0.113	0.098	0.096	0.025	0.023
0.133	6	0.137	0.135	0.118	0.116	0.030	0.028
0.145	6	0.149	0.147	0.128	0.126	0.032	0.030
0.168	6	0.173	0.171	0.150	0.147	0.036	0.033
0.183	6	0.188	0.186	0.163	0.161	0.039	0.037
0.216	6	0.221	0.219	0.190	0.188	0.050	0.048
0.251	6	0.256	0.254	0.221	0.219	0.060	0.058
0.291	6	0.298	0.296	0.254	0.252	0.068	0.066
0.372	6	0.380	0.377	0.319	0.316	0.092	0.089
0.454	6	0.463	0.460	0.386	0.383	0.112	0.109
0.595	6	0.604	0.601	0.509	0.506	0.138	0.134
0.620	6	0.631	0.627	0.535	0.531	0.149	0.145
0.698	6	0.709	0.705	0.604	0.600	0.168	0.164
0.790	6	0.801	0.797	0.685	0.681	0.189	0.185

All dimensions in inches. Source: Extracted from ANSI B18.3-1969, published by the American Society of Mechanical Engineers.

Table 74. Dimensions of Spline Keys.

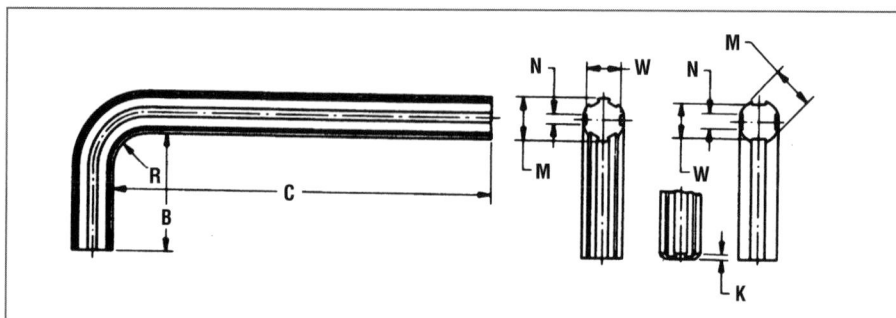

Nom. Socket and Key Size	M Major Dia.		W Minor Dia.		N Width of Indent		B Length of Short Arm	
	Max.	Min.	Max.	Min.	Max.	Min.	Max.	Min.
0.033	0.0330	0.0320	0.0250	0.0240	0.0140	0.0130	0.312	0.125
0.048	0.0480	0.0470	0.0390	0.0380	0.0130	0.0120	0.438	0.250
0.060	0.0600	0.0590	0.0490	0.0480	0.0160	0.0150	0.625	0.438
0.072	0.0720	0.0710	0.0620	0.0610	0.0190	0.0180	0.656	0.469
0.096	0.0960	0.0950	0.0790	0.0775	0.0240	0.0230	0.703	0.516
0.111	0.1110	0.1100	0.0940	0.0925	0.0280	0.0270	0.750	0.562
0.133	0.0330	0.1310	0.1140	0.1120	0.0340	0.0320	0.797	0.609
0.145	0.0450	0.1435	0.1240	0.1225	0.0355	0.0340	0.844	0.656
0.168	0.1680	0.1660	0.1440	0.1420	0.0410	0.0390	0.891	0.703
0.183	0.1830	0.1815	0.1580	0.1565	0.0440	0.0425	0.938	0.750
0.216	0.2160	0.2145	0.1840	0.1825	0.0550	0.0535	1.031	0.844
0.251	0.2510	0.2495	0.2140	0.2125	0.0655	0.0640	1.125	0.938
0.281	0.2910	0.2895	0.2460	0.2445	0.0775	0.0760	1.219	1.031
0.372	0.3720	0.3705	0.3100	0.3085	0.0975	0.0960	1.344	1.156
0.454	0.4540	0.4525	0.3770	0.3755	0.1185	0.1170	1.469	1.281
0.595	0.5950	0.5935	0.5000	0.4975	0.1460	0.1445	1.719	1.531
0.620	0.6200	0.6175	0.5240	0.5215	0.1615	0.1590	1.844	1.656
0.698	0.6980	0.6955	0.5930	0.5905	0.1805	0.1780	1.844	1.656
0.790	0.7900	0.7875	0.6740	0.6715	0.1975	0.1950	1.969	1.781

Nom. Key and Socket Size	C Length of Long Arm				Number of Splines	R Radius of Bend	K Chamfer
	Short Series		Long Series				
	Max.	Min.	Max.	Min.		Min.	Max.
0.033	1.312	1.125	-	-	4	0.062	0.003
0.048	1.312	1.125	-	-	6	0.062	0.004
0.060	1.750	1.562	-	-	6	0.062	0.006
0.072	1.844	1.656	-	-	6	0.062	0.008
0.096	1.969	1.781	-	-	6	0.078	0.008

(Continued)

Table 74. *(Continued)* **Dimensions of Spline Keys.**

| Nom. Key and Socket Size | C Length of Long Arm | | | | Number of Splines | R Radius of Bend | K Chamfer |
| | Short Series | | Long Series | | | | |
	Max.	Min.	Max.	Min.		Min.	Max.
0.111	2.094	1.906	-	-	6	0.094	0.009
0.133	2.219	2.031	3.656	3.469	6	0.125	0.014
0.145	2.344	2.156	3.844	3.656	6	0.125	0.015
0.168	2.469	2.281	4.031	3.844	6	0.156	0.016
0.183	2.594	2.406	4.219	4.031	6	0.156	0.016
0.216	2.844	2.656	4.594	4.406	6	0.188	0.022
0.251	3.094	2.906	4.969	4.781	6	0.219	0.024
0.281	3.344	3.156	5.344	5.156	6	0.250	0.030
0.372	3.844	3.656	6.094	5.906	6	0.312	0.032
0.454	4.344	4.156	6.844	6.656	6	0.375	0.044
0.595	5.344	5.156	8.344	8.156	6	0.500	0.050
0.620	5.844	5.656	9.094	8.906	6	0.500	0.053
0.698	5.844	5.656	-	-	6	0.562	0.055
0.790	6.344	6.156	-	-	6	0.625	0.070

All dimensions in inches. Source: Extracted from ANSI B18.3-1960, published by the American Society of Mechanical Engineers.

Table 75. ISO Tolerances for Metric Fasteners.

Nom. Dia.		Tolerance Zone (External Threads)								
Over	To	h6	h8	h10	h11	h13	h14	h15	h16	js14
0	1	0 −0.006	0 −0.014	0 −0.040	0 −0.060	0 −0.14	−	−	−	−
1	3	0 −0.006	0 −0.014	0 −0.040	0 −0.060	0 −0.14	0 −0.25	0 −0.40	0 −0.60	±0.125
3	6	0 −0.008	0 −0.018	0 −0.048	0 −0.075	0 −0.18	0 −0.30	0 −0.48	0 −0.75	±0.15
6	10	0 −0.009	0 −0.022	0 −0.058	0 −0.090	0 −0.22	0 −0.36	0 −0.58	0 −0.90	±0.18
10	18	0 −0.011	0 −0.027	0 −0.070	0 −0.110	0 −0.27	0 −0.43	0 −0.70	0 −1.10	±0.215
18	30	0 −0.030	0 −0.033	0 −0.084	0 −0.130	0 −0.33	0 −0.52	0 −0.84	0 −1.30	±0.26
30	50	−	−	−	−	0 −0.39	0 −0.62	0 −1.00	0 −1.60	±0.31
50	80	−	−	−	−	0 −0.46	0 −0.74	0 −1.20	0 −1.90	±0.37
80	120	−	−	−	−	0 −0.54	0 −0.87	0 −1.40	0 −2.20	±0.435

(Continued)

Table 75. *(Continued)* **ISO Tolerances for Metric Fasteners.**

Nom. Dia.		Tolerance Zone (External Threads)				Tolerance Zone (Internal Threads)					
Over	To	js15	js16	js17	m6	H7	H8	H9	H11	H13	H14
0	1	-	-	-	+0.002 +0.008	+0.010 0	+0.014 0	+0.025 0	+0.060 0	+0.14 0	-
1	3	±0.20	±0.30	±0.50	+0.002 +0.008	+0.010 0	+0.014 0	+0.025 0	+0.060 0	+0.14 0	+0.25 0
3	6	±0.24	±0.375	±0.60	+0.004 +0.012	+0.012 0	+0.018 0	+0.030 0	+0.075 0	+0.18 0	+0.30 0
6	10	±0.29	±0.45	±0.075	+0.006 +0.015	+0.015 0	+0.022 0	+0.036 0	+0.090 0	+0.22 0	+0.36 0
10	18	±0.35	±0.55	±0.90	+0.007 +0.018	+0.018 0	+0.027 0	+0.043 0	+0.110 0	+0.27 0	+0.43 0
18	30	±0.42	±0.65	±1.05	+0.008 +0.021	+0.021 0	+0.033 0	+0.052 0	+0.130 0	+0.33 0	+0.52 0
30	50	±0.50	±0.80	±1.25	-	-	-	-	-	+0.39 0	+0.62 0
50	80	±0.60	±0.95	±1.50	-	-	-	-	-	+0.46 0	+0.74 0
80	120	±0.70	±1.10	±1.75	-	-	-	-	-	+0.54 0	+0.87 0

All dimensions in millimeters. References: ISO R286, ISO 4759/I, ISO 4759/II, ISO 4759/III. Source: Unbrako/SPS Technologies, Inc.

Table 76. Dimensions of Metric Hexagon Cap Screw (ISO/DIN 4014, DIN 931).

Bolt Size	E Body Dia.		F Width Across Flats		G Width Across Corners	H Head Height		L_T Thread Length			
	Max.	Min.	Max.	Min.	Min.	Max.	Min.	L_T to 125	L_T to 200	L_T over 200
M4	4.0	3.82	7.00	6.78	7.66	2.50	2.40	14.00	-	-
M5	5.0	4.82	8.00	7.78	8.79	3.65	3.35	16.00	-	-
M6	6.0	5.82	10.00	9.78	11.05	4.15	3.85	18.00	24.00	-
M8	8.0	7.78	13.00	12.73	14.38	5.45	5.15	22.00	28.00	-
M10	10.0	9.78	16.00	15.73	18.90	6.58	6.22	26.00	32.00	45.00
M12	12.0	11.73	18.00	17.73	21.10	7.68	7.32	30.00	36.00	49.00
M14	14.0	13.73	21.00	20.67	24.49	8.98	8.62	34.00	40.00	53.00
M16	16.0	15.73	24.00	23.67	26.75	10.18	9.82	38.00	44.00	57.00
M18	18.0	17.73	27.00	26.67	30.14	11.72	11.28	42.00	48.00	61.00

(Continued)

Table 76. *(Continued)* **Dimensions of Metric Hexagon Cap Screw (ISO/DIN 4014, DIN 931).**

Bolt Size	E Body Dia.		F Width Across Flats		G Width Across Corners	H Head Height		L_T Thread Length		
	Max.	Min.	Max.	Min.	Min.	Max.	Min.	L_T to 125	L_T to 200	L_T over 200
M20	20.0	19.67	30.00	29.67	33.53	12.72	12.28	46.00	52.00	65.00
M22	22.0	21.67	34.00	33.38	35.72	14.22	13.78	50.00	56.00	69.00
M24	24.0	23.67	36.00	35.38	39.98	15.22	14.78	54.00	60.00	73.00
M27	27.0	26.67	41.00	40.38	45.20	17.05	16.35	60.00	66.00	79.00
M30	30.0	29.67	46.00	45.00	50.85	19.12	18.28	66.00	72.00	85.00
M33	33.0	32.61	50.00	49.00	55.37	20.92	20.08	72.00	78.00	91.00
M36	36.0	35.61	55.00	53.80	60.79	22.92	22.08	78.00	84.00	97.00

All dimensions in millimeters.

Table 77. Nominal Dimensions of Metric Fine Pitch Hexagon Head Bolt (DIN 960).

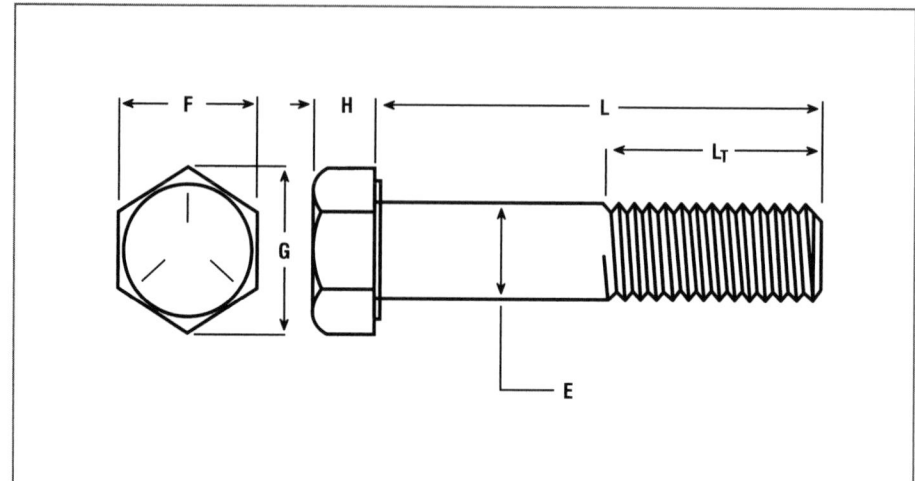

Bolt Size	E Body Dia.	F Width Across Flats	G Width Across Corners	H Head Height	L_T Thread Length		
					L_T to 125	L_T to 200	L_T over 200
M8×1	8.00	13.00	14.38	5.30	22.00	28.00	-
M10×1	10.00	17.00	18.90	6.40	26.00	32.00	-
M10×1.25	10.00	17.00	18.90	6.40	26.00	32.00	-
M12×1.25	12.00	19.00	21.10	7.50	30.00	36.00	-
M12×1.50	12.00	19.00	21.10	7.50	30.00	36.00	-
M14×1.50	14.00	22.00	24.49	8.80	34.00	40.00	-
M16×1.50	16.00	24.00	26.75	10.00	38.00	44.00	-
M18×1.50	18.00	27.00	30.14	11.50	42.00	48.00	-
M20×1.50	20.00	30.00	33.53	12.50	46.00	52.00	-
M20×2.00	20.00	30.00	33.53	12.50	46.00	52.00	-
M24×2.00	24.00	36.00	39.98	15.00	54.00	60.00	-
M27×2.00	27.00	41.00	45.20	17.00	60.00	66.00	79.00
M30×2.00	30.00	46.00	50.85	18.70	66.00	72.00	58.00
M33×2.00	33.00	50.00	55.37	21.00	72.00	78.00	91.00
M36×3.00	36.00	55.00	60.79	22.50	78.00	84.00	97.00

All dimensions in millimeters.

Table 78. Nominal Dimensions of Metric Fine Pitch Hexagon Head Bolt (DIN 961).

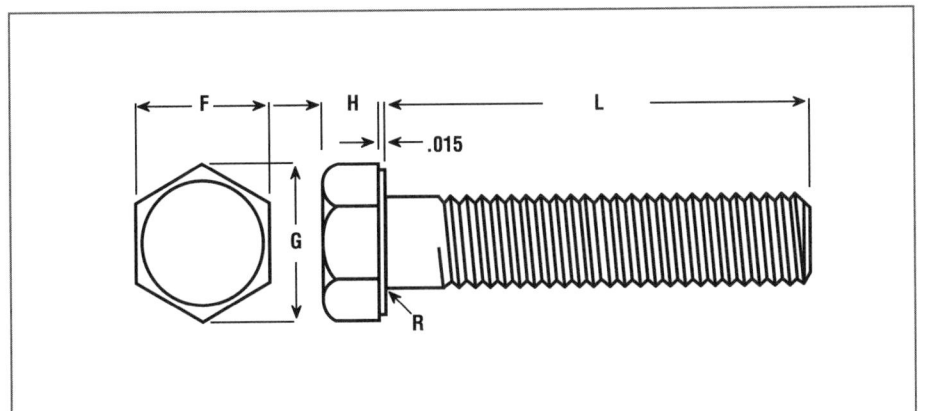

Bolt Size	Body Dia.	F Width Across Flats	G Width Across Corners	H Head Height
M8	8.00	13.00	14.38	5.30
M10	10.00	17.00	18.90	6.40
M12	12.00	19.00	21.10	7.50
M14	14.00	22.00	24.49	8.80
M16	16.00	24.00	26.75	10.00
M18	18.00	27.00	30.14	11.50
M20	20.00	30.00	33.53	12.50
M22	22.00	32.00	35.72	14.00
M24	24.00	36.00	39.98	15.00
M27	27.00	41.00	45.63	17.00
M30	30.00	46.00	51.28	19.00
M33	33.00	50.00	55.80	21.00
M36	36.00	55.00	61.31	23.00
M39	39.00	60.00	66.96	25.00
M42	42.00	65.00	72.61	26.00
M45	45.00	70.00	78.29	28.00
M48	48.00	75.00	83.91	30.00
M52	52.00	80.00	89.56	33.00

All dimensions in inches.

Table 79. Dimensions of Metric Heavy (Structural) Hexagon Head Bolt (DIN 6914). (See Notes.)

Bolt Size	E Body Dia.		F Width Across Flats		G Width Across Corners	H Head Height		L_T Thread Length ≤100	L_T Thread Length 100-200
	Max.	Min.	Max.	Min.	Min.	Max.	Min.		
M16	16.70	15.30	27.00	26.16	29.56	10.75	9.25	26.00	28.00
M20	20.84	19.16	32.00	31.00	35.03	13.90	12.10	31.00	33.00
M22	22.84	21.16	36.00	35.00	39.55	14.90	13.10	32.00	34.00
M24	24.84	23.16	41.00	40.00	45.20	15.90	14.10	34.00	37.00
M27	27.84	26.16	46.00	45.00	50.85	17.90	16.10	37.00	39.00
M30	30.84	29.16	50.00	49.00	55.37	19.75	17.65	40.00	42.00
M36	37.00	35.00	60.00	58.80	66.44	23.55	21.45	48.00	50.00

Notes: These hexagon bolts have increased width across flats for high tensile structural bolting. Use with hex nut specified in DIN 6915 and washer specified in DIN 6916. All dimensions in millimeters.

Table 80. Nominal Dimensions of Metric Hexagon Flange Bolts (DIN 6921).

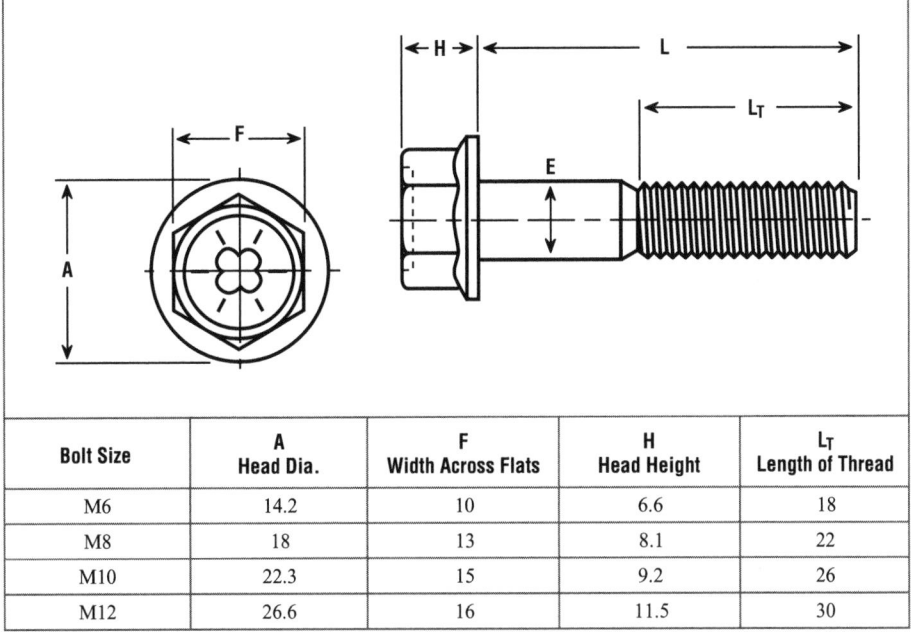

Bolt Size	A Head Dia.	F Width Across Flats	H Head Height	L$_T$ Length of Thread
M6	14.2	10	6.6	18
M8	18	13	8.1	22
M10	22.3	15	9.2	26
M12	26.6	16	11.5	30

All dimensions in millimeters.

Table 81. Maximum Grip Gaging Lengths and Minimum Body Lengths for Metric Hexagon Head Bolts.

Nom. Bolt Length	Nominal Bolt Diameter and Pitch									
	M6 × 1		M8 × 1.25		M10 × 1.5		M12 × 1.75		M14 × 2	
	Max. Grip Length	Min. Body Length	Max. Grip Length	Min. Body Length	Max. Grip Length	Min. Body Length	Max. Grip Length	Min. Body Length	Max. Grip Length	Min. Body Length
30	12.0	7.0								
35	17.0	12.0	13.0	6.8						
40	17.0	12.0	13.0	6.08	14.0	6.5				
45	27.0	22.0	23.0	16.8	19.0	11.5	15.0	6.2		
50	27.0	22.0	23.0	16.8	19.0	11.5	15.0	6.2	16.0	6.0
55	37.0	32.0	33.0	26.8	29.0	21.5	25.0	16.2	21.0	11.0
60	37.0	32.0	33.0	26.8	29.0	21.5	25.0	16.2	21.0	11.0
65			43.0	36.8	39.0	31.5	35.0	26.2	31.0	21.0
70			43.0	36.8	39.0	31.5	35.0	26.2	31.0	21.0
75			53.0	46.8	49.0	41.5	45.0	36.2	41.0	31.0
80			53.0	46.8	49.0	41.5	45.0	36.2	41.0	31.0
85					59.0	51.5	55.0	46.2	51.0	41.0
90					59.0	51.5	55.0	46.2	51.0	41.0
100					74.0	66.5	70.0	61.2	66.0	56.0
110							80.0	71.2	76.0	66.0
120							90.0	81.2	86.0	76.0
130									90.0	80.0
140									100.0	90.0

(Continued)

Table 81. *(Continued)* **Maximum Grip Gaging Lengths and Minimum Body Lengths for Metric Hexagon Head Bolts.**

Nom. Bolt Length	M16 × 2		M20 × 2.5		M24 × 3		M30 × 3.5	
	Max. Grip Length	Min. Body Length	Max. Grip Length	Min. Body Length	Max. Grip Length	Min. Body Length	Max. Grip Length	Min. Body Length
55	17.0	7.0						
60	17.0	7.0						
65	27.0	17.0	19.0	6.5				
70	27.0	17.0	19.0	6.5				
75	37.0	27.0	29.0	16.5				
80	37.0	27.0	29.0	16.5	26.0	11.0		
85	47.0	37.0	39.0	26.5	31.0	16.0		
90	47.0	37.0	39.0	26.5	36.0	21.0		
100	62.0	52.0	54.0	41.5	46.0	31.0	34.0	16.5
110	72.0	62.0	64.0	51.5	56.0	41.0	44.0	26.5
120	82.0	72.0	74.0	61.5	66.0	51.0	54.0	36.5
130	86.0	76.0	78.0	65.5	70.0	55.0	58.0	40.5
140	96.0	86.0	88.0	75.5	80.0	65.0	68.0	50.5
150	106.0	96.0	98.0	85.5	90.0	75.0	78.0	60.5
160	116.0	106.0	108.0	95.5	100.0	85.0	88.0	70.5
170			118.0	105.5	110.0	95.0	98.0	80.5
180			128.0	115.5	120.0	105.0	108.0	90.5
190			138.0	125.5	130.0	115.0	118.0	100.5
200			148.0	135.5	140.0	125.0	128.0	110.5
220					147.0	132.0	135.0	117.5
240					167.0	152.0	155.0	137.5

(Continued)

Table 81. *(Continued)* **Maximum Grip Gaging Lengths and Minimum Body Lengths for Metric Hexagon Head Bolts.**

Nom. Bolt Length	Nominal Bolt Diameter and Pitch							
	M16 × 2		M20 × 2.5		M24 × 3		M30 × 3.5	
	Max. Grip Length	Min. Body Length	Max. Grip Length	Min. Body Length	Max. Grip Length	Min. Body Length	Max. Grip Length	Min. Body Length
260							175.0	157.5
280							195.0	177.5

Notes: Bolts shorter than the sizes shown are threaded full length. All dimensions in millimeters. Source: DOD-B-70331.

Table 82. Dimensions of Metric Round (or Cup) Head Square Neck "Carriage" Bolts (DIN 603).

Bolt Size	E Body Dia.	A Head Dia. Max.	H Head Height Max.	P Square Depth Max.	O Square Width Max.	L_T Thread Length		
						L_T to 125	L_T to 200	L_T over 200
M5	5.00	13.55	3.30	4.10	5.48	16.00	-	-
M6	6.00	16.55	3.88	4.60	6.48	18.00	24.00	-
M8	8.00	20.65	4.88	5.60	8.58	22.00	28.00	-
M10	10.00	24.65	5.38	6.60	10.58	26.00	32.00	45.00
M12	12.00	30.65	6.95	8.75	12.70	30.00	36.00	49.00
M16	16.00	38.8	8.95	12.90	16.70	38.00	44.00	57.00
M20	20.00	46.8	11.05	15.90	-	-	-	46.00

All dimensions in millimeters.

Table 83. Nominal Dimensions of Metric Flat Countersunk Nib Bolts (DIN 604).

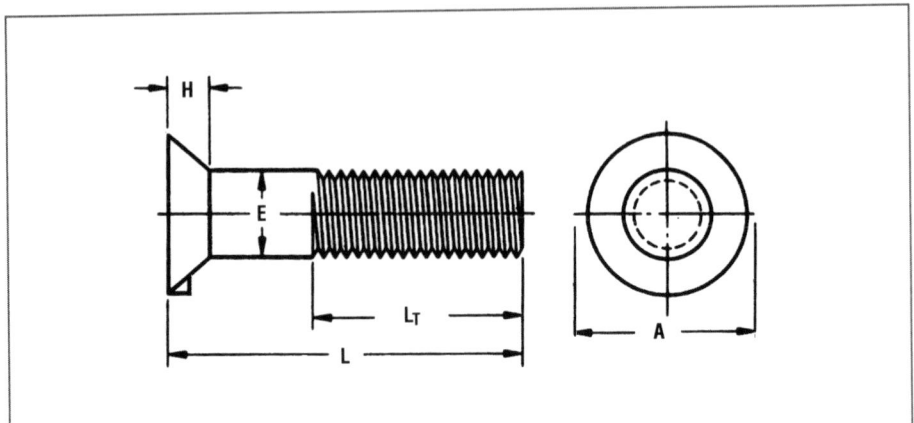

E Bolt Size	A Head Dia.	H Head Height	α Head Angle	L_T Thread Length
M8	16.00	5.00	90°	22.00
M10	19.00	5.50	90°	26.00
M12	24.00	7.00	90°	30.00
M16	32.00	9.00	90°	38.00
M20	32.00	11.50	60°	46.00
M24	38.80	13.00	60°	54.00

All dimensions in millimeters.

Table 84. Dimensions of Metric Flat Countersunk Square Neck Bolts - Long Square (DIN 605).

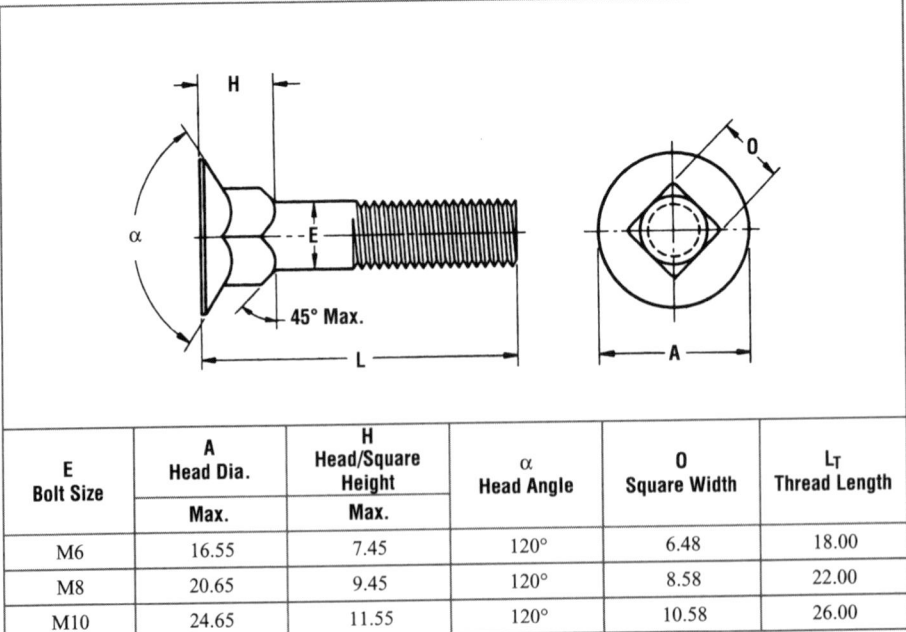

E Bolt Size	A Head Dia. Max.	H Head/Square Height Max.	α Head Angle	O Square Width	L_T Thread Length
M6	16.55	7.45	120°	6.48	18.00
M8	20.65	9.45	120°	8.58	22.00
M10	24.65	11.55	120°	10.58	26.00

All dimensions in millimeters.

Table 85. Nominal Dimensions of Metric Cup Head Nib Bolts (DIN 607).

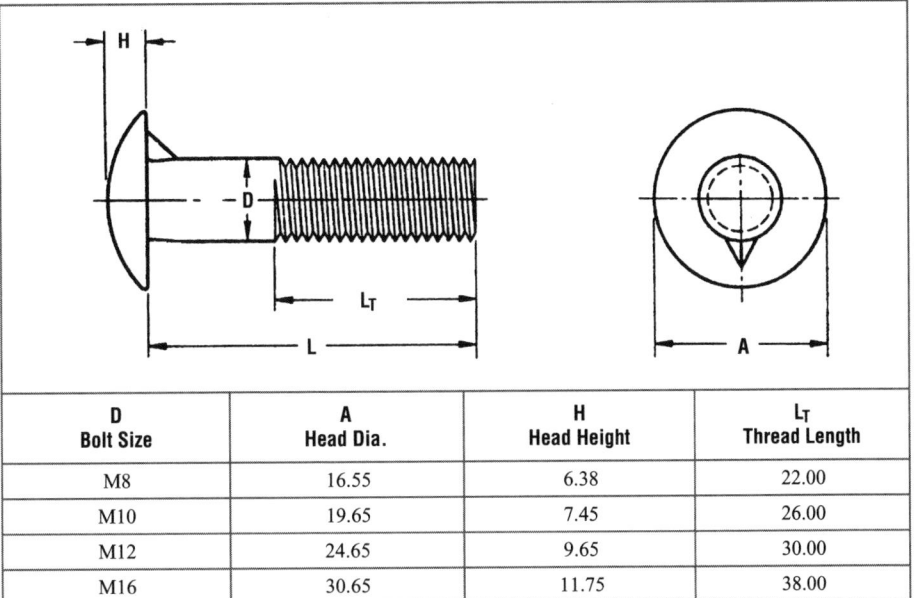

D Bolt Size	A Head Dia.	H Head Height	L$_T$ Thread Length
M8	16.55	6.38	22.00
M10	19.65	7.45	26.00
M12	24.65	9.65	30.00
M16	30.65	11.75	38.00

All dimensions in millimeters.

Table 86. Nominal Dimensions of Metric Hexagon Nuts (ISO/DIN 4032, DIN 934).

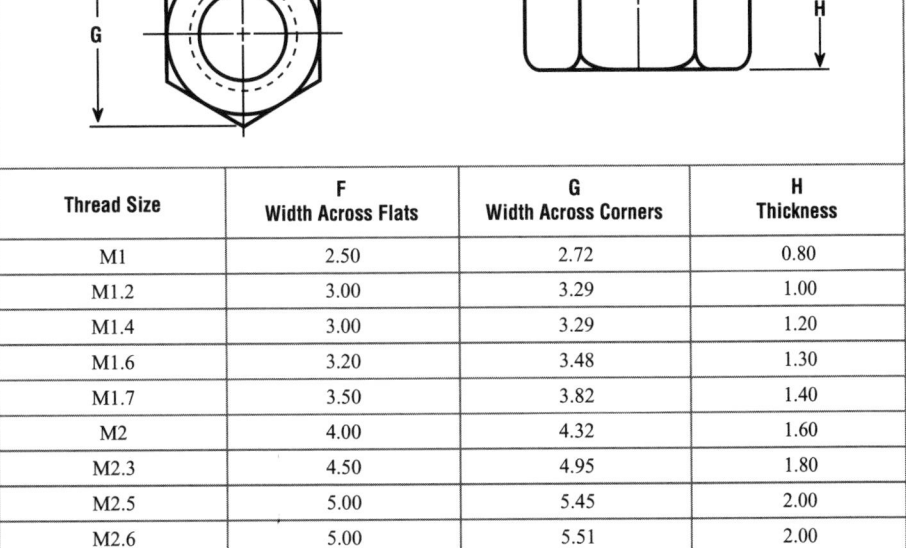

Thread Size	F Width Across Flats	G Width Across Corners	H Thickness
M1	2.50	2.72	0.80
M1.2	3.00	3.29	1.00
M1.4	3.00	3.29	1.20
M1.6	3.20	3.48	1.30
M1.7	3.50	3.82	1.40
M2	4.00	4.32	1.60
M2.3	4.50	4.95	1.80
M2.5	5.00	5.45	2.00
M2.6	5.00	5.51	2.00

(Continued)

Table 86. *(Continued)* Nominal Dimensions of Metric Hexagon Nuts (ISO/DIN 4032, DIN 934).

Thread Size	F Width Across Flats	G Width Across Corners	H Thickness
M3	5.50	6.01	2.40
M3.5	6.00	6.01	2.80
M4	7.00	7.66	3.20
M5	8.00	8.76	4.00
M6	10.00	11.05	5.00
M7	11.00	12.12	5.50
M8	13.00	14.38	6.50
M10	17.00	18.90	8.00
M12	19.00	21.10	10.00
M14	22.00	24.49	11.00
M16	24.00	26.75	13.00
M18	27.00	29.56	15.00
M20	30.00	32.95	16.00
M22	32.00	35.03	18.00
M24	36.00	39.55	19.00
M27	41.00	45.20	22.00
M30	46.00	50.85	24.00
M33	50.00	55.37	26.00
M36	55.00	60.79	29.00
M39	60.00	66.44	31.00
M42	65.00	71.30	34.00
M45	70.00	76.95	36.00
M48	75.00	82.60	38.00
M52	80.00	88.25	42.00
M56	85.00	95.07	45.00
M60	90.00	100.72	48.00
M64	95.00	106.37	51.00
M68	100.00	112.02	54.00
M72	105.00	117.67	58.00
M76	110.00	123.32	61.00
M80	115.00	128.97	64.00
M85	120.00	134.62	68.00
M90	130.00	145.77	72.00
M95	135.00	151.42	76.00
M100	145.00	162.72	80.00
M105	150.00	168.37	84.00
M110	155.00	174.02	88.00
M115	165.00	185.32	92.00
M120	170.00	190.97	96.00

(Continued)

Table 86. *(Continued)* **Nominal Dimensions of Metric Hexagon Nuts (ISO/DIN 4032, DIN 934).**

Thread Size	F Width Across Flats	G Width Across Corners	H Thickness
M125	180.00	202.27	100.00
M130	185.00	207.75	104.00
M135	190.00	213.40	108.00
M140	200.00	224.70	112.00
M145	210.00	236.00	116.00
M150	210.00	236.00	120.00

All dimensions in millimeters.

Table 87. Dimensions of Metric Hexagon Thin Nuts (DIN 439B).

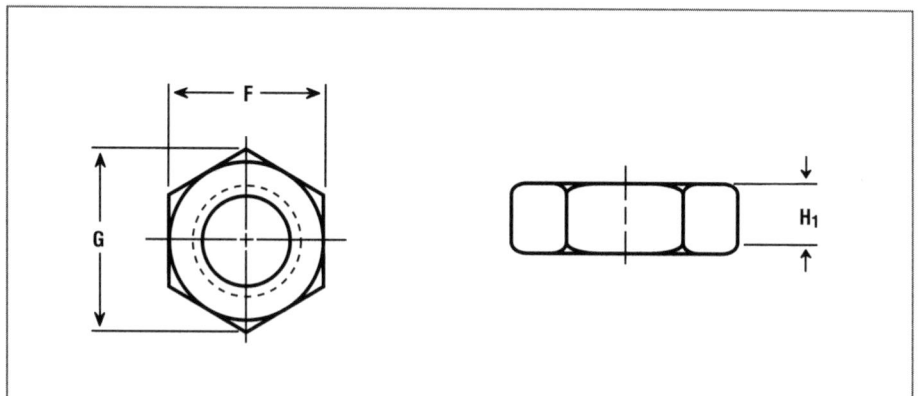

Thread Size	F Width Across Flats		G Width Across Corners	H Thickness	
	Max.	Min.	Min.	Max.	Min.
M2	4.00	-	4.32	1.20	-
M3	5.50	-	6.01	1.80	-
M4	7.00	-	7.66	2.20	-
M5	8.00	7.78	8.79	2.70	2.45
M6	10.00	9.78	11.05	3.20	2.90
M8	13.00	12.73	14.38	4.00	3.70
M10	17.00	15.73	18.90	5.00	4.70
M12	19.00	17.73	21.10	6.00	5.70
M14	22.00	-	24.49	7.00	-
M16	24.00	23.67	26.75	8.00	7.42
M18	27.00	-	29.56	9.00	-
M20	30.00	29.16	32.95	10.00	9.10
M22	32.00	-	35.03	11.00	-
M24	36.00	35.00	39.55	12.00	10.90

(Continued)

Table 87. *(Continued)* **Dimensions of Metric Hexagon Thin Nuts (DIN 439B).**

Thread Size	F Width Across Flats		G Width Across Corners	H Thickness	
	Max.	Min.	Min.	Max.	Min.
M27	41.00	40.00	45.20	13.50	12.40
M30	46.00	45.00	50.85	15.00	13.90
M33	50.00	49.00	55.37	16.50	15.40
M36	55.00	53.80	60.79	18.00	16.90

All dimensions in millimeters.

Table 88. Nominal Dimensions of Metric Structural Hexagon Nuts (DIN 6915). (See Notes.)

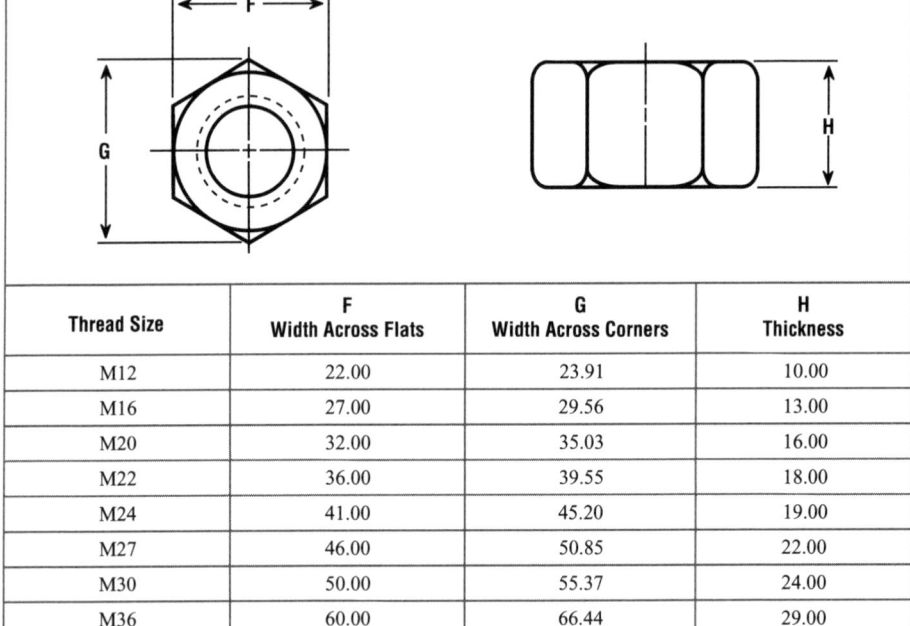

Thread Size	F Width Across Flats	G Width Across Corners	H Thickness
M12	22.00	23.91	10.00
M16	27.00	29.56	13.00
M20	32.00	35.03	16.00
M22	36.00	39.55	18.00
M24	41.00	45.20	19.00
M27	46.00	50.85	22.00
M30	50.00	55.37	24.00
M36	60.00	66.44	29.00

Notes: These hexagon nuts have increased width across flats for high tensile structural bolting. Designed to be used with bolt specified in DIN 6914 and washer specified in DIN 6916. All dimensions in millimeters.

Table 89. Nominal Dimensions of Metric Hexagon Flange Nuts (DIN 6923).

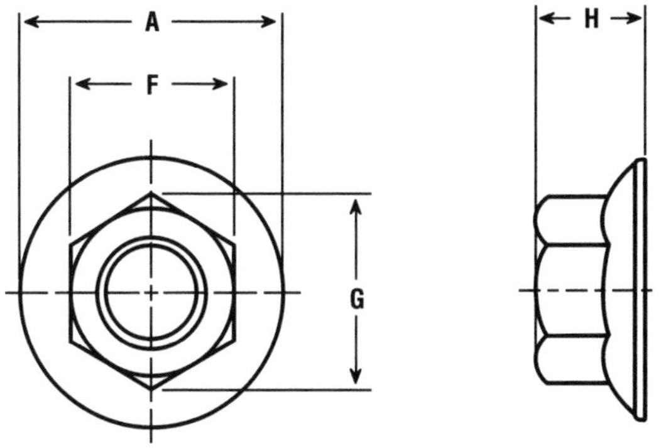

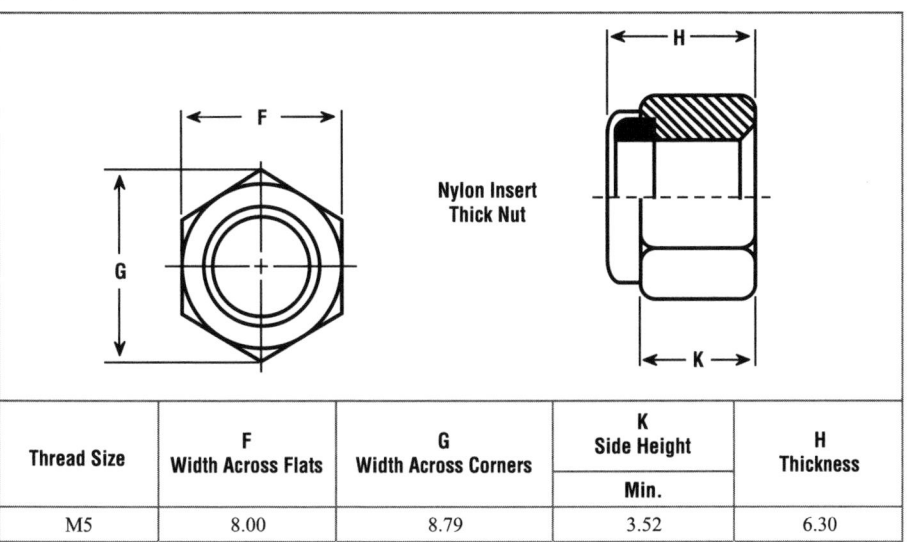

Thread Size	F Width Across Flats	G Width Across Corners	H Thickness	A Width of Flange
M5	8.00	8.79	5.00	11.80
M6	10.00	11.05	6.00	14.20
M8	13.00	14.38	8.00	17.90
M10	15.00	16.64	10.00	21.80
M12	18.00	20.03	12.00	26.00
M14	21.00	23.36	14.00	29.90
M16	24.00	26.75	16.00	34.50
M20	30.00	32.95	20.00	42.80

All dimensions in millimeters.

Table 90. Nominal Dimensions of Metric High-Style Prevailing Torque Hexagon Nut with Nonmetallic Insert (DIN 982).

Thread Size	F Width Across Flats	G Width Across Corners	K Side Height Min.	H Thickness
M5	8.00	8.79	3.52	6.30

Table 90. *(Continued)* **Nominal Dimensions of Metric High-Style Prevailing Torque Hexagon Nut with Nonmetallic Insert (DIN 982).**

Thread Size	F Width Across Flats	G Width Across Corners	K Side Height Min.	H Thickness
M6	10.00	11.05	3.92	8.00
M8	13.00	14.38	5.15	9.50
M10	17.00	18.90	6.43	11.50
M12	19.00	21.10	8.30	14.00
M14	22.00	24.49	9.68	16.00
M16	24.00	26.75	11.28	18.00
M20	30.00	32.95	13.52	22.00
M24	36.00	39.55	16.16	28.00

All dimensions in millimeters.

Table 91. Nominal Dimensions of Metric Low Style Prevailing Torque Hexagon Nuts with Nonmetallic Insert. (DIN 985).

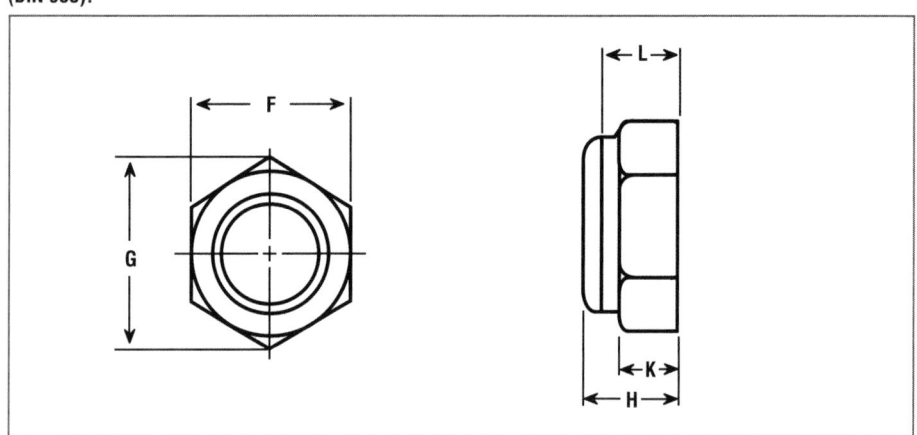

Thread Size	F Width Across Flats	G Width Across Corners	H Thickness	L Side Height Min.	K Thread Length Min.
M2.5/M2.6	5.00	5.51	3.80	2.00	-
M3	5.50	6.01	4.00	2.40	1.65
M4	7.00	7.66	5.00	2.90	2.20
M5	8.00	8.79	5.00	3.20	2.75
M6	10.00	11.05	6.00	4.00	3.30
M8	13.00	14.38	8.00	5.50	4.40
M10	17.00	18.90	10.00	6.50	5.50
M12	19.00	21.10	12.00	8.00	6.60
M14	22.00	24.49	14.00	9.50	7.70
M16	24.00	26.75	16.00	10.50	8.80
M18	27.00	29.56	18.50	13.00	9.90

(Continued)

Table 91. *(Continued)* **Nominal Dimensions of Metric Low Style Prevailing Torque Hexagon Nuts with Nonmetallic Insert. (DIN 985).**

Thread Size	F Width Across Flats	G Width Across Corners	H Thickness	L Side Height Min.	K Thread Length Min.
M20	30.00	32.95	20.00	14.00	11.00
M22	32.00	35.03	22.00	15.00	12.20
M24	36.00	39.55	24.00	15.00	13.20
M27	41.00	45.02	27.00	17.00	14.80
M30	46.00	50.85	30.00	19.00	16.50
M33	50.00	55.37	33.00	22.00	18.20
M36	55.00	60.79	36.00	25.00	19.80
M39	60.00	66.44	39.00	27.00	21.50
M42	65.00	72.09	42.00	29.00	23.10
M45	70.00	76.95	45.00	32.00	24.80
M48	75.00	82.60	48.00	36.00	26.50

All dimensions in millimeters.

Table 92. Nominal Dimensions of Metric Prevailing Torque Hexagon Nuts (DIN 980 Form V).

Thread Size	F Width Across Flats	G Width Across Corners	H Thickness	T Thickness Above Notch
M4	7.00	7.66	3.70	2.32
M5	8.00	8.79	4.50	2.96
M6	10.00	11.05	5.50	3.76
M8	13.00	14.38	7.00	4.91
M10	17.00	18.90	9.00	6.11
M12	19.00	21.10	11.00	7.71
M14	22.00	24.49	12.00	8.24
M16	24.00	26.75	14.00	9.84
M18	27.00	29.56	18.00	9.90
M20	30.00	32.95	20.00	11.00
M22	32.00	35.03	22.00	12.20

(Continued)

Table 92. *(Continued)* **Nominal Dimensions of Metric Prevailing Torque Hexagon Nuts (DIN 980 Form V).**

Thread Size	F Width Across Flats	G Width Across Corners	H Thickness	T Thickness Above Notch
M24	36.00	39.55	24.00	13.20
M27	41.00	45.02	27.00	14.80
M30	46.00	50.85	30.00	16.50
M33	50.00	55.37	33.00	18.20
M36	55.00	60.79	36.00	19.80

All dimensions in millimeters.

Table 93. Nominal Dimensions of Metric Hexagon Castle Nuts (DIN 935).

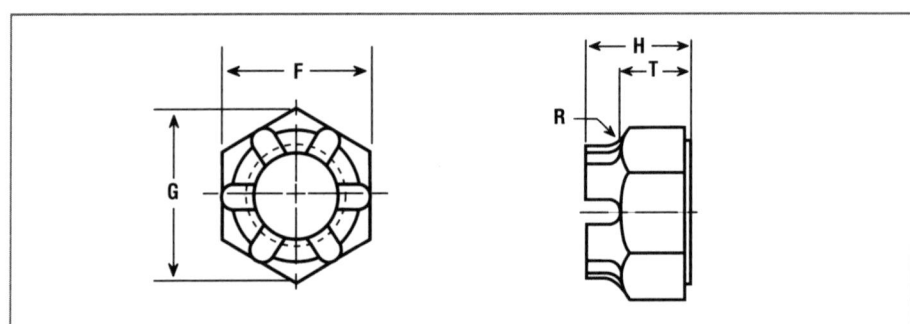

Thread Size	F Width Across Flats	G Width Across Corners	H Thickness	T Unslotted Thickness Min.
M4	7.00	7.74	5.00	3.20
M5	8.00	8.87	6.00	4.00
M6	10.00	11.05	7.50	5.00
M7	11.00	12.12	8.00	5.50
M8	13.00	14.38	9.50	6.50
M10	17.00	18.90	12.00	8.00
M12	19.00	21.10	15.00	10.00
M14	22.00	24.49	16.00	11.00
M16	24.00	26.75	19.00	13.00
M18	27.00	29.56	21.00	15.00
M20	30.00	32.95	22.00	16.00
M22	32.00	35.03	26.00	18.00
M24	36.00	39.55	27.00	19.00
M27	41.00	45.20	30.00	22.00
M30	46.00	50.85	33.00	24.00
M33	50.00	55.37	35.00	26.00
M36	55.00	60.79	38.00	29.00
M39	60.00	66.44	40.00	31.00
M42	65.00	71.30	46.00	34.00

(Continued)

Table 93. *(Continued)* Nominal Dimensions of Metric Hexagon Castle Nuts (DIN 935).

Thread Size	F Width Across Flats	G Width Across Corners	H Thickness	T Unslotted Thickness Min.
M45	70.00	78.26	48.00	36.00
M48	75.00	82.60	50.00	38.00
M52	80.00	88.25	54.00	42.00
M56	85.00	95.07	57.00	45.00
M58	90.00	100.72	63.00	48.00
M60	90.00	100.72	63.00	48.00
M64	95.00	106.37	66.00	51.00
M68	100.00	112.02	69.00	54.00
M72	105.00	117.67	73.00	58.00
M76	110.00	123.32	76.00	61.00
M80	115.00	128.97	79.00	64.00
M85	120.00	134.62	88.00	68.00
M90	130.00	145.77	92.00	72.00
M95	135.00	151.42	96.00	76.00
M100	145.00	162.72	100.00	80.00

Table 94. Nominal Dimensions of Metric Domed Cap Hexagon Nuts (DIN 1587).

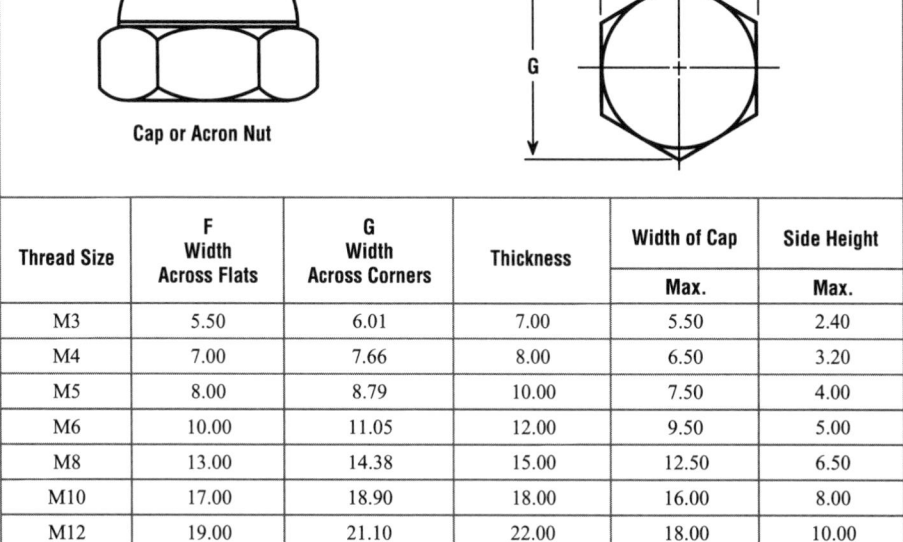

Cap or Acron Nut

Thread Size	F Width Across Flats	G Width Across Corners	Thickness	Width of Cap Max.	Side Height Max.
M3	5.50	6.01	7.00	5.50	2.40
M4	7.00	7.66	8.00	6.50	3.20
M5	8.00	8.79	10.00	7.50	4.00
M6	10.00	11.05	12.00	9.50	5.00
M8	13.00	14.38	15.00	12.50	6.50
M10	17.00	18.90	18.00	16.00	8.00
M12	19.00	21.10	22.00	18.00	10.00
M14	22.00	24.49	25.00	21.00	11.00

(Continued)

Table 94. *(Continued)* Nominal Dimensions of Metric Domed Cap Hexagon Nuts (DIN 1587).

Thread Size	F Width Across Flats	G Width Across Corners	Thickness	Width of Cap Max.	Side Height Max.
M16	24.00	26.75	28.00	23.00	13.00
M20	30.00	33.53	34.00	28.00	16.00
M24	36.00	39.98	42.00	34.00	19.00

All dimensions in millimeters.

Table 95. Nominal Dimensions of Metric Wing Nuts (DIN 315).

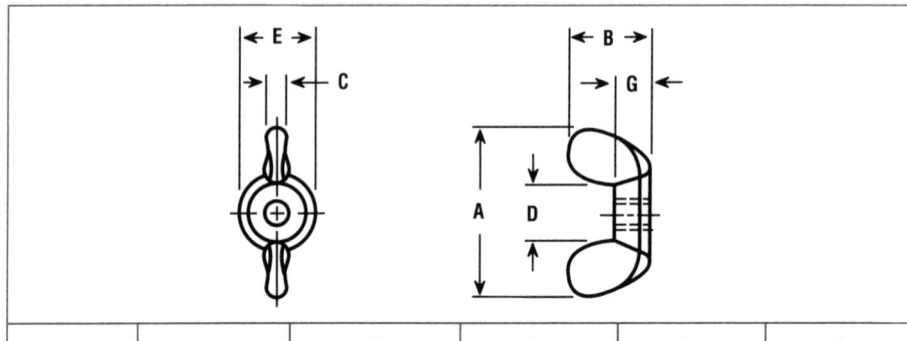

D Thread Size	A Wing Spread	B Wing Height	C Wing Thickness	E Boss Dia.	G Wall Height
M4	18.00	8.50	1.70	6.00	3.20
M5	24.00	11.00	2.30	8.00	4.00
M6	30.00	15.00	2.70	10.00	5.00
M8	36.00	18.00	3.60	13.00	6.50
M10	48.00	23.00	4.60	17.00	8.00
M12	62.00	31.00	5.60	20.00	10.00
M16	70.00	35.00	6.50	26.00	13.00
M20	86.00	44.00	7.00	32.00	16.00
M24	106.00	53.50	8.60	41.00	20.00

All dimensions in millimeters.

Table 96. Nominal Dimensions of Medium (DIN 125) and Grade C (DIN 126) Metric Washers for Hexagon Bolts.

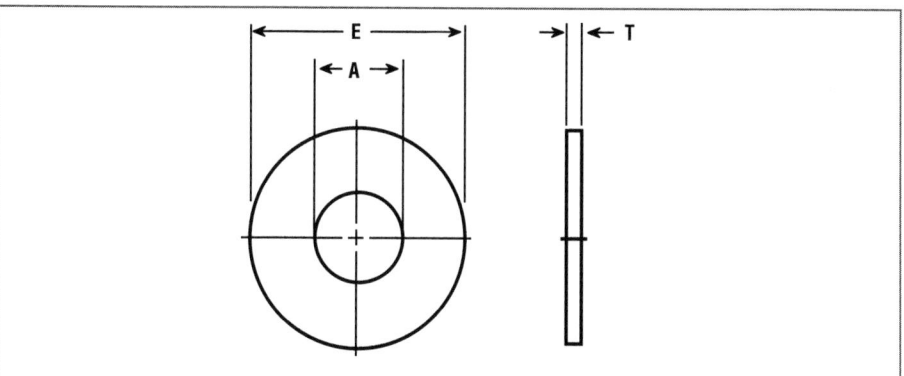

A Inside Dia. DIN 125	A Inside Dia. DIN 126	For Screw Size	E External Dia.	T Thickness
1.7	-	M1.6	4.00	0.30
2.2	-	M2	5.00	0.30
2.7	-	M2.5	6.00	0.50
3.2	-	M3	7.00	0.50
3.7	-	M3.5	8.00	0.50
4.3	-	M4	9.00	0.80
5.3	5.5	M5	10.00	1.00
6.4	6.6	M6	12.00	1.60
7.4	7.6	M7	14.00	1.60
8.4	9.0	M8	16.00	1.60
10.5	11.0	M10	20.00	2.00
13.0	13.5	M12	24.00	2.50
15.0	15.5	M14	28.00	2.50
17.0	17.5	M16	30.00	3.00
19.0	-	M18	34.00	3.00
21.0	22.0	M20	37.00	3.00
23.0	-	M22	39.00	3.00
25.0	26.0	M24	44.00	4.00
28.0	30.0	M27	50.00	4.00
31.0	33.0	M30	56.00	4.00
34.0	36.0	M33	60.00	5.00
37.0	39.0	M36	66.00	5.00
40.0	42.0	M39	72.00	6.00
43.0	45.0	M42	78.00	7.00
-	48.0	M45	85.00	7.00
50.0	52.0	M48	92.00	8.00

(Continued)

Table 96. *(Continued)* **Nominal Dimensions of Medium (DIN 125) and Grade C (DIN 126) Metric Washers for Hexagon Bolts.**

A Inside Dia. DIN 125	A Inside Dia. DIN 126	For Screw Size	E External Dia.	T Thickness
54.0	56.0	M52	98.00	8.00
-	62	M56	105.00	9.00
-	66	M60	110.00	9.00
-	70	M64	115.00	9.00

All dimensions in millimeters.

Table 97. Nominal Dimensions of Metric Split Lock Washers (DIN 127, Form A and B). (See Notes.)

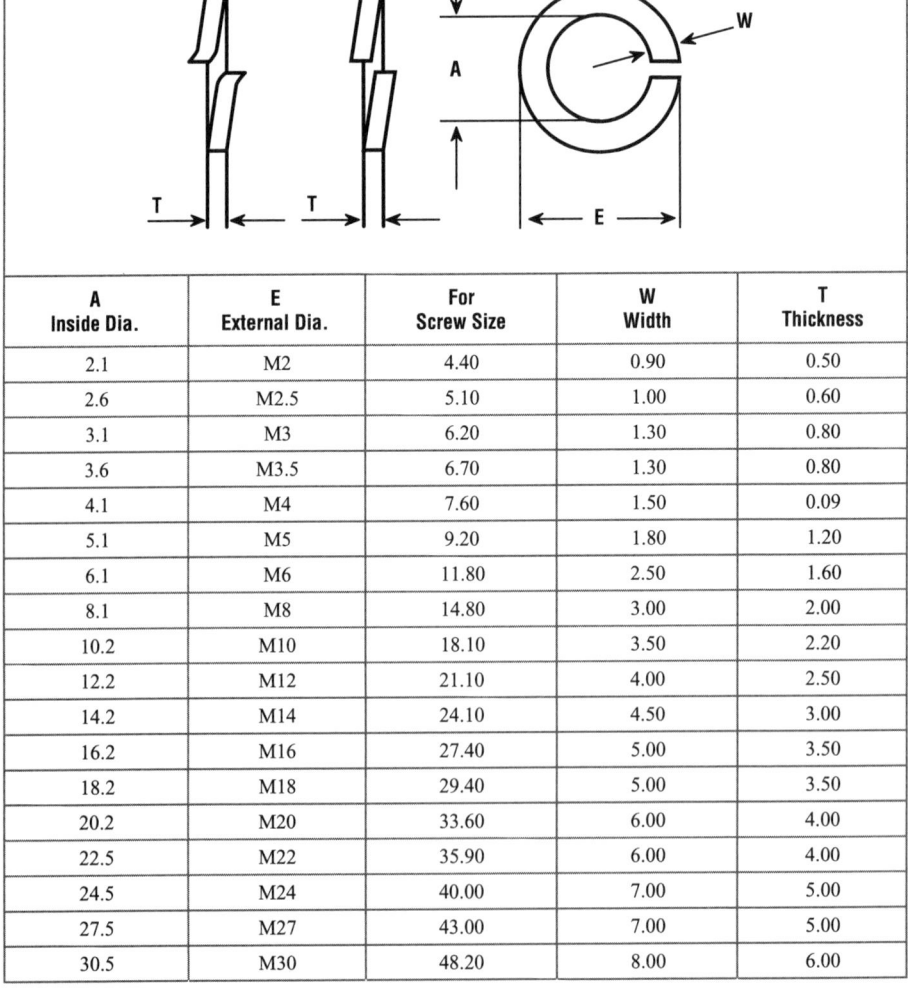

A Inside Dia.	E External Dia.	For Screw Size	W Width	T Thickness
2.1	M2	4.40	0.90	0.50
2.6	M2.5	5.10	1.00	0.60
3.1	M3	6.20	1.30	0.80
3.6	M3.5	6.70	1.30	0.80
4.1	M4	7.60	1.50	0.09
5.1	M5	9.20	1.80	1.20
6.1	M6	11.80	2.50	1.60
8.1	M8	14.80	3.00	2.00
10.2	M10	18.10	3.50	2.20
12.2	M12	21.10	4.00	2.50
14.2	M14	24.10	4.50	3.00
16.2	M16	27.40	5.00	3.50
18.2	M18	29.40	5.00	3.50
20.2	M20	33.60	6.00	4.00
22.5	M22	35.90	6.00	4.00
24.5	M24	40.00	7.00	5.00
27.5	M27	43.00	7.00	5.00
30.5	M30	48.20	8.00	6.00

(Continued)

Table 97. *(Continued)* **Nominal Dimensions of Metric Split Lock Washers (DIN 127, Form A and B).**
(See Notes.)

A Inside Dia.	E External Dia.	For Screw Size	W Width	T Thickness
33.5	M33	55.20	10.00	6.00
36.5	M36	58.20	10.00	6.00
39.0	M38	61.20	10.00	6.00
42.5	M42	68.20	12.00	7.00
45.5	M45	71.20	12.00	7.00
49.0	M48	75.00	12.00	7.00
53.0	M52	86.00	14.00	8.00

Notes: Form A washers have tangs. Form B washers have square ends. All dimensions in millimeters.

Table 98. Nominal Dimensions of External Toothed (Form A) and Internal Toothed (Form J) Metric Lock Washers (DIN 6797).

A Inside Dia.	For Screw Size	E External Dia.	T Thickness
2.20	M2	4.50	0.30
2.50	M2.3	5.00	0.40
2.80	M2.5	5.50	0.40
3.20	M3	6.00	0.40
3.70	M3.5	7.00	0.50
4.30	M4	8.00	0.50
5.10	M5	9.00	0.50
5.30	M5	10.00	0.60
6.40	M6	11.00	0.70
8.20	M8	14.00	0.80
8.40	M8	15.00	0.80
10.50	M10	18.00	0.90
13.00	M12	20.50	1.00

(Continued)

Table 98. *(Continued)* **Nominal Dimensions of External Toothed (Form A) and Internal Toothed (Form J) Metric Lock Washers (DIN 6797).**

A Inside Dia.	For Screw Size	E External Dia.	T Thickness
15.00	M14	24.00	1.00
17.00	M16	26.00	1.20
21.00	M20	33.00	1.40
25.00	M24	38.00	1.50
31.00	M30	48.00	1.60

All dimensions in millimeters.

Table 99. Nominal Dimensions of External (Form A) and Internal (Form J) Serrated Metric Lock Washers (DIN 6798).

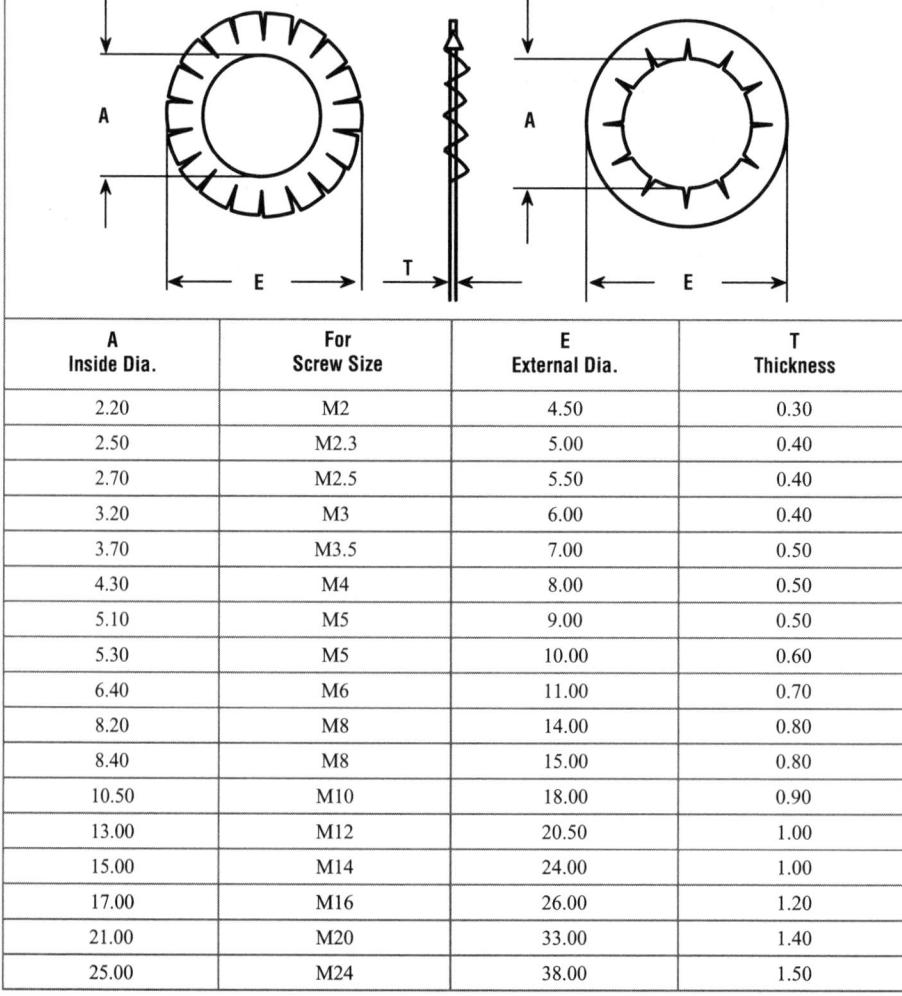

A Inside Dia.	For Screw Size	E External Dia.	T Thickness
2.20	M2	4.50	0.30
2.50	M2.3	5.00	0.40
2.70	M2.5	5.50	0.40
3.20	M3	6.00	0.40
3.70	M3.5	7.00	0.50
4.30	M4	8.00	0.50
5.10	M5	9.00	0.50
5.30	M5	10.00	0.60
6.40	M6	11.00	0.70
8.20	M8	14.00	0.80
8.40	M8	15.00	0.80
10.50	M10	18.00	0.90
13.00	M12	20.50	1.00
15.00	M14	24.00	1.00
17.00	M16	26.00	1.20
21.00	M20	33.00	1.40
25.00	M24	38.00	1.50

All dimensions in millimeters.

Table 100. Nominal Dimensions of Curved (Form A) Spring Metric Lock Washers (DIN 128).

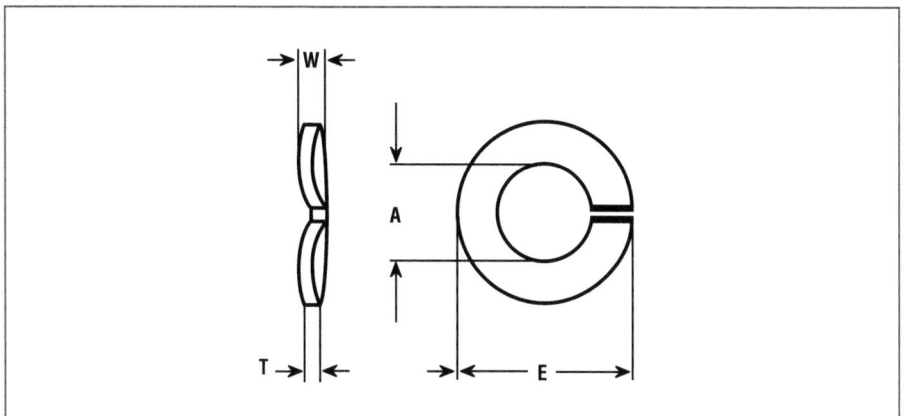

A Inside Dia.	For Screw Size	E External Dia.	T Thickness	W Uncompressed Thickness	
				Max.	Min.
2.1	M2	4.40	0.50	0.90	0.70
2.6	M2.5	5.10	0.60	1.10	0.90
3.1	M3	6.20	0.70	1.30	1.10
3.6	M3.5	6.70	0.70	1.30	1.10
4.1	M4	7.60	0.80	1.40	1.20
5.1	M5	9.20	1.00	1.70	1.50
6.1	M6	11.80	1.30	2.20	2.00
8.1	M8	14.80	1.60	2.75	2.45
10.2	M10	18.10	1.80	3.15	2.85
12.2	M12	21.10	2.10	3.65	3.35
14.2	M14	24.10	2.40	4.30	3.90
16.2	M16	27.40	2.80	5.10	4.50
18.2	M17	29.40	2.80	5.10	4.50
20.2	M20	33.60	3.20	5.90	5.10
22.2	M21	35.90	3.20	5.90	5.10
24.5	M24	40.00	4.00	7.50	6.50
30.5	M30	48.20	6.00	10.50	9.50

All dimensions in millimeters.

Table 101. Nominal Dimensions of Metric Curved (Form A) Spring Washers (DIN 137).

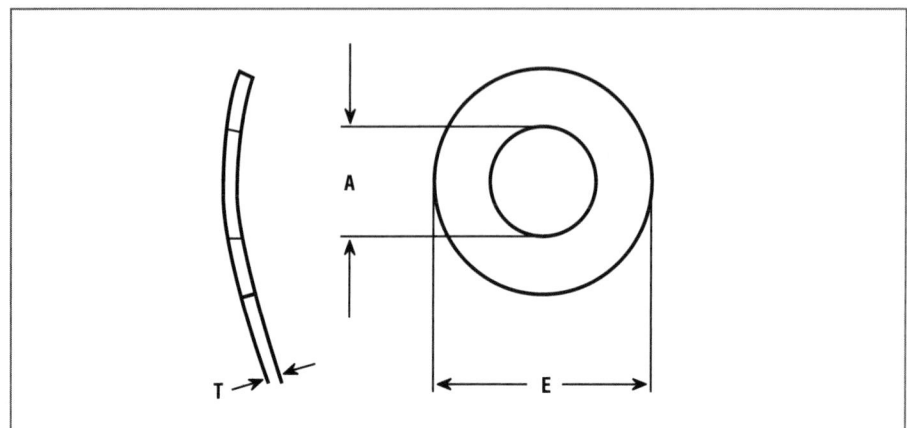

A Inside Dia.	For Screw Size	E External Dia.	T Thickness
1.10	M1	2.50	0.20
1.30	M1.2	3.00	0.20
1.50	M1.4	3.00	0.25
1.80	M1.6/M1.7	4.00	0.25
2.20	M2	4.50	0.30
2.80	M2.5	5.50	0.30
3.20	M3	6.00	0.40
3.70	M3.5	7.00	0.40
4.30	M4	8.00	0.50
5.30	M5	10.00	0.50
6.40	M6	11.00	0.50
7.40	M7	12.00	0.50
8.40	M8	15.00	0.50
10.50	M10	18.00	0.80

All dimensions in millimeters.

Table 102. Nominal Dimensions of Metric Waved (Form B) Spring Washers (DIN 137).

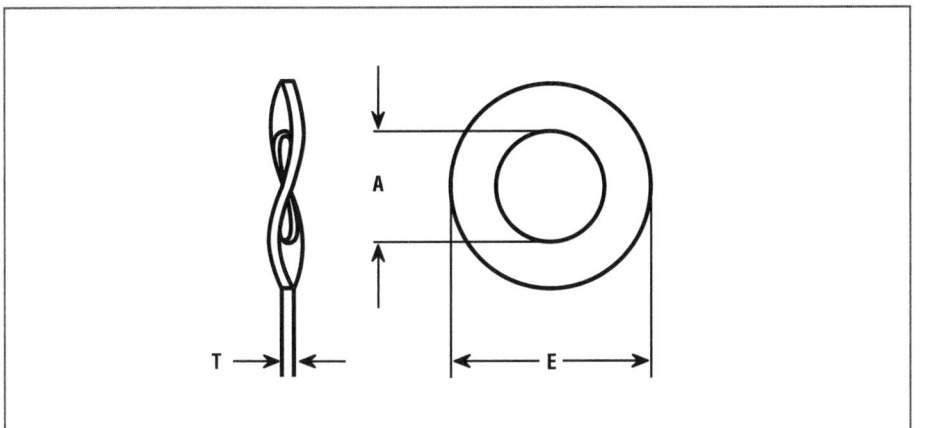

A Inside Dia.	For Screw Size	E External Dia.	T Thickness
3.20	M3	8.00	0.50
3.70	M3.5	8.00	0.50
4.30	M4	9.00	0.50
5.30	M5	11.00	0.50
6.40	M6	12.00	0.50
7.40	M7	14.00	0.80
8.40	M8	15.00	0.80
10.50	M10	21.00	1.00
13.00	M12	24.00	1.20
15.00	M14	28.00	1.60
17.00	M16	30.00	1.60
19.00	M18	34.00	1.60
21.00	M20	36.00	1.60
23.00	M22	40.00	1.80
25.00	M24	44.00	1.80
28.00	M27	50.00	2.00
31.00	M30	56.00	2.20
34.00	M33	60.00	2.20
37.00	M36	68.00	2.50

All dimensions in millimeters.

Table 103. Nominal Dimensions of Metric Disc Spring Washers (DIN 2093).

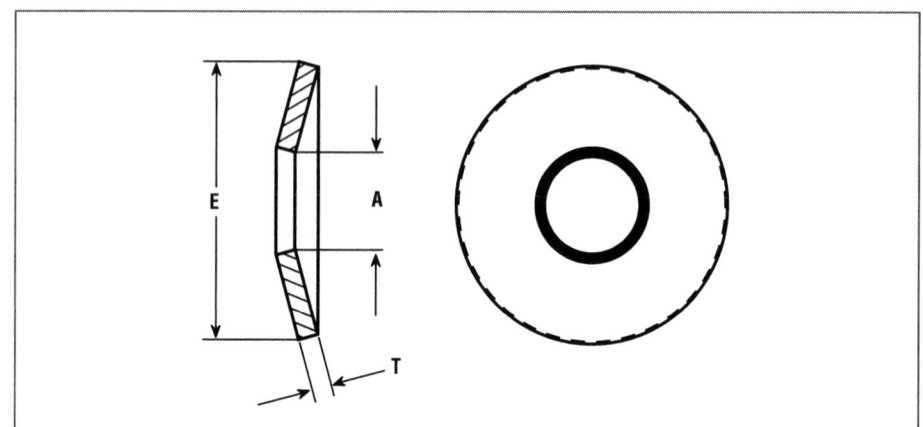

A Inside Dia.	E External Dia.	T Thickness
3.20	8.00	0.20, 0.30, 0.40
4.20	8.00	0.20, 0.30, 0.40
3.20	10.00	0.30, 0.40, 0.50
4.20	10.00	0.40, 0.50
5.20	10.00	0.25, 0.40, 0.50
4.20	12.00	0.40, 0.50, 0.60
5.20	12.00	0.50, 0.60
6.20	12.00	0.50, 0.60
5.20	12.50	0.50
6.20	12.50	0.35, 0.50, 0.70
7.20	14.00	0.35, 0.50, 0.80
5.20	15.00	0.40, 0.50, 0.60, 0.70
6.20	15.00	0.50, 0.60, 0.70
8.20	15.00	0.70, 0.80
8.20	16.00	0.40, 0.60, 0.70, 0.80, 0.90
6.20	18.00	0.40, 0.50, 0.60, 0.70, 0.80
8.20	18.00	0.50, 0.70, 0.80, 1.00
9.20	18.00	0.45, 0.70, 1.00
8.20	20.00	0.60, 0.70, 0.80, 0.90, 1.00
10.20	20.00	0.50, 0.80, 0.90, 1.00, 1.10, 1.25, 1.50
11.20	22.50	0.60, 0.80, 1.25
8.20	23.00	0.70, 0.80, 0.90, 1.00
10.20	23.00	0.90, 1.00, 1.25
12.20	23.00	1.00, 1.25, 1.50
10.20	25.00	1.00
12.20	25.00	0.70, 0.90, 1.00, 1.25, 1.50

(Continued)

Table 103. *(Continued)* Nominal Dimensions of Metric Disc Spring Washers (DIN 2093).

A Inside Dia.	E External Dia.	T Thickness
10.20	28.00	0.80, 1.00, 1.25, 1.50
12.20	28.00	1.00, 1.25, 1.50
14.20	28.00	0.80, 1.00, 1.25, 1.50
12.20	31.50	1.00, 1.25, 1.50
16.30	31.50	0.80, 1.25, 1.50, 1.75, 2.00
12.30	34.00	1.00, 1.25, 1.50
14.30	34.00	1.25, 1.50
16.30	34.00	1.50, 2.00
18.30	35.50	0.90, 1.25, 2.00
14.30	40.00	1.25, 1.50, 2.00
16.30	40.00	1.50, 2.00
18.30	40.00	2.00
20.40	40.00	1.00, 1.50, 2.00, 2.25, 2.50
22.40	45.00	1.25, 1.75, 2.50
18.40	50.00	1.25, 1.50, 2.00, 2.50, 3.00
20.40	50.00	2.00, 2.50
22.40	50.00	2.00, 2.50
25.40	50.00	1.25, 1.50, 2.00, 2.50, 3.00
28.50	56.00	1.50, 2.00, 3.00
20.50	60.00	2.00, 2.50, 3.00
25.50	60.00	2.50, 3.00
30.50	60.00	2.50, 3.00, 3.50
31.00	63.00	1.80, 2.50, 3.00, 3.50
25.50	70.00	2.00
30.50	70.00	2.50, 3.00
35.50	70.00	3.00, 4.00
40.50	70.00	4.00, 5.00
36.00	71.00	2.00, 2.50, 4.00
31.00	80.00	2.50, 3.00, 4.00
36.00	80.00	3.00, 4.00
41.00	80.00	2.25, 3.00, 4.00, 5.00
46.00	90.00	2.50, 3.50, 5.00
41.00	100.00	4.00, 5.00
51.00	100.00	2.70, 3.50, 4.00, 5.00, 6.00
57.00	112.00	3.00, 4.00, 6.00
41.00	125.00	4.00
51.00	125.00	4.00, 5.00, 6.00
61.00	125.00	5.00, 6.00, 8.00

(Continued)

Table 103. *(Continued)* **Nominal Dimensions of Metric Disc Spring Washers (DIN 2093).**

A Inside Dia.	E External Dia.	T Thickness
64.00	125.00	3.50, 5.00, 8.00
71.00	125.00	6.00, 8.00, 10.00
72.00	140.00	3.80, 5.00, 8.00
61.00	150.00	5.00, 6.00
71.00	150.00	6.00, 8.00
81.00	150.00	8.00, 10.00

All dimensions in millimeters.

Table 104. Nominal Dimensions of Metric Heavy Structural Steel Bolting Washer (DIN 6916). (See Notes.)

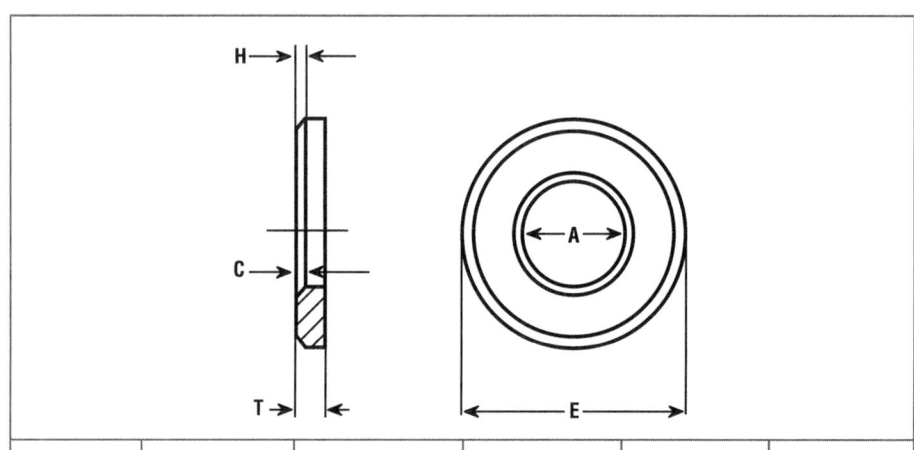

A Inside Dia.	For Thread Size	E External Dia.	T Thickness	H Chamfer Height	C Chamfer Width
13.00	M12	24.00	3.00	1.60	0.50
17.00	M16	30.00	4.00	1.60	1.00
21.00	M20	37.00	4.00	2.00	1.00
23.00	M22	39.00	4.00	2.00	1.00
25.00	M24	44.00	4.00	2.00	1.00
28.00	M27	50.00	5.00	2.50	1.00
31.00	M30	56.00	5.00	2.50	1.00
37.00	M36	66.00	6.00	3.00	1.50

Notes: These washers are intended for high tensile structural bolting. Use with hexagon bolt specified in DIN 9614 and hex nut specified in DIN 6915. All dimensions in millimeters.

Table 105. Nominal Dimensions of Large Size Metric Fender Washers (DIN 9021).

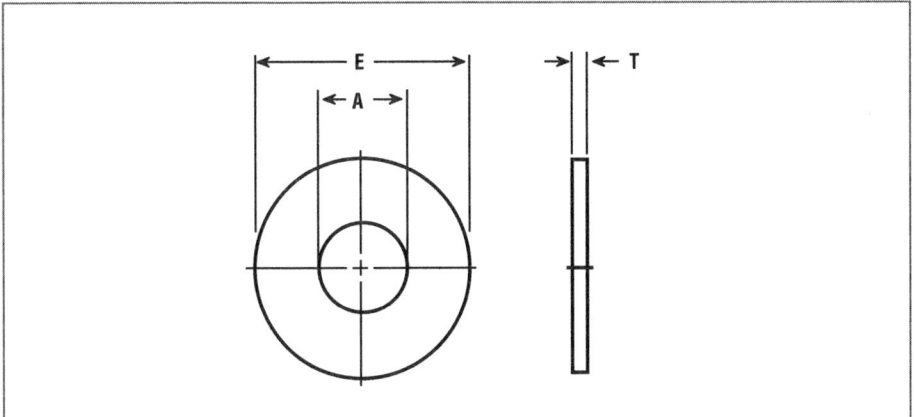

A Inside Dia.	For Screw Size	E External Dia. Max.	T Thickness
2.50	M2.3	7.00	0.80
2.70	M2.5	8.00	0.80
3.20	M3	9.00	0.80
4.30	M4	12.00	1.00
5.30	M5	15.00	1.20
6.40	M6	18.00	1.60
8.40	M8	24.00	2.00
10.50	M10	30.00	2.50
13.00	M12	37.00	3.00
15.00	M14	44.00	3.00
17.00	M16	50.00	3.00
20.00	M18	56.00	4.00
22.00	M20	60.00	4.00

All dimensions in millimeters.

Table 106. Nominal Dimensions of Metric Washers for Cheese Head Screws (DIN 433).

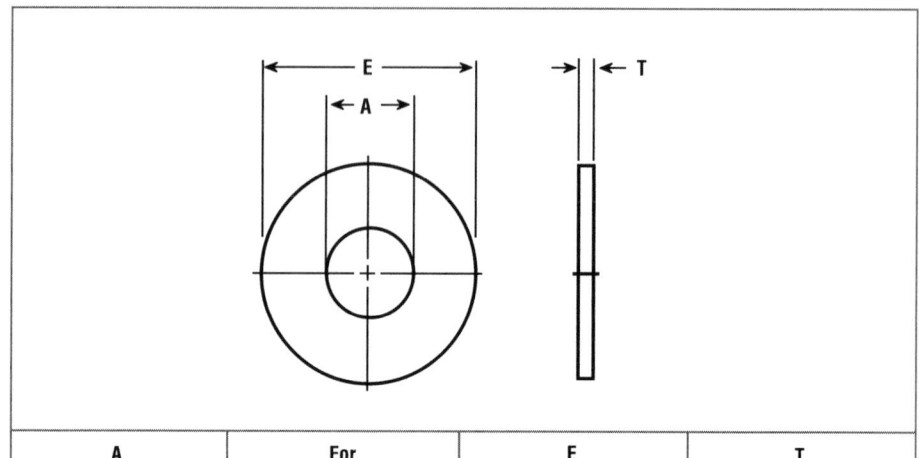

A Inside Dia.	For Screw Size	E External Dia.	T Thickness
1.10	M1	2.50	0.30
1.30	M1.3	3.00	0.30
1.50	M1.5	3.00	0.30
1.90	M1.8	4.00	0.30
2.20	M2	4.50	0.30
2.70	M2.5	5.00	0.50
3.20	M3	6.00	0.50
3.70	M3.5	7.00	0.50
4.30	M4	8.00	0.50
5.30	M5	9.00	1.00
6.40	M6	11.00	1.60
8.40	M8	15.00	1.60
10.50	M10	18.00	1.60
13.00	M12	20.00	2.00
15.00	M14	24.00	2.50
17.00	M16	28.00	2.50
19.00	M18	30.00	2.50
21.00	M20	34.00	3.00

All dimensions in millimeters.

Table 107. Nominal Dimensions of Medium Metric Washers for Clevis Pins (DIN 1440).

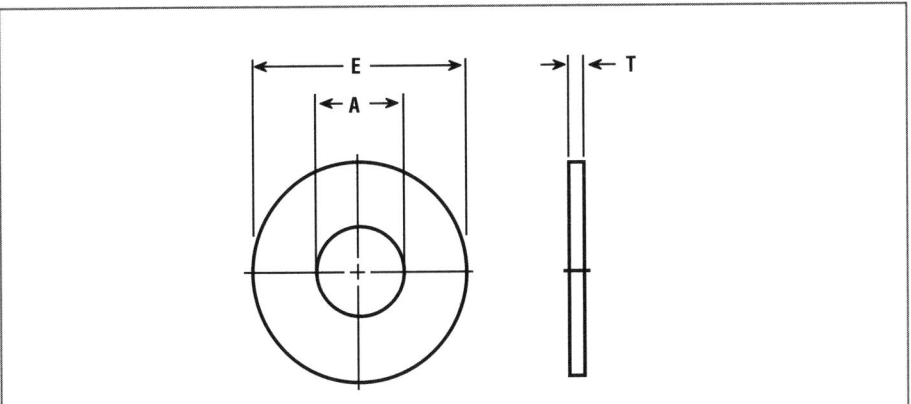

For Pin Size	E External Dia.	T Thickness
M3	6.00	0.80
M4	8.00	0.80
M5	10.00	0.80
M6	12.00	1.60
M7	14.00	1.60
M8	16.00	2.00
M10	20.00	2.50
M12	25.00	3.00
M13	25.00	3.00
M14	28.00	3.00
M16	28.00	3.00
M18	30.00	4.00
M20	32.00	4.00
M22	34.00	4.00
M23	36.00	4.00
M25	40.00	4.00
M26	40.00	5.00
M28	42.00	5.00
M30	45.00	5.00
M32	50.00	5.00
M33	50.00	5.00
M35	52.00	6.00
M36	52.00	6.00
M40	58.00	6.00
M45	62.00	7.00
M50	68.00	8.00

All dimensions in millimeters.

Table 108. Nominal Dimensions of Metric Hexagon Cap Screw (DIN 933).

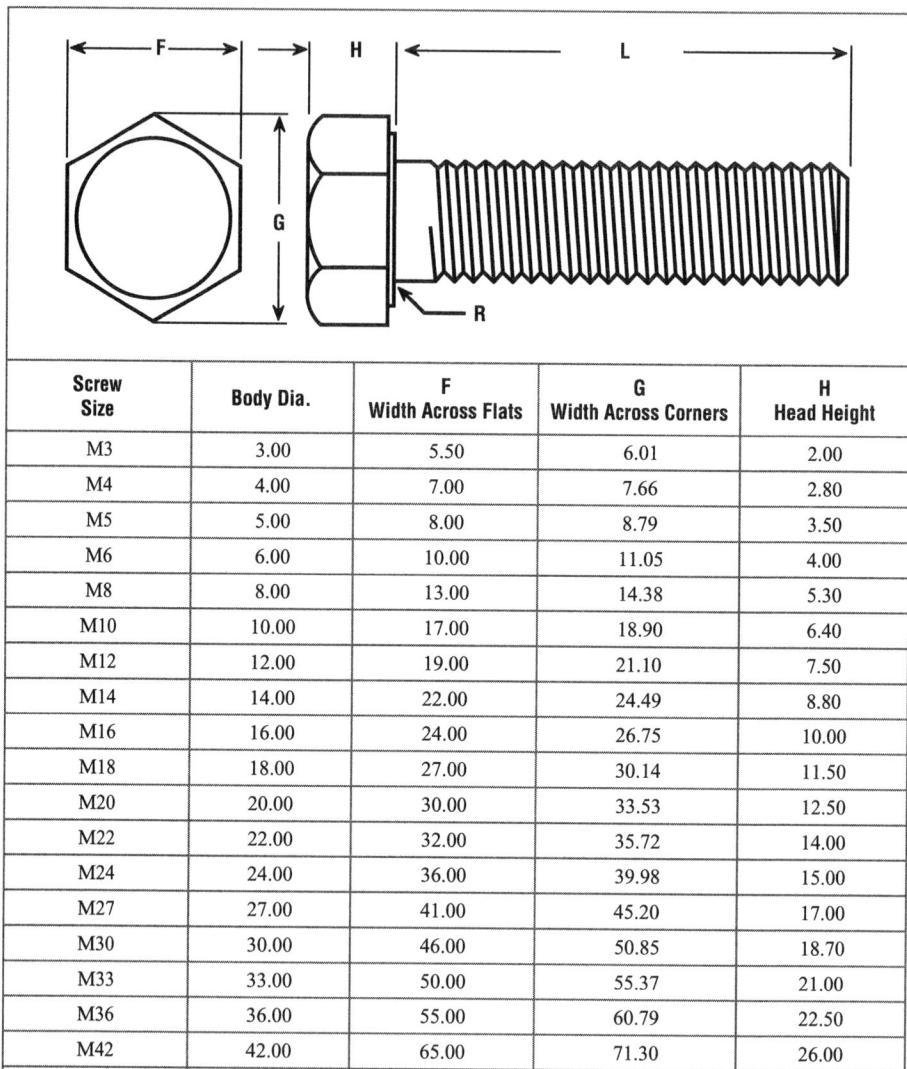

Screw Size	Body Dia.	F Width Across Flats	G Width Across Corners	H Head Height
M3	3.00	5.50	6.01	2.00
M4	4.00	7.00	7.66	2.80
M5	5.00	8.00	8.79	3.50
M6	6.00	10.00	11.05	4.00
M8	8.00	13.00	14.38	5.30
M10	10.00	17.00	18.90	6.40
M12	12.00	19.00	21.10	7.50
M14	14.00	22.00	24.49	8.80
M16	16.00	24.00	26.75	10.00
M18	18.00	27.00	30.14	11.50
M20	20.00	30.00	33.53	12.50
M22	22.00	32.00	35.72	14.00
M24	24.00	36.00	39.98	15.00
M27	27.00	41.00	45.20	17.00
M30	30.00	46.00	50.85	18.70
M33	33.00	50.00	55.37	21.00
M36	36.00	55.00	60.79	22.50
M42	42.00	65.00	71.30	26.00
M48	48.00	75.00	82.60	30.00

All dimensions in millimeters.

Table 109. Dimensions of Metric Lag (or Coach) Screws (DIN 571). (See Notes.)

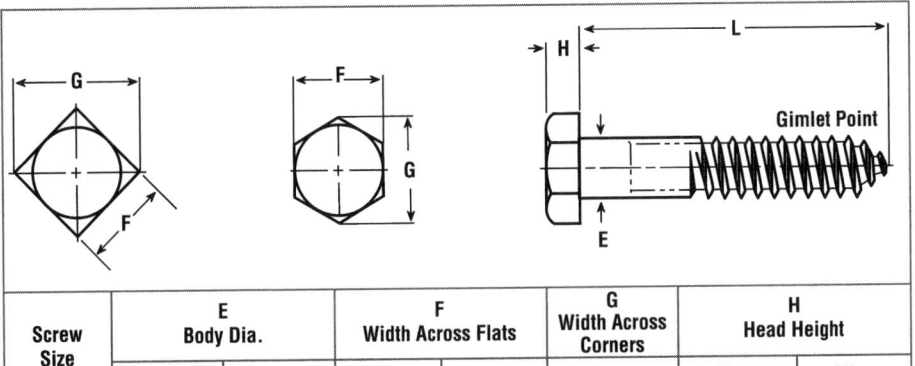

Screw Size	E Body Dia.		F Width Across Flats		G Width Across Corners	H Head Height	
	Max.	Min.	Max.	Min.	Min.	Max.	Min.
M6	6.48	5.52	10.0	9.64	10.89	4.38	3.62
M8	8.58	7.42	13.0	12.57	14.20	5.68	4.92
M10	10.58	9.42	16.0	15.57	18.72	6.85	5.95
M12	12.70	11.30	18.0	17.57	20.88	7.95	7.05
M16	16.70	15.30	24.0	23.16	26.17	10.75	9.25

Notes: L_T (length of thread) = 0.61L mm (minimum). All dimensions in millimeters.

Table 110. Nominal Dimensions of Metric Cross Recessed Raised Cheese Head Screws (DIN 7985).

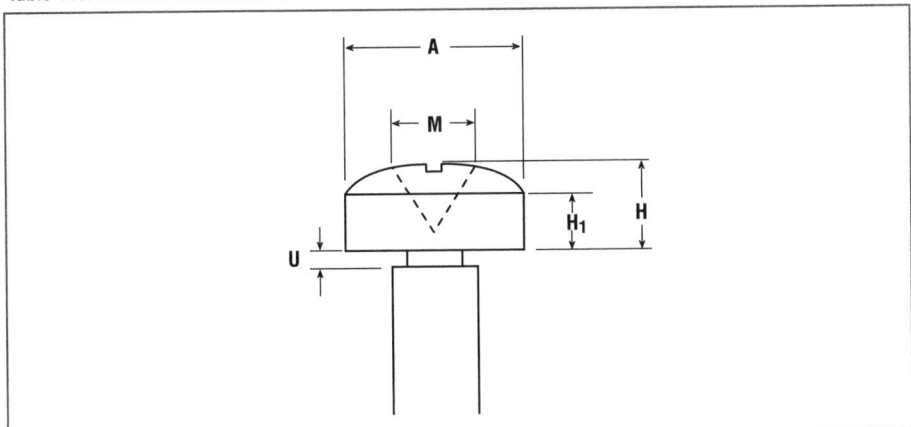

Screw Size	A Head Dia.	H Head Height	H₁ Flat Height	U Neck Relief Width	M Recess Width
	Max.		Approx.		Approx.
M3	6.00	2.40	1.60	1.00	3.10
M4	8.00	3.10	2.00	1.40	4.60
M5	10.00	3.80	2.50	1.60	5.30
M6	12.00	4.60	3.00	2.00	6.80
M8	16.00	6.00	3.70	2.50	9.00

All dimensions in millimeters.

Table 111. Nominal Dimensions of Metric Slotted Cheese Head Screws (DIN 84).

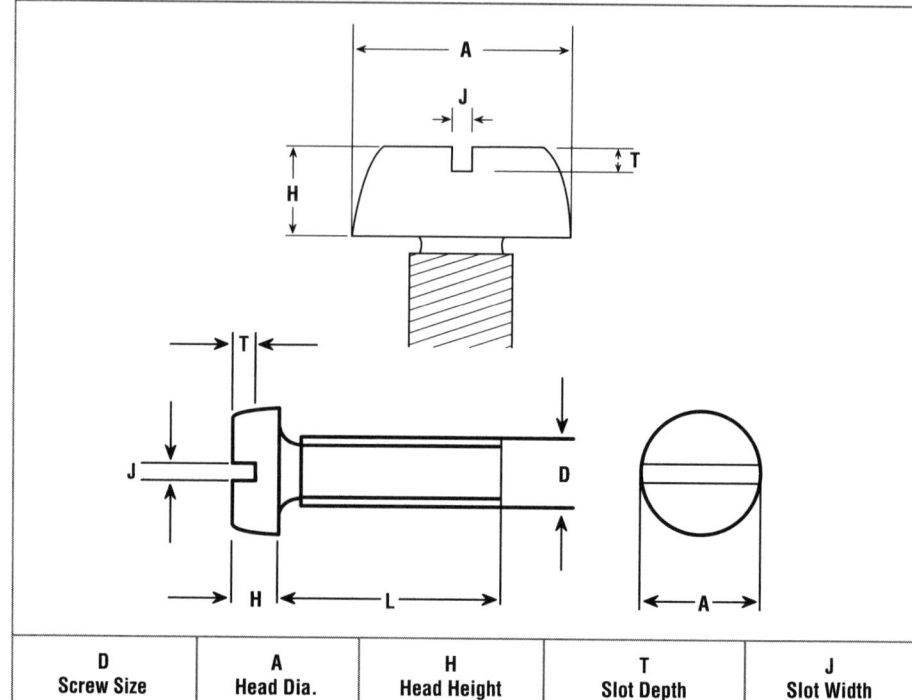

D Screw Size	A Head Dia.	H Head Height	T Slot Depth	J Slot Width
M1.6	3.00	1.00	0.45	0.40
M1.8	3.40	1.20	0.55	0.04
M2	3.80	1.30	0.60	0.50
M2.5	4.50	1.60	0.70	0.60
M3	5.50	2.00	0.85	0.80
M3.5	6.00	2.40	1.00	1.00
M4	7.00	2.60	1.10	1.20
M5	8.50	3.30	1.30	1.20
M6	10.00	3.90	1.60	1.60
M8	13.00	5.00	2.00	2.00
M10	16.00	6.00	2.40	2.50
M12	18.00	7.00	2.40	2.50

All dimensions in millimeters.

Table 112. Nominal Dimensions of Metric Slotted Pan Head Screws (DIN 85).

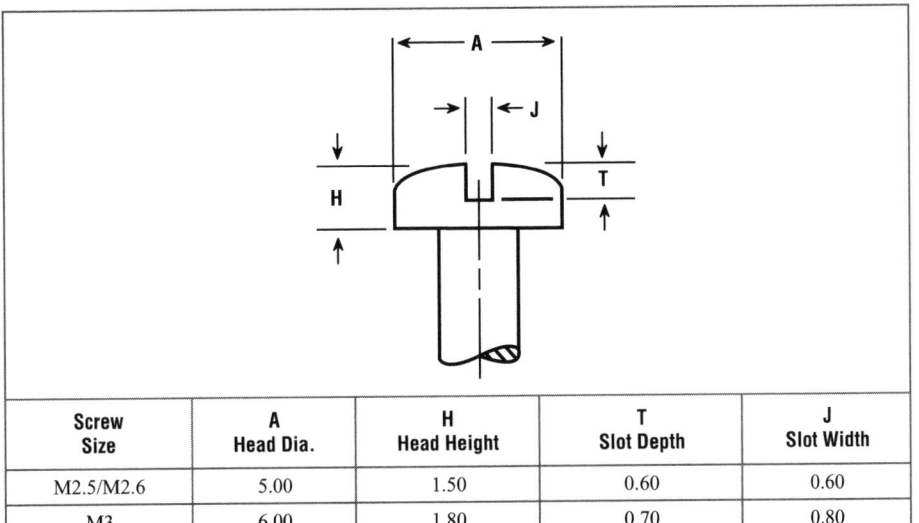

Screw Size	A Head Dia.	H Head Height	T Slot Depth	J Slot Width
M2.5/M2.6	5.00	1.50	0.60	0.60
M3	6.00	1.80	0.70	0.80
M4	8.00	2.40	1.00	1.20
M5	10.00	3.00	1.20	1.20
M6	12.00	3.60	1.40	1.60
M8	16.00	4.80	1.90	2.00
M10	20.00	6.00	2.40	2.50

All dimensions in millimeters.

Table 113. Nominal Dimensions of Metric Slotted Flat Head Countersunk Head Screws (DIN 963).

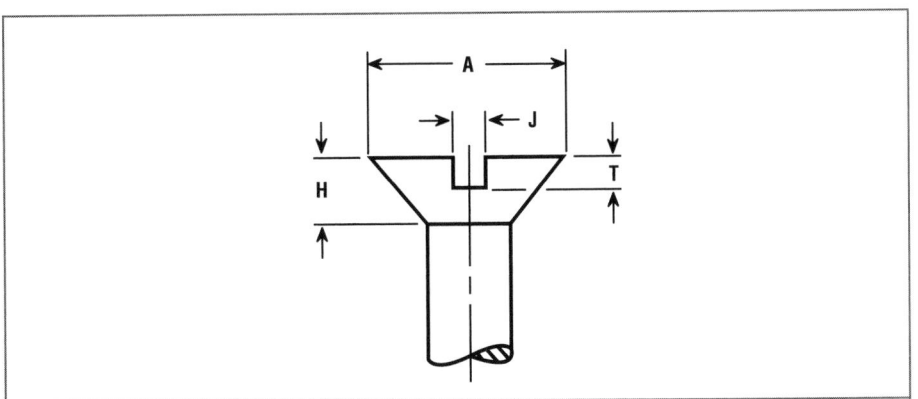

Screw Size	A Head Dia.	H Head Height	T Slot Depth Min.	J Slot Width
M1	1.90	0.60	0.20	0.25
M1.2	2.30	0.72	0.25	0.30
M1.4	2.60	0.84	0.28	0.30
M1.6	3.00	0.96	0.32	0.40

(Continued)

Table 113. *(Continued)* **Nominal Dimensions of Metric Slotted Flat Head Countersunk Head Screws (DIN 963).**

Screw Size	A Head Dia.	H Head Height	T Slot Depth Min.	J Slot Width
M1.8	3.40	1.08	0.35	0.40
M2	3.80	1.20	0.40	0.50
M2.5	4.70	1.50	0.50	0.60
M3	5.60	1.65	0.60	0.80
M3.5	6.50	1.93	0.70	0.80
M4	7.50	2.20	0.80	1.00
M5	9.20	2.50	1.00	1.20
M6	11.00	3.00	1.20	1.60
M8	14.50	4.00	1.60	2.00
M10	18.00	5.00	2.00	2.50
M12	22.00	6.00	2.40	3.00
M14	25.00	7.00	2.80	3.00
M16	29.00	8.00	3.20	4.00
M18	33.00	9.00	3.60	4.00
M20	36.00	10.00	4.00	5.00

All dimensions in millimeters.

Table 114. **Nominal Dimensions of Metric Slotted Oval Countersunk Head Screws (DIN 964).**

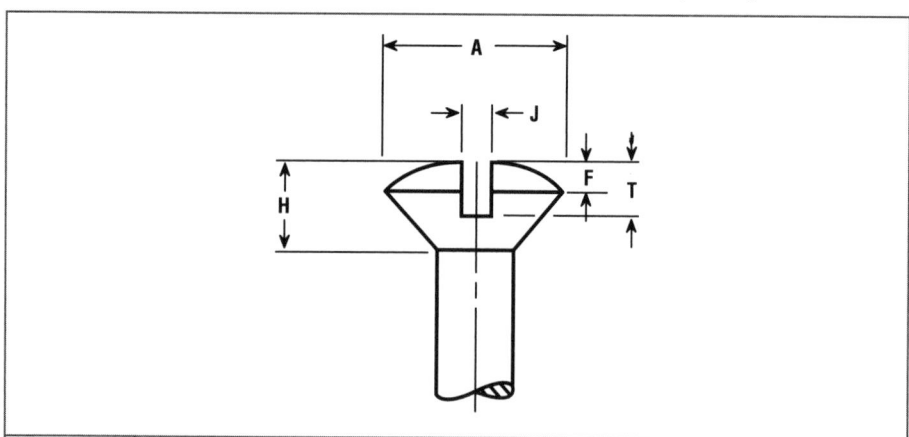

Screw Size	A Head Dia.	H Head Height	F Raised Height	T Slot Depth Min.	J Slot Width
M1	1.90	0.60	0.25	0.40	0.25
M1.2	2.30	0.72	0.30	0.50	0.30
M1.4	2.60	0.84	0.35	0.52	0.30
M1.6	3.00	0.96	0.40	0.65	0.40
M2	3.80	1.20	0.50	0.80	0.50

(Continued)

Table 114. *(Continued)* Nominal Dimensions of Metric Slotted Oval Countersunk Head Screws (DIN 964).

Screw Size	A Head Dia.	H Head Height	F Raised Height	T Slot Depth Min.	J Slot Width
M2.5	4.70	1.50	0.60	1.00	0.60
M3	5.60	1.65	0.75	1.20	0.80
M3.5	6.50	1.93	0.90	1.40	0.80
M4	7.50	2.20	1.00	1.60	1.00
M5	9.20	2.50	1.25	2.00	1.20
M6	11.00	3.00	1.50	2.40	1.60
M8	14.50	4.00	2.00	3.20	2.00
M10	18.00	5.00	2.50	4.00	2.50

All dimensions in millimeters.

Table 115. Nominal Dimensions of Metric Cross Recessed Countersunk Flat Head Screws (DIN 965).

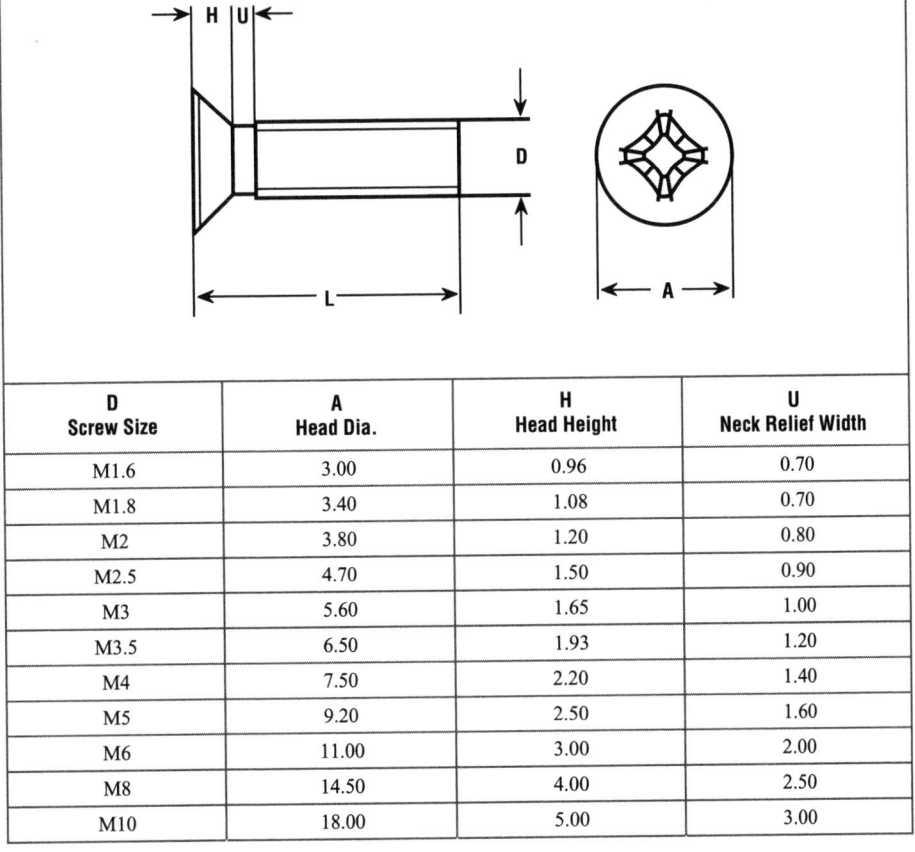

D Screw Size	A Head Dia.	H Head Height	U Neck Relief Width
M1.6	3.00	0.96	0.70
M1.8	3.40	1.08	0.70
M2	3.80	1.20	0.80
M2.5	4.70	1.50	0.90
M3	5.60	1.65	1.00
M3.5	6.50	1.93	1.20
M4	7.50	2.20	1.40
M5	9.20	2.50	1.60
M6	11.00	3.00	2.00
M8	14.50	4.00	2.50
M10	18.00	5.00	3.00

All dimensions in millimeters.

Table 116. Nominal Dimensions of Metric Cross Recessed Raised Countersunk Flat Head Screws (DIN 966).

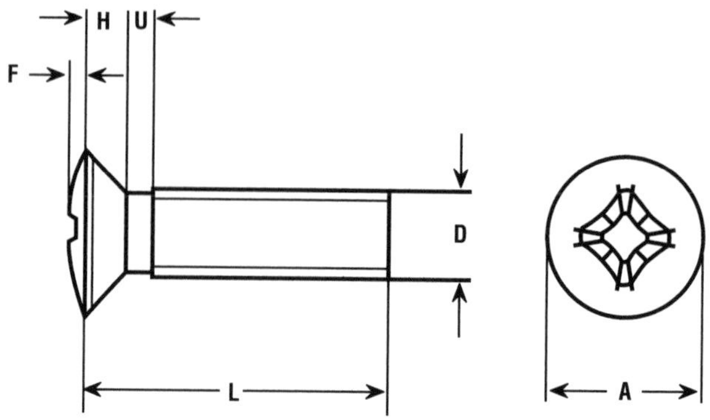

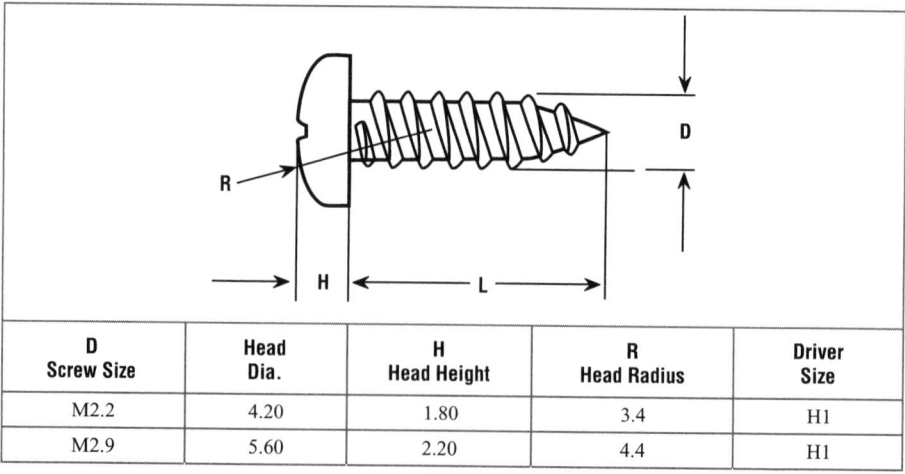

D Screw Size	A Head Dia.	H Head Height	F Raised Height	U Neck Relief Width
M1.6	3.00	0.96	0.40	0.70
M1.8	3.40	1.08	0.45	0.70
M2	3.80	1.20	0.50	0.80
M2.5	4.70	1.50	0.60	0.90
M3	5.60	1.65	0.75	1.00
M3.5	6.50	1.93	0.90	1.20
M4	7.50	2.20	1.00	1.40
M5	9.20	2.50	1.25	1.50
M6	11.00	3.00	1.50	2.00
M8	14.50	4.00	2.00	2.50
M10	18.00	5.00	2.50	3.00

All dimensions in millimeters.

Table 117. Nominal Dimensions of Metric Recessed Pan Head Sheet Metal Tapping Screws (ISO/DIN 7049, DIN 7981).

D Screw Size	Head Dia.	H Head Height	R Head Radius	Driver Size
M2.2	4.20	1.80	3.4	H1
M2.9	5.60	2.20	4.4	H1

(Continued)

Table 117. *(Continued)* **Nominal Dimensions of Metric Recessed Pan Head Sheet Metal Tapping Screws (ISO/DIN 7049, DIN 7981).**

D Screw Size	A Head Dia.	H Head Height	R Head Radius	Driver Size
M3.5	6.90	2.60	5.4	H2
M3.9	7.50	2.80	5.8	H2
M4.2	8.20	3.05	6.2	H2
M4.8	9.50	3.55	7.2	H2
M5.5	10.80	3.95	8.2	H3
M6.3	12.50	4.55	9.5	H3
M8	15.80	5.25	-	-

All dimensions in millimeters.

Table 118. Dimensions of Metric Recessed Flat Head Countersunk Sheet Metal Tapping Screws (ISO/DIN 7050, DIN 7982).

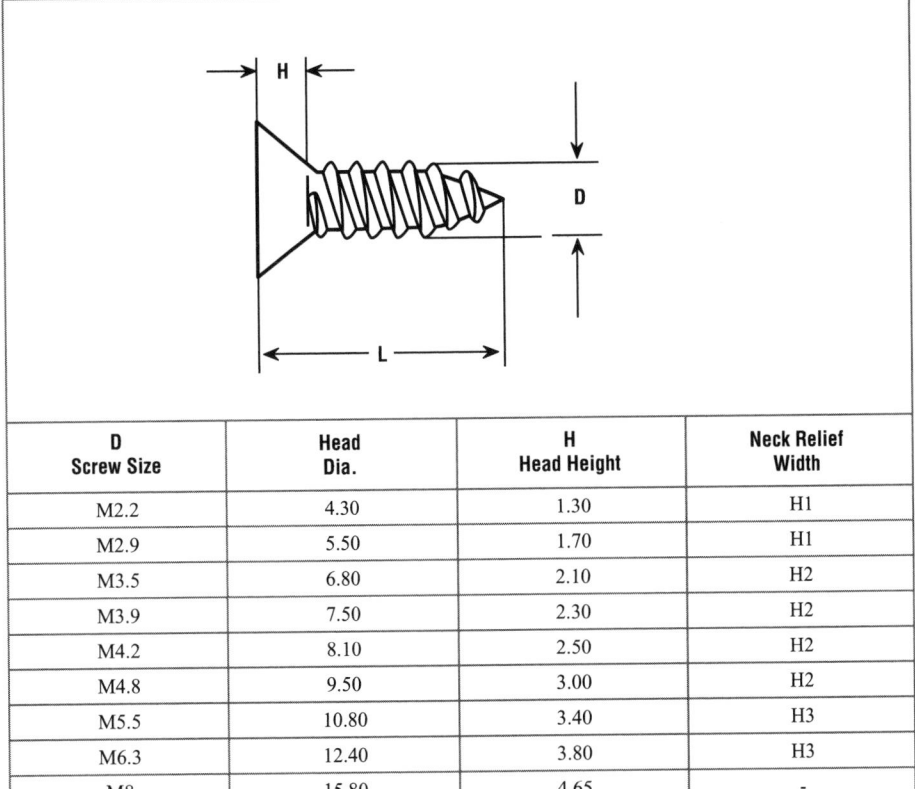

D Screw Size	Head Dia.	H Head Height	Neck Relief Width
M2.2	4.30	1.30	H1
M2.9	5.50	1.70	H1
M3.5	6.80	2.10	H2
M3.9	7.50	2.30	H2
M4.2	8.10	2.50	H2
M4.8	9.50	3.00	H2
M5.5	10.80	3.40	H3
M6.3	12.40	3.80	H3
M8	15.80	4.65	-

All dimensions in millimeters.

Table 119. Nominal Dimensions of Metric Recessed Oval Head Countersunk Sheet Metal Tapping Screws (ISO/DIN 7051, DIN 7983).

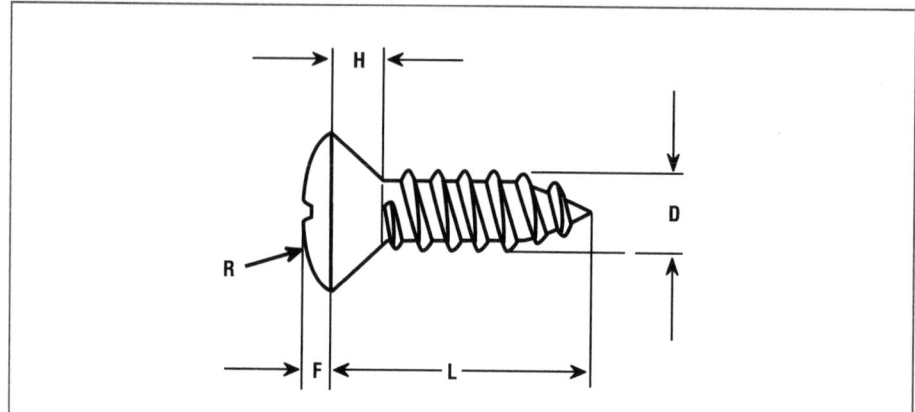

D Screw Size	Head Dia.	F Oval Height	H Countersink Depth	R Head Radius	Driver Size
M2.2	4.30	0.70	1.30	3.8	H1
M2.9	5.50	0.90	1.70	4.6	H1
M3.5	6.80	1.20	2.10	5.4	H2
M3.9	7.50	1.30	2.30	6	H2
M4.2	8.10	1.40	2.50	6.5	H2
M4.8	9.50	1.50	3.00	8.2	H2
M5.5	10.80	1.70	3.40	9.4	H3
M6.3	12.40	2.00	3.80	11.1	H3
M8	15.80	-	4.65	-	-

All dimensions in millimeters.

Table 120. Nominal Dimensions of Metric Square Head Set Screw with Short Dog Point (DIN 479).

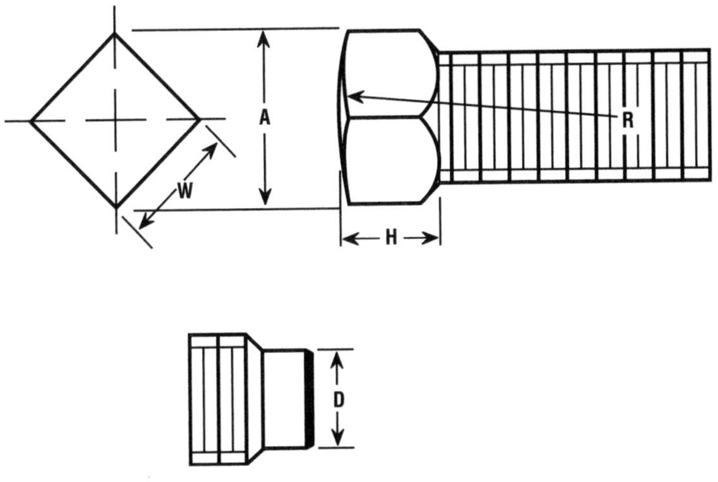

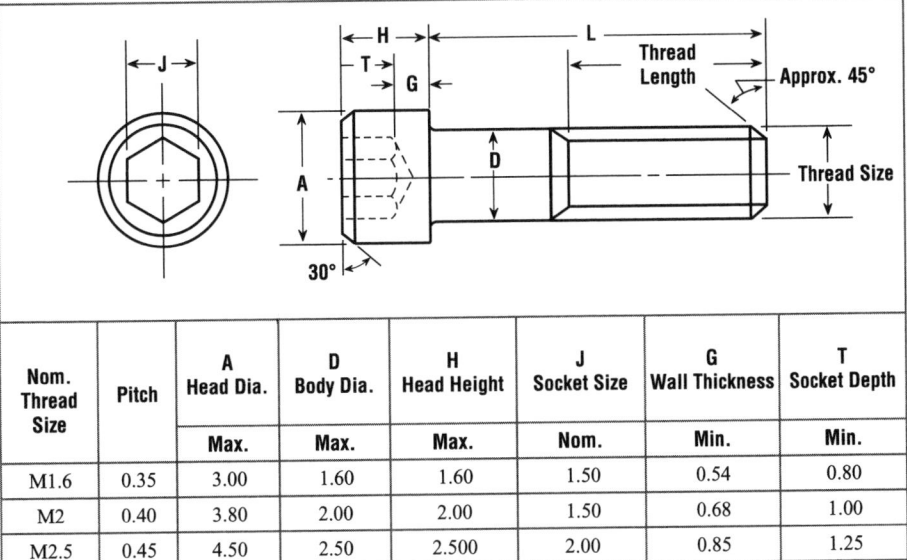

D Screw Size	W Width Across Flats	A Width Across Corners	D Point Dia.	H Head Height
M6	6.00	8.00	4.00	6.00
M8	8.00	10.00	5.50	8.00
M10	10.00	13.00	7.00	10.00
M12	13.00	17.00	8.50	12.00
M16	16.00	21.00	12.00	16.00

All dimensions in millimeters.

Table 121. Dimensions of Metric Socket Head Cap Screws.

Nom. Thread Size	Pitch	A Head Dia.	D Body Dia.	H Head Height	J Socket Size	G Wall Thickness	T Socket Depth
		Max.	Max.	Max.	Nom.	Min.	Min.
M1.6	0.35	3.00	1.60	1.60	1.50	0.54	0.80
M2	0.40	3.80	2.00	2.00	1.50	0.68	1.00
M2.5	0.45	4.50	2.50	2.500	2.00	0.85	1.25

(Continued)

Table 121. *(Continued)* **Dimensions of Metric Socket Head Cap Screws.**

Nom. Thread Size	Pitch	A Head Dia.	D Body Dia.	H Head Height	J Socket Size	G Wall Thickness	T Socket Depth
		Max.	Max.	Max.	Nom.	Min.	Min.
M3	0.50	5.50	3.00	3.00	2.50	1.02	1.50
M4	0.70	7.00	4.00	4.00	3.00	1.52	2.00
M5	0.80	8.50	5.00	5.00	4.00	1.90	2.50
M6	1.00	10.00	6.00	6.00	5.00	2.28	3.00
M8	1.25	13.00	8.00	8.00	6.00	3.20	4.00
M10	1.50	16.00	10.00	10.00	8.00	4.00	5.00
M12	1.75	18.00	12.00	12.00	10.00	4.80	6.00
M14	2.00	21.00	14.00	14.00	12.00	5.60	7.00
M16	2.00	24.00	16.00	16.00	14.00	6.40	8.00
M20	2.50	30.00	20.00	20.00	17.00	8.00	10.00
M24	3.00	36.00	24.00	24.00	19.00	9.60	12.00
M30	3.50	45.00	30.00	30.00	22.00	12.00	15.00
M36	4.00	54.00	36.00	36.00	27.00	14.40	18.00
M42	4.50	63.00	42.00	42.00	32.00	16.80	21.00
M48	5.00	72.00	48.00	48.00	36.00	19.20	24.00

Length tolerances for socket head cap screws					
Size	Up to 16 mm incl.	Over 16 to 50 mm incl.	Over 50 to 120 mm incl.	Over 120 to 200 mm incl.	Over 200 mm
M1.6 thru M10	± 0.3	± 0.4	± 0.7	± 1.0	± 2.0
M12 thru M20	± 0.3	± 0.4	± 1.0	± 1.5	± 2.5
Over M20	-	± 0.7	± 1.5	± 2.0	± 3.0

All dimensions in millimeters. Source: Unbrako/SPS Technologies, Inc.

Table 122. Grip (L_G) and Body (L_B) Lengths for Metric Socket Head Cap Screws. (See Notes.)

Length	M1.6		M2		M2.5		M3		M4		M5		M6	
	L_G	L_B	L_G	L_B	L_G	L_B	L_G	L_B	L_G	L_B	L_G	L_B	L_G	L_B
20	4.8	3.0	4.0	2.0										
25	9.8	8.0	9.0	7.0	8.0	5.7	7.0	4.5						
30	14.8	13.0	14.0	12.0	13.0	10.7	12.0	9.5	10.0	6.5				
35			19.0	17.0	18.0	15.7	17.0	14.5	15.0	11.5	13.0	9.0	11.0	6.0
40			24.0	22.0	23.0	20.7	22.0	19.5	20.0	16.5	18.0	14.0	16.0	11.0
45					28.0	25.7	27.0	24.5	25.0	21.5	23.0	19.0	21.0	16.0
50					33.0	30.7	32.0	29.5	30.0	26.5	28.0	24.0	26.0	21.0
55							37.0	34.5	35.0	31.5	33.0	29.0	31.0	26.0
60							42.0	39.5	40.0	36.5	38.0	34.0	36.0	31.0
65							47.0	44.5	45.0	41.5	43.0	39.0	41.0	36.0
70									50.0	46.5	48.0	44.0	46.0	41.0
80									60.0	56.5	58.0	54.0	56.0	51.0
90											68.0	64.0	66.0	61.0

(Continued)

Table 122. *(Continued)* Grip (L_G) and Body (L_B) Lengths for Metric Socket Head Cap Screws. (See Notes.)

Length	M1.6 L_G	M1.6 L_B	M2 L_G	M2 L_B	M2.5 L_G	M2.5 L_B	M3 L_G	M3 L_B	M4 L_G	M4 L_B	M5 L_G	M5 L_B	M6 L_G	M6 L_B
100											78.0	74.0	76.0	71.0
110													86.0	81.0
120													96.0	91.0

Length	M8 L_G	M8 L_B	M10 L_G	M10 L_B	M12 L_G	M12 L_B	M14 L_G	M14 L_B	M16 L_G	M16 L_B	M20 L_G	M20 L_B	M24 L_G	M24 L_B
45	17.0	10.7												
50	22.0	15.7	18.0	10.5										
55	27.0	20.7	23.0	15.5										
60	32.0	25.7	28.0	20.5	24.0	15.2								
65	37.0	30.7	33.0	25.5	29.0	20.2	25.0	15.0						
70	42.0	35.7	38.0	30.5	34.0	25.2	30.0	20.0	26.0	16.0				
80	52.0	45.7	48.0	40.5	44.0	35.2	40.0	30.0	36.0	26.0				
90	62.0	55.7	58.0	50.5	54.0	45.2	50.0	40.0	46.0	36.0	38.0	25.5		
100	72.0	65.7	68.0	60.5	64.0	55.2	60.0	50.0	56.0	46.0	48.0	35.5	40.0	25.0
110	82.0	75.7	78.0	70.5	74.0	65.2	70.0	60.0	66.0	56.0	58.0	45.5	50.0	35.0
120	92.0	85.7	88.0	80.5	84.0	75.2	80.0	70.0	76.0	66.0	68.0	55.5	60.0	45.0
130	102.0	95.7	98.0	90.5	94.0	85.2	90.0	80.0	86.0	76.0	78.0	65.5	70.0	55.0
140	112.0	105.7	108.0	100.5	104.0	95.2	100.0	90.0	96.0	86.0	88.0	75.5	80.0	65.0
150	122.0	115.7	118.0	110.5	114.0	105.2	110.0	100.0	106.0	96.0	98.0	85.5	90.0	75.0
160	132.0	125.7	128.0	120.5	124.0	115.2	120.0	110.0	116.0	106.0	108.0	95.5	100.0	85.0
180			148.0	140.5	144.0	135.2	140.0	130.0	136.0	126.0	128.0	115.5	120.0	105.0
200			168.0	160.5	164.0	155.2	160.0	150.0	156.0	146.0	148.0	135.5	140.0	125.0
220					184.0	175.2	180.0	170.0	176.0	166.0	168.0	155.5	160.0	145.0

(Continued)

Table 122. *(Continued)* **Grip (L_G) and Body (L_B) Lengths for Metric Socket Head Cap Screws.** (See Notes.)

Length	M8		M10		M12		M14		M16		M20		M24	
	L_G	L_B	L_G	L_B	L_G	L_B	L_G	L_B	L_G	L_B	L_G	L_B	L_G	L_B
240					204.0	195.2	200.0	190.0	196.0	186.0	188.0	175.5	180.0	165.0
260							220.0	210.0	216.0	206.0	208.0	195.5	200.0	185.0
300									256.0	246.0	248.0	235.5	240.0	225.0

Notes: L_G is the maximum grip length and is the distance from the bearing surface to the first complete thread. L_B is the minimum body length and is the length of the unthreaded cylindrical portion of the shank. All dimensions in millimeters. Source: Unbrako/SPS Technologies, Inc.

Table 123. Drill and Counterbore Sizes for Metric Socket Head Cap Screws.

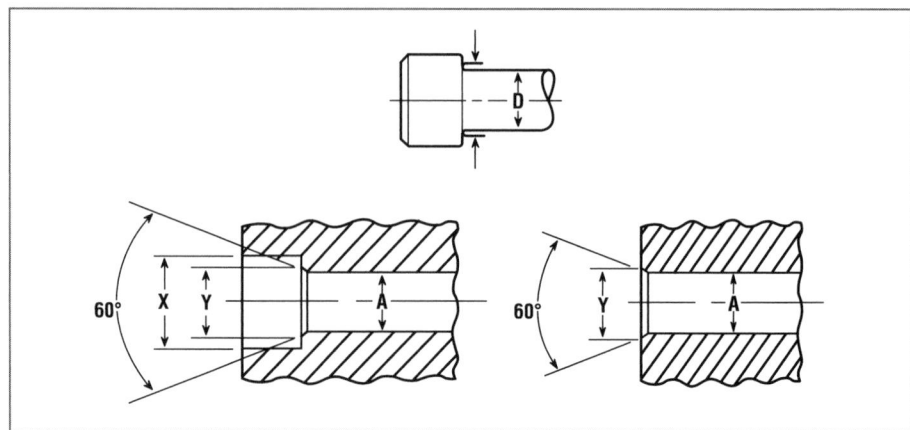

Nom. Size	D Basic Screw Dia.	A		X Counter-bore Dia.	Y Counter-sink Dia.[3]
		Drill Size for Hole A			
		Close Fit[1]	Normal Fit[2]		
M1.6	1.60	1.80	1.95	3.50	2.0
M2	2.00	2.20	2.40	4.40	2.6
M2.5	2.50	2.70	3.00	5.40	3.1
M3	3.00	3.40	3.70	6.50	3.6
M4	4.00	4.40	4.80	8.25	4.7
M5	5.00	5.40	5.80	9.75	5.7
M6	6.00	6.40	6.80	11.25	6.8
M8	8.00	8.40	8.80	14.25	9.2
M10	10.00	10.50	10.80	17.25	11.2
M12	12.00	12.50	12.80	19.25	14.2
M14	14.00	14.50	14.75	22.25	16.2
M16	16.00	16.50	16.75	25.50	18.2
M20	20.00	20.50	20.75	31.50	22.4
M24	24.00	24.50	24.75	37.50	26.4
M30	30.00	30.75	31.75	47.50	33.4
M36	36.00	37.00	37.50	56.50	39.4
M42	42.00	43.00	44.00	66.00	45.6
M48	48.00	49.00	50.00	75.00	52.6

Notes: [1] Close fit is limited to holes for those lengths of screws threaded to the head in assemblies in which: A) only one screw is used; or B) two or more screws are used and the mating holes are produced at assembly or by matched and coordinated tooling. [2] Normal fit is intended for: A) screws of relatively long length; or B) assemblies that involve two or more screws and where the mating holes are produced by conventional tolerancing methods. This fit provides the maximum allowable eccentricity of the longest standard screws and for certain deviations in the parts being fastened. [3] It is considered good practice to chamfer or break the edges of holes that are smaller than the screw diameter plus twice the fillet height (D Max. + 2F Max.) in parts in which the hardness approaches, equals, or exceeds the screw hardness. Dimension "Y" on the Table provides this value. All dimensions in millimeters. Source: Unbrako/SPS Technologies, Inc.

Table 124. ISO Tolerances for Socket Screws.

Nom. Dia.		Tolerance Zone										
over	to	C13	C14	D9	D10	D11	D12	EF8	E11	E12	JS9	K9
0	3	+0.20 +0.06	+0.31 +0.06	+0.045 +0.020	+0.060 +0.020	+0.080 +0.020	+0.12 +0.02	+0.024 +0.010	+0.074 +0.014	+0.100 +0.014	± 0.0125	0 −0.025
3	6	+0.24 +0.06	+0.37 +0.07	+0.060 +0.030	+0.078 +0.030	+0.115 +0.030	+0.15 +0.03	+0.028 +0.014	+0.095 +0.020	+0.140 +0.020	± 0.015	0 −0.030
6	10	-	-	-	-	+0.130 +0.040	+0.19 +0.40	+0.040 +0.018	+0.115 +0.025	+0.115 +0.025	± 0.018	0 −0.036
10	18	-	-	-	-	-	+0.2 +0.05	-	+0.142 +0.032	+0.212 +0.032	-	-
18	30	-	-	-	-	-	+0.275 +0.065	-	-	-	-	-
30	50	-	-	-	-	-	+0.33 +0.08	-	-	-	-	-
50	80	-	-	-	-	-	+0.40 +0.10	-	-	-	-	-

All dimensions in millimeters. References: ISO R286, ISO 4759/I, ISO 4759/II, ISO 4759/III. Source: Unbrako/SPS Technologies, Inc.

Table 125. Dimensions of Metric Socket Flat Head Cap Screws.

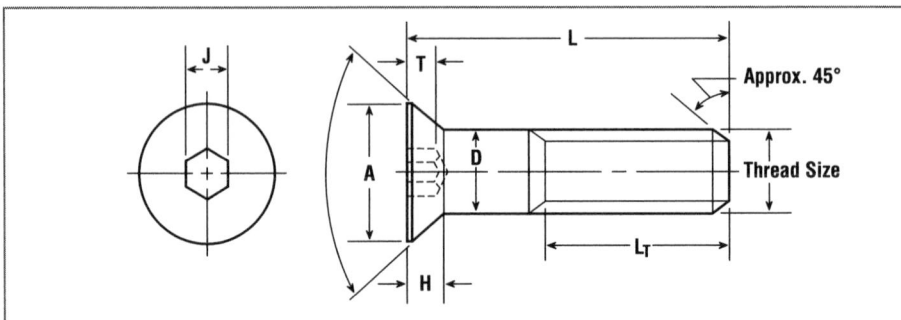

Nom. Thread Size	Pitch	A Head Dia.	D Body Dia.	H Head Height	J Socket Size	L_T Thread Length	T Socket Depth
		Max.	Max.	Ref.	Nom.	Min.	Min.
M3	0.5	6.72	3.00	1.70	2.00	18.00	1.10
M4	0.7	8.96	4.00	2.30	2.50	20.00	1.55
M5	0.8	11.20	5.00	2.80	3.00	22.00	2.05
M6	1.0	13.44	6.00	3.30	4.00	24.00	2.25
M8	1.25	17.92	8.00	4.40	5.00	28.00	3.20
M10	1.50	22.40	10.00	5.50	6.00	32.00	3.80
M12	1.75	26.88	12.00	6.50	8.00	36.00	4.35
M16	2.00	33.60	16.00	7.50	10.00	44.00	4.89
M20	2.50	40.32	20.00	8.50	12.00	52.00	5.45
M24	3.00	40.42	24.00	14.0	14.00	60.00	10.15

Length tolerances for socket flat head cap screws			
Size	Up to 16 mm incl.	Over 16 to 60 mm incl.	Over 60 mm
M3 thru M24	± 0.3	± 0.5	± 0.8

All dimensions in millimeters. Source: Unbrako/SPS Technologies, Inc.

Table 126. Dimensions of Metric Socket Button Head Cap Screws.

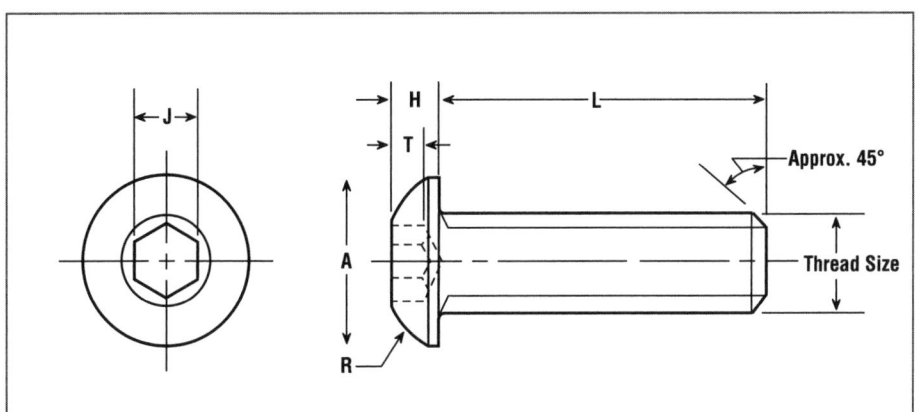

Screw Size	A Head Dia.	Oval Height	H Head Height	T Socket Depth	J Socket Size
M3	0.5	5.70	1.65	1.05	2.0
M4	0.7	7.60	2.20	1.35	2.5
M5	0.8	9.50	2.75	1.92	3.0
M6	1.0	10.50	3.30	2.08	4.0
M8	1.285	14.00	4.40	2.75	5.0
M10	1.50	18.00	5.50	3.35	6.0
M12	1.75	21.00	6.60	4.16	8.0
M16	2.0	28.00	8.60	5.20	10.0

Length tolerances for socket flat head cap screws			
Size	Up to 16 mm incl.	Over 16 to 60 mm incl.	Over 60 mm
M3 thru M24	± 0.3	± 0.5	± 0.8

All dimensions in millimeters. Source: Unbrako/SPS Technologies, Inc.

Table 127. Dimensions of Metric Socket Head Shoulder Screws.

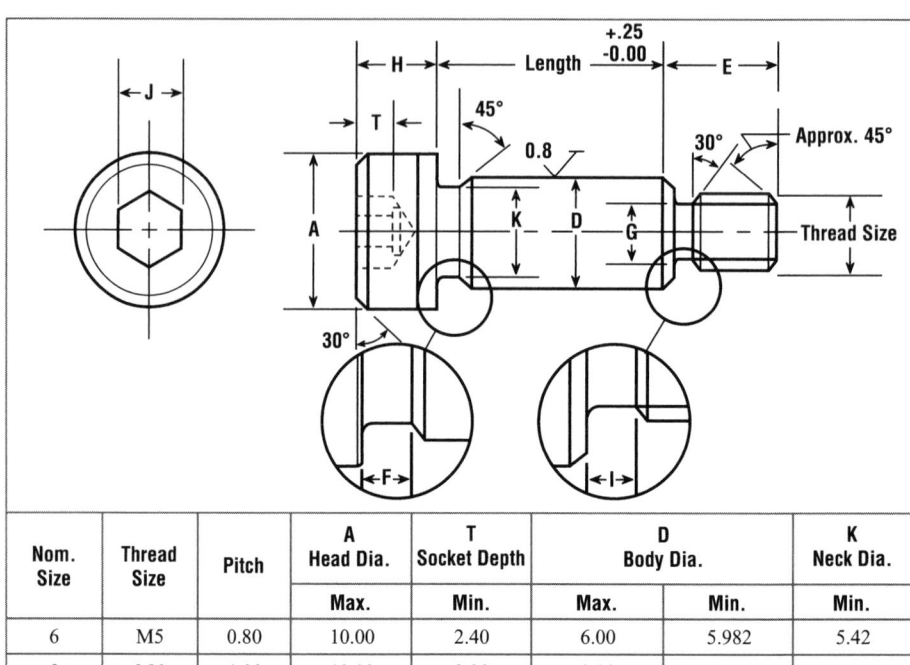

Nom. Size	Thread Size	Pitch	A Head Dia.	T Socket Depth	D Body Dia.		K Neck Dia.
			Max.	Min.	Max.	Min.	Min.
6	M5	0.80	10.00	2.40	6.00	5.982	5.42
8	M6	1.00	13.00	3.30	8.00	7.978	7.42
10	M8	1.25	16.00	4.20	10.00	9.978	9.42
12	M10	1.50	18.00	4.90	12.00	11.973	11.42
16	M12	1.75	24.00	6.60	16.00	15.973	15.42
20	M16	2.00	30.00	8.80	20.00	19.967	19.42
24	M20	2.50	36.00	10.00	24.00	23.967	23.42

Nom. Size	Thread Size	H Head Height	G Thread Neck Dia.	F Neck Width	I Thread Neck Width	E Thread/Neck Length	J Socket Size
		Max.	Min.	Max.	Max.	Max.	Nom.
6	M5	4.50	3.68	2.50	2.40	9.75	3.00
8	M6	5.50	4.40	2.50	2.60	11.25	4.00
10	M8	7.00	6.03	2.50	2.80	13.25	5.00
12	M10	8.00	7.69	2.50	3.00	16.40	6.00
16	M12	10.00	9.35	2.50	4.00	18.40	8.00
20	M16	14.00	12.96	2.50	4.80	22.40	10.00
24	M20	16.00	16.30	3.00	5.60	27.40	12.00

All dimensions in millimeters. Source: Unbrako/SPS Technologies, Inc.

Table 128. Dimensions of Metric Low Head Cap Screws.

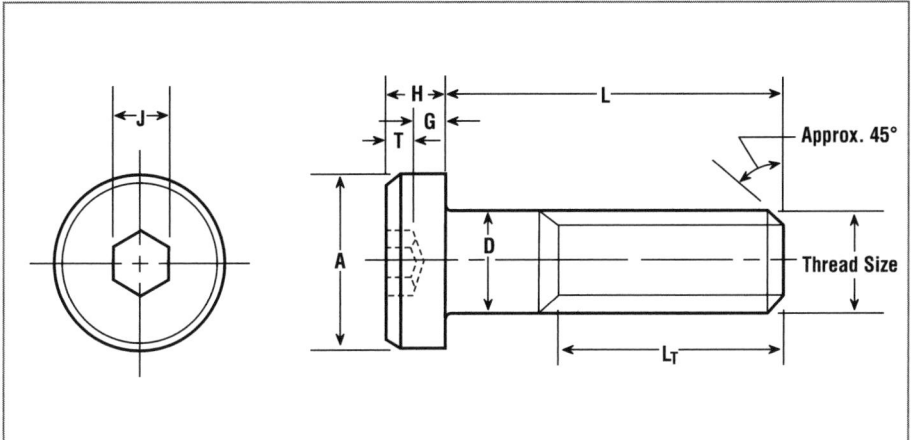

Nom. Thread Size	Pitch	A Head Dia.	D Body Dia.	G Wall Thickness	T Socket Depth	H Head Height	L_T Thread Length	J Socket Size
		Max.	Max.	Min.	Min.	Max.	Min.	Nom.
M4	0.7	7.00	4.00	1.06	1.48	2.8	20.00	3.0
M5	0.8	8.50	5.00	1.39	1.85	3.5	22.00	4.0
M6	1.0	10.00	6.00	1.65	2.09	4.0	24.00	5.0
M8	1.25	13.00	8.00	2.24	2.48	5.0	28.00	6.0
M10	1.5	16.00	10.00	2.86	3.36	6.5	32.00	8.0
M12	1.75	18.00	12.00	3.46	4.26	8.0	36.00	10.0
M16	2.0	24.00	16.00	4.91	4.76	10.0	44.00	12.0
M20	2.5	30.00	20.00	6.10	6.07	12.5	52.00	14.0

All dimensions in millimeters. Source: Unbrako/SPS Technologies, Inc.

Table 129. Dimensions of Metric Socket Set Screws - Knurled Cup Point and Plain Cup Point.

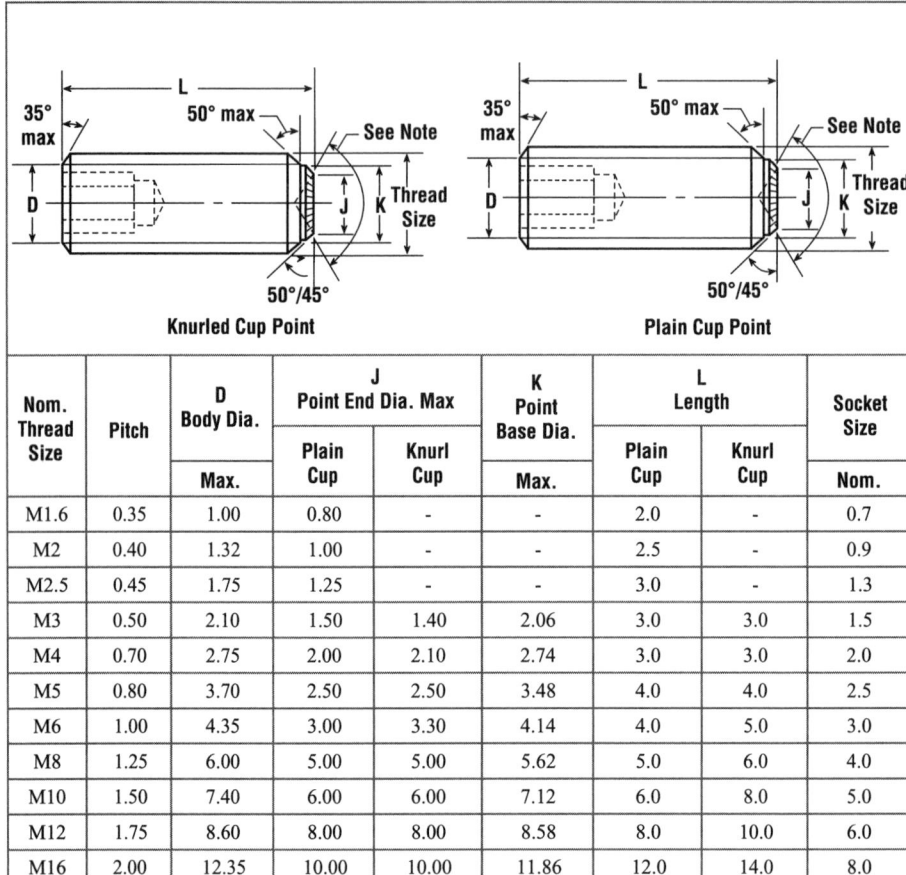

Knurled Cup Point   Plain Cup Point

Nom. Thread Size	Pitch	D Body Dia. Max.	J Point End Dia. Max		K Point Base Dia. Max.	L Length		Socket Size Nom.
			Plain Cup	Knurl Cup		Plain Cup	Knurl Cup	
M1.6	0.35	1.00	0.80	-	-	2.0	-	0.7
M2	0.40	1.32	1.00	-	-	2.5	-	0.9
M2.5	0.45	1.75	1.25	-	-	3.0	-	1.3
M3	0.50	2.10	1.50	1.40	2.06	3.0	3.0	1.5
M4	0.70	2.75	2.00	2.10	2.74	3.0	3.0	2.0
M5	0.80	3.70	2.50	2.50	3.48	4.0	4.0	2.5
M6	1.00	4.35	3.00	3.30	4.14	4.0	5.0	3.0
M8	1.25	6.00	5.00	5.00	5.62	5.0	6.0	4.0
M10	1.50	7.40	6.00	6.00	7.12	6.0	8.0	5.0
M12	1.75	8.60	8.00	8.00	8.58	8.0	10.0	6.0
M16	2.00	12.35	10.00	10.00	11.86	12.0	14.0	8.0
M20	2.50	16.00	14.00	14.00	14.83	16.0	18.0	10.0
M24	3.00	18.95	16.00	16.00	17.80	20.0	20.0	12.0

Note: The cap angle is 135° max. for screw lengths equal to or smaller than screw diameter. For longer lengths, the cup angle will be 124° max.

Length tolerances for Metric socket set screws with plain cup and knurled cup point			
Size	Up to 12 mm incl.	Over 12 to 50 mm incl.	Over 50 mm
M1.6 thru M24	± 0.3	± 0.5	± 0.8

All dimensions in millimeters. Source: Unbrako/SPS Technologies, Inc.

Table 130. Dimensions of Metric Socket Set Screws - Flat Point, Cone Point, and Dog Point. (See Notes.)

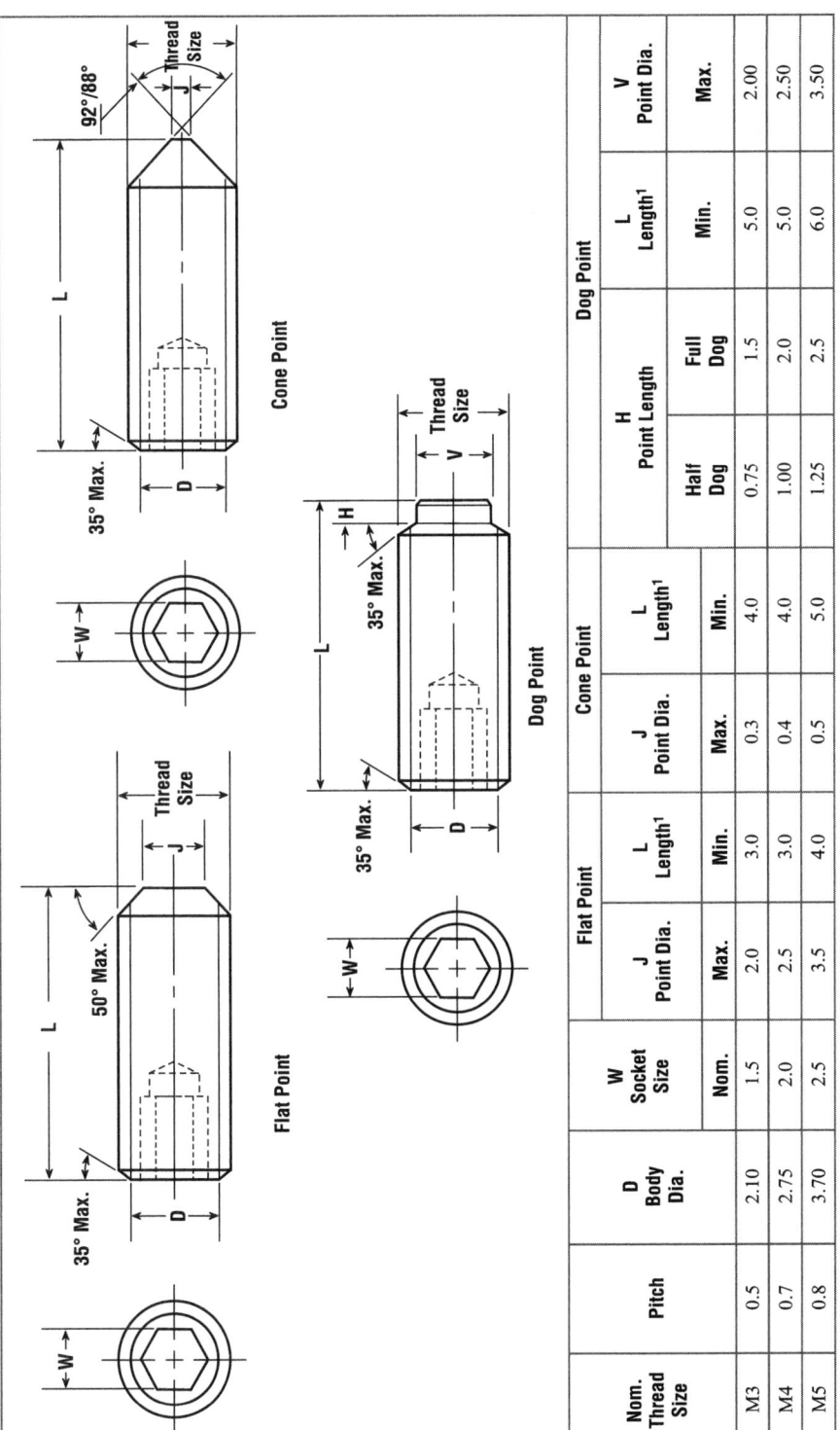

Cone Point

Dog Point

Flat Point

Nom. Thread Size	Pitch	D Body Dia.	W Socket Size	Flat Point		Cone Point			Dog Point				
				J Point Dia.	L Length[1]	J Point Dia.	L Length[1]		H Point Length		L Length[1]	V Point Dia.	
			Nom.	Max.	Min.	Max.	Min.		Half Dog	Full Dog	Min.	Max.	
M3	0.5	2.10	1.5	2.0	3.0	0.3	4.0		0.75	1.5	5.0	2.00	
M4	0.7	2.75	2.0	2.5	3.0	0.4	4.0		1.00	2.0	5.0	2.50	
M5	0.8	3.70	2.5	3.5	4.0	0.5	5.0		1.25	2.5	6.0	3.50	

(Continued)

Table 130. *(Continued)* **Dimensions of Metric Socket Set Screws - Flat Point, Cone Point, and Dog Point.** (See Notes.)

Nom. Thread Size	Pitch	D Body Dia.	W Socket Size Nom.	Flat Point J Point Dia. Max.	Flat Point L Length¹ Min.	Cone Point J Point Dia. Max.	Cone Point L Length¹ Min.	Dog Point H Point Length Half Dog	Dog Point H Point Length Full Dog	Dog Point L Length¹ Min.	Dog Point V Point Dia. Max.
M6	1.00	4.25	3.0	4.0	4.0	1.5	6.0	1.50	3.0	6.0	4.00
M8	1.25	6.00	4.0	5.5	5.0	2.0	6.0	2.00	4.0	8.0	5.50
M10	1.50	7.40	5.0	7.0	6.0	2.5	8.0	2.50	5.0	8.0	7.00
M12	1.75	8.60	6.0	8.5	8.0	3.0	10.0	3.00	6.0	12.0	8.50
M16	2.00	12.35	8.0	12.0	12.0	4.0	14.0	4.00	8.0	16.0	12.00
M20	2.50	16.00	10.0	15.0	14.0	6.0	18.0	5.00	10.0	20.0	15.00
M24	3.00	18.95	12.0	18.0	20.0	8.0	20.0	6.00	12.0	22.0	18.00

Notes: The cap angle is 135° max. for screw lengths equal to or smaller than screw diameter. For longer lengths, the cup angle will be 124° max. ¹ Minimum length is preferred.

Length tolerances for Metric socket set screws with flat point, cone point and dog point

Size	Up to 12 mm, incl.	Over 12 to 50 mm, incl.	Over 50 mm
M1.6 thru M24	± 0.3	± 0.5	± 0.8

All dimensions in millimeters. Source: Unbrako/SPS Technologies, Inc.

Table 131. Metric Hexagon Key Application Chart.

W Size	Socket Head Cap Screws	Flat Head Cap Screws	Button Head Cap Screws	Socket Head Shoulder Screws	Socket Set Screws	Low Head Cap Screws
0.7	-	-	-	-	M1.6	-
0.9	-	-	-	-	M2	-
1.3	-	-	-	-	M2.5	-
1.5	M1.6, M2	-	-	-	M3	-
2.0	M2.5	M3	M3	-	M4	-
2.5	M3	M4	M4		M5	
3.0	M4	M5	M5	M6	M6	M4
4.0	M5	M6	M6	M8	M8	M5
5.0	M6	M8	M8	M10	M10	M6
6.0	M8	M10	M10	M12	M12	M8
8.0	M10	M12	M12	M16	M16	M10
10.0	M12	M16	M16	M20	M20	M12
12.0	M14	M20	-	M24	M24	M16
14.0	M16	M24	-	-	-	M20
17.0	M20	-	-	-	-	-
19.0	M24	-	-	-	-	-
22.0	M30	-	-	-	-	-
27.0	M36	-	-	-	-	-
32.0	M42	-	-	-	-	-
36.0	M48	-	-	-	-	-

All dimensions in millimeters. Source: Unbrako/SPS Technologies, Inc.

Table 132. Dimensions and Mechanical Properties of Metric Hexagon Keys.

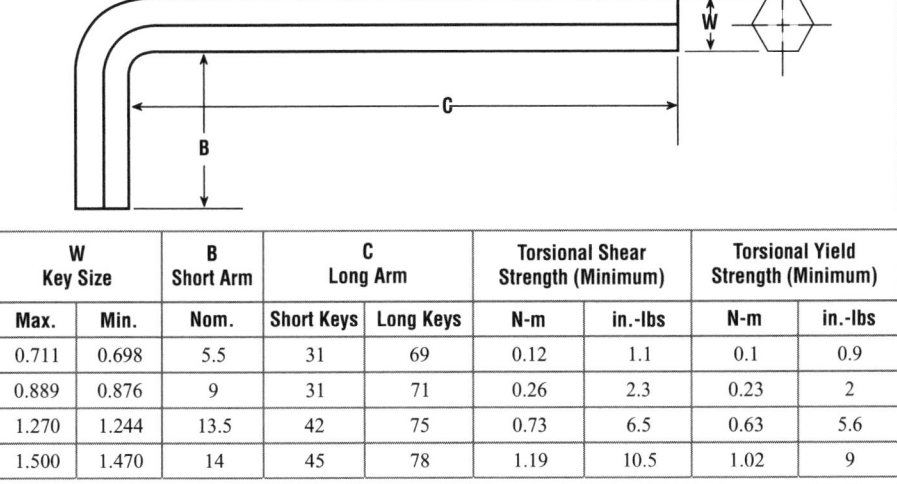

W Key Size		B Short Arm	C Long Arm		Torsional Shear Strength (Minimum)		Torsional Yield Strength (Minimum)	
Max.	Min.	Nom.	Short Keys	Long Keys	N-m	in.-lbs	N-m	in.-lbs
0.711	0.698	5.5	31	69	0.12	1.1	0.1	0.9
0.889	0.876	9	31	71	0.26	2.3	0.23	2
1.270	1.244	13.5	42	75	0.73	6.5	0.63	5.6
1.500	1.470	14	45	78	1.19	10.5	1.02	9

(Continued)

Table 132. *(Continued)* Dimensions and Mechanical Properties of Metric Hexagon Keys.

W Key Size		B Short Arm	C Long Arm		Torsional Shear Strength (Minimum)		Torsional Yield Strength (Minimum)	
Max.	Min.	Nom.	Short Keys	Long Keys	N-m	in.-lbs	N-m	in.-lbs
2.000	1.970	16	50	83	2.9	26	2.4	21
2.500	2.470	18	56	90	5.4	48	4.4	39
3.000	2.960	20	63	100	9.3	82	8	71
4.000	3.960	25	70	106	22.2	196	18.8	166
5.000	4.960	28	80	118	42.7	378	36.8	326
6.000	5.960	32	90	140	74	655	64	566
8.000	7.950	36	100	160	183	1,620	158	1,400
10.000	9.950	40	112	170	345	3,050	296	2,620
12.000	11.950	45	125	212	634	5,610	546	4,830
14.000	13.930	55	140	236	945	8,360	813	7,200
17.000	16.930	60	160	250	1,690	15,000	1,450	12,800
19.000	18.930	70	180	280	2,360	20,900	2,030	18,000
22.000	21.930	80	200	335	3,670	32,500	3,160	28,000
24.000	23.930	90	224	375	4,140	36,600	3,560	31,500
27.000	26.820	100	250	500	5,870	51,900	5,050	44,700
32.000	31.820	125	315	630	8,320	73,600	7,150	63,300
36.000	35.820	140	355	710	11,800	104,000	10,200	90,300

All dimensions in millimeters. Source: Unbrako/SPS Technologies, Inc.

Aircraft Hardware

AN and MS identification nomenclature. Aircraft hardware is normally specified by materials and dimensions set forth in aeronautical and Department of Defense (DoD) standards. AN (Air Force–Navy Aeronautical) standards have often been superceded or supplemented by National Aerospace Standards (NAS) and Military Standards (MS). Together, these standards present a universally accepted method for identification of aircraft hardware. All fasteners are identified with a specification number and a series of letters and/or "dash" numbers identifying their size, type of material, etc. This system presents a reliable method of identifying and cataloging thousands of fasteners and related hardware. Many of these have both an AN number and an MS number that are used interchangeably to identify the exact same item. The most widely used aircraft hardware will be described in this section, which provides alternatives to the more common commercial fasteners that may be inadequate for special applications. Much of the information in this section was provided by Ron Alexander, who has written extensively on the subject of aircraft hardware, and additional material has been extracted from the Federal Aviation Administration's (FAA) Advisory Circular AC 43.13-1B, and applicable standards.

Bolts, nuts, and washers

Aircraft quality bolts, made from alloy steel, stainless or corrosion resistant steel, aluminum alloys and titanium, are used where high strength is needed. Where this strength is not necessary screws are substituted. Aircraft bolts will always have a marking on their head—If there are no markings at all on the head of a bolt, it is probably a commercial grade fastener. The markings on bolts vary according to the manufacturer. A chart of typical bolt heads is presented in (*Figure 1*).

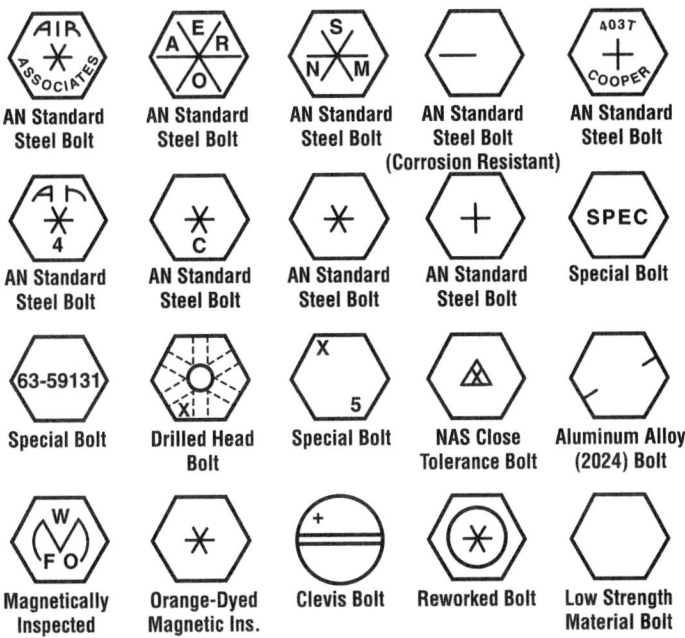

Figure 1. Typical aircraft bolt markings. (Source: AC 43-13-1B.)

NAS bolts have a higher tensile strength (usually about 160,000 PSI) and can be identified by a cupped out head. Close tolerance bolts are machined more accurately than general purpose bolts and they are used in applications requiring a very tight fit. Close tolerance bolts can be either AN or NAS and typically have a head marking consisting of a raised or recessed triangle. NAS-144 through NAS-158 bolts come with UNRF threads and are rated at 160,000 PSI ultimate tensile strength and 96,000 PSI ultimate shear strength. NAS bolts should be used with NAS-143 or MS2002 series chamfered washers. Typical NAS head markings are shown in *Figure 2*.

Head Marking	Designation and Strength	Notes
	Bolt, machine, aircraft. AN73 thru AN71 UTS 125,000 PSI	Available in both UNF and UNC threads. Bolt heads is higher AN3 thry AN20 series but bolts are interchangable.
	Bolt, hex head, close tolerance. NAS 1303 thru NAS 1320 UTS 160,000 PSI	Dimensionally similar to AN3 thru AN20 but shanks are slightly larger in diameter for closer fit. Much stronger than AN3 thru AN20.
	Bolt, .0156 oversize shank, close tolerance. NAS 2903 thru NAS 2920. UTS 160,000 PSI	Special purpose oversize repair bolt. Will not fit in standard holes. Identical to NAS3003 thru NAS3020. Identified by letter "E." Often found surplus.
	Bolt, hex head, close tolerance. Short thread. NAS1103 thru NAS1120. UTS 160,000 PSI	Close tolerance shear bolt. Thin head and short thread. To be used only in double shear applications. Weight saver.
	Bolt, shear, close tolerance. NAS 464-3 thru NAS 464-12. UTS 160,000 PSI	Similar to and interchangeable with NAS 1103 thru NAS 1120.
	Single raised or recessed dash designates a corrosion resistant steel bolt.	To be used only for corrosion resistance, strength, particularly at elevated temperatures, is low. Letters RM are an optional manufacturer's identification.
	Two raised or recessed dashes designate an aluminum alloy bolt.	Lightweight but low strength special purpose bolt. Do not use standard torque tables.
		Bolt, internal wrenching steel. NAS 144 thru NAS158. UTS 160,000 PSI. Aerospace quality internal wrenching bolt. Internal hex may strip with repeat use.
		Bolt, 12 point, external wrenching. NAS 624 thru NAs 644. UTS 180,000 PSI. Most common of the family of NAS "Super Bolts."

Figure 2. Typical NAS bolt markings.

The standard bolts used in aircraft construction are AN3 through AN20. The size, material, etc. of MS and AN bolts is identified by a part number. A breakdown of a $3/8$" dia., 1 $5/64$" long, undrilled AN bolt is shown in *Figure 3*.

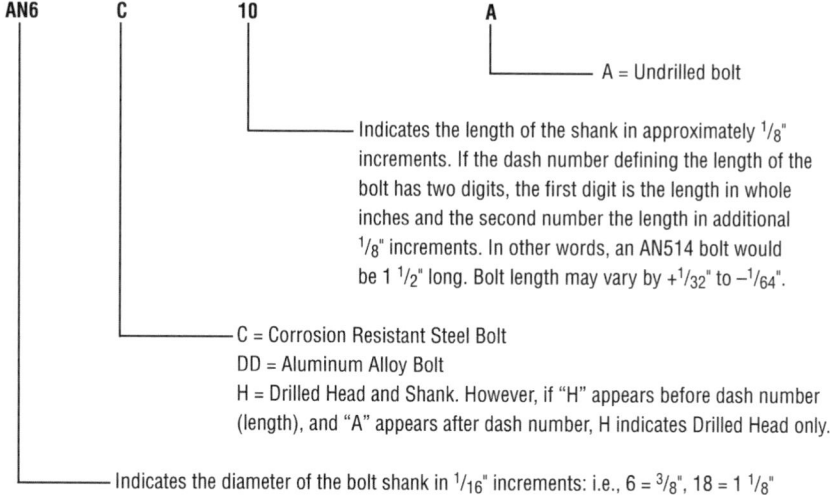

Figure 3. AN, MS, and NAS bolt number designation system.

The length must be sufficient to ensure no more than one thread will remain inside the bolt hole. This is the *grip length* of the bolt and it is measured from the underneath portion of the head to the beginning of the threads. The grip length should be equal to the material thickness that is being held by the bolt, or slightly longer. Washers with a combined maximum height of $1/8$" may be used if the bolt is slightly longer. Approximate threaded lengths for AN 3 through AN 20 bolts are given in **Table 1**.

Table 1. Approximate Threaded Length of AN 3 Through AN 20 Bolts. *(See Notes)*

AN Number	AN 3 & 4	AN 5	AN 6	AN 7	AN 8	AN 9
Threaded Length	13/32	17/32	41/64	21/32	25/32	29/32
AN Number	AN 10	AN 12	AN 14	AN 16	AN 18	AN 20
Threaded Length	61/64	1 3/32	1 1/4	1 3/8	1 1/2	1 11/16

Notes: All dimensions are approximate. Bolt length may vary by $+1/32$" to $-1/16$". To determine approximate grip length, subtract threaded length from the overall length of the fastener. Shank length is equal to grip length plus threaded length. All dimensions in inches.

Nuts. Aircraft nuts usually have no identification markings, but they are made from the same material as bolts. Due to the wide range of vibrations experienced in airframes and engines, nuts must have some form of a locking device to keep them in place. The most common ways of locking are cotter pins used in castle nuts, fiber inserts, lockwashers, and safety wire. Castle nuts (AN310) are used on bolts subjected to tension loads. Castellated shear nuts (AN320) are for applications subjected to shear stress only. Plain nuts (AN315 and AN335) require a locking device such as a checknut (AN 316) or safety wire—they can withstand high tension loading. Light hex nuts (AN340 and AN345) require a locking device and are satisfactory only for nonstructural applications.

Washers. The most common washers are designated AN960 and AN970. The main purposes of a flat washer in aircraft installations are to provide a shim when needed, act as a smooth load-bearing surface, and to adjust the position of castle nuts in relation to the drilled hole in a bolt. Also, plain washers should be used under a lockwasher to prevent damage to the surface of the metal. AN960 washers are the most common. They are manufactured in a regular thickness and a thinner thickness (one half the thickness of regular). The dash number following the AN960 indicates the size bolt for which they are used. The system differs from the one used for bolts. As an example, an AN960-616 is used with a $^3/_8$" bolt. If there is an "L" after the dash number, it is a thin or "light" washer. An AN960C is a stainless washer. The AN970 is a larger area flat washer used mainly for wood applications to protect the surface. Lockwashers (AN935 and AN936) should be used only where self-locking or castellated nuts are not suitable for the application.

Screws. Most screws are not used in structural applications in aircraft, as they have a lower material strength, looser thread fit, and do not have a clearly defined grip (they are threaded over their entire length). However, socket head machine screws made to NAS608 and NAS609 specifications are made from high-tensile strength materials and can be used in structural applications.

Safetying. As defined in AC 43.13-1B, safetying is "securing by various means and nut, bolt, turnbuckle, etc., ... so that vibration will not cause it to loosen during operation." Wire (usually soft brass, aluminum, or steel) and cotter pins are most often used for this purpose. Two different methods of safety wiring are used: The single wire method (*Figure 4*) is used on screws, bolts, and/or nuts in a closely-spaced pattern, and the largest wire size that will fit the holes should be used. The double-twist method (*Figure 5*) usually uses .032" minimum diameter wire and provides a secure means of assuring that critical fasteners do not loosen.

Figure 4. Single wire method. (Source: AC 43.13-1B)

Bolt Heads   **Castle Nuts**

Figure 5. Double-twist method. (Source: AC 43.13-1B)

Solid shank rivets

Sheet metal construction is very popular within several industries, and rivets play a major role in this type of construction—basically, sheet metal parts may be riveted, bolted, or welded together. Rivets are light in weight and cheap in price, and they are not only used to hold steel or aluminum sheets together, but they are also used in securing fittings, nut plates, and spars and ribs used in aircraft construction. Simply defined, a rivet is a bolt without a nut. More precisely, it is a small metal pin that when properly installed provides a very effective and strong bond.

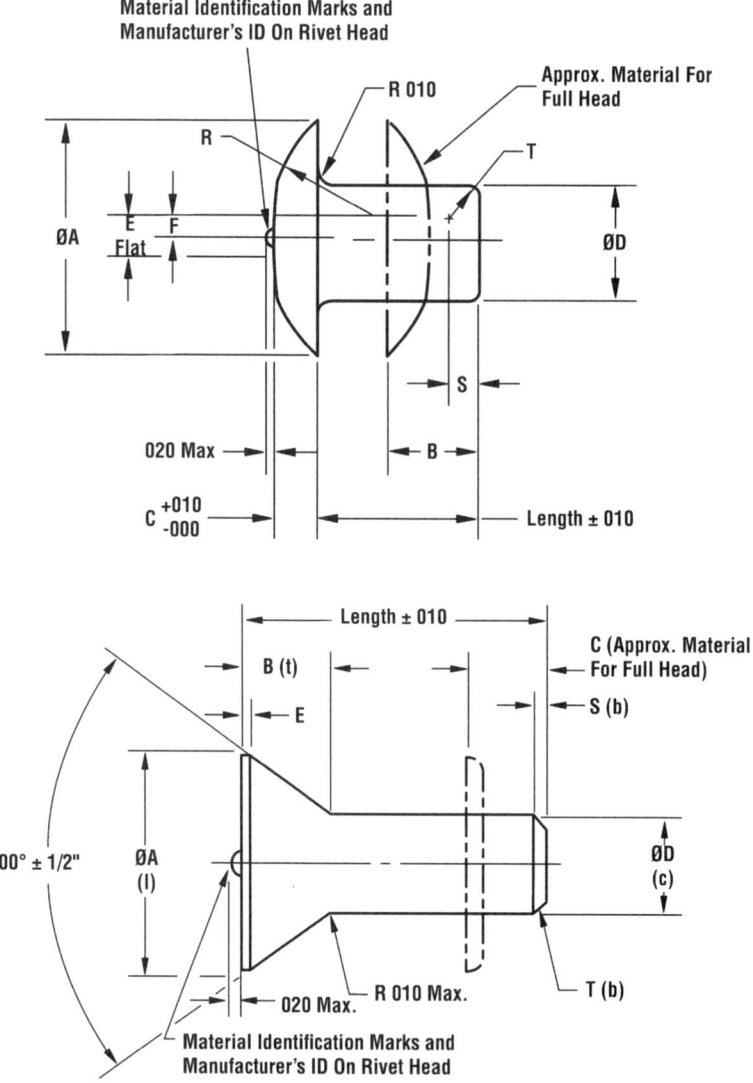

Figure 6. Universal head (MS20470 and MS20515) and 100° countersunk head (MS20426 and MS20427) solid shank rivets. The length of a rivet is measured from the bottom of the rivet head to the end of the shank on a universal head rivet, and from the top of the head to the end of the shank on a countersunk rivet.

A solid rivet consists of a manufacturer's head, a shank, and a shop head (often termed a bucktail). The shop head is formed during the installation process by squeezing the rivet or by using a bucking bar and a rivet gun. The shop head functions much the same as a nut on a bolt. When the rivet is installed, the action of the riveting tool also expands the shank of the rivet slightly, thus providing a very tight fit.

Solid shank rivets are identified by their material, head type, temper condition, and the dimension of their shank. Rivets are most commonly manufactured with either a 100° countersunk head or a universal head. The aluminum and titanium alloy universal head rivet is designated MS20470, and the 100° countersunk head rivet is designated MS20426 (see *Figure 6*). These MS numbers are also interchangeable with AN numbers: AN470 = MS20470 and AN426 = MS20426. Additionally, five alternative materials are available within these designations (see *Figure 7*).

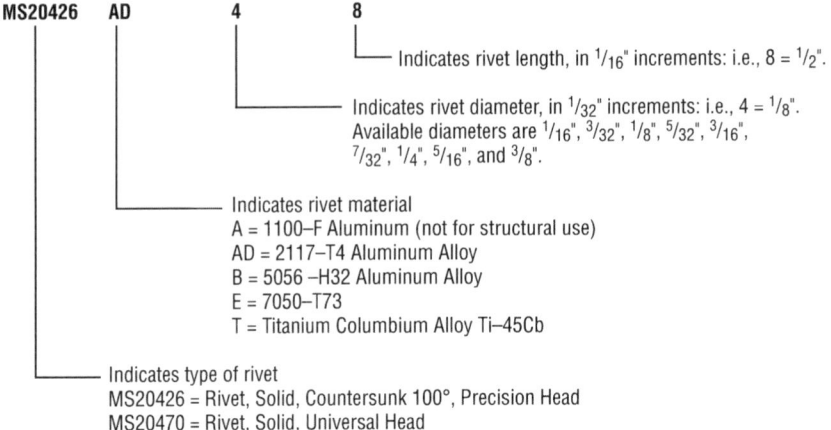

Figure 7. Solid shank rivet number designation system.

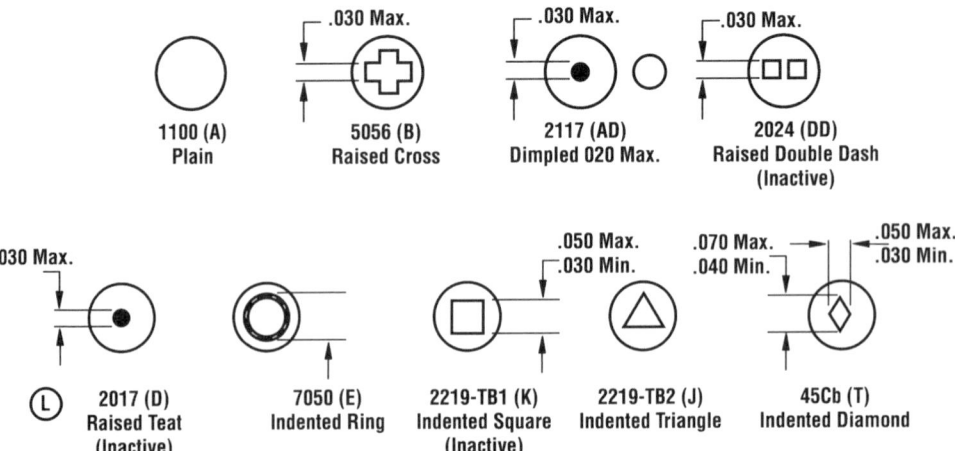

Figure 8. Material identification rivet head marking system.

Two other solid shank rivets should be noted. MS20615 is the designation for brass, copper, and nickel copper alloy universal head rivets, and MS20427 is for 100° flush head carbon steel, corrosion resistant steel, Monel, and copper rivets.

Solid rivet shear strength. Pure aluminum 1100-F has a low shear strength (usually 10,000 PSI). These rivets are used in softer aluminum alloys such as 3003 or 5052 where strength is not a consideration. They use an "A" as identification and are termed soft: i.e., MS20470A. Aluminum alloy 2117-T4 is heat treated for strength, and typically has a minimum shear strength of 26,000 PSI. Their material is also highly resistant to corrosion and rivets made from 2117-T4 use the letters "AD" as an identifier (MS20470AD). AD rivets should be used for all structural applications on sheet aluminum, and are easily identifiable because they are dimpled on their head. This dimple is also used as an aid to drill out defective or improperly installed rivets. The 5056-H32 aluminum alloy and has a minimum shear strength of 24,000 PSI; and rivets made from this material ("B" series) are used for magnesium alloy structures. Minimum shear strengths for other materials are: 7050-T73 = 41,000 PSI; Monel = 49,000 PSI; and Ti-45Cb = 50,000 PSI. Single rivet shear strengths are shown in **Table 2**.

Table 2. Single Shear Strength of Solid Shank Rivets.

Rivet Material	Driven Single Shear Strength (lbs)							
	1/16"	3/32"	1/8"	5/32"	3/16"	1/4"	5/16"	3/8"
2117-T4	106	217	389	596	860	1555	2455	3510
5056-H32	99	203	363	556	802	1450	2290	3275
7050-T73	152	311	558	854	1230	2230	3520	5030
Monel	183	376	674	1030	1490	2695	4260	6085
Ti-45Cb	187	384	687	1050	1515	2745	4340	6200

Source: Mil-Hdbk-5H.

Sources are not in unanimous agreement on the dimensions for rivet hole sizes and head dimensions for the installation of solid rivets, and those provided in **Table 3** are from MIL-HDBK-5H, which provides authoritative recommendations for structural fasteners, and MS20426 and MS20470.

Blind rivets

Blind rivets are headed fastening devices that have a shank that is manually expanded by a means of a separate internal stem or mandrel. Their design allows installation from one side only, which greatly simplifies assembly and also makes it possible to securely join and fasten two items (typically two lengths of sheet metal) even though one side is not accessible. There are a wide variety of blind rivets available, including the common "pop" rivet that can be purchased at any hardware store. However, common rivets should be used only in nonstructural applications, as their shear strength is very poor. The blind rivets that will be examined in this section are structural rivets, which can be obtained from aircraft hardware suppliers. Structural rivets retain part of the mandrel inside the rivet shank after the mandrel (or stem) breaks, and the retained mandrel extends through the shear plane of the work. To be sure that the mandrel contributes to the structural integrity of the rivet, the mandrel must break above the plane of the upper work surface, and it must be securely locked in place.

MS Blind Rivets. The most widely used structural blind rivets are the MS20600 series rivets, which typically have a minimum shear strength of 25,000 PSI for those made of 2117 or 5056 aluminum alloy, and 49,000 PSI for those made of Monel. General guidelines

Table 3. Dimensional Sizes and Bucked Head Dimensions for Solid Shank Rivets.

Dimension		Nominal Shank Diameter [1]					
		1/16"	3/32"	1/8"	5/32"	3/16"	7/32"*
Drill Size		#51	#41	#30	#21	#11	#1
Nominal Hole Dia.		0.067	0.096	0.128	0.159	0.191	0.228
Universal Head Diameter [2]		0.125 □0.006	0.187 □0.009	0.250 □0.012	0.312 □0.016	0.375 □0.019	0.437 □0.022
Universal Head Height +0.010/–0.000 [2]		0.027	0.040	0.054	0.067	0.080	0.093
Countersunk Head Diameter □0.004 [3]		0.114	0.179	0.225	0.288	0.353	0.415
Countersunk Head Height (Reference) [3]		0.022	0.036	0.042	0.055	0.070	0.083
Material Required For Full Head [4]		0.094	0.141	0.188	0.234	0.281	0.328
Minimum Driven Head Diameter [5]		0.081	0.122	0.163	0.203	0.244	0.285
Driven Head Height	Max.	0.040	0.050	0.070	0.092	0.105	0.110
	Min.	0.025	0.038	0.050	0.062	0.075	0.087
Dimension		1/4"	9/32"*	5/16"	11/32"*	3/8"	13/32"*
Drill Size		F	L	P	S	W	Z
Nominal Hole Dia.		0.257	0.290	0.323	0.348	0.386	0.413
Universal Head Diameter [2]		0.500 □0.025	0.562 □0.028	0.625 □0.031	0.687 □0.034	0.750 □0.037	0.812 □0.040
Universal Head Height +0.010/–0.000 [2]		0.107	0.120	0.133	0.147	0.161	0.174
Countersunk Head Diameter □0.004 [3]		0.476	0.524	0.564	0.627	0.694	0.758
Countersunk Head Height (Reference) [3]		0.095	–	0.106	–	0.134	–
Material Required For Full Head [4]		0.375	0.422	0.469	0.516	0.562	0.609
Minimum Driven Head Diameter [5]		0.325	0.365	0.406	0.450	0.488	0.530
Driven Head Height	Max.	0.130	0.140	0.158	0.170	0.190	0.200
	Min.	0.100	0.113	0.125	0.137	0.150	0.165

* Indicates oversize replacement rivet. [1] Tolerance on shaft diameter is +0.003/–0.001. [2] Universal head rivet only. [3] Countersunk head rivet only. [4] The length of shaft projecting beyond the workpiece that will, when bucked, produce a full diameter head. This is sometimes referred to as the "clinch allowance." [5] For rivets made of 7050-T3, upset head diameter should be at least 1.4 times nominal shank diameter. All dimensions, except drill sizes, are in inches. Source: MIL-HDBK-5H, MS20426, and MS20470.

are to use five of these rivets in the same area that would normally require only three solid rivets, or to specify a size $1/32$" larger than the solid rivet that would be used in the same application. However, there are some varieties of blind rivets that are suitable "one for one" replacements for the same size shank solid rivet, and these aerospace structural rivets will be discussed later in this section.

The Department of Defense and other standardizing bodies have assigned "dash" numbers to identify the diameter and grip range of blind rivets. The "dash" numbers are universally recognized as part numbers by domestic rivet manufacturers and distributors, and are commonly used to qualify the specifications and dimensions of blind rivets. An example of the dash numbers for a standard MS20601 100° flush head, pull stem, self-plugging rivet (as explained in *Figure 9*) might be MS20601AD8W5.

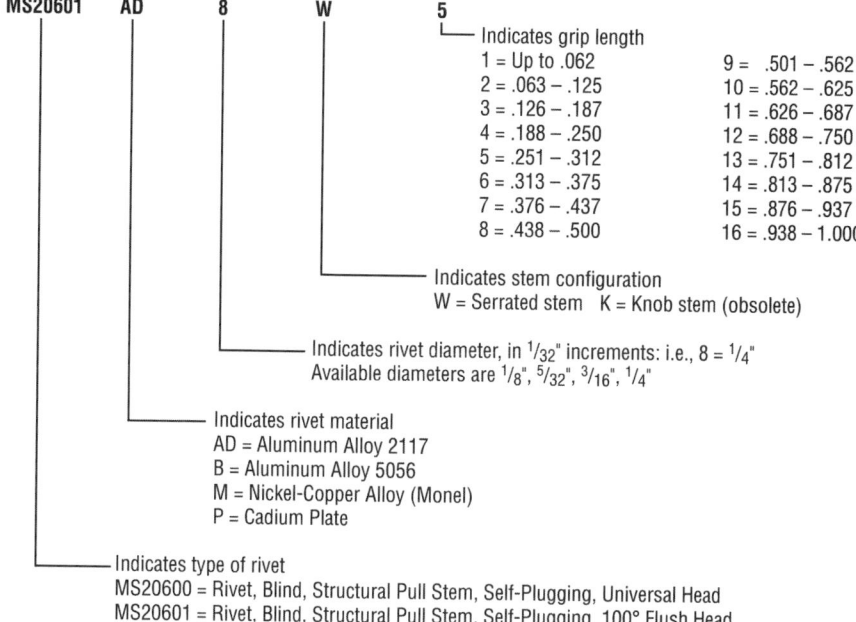

Figure 9. Blind rivet number designation system.

When ordering rivets, it is wise to provide the complete designation, with identifying numbers and letters, but it should be kept in mind that not all rivets are available in the range of diameters and grip lengths specified above.

DoD specification rivets are manufactured by several companies, two of the most widely available are made by the Avdel and by Cherry Aerospace Fasteners; both divisions of Textron, Inc. The "Cherry MS" series part numbers correspond to DoD designations as shown below. Diameters and grip lengths are identified by the same dash number system used for the DoD series rivets.

Material Designation	MS20600	MS20601
	Cherry Number	
AD	CR9163	CR9162
B	CR9157	CR9156
M	CR9563M	CR9562M
MP	CR9563	CR9562

Aerospace Structural Rivets. Blind rivets can actually be stronger than a solid alumi-num rivet of the same diameter, and can sometimes be used in their place. However, the FAAs Advisory Circular AC 43.13-1B clearly specifies when blind rivets may be used in aircraft construction: "Blind Rivets in the MS20600 through MS20603 series rivets and the mechanically locked stem NAS 1389, 1399, 1738, and 1739 rivets sometimes may be substituted for solid rivets. They should not be used where the looseness or failure of a few rivets will impair the airworthiness of the aircraft." Two examples of rivets that meet the NAS standards are Cherrylock® (NAS 1738 and 1739) and CherryMAX AB® (NAS 1398 and 1399) wiredrawn rivets (both from Cherry Aerospace Fasteners), which are suitable replacements for the same size solid shank rivet. When these rivets are pulled, they form a large bulb that appears very similar to a driven solid shank rivet. They are available with a universal head and with a 100° countersunk head. Cherrylock and CherryMAX rivets are expensive compared to MS20600 series rivets, but as can be seen in **Tables 4** and **5**, their strength often justifies the additional cost.

Table 4. Materials and Strength of Cherrylock® Rivets.

Materials			
Rivet Material		**Ultimate Rivet Shear Strength (PSI)[1]**	**Typical Maximum Temperature in Use (°F)**
Sleeve	**Stem**		
5056 Al	Alloy Steel Inconel 600	50,000	250
Monel	Inconel 600	55,000	900
Inconel 600	A-286 Cres	75,000	1400

Minimum Ultimate Rivet Shear and Tensile Strength (lbs)[2]						
Rivet Material	**Single Shear**			**Tensile**		
	1/8"	5/32"	3/16"	1/8"	5/32"	3/16"
Aluminum	619	935	1260	345	530	710
Monel	895	1353	1823	490	740	1000
Cres.	1221	1845	2488	570	860	1160

[1] At room temperature. [2] Test method per MIL-STD-1312-8&-20. Values are for fastener capabilities only. Source: Cherry Aerospace Fasteners.

Table 5. Materials and Strength of CherryMAX AB® Rivets.

Materials			
Rivet Material		**Ultimate Rivet Shear Strength (PSI)[1]**	**Typical Maximum Temperature in Use (°F)**
Sleeve	**Stem**		
2017 Al	7075 Al	38,000	250
Monel	A-286 Cres	55,000	900
A-286 Cres	A-286 Cres	75,000	1400

Minimum Ultimate Rivet Shear and Tensile Strength (lbs)[2]								
Rivet Material	**Single Shear**				**Tensile**			
	1/8"	5/32"	3/16"	1/4"	1/8"	5/32"	3/16"	1/4"
Aluminum	494	755	1090	1970	230	375	540	1000
Monel	710	1090	1590	2840	340	550	780	1450
Cres.	970	1490	2150	3890	640	1000	1500	2700

[1] At room temperature. [2] Test method per MIL-STD-1312-8&-20. Values are for fastener capabilities only. Note: The above tensile strengths do not apply to CherryMAX part number CR4626, which has a slightly lower tensile strength than that stated for corrosion resistant steel rivets in the Table. Source: Cherry Aerospace Fasteners.

Rivet material compatibility

The surfaces being joined must be considered when selecting a rivet, as some materials, when placed in proximity, are incompatible and will instigate corrosion where the rivet is in contact with the work surface. The following guidelines should be followed. *Aluminum rivets*: 1100 (pure aluminum) rivets are used in soft aluminum alloys such as 1100, 3003, and 5052. 2117-T4 rivets are most commonly used for aluminum alloy structures. 2017-T3, 2017-T31, and 2024-T4 rivets are used in aluminum alloys where high strength is required. 5056-H32 rivets are specified for joining magnesium alloys. *Mild steel rivets*: Used primarily for riveting steel parts. *Corrosion resistant steel* (*Cres*) *rivets*: Used for corrosion resistant (stainless) steel parts. *Monel rivets*: Used for riveting high-nickel steel alloys and nickel alloys, and for stainless steel. *Copper rivets* can be used for copper and for nonmetallic materials.

Rivet installation

All references to dimensions in this section refer to MS20426 and MS20470 solid shank rivets. General guidelines are harder to establish for blind rivets because their quality ranges from very poor (pop rivets) to excellent (aerospace rivets). A rule of thumb, often stated, for MS20600 and MS20601 blind rivets is as follows: One bolt is equal in strength to three solid rivets or five blind rivets. When using aerospace structural rivets, this guideline is obsoleted by their superior quality, and the manufacturer's installation guidelines should be followed.

Several dimensions are required when laying out a riveting operation. First, the proper diameter rivet must be selected, which involves two considerations. First, the rivet diameter should be 1.5 to 2 times the total workpiece thickness, and it should be 2.5 to 3 times the thickness of the thickest sheet being riveted. Next, "pitch" and "row spacing" must be established (*Figure 10*). Pitch is defined as the distance between the hole centers of adjacent rivets in a row, and minimum pitch distances are given in **Table 6**. Row spacing is the distance between rows of adjacent rivets, and minimum row spacing is .866 times pitch. Note that these are minimums: normal pitch can run eight to ten times rivet diameter, and normal row spacing is four to six times rivet diameter. Edge margin is the distance between hole centers and the edge of the material, and may be either visible or invisible. When viewed from the outside (the side from which the rivet is inserted), the material edge in view is the visible edge, and the edge hidden is the invisible edge. **Table 7** provides edge margin minimums.

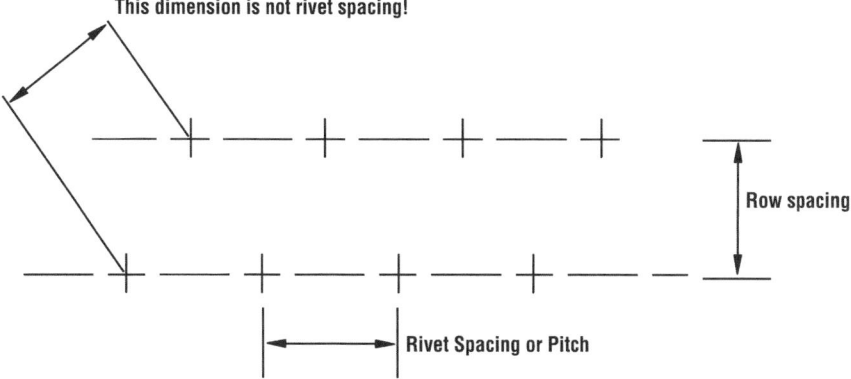

Figure 10. Measuring rivet spacing and row spacing.

Table 6. Minimum Pitch Spacing for Solid Shank Rivets.

Rivet Dia.	Dimpled Csk Rivets	Machine Csk Rivets	Universal Head Rivets	Rivet Dia.	Dimpled Csk Rivets	Machine Csk Rivets	Universal Head Rivets
1/16	0.375	0.313	0.250	1/4	0.938	1.375	0.875
3/32	0.563	0.438	0.375	9/32*	1.125	1.125	1.000
1/8	0.625	0.531	0.500	5/16	1.250	1.250	1.125
5/32	0.688	0.625	0.563	11/32	1.375	1.375	1.250
3/16	0.750	0.750	0.688	3/8	1.500	1.500	1.344
7/32*	0.875	0.875	0.781	13/32*	1.625	1.625	1.469

* Indicates sizes normally used only for oversize replacement. All dimensions in inches. See Table 7 for examples of machine countersinks and dimpled countersinks. Source: MIL-R-47196A.

Table 7. Minimum Edge Margins for Solid Shank Rivets.

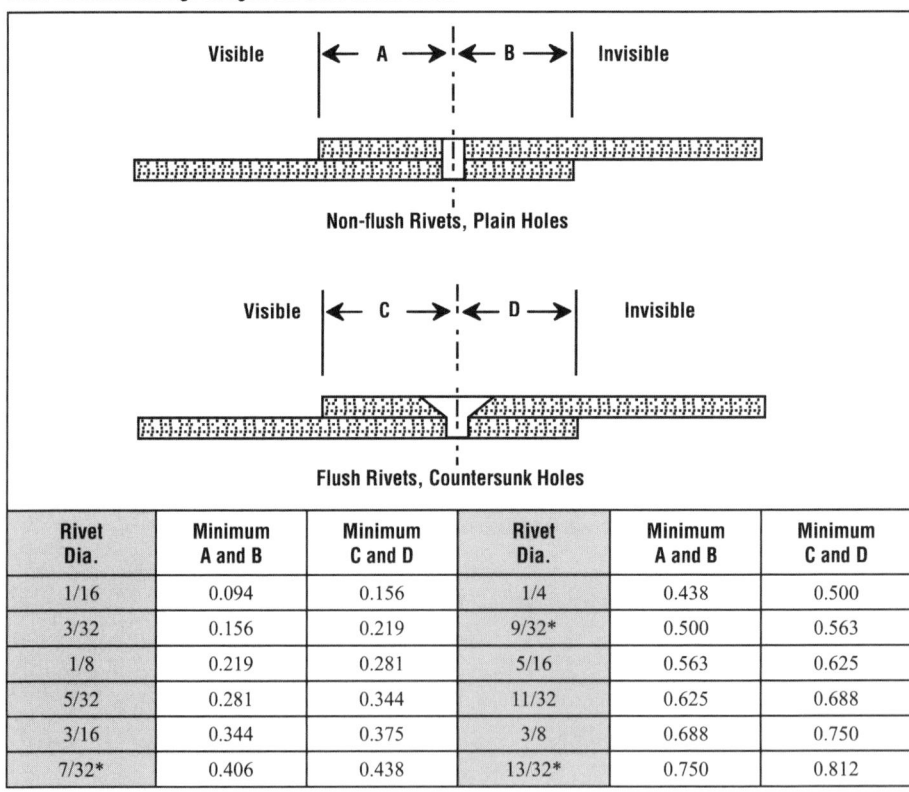

Rivet Dia.	Minimum A and B	Minimum C and D	Rivet Dia.	Minimum A and B	Minimum C and D
1/16	0.094	0.156	1/4	0.438	0.500
3/32	0.156	0.219	9/32*	0.500	0.563
1/8	0.219	0.281	5/16	0.563	0.625
5/32	0.281	0.344	11/32	0.625	0.688
3/16	0.344	0.375	3/8	0.688	0.750
7/32*	0.406	0.438	13/32*	0.750	0.812

* Indicates sizes normally used only for oversize replacement. All dimensions in inches. Source: MIL-R-47196A.

Countersinking and Dimpling. The most widely used method for preparing sheet metal for countersunk rivets is done by chamfering a conical depression around the top of an existing hole for receiving the head of a fastener. The use of this method is not acceptable for unacceptably thin panels because the bearing strength of the countersunk head is marginalized if the thickness of the sheet is less than the depth of the countersink. Dimpling is

Table 8. Hole Preparation for Countersunk Solid Shank Rivets (MS20426).

100° ± .5°
100° ± .5°
Dimple Upper Dia.
Outer Sheet Thickness
Nominal Rivet Head
H
Dimple Lower Dia.
Middle Sheet Thickness
X
Nominal Dimple
G
Inner Sheet Thickness
G = Gap
X = Thickness of Sheet
H = Head Height
X

Dimensions		Nominal Shank Diameter			
		1/16"	3/32"	1/8"	5/32"
Hole Size Before Dimpling	Min.	0.064	0.096	0.128	0.159
Final Hole Size	Min.	0.067	0.098	0.129	0.061
	Max	0.071	0.106	0.139	0.172
Sheet Thickness For Dimpling	Min.	0.016	0.016	0.016	0.016
	Max.	0.040	0.051	0.064	0.072
Sheet Thickness For Countersinking	Min.	0.032	0.040	0.051	0.064
Countersink Diameter [1]	Min.	0.095	0.160	0.206	0.267
	Max.	0.105	0.170	0.216	0.277
Countersink Diameter [2]	Min.	0.117	0.167	0.215	0.285
	Max.	0.127	0.177	0.225	0.293
"G" Gap	Max.	0.005	0.010		

Dimensions		Nominal Shank Diameter			
		3/16"	1/4"	5/16"	3/8"
Hole Size Before Dimpling	Min.	0.191	0.250	0.312	0.375
Final Hole Size	Min.	0.191	0.257	0.320	0.386
	Max	0.204	0.266	0.330	0.404
Sheet Thickness For Dimpling	Min.	0.016	0.016	0.016	0.016
	Max.	0.091	0.125	0.156	0.188
Sheet Thickness For Countersinking	Min.	0.072	0.102	0.125	0.156
Countersink Diameter [1]	Min.	0.334	0.457	0.545	0.675
	Max.	0.344	0.467	0.555	0.685
Countersink Diameter [2]	Min.	0.349	0.471	–	–
	Max.	0.359	0.481	–	–
"G" Gap	Max.	0.010			

[1] Dimension to be used when countersinking is the only method of hole preparation used. [2] Dimensions to be used when combination predimple and countersink method of hole preparation is used. The countersink dimensions on these Tables will result in the rivet head being "flush" to "high" after driving, which may necessitate shaving the protruding head depending on surface requirements. Source: MIL-STD-403C.

the process of stretching a relatively small, shallow indentation into sheet metal to accept a countersunk rivet or screw. Dimpling is substantially stronger than machine countersinking, and should always be used on relatively thin panels. When dimpling, a pilot hole should be drilled, then the material is dimpled and the hole size should be finalized by drilling or reaming. Most aluminum may be dimpled at ambient temperature, but 7075-T6 and 2024-T6, and magnesium, should be dimpled at temperatures between 149° and 204°F (65° to 96°C). Titanium sheets should be dimpled at temperatures between 316° and 426°F (157° to 219°C). Combination dimple and countersinking is advisable when the outer sheet is less than the least thickness for countersinking, and the inner sheet is greater than the greatest thickness for dimpling. **Table 8** provides minimum sheet thicknesses for countersinking, and minimum and maximum sheet thicknesses for dimpling when using solid shank rivets, and **Table 9** gives the same information for blind rivets. Note that the countersink dimensions on these Tables will result in the rivet head being "flush" to "high" after driving, which may necessitate shaving the protruding head depending on surface requirements.

120° Countersunk rivets. In addition to standard precision head 100° degree rivets, countersunk rivets are also available with 120° heads. These wide head rivets are specified for aerospace work when working with composite workpieces. They can also be useful when attaching thin sheets of aluminum, but only with the understanding that, for structural purposes, their integrity will be sacrificed if the sheet thickness is not sufficient to provide acceptable bearing area.

Table 9. Hole Preparation for Countersunk Blind Rivets (MS20601).

Dimensions		Nominal Shank Diameter			
		1/8"	5/32"	3/16"	1/4"
Hole Size Before Dimpling	Min.	0.120	0.128	0.161	0.234
Final Hole Size	Min.	0.128	0.160	0.192	0.256
	Max	0.132	0.164	0.196	0.261
Sheet Thickness For Dimpling	Min.	0.016	0.016	0.016	0.016
	Max.	0.051	0.072	0.091	0.125
Sheet Thickness For Countersinking	Min.	0.051	0.064	0.072	0.102
Countersink Diameter [1]	Min.	0.206	0.267	0.334	0.457
	Max.	0.216	0.277	0.344	0.467
Countersink Diameter [2]	Min.	0.215	0.283	0.349	0.471
	Max.	0.225	0.293	0.359	0.481
"G" Gap (See Table 8)	Max.	0.010			

[1] Dimension to be used when countersinking is the only method of hole preparation used. [2] Dimensions to be used when combination predimple and countersink method of hole preparation is used. The countersink dimensions on these Tables will result in the rivet head being "flush" to "high" after driving, which may necessitate shaving the protruding head depending on surface requirements. Source: MIL-STD-403C.

Thread Measurement With The Three-Wire Method

Several gages are available to measure the pitch diameter of threads, including the thread micrometer, which uses a pointed anvil and spindle to measure the minor diameter of an external thread. Another method provides for locating segments of wire of a known diameter at three places on a thread and using a micrometer to measure the diameter over the three points of the wire, as shown in *Figure 1*. The key to an exact reading is in selecting the proper size wire for the measurement. The best size wires are those that touch the thread at the pitch diameter, and the size will obviously vary with the angle of the thread. The best size wire can be identified by using the following equation to identify a constant for the angle, and then dividing the constant by the pitch (number of threads per inch).

$$\text{Constant} = \sec \tfrac{1}{2} \text{ angle of the thread} \times 0.5.$$

For 60° threads, this means that the best wire size is $0.57735 \div$ pitch; and for 55° threads, the best wire size is $0.56369 \div$ pitch.

Therefore, the best wire size for measuring a $^7/_{16} \times 24$ thread would be found by dividing the pitch by the constant, or $0.57735 \div 24 = 0.02405625$. Best wire sizes and constants for Unified threads are given in **Table 1**, and for Metric threads in **Table 2**. Metric constants are derived from a different formula based on pitch and the wire diameter rather than the number of threads per inch. Wires used in thread measurement should be hardened and lapped steel wires with an accuracy of 0.0002 inch.

After three wires of equal diameter have been selected, they are positioned in the thread grooves as shown in *Figure 1*. The anvil and spindle of an ordinary micrometer are then placed against the three wires and a measurement is taken. To determine what the reading of the micrometer *should* be if the thread is the correct finish size, use one of the following formulas.

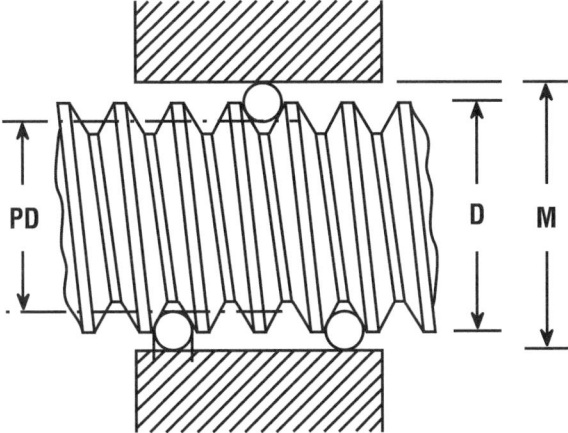

Figure 1. Measuring threads with the three-wire method.

For Unified threads $$M = (D + 3W) - (1.5155 \div P)$$

where M = Measurement over the best wire size, in inches
D = Outside diameter of the thread, in inches
W = Diameter of the best wire size
P = Pitch (number of threads per inch).

For Metric threads $$M = P + C$$

where M = Measurement over the best wire size, in millimeters
P = Pitch, in millimeters
C = Constant (from **Table 2**).

Note that **Table 2** has pitch, best wire size, and the constant stated in millimeters and converted to decimal inches, so M can be determined using the inch conversions in the Table. When the conversions are used, the value of M will be in inches, rather than millimeters.

Table 1. Three-Wire Measurement for Unified Threads.

Pitch	Best Wire Size	Pitch	Best Wire Size
4	0.144338	20	0.028868
4.5	0.128300	22	0.026243
5	0.115470	24	0.024056
5.5	0.104973	26	0.022205
6	0.096225	27	0.021383
7	0.082478	28	0.020619
7.5	0.076980	30	0.019245
8	0.072169	32	0.018042
9	0.064150	36	0.016037
10	0.057735	40	0.014434
11	0.052538	44	0.013121
11.5	0.050204	48	0.012028
12	0.048112	50	0.011547
13	0.044411	56	0.010310
14	0.041239	64	0.090211
16	0.036084	72	0.008018
18	0.032075	80	0.007217

All dimensions in inches.

Table 2. Three-Wire Thread Measurement for Metric Threads.

Pitch		Best Wire Size		Constant	
Millimeters	Inches	Millimeters	Inches	Millimeters	Inches
0.35	0.01378	0.2021	0.00796	0.3031	0.01193
0.40	0.01576	0.2309	0.00909	0.3464	0.01364
0.45	0.01772	0.2598	0.01023	0.3897	0.01534
0.50	0.01989	0.2887	0.01137	0.4330	0.01705
0.60	0.02362	0.3464	0.01364	0.5195	0.02046
0.70	0.02756	0.4041	0.01591	0.6062	0.02387
0.80	0.03150	0.4619	0.01818	0.6928	0.02728
1.00	0.03937	0.5774	0.02273	0.8660	0.03410
1.25	0.04921	0.7217	0.02841	1.0815	0.04262
1.50	0.05906	0.8660	0.03410	1.2990	0.05114
1.75	0.06890	1.0104	0.03978	1.5155	0.05967
2.00	0.07874	1.1547	0.04646	1.7321	0.06819
2.50	0.09843	1.4434	0.05688	2.1651	0.08524
3.00	0.11811	1.7321	0.06819	2.5981	0.10229
3.50	0.13780	2.0207	0.07956	3.0311	0.11933
4.00	0.15748	2.3094	0.09092	3.4641	0.13638
4.50	0.17717	2.5981	0.10229	3.8971	0.15343
5.00	0.19685	2.8868	0.11365	4.3301	0.17048
5.50	0.21654	3.1754	0.12602	4.7631	0.18753
6.00	0.23622	3.4641	0.13638	5.1962	0.20457

Fasteners for Wood

Nails

Nails, of many types, forms, and shapes, are the most common fasteners for wood construction. Nails in use must resist withdrawal and/or lateral loads, and their resistance is affected by the species of wood and the type of nail. Sizes of readily available nails are given in **Table 1**. Because international nail producers do not always adhere to the dimensions in the Table, it is advisable to specify length and diameters when purchasing nails.

The resistance of a nail to withdraw from a piece of wood is dependent on the density of the wood, the diameter of the nail, and the depth of penetration. For bright common wire nails driven into the side grain of seasoned wood, or unseasoned wood that remains wet, the maximum withdrawal load can be obtained with the following equation.

For withdrawal load in pounds $p = 7,850 \times G^{5/2} \times D \times L$

For withdrawal load in Newtons $p = 54.12 \times G^{5/2} \times D \times L$

> where p = maximum load in pounds or Newtons
> G = specific gravity of the wood at 12% moisture content
> D = diameter of nail in inches or millimeters
> L = depth of penetration of the nail in inches or millimeters.

The resistance of nails to withdrawal is generally greatest when they are driven perpendicular to the grain of the wood. When the nail is driven parallel to the wood fibers (into the end of the piece), withdrawal resistance in the softer woods drops by 25 to 50% of the resistance obtained when the nail is driven perpendicular to the grain. The difference between side- and end-grain withdrawal loads is less for dense woods than for softer woods. Withdrawal resistance is also affected by other factors including the type of nail point, type of shank, time the nail remains in the wood, surface coatings of the nail, and moisture content changes in the wood. Another factor to consider when nailing wood is that the less dense species do not split as readily as the denser ones, thus offering the opportunity to increase the diameter, length, and number of nails to compensate for the wood's lower resistance to nail withdrawal. Withdrawal resistance in plywood is approximately 15 to 30% less than solid wood of the same thickness, primarily because the fiber distortion is less uniform in plywood.

Wood Screws

The common types of wood screws have flat, oval, or round heads, as shown in *Figure 1*. For flush surfaces, flat heads are essential, but round head screws are required when countersinking is objectionable. The root diameter for most screws averages about two-thirds the shank diameter. Resistance of wood screw shanks to withdrawal varies directly with the square of the specific gravity of the wood. Within limits, the withdrawal load varies directly with the depth of penetration of the threaded portion and the diameter of the screw, provided that the screw does not fail in tension. Failure in tension results when the screw's strength is exceeded by the withdrawal strength from the wood, and the limiting length to cause a tension failure decreases as the density of the wood increases because the withdrawal strength of the wood increases with density. The longer lengths of standard screws are therefore superfluous in dense hardwoods.

Table 1. Dimensions of Commonly Available Nails.

Size	Gauge	Length		Diameter	
		inch	mm	inch	mm
Bright Common Wire Nails (Penny Nails)					
6d	11-1/2	2	50.8	0.113	2.87
8d	10-1/4	2-1/2	63.5	0.131	3.33
10d	9	3	76.2	0.148	3.76
12d	9	3-1/4	82.6	0.148	3.76
16d	8	3-1/2	88.9	0.162	4.11
20d	6	4	101.6	0.192	4.88
30d	5	4-1/2	114.3	0.207	5.26
40d	4	5	127.0	0.225	5.72
50d	3	5-1/2	139.7	0.244	6.20
60d	2	6	152.4	0.262	6.65
Smooth Box Nails					
3d	14-1/2	1-1/4	31.8	0.076	1.93
4d	14	1-1/2	38.1	0.080	2.03
5d	14	1-3/4	44.5	0.080	2.03
6d	12-1/2	2	50.8	0.098	2.49
7d	12-1/2	2-1/4	57.2	0.098	2.49
8d	11-1/2	2-1/2	63.5	0.113	2.87
10d	10-1/2	3	76.2	0.128	3.25
16d	10	3-1/2	88.9	0.135	3.43
20d	9	4	101.6	0.148	3.76

Size	Length		Diameter	
	inch	mm	inch	mm
Helical and Annularly Threaded Nails				
6d	2	50.8	0.120	3.05
8d	2-1/2	63.5	0.120	3.05
10d	3	76.2	0.135	3.43
12d	3-1/4	82.6	0.135	3.43
16d	3-1/2	88.9	0.148	3.76
20d	4	101.6	0.177	4.50
30d	4-1/2	114.3	0.177	4.50
40d	5	127.0	0.177	4.50
50d	5-1/2	139.7	0.177	4.50
60d	6	152.4	0.177	4.50
70d	7	177.8	0.207	5.26
80d	8	203.2	0.207	5.26
90d	9	228.6	0.207	5.26

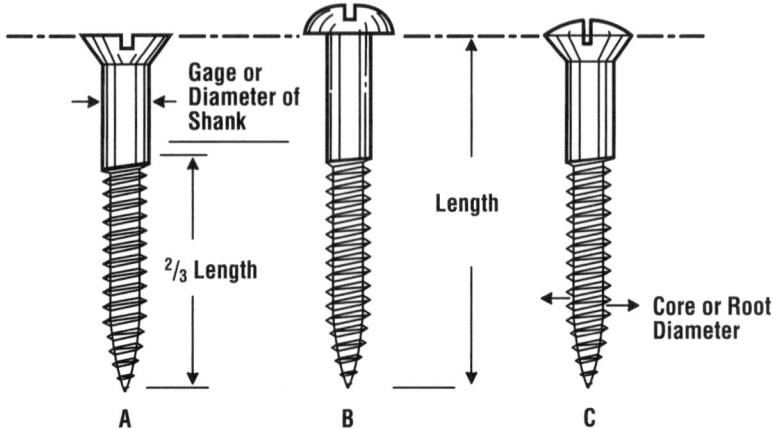

Figure 1. Common wood screws: A) flathead, B) roundhead, C) ovalhead.

Ultimate test values for withdrawal loads of wood screws inserted into the side grain of seasoned wood may be obtained with the following equation.

For withdrawal load in pounds $\quad p = 15{,}700 \times G^2 \times D \times L$

For withdrawal load in Newtons $\quad p = 108.25 \times G^2 \times D \times L$

where p = maximum load in pounds or Newtons
G = specific gravity of the wood at 12% moisture content
D = shank diameter of screw in inches or millimeters
L = depth of penetration of the threaded part of the screw in inches or millimeters.

This equation is applicable when the prebored screw lead hole has a diameter of about 70% of the root diameter of the threads in softwoods, and about 90% in hardwoods. The equation values are applicable to the screw sizes listed in **Table 2**, and when lengths and gages are outside the indicated limits, the actual values are likely to be less than the equation values. The withdrawal loads of screws inserted in the end grain of wood are somewhat erratic, but when splitting is avoided, they should average 75% of the load sustained by screws inserted in the side grain.

Withdrawal resistance of tapping screws (screws that have threads the full length of the shank–commonly known as sheet metal screws) is in general about 10% greater than that for wood screws of comparative diameter and length of threaded portion. The ratio between the withdrawal resistance of tapping screws and wood screws varies from 1.16 in denser woods such as oak, to 1.05 in lighter woods such as redwood.

Lag Screws
Lag screws are normally used where it would be difficult to fasten a bolt, or in instances where a nut on the surface would be objectionable. Commonly available lag screws range from about 0.2 to one inch (5.1 to 25.4 mm) in diameter, and from 1 to 16 inches (25.4 to 406 mm) in length. The thread length varies with screw length and ranges from $^3/_4$ inch (19 mm) in lengths up to 1 $^1/_4$ inch (31.8 mm), to one-half the overall screw length in lengths 10 inches (254 mm) or more. The head on lag screws is hexagonal shaped to provide for

Table 2. Screw Sizes and Gage Diameters.

Screw Length		Gage Limits	Gage Conversion to Diameter		
inch	mm		Gage Number	Dia. inch	Dia. mm
1/2	12.7	4 to 6	4	0.112	2.84
			5	0.125	3.18
3/4	19.0	4 to 11	6	0.138	3.51
			7	0.151	3.84
1	25.4	4 to 12	8	0.164	4.17
			9	0.177	4.50
1 1/2	38.1	5 to 14	10	0.190	4.83
			11	0.203	5.16
2	50.8	7 to 16	12	0.216	5.49
			14	0.242	6.15
2 1/2	63.5	9 to 18	16	0.268	6.81
			18	0.294	7.47
3	76.2	12 to 20	20	0.320	8.13
			24	0.372	9.45

tightening by wrench. The equation for withdrawal resistance of lag screws, which follows, is based on the screw material having a tensile yield strength of 45 ksi (310 MPa) and an average ultimate tensile strength of 77 ksi (531 MPa).

For withdrawal load in pounds $p = 125.4 \times G^{3/2} \times D^{3/4} \times L$

For withdrawal load in Newtons $p = 108.25 \times G^{3/2} \times D^{3/4} \times L$

where p = maximum load in pounds or Newtons
G = specific gravity of the wood at 12% moisture content
D = shank diameter of screw in inches or millimeters
L = depth of penetration of the threaded part of the screw in inches or millimeters.

Lag screws, like wood screws, require prebored lead holes, and the diameter of the lead hole for the threaded section varies with the density of the wood. For low density softwoods, such as cedars and white pines, the lead hole should be 40 to 70% of the shank diameter; for Douglas fir and Southern pine, 60 to 75%; and for dense hardwoods, 65 to 85%. The smaller percentage in each range applies to lag screws of the smaller diameters, and the larger percentage to lag screws with larger diameters. A lead hole should also be prebored to the depth of penetration of the shank, equal to the diameter of the shank.

Definitions of Threaded Fasteners (*MIL-STD-1251A*)

Bolt, Clevis. An externally threaded fastener whose threaded and unthreaded portions are of one nominal diameter and are separated by a narrow circumferential groove. The head has a recess for holding or driving.

Bolt, Close Tolerance. An externally threaded fastener whose unthreaded portion is of a specified grip length and is machined to a tolerance of 0.001 inch or less. Items over 1.000 inch in diameter shall have a tolerance of 0.0015 inch or less. The nominal major diameter of the threads shall be at least 0.001 inch below the minimum shank diameter, but not below the minimum major diameter for applicable class of fit. The head is designed for external wrenching. The minimum tensile strength shall be less than 160,000 PSI.

Bolt, Eye. An externally threaded device whose threaded portion is of one nominal diameter, without a head, but with the unthreaded end bent more than 225°, or cast, forged, or punched to resemble an eye.

Bolt, Hook. An externally threaded device whose threaded portion is of one nominal diameter, without a head, but with the unthreaded end bent to less that 225°.

Bolt, Internal Wrenching. An externally threaded fastener whose threaded portion is of one nominal diameter. The head is beveled (conical) in shape, and has an internal socket for internal wrenching.

Bolt, Lag. An externally threaded fastener having a square or hexagon head and with a continuous thread (wood screw or fetter drive type) extending from a gimlet or cone point for a distance of slightly more than one-half the length of the bolt.

Bolt, Machine. An externally threaded fastener whose threaded and unthreaded portions are each of one nominal diameter, 0.190 inch or larger. The length of the unthreaded portion (of hexagon head fasteners) is controlled and is machined to a tolerance greater than that specified for "Bolt, Close Tolerance." The head is designed for external wrenching only. If the head contains a recess, slot, or socket, see "Screw, Machine."

Bolt, Self-Locking. A "Bolt, Machine," or "Screw, Cap, Hexagon Head" with the added characteristic of a locking feature incorporated in the design of the head or in the threads.

Bolt, Shear. A "Bolt, Close Tolerance" except that item is fabricated from material having a minimum tensile strength of 160,000 PSI or greater.

Bolt, Shoulder. A "Bolt, Machine," or "Screw, Cap Hexagon Head" that has a round unthreaded neck or shank, all or part of which is of greater diameter than the threaded portion.

Bolt, Square Neck. A headed externally threaded fastener whose threaded portion is of one nominal diameter, with a square neck directly beneath the head.

Bolt, Tee Head. An externally threaded fastener whose threaded portion is one nominal diameter and with a head specifically designed to fit in a slot and hold against turning.

Bolt, U. An externally threaded fastener bent approximately 180° in the shape of the letter "u" and with both ends threaded.

Screw, Cap, Hexagon Head. A "Bolt, Machine (Hexagon Head)," except that the length of the unthreaded portion is not controlled.

Screw, Cap, Socket Head. An externally threaded fastener whose threaded portion is of one nominal diameter. The head is cylindrical in shape, and has an internal socket or multiple spline for use with an inserted driver. Excludes items with bevel (conical) heads.

Screw, Close Tolerance. A "Bolt, Close Tolerance" except that the head has an internal socket, recess, or slot, and the minimum tensile strength may be of any value.

Screw, Drive. A hardened cylindrical fastener with multiple spiral flutes on its shank. It also has an end smaller in diameter than the outside diameter of the spiral flutes, which acts as a pilot when driven into a drilled hole.

Screw, Eye. A fastening device with one end formed in the shape of an eye, and the other threaded with a lag or wood screw type of head.

Screw, Instrument. A "Screw, Machine," except that the thread diameter is less than 0.060 inch.

Screw, Machine. An externally threaded fastener whose threaded portion is of one nominal diameter. The unthreaded portion has a tolerance greater than that specified for "Bolt, Close Tolerance." For thread sizes 0.060 through 0.164 inch, any head may be used except "Screw, Cap, Socket Head" or "Bolt, Internal Wrenching." For thread sizes 0.190 inch and larger, any recess, slot, or socket head may be used except "Screw, Cap, Socket Head" or "Bolt, Internal Wrenching."

Screw, Self-Locking. A "Screw, Machine" or "Screw, Cap, Socket Head" with the added characteristic of a locking feature incorporated in the design of the head or in the threads.

Screw, Shoulder. A "Screw, Machine" except that it has a round unthreaded neck or shank, all or part of which is of greater diameter than the threaded portion.

Screw, Tapping, Thread Cutting. A hardened externally threaded fastener whose thread extends from a tapered end to the bearing surface of the head and is interrupted by flutes or slots to permit cutting its own mating thread.

Screw, Tapping Thread Forming. A hardened externally threaded fastener whose thread usually extends from a gimlet or dog type point to the bearing surface of the head and designed to form its own mating thread.

Screw, Wood. An unhardened externally threaded fastener whose continuous thread extends from a gimlet point for a distance of approximately two-thirds of the length of the screw, and which is designed to be driven with an inserted driver.

Setscrew. An externally threaded device whose threaded portion is one of nominal diameter with or without a head and having a cup, cone, or other type of machined point to prevent or restrict relative movement of parts and designed to be driven with either a wrench or inserted driver.

Thumbscrew. An externally threaded fastener whose threaded portion is of one nominal diameter. It may have an unthreaded portion with a diameter less than, equal to, or grater than the diameter of the threaded portion. It has either a vertically flattened, circular knurled, or wing type head, all of which are designed for rotation by the thumb and fingers.

Conversion Factors for Commonly Used Units of Measurement.

To Convert... U.S. System	To... Metric System	Multiply by...	To Convert... Metric System	To... U.S. System	Multiply by...
			Length		
mil	millimeter	0.0254	millimeter	mil	39.37
inch	μm	25,400	mm	inch	0.00003937
inch	millimeter	25.4	millimeter	inch	0.03937
inch	centimeter	2.54	centimeter	inch	0.3937
inch	meter	0.0254	meter	inch	39.37
foot	millimeter	304.8	millimeter	foot	0.0032808
foot	centimeter	30.48	centimeter	foot	0.032808
foot	meter	0.3048	meter	foot	3.2808
yard	millimeter	914.1	millimeter	yard	0.0010936
yard	centimeter	91.44	centimeter	yard	0.010936
yard	meter	0.9144	meter	yard	1.0936
mile	meter	1609.344	meter	mile	0.000621
mile	kilometer	1.609344	kilometer	mile	0.621
			Area		
inch^2	millimeter^2	645.16	millimeter^2	inch^2	0.00155
inch^2	centimeter^2	6.4516	centimeter^2	inch^2	0.155
foot^2	millimeter^2	92903.04	millimeter^2	foot^2	0.0000107639
foot^2	centimeter^2	929.0304	centimeter^2	foot^2	0.001076391
foot^2	meter^2	0.0929	meter^2	foot^2	10.76391
yard^2	millimeter^2	836127.36	millimeter^2	yard^2	0.0000011960
yard^2	centimeter^2	8361.2736	centimeter^2	yard^2	0.000119599
yard^2	meter^2	0.83612736	meter^2	yard^2	1.19599
acre	meter^2	4046.8726	meter^2	acre	0.0002471
			Volume		
fluidounce	centimeter^3	29.57352956	centimeter^3	fluidounce	0.0338
quart	liter	0.946352946	liter	quart	1.056688
gallon	liter	3.785411748	liter	gallon	0.264172
gallon	meter^3	0.003785412	meter^3	gallon	0.0264172
inch^3	centimeter^3	16.387064	centimeter^3	inch^3	0.06102374
foot^3	centimeter^3	28316.85185	centimeter^3	foot^3	0.000035315
foot^3	liter	28.31685185	liter	foot^3	0.235315
foot^3	meter^3	0.028316852	meter^3	foot^3	35.315
yard^3	liter	764.555	liter	yard^3	0.00130795
yard^3	meter^3	0.764555	meter^3	yard^3	1.30795
inch^3/pound	meter^3/kilogram	0.000036	meter^3/kilogram	inch^3/pound	27.680
foot^3/pound	meter^3/kilogram	0.0624	meter^3/kilogram	inch^3/pound	16.018
			Mass		
ounce(avdp.)	gram	28.34952313	gram	ounce	0.03527396
ounce(avdp.)	kilogram	0.028349523	kilogram	ounce	35.27396
pound	gram	453.59237	gram	pound	0.002204622
pound	kilogram	0.45359237	kilogram	pound	2.204623
ton(U.S.short)	metricton	0.09072	metricton	ton(U.S.)	1.1023
			Density		
pound/inch^3	gram/centimeter^3	27.67990	gram/centimeter^3	pound/inch^3	0.0361273
pound/foot^3	gram/centimeter^3	0.160185	gram/centimeter^3	pound/foot^3	62.4279744
pound/foot^3	kilogram/meter^3	16.01846	kilogram/meter^3	pound/foot^3	0.0624280

Conversion Factors for Commonly Used Units of Measurement. *(continued)*

To Convert... U.S. System	To... Metric System	Multiply by...	To Convert... Metric System	To... U.S. System	Multiply by...
Force					
poundforce	Newton	4.448222	Newton	poundforce	0.2248089
poundfoot	Newtonmeter	1.355818	Newtonmeter	poundfoot	0.7375612
pound/foot	Newton/meter	175.1268	Newton/meter	pound/foot	0.06852178
Pressure					
pound/inch^2	pascal	6894.76	pascal	pound/inch^2	0.000145
pound/inch^2	megapascal	0.00689476	megapascal	pound/inch^2	145.04
pound/inch^2	bar	0.0689476	bar	pound/inch^2	14.504
atmosphere	pascal	101.325	pascal	atmosphere	0.9869
Energy					
foot pound force	joule	1.355818	joule	foot pound force	0.737562
BTU	joule	1055.0559	joule	BTU	0.0009478
calorie	joule	4.1868	joule	calorie	0.23885
watt hour	joule	3600	joule	watt hour	0.0002778
Power					
horsepower	kilowatt	0.7456999	kilowatt	horsepower	1.341022
footpound/minute	watt	0.0225969	watt	footpound/minute	44.25372
Velocity					
foot/second	meter/second	0.3048	meter/second	foot/second	3.28084
foot/minute	meter/second	0.00508	meter/second	foot/minute	196.8504
foot/second	kilometer/hour	1.09728	kilometer/hour	foot/second	0.911344
mile/hour	kilometer/hour	1.609344	kilometer/hour	mile/hour	0.6213712
knots	kilometer/hour	1.852	kilometer/hour	knots	0.54
Temperature					
°Fahrenheit	°Celsius	(°F−32)/(1.8)	°Celsius	°Fahrenheit	(1.8°C)+32
°Fahrenheit	°Kelvin	(°F+459.67)/(1.8)	°Kelvin	°Fahrenheit	1.8°K−459.67
°Kelvin	°Celsius	°K−273.15	°Celsius	°Kelvin	°C+273.15

(Entries in bold type refer to Tables)

(Entries in bold type refer to Tables)

(Entries in bold type refer to Tables)